Stability of Nonlinear Waves in Hamiltonian Dynamical Systems

Mathematical
Surveys
and
Monographs

Volume 288

Stability of Nonlinear Waves in Hamiltonian Dynamical Systems

Anna Geyer
Dmitry E. Pelinovsky

AMS AMERICAN MATHEMATICAL SOCIETY
Providence, Rhode Island

2020 *Mathematics Subject Classification*. Primary 37J25, 37K45, 35C07, 35B10, 35C08, 37K10, 35Q55, 35Q53.

Library of Congress Cataloging-in-Publication Data

Names: Geyer, Anna, 1985- author. | Pelinovsky, Dmitry, author.
Title: Stability of nonlinear waves in hamiltonian dynamical systems / Anna Geyer, Dmitry E. Pelinovsky, authors.
Description: Providence, Rhode Island : American Mathematical Society, [2025] | Series: Mathematical surveys and monographs, 0076-5376 ; volume 288 | Includes bibliographical references.
Identifiers: LCCN 2025005090 | ISBN 9781470475529 (paperback) | ISBN 9781470480806 (ebook)
Subjects: LCSH: Nonlinear waves–Mathematical models. | Hamiltonian systems – Mathematical models. | AMS: Dynamical systems and ergodic theory – Finite-dimensional Hamiltonian, Lagrangian, contact, and nonholonomic systems – Stability problems. | Dynamical systems and ergodic theory – Infinite-dimensional Hamiltonian systems – Stability problems. | Partial differential equations – Representations of solutions – Traveling wave solutions. | Partial differential equations – Qualitative properties of solutions – Periodic solutions. | Partial differential equations – Representations of solutions – Soliton solutions. | Dynamical systems and ergodic theory – Infinite-dimensional Hamiltonian systems – Completely integrable systems, integrability tests, bi-Hamiltonian structures, hierarchies (KdV, KP, Toda, etc.). | Partial differential equations – Equations of mathematical physics and other areas of application – NLS-like equations (nonlinear Schroödinger). | Partial differential equations – Equations of mathematical physics and other areas of application – KdV-like equations (Korteweg-de Vries).
Classification: LCC QC174.24.N6 G494 2025 | DDC 515/.39–dc23/eng/20250618
LC record available at https://lccn.loc.gov/2025005090

Mathematical Surveys and Monographs ISSN: 0076-5376 (print); 2331-7159 (online)

DOI: https://doi.org/10.1090/surv/288

∞ The paper used in this book is acid-free and falls within the guidelines
established to ensure permanence and durability.
Visit the AMS home page at `https://www.ams.org/`

10 9 8 7 6 5 4 3 2 1   30 29 28 27 26 25

Contents

Preface

Nonlinear dispersive wave equations arise as reduced mathematical models from governing equations of mathematical physics, such as the Navier–Stokes and Maxwell equations. These reduced models combine the leading-order balance between nonlinear and dispersive effects present in wave propagation. Questions concerning the well-posedness of the time evolution, the existence and stability of coherent structures such as traveling, standing, or time-periodic solutions, and the long-time dynamics near these coherent structures are of paramount importance for applications.

This monograph addresses the question of stability of nonlinear waves, i.e., solutions of nonlinear wave equations, in the particular case when the nonlinear dispersive wave equations are defined by Hamiltonian systems. Stability in finite-dimensional Hamiltonian systems has been studied since the groundbreaking works of A. M. Lyapunov (1857–1918), L. S. Pontryagin (1908–1988), and M. G. Krein (1907–1989). Following the remarkable discovery of solitary waves, breathers, and other coherent structures, the stability of these solutions has been a central topic of investigation in infinite-dimensional Hamiltonian systems, attracting the attention of a vast number of researchers.

The objective of this monograph is to provide a comprehensive overview of standard methods in the stability theory of nonlinear waves in Hamiltonian systems, and to present recent advances that address situations where these methods cannot be applied.

The book is structured as follows: Chapter 1 comprises material for a basic graduate course on dynamical systems and reviews energy and linearization methods for stability of equilibrium points and periodic orbits in finite-dimensional systems. Particular attention is given to the stability of saddle points in Hamiltonian systems and to the orbital stability of periodic orbits.

Chapters 2 and 3 can be viewed as a specialized graduate course on the stability of nonlinear waves (either spatially decaying or spatially periodic) in infinite-dimensional Hamiltonian systems. Basic concepts related to orbital stability, such as the Vakhitov–Kolokolov slope criterion, the Grillakis–Shatah–Strauss approach, and the Weinstein variational methods, are rigorously developed and applied to standard examples such as the nonlinear Schrödinger equation, the Korteweg–de Vries equation and the nonlinear wave equation. In cases where the orbital stability theory is not applicable, linearization methods are reviewed to study, for instance, the spectral stability of solutions corresponding to saddle points of the energy functional.

After the first three chapters, the book turns into a monograph which reviews special topics of interest based on previous works of the authors.

Chapters 4 and 5 cover the stability theory for nonlinear waves in integrable Hamiltonian systems. In this case, the presence of infinitely many conserved quantities, the existence of the Bäcklund–Darboux transformation between solutions of the same equation, and the relation between the nonlinear wave equation and a system of linear equations involving a spectral parameter, help to overcome difficulties met when trying to apply conventional methods of stability analysis.

Chapter 6 deals with peaked traveling waves, which are commonly found in models for water wave propagation. Methods of linear and nonlinear stability analysis must be modified for solutions of limited smoothness in order to capture the intrinsic instability of peaked traveling waves even when the conventional methods based on energy functions suggest their orbital stability.

Chapter 7 covers solitary waves over a nonzero background when the conventional energy methods of stability analysis fail. One needs to use exponentially weighted spaces in order to apply the energy methods to solitary waves such as dark solitons and domain walls.

Only very few books have been published to this day on the subject of stability of nonlinear waves, despite the large number of original research publications on the subject, which has been steadily high in the last twenty years. The AMS monograph by J. Angulo Pava [**14**] gives the first account of the Grillakis–Shatah–Strauss approach in application to solitary and periodic waves of classical nonlinear dispersive wave equations. The monograph by K. Promislow and T. Kapitula [**152**] includes a chapter on stability in Hamiltonian systems but specializes on the stability of solitary waves in dissipative systems. The monographs of J. Yang [**249**] and D. Pelinovsky [**216**] are focused on the particular context of the nonlinear Schrödinger equation with spatially extended (e.g., periodic) potentials and cover the stability of nonlinear waves only to a limited degree. With this monograph, we aim to highlight the analysis of stability of nonlinear waves in Hamiltonian systems and to bring it to the attention of a wider audience.

Different chapters of this book have been used for specialized mini-courses and lecture series, e.g., at the University des Antilles, Guadeloupe, France (December 2019), Delft University, The Netherlands (April–June 2022, 2024, and 2025), Karlsruhe Institute of Technology, Germany (September 2021 and June–July 2022), McMaster University, Canada (September–November, 2022), and Southeast University, China (August, 2023). The authors thank their students for many corrections which improved the presentation of the book.

Needless to say that many results reviewed in this monograph are based on the collaborative research with various colleagues, and the authors appreciate very much the collective efforts made to obtain these original results.

Anna Geyer
Dmitry E. Pelinovsky
January 2025

CHAPTER 1

Stability in finite-dimensional systems

This chapter presents a graduate-level introduction to the stability of equilibrium points and periodic orbits in finite-dimensional dynamical systems.

We introduce the basic definitions and examples of stable, asymptotically stable, and unstable equilibrium points (Section 1.1). Then, we describe two main approaches to analyze the local stability of equilibrium points. The first approach is based on linearizing a nonlinear dynamical system at its equilibrium points (Section 1.2). The second approach is based on the construction of energy functions which allow us to control the local behavior of a nonlinear dynamical systems near its equilibrium points (Section 1.3). In the framework of energy methods, we introduce the concept of gradient and Hamiltonian systems and analyze the stability of critical points of the corresponding energy functions. Saddle points of Hamiltonian systems are excluded from the general stability and instability theorems and have to be studied separately by using the concept of Krein matrices and the Hamiltonian–Krein theorem (Section 1.4). These two methods of stability analysis are then extended to periodic orbits where the distance to the periodic orbit is controlled in the time evolution according to the definition of orbital stability, asymptotic stability, and instability (Section 1.5).

1.1. Stability of equilibrium points

Let us consider a finite-dimensional dynamical system written in the evolution form:

$$(1.1)\qquad \begin{cases} \dfrac{dx}{dt} = f(x), & t > 0, \\ x(0) = x_0, \end{cases}$$

where $x_0 \in \mathbb{R}^n$ and $f(x) : \mathbb{R}^n \mapsto \mathbb{R}^n$. For the existence and uniqueness of local solutions $x \in C^1([0,\tau_0),\mathbb{R}^n)$ to the initial-value problem (1.1) for some $\tau_0 > 0$, we require that f be at least Lipschitz continuous on an open subset of $\mathbb{R}^n$ that contains x_0. Under the same condition, the solution $x \in C^1([0,\tau_0),\mathbb{R}^n)$ is Lipschitz continuous with respect to initial data x_0 as well as with respect to parameters of the vector field f.

The point $x_* \in \mathbb{R}^n$ is called **an equilibrium point** if $f(x_*) = 0$. The stability of equilibrium points is an important problem both for understanding local and global dynamics of the initial-value problem (1.1) in the phase space $\mathbb{R}^n$, as well as for practical applications in which physical models are described by this dynamical system.

We are concerned with the local stability in an open neighborhood of the equilibrium point x_*. For simplicity, we work with an open ball $B_\epsilon(x_*) \subset \mathbb{R}^n$ of radius

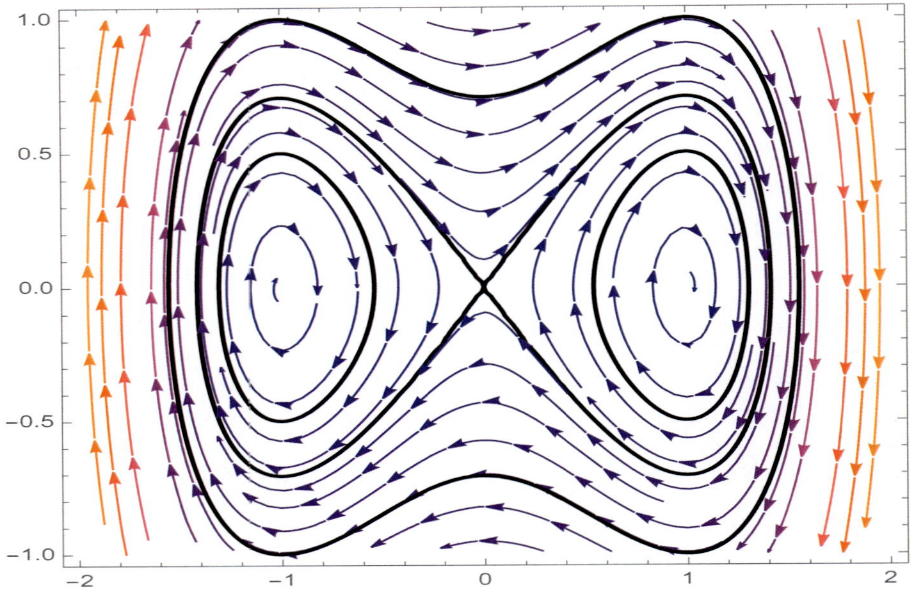

FIGURE 1.1. Phase portrait of system (1.3) with the saddle point $(0,0)$ and the center points $(\pm 1, 0)$ on the (x,y) plane.

$\epsilon > 0$ centered at the point x_*:

$$B_\epsilon(x_*) := \{x \in \mathbb{R}^n : \quad |x - x_*| < \epsilon\},$$

where $|x|$ is the vector (Euclidean) norm of $x \in \mathbb{R}^n$. The boundary of the ball (the sphere of radius $\epsilon > 0$) is denoted by $\partial B_\epsilon(x_*)$ and the closure of the ball is denoted by $\overline{B}_\epsilon(x_*) = B_\epsilon(x_*) \cup \partial B_\epsilon(x_*)$.

DEFINITION 1.1. We say that the equilibrium point x_* is **stable** if for every $\epsilon > 0$ (small) there exists some $\delta > 0$ (smaller) such that for every $x_0 \in B_\delta(x_*)$, the unique solution $x \in C^1(\mathbb{R}_+, \mathbb{R}^n)$ of the initial-value problem (1.1) satisfies $x(t) \in B_\epsilon(x_*)$ for every $t > 0$. In addition, if $\lim_{t\to\infty} x(t) = x_*$, the equilibrium point x_* is **asymptotically stable**.

DEFINITION 1.2. We say that the equilibrium point x_* is **unstable** if it is not stable. This implies that there is some $\epsilon > 0$ (finite) such that for every $\delta > 0$ (small) there exists some $x_0 \in B_\delta(x_*)$ such that the unique solution $x \in C^1([0,\tau_0), \mathbb{R}^n)$ of the initial-value problem (1.1) satisfies $x(t_0) \notin B_\epsilon(x_*)$ for some $t_0 \in (0, \tau_0)$.

Stable, asymptotically stable, and unstable equilibrium points are illustrated in the following two examples.

EXAMPLE 1.1. A typical example illustrating Definitions 1.1 and 1.2 is the Duffing oscillator modelled by the second-order differential equation

$$\frac{d^2x}{dt^2} - x + x^3 = 0. \tag{1.2}$$

The second-order equation can be cast as a system of two first-order equations

$$\begin{cases} \dfrac{dx}{dt} = y, \\ \dfrac{dy}{dt} = x - x^3. \end{cases} \tag{1.3}$$

The system (1.3) is integrable with the conserved quantity

$$V(x,y) = \frac{1}{2}y^2 - \frac{1}{2}x^2 + \frac{1}{4}x^4, \tag{1.4}$$

which has the physical meaning of total energy of the Duffing oscillator. The quantity (1.4) is conserved in the sense that

$$\frac{d}{dt}V(x(t),y(t)) = y\frac{dy}{dt} + (x^3 - x)\frac{dx}{dt} = 0$$

for every solution $(x,y) \in C^1([0,\tau_0),\mathbb{R}^2)$ to the system (1.3). As a result, all orbits of the system (1.3) are represented on the phase plane (x,y) as the level curves $V(x,y) = C$ for some $C \in \mathbb{R}$. The phase portrait of the system (1.3) is shown in Figure 1.1. The system (1.3) has three equilibrium points, at $(0,0)$ and $(\pm 1,0)$. The equilibrium point $(0,0)$ is a saddle point of V and the orbits corresponding to the level $C = 0$ diverge from the origin (with the exception of the stable manifolds of the saddle point, which are part of the homoclinic orbits). Therefore, the saddle point is unstable in the sense of Definition 1.2. The equilibrium points $(\pm 1, 0)$ are the global minima of V for $C = -\frac{1}{4}$ and the orbits for $C \in \left(-\frac{1}{4}, 0\right)$ surround the minima, which are referred to as the center points. Therefore, the center points are stable but not asymptotically stable in the sense of Definition 1.1.

EXAMPLE 1.2. Another illustrative example is formulated in polar coordinates

$$\begin{cases} x = r\cos\theta, \\ y = r\sin\theta, \end{cases} \qquad r > 0, \quad \theta \in [0, 2\pi],$$

as the following system of two first-order equations:

$$\begin{cases} \dfrac{dr}{dt} = r(1-r), \\ \dfrac{d\theta}{dt} = \sin^2\left(\dfrac{\theta}{2}\right). \end{cases} \tag{1.5}$$

The phase portrait of the system (1.5) is shown in Figure 1.2. Two equilibrium points exist on the (x,y) plane: $(0,0)$ and $(1,0)$. It follows from the first equation of system (1.5) that $\lim_{t\to+\infty} r(t) = 1$ for every $r_0 > 0$. Therefore, $(0,0)$ is unstable in the sense of Definition 1.2. On the other hand, the second equation of the system (1.5) implies that $\theta(t)$ is monotonically increasing so that $\lim_{t\to+\infty}\theta(t) = 2\pi$ for every $\theta_0 \in (0,2\pi]$. For every initial point $(x_0,y_0) \neq (0,0)$, the solution (x,y) approaches the equilibrium point $(1,0)$ as $t \to +\infty$. This may suggest that $(1,0)$ is asymptotically stable in the sense of Definition 1.1. However, orbits starting from (x_0,y_0) with $x_0 \approx 1$ and $y_0 > 0$ move out the local neighborhood of $(1,0)$, hence $(1,0)$ is unstable in the sense of Definition 1.2.

REMARK 1.1. Example 1.2 shows that the last condition in Definition 1.1 does not guarantee asymptotic stability of the equilibrium point unless it is stable.

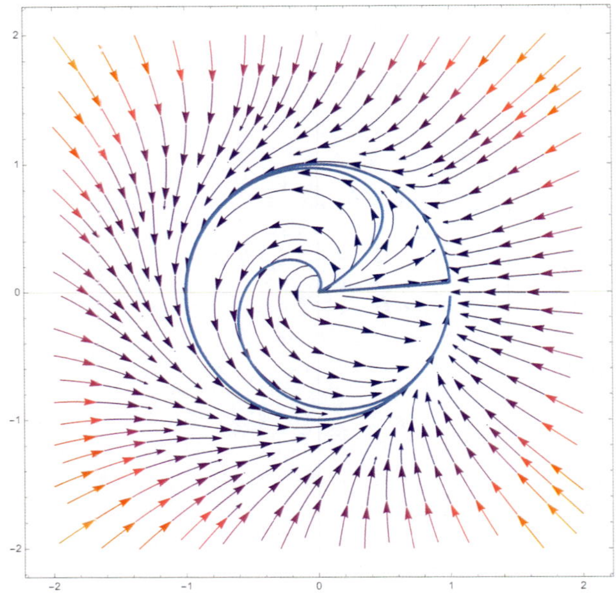

FIGURE 1.2. Phase portrait of system (1.5) in polar coordinates with the unstable points $(0,0)$ and $(1,0)$ on the (x,y) plane. The orbits of three particular solutions with different initial conditions are also depicted.

1.2. Method of linearization

We assume that $f(x) : \mathbb{R}^n \mapsto \mathbb{R}^n$ is a C^1 function at the equilibrium point x_*, so that we can use the Taylor expansion

$$f(x) = \underbrace{f(x_*)}_{=0} + \underbrace{D_x f(x_*)}_{=:A}(x - x_*) + R(x), \tag{1.6}$$

where $A \in \mathbb{M}^{n\times n}$ is constant and R is a superlinear remainder term satisfying the following limit:

$$\lim_{|x-x_*|\to 0} \frac{|R(x)|}{|x - x_*|} = 0.$$

Many computations become simplified if we assume that the remainder term actually satisfies the following estimate: for every $\epsilon \in (0, \epsilon_0)$ with some $\epsilon_0 > 0$, there exists $\nu > 0$ and $k > 0$ such that

$$|R(x)| \le k|x - x_*|^{1+\nu}, \quad x \in B_\epsilon(x_*). \tag{1.7}$$

For instance, if f is a C^2 function at x_*, then the bound (1.7) is satisfied with $\nu = 1$.

By introducing the perturbation $\xi(t)$ to the equilibrium point x_* as $\xi(t) := x(t) - x_*$, we can rewrite the dynamical system (1.1) as

$$\begin{cases} \dfrac{d\xi}{dt} = A\xi + R(\xi), & t > 0, \\ \xi(0) = \xi_0, \end{cases} \tag{1.8}$$

where $\xi_0 \in B_\delta(0)$ is defined in a local neighborhood of 0 for some $\delta > 0$ (small). The equilibrium point x_* of system (1.1) now corresponds to the zero equilibrium point of system (1.8). The remainder term $R(\xi)$ has been redefined from $R(x)$ in (1.6) and it follows from (1.7) that $R(\xi)$ satisfies

$$|R(\xi)| \leq k|\xi|^{1+\nu}, \quad \xi \in B_\epsilon(0). \tag{1.9}$$

The method of linearization relies on computations of eigenvalues of A in the spectral problem

$$(A - \lambda I)u = 0, \quad u \in \mathbb{C}^n,$$

where I denotes the identity matrix in $\mathbb{M}^{n\times n}$. The characteristic equation $\det(A - \lambda I) = 0$ is given by a polynomial of degree n, which determines n eigenvalues in the spectrum of A denoted by $\sigma(A) = \{\lambda_1, \lambda_2, \dots, \lambda_n\}$. We recall the following two definitions from linear algebra.

DEFINITION 1.3. A root $\lambda = \lambda_0$ of the characteristic equation $\det(A - \lambda I) = 0$ is called an **eigenvalue** of A. The multiplicity of the root λ_0 is called **algebraic multiplicity** of the eigenvalue λ_0. The number of linearly independent solutions of $(A - \lambda_0 I)u = 0$ for $u \in \mathbb{C}^n$ is called **geometric multiplicity** of the eigenvalue λ_0. The eigenvalue λ_0 is called **semi-simple** if its algebraic and geometric multiplicities coincide. It is called **defective** if the algebraic multiplicity exceeds the geometric multiplicity and the difference between algebraic and geometric multiplicities is called the defect of the eigenvalue λ_0.

DEFINITION 1.4. Let $\lambda = \lambda_0$ be an eigenvalue of A of geometric multiplicity 1 and algebraic multiplicity m_0. A canonical basis $\{u_j\}_{j=1}^{m_0}$ of generalized eigenvectors of A is defined by the Jordan chain of equations:

$$Au_j = \lambda_0 u_j + u_{j-1}, \quad 1 \leq j \leq m_0, \tag{1.10}$$

where $u_0 = 0$. In this basis of generalized eigenvectors of A, the matrix A is reduced to the Jordan canonical form of a block-diagonal matrix with upper triangular blocks.

REMARK 1.2. If $\lambda_0 \in \mathbb{C}\backslash\mathbb{R}$ is an eigenvalue of A, then the basis of generalized eigenvectors of A belongs to $\mathbb{C}^n$ rather than to $\mathbb{R}^n$. If $x \in \mathbb{R}^n$ and A is a real matrix, then it is necessary either to restrict coordinates over the generalized eigenvectors to complex-conjugate coordinates or to take real and imaginary parts of the generalized eigenvectors in which case the Jordan canonical form is no longer upper triangular (see Remark 1.4).

Depending on the multiplicity of an eigenvalue λ_0 in Definition 1.3, the time evolution of solutions of the linear equation $\frac{d\xi}{dt} = A\xi$ in the subspace spanned by corresponding generalized eigenvectors of A is described as follows.

- If λ_0 is a semi-simple eigenvalue, the corresponding block in the Jordan canonical form is diagonal and all linearly independent solutions of the linear equation associated to the eigenvalue λ_0 behave like $e^{\lambda_0 t}$.

- If λ_0 is a defective eigenvalue, the corresponding block in the Jordan canonical form is an upper triangular matrix and the linearly independent solutions of the linear equation associated to the eigenvalue λ_0 behave like $p_k(t)e^{\lambda_0 t}$, where $p_k(t)$ is a polynomial in t of degree k with k being the defect of λ_0.

The following theorem gives the precise conditions for stability, asymptotic stability, and instability of the zero equilibrium point in the linear equation $\frac{d\xi}{dt} = A\xi$ according to Definitions 1.1 and 1.2.

THEOREM 1.1. *Let* $\sigma(A) = \{\lambda_1, \lambda_2, \ldots, \lambda_n\}$. *The zero equilibrium point in the linear equation* $\frac{d\xi}{dt} = A\xi$ *is*

- *stable if and only if* $\mathrm{Re}(\lambda_j) \le 0$ *for all* j *and each eigenvalue* λ_j *with* $\mathrm{Re}(\lambda_j) = 0$ *is semi-simple;*
- *asymptotically stable if and only if* $\mathrm{Re}(\lambda_j) < 0$ *for all* j*;*
- *unstable if and only if either* $\mathrm{Re}(\lambda_j) > 0$ *for at least one* j *or* $\mathrm{Re}(\lambda_j) \le 0$ *for all* j *but at least one eigenvalue* λ_j *with* $\mathrm{Re}(\lambda_j) = 0$ *is defective.*

PROOF. The unique solution of the linear equation $\frac{d\xi}{dt} = A\xi$ subject to the initial condition $\xi(0) = \xi_0$ is given by $\xi(t) = e^{tA}\xi_0$ written in terms of the matrix exponential e^{tA}. The matrix exponential e^{tA} is understood in the sense of the power series

$$e^{tA} = \sum_{n=0}^{\infty} \frac{t^n}{n!} A^n,$$

which converges uniformly and absolutely for every matrix A with a finite matrix norm $\|A\| < \infty$ defined by

$$\|A\| := \sup_{x \in \mathbb{R}^n} \{|Ax| : \quad |x| = 1\}.$$

The matrix exponential e^{tA} can be analyzed in the Jordan canonical form since if the invertible transformation $S \in \mathbb{M}^{n\times n}$ reduces A to the upper triangular matrix J by $S^{-1}AS = J$, then e^{tA} is reduced to the upper triangular matrix e^{tJ} by the same transformation:

$$\begin{aligned} S^{-1}e^{tA}S &= S^{-1}\left(\sum_{n=0}^{\infty} \frac{t^n}{n!} A^n\right) S \\ &= \sum_{n=0}^{\infty} \frac{t^n}{n!} S^{-1}A^n S \\ &= \sum_{n=0}^{\infty} \frac{t^n}{n!} (S^{-1}AS)^n \\ &= \sum_{n=0}^{\infty} \frac{t^n}{n!} J^n = e^{tJ}, \end{aligned}$$

where each power series converges uniformly and absolutely, which justifies the algebraic manipulations. We are now using the fact that e^{tJ} contains diagonal entries $e^{\lambda_0 t}$ for the semi-simple eigenvalue λ_0 and $p_k(t)e^{\lambda_0 t}$ for the defective eigenvalue λ_0 with defect k, where $p_k(t)$ is a polynomial in t of degree k. It follows from the Jordan canonical form e^{tJ} that

- if $\mathrm{Re}(\lambda_j) \le 0$ for all j and each eigenvalue λ_j with $\mathrm{Re}(\lambda_j) = 0$ is semi-simple, then there exists $c > 0$ such that

$$\|e^{tA}\| \le c, \quad t > 0.$$

Hence, $|\xi(t)| \le c|\xi_0|$ for every $\xi_0 \in \mathbb{R}^n$, so that the zero equilibrium point is stable.

- if $\text{Re}(\lambda_j) < 0$ for all j, then there exist some fixed constants $a > 0$ and $c > 0$ such that
$$\|e^{tA}\| \le ce^{-at}, \quad t > 0.$$
Hence, $|\xi(t)| \le ce^{-at}|\xi_0|$ for every $\xi_0 \in \mathbb{R}^n$, so that the zero equilibrium point is asymptotically stable.

- if $\text{Re}(\lambda_j) > 0$ for at least one j, then there is $c > 0$ and $\xi_0 \in \mathbb{R}^n$ spanned by (generalized) eigenvectors for the eigenvalue λ_j such that
$$|\xi(t)| \ge ct^{k_j} e^{\text{Re}(\lambda_j)t}|\xi_0|, \quad t > 0,$$
where k_j is the defect of the eigenvalue λ_j, so that the zero equilibrium point is unstable.

- if $\text{Re}(\lambda_j) \le 0$ for all j but at least one eigenvalue λ_j with $\text{Re}(\lambda_j) = 0$ is defective, then there is $c > 0$ and $\xi_0 \in \mathbb{R}^n$ spanned by (generalized) eigenvectors for the eigenvalue λ_j such that
$$|\xi(t)| \ge ct^{k_j}|\xi_0|, \quad t > 0,$$
where k_j is the defect of the eigenvalue λ_j, so that the zero equilibrium point is unstable.

This list covers all possible cases in $\sigma(A) = \{\lambda_1, \lambda_2, \dots, \lambda_n\}$. □

REMARK 1.3. Let λ_0 be an eigenvalue of A of geometric multiplicity 1 and algebraic multiplicity m_0. Let $S \in \mathbb{M}^{n\times n}$ be the invertible transformation that contains generalized eigenvectors defined by the Jordan chain of equations (1.10) as in Definition 1.4. Then the m_0-by-m_0 upper triangular block of the matrix $J = S^{-1}AS$ is given in the form
$$J_0 = \begin{bmatrix} \lambda_0 & 1 & 0 & 0 & \dots & 0 \\ 0 & \lambda_0 & 1 & 0 & \dots & 0 \\ 0 & 0 & \lambda_0 & 1 & \dots & 0 \\ 0 & 0 & 0 & \lambda_0 & \dots & 0 \\ \vdots & \vdots & \vdots & \vdots & \ddots & \vdots \\ 0 & 0 & 0 & 0 & \dots & \lambda_0 \end{bmatrix} = \lambda_0 I_{m_0} + N_{m_0},$$
where N_{m_0} is the nilpotent matrix of order m_0 satisfying $N_{m_0}^{m_0} = O_{m_0}$ with $O_{m_0} \in \mathbb{M}^{m_0\times m_0}$ being the zero matrix.

REMARK 1.4. Let $\lambda_0 = \alpha_0 + i\beta_0$ be a simple complex eigenvalue and A be a real matrix. Then $\bar{\lambda}_0 = \alpha_0 - i\beta_0$ is also a simple complex eigenvalue of A. The complex transformation matrix $S \in \mathbb{M}^{n\times n}$ has complex-conjugate eigenvectors u_0 and $\bar{u}_0$ in order to reduce A to the complex upper-triangular matrix $J = S^{-1}AS$. It might be more convenient to use the real transformation matrix $S \in \mathbb{M}^{n\times n}$ with the real and imaginary parts of the eigenvector $u_0 = v_0 + iw_0$ satisfying
$$\begin{cases} Av_0 = \alpha_0 v_0 - \beta_0 w_0, \\ Aw_0 = \beta_0 v_0 + \alpha w_0, \end{cases}$$

and reduce A to the real block-diagonal matrix $J = S^{-1}AS$ with the corresponding 2-by-2 block given by

$$J_0 = \begin{bmatrix} \alpha_0 & \beta_0 \\ -\beta_0 & \alpha_0 \end{bmatrix} = \alpha_0 I_2 + \beta_0 R_2\left(\frac{\pi}{2}\right),$$

where

$$R_2(\theta) := \begin{bmatrix} \cos(\theta) & \sin(\theta) \\ -\sin(\theta) & \cos(\theta) \end{bmatrix}$$

is the 2-by-2 matrix of clockwise rotations in the plane with angle θ.

EXAMPLE 1.3. For $n = 2$, every real matrix $A \in \mathbb{M}^{2\times 2}$ can be transformed to one of the following three real Jordan canonical forms:

$$\begin{bmatrix} \lambda_1 & 0 \\ 0 & \lambda_2 \end{bmatrix}, \quad \begin{bmatrix} \lambda_1 & 1 \\ 0 & \lambda_1 \end{bmatrix}, \quad \begin{bmatrix} \alpha_1 & \beta_1 \\ -\beta_1 & \alpha_1 \end{bmatrix}.$$

For $n = 3$, every real matrix $A \in \mathbb{M}^{3\times 3}$ can be transformed to one of the following four real Jordan canonical forms:

$$\begin{bmatrix} \lambda_1 & 0 & 0 \\ 0 & \lambda_2 & 0 \\ 0 & 0 & \lambda_3 \end{bmatrix}, \quad \begin{bmatrix} \lambda_1 & 1 & 0 \\ 0 & \lambda_1 & 0 \\ 0 & 0 & \lambda_2 \end{bmatrix}, \quad \begin{bmatrix} \lambda_1 & 1 & 0 \\ 0 & \lambda_1 & 1 \\ 0 & 0 & \lambda_1 \end{bmatrix}, \quad \begin{bmatrix} \lambda_1 & 0 & 0 \\ 0 & \alpha_2 & \beta_2 \\ 0 & -\beta_2 & \alpha_2 \end{bmatrix}.$$

The number of irreducible real Jordan canonical forms grows rapidly with larger n. For $n = 4$, nine real Jordan forms of a real matrix $A \in \mathbb{M}^{4\times 4}$ are possible.

It is natural to wonder if the results of Theorem 1.1 can be transferred to the dynamical system (1.8) near the zero equilibrium point, or in other words, if the linear stability or instability of equilibrium points persists in the dynamics of the full nonlinear system. By using the integral formulation of the initial-value problem (1.8), we will show in the following two theorems that the answer to this question is positive if either $\mathrm{Re}(\lambda_j) < 0$ for all j or there is at least one j such that $\mathrm{Re}(\lambda_j) > 0$. The proof of Theorem 1.2 can be found in [**19**]. The proof of Theorem 1.3 is inspired by [**229**].

THEOREM 1.2. *Let the zero equilibrium point be asymptotically stable in the linear equation* $\frac{d\xi}{dt} = A\xi$. *Then,* x_* *is asymptotically stable in the nonlinear equation* $\frac{dx}{dt} = f(x)$, *where* $f \in C^1(\mathbb{R})$ *satisfies* (1.6) *and* (1.7).

PROOF. By Theorem 1.1, if the zero equilibrium point is asymptotically stable in the linear equation $\frac{d\xi}{dt} = A\xi$, then the matrix exponential e^{tA} satisfies the bound

$$\|e^{tA}\| \leq ce^{-at}, \quad t > 0. \tag{1.11}$$

We will prove that there exist fixed constants $b > 0$ and $c > 0$ such that for every $\xi_0 \in B_\delta(0)$ with every $\delta > 0$ (small), the unique solution $\xi \in C^1(\mathbb{R}_+, \mathbb{R}^n)$ of the perturbed system (1.8) satisfies

$$|\xi(t)| \leq ce^{-bt}|\xi_0|, \quad t > 0. \tag{1.12}$$

Thus, $\xi(t) \in B_\epsilon(0)$ for every $t > 0$ with $\epsilon = c\delta$ and $|\xi(t)| \to 0$ as $t \to \infty$, so that 0 is asymptotically stable in the perturbed system (1.8) and $\delta := \epsilon/c$ is uniquely determined in terms of a given $\epsilon > 0$.

In order to prove the bound (1.12) from the bound (1.11), we use the variation of constants formula and write the unique solution to the initial-value problem (1.8) in the integral form

$$\xi(t) = e^{tA}\xi_0 + \int_0^t e^{(t-\tau)A} R(\xi(\tau)) d\tau. \tag{1.13}$$

It follows from the bound (1.11) and the triangle inequality that

$$|\xi(t)| \leq c e^{-at}|\xi_0| + c\int_0^t e^{-a(t-\tau)} |R(\xi(\tau))| d\tau. \tag{1.14}$$

For every $\epsilon \in (0, \epsilon_0)$ with some $\epsilon_0 > 0$, it follows from the bound (1.9) that

$$|R(\xi)| \leq k\epsilon_0^{\nu} |\xi|, \quad \xi \in B_\epsilon(0). \tag{1.15}$$

Multiplying the bound (1.14) by e^{at} and using bound (1.15), we rewrite the estimate in the form:

$$e^{at}|\xi(t)| \leq c|\xi_0| + ck\epsilon_0^{\nu} \int_0^t e^{a\tau} |\xi(\tau)| d\tau.$$

By the well-known Gronwall's lemma, this inequality implies that

$$e^{at}|\xi(t)| \leq c|\xi_0| e^{ck\epsilon_0^{\nu} t}$$

or equivalently that

$$|\xi(t)| \leq c|\xi_0| e^{-(a - ck\epsilon_0^{\nu})t}.$$

If $\epsilon_0 > 0$ is sufficiently small, then we obtain the estimate (1.12) with $b := a - ck\epsilon_0^{\nu} > 0$ and $c > 0$ as in (1.11). ∎

THEOREM 1.3. *Assume that among the eigenvalues of A there exists either a maximal real eigenvalue $\lambda_0 > 0$ or a pair of complex conjugate eigenvalues with maximal real part $\lambda_0 > 0$. Then, x_* is unstable in the nonlinear equation $\frac{dx}{dt} = f(x)$, where $f \in C^1(\mathbb{R})$ satisfies (1.6) and (1.7).*

PROOF. For simplicity, we assume that the maximal real eigenvalue $\lambda_0 > 0$ or the pair of complex conjugate eigenvalues with maximal real part $\lambda_0 > 0$ is simple. This implies the upper bound on the matrix exponential e^{tA}:

$$\|e^{tA}\| \leq c e^{\lambda_0 t}, \quad t > 0, \tag{1.16}$$

for some fixed constant $c \geq 1$, and the existence of $\xi_0 \in \mathbb{R}^n$ such that

$$|e^{tA}\xi_0| \geq c_0 e^{\lambda_0 t}|\xi_0|, \quad t > 0, \tag{1.17}$$

for some $c_0 \in (0, c]$. For instance, if $\xi_0 \in \mathbb{R}^n$ is an eigenvector of A for a simple real eigenvalue λ_0, then $e^{tA}\xi_0 = e^{\lambda_0 t}\xi_0$ so that the bound (1.17) is attained with $c_0 = 1$. The bound (1.17) holds for every $\xi_0 \in \mathbb{R}^n$ which has a nonzero projection to the corresponding eigenvector. To show that x_* is unstable as in Definition 1.2, we will prove that there exists $\epsilon > 0$ (finite) such that for every $\delta > 0$ (small), there is some $\xi_0 \in \mathbb{R}^n$ with $|\xi_0| = \delta$ such that the unique solution $\xi \in C^1([0, \tau_0), \mathbb{R}^n)$ of the perturbed system (1.8) satisfies

$$|\xi(t_0)| \geq \epsilon, \quad \text{for some } t_0 \in (0, \tau_0), \tag{1.18}$$

where $\tau_0 > 0$ is the maximal (either finite or infinite) existence time.

Let ϵ be fixed in $(0, \epsilon_0)$, where

$$\epsilon_0 := \frac{c_0}{2}\left(\frac{c_0 \nu \lambda_0}{k(2c)^{2+\nu}}\right)^{1/\nu},$$

and define a time $T_\delta > 0$ that depends on δ according to the following bound

$$\frac{2}{c_0}\epsilon \le \delta e^{\lambda_0 T_\delta} \le \left(\frac{c_0 \nu \lambda_0}{k(2c)^{2+\nu}}\right)^{1/\nu}. \tag{1.19}$$

Using again the integral form (1.13), we will control the convolution term during the time span $[0, T]$, where

$$T := \sup\left\{t > 0 : \quad |\xi(t) - e^{tA}\xi_0| < \frac{1}{2}c_0 e^{\lambda_0 t}\delta\right\}. \tag{1.20}$$

If $T \ge T_\delta$, then it follows from the reversed triangle inequality with the help of the bounds (1.17) with $|\xi_0| = \delta$ and (1.20) that

$$\begin{aligned} |\xi(T_\delta)| &\ge |e^{T_\delta A}\xi_0| - |\xi(T_\delta) - e^{T_\delta A}\xi_0| \\ &\ge c_0 e^{\lambda_0 T_\delta}\delta - \frac{1}{2}c_0 e^{\lambda_0 T_\delta}\delta \\ &= \frac{1}{2}c_0 e^{\lambda_0 T_\delta}\delta. \end{aligned}$$

Thanks to the lower bound in (1.19), this implies that $|\xi(T_\delta)| \ge \epsilon$, which is the bound (1.18) for $t_0 = T_\delta$. Hence, we only need to show that $T \ge T_\delta$.

Let us assume to the contrary that $T < T_\delta$. For every $t \in [0, T]$, we can use the triangle inequality with the help of the bounds (1.16) and (1.20) to get

$$\begin{aligned} |\xi(t)| &\le |e^{tA}\xi_0| + |\xi(t) - e^{tA}\xi_0| \\ &\le c e^{\lambda_0 t}\delta + \frac{1}{2}c_0 e^{\lambda_0 t}\delta \\ &\le 2c e^{\lambda_0 t}\delta. \end{aligned}$$

Then, for every $t \in [0, T]$, we analyze the convolution term in (1.13) with the bounds (1.9) and (1.16) to find that

$$\begin{aligned} |\xi(t) - e^{tA}\xi_0| &\le ck \int_0^t e^{\lambda_0(t-\tau)}|\xi(\tau)|^{1+\nu} d\tau \\ &\le ck(2c\delta)^{1+\nu} \int_0^t e^{\lambda_0 t + \nu\lambda_0\tau} d\tau \\ &\le ck(2c\delta)^{1+\nu}\nu^{-1}\lambda_0^{-1} e^{\lambda_0(1+\nu)t}. \end{aligned}$$

Since $T < \infty$, it follows from (1.20) that

$$|\xi(T) - e^{TA}\xi_0| = \frac{1}{2}c_0 e^{\lambda_0 T}\delta,$$

so that we obtain at $t = T$ that

$$ck(2c\delta)^{1+\nu}\nu^{-1}\lambda_0^{-1} e^{\lambda_0(1+\nu)T} \ge |\xi(T) - e^{TA}\xi_0| = \frac{1}{2}c_0 e^{\lambda_0 T}\delta.$$

Together with the upper bound in (1.19) this yields

$$\delta e^{\lambda_0 T} \ge \left(\frac{c_0 \nu \lambda_0}{k(2c)^{2+\nu}}\right)^{1/\nu} \ge \delta e^{\lambda_0 T_\delta},$$

which contradicts our assumption $T < T_\delta$. Hence, $T \ge T_\delta$ and the estimate (1.18) is proven. □

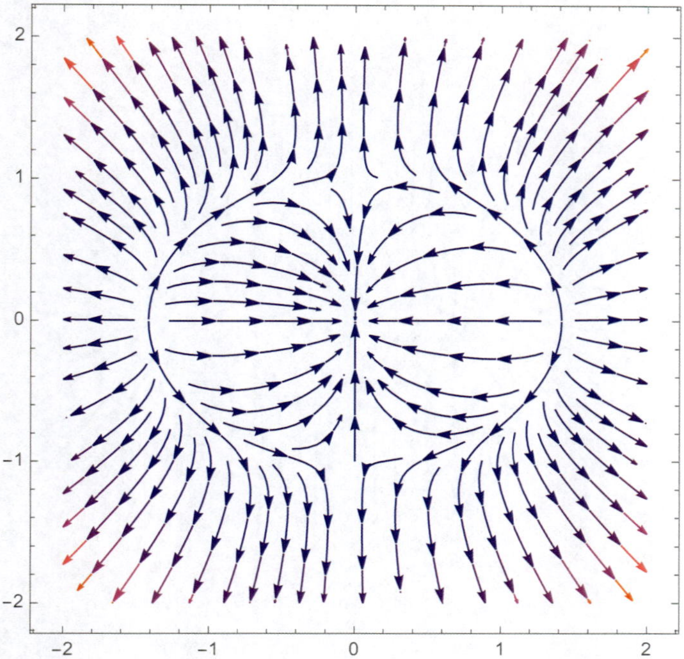

FIGURE 1.3. Phase portrait of system (1.21) with the asymptotically stable point $(0,0)$ and the unstable points $(\pm\sqrt{2},0)$ and $(0,\pm 1)$ on the (x,y) plane.

EXAMPLE 1.4. Theorems 1.2 and 1.3 can be applied to determine the stability of equilibrium points of the system

$$\left\{\begin{aligned} \frac{dx}{dt} &= -2x + x^3 + xy^2, \\ \frac{dy}{dt} &= -y + x^2y + y^3. \end{aligned}\right. \tag{1.21}$$

The matrix of the linearization of the vector field at the equilibrium point at $(0,0)$ is given by $\mathrm{diag}(-2,-1)$, hence $(0,0)$ is asymptotically stable by Theorem 1.2. There exist two pairs of equilibrium points $(\pm\sqrt{2},0)$ and $(0,\pm 1)$ with the matrices of linearization given by $\mathrm{diag}(4,1)$ and $\mathrm{diag}(-1,2)$, respectively. Both pairs of equilibrium points are unstable by Theorem 1.3. Figure 1.3 illustrates these conclusions and shows that the orbits blow up to infinity far away from the local neighborhood of the asymptotically stable point $(0,0)$.

EXAMPLE 1.5. Another example is given by the system

$$\left\{\begin{aligned} \frac{dx}{dt} &= x - x^3 - xy^2, \\ \frac{dy}{dt} &= -3y - x^2y - y^3. \end{aligned}\right. \tag{1.22}$$

The matrix of the linearization of the vector field at the equilibrium point at $(0,0)$ is given by $\mathrm{diag}(1,-3)$, hence $(0,0)$ is an unstable equilibrium by Theorem 1.3. There

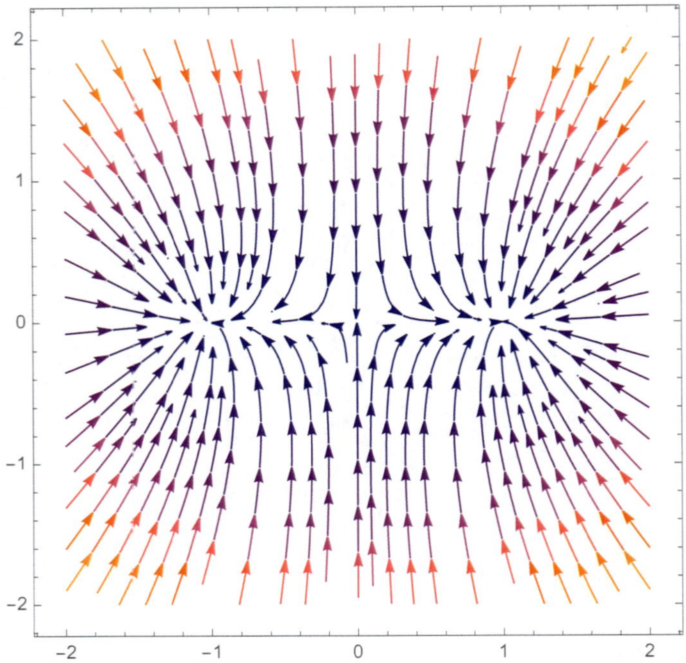

FIGURE 1.4. Phase portrait of system (1.22) with the unstable point $(0,0)$ and the asymptotically stable points $(\pm 1, 0)$ on the (x, y) plane.

exists a pair of equilibrium points $(\pm 1, 0)$ with the matrix of linearization given by $\mathrm{diag}(-2,-4)$. These equilibrium points are asymptotically stable by Theorem 1.2. Figure 1.4 illustrates these conclusions and shows that the orbits do not blow up to infinity away from the local neighborhood of the unstable point $(0,0)$ but instead approach one of the two asymptotically stable points $(\pm 1, 0)$.

1.3. Method of energy functions

The method of linearization is not conclusive if the eigenvalues in $\sigma(A) = \{\lambda_1, \lambda_2, \ldots, \lambda_n\}$ satisfy $\mathrm{Re}(\lambda_j) \leq 0$ for all j and $\mathrm{Re}(\lambda_j) = 0$ for some j. By Theorem 1.1, the zero equilibrium point in $\frac{d\xi}{dt} = A\xi$ is linearly stable if each eigenvalue λ_j with $\mathrm{Re}(\lambda_j) = 0$ is semi-simple and linearly unstable if there is at least one defective eigenvalue λ_j with $\mathrm{Re}(\lambda_j) = 0$. Depending on the remainder term R, the zero equilibrium point may be stable or asymptotically stable or unstable in the perturbed system (1.8).

A useful approach to conclude on stability of equilibrium points in such marginal cases is based on the center manifold theory and computations of normal forms [**138**]. Since this approach is difficult to apply for infinite-dimensional Hamiltonian systems [**122**], we will not review details of the center manifold theory and normal forms, but will focus on the method of energy functions.

The method of energy functions was pioneered by A. M. Lyapunov [**190**]. We will review two of Lyapunov's theorems on stability and instability of equilibrium points in the dynamical system (1.1). The energy functions introduced in these theorems are often called **Lyapunov functions**.

The first Lyapunov theorem gives stability and asymptotic stability of an equilibrium point in the dynamical system (1.1).

THEOREM 1.4. *Let x_* be an equilibrium point of $\frac{dx}{dt} = f(x)$. Assume that there is an open set $D \subset \mathbb{R}^n$ containing x_* and a C^1 function $V(x) : D \mapsto \mathbb{R}$ such that*

(S1) *$V(x_*) = 0$ and $V(x) > 0$ for every $x \in D \backslash \{x_*\}$;*

(S2) *$\nabla V(x) \cdot f(x) \leq 0$ for every $x \in D$.*

Then, x_* is stable in the sense of Definition 1.1. *Moreover, if the condition (S2) is replaced by the stricter condition*

(S2') *$\nabla V(x) \cdot f(x) < 0$ for every $x \in D \backslash \{x_*\}$,*

then x_* is asymptotically stable.

PROOF. In order to show the stability of x_*, let us consider a ball $B_\epsilon(x_*) \subset D \subset \mathbb{R}^n$ for an arbitrary $\epsilon > 0$ (small). Since the function $V(x)$ is continuous, the property (S1) implies that there exists

$$m_\epsilon := \min_{x \in \partial B_\epsilon(x_*)} V(x) > 0,$$

such that there exists a closed loop inside $\overline{B}_\epsilon(x_*)$ on which $V(x) = m_\epsilon$ and which touches $\partial B_\epsilon(x_*)$ at one or more points, see Figure 1.5. By continuity of V, there exists $\delta > 0$ (smaller) such that

$$0 < \max_{x \in \overline{B_\delta(x_*)}} V(x) < m_\epsilon.$$

As is illustrated on Figure 1.5, δ can be any number smaller than the shortest distance from x_* to the level surface $V(x) = m_\epsilon$. For every $x_0 \in B_\delta(x_*)$, there exists a unique solution $x \in C^1([0, \tau_0), \mathbb{R}^n)$ to the initial-value problem (1.1) with some $\tau_0 > 0$. Since

$$\frac{d}{dt} V(x(t)) = \nabla V(x(t)) \cdot f(x(t)) \leq 0, \quad t \in [0, \tau_0)$$

by property (S2), the map $t \mapsto V(x(t))$ is nonincreasing along the solution at least for small $t > 0$. Hence $V(x(t)) \leq V(x_0) < m_\epsilon$, which implies that $x(t) \in B_\epsilon(x_*)$ for $t \in (0, \tau_0)$. Indeed, if we assume to the contrary that there exists $t_1 \in (0, \tau_0)$ such that $x(t_1) \notin \overline{B_\epsilon(x_*)}$, then there exists $t_2 \in (0, t_1)$ such that $x(t_2) \in \partial B_\epsilon(x_*)$. Since $\min_{x \in \partial B_\epsilon(x_*)} V(x) = m_\epsilon$ and $x(t_2) \in \partial B_\epsilon(x_*)$ we have that $V(x(t_2)) \geq m_\epsilon$ which is a contradiction. Since $x(t)$ stays in a bounded subset of $D \subset \mathbb{R}^n$, the local solution can be continued for all $t > 0$ and $x(t) \in B_\epsilon(x_*)$ for all $t > 0$. Hence, x_* is stable in the sense of Definition 1.1.

In order to show the asymptotic stability of x_*, we now use the stricter condition (S2') which implies that the map $t \mapsto V(x(t))$ is strictly decreasing along the solution for small $t > 0$. We need to prove that $x(t) \to x_*$ as $t \to +\infty$ since the solution $x(t)$ cannot reach x_* in finite time as this would contradict the uniqueness of solutions of the initial-value problem (1.1). Since $x(t) \in B_\epsilon(x_*)$ for all $t > 0$, the solution stays in a bounded and compact subset of D so that there is a sequence of times $\{t_n\}_{n \in \mathbb{N}}$ and $y \in D$ such that $t_n \to +\infty$ and $x(t_n) \to y$ as $n \to \infty$. Assuming that $y \neq x_*$ yields a contradiction. Indeed, since V is continuous, we have $V(y) = \lim_{n\to\infty} V(x(t_n))$. Since the map $t \mapsto V(x(t))$ is strictly decreasing, we have $V(y) < V(x(t_n))$ for every $n \in \mathbb{N}$. If $y \neq x_*$, the solution can be continued

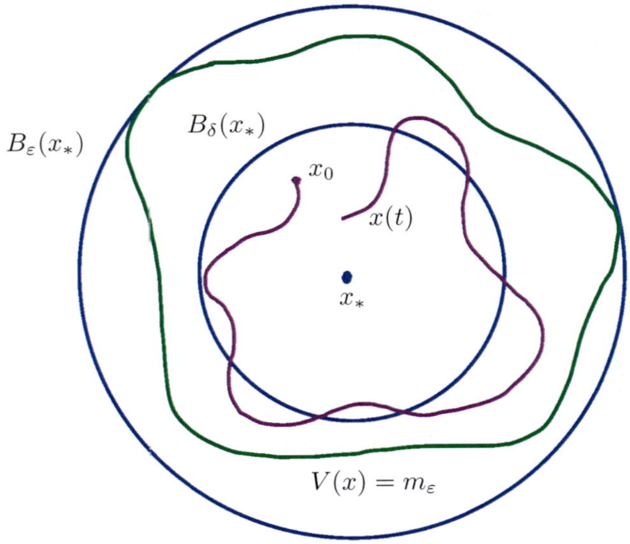

FIGURE 1.5. Illustration of the proof of Theorem 1.4.

in new time s from the initial condition y. Let us denote this solution by $\tilde{x}(s)$, $s > 0$ such that $\tilde{x}(0) = y$. By the monotonicity of the map $s \mapsto V(\tilde{x}(s))$, we have

$$V(\tilde{x}(s)) < V(y) \quad \text{for all } \ s > 0.$$

Consider $x(t_n)$ for large values of n such that $|x(t_n) - y|$ is small. Let us fix $s > 0$. By continuous dependence, $x(t_n + s)$ can be made as close to $\tilde{x}(s)$ as desired by taking n sufficiently large. By the continuity of $V(x)$, this implies that $|V(x(t_n + s)) - V(\tilde{x}(s))|$ is small and since $V(\tilde{x}(s)) < V(y)$, we have $V(x(t_m)) < V(y)$ for some m. However, this is a contradiction since $V(y) < V(x(t_m))$ for every $m \in \mathbb{N}$. Hence, $y = x_*$ so that $x(t_n) \to x_*$ as $t_n \to +\infty$, irrespective of the choice of the sequence $\{t_n\}_{n \in \mathbb{N}}$. Thus, x_* is asymptotically stable in the sense of Definition 1.1. □

EXAMPLE 1.6. Theorem 1.4 can be applied to determine the stability of equilibrium points of the system

$$\begin{cases} \dfrac{dx}{dt} = -2x^3 - y^3, \\ \dfrac{dy}{dt} = -x^3 - y^3. \end{cases} \tag{1.23}$$

The only equilibrium point of this system is at $(0,0)$ and the matrix of linearization $A = O_2$ has a double zero eigenvalue. Hence, the methods based on linearization are not conclusive to study the stability of $(0,0)$ in the system (1.23). On the other hand, the function

$$V(x, y) = \frac{1}{4}(x^4 + y^4)$$

can serve as a Lyapunov function in Theorem 1.4. It clearly satisfies property (S1). In addition, computing the time derivative along the solutions of system (1.23), we

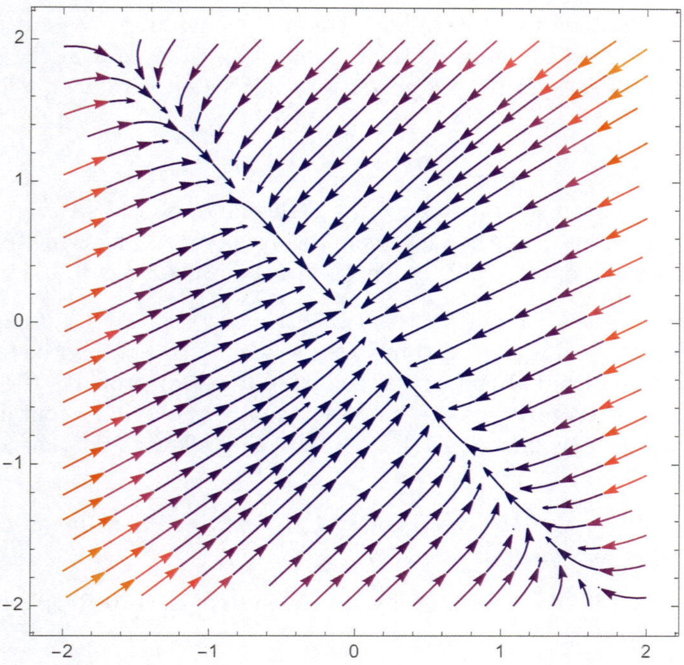

FIGURE 1.6. Phase portrait of system (1.23) with the asymptotically stable point $(0,0)$ on the (x,y) plane.

verify property (S2'):

$$\begin{aligned}\frac{d}{dt}V(x(t),y(t)) &= x^3\frac{dx}{dt} + y^3\frac{dy}{dt}\\ &= -2x^6 - 2x^3y^3 - y^6\\ &= -x^6 - (x^3+y^3)^2,\end{aligned}$$

which is strictly negative for every $(x,y) \neq (0,0)$. Hence $(0,0)$ is asymptotically stable by Theorem 1.4. Figure 1.6 shows that the orbits all converge to $(0,0)$ as $t \to +\infty$.

Example 1.6 is special because the function $V(x,y) = \frac{1}{4}(x^4+y^4)$ is strictly decreasing along nonzero solutions to system (1.23). Generally, it is easy to construct functions V satisfying property (S1) but it is not so easy (and sometimes impossible) to satisfy property (S2) or its stricter version (S2') with such V. Hence, it is nice to know that property (S2) (respectively, (S2')) is always satisfied for Hamiltonian (respectively, gradient) dynamical systems.

DEFINITION 1.5. We say that a dynamical system is **gradient** if it can be written in the form

$$\frac{dx}{dt} = -\nabla V(x), \tag{1.24}$$

where $V(x) : \mathbb{R}^n \mapsto \mathbb{R}$ is at least C^1.

REMARK 1.5. The name "gradient system" comes from the steepest descent method, which is used to minimize the function $V(x) : \mathbb{R}^n \mapsto \mathbb{R}$ along the direction opposite to the direction of the gradient vector $\nabla V(x)$. Gradient systems satisfy the property

$$\frac{d}{dt}V(x(t)) = -|\nabla V(x(t))|^2,$$

for solutions of system (1.24), where the right-hand side is strictly negative for every $x \in \mathbb{R}^n$ different from the set of equilibrium points x_* for which $\nabla V(x_*) = 0$. Hence the stricter property (S2') of Theorem 1.4 is satisfied.

Property (S1) of Theorem 1.4 is satisfied for every isolated minimum of $V(x)$. If $V(x_*) \neq 0$ at the minimum x_*, one can consider $V(x) - V(x_*)$ instead of $V(x)$. Note that the equilibrium point x_* has to be **isolated** for property (S1) to be true, which means that there is some $\epsilon > 0$ such that $B_\epsilon(x_*)\backslash\{x_*\}$ contains no other equilibrium points. By Theorem 1.4, we obtain the following corollary for gradient systems.

COROLLARY 1.1. *Let x_* be an isolated minimum of V in the gradient system* (1.24). *Then, x_* is asymptotically stable.*

DEFINITION 1.6. We say that a dynamical system is **Hamiltonian** if it can be written in the form

$$\frac{dx}{dt} = J\nabla V(x), \tag{1.25}$$

where $V(x) : \mathbb{R}^n \mapsto \mathbb{R}$ is at least C^1 and $J \in \mathbb{M}^{n\times n}$ is an invertible and skew-symmetric matrix with constant entries satisfying $J^T = -J$. Since J is invertible and skew-symmetric, the dimension n is necessarily even and the integer $m := \frac{n}{2}$ is referred as **the degree** of the Hamiltonian system (1.25).

REMARK 1.6. The name "Hamiltonian system" comes from classical mechanics, where such systems are generated by the Hamiltonian function $V(x) : \mathbb{R}^{2m} \mapsto \mathbb{R}$ which can be interpreted as the conserved energy of a mechanical system of m particles whose motion is described by the system (1.25). Hamiltonian systems satisfy the property

$$\frac{d}{dt}V(x(t)) = \nabla V(x(t)) \cdot (J\nabla V(x(t))) = 0,$$

for solutions of system (1.25), since for every real-valued $y \in \mathbb{R}^n$, we have $y \cdot Jy = J^T y \cdot y = -Jy \cdot y = -y \cdot Jy$ implying $y \cdot Jy = 0$. Hence property (S2) of Theorem 1.4 is satisfied.

Property (S1) of Theorem 1.4 is satisfied again for every isolated minimum of $V(x) - V(x_*)$. Moreover, for every isolated maximum of $V(x)$, we can take $-V(x) + V(x_*)$ in Theorem 1.4 since $-V(x(t))$ is also constant for every solution to the Hamiltonian system (1.25). By Theorem 1.4, we obtain the following corollary.

COROLLARY 1.2. *Let x_* be an isolated minimum or maximum of V in the Hamiltonian system* (1.25). *Then, x_* is stable.*

Complementary to the definite results in Corollaries 1.1 and 1.2 on isolated equilibrium points, we will consider stability or instability of isolated maxima and saddle points in the gradient system (1.24) and saddle points in the Hamiltonian system (1.25). For this purpose we need the second Lyapunov theorem which ensures instability of equilibrium points in the dynamical system (1.1).

THEOREM 1.5. *Let x_* be an equilibrium point of $\frac{dx}{dt} = f(x)$. Assume that there is an open set $D \subset \mathbb{R}^n$ containing x_*, an open subset $D_1 \subset D$ with $x_* \in \partial D_1$, and a C^1 function $V(x) : D \mapsto \mathbb{R}$ such that*

(I1) *$V(x_*) = 0$ and $V(x) > 0$ for every $x \in D_1$;*
(I2) *$\nabla V(x) \cdot f(x) > 0$ for every $x \in D_1$.*

Then, x_* is unstable in the sense of Definition 1.2.

PROOF. For every $\delta > 0$ (small), let us consider the ball $B_\delta(x_*)$ and pick the initial point $x_0 \in D_1 \cap B_\delta(x_*)$. Then, we have $V(x_0) > 0$ from property (I1) and

$$\frac{d}{dt}V(x(t)) = \nabla V(x(t)) \cdot f(x(t)) > 0,$$

for solutions of the initial-value problem (1.1), at least for small $t > 0$ from property (I2). Therefore, the map $t \mapsto V(x(t))$ is strictly increasing and $x(t)$ remains in D_1 at least for small $t > 0$ with $V(x(t)) > V(x_0) > 0$.

Fix $\epsilon > \delta$ and consider $D_1 \cap B_\epsilon(x_*)$. It follows from the previous discussion that $x(t) \in D_1 \cap B_\epsilon(x_*)$ for small $t > 0$ and that there is a subset $D_1' \subset D_1$ with $x_* \in \partial D_1'$ such that $x(t) \notin D_1'$ for all $t > 0$ for which $x(t)$ exists in $B_\epsilon(x_*)$. The subsets are illustrated in Figure 1.7. By property (I1), we can define

$$m := \min_{x \in D_1 \cap B_\epsilon(x_*) \backslash D_1'} \nabla V(x(t)) \cdot f(x(t)) > 0,$$

so that the solution $x(t) \in D_1 \cap B_\epsilon(x_*)$ for small $t > 0$ satisfies

$$\frac{d}{dt}V(x(t)) = \nabla V(x(t)) \cdot f(x(t)) \geq m.$$

Integrating this inequality in time, we obtain $V(x(t)) \geq V(x_0) + mt$ and since the lower bound grows linearly in t, it is true for some $t_* > 0$ that

$$V(x(t_*)) \geq V(x_0) + mt_* > \max_{x \in D_1 \cap B_\epsilon(x_*)} V(x).$$

Hence, $x(t_*) \notin B_\epsilon(x_*)$ so that x_* is unstable in the sense of Definition 1.2. □

EXAMPLE 1.7. Theorem 1.5 can be applied to determine the stability of equilibrium points of the system

$$\begin{cases} \dfrac{dx}{dt} = x^3 - y^3, \\ \dfrac{dy}{dt} = -x^3 - y^3. \end{cases} \tag{1.26}$$

Similarly to Example 1.6, the only equilibrium point is at $(0, 0)$ and the matrix of linearization $A = O_2$ has a double zero eigenvalue. The function

$$V(x, y) = \frac{1}{4}(x^4 - y^4)$$

can serve as a Lyapunov function in Theorem 1.5. It clearly satisfies property (I1) for $x^4 > y^4$, e.g., in the cone $D_1 = \{x > 0, \ -x < y < x\}$. In addition, computing

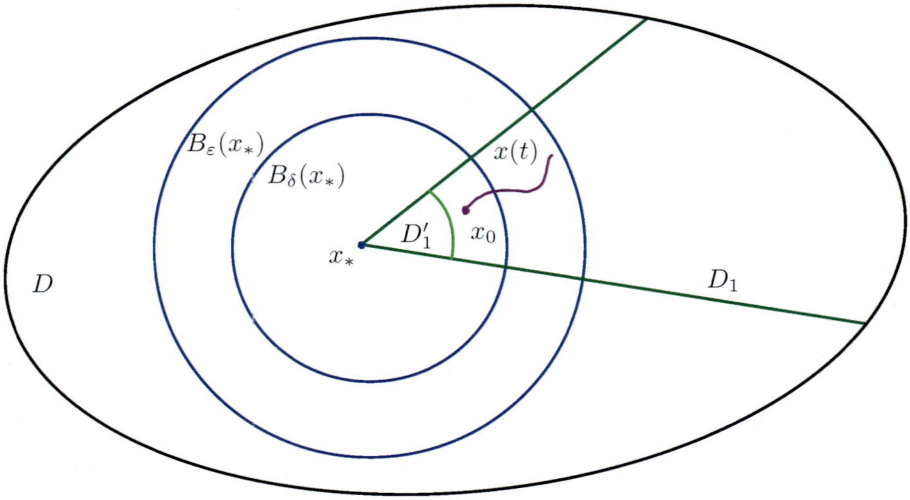

FIGURE 1.7. Illustration of the proof of Theorem 1.5.

the time derivative along the solutions of system (1.26), we verify property (I2):

$$\begin{aligned}\frac{d}{dt}V(x(t),y(t)) &= x^3\frac{dx}{dt} - y^3\frac{dy}{dt}\\ &= x^6 + y^6,\end{aligned}$$

which is strictly positive for every (x, y) in D_1. Hence $(0, 0)$ is unstable by Theorem 1.5. Figure 1.8 shows that all orbits diverge to infinity.

REMARK 1.7. By replacing $V(x)$ by $-V(x)$, the same proof of instability as in Theorem 1.5 can be developed for the function $V(x)$ satisfying the properties

(I1') $V(x_*) = 0$ and $V(x) < 0$ for every $x \in D_2$;
(I2') $\nabla V(x) \cdot f(x) < 0$ for every $x \in D_2$,

in another subset $D_2 \subset D$ with $x_* \in \partial D_2$.

Since the modified property (I2') in Remark 1.7 is satisfied for V in the gradient dynamical system (1.24), this equivalent version of Theorem 1.5 proves the instability of isolated maxima and saddle points of V with either $D_2 \equiv D\backslash\{x_*\}$ for an isolated maximum or with D_2 being a proper subset of D for the isolated saddle point. By Theorem 1.5, we obtain the following corollary.

COROLLARY 1.3. *Let x_* be an isolated maximum or an isolated saddle point of V in the gradient system* (1.24). *Then, x_* is unstable.*

Corollaries 1.1 and 1.3 clarify stability or instability of every isolated equilibrium point of the gradient dynamical system (1.24). For the Hamiltonian dynamical system (1.25), stability of isolated minimum and maximum points is given by Corollary 1.2. It is natural to anticipate instability of isolated saddle points of V. The following example shows that saddle points are unstable in degree-one Hamiltonian systems. This is essentially due to the fact that in a two-dimensional phase space, there is not enough room for nonintersecting orbits, given by the level curves of a

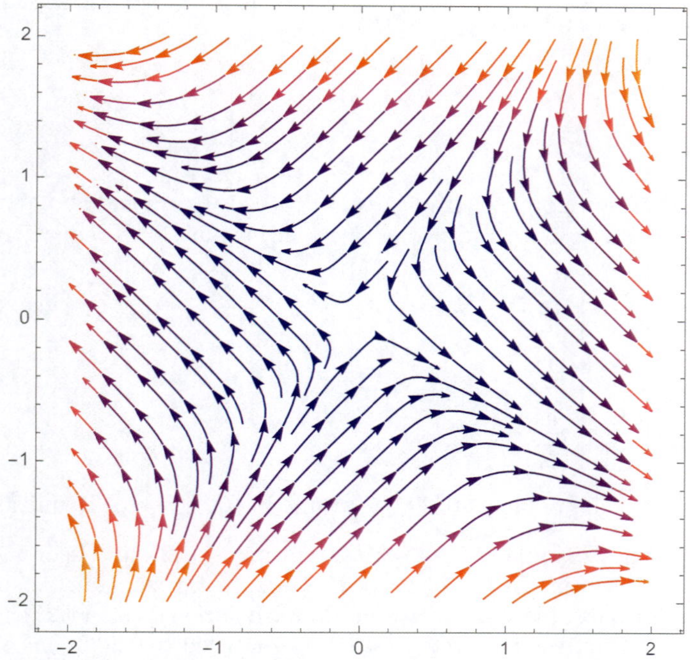

FIGURE 1.8. Phase portrait of system (1.26) with the unstable point $(0,0)$ on the (x,y) plane.

sign-indefinite Hamiltonian function V, to remain in a neighborhood of a saddle point.

EXAMPLE 1.8. Consider the Duffing oscillator (1.2) in Example 1.1. The equilibrium point $(0,0)$ is a saddle point of the Hamiltonian function $V(x,y)$ in (1.4) and it is unstable because the level curves of $V(x,y)$ diverge from $(0,0)$, see Figure 1.1. In comparison, the center points $(\pm 1, 0)$ are minima of $V(x,y)$. By Corollary 1.2, the center points $(\pm 1, 0)$ are stable and the level curves of $V(x,y)$ surrounds $(\pm 1, 0)$, see Figure 1.1.

In contrast to degree-one Hamiltonian systems, saddle points of a degree-m Hamiltonian system are not necessarily unstable as the orbits along the level surface have extra space in $\mathbb{R}^{2m}$ to avoid self-intersections and to stay in a local neighborhood of the saddle point if $m \geq 2$. The following example illustrates stability of saddle points for the degree-two Hamiltonian system.

EXAMPLE 1.9. Consider the Hamiltonian system (1.25) with the conserved energy

$$V(x_1, y_1, x_2, y_2) = \frac{1}{2}(x_1^2 + y_1^2) - \frac{1}{2}(x_2^2 + y_2^2) \tag{1.27}$$

and the skew-symmetric matrix J in the form

$$J = \begin{bmatrix} 0 & 1 & 0 & 0 \\ -1 & 0 & 0 & 0 \\ 0 & 0 & 0 & 1 \\ 0 & 0 & -1 & 0 \end{bmatrix}. \tag{1.28}$$

The only equilibrium point $(x_1, y_1, x_2, y_2) = (0, 0, 0, 0)$ is a saddle point of V since its Hessian has two positive and two negative eigenvalues. The equations of motion

$$\begin{cases} \dfrac{dx_1}{dt} = y_1 \\ \dfrac{dy_1}{dt} = -x_1 \\ \dfrac{dx_2}{dt} = -y_2 \\ \dfrac{dy_2}{dt} = x_2 \end{cases} \quad \Rightarrow \quad \begin{cases} \dfrac{d^2x_1}{dt^2} + x_1 = 0 \\ \dfrac{d^2x_2}{dt^2} + x_2 = 0 \end{cases} \tag{1.29}$$

imply that the saddle point $(0, 0, 0, 0)$ is stable in the sense of Definition 1.1.

REMARK 1.8. Example 1.9 shows that property (I2) in Theorem 1.5 cannot be relaxed to the form $\nabla V(x) \cdot f(x) \geq 0$ because the constant m in the proof may be zero. In particular, it is zero for Hamiltonian systems (1.25). As a result, saddle points of Hamiltonian systems are not governed by the definite stability or instability conclusions obtained with energy function methods.

1.4. Saddle points of Hamiltonian systems

The main focus of this monograph is the stability analysis of nonlinear waves, that is, solutions to nonlinear wave equations which can be cast as equilibrium points of infinite-dimensional Hamiltonian systems. With a rare exception, these nonlinear waves are saddle points of an associated energy functional. As a preparatory step, it is therefore important to first clarify the linear and nonlinear stability of saddle points in finite-dimensional Hamiltonian systems.

We start with the linear stability of saddle points where some results can be obtained with the Hamiltonian–Krein theorem. We then turn to the nonlinear stability of saddle points for which no definite conclusions can be drawn without specific analysis of nonlinear terms of the Hamiltonian system.

1.4.1. Linear stability. Let x_* be a critical point of the Hamiltonian system (1.25) with $V \in C^2(\mathbb{R}^n, \mathbb{R})$ and consider a perturbation $\xi := x - x_*$. Taylor expansion of the Hamiltonian function gives

$$V(x) = V(x_*) + \frac{1}{2}\langle \mathcal{L}\xi, \xi \rangle + V_R(\xi), \tag{1.30}$$

where the symmetric matrix $\mathcal{L} = D^2V(x_*)$ is the Hessian of V at x_*, $V_R(\xi)$ is a superquadratic remainder term near $\xi = 0$, and we have used that $\nabla V(x_*) = 0$. It follows from (1.25) and (1.30) that the linear evolution is described by the system

$$\frac{d\xi}{dt} = J\mathcal{L}\xi, \tag{1.31}$$

where J is an invertible and skew-symmetric matrix such that $J^T = -J$. Note that $\langle x, y \rangle := \sum_{j=1}^n x_j \bar{y}_j$ is the inner product on $\mathbb{C}^n$. Although $\xi \in \mathbb{R}^n$ and $J, \mathcal{L} \in$

$\mathbb{M}^{n\times n}$ are real matrices, the extension of the inner product to $\mathbb{C}^n$ is needed for understanding the eigenvectors of $J\mathcal{L}$.

The stability of the equilibrium point 0 in the linear Hamiltonian system (1.31) is determined by the eigenvalues $\sigma(J\mathcal{L}) = \{\lambda_1, \lambda_2, \cdots, \lambda_n\}$ of $J\mathcal{L}$ according to Theorem 1.1 with $A := J\mathcal{L}$. Eigenvalues in the linear Hamiltonian system (1.31) are symmetric about the real and imaginary axes, as described in the following lemma.

LEMMA 1.1. *If $\lambda_0 \in \mathbb{R} \cup (i\mathbb{R})\backslash\{0\}$ is an eigenvalue, then $-\lambda_0$ is also an eigenvalue of the same algebraic multiplicity. If $\lambda_0 \in \mathbb{C} \setminus \{\mathbb{R} \cup (i\mathbb{R})\}$ is a complex eigenvalue, then $\bar{\lambda}_0$, $-\lambda_0$, and $-\bar{\lambda}_0$ are also eigenvalues of the same algebraic multiplicity.*

PROOF. Since J and $\mathcal{L}$ are real matrices and eigenvalues of $J\mathcal{L}$ are roots of the characteristic polynomial

$$P(\lambda) := \det(J\mathcal{L} - \lambda I)$$

with real coefficients, eigenvalues of the same algebraic multiplicity are symmetric about the real axis, so that both λ and $\bar{\lambda}$ are roots of the characteristic polynomial P.

Since roots of $\det(J\mathcal{L} - \bar{\lambda}I) = 0$ and $\det((J\mathcal{L})^T - \bar{\lambda}I) = 0$ coincide and $(J\mathcal{L})^T = -\mathcal{L}J$, eigenvalues of $J\mathcal{L}$ can equivalently be found from roots of $\det(\mathcal{L}J + \bar{\lambda}I)$. Since J is invertible and

$$\det(\mathcal{L}J + \bar{\lambda}I) = \det(J^{-1}(J\mathcal{L} + \bar{\lambda}I)J) = \det(J\mathcal{L} + \bar{\lambda}I),$$

eigenvalues of the same algebraic multiplicity are symmetric about the imaginary axis so that both λ and $-\bar{\lambda}$ are roots of the characteristic polynomial P. Combining the two symmetries yields the assertion of the lemma. □

Theorem 1.1 and Lemma 1.1 give the following result for linear Hamiltonian systems.

THEOREM 1.6. *Let $\{\lambda_1, \lambda_2, \ldots, \lambda_n\}$ be the eigenvalues of $J\mathcal{L}$. The zero equilibrium point in the linear Hamiltonian system* (1.31) *is stable if and only if $\lambda_j \in i\mathbb{R}$ for all j and each eigenvalue λ_j is semi-simple. It is unstable if and only if either $\lambda_j \notin i\mathbb{R}$ for at least one j or $\lambda_j \in i\mathbb{R}$ for all j but at least one eigenvalue λ_j is defective.*

The symmetric matrix $\mathcal{L}$ is the Hessian of the energy function V at the equilibrium point x_*, see (1.30). If x_* is an isolated minimum or maximum of V, then $\mathcal{L}$ is positive- or negative-definite, respectively. Hence, the origin is stable by Corollary 1.2.

Suppose now that x_* is an isolated saddle point of V, so that the matrix $\mathcal{L}$ has $n(\mathcal{L})$ negative and $p(\mathcal{L})$ positive eigenvalues, taking into account their multiplicities. We assume that $\mathcal{L}$ has no zero eigenvalue so that $n(\mathcal{L}) + p(\mathcal{L}) = 2m$, where $m := \frac{n}{2}$ is the degree of the linear Hamiltonian system (1.31).

To determine the stability of the equilibrium point x_* of the linear Hamiltonian system (1.31) we need to analyze the eigenvalues of $J\mathcal{L}$. Since the eigenvalues of $\mathcal{L}$ are known, the main question is whether there exists a relation between the real eigenvalues of $\mathcal{L}$ and the complex eigenvalues of $J\mathcal{L}$. In order to answer this question, we explain the concept of Krein signatures named after M. G. Krein [**111, 137**]. This concept was introduced by R. MacKay [**192**] in the context of stability of the linear Hamiltonian systems.

DEFINITION 1.7. Let $\lambda_0 \in \sigma(J\mathcal{L})$ be an eigenvalue of algebraic multiplicity m_0 and $\{u_j\}_{j=1}^{m_0}$ be the basis of generalized eigenvectors of $J\mathcal{L} - \lambda_0 I$. The matrix $K(\lambda_0) \in \mathbb{M}^{m_0 \times m_0}$ with elements given by

$$[K(\lambda_0)]_{ij} := \langle \mathcal{L} u_i, u_j \rangle, \quad 1 \le i, j \le m_0,$$

is called the **Krein matrix** associated with the eigenvalue λ_0.

The following lemma describes the main properties of the Krein matrix $K(\lambda_0)$. In order to simplify the arguments, we assume that the defective eigenvalue $\lambda_0 \in \sigma(J\mathcal{L})$ has geometric multiplicity one and algebraic multiplicity m_0, so that we can use the Jordan chain of equations (1.10) in Definition 1.4.

LEMMA 1.2. *Let $\lambda_0 \in \sigma(J\mathcal{L})$ be an eigenvalue of geometric multiplicity one and algebraic multiplicity m_0. Then, we have:*

(K1) $K(\lambda_0) = 0$ *if* $\lambda_0 \in \mathbb{C}\backslash\{i\mathbb{R}\}$.
(K2) $K(\lambda_0)$ *is Hermitian and invertible if* $\lambda_0 \in i\mathbb{R}\backslash\{0\}$.
(K3) *If* m_0 *is even, then* $p_{\lambda_0} = n_{\lambda_0} = \frac{m_0}{2}$.
(K4) *If* m_0 *is odd, then*

$$p_{\lambda_0} = n_{\lambda_0} + 1 = \frac{m_0 + 1}{2} \quad \text{if } \langle \mathcal{L} u_{\frac{m_0+1}{2}}, u_{\frac{m_0+1}{2}} \rangle > 0 \tag{1.32}$$

or

$$p_{\lambda_0} = n_{\lambda_0} - 1 = \frac{m_0 - 1}{2} \quad \text{if } \langle \mathcal{L} u_{\frac{m_0+1}{2}}, u_{\frac{m_0+1}{2}} \rangle < 0, \tag{1.33}$$

where p_{λ_0} (resp. n_{λ_0}) denotes the number of positive (resp. negative) eigenvalues of $K(\lambda_0)$.

PROOF. In agreement with Definition 1.4, we define the canonical basis $\{u_j\}_{j=1}^{m_0}$ of the generalized eigenvectors of $J\mathcal{L} - \lambda_0 I$ by the Jordan chain of equations

$$J\mathcal{L} u_j = \lambda_0 u_j + u_{j-1}, \quad 1 \le j \le m_0, \tag{1.34}$$

where $u_0 = 0$. It follows from (1.34) that

$$\mathcal{L} J \mathcal{L} u_j = \lambda_0 \mathcal{L} u_j + \mathcal{L} u_{j-1}, \quad 1 \le j \le m_0, \tag{1.35}$$

so that

$$\langle \mathcal{L} J \mathcal{L} u_i, u_j \rangle = \lambda_0 \langle \mathcal{L} u_i, u_j \rangle + \langle \mathcal{L} u_{i-1}, u_j \rangle, \quad 1 \le i, j \le m_0. \tag{1.36}$$

Since $\mathcal{L}^T = \mathcal{L}$ and $J^T = -J$, we obtain

$$\langle \mathcal{L} J \mathcal{L} u_i, u_j \rangle = -\langle u_i, \mathcal{L} J \mathcal{L} u_j \rangle = -\overline{\langle \mathcal{L} J \mathcal{L} u_j, u_i \rangle},$$

which yields with the help of (1.36) that

$$(\lambda_0 + \bar{\lambda}_0) \langle \mathcal{L} u_i, u_j \rangle + \langle \mathcal{L} u_{i-1}, u_j \rangle + \langle \mathcal{L} u_i, u_{j-1} \rangle = 0, \quad 1 \le i, j \le m_0. \tag{1.37}$$

These prerequisites allow us to prove properties (K1)–(K4).

PROOF OF (K1). If $\lambda_0 \in \mathbb{C}\backslash\{i\mathbb{R}\}$, then $\lambda_0 + \bar{\lambda}_0 \neq 0$. We obtain from (1.37) by induction that

$$\langle \mathcal{L} u_i, u_j \rangle = 0, \quad 1 \le i, j \le m_0. \tag{1.38}$$

Indeed, if $i = j = 1$, then $\langle \mathcal{L} u_1, u_1 \rangle = 0$ since $u_0 = 0$. Incrementing i and j in (1.37) by one gives $\langle \mathcal{L} u_2, u_1 \rangle = \langle \mathcal{L} u_1, u_2 \rangle = 0$, and so on for all $1 \le i, j \le m_0$. Hence, $K(\lambda_0) = 0$ and property (K1) is proven. □

PROOF OF (K2). If $\lambda_0 \in i\mathbb{R}\backslash\{0\}$, then $\lambda_0 + \bar{\lambda}_0 = 0$. We obtain from (1.37) by induction that

$$\langle \mathcal{L}u_i, u_j \rangle = 0, \quad 2 \le i + j \le m_0. \tag{1.39}$$

Indeed, if $j = 1$, then $\langle \mathcal{L}u_i, u_1 \rangle = 0$ for all $1 \le i \le m_0 - 1$ since $u_0 = 0$ and $\lambda_0 + \bar{\lambda}_0 = 0$. Incrementing j in (1.37) by one gives (1.39). Hence, $K(\lambda_0)$ contains zero entries above the main antidiagonal:

$$K(\lambda_0) = \begin{bmatrix} 0 & 0 & \cdots & 0 & \langle \mathcal{L}u_1, u_{m_0} \rangle \\ 0 & 0 & \cdots & \langle \mathcal{L}u_2, u_{m_0-1} \rangle & \langle \mathcal{L}u_2, u_{m_0} \rangle \\ \vdots & \vdots & \vdots & \vdots & \vdots \\ 0 & \langle \mathcal{L}u_{m_0-1}, u_2 \rangle & \cdots & \langle \mathcal{L}u_{m_0-1}, u_{m_0-1} \rangle & \langle \mathcal{L}u_{m_0-1}, u_{m_0} \rangle \\ \langle \mathcal{L}u_{m_0}, u_1 \rangle & \langle \mathcal{L}u_{m_0}, u_2 \rangle & \cdots & \langle \mathcal{L}u_{m_0}, u_{m_0-1} \rangle & \langle \mathcal{L}u_{m_0}, u_{m_0} \rangle \end{bmatrix}.$$

Furthermore, it follows from (1.37) that

$$a_0 := \langle \mathcal{L}u_1, u_{m_0} \rangle = -\langle \mathcal{L}u_2, u_{m_0-1} \rangle = \cdots = (-1)^{m_0-1} \langle \mathcal{L}u_{m_0}, u_1 \rangle \tag{1.40}$$

and similar relations hold for the other nonzero elements of the anti-diagonals of $K(\lambda_0)$. We claim that $a_0 \neq 0$. Indeed, for every $f \in \mathbb{C}^n$ we have

$$\begin{aligned} 0 &= -\langle (J\mathcal{L} - \lambda_0 I)u_1, \mathcal{L}f \rangle \\ &= \langle \mathcal{L}u_1, J\mathcal{L}f \rangle + \lambda_0 \langle \mathcal{L}u_1, f \rangle \\ &= \langle \mathcal{L}u_1, (J\mathcal{L} - \lambda_0 I)f \rangle, \end{aligned}$$

due to $\lambda_0 + \bar{\lambda}_0 = 0$. Hence, $\mathrm{Ran}(J\mathcal{L} - \lambda_0 I)$ is orthogonal to $\mathcal{L}u_1$. If $a_0 = \langle \mathcal{L}u_1, u_{m_0} \rangle = 0$, u_{m_0} is also orthogonal to $\mathcal{L}u_1$ and since the subspace of $(J\mathcal{L} - \lambda_0 I)$ is one-dimensional (geometric multiplicity of λ_0 is one), then $u_{m_0} \in \mathrm{Ran}(J\mathcal{L} - \lambda_0 I)$ by the Fredholm Alternative Theorem B.2. Hence, there exists $u_{m_0+1} \in \mathbb{C}^n$ such that

$$J\mathcal{L}u_{m_0+1} = \lambda_0 u_{m_0+1} + u_{m_0},$$

i.e., relation (1.34) holds also for $j = m_0 + 1$, which implies that the algebraic multiplicity of λ_0 exceeds m_0, a contradiction. Therefore, $a_0 \neq 0$ so that $K(\lambda_0)$ is invertible. Since $\mathcal{L}$ is symmetric, $K(\lambda_0)$ is Hermitian, so that property (K2) is proven. □

PROOF OF (K3) AND (K4). It remains to compute n_{λ_0} and p_{λ_0} satisfying $n_{\lambda_0} + p_{\lambda_0} = m_0$. Consider the canonical basis $\{e_1, e_2, \ldots, e_{m_0}\}$ in $\mathbb{R}^{m_0}$. Let Π_0 be a projection matrix from $\mathbb{R}^{m_0}$ to $\mathrm{Span}\{e_1, e_2, \ldots, e_{d_0}\}$, where $d_0 = \frac{m_0}{2}$ if m_0 is even and $d_0 = \frac{m_0-1}{2}$ if m_0 is odd. Since $\Pi_0 K(\lambda_0) \Pi_0 = 0$, all d_0 eigenvalues of $\Pi_0 K(\lambda_0) \Pi_0$ are zero. By the Cauchy interlacing theorem (Theorem B.6), at least d_0 eigenvalues of $K(\lambda_0)$ are nonpositive. Since $K(\lambda_0)$ is invertible, zero is not an eigenvalue. Hence at least d_0 eigenvalue of $K(\lambda_0)$ are strictly negative, so that $n_{\lambda_0} \ge d_0$. The same argument with $-K(\lambda_0)$ shows that $p_{\lambda_0} \ge d_0$.

If m_0 is even, then n_{λ_0} and p_{λ_0} cannot exceed d_0 since otherwise $n_{\lambda_0} + p_{\lambda_0} > 2d_0 = m_0$ but $n_{\lambda_0} + p_{\lambda_0} = m_0$. Hence $n_{\lambda_0} = p_{\lambda_0} = d_0 = \frac{m_0}{2}$ and property (K3) is proven.

If m_0 is odd, then it only follows from this count that $n_{\lambda_0}, p_{\lambda_0} \ge \frac{m_0-1}{2}$. In order to make it more precise, let us compute $\det K(\lambda_0)$ which we can do for both even and odd m_0.

Since $\mathcal{L}$ is symmetric, $a_0 \in i\mathbb{R}$ if m_0 is even and $a_0 \in \mathbb{R}$ if m_0 is odd. It follows from (1.40) that if m_0 is even, then

$$\langle \mathcal{L}u_1, u_{m_0}\rangle = -\langle \mathcal{L}u_{m_0}, u_1\rangle = -\overline{\langle \mathcal{L}u_1, u_{m_0}\rangle},$$

whereas if m_0 is odd, then

$$\langle \mathcal{L}u_1, u_{m_0}\rangle = \langle \mathcal{L}u_{m_0}, u_1\rangle = \overline{\langle \mathcal{L}u_1, u_{m_0}\rangle}.$$

By the Laplace cofactor expansion, we compute $\det K(\lambda_0)$ separately in the two cases and find that

$$\det K(\lambda_0) = \begin{cases} |a_0|^{m_0}(-1)^{\frac{m_0(m_0+3)}{2}}, & m_0 \text{ is even}, \\ a_0^{m_0}(-1)^{\frac{m_0-1}{2}}(-1)^{\frac{m_0(m_0+3)}{2}}, & m_0 \text{ is odd}. \end{cases} \tag{1.41}$$

If m is odd, the sign of $\det K(\lambda_0)$ is $(-1)^{n_{\lambda_0}}$. Comparison with (1.41) shows that $n_{\lambda_0} = d_0 = \frac{m_0-1}{2}$ and $p_{\lambda_0} = m_0 - d_0 = \frac{m_0+1}{2}$ if the sign of $a_0(-1)^{\frac{m_0+3}{2}}$ is positive, whereas $n_{\lambda_0} = d_0 + 1 = \frac{m_0+1}{2}$ and $p_{\lambda_0} = m_0 - d_0 = \frac{m_0-1}{2}$ if the sign of $a_0(-1)^{\frac{m_0+3}{2}}$ is negative. However, the sign of $a_0(-1)^{\frac{m_0+3}{2}}$ coincides with the sign of

$$\langle \mathcal{L}u_{\frac{m_0+1}{2}}, u_{\frac{m_0+1}{2}}\rangle.$$

This yields (1.32) and (1.33) so that property (K4) is proven. □

EXAMPLE 1.10. If $m_0 = 1$, the eigenvalue λ_0 is simple and u_1 is the eigenvector satisfying $J\mathcal{L}u_1 = \lambda_0 u_1$. Then, $K(\lambda) = \langle \mathcal{L}u_1, u_1\rangle$ is either positive ($p_{\lambda_0} = 1$, $n_{\lambda_0} = 0$) or negative ($p_{\lambda_0} = 0$, $n_{\lambda_0} = 1$). Accordingly, the Krein signature for the eigenvalue λ_0 is positive in the former case and negative in the latter case.

EXAMPLE 1.11. If $m_0 = 2$, then $p_{\lambda_0} = n_{\lambda_0} = 1$ independenly of the sign of the nonzero $\langle \mathcal{L}u_2, u_1\rangle$, where u_2 is the generalized eigenvector satisfying $J\mathcal{L}u_2 = \lambda_0 u_2 + u_1$.

The following theorem presents the completeness relations between eigenvalues of $\mathcal{L}$ and eigenvalues of $J\mathcal{L}$. The result is usually referred as the **Hamiltonian–Krein theorem**. This theory was pioneered by L. S. Pontryagin [**225**] and later elaborated in the scientific school of M. G. Krein [**111, 137**]. The latest applications of the Hamiltonian–Krein theorem were inspired by particular examples in [**76, 123, 150, 213**] and were generalized in [**55, 151, 180**].

THEOREM 1.7. *Assume that $\mathcal{L}$ is invertible with $p(\mathcal{L})$ positive and $n(\mathcal{L})$ negative eigenvalues satisfying $p(\mathcal{L}) + n(\mathcal{L}) = 2m$. Eigenvalues of $J\mathcal{L}$ satisfy the following completeness relations:*

$$\begin{cases} p(\mathcal{L}) = N_{\text{real}} + 2N_{\text{comp}} + 2\sum_{\lambda_0 \in i\mathbb{R}_+} p_{\lambda_0}, \\ n(\mathcal{L}) = N_{\text{real}} + 2N_{\text{comp}} + 2\sum_{\lambda_0 \in i\mathbb{R}_+} n_{\lambda_0}, \end{cases} \tag{1.42}$$

where N_{real} is the number of real positive eigenvalues of $J\mathcal{L}$, $2N_{\text{comp}}$ is the (even) number of complex eigenvalues of $J\mathcal{L}$ with positive real part, and p_{λ_0} (resp. n_{λ_0}) is the number of positive (resp. negative) eigenvalues of the Krein matrix $K(\lambda_0)$ for each eigenvalue $\lambda_0 \in i\mathbb{R}_+$ of $J\mathcal{L}$, taking into account algebraic multiplicities.

PROOF. We again simplify the proof by assuming that each eigenvalue of $J\mathcal{L}$ has geometric multiplicity one. The main idea of the proof is to compute the number of positive and negative eigenvalues of the coefficient matrix obtained from the quadratic form $\langle \mathcal{L}u, u\rangle$ with $u \in \mathbb{R}^n$ represented over the basis of (generalized) eigenvectors of $J\mathcal{L}$.

Let $\{u_i\}_{i=1}^{m_0}$ and $\{v_j\}_{j=1}^{k_0}$ be the basis of (generalized) eigenvectors of $J\mathcal{L}$ for the eigenvalue λ_0 and μ_0, respectively, satisfying (1.34). Repeating the same computation that lead to (1.37) yields

$$(\lambda_0 + \bar{\mu}_0)\langle \mathcal{L}u_i, v_j\rangle + \langle \mathcal{L}u_{i-1}, v_j\rangle + \langle \mathcal{L}u_i, v_{j-1}\rangle = 0, \quad 1 \le i \le m_0, \ 1 \le j \le k_0.$$

If $\lambda_0 + \bar{\mu}_0 \neq 0$, we obtain similarly to (1.38) that

$$\langle \mathcal{L}u_i, v_j\rangle = 0, \quad 1 \le i \le m_0, \ 1 \le j \le k_0.$$

Hence, the representation of $\langle \mathcal{L}u, u\rangle$ with $u \in \mathbb{R}^n$ over the basis of (generalized) eigenvectors of $J\mathcal{L}$ is block-diagonalized and the nonzero blocks correspond to either a pair of eigenvalues λ_0 and $\mu_0 = -\bar{\lambda}_0$ or an eigenvalue $\lambda_0 \in i\mathbb{R}$. Since J and $\mathcal{L}$ are invertible, no zero eigenvalues exist and therefore $\lambda_0 \neq 0$.

For each eigenvalue $\lambda_0 \in i\mathbb{R}\backslash\{0\}$, the numbers of positive and negative eigenvalues of $K(\lambda_0)$ are given by properties (K2), (K3), and (K4) in Lemma 1.2.

For each eigenvalue $\lambda_0 \in \mathbb{C}\backslash\{i\mathbb{R}\}$, property (K1) in Lemma 1.2 states that $K(\lambda_0) = 0$. However, by Lemma 1.1, $\mu_0 = -\bar{\lambda}_0$ is also an eigenvalue of $J\mathcal{L}$ of the same algebraic multiplicity as λ_0. In addition, if $\lambda \in \mathbb{C}\backslash\{\mathbb{R}\cup i\mathbb{R}\}$, then $\bar{\lambda}_0$ and $-\bar{\lambda}_0$ are also eigenvalues of the same algebraic multiplicities. For simplicity, we consider the case $\lambda_0 \in \mathbb{R}$ only; the case $\lambda_0 \in \mathbb{C}\backslash\{\mathbb{R}\cup i\mathbb{R}\}$ can be considered similarly with longer computations.

For $\lambda_0 \in \mathbb{R}$, let $\{u_i\}_{i=1}^{m_0}$ be the basis of (generalized) eigenvectors in $J\mathcal{L}$ for the eigenvalue λ_0 satisfying (1.34) and let $\{u_i^*\}_{i=1}^{m_0}$ be the respective basis for the eigenvalue $\mu_0 = -\lambda_0$ satisfying

$$J\mathcal{L}u_j^* = -\lambda_0 u_j^* - u_{j-1}^*, \quad 1 \le j \le m_0.$$

By taking linear combinations, we compute

$$J\mathcal{L}(u_j \pm u_j^*) = \lambda_0(u_j \mp u_j^*) + (u_{j-1} \mp u_{j-1}^*), \quad 1 \le j \le m_0.$$

Repeating computations as in (1.35) gives for $1 \le i, j \le m_0$:

$$\begin{aligned} &\lambda_0 \left[\langle \mathcal{L}(u_i \pm u_i^*), (u_j \pm u_j^*)\rangle + \langle \mathcal{L}(u_i \mp u_i^*), (u_j \mp u_j^*)\rangle\right] \\ &\quad + \langle \mathcal{L}(u_{i-1} \pm u_{i-1}^*), (u_j \pm u_j^*)\rangle + \langle \mathcal{L}(u_i \pm u_i^*), (u_{j-1} \pm u_{j-1}^*)\rangle = 0, \end{aligned}$$

and

$$\begin{aligned} &\lambda_0 \left[\langle \mathcal{L}(u_i \pm u_i^*), (u_j \mp u_j^*)\rangle + \langle \mathcal{L}(u_i \mp u_i^*), (u_j \pm u_j^*)\rangle\right] \\ &\quad + \langle \mathcal{L}(u_{i-1} \pm u_{i-1}^*), (u_j \mp u_j^*)\rangle + \langle \mathcal{L}(u_i \pm u_i^*), (u_{j-1} \mp u_{j-1}^*)\rangle = 0. \end{aligned}$$

Since $\lambda_0 \in \mathbb{R}$, $u_j, u_j^* \in \mathbb{R}^n$ for all j. By induction, it follows from these relations that

$$\langle \mathcal{L}(u_i + u_i^*), (u_j + u_j^*)\rangle = -\langle \mathcal{L}(u_i - u_i^*), (u_j - u_j^*)\rangle, \quad 1 \le i, j \le m_0$$

and

$$\langle \mathcal{L}(u_{i-1} \pm u_{i-1}^*), (u_j \mp u_j^*)\rangle = 0, \quad 1 \le i, j \le m_0.$$

As a result, $\langle \mathcal{L}u, u\rangle$ becomes block-diagonal in vectors $\{u_j + u_j^*, u_j - u_j^*\}_{j=1}^{m_0}$ and the number of positive and negative eigenvalues of $\langle \mathcal{L}u, u\rangle$ over $\{u_j + u_j^*, u_j - u_j^*\}_{j=1}^{m_0}$ coincide. Moreover, since $\{u_1, u_2, \dots, u_{m_0}, u_1^*, u_2^*, \dots, u_{m_0}^*\}$ is a basis of (generalized) eigenvectors of $J\mathcal{L}J\mathcal{L}$ for the eigenvalue λ_0^2, the matrix with the elements

$$\langle \mathcal{L}(u_i + u_i^*), (u_j + u_j^*)\rangle, \quad 1 \le i, j \le m_0$$

is invertible. Indeed, if it had a zero eigenvalue, then the algebraic multiplicity of eigenvalue λ_0^2 for $J\mathcal{L}J\mathcal{L}$ would exceed $2m_0$. Thus, $\langle \mathcal{L}u, u\rangle$ on the basis $\{u_j + u_j^*, u_j - u_j^*\}_{j=1}^{m_0}$ has exactly m_0 positive and m_0 negative eigenvalues.

We are now ready to complete the proof of Theorem 1.7. There exist two orthogonal bases for $\langle \mathcal{L}u, u\rangle$ for $u \in \mathbb{R}^n$:

- One basis is given by the eigenvectors of $\mathcal{L}$ for $p(\mathcal{L})$ positive and $n(\mathcal{L})$ negative eigenvalues of $\mathcal{L}$.
- Another basis is given by (generalized) eigenvectors of $J\mathcal{L}$ after using the Gram–Schmidt orthogonalization procedure with respect to $\langle \mathcal{L}\cdot, \cdot\rangle$.

By Sylvester's inertia theorem (Theorem B.11), the total number of positive and negative terms in the two equivalent diagonalizations of the quadratic form $\langle \mathcal{L}u, u\rangle$ must coincide. Counting all positive and negative eigenvalues in the Krein matrix $K(\lambda_0)$ for $\lambda_0 \in i\mathbb{R}$ and the extended Krein matrix $K(\lambda_0)$ over $\{u_j + u_j^*, u_j - u_j^*\}_{j=1}^{m_0}$ for $\lambda_0 \in \mathbb{R}$ (and similarly for $\lambda_0 \in \mathbb{C}\backslash\{i\mathbb{R} \cup \mathbb{R}\}$) gives the completeness relations (1.42). □

REMARK 1.9. Since all numbers in the completeness relations (1.42) are positive and the eigenvalues are symmetric by Lemma 1.1, if either $p(\mathcal{L}) = 0$ or $n(\mathcal{L}) = 0$, then $N_{\text{real}} = 0$ and $N_{\text{comp}} = 0$, and therefore, $\lambda_j \in i\mathbb{R}$ for each eigenvalue λ_j of $J\mathcal{L}$. Moreover, $p_{\lambda_0} = 0$ if $p(\mathcal{L}) = 0$ and $n_{\lambda_0} = 0$ if $n(\mathcal{L}) = 0$ for every $\lambda_0 \in i\mathbb{R}_+$, which implies by statement (K4) of Lemma 1.2 that each eigenvalue λ_0 with $\text{Re}(\lambda_0) = 0$ is semi-simple. Hence, the zero equilibrium point is stable in the linear Hamiltonian system (1.31) if either $p(\mathcal{L}) = 0$ or $n(\mathcal{L}) = 0$ by Theorem 1.1. This conclusion agrees with Corollary 1.2 since $n(\mathcal{L}) = 0$ (resp. $p(\mathcal{L}) = 0$) corresponds to x_* being an isolated minimum (resp. maximum) of the energy function V in (1.30).

Assume that x_* is an isolated saddle point of V. Since $p(\mathcal{L}) + n(\mathcal{L}) = n = 2m$ is even, $p(\mathcal{L})$ and $n(\mathcal{L})$ can either both be odd or both be even. If they are both odd, then $N_{\text{real}} \ge 1$ in the count (1.42). This allows us to formulate the following corollary from Theorems 1.3 and 1.7.

COROLLARY 1.4. *Let x_* be an isolated saddle point of $V(x)$ in the Hamiltonian system* (1.25). *Then, x_* is unstable if $p(\mathcal{L})$ and $n(\mathcal{L})$ are odd.*

The only inconclusive scenario is the case of an isolated saddle point of V if $p(\mathcal{L})$ and $n(\mathcal{L})$ are even. Example 1.9 suggests that such an isolated saddle point can be linearly stable. The next example shows that such an isolated saddle point can also be linearly unstable and that the completeness relations (1.42) can be satisfied with three posible cases if $p(\mathcal{L}) = n(\mathcal{L}) = 2$.

EXAMPLE 1.12. Example 1.9 corresponds to the choice $n(\mathcal{L}) = p(\mathcal{L}) = 2$, for which $N_{\text{real}} = 0$ and $N_{\text{comp}} = 0$, and $p_{\lambda_0} = n_{\lambda_0} = 1$ for the double eigenvalue $\lambda_0 = i$. It is easy to construct two more examples to exemplify the completeness relations (1.42).

If the conserved energy with $n(\mathcal{L}) = p(\mathcal{L}) = 2$ is

$$V(x_1, y_1, x_2, y_2) = \frac{1}{2}(x_1^2 - y_1^2) + \frac{1}{2}(x_2^2 - y_2^2) \tag{1.43}$$

and the skew-symmetric matrix J is the same as in (1.28), then the equations of motion are given by

$$\begin{cases} \frac{dx_1}{dt} = -y_1, \\ \frac{dy_1}{dt} = -x_1, \\ \frac{dx_2}{dt} = -y_2, \\ \frac{dy_2}{dt} = -x_2, \end{cases} \quad \Rightarrow \quad \begin{cases} \frac{d^2x_1}{dt^2} - x_1 = 0, \\ \frac{d^2x_2}{dt^2} - x_2 = 0, \end{cases}$$

hence the saddle point $(0, 0, 0, 0)$ is unstable with $N_{\text{real}} = 2$, $N_{\text{comp}} = 0$, and no eigenvalues $\lambda_0 \in i\mathbb{R}$.

If the conserved energy with $n(\mathcal{L}) = p(\mathcal{L}) = 2$ is

$$V(x_1, y_1, x_2, y_2) = x_1 x_2 + \frac{1}{2}(y_1^2 - y_2^2) \tag{1.44}$$

and the skew-symmetric matrix J is the same as in (1.28), then the equations of motion are given by

$$\begin{cases} \frac{dx_1}{dt} = y_1, \\ \frac{dy_1}{dt} = -x_2, \\ \frac{dx_2}{dt} = -y_2, \\ \frac{dy_2}{dt} = -x_1, \end{cases} \quad \Rightarrow \quad \begin{cases} \frac{d^2x_1}{dt^2} + x_2 = 0, \\ \frac{d^2x_2}{dt^2} - x_1 = 0, \end{cases} \quad \Rightarrow \quad \frac{d^4x_1}{dt^4} + x_1 = 0,$$

hence the saddle point $(0, 0, 0, 0)$ is unstable with $N_{\text{comp}} = 1$, $N_{\text{real}} = 0$, and no eigenvalues $\lambda_0 \in i\mathbb{R}$.

1.4.2. Linear stability in canonical variables. A canonical form of the linear Hamiltonian system (1.31) with (even) $n = 2m$ is represented by the system

$$J = \begin{bmatrix} 0_m & -I_m \\ I_m & 0_m \end{bmatrix}, \quad \mathcal{L} = \begin{bmatrix} \mathcal{L}_+ & 0_m \\ 0_m & \mathcal{L}_- \end{bmatrix}, \tag{1.45}$$

where $\mathcal{L}_\pm \in \mathbb{M}^{m \times m}$ are symmetric matrices and 0_m and I_m are the zero and identity matrices in $\mathbb{M}^{m \times m}$, respectively. If $\xi = (p, q)^T \in \mathbb{R}^n$, then the quadratic form in the expansion (1.30) is given by the sum of the two terms

$$\langle \mathcal{L}\xi, \xi \rangle = \langle \mathcal{L}_+ p, p \rangle + \langle \mathcal{L}_-, q, q \rangle, \tag{1.46}$$

where the first term represents the kinetic energy written in the momentum variables $p \in \mathbb{R}^m$ and the second term represents the potential energy written in the position variables $q \in \mathbb{R}^m$. The equations of motion obtained from (1.45) and (1.46) can be written in the form

$$\frac{dq}{dt} = \mathcal{L}_+ p, \quad \frac{dp}{dt} = -\mathcal{L}_- q, \tag{1.47}$$

and the eigenvalues of $J\mathcal{L}$ are defined by the spectral problem

$$\mathcal{L}_+ p = \lambda q, \quad \mathcal{L}_- q = -\lambda p. \tag{1.48}$$

For the spectral problem (1.48) a sharper count of positive and negative eigenvalues compared to the completeness relations (1.42) was derived in [**213**]. Methods relying on the count of eigenvalues in the generalized eigenvalue problem were later generalized in [**55, 215**].

The following theorem gives the sharper relations between the eigenvalues of $\mathcal{L}_\pm$ and eigenvalues of $J\mathcal{L}$ defined by the spectral stability problem (1.48).

THEOREM 1.8. *Assume that $\mathcal{L}_\pm$ are invertible with $p(\mathcal{L}_\pm)$ positive and $n(\mathcal{L}_\pm)$ negative eigenvalues satisfying $p(\mathcal{L}_\pm) + n(\mathcal{L}_\pm) = m$. Eigenvalues of the spectral stability problem* (1.48) *satisfy the following completeness relations:*

$$\begin{cases} p(\mathcal{L}_\pm) = N^\pm_{\text{real}} + N_{\text{comp}} + \sum_{\lambda_0 \in i\mathbb{R}_+} p_{\lambda_0}, \\ n(\mathcal{L}_\pm) = N^\mp_{\text{real}} + N_{\text{comp}} + \sum_{\lambda_0 \in i\mathbb{R}_+} n_{\lambda_0}, \end{cases} \tag{1.49}$$

where N_{comp}, p_{λ_0}, n_{λ_0} are the same as in Theorem 1.7 *and N^+_{real} (resp., N^-_{real}) is the number of positive (resp. negative) eigenvalues of Krein matrices $K_+(\lambda_0)$ related to $\mathcal{L}_+$ for the eigenvalue $\lambda_0 \in \mathbb{R}_+$ of $J\mathcal{L}$, taking into account algebraic multiplicities.*

PROOF. Since both $\mathcal{L}_+$ and $\mathcal{L}_-$ are invertible, we set $z := -\lambda^2$ and rewrite the spectral problem (1.48) as the two equivalent generalized eigenvalue problems

$$\mathcal{L}_- p = z(\mathcal{L}_-)^{-1} p, \qquad \mathcal{L}_- q = z(\mathcal{L}_+)^{-1} q. \tag{1.50}$$

The mapping $z = -\lambda^2$ has the following properties:

- A pair of real eigenvalues $\{\lambda_0, -\lambda_0\}$ is mapped into one negative eigenvalue z_0.
- A pair of purely imaginary eigenvalues $\{\lambda_0, \bar{\lambda}_0\}$ is mapped into one positive eigenvalue z_0.
- A quadruplet of complex eigenvalues $\{\lambda_0, \bar{\lambda}_0, -\lambda_0, -\bar{\lambda}_0\}$ is mapped into a pair of complex-conjugate eigenvalues $\{z_0, \bar{z}_0\}$, where $z_0 = -\lambda_0^2$.

Since both $\mathcal{L}_\pm$ and $(-\mathcal{L}_\pm)^{-1}$ are symmetric, the proof of Theorem 1.8 follows from Sylvester's inertia theorem (Theorem B.11) applied to the generalized eigenvalue problems (1.50), as we show below.

In order to simplify the computational details, let us assume here that each eigenvalue λ_0 in the spectral problem (1.48) is simple. A more general case of defective eigenvalues can be handled with Jordan block decomposition as in the proof of Theorem 1.7.

The advantage of the generalized eigenvalue problems (1.50) is that we do not need to worry about the symmetric pairs of eigenvalues $\{\lambda_0, -\bar{\lambda}_0\}$. Indeed, if $\lambda_0 \in i\mathbb{R} \cup \mathbb{R}$ is simple, then the corresponding eigenvectors of the spectral problem (1.48) are $(p, \pm q)^T$ for the eigenvalues $\pm\lambda_0$, hence the eigenvalue $z_0 = -\lambda_0^2$ is simple. If $\lambda_0 \in \mathbb{C}\backslash(i\mathbb{R} \cup \mathbb{R})$ are simple, then the same symmetry gives a pair of simple eigenvalues $\{z_0, \bar{z}_0\}$ with $z_0 = -\lambda_0^2$.

It follows from (1.50) that

$$\langle \mathcal{L}_+ p_0, p_0 \rangle = z_0 \langle (\mathcal{L}_-)^{-1} p_0, p_0 \rangle, \qquad \langle \mathcal{L}_- q_0, q_0 \rangle = z_0 \langle (\mathcal{L}_+)^{-1} q_0, q_0 \rangle, \tag{1.51}$$

where $(p_0, q_0)^T$ is the eigenvector of the spectral problem (1.48) for the eigenvalue λ_0. If $\{p_1, p_2, \dots, p_m\}$ is the basis of eigenvectors of the first generalized eigenvalue problem in (1.50), then it is orthogonal with respect to both $\mathcal{L}_+$ and $(\mathcal{L}_-)^{-1}$. Similarly, if $\{q_1, q_2, \dots, q_m\}$ is the basis of eigenvectors of the second generalized eigenvalue problem in (1.50), then it is orthogonal with respect to both $\mathcal{L}_-$ and

$(\mathcal{L}_+)^{-1}$. The first problem is sufficient since by setting $\tilde{q}_0 := (\mathcal{L}_-)^{-1}p_0$, we can rewrite (1.51) in the form:

$$\langle \mathcal{L}_+ p_0, p_0\rangle = z_0 \langle \mathcal{L}_- \tilde{q}_0, \tilde{q}_0\rangle, \tag{1.52}$$

since each quadratic form is real-valued. Hence, the basis $\{q_1, q_2, \dots, q_m\}$ can be equivalently written in terms of $\{\tilde{q}_1, \tilde{q}_2, \dots, \tilde{q}_m\}$ with $\tilde{q}_j = (\mathcal{L}_-)^{-1}p_j$ being uniquely expressed in terms of the basis $\{p_1, p_2, \dots, p_m\}$.

By Fredholm's theorem (Theorem B.2), if $z_0 \in \mathbb{R}\backslash\{0\}$ is simple, then

$$\langle \mathcal{L}_+ p_0, p_0\rangle \neq 0,$$

which implies in view of (1.52) that $\langle \mathcal{L}_- \tilde{q}_0, \tilde{q}_0\rangle \neq 0$. If $z_0 > 0$ ($\lambda_0 \in i\mathbb{R}$), then both quadratic forms are of the same sign. If $z_0 < 0$ ($\lambda_0 \in \mathbb{R}$), then the two quadratic forms are of opposite signs.

If $z_0 \in \mathbb{C}\backslash\mathbb{R}$, then both quadratic forms in (1.52) are zero and we can introduce $z_0 = \mathrm{Re}(z_0) + i\,\mathrm{Im}(z_0)$ and $p_0 = \mathrm{Re}(p_0) + i\,\mathrm{Im}(p_0)$ according to the generalized eigenvalue problem

$$\begin{cases} \mathcal{L}_+ \mathrm{Re}(p_0) = \mathrm{Re}(z_0)(\mathcal{L}_-)^{-1}\mathrm{Re}(p_0) - \mathrm{Im}(z_0)(\mathcal{L}_-)^{-1}\mathrm{Im}(p_0), \\ \mathcal{L}_+ \mathrm{Im}(p_0) = \mathrm{Re}(z_0)(\mathcal{L}_-)^{-1}\mathrm{Im}(p_0) + \mathrm{Im}(z_0)(\mathcal{L}_-)^{-1}\mathrm{Re}(p_0). \end{cases} \tag{1.53}$$

Since $\langle \mathcal{L}_+ p_0, p_0\rangle = 0$, we have

$$\begin{cases} \langle \mathcal{L}_+ \mathrm{Re}(p_0), \mathrm{Re}(p_0)\rangle + \langle \mathcal{L}_+ \mathrm{Im}(p_0), \mathrm{Im}(p_0)\rangle = 0, \\ \langle \mathcal{L}_+ \mathrm{Re}(p_0), \mathrm{Im}(p_0)\rangle - \langle \mathcal{L}_+ \mathrm{Im}(p_0), \mathrm{Re}(p_0)\rangle = 0, \end{cases}$$

where each term is nonzero if z_0 is a simple eigenvalue. Hence, the quadratic form $\langle \mathcal{L}_+ p, p\rangle$ over the basis of $\{\mathrm{Re}(p_0), \mathrm{Im}(p_0)\}$ has one positive and one negative eigenvalue. Similarly, the quadratic form $\langle \mathcal{L}_- \tilde{q}, \tilde{q}\rangle$ over the basis of $\{\mathrm{Re}(\tilde{q}_0), \mathrm{Im}(\tilde{q}_0)\}$ has one positive and one negative eigenvalue.

Summarizing, there exist two orthogonal bases for $\langle \mathcal{L}_+ p, p\rangle$ and $\langle \mathcal{L}_- q, q\rangle$ for $p, q \in \mathbb{R}^m$:

- One basis is given by the eigenvectors of $\mathcal{L}_\pm$ for $p(\mathcal{L}_\pm)$ positive and $n(\mathcal{L}_\pm)$ negative eigenvalues of $\mathcal{L}_\pm$.
- Another basis is given by eigenvectors of the generalized eigenvalue problems (1.50) and (1.53).

By Sylvester's inertia theorem (Theorem B.11), the total number of positive and negative terms in the two equivalent diagonalizations of the quadratic forms $\langle \mathcal{L}_+ p, p\rangle$ and $\langle \mathcal{L}_- q, q\rangle$ must coincide. Counting all positive, negative, and complex-conjugate eigenvalues gives the completeness relations (1.49). □

REMARK 1.10. If $\lambda_0 \in \mathbb{R}_+$, then $z_0 = -\lambda_0^2 \in \mathbb{R}_-$. It follows from (1.50) that the number N_{real}^+ (resp. N_{real}^-) of positive (resp. negative) eigenvalues of Krein matrices $K_+(\lambda_0)$ related to $\mathcal{L}_+$ give the number of negative (resp. positive) eigenvalues of Krein matrices $K_-(\lambda_0)$ related to $\mathcal{L}_-$.

REMARK 1.11. Since $N_{\mathrm{real}}^+ + N_{\mathrm{real}}^- = N_{\mathrm{real}}$, adding

$$p(\mathcal{L}) = p(\mathcal{L}_+) + p(\mathcal{L}_-), \quad n(\mathcal{L}) = n(\mathcal{L}_+) + n(\mathcal{L}_-)$$

in (1.49) yields (1.42).

REMARK 1.12. Subtracting $p(\mathcal{L}_+) - p(\mathcal{L}_-)$ and $n(\mathcal{L}_+) - n(\mathcal{L}_-)$ in (1.49), we obtain

$$p(\mathcal{L}_+) - p(\mathcal{L}_-) = N^+_{\text{real}} - N^-_{\text{real}} = n(\mathcal{L}_-) - n(\mathcal{L}_+).$$

The saddle point is unstable if

$$n(\mathcal{L}_+) \neq n(\mathcal{L}_-), \tag{1.54}$$

since either N^+_{real} or N^-_{real} or both are nonzero. If $n(\mathcal{L}) = n(\mathcal{L}_+) + n(\mathcal{L}_-)$ is odd, as in Corollary 1.4, then the condition (1.54) is satisfied so that the instability criterion (1.54) is a sharper criterion for instability of isolated saddle points than the condition that $n(\mathcal{L})$ is odd.

REMARK 1.13. If the operator $\mathcal{L}_+$ corresponding to the kinetic energy is positive definite, as for instance in classical mechanical systems, then $n(\mathcal{L}_+) = 0$, so that $N^-_{\text{real}} = N_{\text{comp}} = n_{\lambda_0} = 0$ and hence $n(\mathcal{L}_-) = N^+_{\text{real}}$ follows from the completeness relations (1.49). This implies that an isolated saddle point with $n(\mathcal{L}_+) = 0$ and $n(\mathcal{L}_-) \neq 0$ is unstable. This also agrees with the condition (1.54).

REMARK 1.14. The count of unstable eigenvalues in the linear Hamiltonian system (1.47) with positive $\mathcal{L}_+$ is the same as in linear gradient systems

$$\frac{d\xi}{dt} = -\mathcal{L}\xi,$$

where $\mathcal{L}$ is a symmetric matrix. The number of unstable (real and positive) eigenvalues in both systems is given by the number of negative eigenvalues of the matrix $\mathcal{L}_-$ or $\mathcal{L}$, respectively. The only difference is that every real positive eigenvalue λ has a corresponding symmetric negative eigenvalue $-\lambda$ in the spectral problem (1.48), whereas no such symmetry exists in the spectral problem $\lambda\xi = -\mathcal{L}\xi$.

EXAMPLE 1.13. Examples 1.9 and 1.12 represent the linear Hamiltonian system (1.47) with $q = x \in \mathbb{R}^2$ and $p = y \in \mathbb{R}^2$.

For (1.27), we have

$$\mathcal{L}_+ = \begin{bmatrix} 1 & 0 \\ 0 & -1 \end{bmatrix}, \quad \mathcal{L}_- = \begin{bmatrix} 1 & 0 \\ 0 & -1 \end{bmatrix}$$

and $N^+_{\text{real}} = N^-_{\text{real}} = N_{\text{comp}} = 0$ and $p_{\lambda_0} = n_{\lambda_0} = 1$ for the double eigenvalue $\lambda_0 = i$.

For (1.43), we have

$$\mathcal{L}_+ = \begin{bmatrix} -1 & 0 \\ 0 & -1 \end{bmatrix}, \quad \mathcal{L}_- = \begin{bmatrix} 1 & 0 \\ 0 & 1 \end{bmatrix}$$

and $N^+_{\text{real}} = N_{\text{comp}} = 0$ and $N^-_{\text{real}} = 2$ with no eigenvalues $\lambda_0 \in i\mathbb{R}$.

For (1.44), we have

$$\mathcal{L}_+ = \begin{bmatrix} 1 & 0 \\ 0 & -1 \end{bmatrix}, \quad \mathcal{L}_- = \begin{bmatrix} 0 & 1 \\ 1 & 0 \end{bmatrix}$$

and $N^+_{\text{real}} = N^-_{\text{real}} = 0$ and $N_{\text{comp}} = 1$ with no eigenvalues $\lambda_0 \in i\mathbb{R}$.

1.4.3. Nonlinear stability. Persistence of the stability conclusions from linear to nonlinear Hamiltonian systems depends on the actual form of the remainder term $V_R(\xi)$ in the expansion (1.30). There are no definite results in this direction and every model may bring new discoveries.

We illustrate this unpleasant situation with two examples. Each example starts with the linear Hamiltonian system in Example 1.9, where the isolated saddle point $(0,0,0,0)$ is stable. Then we add different nonlinear terms to the Hamiltonian function and check whether the saddle point is nonlinearly stabile or unstable.

EXAMPLE 1.14. If the conserved energy is

$$V(x_1,y_1,x_2,y_2) = \frac{1}{2}(x_1^2+y_1^2) - \frac{1}{2}(x_2^2+y_2^2) + \frac{1}{4}(x_1+x_2)^4,$$

then the equations of motion are given by

$$\left\{\begin{aligned} &\frac{d^2x_1}{dt^2} + x_1 + (x_1+x_2)^3 = 0, \\ &\frac{d^2x_2}{dt^2} + x_2 - (x_1+x_2)^3 = 0. \end{aligned}\right. \tag{1.55}$$

We show that the saddle point $(0,0,0,0)$ is unstable in the nonlinear system (1.55), even though it is stable in the linear system (1.29). The coordinates $x_\pm := x_1 \pm x_2$ satisfy the system of differential equations

$$\left\{\begin{aligned} &\frac{d^2x_+}{dt^2} + x_+ = 0, \\ &\frac{d^2x_-}{dt^2} + x_- + 2x_+^3 = 0, \end{aligned}\right.$$

with the exact solution

$$\left\{\begin{aligned} &x_+(t) = ce^{it} + \bar{c}e^{-it}, \\ &x_-(t) = \tfrac{1}{4}(c^3e^{3it} + \bar{c}^3e^{-3it}) + 3it|c|^2(ce^{it} - \bar{c}e^{-it}) + de^{it} + \bar{d}e^{-it}, \end{aligned}\right.$$

where c and d are arbitrary complex coefficients which preserve reality of x_+ and x_-. Because $|x_-(t)|$ grows linearly in t, the zero equilibrium point is unstable in the nonlinear system (1.55) in the sense of Definition 1.2.

EXAMPLE 1.15. If the conserved energy is

$$V(x_1,y_1,x_2,y_2) = \frac{1}{2}(x_1^2+y_1^2) - \frac{1}{2}(x_2^2+y_2^2) + \frac{1}{2}x_1^2x_2^2,$$

then the equations of motion are given by

$$\left\{\begin{aligned} &\frac{d^2x_1}{dt^2} + x_1 + x_1x_2^2 = 0, \\ &\frac{d^2x_2}{dt^2} + x_2 - x_1^2x_2 = 0. \end{aligned}\right. \tag{1.56}$$

This example was given in [**195**], where it was claimed that the zero equilibrium point is stable in the nonlinear Hamiltonian system (1.56). We verify this claim by reducing the system to the normal form with normal coordinates

$$x_1(t) = Q_1(t)e^{it} + \bar{Q}_1(t)e^{-it}, \quad x_2(t) = Q_2(t)e^{it} + \bar{Q}_2(t)e^{-it}.$$

By using the method of variation of constants, we obtain

$$\begin{aligned}
&\frac{dQ_1}{dt}e^{it} + \frac{d\bar{Q}_1}{dt}e^{-it} = 0,\\
&\frac{dQ_1}{dt}e^{it} - \frac{d\bar{Q}_1}{dt}e^{-it} = i(Q_1e^{it} + \bar{Q}_1e^{-it})(Q_2e^{it} + \bar{Q}_2e^{-it})^2,\\
&\frac{dQ_2}{dt}e^{it} + \frac{d\bar{Q}_2}{dt}e^{-it} = 0,\\
&\frac{dQ_2}{dt}e^{it} - \frac{d\bar{Q}_2}{dt}e^{-it} = -i(Q_1e^{it} + \bar{Q}_1e^{-it})^2(Q_2e^{it} + \bar{Q}_2e^{-it}).
\end{aligned}$$

The normal form arises when $e^{\pm it}$ are grouped together and $e^{\pm 3it}$ are removed by the formal procedure:

$$\begin{aligned}
2\frac{dQ_1}{dt} &= i\int_0^{2\pi} e^{-it}(Q_1e^{it} + \bar{Q}_1e^{-it})(Q_2e^{it} + \bar{Q}_2e^{-it})^2 dt,\\
2\frac{dQ_2}{dt} &= -i\int_0^{2\pi} e^{-it}(Q_1e^{it} + \bar{Q}_1e^{-it})^2(Q_2e^{it} + \bar{Q}_2e^{-it}) dt.
\end{aligned}$$

This gives the following reduced system:

$$\text{(1.57)} \qquad \begin{cases} 2\dfrac{dQ_1}{dt} = i(2|Q_2|^2Q_1 + Q_2^2\bar{Q}_1),\\[2mm] 2\dfrac{dQ_2}{dt} = -i(2|Q_1|^2Q_2 + Q_1^2\bar{Q}_2).\end{cases}$$

The reduced system (1.57) is now transformed to a degree-one Hamiltonian system. By using the polar coordinates $Q_{1,2}(t) = q_{1,2}(t)e^{i\varphi_{1,2}(t)}$, the system (1.57) splits into a system of four real-valued equations:

$$\text{(1.58)} \qquad \begin{cases} \dfrac{dq_1}{dt} = \dfrac{1}{2}q_1q_2^2\sin 2(\varphi_1 - \varphi_2),\\[2mm] \dfrac{dq_2}{dt} = \dfrac{1}{2}q_1^2q_2\sin 2(\varphi_1 - \varphi_2),\\[2mm] \dfrac{d\varphi_1}{dt} = q_2^2 + \dfrac{1}{2}q_2^2\cos 2(\varphi_1 - \varphi_2),\\[2mm] \dfrac{d\varphi_2}{dt} = -q_1^2 - \dfrac{1}{2}q_1^2\cos 2(\varphi_1 - \varphi_2).\end{cases}$$

It is clear that $C := q_1^2 - q_2^2$ is a constant of motion and that the system depends on $\varphi_1 - \varphi_2$ but not on $\varphi_1 + \varphi_2$. Hence, we introduce the coordinates $\rho := q_2^2$ and $\Delta := 2(\varphi_1 - \varphi_2)$ and reduce (1.58) to a closed system of two equations:

$$\text{(1.59)} \qquad \begin{cases} \dfrac{d\rho}{dt} = (C + \rho)\rho\sin\Delta,\\[2mm] \dfrac{d\Delta}{dt} = (C + 2\rho)(2 + \cos\Delta).\end{cases}$$

This system admits the conserved quantity

$$E := \rho(C + \rho)(2 + \cos\Delta).$$

Since $1 \le 2 + \cos\Delta \le 3$ and $E > 0$, orbits of the degree-one Hamiltonian system (1.59) remain within the bounds

$$\frac{E}{3} \le q_1^2 q_2^2 \le E,$$

where we have recalled $\rho = q_2^2$ and $C + \rho = q_1^2$. For every $\epsilon > 0$ (small), there exists $\delta > 0$ (smaller) such that if $q_1(0)^2 + q_2(0)^2 < \delta$, then $E < 3\delta^2$ and $|C| < 2\delta$ imply

$$q_1(t)^4 + q_2(t)^4 = C^2 + 2q_1^2 q_2^2 < 10\delta^2, \qquad t > 0$$

so that $q_1(t)^2 + q_2(t)^2 < 4\delta =: \epsilon$. Hence, the zero equilibrium point in the system (1.57) is stable in the sense of Definition 1.1. This suggests by using the normal form theory [**138**] that the zero equilibrium point in the system (1.56) is also stable.

1.5. Stability of periodic orbits

Periodic orbits are as important for understanding the dynamical behavior of the initial-value problem (1.1) as equilibrium points.

A **periodic orbit** is a curve γ in the phase space $\mathbb{R}^n$ parameterized by time t. The curve γ is formed by the image of a periodic solution x_γ with some minimal period T. To be more precise, there exists a smallest $T \in (0, \infty)$ such that the periodic solution $x_\gamma \in C^1_{\text{per}}(\mathbb{R}, \mathbb{R}^n)$ of the initial-value problem (1.1) satisfies

$$x_\gamma(t + T) = x_\gamma(t), \qquad \text{for every} \quad t \in \mathbb{R}.$$

The corresponding periodic orbit γ is given by

$$\gamma := \{x_\gamma(t), \ \ t \in [0, T]\} \subset \mathbb{R}^n.$$

The curve γ is C^1 since $x_\gamma \in C^1_{\text{per}}(\mathbb{R}, \mathbb{R}^n)$. It is bounded since T is finite and closed due to the periodicity of $x_\gamma(t)$. The curve γ does not contain equilibrium points because it would take infinite time for an orbit of the dynamical system with $f \in C^1(\mathbb{R}^n, \mathbb{R}^n)$ to reach an equilibrium point.

Because the solution $x_\gamma \in C^1_{\text{per}}(\mathbb{R}, \mathbb{R}^n)$ depends on time t, we have to revise definitions of stable, asymptotically stable, and unstable periodic orbits compared to those for equilibrium points. The periodic orbit γ occupies a compact region in $\mathbb{R}^n$. If the initial point $x_0 \in \mathbb{R}^n$ is defined near γ but not at γ, the actual distance between the solution $x \in C^1(\mathbb{R}, \mathbb{R}^n)$ passing through x_0 at $t = 0$ and the periodic solution $x_\gamma \in C^1_{\text{per}}(\mathbb{R}, \mathbb{R}^n)$ with $x_\gamma(0)$ near x_0 may grow in time but $x(t)$ may still be located in a local neighborhood of γ for all $t > 0$. Based on this geometric observation, illustrated in Figure 1.9, we introduce the concept of orbital stability, asymptotic stability, and instability of the periodic orbit γ.

DEFINITION 1.8. We say that a periodic orbit γ is **orbitally stable** if for every $\epsilon > 0$ (small) there exists some $\delta > 0$ (smaller) such that for every $x_0 \in \mathbb{R}^n$ satisfying

$$\inf_{x_\gamma \in \gamma} |x_0 - x_\gamma| < \delta,$$

the unique solution $x \in C^1(\mathbb{R}_+, \mathbb{R}^n)$ of the initial-value problem (1.1) satisfies

$$\inf_{x_\gamma \in \gamma} |x(t) - x_\gamma| < \epsilon, \qquad \text{for every } t > 0. \tag{1.60}$$

It is called **orbitally asymptotically stable** if it is orbitally stable and

$$\lim_{t \to +\infty} \inf_{x_\gamma \in \gamma} |x(t) - x_\gamma| = 0.$$

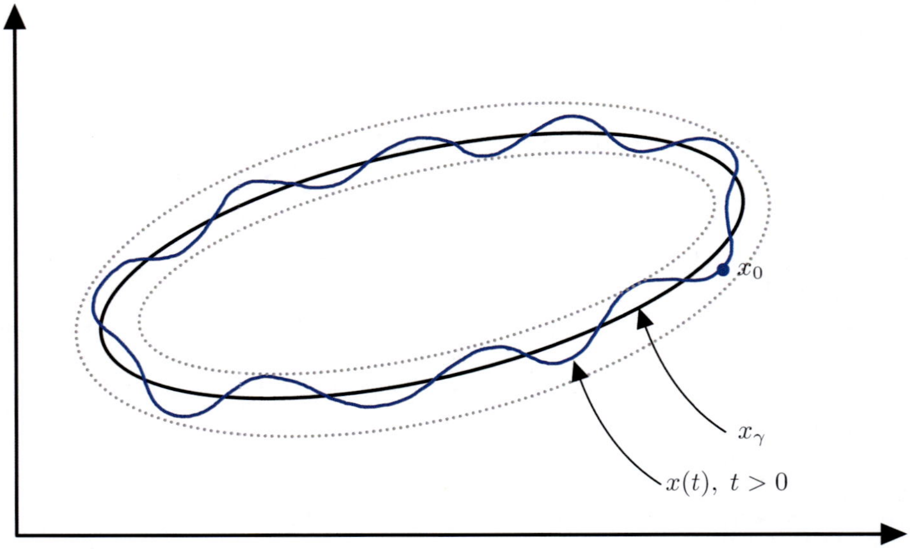

FIGURE 1.9. A sketch of a periodic orbit γ corresponding to a periodic solution x_γ and a nearby solution $x(t)$ with initial condition x_0. If the solution stays within a tubular neighborhood of γ for all $t > 0$, then the periodic orbit γ is orbitally stable.

It is called **orbitally unstable** if it is not orbitally stable.

The following example illustrates the concept of orbital stability versus stability.

EXAMPLE 1.16. Consider the anharmonic oscillator modelled by Newton's equation

$$\frac{d^2x}{dt^2} + |x|^{p-1}x = 0, \tag{1.61}$$

where $p > 1$ is fixed. The equilibrium point at $(0,0)$ is the minimum of the conserved energy

$$V(x,y) = \frac{1}{2}y^2 + \frac{1}{p+1}|x|^{p+1}, \quad y = \frac{dx}{dt}. \tag{1.62}$$

Figure 1.10 shows that the origin is surrounded by a family of periodic orbits $\gamma_{\mathcal{E}}$ parametrized by the energy levels $V(x,y) = \mathcal{E}$. The phase portrait suggests that the periodic orbits are orbitally stable in the sense of Definition 1.8 because a small perturbation of one periodic orbit $\gamma_{\mathcal{E}_1}$ yields another periodic orbit $\gamma_{\mathcal{E}_2}$ for a small perturbation of the energy level $\mathcal{E}_2 \neq \mathcal{E}_1$. The two periodic orbits are located near each other, hence, the distance between the two orbits remains small for every $t > 0$. At the same time, the distance between the two solutions grows in time due to the anharmonicity of the oscillator. Indeed, it follows from (1.62) that the

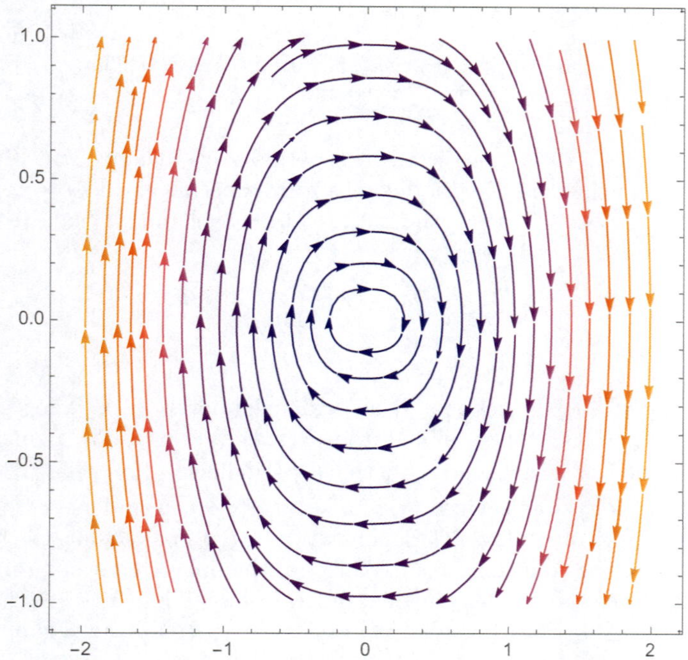

FIGURE 1.10. Phase portrait of the anharmonic oscillator (1.61) with the center point $(0,0)$ on the (x,y) plane.

periodic orbit for the energy level $V(x,y) = \mathcal{E}$ has the period

$$T = 4\int_0^{[(p+1)\mathcal{E}]^{\frac{1}{p+1}}} \frac{dx}{\sqrt{2\left(\mathcal{E} - \frac{x^{p+1}}{p+1}\right)}},$$

due to the symmetry of the periodic orbit about the axes $x = 0$ and $y = 0$. With the change of variables $x = [(p+1)\mathcal{E}]^{\frac{1}{p+1}} z$, we reduce the integral to the form

$$T = 2\sqrt{2}(p+1)^{\frac{1}{p+1}} \mathcal{E}^{\frac{1-p}{2(1+p)}} \int_0^1 \frac{dz}{\sqrt{1 - z^{p+1}}}. \tag{1.63}$$

Since $p > 1$, $T = T(\mathcal{E})$ is strictly monotonically decreasing in $\mathcal{E}$ so that the periodic orbits for two energy levels $\mathcal{E}_1 < \mathcal{E}_2$ have two different periods $T_1 > T_2$. Hence $|x_{\gamma_1}(t) - x_{\gamma_2}(t)|$ cannot be bounded by a small constant for all values of $t > 0$, where $x_{\gamma_{1,2}}$ corresponds to $\gamma_{\mathcal{E}_{1,2}}$. We conclude that each periodic orbit of the anharmonic oscillator (1.61) is orbitally stable in the sense of Definition 1.8 but it is not stable in the sense of Definition 1.1.

The main approach to proving orbital stability of periodic orbits is based on the methods of linearization and Poincaré mappings. For the method of linearization (see Section 1.2), we assume that f is a C^1 function at $x_\gamma \in \gamma$ and use the Taylor expansion

$$f(x) = \underbrace{f(x_\gamma)}_{=\frac{dx_\gamma}{dt}} + \underbrace{D_x f(x_\gamma)}_{=A(t)}(x - x_\gamma) + R(x,t), \tag{1.64}$$

where $A(t) \in \mathbb{M}^{n\times n}$ is a matrix with T-periodic coefficients and $R(x,t)$ is the remainder term, which is T-periodic in t and superlinear in $(x - x_\gamma)$. For the perturbation $\xi(t) := x(t) - x_\gamma(t)$ the initial-value problem (1.1) can be rewritten in the form

$$\begin{cases} \dfrac{d\xi}{dt} = A(t)\xi + R(\xi,t), \quad t > 0, \\ \xi(0) = \xi_0, \end{cases} \tag{1.65}$$

where $\xi_0 \in B_\delta(x_\gamma(0))$ is defined in a local neighborhood of 0 for some $\delta > 0$ (small) and $R(\xi,t)$ has been redefined from $R(x,t)$ in (1.64). The truncated system $\frac{d\xi}{dt} = A(t)\xi$ is the linearized system at the T-periodic solution $x_\gamma \in \gamma$.

The stability of the linearized system $\frac{d\xi}{dt} = A(t)\xi$ can no longer be determined from the eigenvalues of $A(t)$, which depend on time t. In particular, the following example shows that the solution $\xi \in C^1(\mathbb{R},\mathbb{R}^n)$ of the linearized system may grow exponentially in time even if all eigenvalues of $A(t)$ denoted by $\{\lambda_1(t), \lambda_2(t), \dots, \lambda_n(t)\}$ satisfy $\mathrm{Re}(\lambda_j(t)) \le 0$. Such behavior is impossible for equilibrium points, for which A is time-independent, as this would contradict Theorem 1.1.

EXAMPLE 1.17. Consider the 4-by-4 matrix $A(t)$ in the form

$$A(t) = \begin{bmatrix} -1 + \frac{3}{2}\cos^2(t) & 1 - \frac{3}{2}\sin(t)\cos(t) & 0 & 0 \\ -1 - \frac{3}{2}\sin(t)\cos(t) & -1 + \frac{3}{2}\sin^2(t) & 0 & 0 \\ 0 & 0 & 0 & 2 \\ 0 & 0 & -2 & 0 \end{bmatrix}.$$

Eigenvalues of $A(t)$ are t-independent and can be computed exactly. Eigenvalues form two complex-conjugate pairs $\lambda_{1,2} = -\frac{1}{4} \pm \frac{\sqrt{7}}{4}i$ and $\lambda_{3,4} = \pm 2i$ so that $\mathrm{Re}(\lambda_{1,2,3,4}) \le 0$. On the other hand, the linearly independent solutions of $\frac{d\xi}{dt} = A(t)\xi$ are given by

$$\xi_1(t) = e^{\frac{1}{2}t}\begin{bmatrix} \cos(t) \\ -\sin(t) \\ 0 \\ 0 \end{bmatrix}, \quad \xi_2(t) = e^{-t}\begin{bmatrix} \sin(t) \\ \cos(t) \\ 0 \\ 0 \end{bmatrix},$$

$$\xi_3(t) = \begin{bmatrix} 0 \\ 0 \\ \cos(2t) \\ -\sin(2t) \end{bmatrix}, \quad \xi_4(t) = \begin{bmatrix} 0 \\ 0 \\ \sin(2t) \\ \cos(2t) \end{bmatrix},$$

where the first solution grows exponentially in t. Therefore the zero equilibrium in the linearized system $\frac{d\xi}{dt} = A(t)\xi$ for A given above is not stable, despite the fact that all eigenvalues have nonnegative real part.

In order to understand the dynamics of the perturbation $\xi(t)$ in the linearized system $\frac{d\xi}{dt} = A(t)\xi$, we have to analyze the solutions of the linear system of differential equations with T-periodic coefficients. The linearly independent solutions to the linear system can be combined as column vectors in an invertible matrix

$\Phi \in C^1(\mathbb{R}, \mathbb{M}^{n\times n})$. We say that $\Phi(t)$ is the **fundamental solution** if it satisfies the initial-value problem

$$\begin{cases} \dfrac{d\Phi}{dt} = A(t)\Phi, \\ \Phi(0) = I, \end{cases} \quad . \tag{1.66}$$

where I is the identity matrix in $\mathbb{M}^{n\times n}$. It follows from Liouville's formula that

$$\det \Phi(t) = e^{\int_0^t \mathrm{tr}(A)(t')dt'} \det \Phi(0), \tag{1.67}$$

where $\mathrm{tr}(A)$ denotes the trace of the matrix $A(t)$ and $\det \Phi(0) = \det I = 1$. Therefore, $\Phi(t)$ is invertible for every $t \in \mathbb{R}$ if $\mathrm{tr}(A)(t)$ is bounded for every $t \in \mathbb{R}$.

Floquet theory for linear differential equations with time-periodic coefficients transforms every T-periodic linear system into a linear system with constant coefficients. The following lemma gives a technical result used in the proof of Floquet's theorem (see Theorem 1.9).

LEMMA 1.3. *Let $M \in \mathbb{M}^{n\times n}$ be invertible and $T > 0$. There exists $B \in \mathbb{M}^{n\times n}$ such that $e^{TB} = M$.*

PROOF. Let $S \in \mathbb{M}^{n\times n}$ be an invertible transformation of M to the Jordan canonical form $J = S^{-1}MS$, where each upper triangular block of J is invertible because M is invertible. Therefore, we can define a matrix $K \in \mathbb{M}^{n\times n}$ from the matrix equation $e^{TK} = J$ separately for each block of J.

For each diagonal entry of J given by $\{\mu_0\}$, where $\mu_0 \neq 0$ is a semi-simple eigenvalue of M, the diagonal entry of TK is given by $\log(\mu_0)$. For each upper triangular block of J given by $\mu_0 I_{m_0} + N_{m_0}$, where m_0 is the defect of the multiple eigenvalue $\mu_0 \neq 0$ and N_{m_0} is the nillpotent matrix of order m_0, the upper triangular block of TK is given by

$$\log(\mu_0) I_{m_0} + \sum_{j=1}^{m_0-1} \frac{(-1)^{j+1}}{j\mu_0^j} N_{m_0}^j,$$

which is the Taylor series expansion of $\log(\mu_0 I_{m_0} + N_{m_0})$ truncated by finitely many terms due to the property $N_{m_0}^{m_0} = O_{m_0}$. The matrix K constructed in this way represents the Jordan canonical form of B obtained by using the same transformation S such that $K = S^{-1}BS$ because $M = SJS^{-1} = Se^{TK}S^{-1} = e^{TSKS^{-1}} =: e^{TB}$, where we have used

$$Se^{TK}S^{-1} = S\left(\sum_{n=0}^{\infty} \frac{T^n}{n!} K^n\right) S^{-1} = \sum_{n=0}^{\infty} \frac{T^n}{n!} (SKS^{-1})^n = e^{TSKS^{-1}}.$$

The matrix B is determined by K as $B = SKS^{-1}$. □

THEOREM 1.9. *Let Φ be the fundamental solution satisfying* (1.66). *There exist a T-periodic, invertible matrix $P \in C^1_{\mathrm{per}}(\mathbb{R}, \mathbb{M}^{n\times n})$ and a constant matrix $B \in \mathbb{M}^{n\times n}$ such that*

$$\Phi(t) = P(t)e^{tB}, \quad t \in \mathbb{R}. \tag{1.68}$$

If $\xi \in C^1(\mathbb{R}, \mathbb{R}^n)$ is a solution of $\frac{d\xi}{dt} = A(t)\xi$ and $\xi(t) = P(t)\eta(t)$, then $\eta \in C^1(\mathbb{R}, \mathbb{R}^n)$ is a solution of $\frac{d\eta}{dt} = B\eta$.

PROOF. Let $\tilde{\Phi}(t) := \Phi(t+T)$. It satisfies the same system of differential equations

$$\frac{d}{dt}\tilde{\Phi}(t) = \frac{d}{dt}\Phi(t+T) = A(t+T)\Phi(t+T) = A(t)\tilde{\Phi}(t),$$

due to the periodicity of $A(t+T) = A(t)$. The initial condition for $\tilde{\Phi}(t)$ is $\tilde{\Phi}(0) = \Phi(T)$. By uniqueness of solutions and the linear superposition principle, $\tilde{\Phi}(t) = \Phi(t)\Phi(T)$ so that

$$\Phi(t+T) = \Phi(t)\Phi(T).$$

Since $\det\Phi(0) = 1$, it follows from (1.67) that $\det\Phi(t) \neq 0$ for every $t \in \mathbb{R}$ so that $\Phi(t)$ is invertible for every $t \in \mathbb{R}$. Hence, $M := \Phi(T)$ is an invertible matrix in $\mathbb{M}^{n\times n}$.

By Lemma 1.3, there exists $B \in \mathbb{M}^{n\times n}$ such that $e^{TB} = M$. Hence we have

$$\Phi(t+T) = \Phi(t)e^{TB}.$$

Define an invertible matrix $P(t) := \Phi(t)e^{-tB}$, where $\Phi(t)$ and e^{-tB} are invertible. It follows from $\Phi \in C^1(\mathbb{R}, \mathbb{M}^{n\times n})$ that $P \in C^1(\mathbb{R}, \mathbb{M}^{n\times n})$. In addition,

$$P(t+T) = \Phi(t+T)e^{-TB-tB} = \Phi(t)e^{TB}e^{-TB}e^{-tB} = \Phi(t)e^{-tB} = P(t)$$

and

$$\begin{aligned} P'(t+T) &= \Phi'(t+T)e^{-TB-tB} - \Phi(t+T)e^{-TB-tB}B \\ &= \Phi'(t)e^{-tB} - \Phi(t)e^{-tB}B \\ &= P'(t), \end{aligned}$$

so that $P(t) \in C^1_{\text{per}}(\mathbb{R}, \mathbb{M}^{n\times n})$.

Finally, we observe from (1.66) and (1.68) that

$$\frac{dP}{dt} + P(t)B = A(t)P(t).$$

We substitute $\xi(t) = P(t)\eta(t)$ into $\frac{d\xi}{dt} = A(t)\xi$ and obtain

$$\frac{dP}{dt}\eta + P(t)\frac{d\eta}{dt} = A(t)P(t)\eta$$

from which we conclude that

$$P(t)\frac{d\eta}{dt} = P(t)B\eta.$$

Since $P(t)$ is an invertible matrix, the linear system with periodic coefficients $\frac{d\xi}{dt} = A(t)\xi$ is equivalent to the linear system with constant coefficients $\frac{d\eta}{dt} = B\eta$. □

REMARK 1.15. If $\mu_0 < 0$ in the proof of Theorem 1.9, then $\log(\mu_0) = \log|\mu_0| + \pi i$. If $\mu_0 \in \mathbb{C}$, then $\log(\mu_0) = \log|\mu_0| + i\arg(\mu_0)$. The choice of B is not unique since $e^{2\pi i I_n} = I_n$.

DEFINITION 1.9. Eigenvalues of B are called **characteristic exponents** for solutions of the linear system $\frac{d\xi}{dt} = A(t)\xi$. The matrix $M := \Phi(T)$ is called **the monodromy matrix**. Eigenvalues of M are called **Floquet multipliers**.

Recall from Theorem 1.1 that the spectrum $\sigma(B) = \{\sigma_1, \sigma_2, \dots, \sigma_n\}$ of the characteristic exponents determines the stability of the zero equilibrium in the linear equation $\frac{d\eta}{dt} = B\eta$. On the other hand, the unique solution of the initial-value problem for the linear equation $\frac{d\xi}{dt} = A(t)\xi$ with $\xi(0) = \xi_0$ is given by $\xi(t) = \Phi(t)\xi_0$.

The growth or decay of $\xi(t)$ is determined by $M = \Phi(T)$ since the T-periodicity of the coefficients of $A(t)$ implies that $\xi(NT) = M^N \xi_0$ for any $N \in \mathbb{N}$. Therefore, the growth or decay of $\xi(t)$ is determined by the spectrum $\sigma(M) = \{\mu_1, \mu_2, \dots, \mu_n\}$ of Floquet multipliers.

The following theorem gives the stability results for the zero equilibrium point in the linear system $\frac{d\xi}{dt} = A(t)\xi$ in the sense of Definition 1.1. These results do not determine yet the linear orbital stability of the periodic orbits in the nonlinear system (1.65) because they are not using the distance (1.60) in Definition 1.8.

THEOREM 1.10. *Let $\sigma(M) = \{\mu_1, \mu_2, \dots, \mu_n\}$ be the spectrum of Floquet multipliers. The zero equilibrium point in the linear equation $\frac{d\xi}{dt} = A(t)\xi$ with T-periodic matrix $A(t)$ is*

- *stable if and only if $|\mu_j| \le 1$ for all j and each eigenvalue μ_j with $|\mu_j| = 1$ is semi-simple;*
- *asymptotically stable if and only if $|\mu_j| < 1$ for all j;*
- *unstable if and only if either $|\mu_j| > 1$ for at least one j or $|\mu_j| \le 1$ for all j but at least one eigenvalue μ_j with $|\mu_j| = 1$ is defective.*

PROOF. By Theorem 1.9 the unique solution $\xi(t) = P(t)e^{tB}\xi_0$ of $\frac{d\xi}{dt} = A(t)\xi$ is expressed in terms of the unique solution $\eta(t) = e^{tB}\eta_0$ of $\frac{d\eta}{dt} = B\eta$ with $\xi_0 = \eta_0$, where $P(t)$ is an invertible T-periodic matrix.

Since $M = e^{TB}$ and the Jordan block diagonalization of B into upper triangular blocks yields the Jordan block diagonalization of $M = e^{TB}$ with the same invertible matrix S since $S^{-1}MS = e^{TS^{-1}BS}$, we obtain the spectral mapping:

$$\mu_j = e^{T\sigma_j}, \quad 1 \le j \le n, \tag{1.69}$$

such that multiplicity and defectivity of the Floquet multiplier μ_j coincide with those of the characteristic exponent σ_j. Since $P(t)$ is an invertible T-periodic matrix, the solution $\xi(t) = P(t)e^{tB}\xi_0$ remains bounded, decays, or grows as $t \to +\infty$ if and only if the solution $\eta(t) = e^{tB}\xi_0$ remains bounded, decays, or grows as $t \to +\infty$. The results of Theorem 1.10 now follow from the results of Theorem 1.1 because the conditions on $\sigma(M)$ in Theorem 1.10 are identical to the conditions on $\sigma(B)$ in Theorem 1.1. ☐

COROLLARY 1.5. *For each eigenvalue σ_j of geometric multiplicity k and algebraic multiplicity $m \ge k$, there exist k solutions of $\frac{d\xi}{dt} = A(t)\xi$ of the form*

$$\xi(t) = p_{j_s}(t)e^{t\sigma_j}, \quad p_{j_s}(t+T) = p_{j_s}(t), \quad 1 \le j_s \le k \tag{1.70}$$

and $m - k$ solutions of the form

$$\xi(t) = q_{j_s}(t)e^{t\sigma_j}, \quad k+1 \le j_s \le m, \tag{1.71}$$

where $q_{j_s}(t)$ is a sum of powers (not exceeding t^{m-k}) multiplied by T-periodic functions of t.

PROOF. Due to the Jordan block decomposition $S^{-1}MS = e^{TS^{-1}BS}$, vector columns of S are eigenvectors and generalized eigenvectors of B and simultaneously, eigenvectors and generalized eigenvectors of M. The periodic eigenvectors in (1.70) are obtained from $\xi(t) = \Phi(t)s_{j_s} = P(t)e^{tB}s_{j_s} = (P(t)s_{j_s})e^{t\sigma_j}$, where s_{j_s} is the eigenvector of B for the eigenvalue σ_j. The nonperiodic eigenvectors in (1.71) are obtained from $\xi(t) = \Phi(t)s_{j_s}$, where s_{j_s} is spanned by generalized eigenvectors of B for the eigenvalue σ_j. ☐

REMARK 1.16. The characteristic exponents $\sigma(B)=\{\sigma_1,\sigma_2,\ldots,\sigma_n\}$ are uniquely determined in the semi-open strip

$$-\frac{\pi}{T} < \mathrm{Im}(\sigma) \leq \frac{\pi}{T},$$

because the exponential function $\mu = e^{T\sigma}$ is multi-valued for the same value of Floquet multipliers in $\sigma(M) = \{\mu_1, \mu_2, \ldots, \mu_n\}$.

REMARK 1.17. The Floquet multipliers $\sigma(M) = \{\mu_1, \mu_2, \ldots, \mu_n\}$ are independent of the initial point of the monodromy matrix. Indeed, picking any $t_0 \in [0,T]$ and using

$$\Phi(t_0 + t) = \Phi(t)\Phi(t_0),$$

by uniqueness of solutions and the linear superposition principle, we obtain

$$\begin{aligned}\Phi(t_0 + t + T) &= \Phi(t+T)\Phi(t_0) = \Phi(t)\Phi(T)\Phi(t_0)\\ &= \Phi(t)\Phi(t_0)[\Phi(t_0)]^{-1}\Phi(T)\Phi(t_0) = \Phi(t_0+t)\tilde{M},\end{aligned}$$

where $\tilde{M} := [\Phi(t_0)]^{-1}M\Phi(t_0)$ is the monodromy matrix relative to the initial point t_0. Because $\Phi(t_0)$ is invertible, $\sigma(\tilde{M}) = \sigma(M)$ since $\tilde{M}$ and M are related by a similarity transformation.

EXAMPLE 1.18. In the scalar case ($n = 1$), the unique solution of $\frac{d\xi}{dt} = A(t)\xi$ with $\xi(0) = \xi_0$ is given explicitly by $\xi(t) = e^{\int_0^t A(t')dt'}\xi_0$, from which it follows that

$$M = e^{\int_0^T A(t)dt} = \mu_1 \quad \text{and} \quad \sigma_1 = \frac{1}{T}\int_0^T A(t)dt,$$

the mean value of the T-periodic function $A \in C^0_{\mathrm{per}}(\mathbb{R},\mathbb{R})$. Consequently,

$$P(t) = e^{\int_0^t A(t')dt' - \frac{t}{T}\int_0^T A(t')dt'}$$

is the T-periodic function $P \in C^1_{\mathrm{per}}(\mathbb{R},\mathbb{R})$.

EXAMPLE 1.19. We compute the Floquet multipliers explicitly for the system of differential equations

$$\begin{cases} x' = a^2 x - y - (x^2+y^2)x, \\ y' = x + a^2 y - (x^2+y^2)y, \end{cases} \tag{1.72}$$

where $a^2 > 0$ is fixed parameter. The system is the normal form for Hopf bifurcation of periodic orbits [**138**]. It is transformed to the following system of uncoupled equations with the use of polar coordinates $x = r\cos\theta$ and $y = r\sin\theta$:

$$\begin{cases} \dfrac{dr}{dt} = r(a^2 - r^2), \\ \dfrac{d\theta}{dt} = 1. \end{cases}$$

The angular dynamics is a counterclockwise rotation with constant speed, whereas the radial dynamics shows that the equilibrium point $(0,0)$ is a spiral unstable point and the circle $\gamma_a := \{(x,y) : x^2+y^2 = a^2\}$ is an orbitally asymptotically stable periodic orbit of the system in the sense of Definition 1.8. Figure 1.11 shows the phase portrait for the system (1.72). Parameterizing the solution with

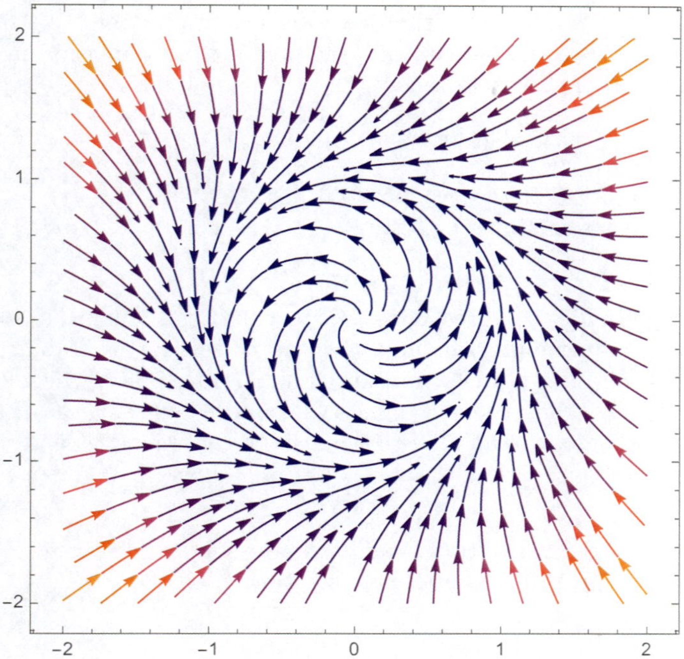

FIGURE 1.11. Phase portrait of system (1.72) with the unstable spiral point $(0,0)$ and the asymptotically stable periodic orbit γ_a in the (x,y) plane.

$x_{\gamma_a}(t) = a\cos(t)$ and $y_{\gamma_a}(t) = a\sin(t)$ and linearizing (1.72) yield the linearized system

$$\begin{cases} \xi' = -\eta - a^2(1+\cos(2t))\xi - a^2\sin(2t)\eta, \\ \eta' = \xi - a^2\sin(2t)\xi - a^2(1-\cos(2t))\eta. \end{cases} \tag{1.73}$$

The fundamental matrix solution of the linearized system (1.73) is explicitly given by

$$\Phi(t) = \begin{bmatrix} e^{-2a^2t}\cos(t) & -\sin(t) \\ e^{-2a^2t}\sin(t) & \cos(t) \end{bmatrix}.$$

Since the fundamental period of the periodic orbit γ_a is $T = 2\pi$, the monodromy matrix can be computed directly

$$M = \Phi(2\pi) = \begin{bmatrix} e^{-4\pi a^2} & 0 \\ 0 & 1 \end{bmatrix},$$

which has the two Floquet multipliers $\mu_1 = e^{-4\pi a^2}$ and $\mu_2 = 1$. According to Theorem 1.10, the zero equilibrium of the linearized system (1.73) is stable but not asymptotically stable in the sense of Definition 1.1.

Example 1.19 shows that the result of Theorem 1.10 is not sufficient for understanding orbital stability, asymptotic stability, and instability of periodic orbits in the sense of Definition 1.8. Since the dynamical system (1.1) is autonomous with time-independent $f(x)$, it is true that if $x(t)$ is a solution, then $x(t+t_0)$ is another solution for any $t_0 \in \mathbb{R}$. Differentiating the periodic solution $x_\gamma(t)$ in t yields one particular solution of the linearized system

$$\frac{d\xi}{dt} = A(t)\xi \quad \text{with } A(t) := D_x f(x_\gamma(t)). \tag{1.74}$$

We place this solution at the last position in the set $\{\xi_1(t), \xi_2(t), \dots, \xi_n(t)\}$ of n linearly independent solutions by setting

$$\xi_n(t) := x_\gamma'(t) = f(x_\gamma(t)). \tag{1.75}$$

Since ξ_n is T-periodic, it corresponds to $\sigma_n = 0$ or $\mu_n = 1$ in (1.69). It is then clear that the Floquet multiplier $\mu_n = 1$ plays no role in the stability analysis in the sense of Definition 1.8 due to translation of $x_\gamma(t)$ in time t, which leaves system (1.1) invariant. The translational symmetry is removed by using the distance (1.60). Since x_γ is T-periodic, we can rewrite the distance between $x(t)$ and γ in the form

$$\text{dist}(x(t), \gamma) := \inf_{t_0 \in [0,T]} |x(t) - x_\gamma(t+t_0)|.$$

For $t = 0$, the first derivative test to minimize the squared distance $|x(0) - x_\gamma(t_0)|^2$ with respect to t_0 gives us the orthogonality condition

$$[x(0) - x_\gamma(t_0)] \cdot x_\gamma'(t_0) = 0. \tag{1.76}$$

This imposes the constraint $\xi(t_0) \cdot f(x_\gamma(t_0)) = 0$, where $\xi(t+t_0) := x(t) - x_\gamma(t+t_0)$ is the perturbation to the periodic solution x_γ satisfying the linearized equation $\frac{d\xi}{dt} = A(t+t_0)\xi$. Without loss of generality, we can set $t_0 = 0$. This setup introduces a decomposition of the phase space $\mathbb{R}^n$ for the linearized system (1.74) into the orthogonal sum $f(x_\gamma(0)) \oplus X_0$, where

$$X_0 := \{\xi \in \mathbb{R}^n : \ \xi \cdot f(x_\gamma(0)) = 0\}. \tag{1.77}$$

REMARK 1.18. Since $A(0)$ is typically not a symmetric matrix, the orthogonal decomposition $\mathbb{R}^n = f(x_\gamma(0)) \oplus X_0$ is not preserved in the time evolution of the linearized equation $\frac{d\xi}{dt} = A(t)\xi$. In other words, it is not generally true that $\xi(t) = f(x_\gamma(t)) + \tilde{\xi}(t)$ with $\tilde{\xi}(t) \perp f(x_\gamma(t))$ for $t \neq 0$ even if $\tilde{\xi}(0) \perp f(x_\gamma(0))$. Nevertheless, one can work with the orthogonal decomposition $\mathbb{R}^n = f(x_\gamma(0)) \oplus X_0$ in a countable sequence of times $\{Tk\}_{k\in\mathbb{N}}$ and obtain the linear orbital stability of the periodic orbits as is done in the proof of Theorem 1.11.

REMARK 1.19. In the Hamiltonian system (1.25), the matrix $A(t)$ has the structure $A(t) = J\mathcal{L}(t)$, where $J^T = -J$ and $\mathcal{L}(t)^T = \mathcal{L}(t)$. In this case, one can introduce the symplectically orthogonal decomposition $\mathbb{R}^n = f(x_\gamma(0)) \oplus X_0^*$, where

$$X_0^* := \{\xi \in \mathbb{R}^n : \ \xi \cdot J^{-1} f(x_\gamma(0)) = 0\}.$$

The symplectically orthogonal decomposition is preserved in the time evolution of the linearized equation $\frac{d\xi}{dt} = J\mathcal{L}(t)\xi$ so that $\xi(t) = f(x_\gamma(t)) + \tilde{\xi}(t)$ with $\tilde{\xi}(t) \perp J^{-1} f(x_\gamma(t))$ for every $t \neq 0$ if $\tilde{\xi}(0) \perp J^{-1} f(x_\gamma(0))$. Indeed, it follows from the linearized system that

$$\begin{aligned}\frac{d}{dt} J^{-1} f(x_\gamma(t)) \cdot \xi(t) &= J^{-1} J\mathcal{L}(t) x_\gamma'(t) \cdot \xi(t) + J^{-1} f(x_\gamma(t)) \cdot J\mathcal{L}(t)\xi(t) \\ &= \mathcal{L}(t) x_\gamma'(t) \cdot \xi(t) - f(x_\gamma(t)) \cdot \mathcal{L}(t)\xi(t) = 0,\end{aligned}$$

so that $J^{-1}f(x_\gamma(t))\cdot\xi(t)$ is independent of t. This property is useful for symplectic integration of the Hamiltonian systems but will not be used in this monograph.

Definition 1.8 implies that the growth or decay of the projection of $\xi(t)$ onto the subspace X_0 defined in (1.77) is important for orbital stability of the periodic orbits, whereas the growth or decay of the projection of $\xi(t)$ onto $X_0^\perp = \text{Span}\{f(x_\gamma(0))\}$ is not important. The following theorem makes this point precise and gives the linear orbital stability results for the periodic orbit γ.

THEOREM 1.11. *Let $\sigma(M) = \{\mu_1, \mu_2, \dots, \mu_n\}$ be the spectrum of Floquet multipliers with $\mu_n = 1$. The periodic orbit γ related to the periodic solution x_γ is*

- *linearly orbitally stable if and only if $|\mu_j| \le 1$ for all $1 \le j \le n-1$ and each eigenvalue μ_j with $|\mu_j| = 1$ for $1 \le j \le n-1$ is semi-simple;*
- *linearly orbitally asymptotically stable if and only if $|\mu_j| < 1$ for all $1 \le j \le n-1$;*
- *linearly orbitally unstable if and only if either $|\mu_j| > 1$ for at least one $1 \le j \le n-1$ or $|\mu_j| \le 1$ for all $1 \le j \le n-1$ but at least one eigenvalue μ_j with $|\mu_j| = 1$ for $1 \le j \le n-1$ is defective.*

PROOF. We start by noting that $\mu_n = 1$ due to the eigenvector (1.75). Due to the periodicity of the solution we have $f(x_\gamma(t+T)) = f(x_\gamma(t))$ and hence also $A(t+T) = A(t)$ in (1.74). Therefore, we can consider the perturbation $\xi(t)$ at the countable sequence of times $\{Tk\}_{k\in\mathbb{N}}$ such that $\xi(Tk) = M^k\xi(0)$.

Let $\mathbb{R}^n = f(x_\gamma(0)) \oplus X_0$ and let $\Pi_0 : \mathbb{R}^n \mapsto X_0$ be the orthogonal projection operator given by

$$\Pi_0\xi = \xi - \frac{f(x_\gamma(0))\cdot\xi}{|f(x_\gamma(0))|^2} f(x_\gamma(0)).$$

Writing the monodromy matrix M in terms of the basis induced by the orthogonal decomposition $\mathbb{R}^n = f(x_\gamma(0)) \oplus X_0$, we find that it takes a lower-triangular form:

$$\begin{bmatrix} \Pi_0 M \Pi_0 & 0 \\ (I-\Pi_0)M\Pi_0 & 1 \end{bmatrix}.$$

Indeed, the upper triangular entries are zero because

$$M(I-\Pi_0)\xi_0 = \alpha_0 M f(x_\gamma(0)) = \alpha_0 f(x_\gamma(0)), \quad \alpha_0 := \frac{f(x_\gamma(0))\cdot\xi_0}{|f(x_\gamma(0))|^2}, \quad \xi_0 := \xi(0),$$

in view of the eigenvector (1.75), and hence $\Pi_0 M(I-\Pi_0)\xi_0 = 0$. Iterating the decomposition for $k \in \mathbb{N}$ yields

$$\xi(Tk) = \alpha_k f(x_\gamma(0)) + (\Pi_0 M \Pi_0)^k \xi_0,$$

with

$$\alpha_k = \alpha_0 + \sum_{j=1}^{k} \frac{f(x_\gamma(0))\cdot M\Pi_0(\Pi_0 M\Pi_0)^{j-1}\xi_0}{|f(x_\gamma(0))|^2}. \tag{1.78}$$

Since $M = e^{TB}$, the matrix B can be similarly represented in the lower-triangular form

$$\begin{bmatrix} \Pi_0 B \Pi_0 & 0 \\ (I-\Pi_0)B\Pi_0 & 0 \end{bmatrix},$$

due to the spectral mapping (1.69). Hence, the projection of $\xi(Tk)$ onto X_0 remains bounded, decays, or grows as $k \to \infty$ if and only if the projection of $\eta(Tk) = e^{kTB}\xi_0$

onto X_0 remains bounded, decays, or grows as $k \to \infty$. The result of Theorem 1.11 follows from the results of Theorem 1.1 and is determined by the $(n-1)$ Floquet multipliers

$$\sigma(\Pi_0 M \Pi_0) = \{\mu_1, \mu_2, \ldots, \mu_{n-1}\}$$

or, equivalently, by the $(n-1)$ characteristic exponents

$$\sigma(\Pi_0 B \Pi_0) = \{\sigma_1, \sigma_2, \ldots, \sigma_{n-1}\}.$$

The actual behavior of the solution $\xi(t)$ between the countable sequence of times $\{Tk\}_{k\in\mathbb{N}}$ does not affect the boundedness, decay, or growth of the perturbation because $P(t)$ is an invertible T-periodic matrix which determines the solution $\xi(t) = P(t)e^{tB}\xi_0$ for $t > 0$. □

REMARK 1.20. The projection α_k onto $f(x_\gamma(0))$ may grow in k according to the solution (1.78), and hence the perturbation to $x_\gamma(Tk) = x_\gamma(0)$ may not be bounded. However, this growth does not affect the linear orbital stability according to Definition 1.8, since if the perturbation $\alpha_k x'_\gamma(Tk)$ is added to $x_\gamma(Tk)$, then the value of the time is changed in $x_\gamma(Tk + \alpha_k)$ up to the linearized terms. The linearized equation $\frac{d\xi}{dt} = A(t)\xi$ is used at the fixed $t_0 = 0$ but it can be translated to the time $t_0 \neq 0$ since the Floquet multipliers of the monodromy matrix M do not depend on the initial time t_0 in the matrix $A(t+t_0)$. Therefore, the definition of t_0 in the orthogonality condition (1.76) can be changed after each period. This is why we can ignore the growth of the projection α_k while solving the linearized equation $\frac{d\xi}{dt} = A(t)\xi$.

EXAMPLE 1.20. Continuing Example 1.19, we identify

$$X_0 = \{(\xi, \eta) \in \mathbb{R}^2 : \quad \eta = 0\}$$

for which $\Pi_0^* M \Pi_0 = e^{-4\pi a^2} = \mu_1$. Since $\mu_1 < 1$, the periodic orbit γ_a is linearly orbitally asymptotically stable by Theorem 1.11. This agrees with the phase portrait for the nonlinear system (1.72) in Figure 1.11.

EXAMPLE 1.21. Continuing Example 1.16, we introduce the family of periodic orbits

$$\gamma_{\mathcal{E}} := \{(x, y) \in \mathbb{R}^2 : \quad V(x, y) = \mathcal{E}\}$$

parameterized by the fixed energy level $V(x, y) = \mathcal{E}$, where $V(x, y)$ is given by (1.62). The linearized system for the anharmonic oscillator (1.61) is given by

$$\begin{cases} \xi' = \eta, \\ \eta' = -p|x_{\mathcal{E}}|^{p-1}\xi, \end{cases} \tag{1.79}$$

where $x_{\mathcal{E}}(t+T) = x_{\mathcal{E}}(t)$ is the solution for the T-periodic orbit $\gamma_{\mathcal{E}}$, where the period T depends on $\mathcal{E}$ due to (1.63). Let us parameterize time such that

$$x_{\mathcal{E}}(0) = a(\mathcal{E}) := [(p+1)\mathcal{E}]^{\frac{1}{p+1}}, \qquad x'_{\mathcal{E}}(0) = 0.$$

Two linearly independent solutions of the linearized system (1.79) are given by the derivatives of $x_{\mathcal{E}}(t)$ in $\mathcal{E}$ and t. Therefore, the fundamental matrix solution is given by

$$\Phi(t) = \begin{bmatrix} [a'(\mathcal{E})]^{-1} \partial_{\mathcal{E}} x_{\mathcal{E}}(t) & [x''_{\mathcal{E}}(0)]^{-1} x'_{\mathcal{E}}(t) \\ [a'(\mathcal{E})]^{-1} \partial_{\mathcal{E}} x'_{\mathcal{E}}(t) & [x''_{\mathcal{E}}(0)]^{-1} x''_{\mathcal{E}}(t) \end{bmatrix},$$

where both $a'(\mathcal{E})$ and $x''_{\mathcal{E}}(0)$ are nonzero. The dependence of $x_{\mathcal{E}}$ and T on $\mathcal{E} \in (0, \infty)$ is C^1 if $p > 1$. It follows by differentiating $x_{\mathcal{E}}(T) = a(\mathcal{E})$ and $x'_{\mathcal{E}}(T) = 0$ in $\mathcal{E}$ that

$\partial_{\mathcal{E}} x_{\mathcal{E}}(T) = a'(\mathcal{E})$ and $\partial_{\mathcal{E}} x'_{\mathcal{E}}(T) = -T'(\mathcal{E}) x''_{\mathcal{E}}(0)$, from which the monodromy matrix is computed in the form

$$\Phi(T) = \begin{bmatrix} 1 & 0 \\ -T'(\mathcal{E}) \left[a'(\mathcal{E})\right]^{-1} x''_{\mathcal{E}}(0) & 1 \end{bmatrix}.$$

Since $T'(\mathcal{E}) \neq 0$ by (1.63), the Floquet multiplier is double with $\mu_1 = \mu_2 = 1$, which is a standard property of Hamiltonian systems. Since the double multiplier is defective, the perturbations actually grow linearly in time. Therefore, according to Theorem 1.10, the zero equilibrium of the linearized system (1.79) is unstable in the sense of Definition 1.2 but the phase portrait in Figure 1.10 does not show this instability. In order to get the correct conclusion, we again identify

$$X_0 = \{(\xi, \eta) \in \mathbb{R}^2 : \ \eta = 0\}$$

for which $\Pi_0^* M \Pi_0 = 1 = \mu_1$. By Theorem 1.11, the periodic orbit $\gamma_{\mathcal{E}}$ for fixed $\mathcal{E} \in (0, \infty)$ is linearly orbitally stable in the sense of Definitiion 1.8, in agreement with the phase portrait in Figure 1.10.

REMARK 1.21. Since it follows from (1.67) that $\det(M) = e^{\int_0^T \operatorname{tr} A(t) dt}$, the product of Floquet multipliers is positive. Moreover, since $\operatorname{tr}(A) = \operatorname{tr}(J\mathcal{L}) = 0$ in linear Hamiltonian systems and $\mu_n = 1$, then $\mu_1 \cdots \mu_{n-1} = 1$. If $n = 2$, then $\mu_1 = \mu_2 = 1$, in agreement with Example 1.21.

The proof of Theorem 1.11 can be expanded to nonlinear orbital stability of the periodic orbit γ in the sense of Definition 1.8 with the method of **Poincaré's return maps**. An arbitrary point of the periodic orbit, e.g., $x_\gamma(0)$, is picked and the phase space $\mathbb{R}^n$ is decomposed into the orthogonal sum $f(x_\gamma(0)) \oplus X_0$, where X_0 is now defined by

$$X_0 := \{x \in \mathbb{R}^n : \quad (x - x_\gamma(0)) \cdot f(x_\gamma(0)) = 0\}. \tag{1.80}$$

Although the perturbation $\xi(t) := x(t) - x_\gamma(t)$ satisfies the perturbed system (1.65), we will now work with the perturbation relative to the point $x_\gamma(0)$, for which we will use the original dynamical system (1.1).

The following theorem formulates a sharp criterion for nonlinear orbital stability of periodic orbits.

THEOREM 1.12. *Let $x(t)$ be the solution to the dynamical system (1.1) with $x(0) = x_0$. For every (small) $\delta > 0$ and every $x_0 \in X_0$ satisfying $|x_0 - x_\gamma(0)| < \delta$, there exists a return time $t_1 > 0$ and $C_1 > 0$ such that $x_1 := x(t_1) \in X_0$ with*

$$|t_1 - T| + |x_1 - x_\gamma(0)| < C_1 \delta.$$

The periodic orbit γ is orbitally stable, asymptotically stable, or unstable if and only if the mapping $X_0 \ni x_0 \mapsto x_1 \in X_0$ is stable, asymptotically stable, or unstable at the fixed point $x_\gamma(0)$.

PROOF. The existence of the return time $t_1 > 0$ near T follows by continuous dependence of the dynamical system (1.1) on initial conditions. Let $x(t)$ be a solution of (1.1) for $x(0) = x_0 \in X_0$, where X_0 is given by (1.80). The return time $t_1 > 0$ is found from the root of the nonlinear function

$$F(t_1, x_0) := (x(t_1) - x_\gamma(0)) \cdot f(x_\gamma(0)),$$

where the dependence of $F(t_1, x_0)$ on x_0 follows from the dependence of $x(t)$ on $x(0) = x_0 \in X_0$. If $x_0 = x_\gamma(0)$, then $t_1 = T$ is a root of $F(t_1, x_0)$ since x_γ is T-periodic. Hence, we write

$$F(t_1, x_0) = (x_\gamma(t_1) - x_\gamma(0)) \cdot f(x_\gamma(0)) + (x(t_1) - x_\gamma(t_1)) \cdot f(x_\gamma(0)).$$

The function $F(t_1, x_0) : \mathbb{R} \times \mathbb{R} \mapsto \mathbb{R}$ is C^1 with $F(T, x_\gamma(0)) = 0$ and

$$\partial_{t_1} F(T, x_\gamma(0)) = x_\gamma'(T) \cdot f(x_\gamma(0)) = |f(x_\gamma(0))|^2 > 0.$$

By the implicit function theorem, for every $x_0 \in X_0$ satisfying $|x_0 - x_\gamma(0)| < \delta$ with small $\delta > 0$, there exists $t_1 > 0$ such that $|t_1 - T| < C\delta$ for some $C > 0$ and $x(t_1) \in X_0$. Moreover, by continuous dependence, x_1 satisfies $|x_1 - x_\gamma(T)| < C\delta$ for some $C > 0$. Orbital stability, asymptotic stability, or instability of the periodic orbit γ are then equivalent to the boundedness, decay, or growth of $|x_1 - x_\gamma(0)|$ relative to $|x_0 - x_\gamma(0)|$. □

REMARK 1.22. Since the Floquet multipliers $\{\mu_1, \mu_2, \dots, \mu_{n-1}\}$ are eigenvalues of the Poincaré's return map $X_0 \ni x_0 \mapsto x_1 \in X_0$ [**51**], the periodic orbit γ is orbitally asymptotically stable if $|\mu_j| < 1$ for all $1 \le j \le n-1$ and is orbitally unstable if there is $|\mu_j| > 1$ for at least one $1 \le j \le n-1$. These results are similar to Theorems 1.2 and 1.3. On the other hand, the nonlinear stability cannot be concluded if $|\mu_j| \le 1$ for all $1 \le j \le n-1$ with some $|\mu_j| = 1$.

EXAMPLE 1.22. For Example 1.19, Poincaré's return map can be computed explicitly. By using $X_0 := \{(x, y) \in \mathbb{R}^2 : x > 0, \quad y = 0\}$, we solve the radial equation $\frac{dr}{dt} = r(a^2 - r^2)$ explicitly from $r(0) = x_0$ to $r(2\pi) = x_1$ to obtain

$$x_0 \mapsto x_1 = P(x_0) := \frac{a x_0}{\sqrt{x_0^2 + (a^2 - x_0^2) e^{-4\pi a^2}}}.$$

Since $P(a) = a$, $x_0 = a$ is the fixed point of the return map. Since $P'(a) = e^{-4\pi a^2} < 1$, the fixed point is asymptotically stable in iterations of the return map. The value of $P'(a) = e^{-4\pi a^2}$ coincides with the Floquet multiplier $\mu_1 = e^{-4\pi a^2}$ in Example 1.20.

REMARK 1.23. Nonlinear stability of periodic orbits is a difficult problem in the context of Hamiltonian dynamical systems of degree two and higher, partly because Lyapunov functions are no longer applicable to analyze the dynamics of perturbations of the periodic orbits. Dynamics of perturbations may become extremely complicated and chaotic due to resonances. However, the analysis of resonances is not the subject of this monograph.

REMARK 1.24. In the context of nonlinear waves in infinite-dimensional Hamiltonian systems, we get extra help from separation of spatial and temporal variables. Nonlinear waves are decaying or periodic with respect to the spatial coordinate whereas their stability is considered with respect to the time coordinate. As a result, spatially decaying or periodic nonlinear waves can still be cast as the equilibrium points of infinite-dimensional systems, although some methods useful for periodic orbits (such as Floquet theory and the period function) remain useful tools for the analysis of stability of spatially periodic waves in infinite-dimensional Hamiltonian systems. In addition, since some Hamiltonian systems have translational and rotational symmetries, the concept of orbital stability is useful in the context of stability of both spatially decaying and periodic waves.

CHAPTER 2

Stability of solitary waves

In the analysis of nonlinear dispersive wave equations one of the central objects of studies is the class of spatially decaying solutions known as **solitary waves**. Solitary waves are defined in a Banach space X, which is a dense subspace of a Hilbert space $\mathcal{H}$ subject to decaying boundary conditions at infinity. For instance, for one-dimensional solitary waves the Sobolev spaces $H^k(\mathbb{R})$, $k \in \mathbb{N}$, form dense subspaces of the Hilbert space $L^2(\mathbb{R})$, and functions in $H^k(\mathbb{R})$ together with their derivatives up to $(k-1)^{\text{th}}$-order are continuous, bounded, and decaying to zero at infinity.

To analyze the stability of solitary waves, one can recast the nonlinear dispersive wave equation as a Hamiltonian system in which the solitary wave $u_* \in X$ is an equilibrium point. In Chapter 1, Definitions 1.1 and 1.2 introduced the notions of stability, asymptotic stability, and instability for equilibrium points $x_* \in \mathbb{R}^n$ in finite-dimensional Hamiltonian systems. These concepts extend naturally to the infinite-dimensional setting, where the role of the state space $\mathbb{R}^n$ is taken over by a Banach space $X \subset \mathcal{H}$, and where the classical solution $x \in C^1([0,\tau_0),\mathbb{R}^n)$ is replaced by a suitable weak solution $u \in C^0([0,\tau_0),X)$ of the infinite-dimensional Hamiltonian system.

In this chapter we describe how to adapt the approaches based on linearization and energy functions that were introduced in the previous chapter for finite-dimensional systems to the stability analysis of solitary waves in infinite-dimensional Hamiltonian systems. The basic setup and examples of solitary waves as critical points of energy are described in Section 2.1. Stability of minimizers of energy is considered in Section 2.2. Instability of saddle points of energy is studied in Section 2.3. Saddle points of energy may become constrained minimizers of energy due to the existence of additional conserved quantities. Orbital stability of constrained energy minimizers is studied in Section 2.4. Degenerate critical points of energy are considered in Section 2.5.

2.1. Hamiltonian systems with solitary waves

2.1.1. Hamiltonian form of the evolution problem. Let $\mathcal{H}$ be a Hilbert space equipped with the inner product $\langle \cdot,\cdot \rangle$. Let X be a Banach space and X^* be its dual in $\mathcal{H}$ satisfying $X \subset \mathcal{H} \subset X^*$. The norm in X is denoted by $\| \cdot \|_X$ and the induced norm in $\mathcal{H}$ is denoted by $\| \cdot \|$.

We say that an evolution problem is cast in Hamiltonian form if there exists a C^1 energy function $H(u) : X \subset \mathcal{H} \to \mathbb{R}$ and a bounded, densely defined, skew-adjoint operator $J : \mathcal{H} \to \mathcal{H}$ with a bounded inverse J^{-1} such that

$$\begin{cases} \dfrac{du}{dt} = J\nabla H(u), & t > 0, \\ u|_{t=0} = u_0 \in X. \end{cases} \tag{2.1}$$

The skew-adjoint property of J can be expressed in the form:

$$\langle Jf, g\rangle = -\langle f, Jg\rangle, \quad f, g \in \mathcal{H}$$

and we assume that the operator $J : \mathcal{H} \to \mathcal{H}$ can be extended as operators $J : X \to X$ and $J : X^* \to X^*$. The gradient $\nabla H(u) : X \to X^*$ is defined in the sense of the variational derivative

$$\langle \nabla H(u), \varphi\rangle := \lim_{\epsilon\to 0} \frac{H(u + \epsilon\varphi) - H(u)}{\epsilon} \quad \text{for all } \varphi \in X.$$

Finally, we define the domain of $\nabla H(u)$ by $\mathcal{D} := \{u \in X : \nabla H(u) \in \mathcal{H}\}$.

REMARK 2.1. The stability theory for the Hamiltonian system (2.1) can also be extended to gradient systems in the form:

$$\begin{cases} \dfrac{du}{dt} = -\nabla H(u), & t > 0, \\ u|_{t=0} = u_0 \in X. \end{cases} \tag{2.2}$$

The following two examples of Hamiltonian systems are used throughout this chapter.

EXAMPLE 2.1. Consider the nonlinear wave (NW) equation in the scalar form

$$u_{tt} - u_{xx} + \gamma u^m = 0,$$

where $(t, x) \in \mathbb{R} \times \mathbb{R}$, $u \in \mathbb{R}$, $\gamma \in \mathbb{R}$, and $m \in \mathbb{N}$. The NW equation can be written in the Hamiltonian form (2.1) by defining $v = u_t$ and extending the scalar form to the vector form as

$$\frac{d}{dt}\begin{bmatrix} u \\ v \end{bmatrix} = \begin{bmatrix} 0 & 1 \\ -1 & 0 \end{bmatrix}\begin{bmatrix} -u_{xx} + \gamma u^m \\ v \end{bmatrix} = J\begin{bmatrix} \nabla_u \\ \nabla_v \end{bmatrix} H(u, v), \tag{2.3}$$

where

$$J = \begin{bmatrix} 0 & 1 \\ -1 & 0 \end{bmatrix},$$

and

$$H(u, v) = \int_{\mathbb{R}} \left[\frac{1}{2}u_x^2 + \frac{\gamma}{m+1}u^{m+1} + \frac{1}{2}v^2\right] dx.$$

Note that $J^* = J^{-1} = -J$. The energy space is given by

$$X = \{(u, v) : \ u \in H^1(\mathbb{R}), \ v \in L^2(\mathbb{R})\}$$

as a subspace of the Hilbert space $\mathcal{H} = L^2(\mathbb{R}) \times L^2(\mathbb{R})$, and we recall from Sobolev's embedding of $H^1(\mathbb{R})$ into $L^\infty(\mathbb{R})$ that the second term of $H(u, v)$ is bounded in X. Expanding

$$H(u + \epsilon\varphi, v + \epsilon\psi) - H(u, v) = \epsilon\left[\langle u_x, \varphi_x\rangle + \gamma\langle u^m, \varphi\rangle + \langle v, \psi\rangle\right] + \mathcal{O}(\epsilon^2)$$

we obtain

$$\begin{bmatrix} \nabla_u \\ \nabla_v \end{bmatrix} H(u, v) = \begin{bmatrix} -u_{xx} + \gamma u^m \\ v \end{bmatrix}.$$

This expression is well-defined in the dual space

$$X^* = \left\{(u,v): \ \ u \in H^{-1}(\mathbb{R}), \ \ v \in L^2(\mathbb{R})\right\},$$

where $H^{-1}(\mathbb{R})$ is the dual of $H^1(\mathbb{R})$ in $L^2(\mathbb{R})$, which can be defined using the Fourier transform and Plancherel's Theorem.

EXAMPLE 2.2. The nonlinear Schrödinger (NLS) equation can be written in the complex form:

$$i\psi_t = -\psi_{xx} + \gamma|\psi|^{2m}\psi,$$

where $(t,x) \in \mathbb{R}\times\mathbb{R}$, $\psi \in \mathbb{C}$, $\gamma \in \mathbb{R}$, and $m \in \mathbb{N}$. It is rather confusing to cast the NLS equation in the Hamiltonian form (2.1) by using complex-valued variables. It is more transparent to use the decomposition $\psi = u + iv$ with real-valued variables u and v. The NLS equation can be written in the Hamiltonian form (2.1) as

$$\frac{d}{dt}\begin{bmatrix} u \\ v \end{bmatrix} = \begin{bmatrix} 0 & 1 \\ -1 & 0 \end{bmatrix}\begin{bmatrix} -u_{xx} + \gamma(u^2+v^2)^m u \\ -v_{xx} + \gamma(u^2+v^2)^m u \end{bmatrix} = J\begin{bmatrix} \nabla_u \\ \nabla_v \end{bmatrix} H(u,v),$$

where J is the same as in Example 2.1 and

$$H(u,v) = \int_{\mathbb{R}}\left[\frac{1}{2}u_x^2 + \frac{1}{2}v_x^2 + \frac{\gamma}{2(m+1)}(u^2+v^2)^{m+1}\right]dx.$$

The energy space is given by $X = H^1(\mathbb{R}) \times H^1(\mathbb{R})$ as a subspace of the Hilbert space $\mathcal{H} = L^2(\mathbb{R}) \times L^2(\mathbb{R})$. In the original complex-valued coordinate $\psi = u + iv$, $H(u,v)$ can be rewritten in the equivalent form:

$$\tilde{H}(\psi,\bar{\psi}) = 2H(u,v) = \int_{\mathbb{R}}\left[|\psi_x|^2 + \frac{\gamma}{m+1}|\psi|^{2m+2}\right]dx.$$

The NLS equation is cast in the complex-valued Hamiltonian form as

$$\frac{d}{dt}\begin{bmatrix} \psi \\ \bar{\psi} \end{bmatrix} = \begin{bmatrix} 0 & -i \\ i & 0 \end{bmatrix}\begin{bmatrix} -\bar{\psi}_{xx} + W'(|\psi|^2)\bar{\psi} \\ -\psi_{xx} + W'(|\psi|^2)\psi \end{bmatrix} = \tilde{J}\begin{bmatrix} \nabla_\psi \\ \nabla_{\bar{\psi}} \end{bmatrix}\tilde{H}(\psi,\bar{\psi}),$$

where

$$\tilde{J} = \begin{bmatrix} 0 & -i \\ i & 0 \end{bmatrix}.$$

The energy function $\tilde{H}(\psi,\bar{\psi})$ is defined in the energy space $X = H^1(\mathbb{R},\mathbb{C})$. The matrix $\tilde{J}$ is skew-symmetric but not skew-adjoint in the sense that $\tilde{J}^T = -\tilde{J}$ but $\bar{\tilde{J}}^T = \tilde{J} \neq -\tilde{J}$. As a result, the complex-valued Hamiltonian form of the NLS equation requires a more special treatment in a Hilbert space $\mathcal{H} = L^2(\mathbb{R},\mathbb{C})$.

2.1.2. Local solutions of the evolution problem. Once the evolution problem (2.1) is formulated, it becomes relevant to identify its possible solutions. Two types of local solutions will be needed for the stability theory, which are typically referred to as **weak solutions** and **strong solutions**.

DEFINITION 2.1. We say that $u \in C^0([0,\tau_0),X) \cap C^1((0,\tau_0),X^*)$ is a weak local solution to the initial-value problem (2.1) with $u_0 \in X$ if

$$\langle\frac{du}{dt}(t,\cdot) - J\nabla H(u(t,\cdot)), \varphi(t,\cdot)\rangle = 0$$

holds for all $\varphi \in C^0([0,\tau_0],X)$ and every $t \in (0,\tau_0)$.

DEFINITION 2.2. We say that $u \in C^0([0,\tau_0),\mathcal{D}) \cap C^1((0,\tau_0),\mathcal{H})$ is a strong local solution to the initial-value problem (2.1) with $u_0 \in \mathcal{D} \subset X$ if

$$\left\| \frac{du}{dt}(t,\cdot) - J\nabla H(u(t,\cdot)) \right\| = 0$$

holds for every $t \in (0,\tau_0)$.

REMARK 2.2. Local solutions to nonlinear dispersive equations can be obtained in many cases by using Duhamel's principle and contraction mapping arguments. For Examples 2.1 and 2.2, local well-posedness of the Hamiltonian system (2.1) has been proven in the energy and domain spaces [**40, 181**]. Since analysis of local well-posedness is not the subject of this monograph, we will always assume that the nonlinear dispersive wave equations that we study in our examples are locally well-posed.

If $u_* \in X$ is a critical point of the energy function $H(u)$, then $\nabla H(u_*) = 0$ and $u(t,\cdot) = u_*$ is independent of time $t \in \mathbb{R}$. Stability, asymptotic stability, and instability of the critical point $u_* \in X$ can be defined similarly to Definitions 1.1 and 1.2.

DEFINITION 2.3. Let $u_* \in X$ be a critical point of $H(u)$. We say that u_* is **stable** if for every $\epsilon > 0$ (small) there exists some $\delta > 0$ such that for every $u_0 \in X$ satisfying

$$\|u_0 - u_*\|_X < \delta,$$

the unique weak solution $u \in C^0(\mathbb{R}_+, X)$ of the Hamiltonian system (2.1) satisfies

$$\|u(t,\cdot) - u_*\|_X < \epsilon \quad \text{for every } t > 0.$$

If in addition, $\lim_{t\to\infty} \|u(t,\cdot) - u_*\|_X = 0$, we say that u_* is **asymptotically stable**. If it is not stable, then it is **unstable**.

2.1.3. Symmetries of the Hamiltonian systems. The autonomous Hamiltonian system (2.1) is invariant under translations in time t:

$$u(t,\cdot) \mapsto u(t+t_0,\cdot), \quad \text{for arbitrary } t_0 \in \mathbb{R}.$$

This translation symmetry is associated with the fact that the energy function $H(u)$ is conserved along every strong solution $u \in C^0([0,\tau_0),\mathcal{D})$ as in Definition 2.2. Indeed,

$$\frac{d}{dt}H(u(t,\cdot)) = \langle \nabla H(u(t,\cdot)), J\nabla H(u(t,\cdot))\rangle = 0,$$

since $\nabla H(u) \in \mathcal{H}$ and

$$\langle \nabla H(u), J\nabla H(u)\rangle = -\langle J\nabla H(u), \nabla H(u)\rangle = -\langle \nabla H(u), J\nabla H(u)\rangle.$$

By an approximation argument for a sequence of strong local solutions converging to a weak local solution of the Hamiltonian system (2.1), conservation of energy

$$H(u(t,\cdot)) = H(u_0), \quad t \in [0,\tau_0)$$

is extended to every weak solution $u \in C^0([0,\tau_0),X)$ as in Definition 2.1.

In many physical problems, the Hamiltonian system (2.1) has additional symmetries such as translations in space and phase rotation.

We represent each additional symmetry with the symmetry operator $T(\theta) : \mathcal{H} \to \mathcal{H}$, where $\theta \in \mathbb{R}$ is a parameter and assume that the symmetry operator can be extended as $T(\theta) : X \to X$. We define the infinitesimal generator of T as the linear operator $T' : \mathrm{Dom}(T') \subset \mathcal{H} \to \mathcal{H}$,

$$T'u := \lim_{\theta \to 0} \frac{T(\theta)u - u}{\theta},$$

for every $u \in \mathrm{Dom}(T') \subset \mathcal{H}$ for which the limit exists. If we set $T'(0) = T'$ and $T(0) = \mathrm{id}_{\mathcal{H}}$, then we can equally define $T'(\theta) = T(\theta)T' = T'T(\theta)$ for every $\theta \in \mathbb{R}$. We assume that if $u \in X$, then $J^{-1}T'u \in X^*$.

The following definition collects the five most important properties of $T(\theta)$ and T'.

DEFINITION 2.4. We say that the bounded linear operator $T(\theta) : \mathcal{H} \to \mathcal{H}$, $\theta \in \mathbb{R}$, is a **symmetry operator** for the Hamiltonian system (2.1) if it satisfies the following properties:

(a) **Group**: $T(0) = \mathrm{id}_{\mathcal{H}}$ is the identity on $\mathcal{H}$ and

$$T(\theta_1 + \theta_2) = T(\theta_1)T(\theta_2) = T(\theta_2)T(\theta_1), \quad \theta_1, \theta_2 \in \mathbb{R}.$$

This implies that $T(\theta)$ is invertible with inverse $[T(\theta)]^{-1} = T(-\theta)$.

(b) **Isometry**: $\mathcal{H}$ is invariant under $T(\theta)$ such that

$$\langle T(\theta)f, T(\theta)g\rangle = \langle f, g\rangle, \quad f, g \in \mathcal{H}.$$

(c) **Invariance**: $H(T(\theta)u) = H(u)$ for all $u \in X$.

(d) **Commutativity**: $JT(\theta) = T(\theta)J$.

(e) **Skew-adjointness**:

$$\langle T'f, g\rangle = -\langle f, T'g\rangle, \quad f, g \in \mathrm{Dom}(T') \subset \mathcal{H}.$$

Note that if $\mathrm{Dom}(T') = \mathcal{H}$, then skew-adjointness of T' follows from

$$0 = \partial_\theta \langle T(\theta)u, T(\theta)u\rangle\big|_{\theta=0} = \langle T'u, u\rangle + \langle u, T'u\rangle,$$

where we have used the isometry property (b).

Associated to each symmetry operator $T(\theta)$ there exists an additional conserved quantity denoted by $Q(u)$. The following lemma explains why $T(\theta)$ is a symmetry operator for the Hamiltonian system (2.1) and how the associated conserved quantity can be obtained.

LEMMA 2.1. *Assume that $T(\theta) : \mathcal{H} \to \mathcal{H}$ is the symmetry operator in Definition 2.4 and define $Q(u) : X \to \mathbb{R}$ by*

$$Q(u) := \frac{1}{2}\langle J^{-1}T'u, u\rangle. \tag{2.4}$$

If $u \in C^0([0, \tau_0), X)$ is a weak solution of the Hamiltonian system (2.1), then so is $T(\theta)u$ for every fixed $\theta \in \mathbb{R}$. For the generator, we obtain that

$$T'u = J\nabla Q(u). \tag{2.5}$$

Moreover, Q is invariant with respect to $T(\theta)$ and conserved in time, that is

$$Q(T(\theta)u) = Q(u)$$

for every $\theta \in \mathbb{R}$ and $u \in X$, and

$$Q(u(t, \cdot)) = Q(u_0), \quad t \in [0, \tau_0),$$

for every strong solution $u \in C^0([0, \tau_0), \mathcal{D})$.

PROOF. By using the group property (a), isometry (b), and invariance (c) we obtain that

$$\begin{aligned}\langle \nabla H(T(\theta)u), \phi\rangle &= \lim_{\varepsilon\to 0}\frac{1}{\varepsilon}\left(H(T(\theta)u+\varepsilon\phi) - H(T(\theta)u)\right)\\ &= \lim_{\varepsilon\to 0}\frac{1}{\varepsilon}\left(H(T(\theta)(u+T(-\theta)\varepsilon\phi) - H(T(\theta)u)\right)\\ &= \lim_{\varepsilon\to 0}\frac{1}{\varepsilon}\left(H(u+T(-\theta)\varepsilon\phi) - H(u)\right)\\ &= \langle \nabla H(u), T(-\theta)\phi\rangle\\ &= \langle T(\theta)\nabla H(u), \phi\rangle.\end{aligned}$$

Therefore, using commutativity (d), we find that $J\nabla H(T(\theta)u) = T(\theta)J\nabla H(u)$. Since $T(\theta)$ is continuous, $T(\theta)u \in C^0([0,\tau_0],X)$ for every $u \in C^0([0,\tau_0],X)$. This implies that $T(\theta)u$ is a weak solution of (2.1) if u is a weak solution in the sense of Definition 2.1, since using the group property (a) and isometry (b) we find that

$$\begin{aligned}\langle \frac{d}{dt}T(\theta)u, \varphi\rangle &= \langle T(\theta)\frac{d}{dt}u, \varphi\rangle = \langle \frac{d}{dt}u, T(-\theta)\varphi\rangle = \langle J\nabla H(u), T(-\theta)\varphi\rangle\\ &= \langle T(\theta)J\nabla H(u), \varphi\rangle = \langle J\nabla H(T(\theta)u), \varphi\rangle,\end{aligned}$$

for every fixed $\theta \in \mathbb{R}$ and every $\varphi \in C^0([0,\tau_0],X)$. Furthermore, in view of property (d), J commutes with T'. Moreover, it follows from property (d) that J^{-1} commutes with T, and hence with T'. Therefore, the composition $J^{-1}T'$ of two skew-adjoint and commuting operators J^{-1} and T' is self-adjoint. It follows that the first variation of Q gives $\nabla Q(u) = J^{-1}T'u$, which implies (2.5). Using (2.5) we can prove the invariance of $Q(u)$ with respect to time t using the skew-adjointness of J, properties (b) and (c):

$$\begin{aligned}\frac{d}{dt}Q(u) &= \langle \nabla Q(u), J\nabla H(u)\rangle = \langle J^{-1}T'u, J\nabla H(u)\rangle = -\langle T'u, \nabla H(u)\rangle\\ &= -\langle T(\theta)T'u, T(\theta)\nabla H(u)\rangle = -\langle T'(\theta)u, \nabla H(T(\theta)u)\rangle\\ &= -\frac{d}{d\theta}H(T(\theta)u) = -\frac{d}{d\theta}H(u) = 0,\end{aligned}$$

where all inner products are well-defined for the strong solution $u \in C^0([0,\tau_0),\mathcal{D})$. The invariance of $Q(u)$ with respect to $T(\theta)$ follows from the commutativity of J^{-1} and T' and the isometry property (b):

$$\begin{aligned}Q(T(\theta)u) &= \frac{1}{2}\langle J^{-1}T'T(\theta)u, T(\theta)u\rangle\\ &= \frac{1}{2}\langle T(\theta)J^{-1}T'u, T(\theta)u\rangle\\ &= \frac{1}{2}\langle J^{-1}T'u, u\rangle\\ &= Q(u),\end{aligned}$$

for every $u \in X$. □

REMARK 2.3. Similar to the conservation of energy $H(u)$, conservation of $Q(u)$ is extended to every weak solution $u \in C^0([0,\tau_0),X)$.

REMARK 2.4. Since T' is independent of u, the conserved quantity $Q(u)$ in (2.4) is quadratic in u. A linear conserved quantity $M(u) = \langle 1, u\rangle$ arises in Hamiltonian systems with an unbounded skew-adjoint operator J (Example 3.2). Other conserved quantities with polynomial dependence on u may also arise, e.g., in integrable Hamiltonian systems (Section 4.1).

2.1.4. Relative equilibria. The equilibrium $u(t,\cdot) = u_*$ given by a critical point of $H(u)$ is extended with an arbitrary parameter $\omega \in \mathcal{I} \subset \mathbb{R}$ to a family $u(t,\cdot) = T(\omega t)u_\omega$ of **relative equilibria** of the Hamiltonian system (2.1), where $\mathcal{I}$ is a suitable interval (bounded or unbounded) and u_ω is the spatial profile for $\omega \in \mathcal{I}$. The relative equilibria are also known as **stationary states** of the Hamiltonian system (2.1).

LEMMA 2.2. *Let $u(t,\cdot) = T(\omega t)u_\omega$, $t \in \mathbb{R}$, be a stationary state of the Hamiltonian system* (2.1) *in X. Then, $u_\omega \in X$ is a critical point of the* ***augmented energy***

$$\Lambda_\omega(u) := H(u) - \omega Q(u),$$

so that $\nabla \Lambda_\omega(u_\omega) = \nabla H(u_\omega) - \omega \nabla Q(u_\omega) = 0$.

PROOF. If $u(t,\cdot) = T(\omega t)u_\omega$, $t \in \mathbb{R}$ is a weak solution of the Hamiltonian system (2.1), then $u_\omega \in X$ satisfies

$$T'(\omega t)\omega u_\omega = J\nabla_{T(\omega t)u_\omega} H(T(\omega t)u_\omega) = JT(\omega)\nabla H(T(\omega t)u_\omega) = T(\omega t)J\nabla H(u_\omega).$$

Since $T'(\omega t) = T(\omega t)T'$, it follows that u_ω satisfies

$$\omega T' u_\omega = J\nabla H(u_\omega).$$

It follows from (2.5) that $u_\omega \in X$ is a critical point of $\Lambda_\omega(u)$ since J is invertible. □

For the stationary state $u(t,\cdot) = T(\omega t)u_\omega$, the concept of stability can be extended to the concept of orbital stability, similar to Definition 1.8 for periodic orbits in finite-dimensional dynamical systems.

DEFINITION 2.5. Fix $\omega \in \mathcal{I} \subset \mathbb{R}$ and let $u_\omega \in X$ be a critical point of $\Lambda_\omega(u)$. We say that the critical point u_ω is **orbitally stable in** X if for every $\epsilon > 0$ (small) there exists some $\delta > 0$ such that for every $u_0 \in X$ satisfying

$$\|u_0 - u_\omega\|_X < \delta,$$

the unique weak solution $u \in C^0(\mathbb{R}_+, X)$ satisfies

$$\inf_{\theta \in \mathbb{R}} \|u(t,\cdot) - T(\theta)u_\omega\|_X < \epsilon, \qquad \text{for every } t > 0.$$

It is called **orbitally unstable** if it is not orbitally stable.

REMARK 2.5. The argument of $\inf_{\theta \in \mathbb{R}} \|u(t,\cdot) - T(\theta)u_\omega\|_X$, say $\theta_*(t)$, depends on time t. However, Definition 2.5 does not require us to control the evolution of θ_* in t.

REMARK 2.6. If u_* is a critical point of $H(u)$, then Definition 2.5 can be used for orbital stability of the equilibrium u_* in the particular case $\omega = 0$.

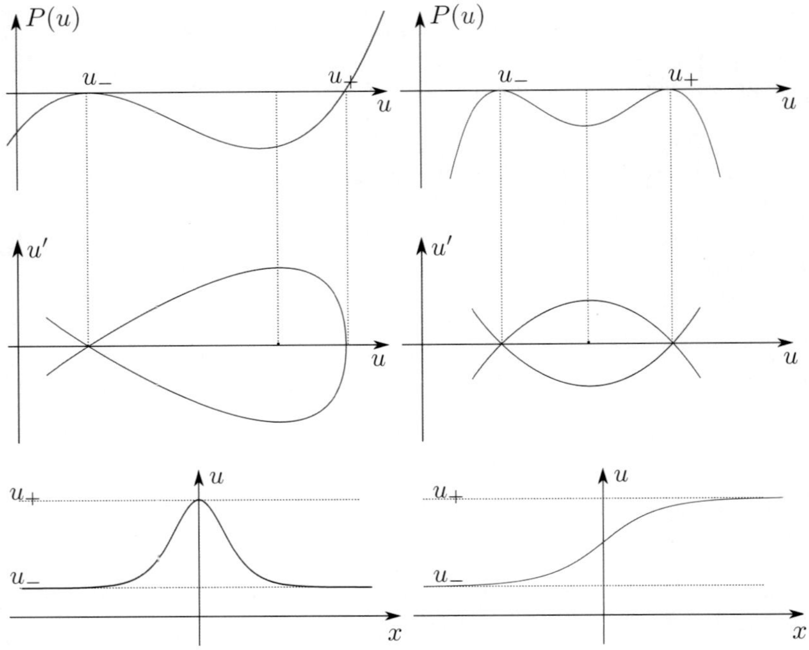

FIGURE 2.1. Cases (i) (left) and (ii) (right) of Lemma 2.3. Top: potential energy P versus u. Middle: phase portraits of Newton's system (2.8). Bottom: homoclinic and heteroclinic orbits versus x.

2.1.5. Examples of solitary waves. In order to be able to classify the solitary waves, let us first give an elementary result on the existence of homoclinic and heteroclinic orbits in Newton's equation:

$$\frac{d^2u}{dx^2} + P'(u) = 0, \tag{2.6}$$

where $P \in C^2(\mathbb{R})$ and $u = u(x)$. This degree-one Hamiltonian system is integrable with the energy function

$$E(u,v) = \frac{1}{2}v^2 + P(u), \quad v := \frac{du}{dx}, \tag{2.7}$$

where $E(u,v) = \mathcal{E}$ is constant in x along solutions. Newton's equation (2.6) describes the dynamics of a particle with position u and velocity v in a one-dimensional potential field, for which $P(u)$ is the potential energy, $\frac{1}{2}v^2$ is the kinetic energy, and $\mathcal{E}$ is the total energy.

The following lemma gives a technical result on the existence of homoclinic and heteroclinic orbits of Newton's equation (2.6). The two cases are illustrated in Figure 2.1.

LEMMA 2.3. *Assume* $P \in C^2(\mathbb{R})$.

(i) *If there exist* $\mathfrak{u}_- < \mathfrak{u}_+$ *such that*
- $P(\mathfrak{u}_-) = 0,\ P'(\mathfrak{u}_-) = 0,\ P''(\mathfrak{u}_-) < 0,$
- $P(\mathfrak{u}_+) = 0,\ P'(\mathfrak{u}_+) > 0,$
- $P(u) < 0$ *for* $u \in (\mathfrak{u}_-, \mathfrak{u}_+)$,

then there exists a homoclinic orbit satisfying $u(x) \in (\mathfrak{u}_-, \mathfrak{u}_+]$ *for* $x \in \mathbb{R}$ *and* $u(x) \to \mathfrak{u}_-$ *as* $x \to \pm\infty$.

(ii) *If there exist* $\mathfrak{u}_- < \mathfrak{u}_+$ *such that*
- $P(\mathfrak{u}_\pm) = 0$, $P'(\mathfrak{u}_\pm) = 0$, $P''(\mathfrak{u}_\pm) < 0$,
- $P(u) < 0$ *for* $u \in (\mathfrak{u}_-, \mathfrak{u}_+)$,

then there exist two heteroclinic orbits satisfying $u(x) \in (\mathfrak{u}_-, \mathfrak{u}_+)$ *for* $x \in \mathbb{R}$ *and either* $u(x) \to \mathfrak{u}_\pm$ *as* $x \to \pm\infty$ *or* $u(x) \to \mathfrak{u}_\mp$ *as* $x \to \pm\infty$.

PROOF. Newton's equation (2.6) can be written as the planar system

$$\begin{cases} u' = v, \\ v' = -P'(u). \end{cases} \tag{2.8}$$

A straightforward computation shows that system (2.8) possesses a first integral of the form (2.7). Orbits of the planar system (2.8) correspond to the level curves $E(u, v) = \mathcal{E}$ for a fixed value of $\mathcal{E}$. Since the energy $E(u, v)$ is quadratic in v, the orbits in the (u, v) phase plane are symmetric with respect to the horizontal axis.

If the conditions $P(\mathfrak{u}_-) = 0$, $P'(\mathfrak{u}_-) = 0$, $P''(\mathfrak{u}_-) < 0$ are satisfied as in (i), then $(\mathfrak{u}_-, 0)$ is a saddle point of $E(u, v)$ corresponding to the level $\mathcal{E} = 0$ and the orbit on the phase plane exists whenever $P(u) \leq 0$. If the conditions $P(\mathfrak{u}_+) = 0$, $P'(\mathfrak{u}_+) > 0$, and $P(u) < 0$ for $u \in (\mathfrak{u}_-, \mathfrak{u}_+)$ are satisfied, the orbit reaches the point $(\mathfrak{u}_+, 0)$ but since this point is not a critical point of the second-order equation (2.6), the orbit turns and returns back to the saddle point $(\mathfrak{u}_-, 0)$ symmetrically in the (u, v) plane. The homoclinic orbit arises at the intersection of the unstable and stable curves connected at the saddle point $(\mathfrak{u}_-, 0)$, which is reached in the limits $x \to \pm\infty$ whereas the turning point $(\mathfrak{u}_+, 0)$ is reached for a finite x. Hence, the homoclinic orbit satisfies the properties claimed in (i).

If the conditions $P(\mathfrak{u}_\pm) = 0$, $P'(\mathfrak{u}_\pm) = 0$, $P''(\mathfrak{u}_\pm) < 0$, are satisfied as in (ii), then $(\mathfrak{u}_-, 0)$ and $(\mathfrak{u}_+, 0)$ are saddle points of $E(u, v)$ corresponding to the level $\mathcal{E} = 0$. The orbits on the phase plane exist whenever $P(u) \leq 0$. If the condition $P(u) < 0$ for $u \in (\mathfrak{u}_-, \mathfrak{u}_+)$ is satisfied, the orbits reach the saddle points $(\mathfrak{u}_-, 0)$ and $(\mathfrak{u}_+, 0)$ in the limits $x \to \pm\infty$. The heteroclinic orbits arise at the intersection of the unstable curve of one saddle point and the stable curve of the other saddle point and satisfy the properties claimed in (ii). ∎

Solitary waves of the nonlinear dispersive wave equations in Examples 2.1 and 2.2 are obtained from the homoclinic and heteroclinic orbits of Lemma 2.3. These solutions have different physical meanings and different names, as we will see below.

EXAMPLE 2.3. Let us generalize Example 2.1 and consider a generalized NW equation

$$u_{tt} - u_{xx} + V'(u) = 0, \tag{2.9}$$

where $V(u) : \mathbb{R} \to \mathbb{R}$. The equilibrium solution $u(t, x) = u_*(x)$ satisfies Newton's equation

$$-u_*'' + V'(u_*) = 0,$$

which corresponds to (2.6) with $P(u) = -V(u)$. Assuming $V \in C^2(\mathbb{R})$ we have by Lemma 2.3 with $P(u) = -V(u)$ the existence of a heteroclinic orbit satisfying $u_*(x) \to u_\pm$ as $x \to \pm\infty$ if there exist $u_- < u_+$ such that

- $V(u_\pm) = 0$, $V'(u_\pm) = 0$, $V''(u_\pm) > 0$,
- $V(u) > 0$ for $u \in (u_-, u_+)$.

The heteroclinic orbit is known as a **kink** of the NW equation. We also have the existence of a homoclinic orbit satisfying $u_*(x) \to 0$ as $x \to \pm\infty$ if there exists $u_+ > 0$ such that

- $V(0) = 0$, $V'(0) = 0$, $V''(0) > 0$,
- $V(u_+) = 0$, $V'(u_+) < 0$,
- $V(u) > 0$ for $u \in (0, u_+)$.

The homoclinic orbit is known as a **pulse** of the NW equation. Since the NW equation is autonomous in (t, x), the Hamiltonian system admits the conserved energy

$$H(u, v) = \int_{\mathbb{R}} \left[\frac{1}{2} u_x^2 + V(u) + \frac{1}{2} v^2 \right] dx \tag{2.10}$$

and the translation symmetry $T(\theta) : \mathcal{H} \mapsto \mathcal{H}$ with $\mathcal{H} = L^2(\mathbb{R}) \times L^2(\mathbb{R})$ given by

$$T(\theta) \begin{bmatrix} u \\ v \end{bmatrix} (x) := \begin{bmatrix} u \\ v \end{bmatrix} (x + \theta), \quad \theta \in \mathbb{R}. \tag{2.11}$$

By Lemma 2.1 with J in (2.3) and $T' = \partial_x$ with $\mathrm{Dom}(T') = H^1(\mathbb{R}) \times H^1(\mathbb{R})$, the translation symmetry T is related to the additional conserved quantity called **momentum**:

$$P(u, v) = \frac{1}{2} \int_{\mathbb{R}} (v u_x - v_x u) dx. \tag{2.12}$$

Stationary states of the generalized NW equation (2.9) are given by

$$\begin{bmatrix} u \\ v \end{bmatrix} (t, x) = T(ct) \begin{bmatrix} u_c \\ v_c \end{bmatrix} (x) = \begin{bmatrix} u_c \\ v_c \end{bmatrix} (x + ct),$$

which are also called **traveling wave solutions**. By using the energy (2.10) and the momentum (2.12), the augmented energy is defined by

$$\Lambda_c(u, v) := H(u, v) - cP(u, v). \tag{2.13}$$

The critical points (u_c, v_c) of $\Lambda_c(u, v)$ are obtained as solutions of the Euler–Lagrange equations

$$\begin{bmatrix} \nabla_u \\ \nabla_v \end{bmatrix} \Lambda_c(u, v) = \begin{bmatrix} -u_{xx} + V'(u) + c v_x \\ v - c u_x \end{bmatrix} = \begin{bmatrix} 0 \\ 0 \end{bmatrix},$$

from which we find

$$v_c = c u_c', \qquad -(1 - c^2) u_c'' + V'(u_c) = 0. \tag{2.14}$$

By Lemma 2.3 with $P(u) = -(1 - c^2)^{-1} V(u)$, if a solution u_* of (2.14) exists for $c = 0$, then a solution u_c of (2.14) exists for every $c \in (-1, 1)$.

REMARK 2.7. The energy space for the generalized NW equation (2.9) is defined by boundedness of each term in the energy $H(u, v)$ given by (2.10). In the case of kinks, it is easier to set up the energy space if the perturbations to the kink are required to decay to zero at infinity. In the case of pulses, both the pulse and its perturbation are spatially decaying to zero at infinity.

EXAMPLE 2.4. Let us generalize Example 2.2 and consider a generalized NLS equation

$$i\psi_t = -\psi_{xx} + W'(|\psi|^2)\psi, \tag{2.15}$$

where $W(|\psi|^2) : \mathbb{R}_+ \to \mathbb{R}$. The equilibrium solution $\psi(t,x) = \psi_*(x)$ with real ψ_* satisfies the Newton's equation

$$-\psi_*'' + W'(\psi_*^2)\psi_* = 0,$$

which corresponds to (2.6) with $P(u) = -W(u^2)$. Assuming $W \in C^2(\mathbb{R}_+)$ we have by Lemma 2.3 with $P(u) = -W(u^2)$ the existence of a homoclinic orbit satisfying $\psi_*(x) \to 0$ as $x \to \pm\infty$ if there exists $\psi_+ > 0$ such that

- $W(0) = 0$, $W'(0) > 0$,
- $W(\psi_+^2) = 0$, $W'(\psi_+^2) < 0$,
- $W(\psi^2) > 0$ for $\psi \in (0, \psi_+)$.

The homoclinic orbit is known as a **bright soliton** of the NLS equation. We also have the existence of a heteroclinic orbit satisfying $\psi_*(x) \to \pm\psi_+$ as $x \to \pm\infty$ if there exists $\psi_+ > 0$ such that

- $W(\psi_+^2) = 0$, $W'(\psi_+^2) = 0$, $W''(\psi_+^2) > 0$,
- $W(\psi^2) > 0$ for $\psi \in (-\psi_+, \psi_+)$.

The heteroclinic orbit is known as a **black soliton** of the NLS equation. The complex-valued Hamiltonian system with the conserved energy

$$H(\psi, \bar{\psi}) = \int_{\mathbb{R}} \left[|\psi_x|^2 + W(|\psi|^2) \right] dx$$

admits two symmetry operators:

- the translation symmetry T_1,

$$\psi(t,x) \mapsto T_1(\xi)\psi(t,x) := \psi(t, x+\xi), \qquad \xi \in \mathbb{R}, \tag{2.16}$$

- the phase rotation symmetry T_2,

$$\psi(t,x) \mapsto T_2(\theta)\psi(t,x) := e^{-i\theta}\psi(t,x), \qquad \theta \in \mathbb{R}, \tag{2.17}$$

where we note that $T_2(\theta)(\psi, \bar{\psi}) = (e^{-i\theta}\psi, e^{i\theta}\bar{\psi})$. The two symmetry operators are related to the two conserved quantities called **momentum** and **mass**, respectively:

$$P(\psi, \bar{\psi}) = \frac{i}{2} \int_{\mathbb{R}} (\bar{\psi}\psi_x - \bar{\psi}_x \psi) dx$$

and

$$Q(\psi, \bar{\psi}) = \int_{\mathbb{R}} |\psi|^2 dx.$$

These conserved quantities can be obtained from Lemma 2.1 by using the complex-valued Hamiltonian form

$$\frac{d}{dt} \begin{bmatrix} \psi \\ \bar{\psi} \end{bmatrix} = \tilde{J} \begin{bmatrix} \nabla_\psi \\ \nabla_{\bar{\psi}} \end{bmatrix} H(\psi, \bar{\psi}), \quad \tilde{J} = \begin{bmatrix} 0 & -i \\ i & 0 \end{bmatrix}.$$

As is mentioned in Example 2.2, $\tilde{J} = \tilde{J}^{-1}$ is skew-symmetric but not skew-adjoint, hence the inner products in (2.4) should be used without complex conjugation. Hence we obtain

$$\frac{1}{2} \langle \tilde{J}^{-1} T_1' \begin{bmatrix} \psi \\ \bar{\psi} \end{bmatrix}, \begin{bmatrix} \psi \\ \bar{\psi} \end{bmatrix} \rangle = \frac{1}{2} \int_{\mathbb{R}} \begin{bmatrix} -i\bar{\psi}_x \\ i\psi_x \end{bmatrix} \cdot \begin{bmatrix} \psi \\ \bar{\psi} \end{bmatrix} dx = P(\psi, \bar{\psi})$$

and

$$\frac{1}{2}\langle \tilde{J}^{-1}T_2' \begin{bmatrix} \psi \\ \bar{\psi} \end{bmatrix}, \begin{bmatrix} \psi \\ \bar{\psi} \end{bmatrix} \rangle = \frac{1}{2}\int_{\mathbb{R}} \begin{bmatrix} \bar{\psi} \\ \psi \end{bmatrix} \cdot \begin{bmatrix} \psi \\ \bar{\psi} \end{bmatrix} dx = Q(\psi, \bar{\psi})$$

Stationary states of the generalized NLS equation (2.15), also called traveling wave solutions, are given by

$$\psi(t,x) = T_1(ct)T_2(\omega t)\psi_{c,\omega}(x) = \psi_{c,\omega}(x+ct)e^{-i\omega t}.$$

By defining the augmented energy

$$\Lambda_{c,\omega}(\psi,\bar{\psi}) := H(\psi,\bar{\psi}) - cP(\psi,\bar{\psi}) - \omega Q(\psi,\bar{\psi}), \tag{2.18}$$

we obtain the Euler–Lagrange equation in the form

$$\nabla_{\bar{\psi}}\Lambda_{c,\omega}(\psi,\bar{\psi}) = -\psi_{xx} + W'(|\psi|^2)\psi - ic\psi_x - \omega\psi = 0.$$

Critical points $\psi_{c,\omega}$ of $\Lambda_{c,\omega}(\psi,\bar{\psi})$ satisfy

$$\psi_{c,\omega}'' - W'(|\psi_{c,\omega}|^2)\psi_{c,\omega} + ic\psi_{c,\omega}' + \omega\psi_{c,\omega} = 0. \tag{2.19}$$

By an elementary transformation (see Theorem 2.9), one can show that if a solution ψ_* of (2.19) exists for $(c,\omega) = (0,0)$, then a solution $\psi_{c,\omega}$ of (2.19) exists for (c,ω) in an open region $\mathcal{I} \subset \mathbb{R}^2$ containing $(0,0)$.

REMARK 2.8. The energy space for the generalized NLS equation (2.15) is defined by boundedness of each term in the energy $H(\psi)$. In the case of bright solitons, both the bright soliton and its perturbation are spatially decaying. However, in the case of black solitons, both the black soliton and its perturbation might not be spatially decaying due to the complex-valued structure of ψ in $W(|\psi|^2)$. As a result, the stability of black solitons is a delicate problem, which is considered separately in Chapter 7.

2.2. Minimizers of energy

Since solitary waves are critical points of energy, we can adopt the energy function methods from Section 1.3 in order to prove that minimizers of energy are stable in the sense of Definition 2.3.

One necessary ingredient of the theory is the existence of weak solutions in the sense of Definition 2.1 for which the energy is preserved in time. Local well-posedness for many Hamiltonian systems such as those in Examples 2.3 and 2.4 has been proven from a standard contraction principle (Banach fixed-point theorem) and its extensions [**40, 181**]. Since this monograph does not deal with the local well-posedness theory for partial differential equations, we impose it as the following assumption.

ASSUMPTION 2.1. Assume that the Hamiltonian system (2.1) admits a unique local weak solution $u \in C^0([0,\tau_0), X) \cap C^1((0,\tau_0), X^*)$ in the sense of Definition 2.1 such that $H(u(t,\cdot)) = H(u_0)$ for every $t \in [0,\tau_0)$. Moreover, the solution depends continuously on the initial data $u_0 \in X$.

REMARK 2.9. If the Hamiltonian system (2.1) admits a strong local solution $u \in C^0([0,\tau_0),\mathcal{D}) \cap C^1((0,\tau_0),\mathcal{H})$, then the skew-adjointness of the operator J implies that the energy $H(u(t,\cdot)) = H(u_0)$ is conserved in time. In general, however, this property is not true and there are cases where one may only have that the energy is nonincreasing, see for instance [**39**, Theorem 1.2].

Lyapunov's stability theory (Theorem 1.4) implies that every isolated minimum of $H \in C^1(X, \mathbb{R})$ in X with $\dim(X) < \infty$ is stable. Two difficulties arise when the result is extended to the Banach space X with $\dim(X) = \infty$. Let $B_\epsilon(u_*)$ be a ball in X of radius $\epsilon > 0$ centered at $u_* \in X$ and define

$$m_\epsilon := \inf_{u \in \partial B_\epsilon(u_*)} H(u) \quad \text{and} \quad M_\delta := \sup_{u \in \overline{B_\delta(u_*)}} H(u).$$

First, since X is infinite-dimensional, it cannot be taken for granted that

- m_ϵ is necessarily positive, and that
- for every $\epsilon > 0$ there exists $\delta > 0$ such that $0 < M_\delta < m_\epsilon$.

Second, due to the symmetries of the Hamiltonian system, the minimizers of energy are usually degenerate because they are invariant with respect to the symmetry group of translations. We will show here how to overcome both difficulties for infinite-dimensional Hamiltonian systems.

One powerful way to control perturbations of the critical point u_* of $H(u)$ in X is to construct the second variation of $H(u)$ at u_* provided that $H \in C^2(X, \mathbb{R})$. The assumption that $\nabla H(u) : X \to X^*$ is C^1 is equivalent to the assumption used in the linearization methods of Section 1.2 and is more restrictive compared to the assumption used in the energy function methods of Section 1.3.

Since u_* is a critical point of $H(u)$ and $H \in C^2(X, \mathbb{R})$, we have the expansion of energy for the perturbation $w := u - u_*$ of u_*,

$$H(u_* + w) = H(u_*) + \underbrace{\langle \nabla H(u_*), w \rangle}_{=0} + \frac{1}{2} \langle \mathcal{L} w, w \rangle + R_H(w), \quad w \in X, \tag{2.20}$$

where $\mathcal{L} := \nabla^2 H(u_*) : X \to X^*$ is a self-adjoint operator in the Hilbert space $\mathcal{H}$ with $X \subset \mathcal{H} \subset X^*$ and $R_H \in C^2(X, \mathbb{R})$ is a superquadratic remainder term satisfying

$$\lim_{\|w\|_X \to 0} \frac{|R_H(w)|}{\|w\|_X^2} = 0.$$

This implies that there exists a (small) $C_R > 0$ such that

$$|R_H(w)| \le C_R \|w\|_X^2, \quad \forall w \in X : \ \|w\|_X \le R, \tag{2.21}$$

where $C_R \to 0$ as $R \to 0$. The quadratic form of $\mathcal{L}$ is bounded in X with the bound

$$|\langle \mathcal{L} w, w \rangle| \le C \|w\|_X^2, \quad \forall w \in X, \tag{2.22}$$

with some constant $C > 0$. If a symmetry operator $T(\theta)$ is present in the Hamiltonian system (2.1) according to Definition 2.4 and $T' u_* \in X$, then $\mathcal{L}$ has a nontrivial kernel since

$$\mathcal{L} u_{\text{sym}} = 0, \quad u_{\text{sym}} := T' u_*. \tag{2.23}$$

Indeed, since u_* is a critical point of $H(u)$ and H is invariant under T by property (c) of Definition 2.4, it follows by differentiating $\nabla H(T(\theta) u_*) = 0$ with respect to θ that

$$0 = \frac{d}{d\theta} \nabla H(T(\theta) u_*) \bigg|_{\theta = 0} = \mathcal{L} u_{\text{sym}}.$$

The following theorem rephrases the conditions of Theorem 1.4 in terms of the second variation of H in (2.20) and provides the stability of a strict local minimizer of energy in the Hamiltonian system (2.1). The result is equivalent to Corollary 1.2

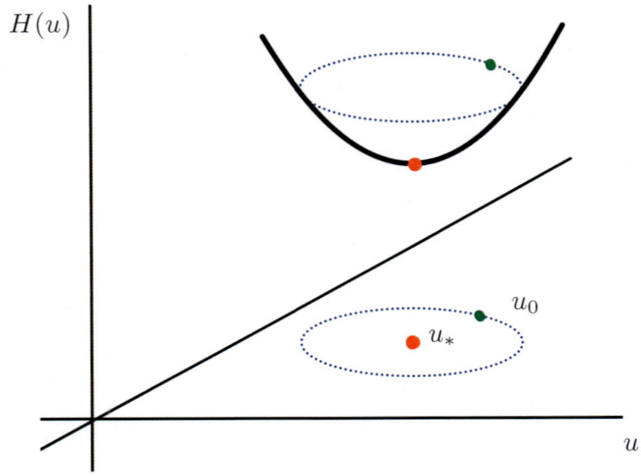

FIGURE 2.2. A sketch of a bounded level set $H(u) = H(u_0)$ in the u-plane near the local minimum u_*.

in the finite-dimensional setting. Figure 2.2 illustrates the level set of $H(u)$ with a minimum at u_*.

THEOREM 2.1. *Let Assumption* 2.1 *be satisfied. Let $u_* \in X$ be a critical point of $H(u) : X \to \mathbb{R}$ such that $H \in C^2(X, \mathbb{R})$. If u_* is a strict local minimizer of $H(u)$, i.e., there exists $C_0 > 0$ such that the linear operator $\mathcal{L} := \nabla^2 H(u_*) : X \to X^*$ satisfies*

$$\langle \mathcal{L}w, w \rangle \geq C_0 \|w\|_X^2, \quad \forall w \in X, \tag{2.24}$$

then u_* is stable in the sense of Definition 2.3.

PROOF. By Assumption 2.1, the Hamiltonian system (2.1) is locally well-posed and $H(u)$ is conserved in time. Let $u \in C^0([0, \tau_0), X)$ be the unique weak solution in X with initial data $u_0 \in X$ for some $\tau_0 > 0$ and set $w(t, \cdot) := u(t, \cdot) - u_* \in X$ such that $w(0, \cdot) = w_0 := u_0 - u_*$. Then, since $H(u) \in C^2(X, \mathbb{R})$ and u_* is a critical point of $H(u)$, we have the expansion (2.20). It follows from (2.21) and (2.22) that there exist $\delta_0 > 0$ and $C_+ > 0$ such that

$$H(u_0) - H(u_*) \leq C_+ \|w_0\|_X^2, \tag{2.25}$$

for every $w_0 \in X$ satisfying $\|w_0\|_X < \delta_0$. On the other hand, it follows from (2.20), (2.21), and (2.24) that there exists $\epsilon_0 > 0$ and $C_- > 0$ such that

$$H(u) - H(u_*) \geq C_- \|w\|_X^2, \tag{2.26}$$

for every $w \in X$ satisfying $\|w\|_X < \epsilon_0$. Since $H(u(t, \cdot)) = H(u_0)$, combining (2.25) and (2.26) we obtain for $\|w_0\|_X < \delta_0$ and $\|w(t, \cdot)\|_X < \epsilon_0$ that

$$\|w(t, \cdot)\|_X^2 \leq \frac{H(u(t, \cdot)) - H(u_*)}{C_-} = \frac{H(u_0) - H(u_*)}{C_-} \leq \frac{C_+}{C_-} \|w_0\|_X^2, \quad t \in [0, \tau_0).$$

For every $\epsilon \in (0, \epsilon_0)$ we can choose $\delta := \min(\delta_0, \sqrt{C_-/C_+}\epsilon)$ such that if $\|w_0\|_X < \delta$, then

$$\|w(t, \cdot)\|_X < \sqrt{C_+/C_-}\delta \leq \epsilon, \quad t \in [0, \tau_0). \tag{2.27}$$

Since $\|u(\tau_0,\cdot)\|_X$ is bounded, local solutions can be continued for $t>\tau_0$ and since the estimate (2.27) is independent of t, it holds for all $t>0$. Hence, u_* is stable in the sense of Definition 2.3. □

REMARK 2.10. Condition (2.24) is referred to as coercivity of the quadratic form $\langle \mathcal{L}w, w\rangle$ associated with the linear operator $\mathcal{L} : X \to X^*$.

REMARK 2.11. Theorem 2.1 also holds for gradient systems (2.2) for which

$$H(u(t,\cdot)) \leq H(u_0), \quad t\geq 0.$$

Theorem 2.1 can be used to prove stability of kinks in the generalized NW equation (2.9) of Example 2.3 in the special case when $V'(u)$ is odd in u.

THEOREM 2.2. *Assume that* $V(u)\in C^2(\mathbb{R})$ *satisfies* $V'(-u) = -V'(u)$ *and there exists* $u_+>0$ *such that*

- $V(u_+) = 0$, $V'(u_+) = 0$, $V''(u_+)>0$,
- $V(u)>0$ *for* $u\in(-u_+,u_+)$.

Then there exists a kink $u_*\in C^3(\mathbb{R})$ *satisfying* $u_*(-x) = -u_*(x)$ *and* $u_*(x)\to\pm u_+$ *as* $x\to\pm\infty$ *such that* $(u,v) = (u_*,0)$ *is a critical point of* $H(u,v)$ *defined in* (2.10). *The kink is stable in the sense of Definition* 2.3 *with respect to perturbations* $(w,v)\in X$, *where*

$$X := \left\{w\in H^1(\mathbb{R}),\ \ v\in L^2(\mathbb{R}),\ \ w(-x) = -w(x),\ \ v(-x) = -v(x),\ \ x\in\mathbb{R}\right\}. \tag{2.28}$$

PROOF. Assumption 2.1 is verified by the local well-posedness theory for the generalized NW equation (2.9), see [**181**]. The existence of u_* follows from Lemma 2.3 applied to

$$-u_*'' + V'(u_*) = 0,$$

as in Example 2.3. Since the differential equation has the translational symmetry and $V'(-u) = -V'(u)$, one can choose u_* to be odd such that $u_*(-x) = -u_*(x)$. Consider $(u,v) = (u_*,0)+(w,v)$ with the perturbation $(w,v)\in X$, where X is the energy space with odd symmetry given by (2.28). The perturbation satisfies the evolution equation

$$\frac{d}{dt}\begin{bmatrix} w \\ v\end{bmatrix} = \begin{bmatrix} v \\ w_{xx} + V'(u_*) - V'(u_*+w)\end{bmatrix}.$$

Expanding $H(u,v)$ about the critical point $(u_*,0)$ gives (2.20) with

$$\mathcal{L} := \begin{bmatrix} \mathcal{L}_0 & 0 \\ 0 & 1\end{bmatrix}, \quad \mathcal{L}_0 := -\partial_x^2 + V''(u_*). \tag{2.29}$$

and

$$R_H := \begin{bmatrix} V'(u_*+w) - V'(u_*) - V''(u_*)w \\ 0\end{bmatrix}.$$

Note that $\mathcal{L}_0$ is an unbounded self-adjoint Schrödinger operator in $L^2(\mathbb{R})$ with dense domain $H^2(\mathbb{R})\subset L^2(\mathbb{R})$. By the translational symmetry (2.11), it follows from (2.23) that

$$\mathcal{L}u_{\rm sym} = 0, \qquad \text{where } u_{\rm sym} := \begin{bmatrix}\partial_x u_* \\ 0\end{bmatrix}.$$

Since u_* is monotonically increasing in x, we have $\partial_x u_*(x) > 0$ for all $x\in\mathbb{R}$. It follows from Sturm's oscillation theorem (Theorem B.9) and positivity of $\partial_x u_*$ on $\mathbb{R}$ that $\lambda = 0$ is the first eigenvalue of $\mathcal{L}_0$, i.e., $\sigma(\mathcal{L}_0)\geq 0$. Moreover, it is a

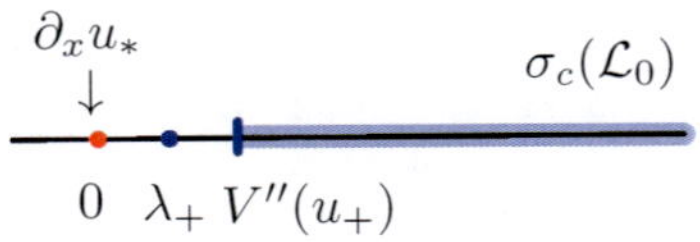

FIGURE 2.3. Sketch of the spectrum of the operator $\mathcal{L}_0$ in the proof of Theorem 2.2.

simple eigenvalue as a second-order differential equation with constant Wronskian cannot have two linearly independent solutions decaying to zero at infinity. Since $V''(u_*) \to V''(u_+)$ as $|x| \to \infty$ exponentially fast, Weyl's theorem (Theorem B.7) implies that $\sigma_c(\mathcal{L}_0) = [V''(u_+), \infty)$ with $V''(u_+) > 0$, so that the simple eigenvalue $\lambda = 0$ is isolated from the rest of $\sigma(\mathcal{L}_0)$.

When the operator $\mathcal{L}_0$ is restricted to the subspace $L^2_{\rm odd} \subset L^2(\mathbb{R})$ of functions with odd symmetry, Theorem B.5 implies that $\sigma(\mathcal{L}_0|_{L^2_{\rm odd}}) \geq \sigma(\mathcal{L}_0)$. Since $\partial_x u_*$ is even in x, the simple eigenvalue $\lambda = 0$ is no longer in the spectrum of $\sigma(\mathcal{L}_0|_{L^2_{\rm odd}})$. Hence there exists $\lambda_+ > 0$ such that $\sigma(\mathcal{L}_0|_{L^2_{\rm odd}}) \geq \lambda_+$, see Figure 2.3, where λ_+ is the second eigenvalue of $\mathcal{L}_0$ corresponding to the eigenfunction with odd symmetry. This implies the coercivity bound in $L^2(\mathbb{R})$:

$$\langle \mathcal{L}_0|_{L^2_{\rm odd}} w, w\rangle \geq \lambda_+ \|w\|^2_{L^2}, \quad \forall w \in H^1(\mathbb{R}).$$

To obtain coercivity of $\langle \mathcal{L}_0|_{L^2_{\rm odd}} w, w\rangle$ in $H^1(\mathbb{R})$, we define a perturbed operator

$$\mathcal{L}_{0,\varepsilon} := -\partial_x^2 + \frac{1}{1-\varepsilon} V''(u_*)$$

with some small $\varepsilon > 0$ and observe by continuity that there exists $\lambda_{+,\varepsilon} > 0$ such that

$$\langle \mathcal{L}_{0,\varepsilon}|_{L^2_{\rm odd}} w, w\rangle \geq \lambda_{+,\varepsilon} \|w\|^2_{L^2} \quad \forall w \in H^1(\mathbb{R}).$$

This implies that

$$\begin{aligned}
\langle \mathcal{L}_0|_{L^2_{\rm odd}} w, w\rangle &= \|w_x\|^2_{L^2} + \langle V''(u_*) w, w\rangle \\
&= \varepsilon \|w_x\|^2_{L^2} + (1-\varepsilon)\left(\|w_x\|^2_{L^2} + \frac{1}{1-\varepsilon}\langle V''(u_*) w, w\rangle \right) \\
&\geq \varepsilon \|w_x\|^2_{L^2} + (1-\varepsilon)\lambda_{+,\varepsilon} \|w\|^2_{L^2} \\
&\geq C_0 \|w\|^2_{H^1}, \qquad \forall w \in H^1(\mathbb{R}),
\end{aligned}$$

where $C_0 := \min\{\varepsilon, (1-\varepsilon)\lambda_{+,\varepsilon}\}$ with fixed $\varepsilon > 0$. Thus, the linear operator $\mathcal{L}|_{L^2_{\rm odd}}$ is coercive in the energy space X with odd symmetry given by (2.28) so that the coercivity bound (2.24) holds. By Theorem 2.1, the kink $(u_*, 0)$ is stable with respect to odd perturbations $(w, v) \in X$. □ □

REMARK 2.12. If $V(u) = \frac{1}{4}(1-u^2)^2$ with $V'(u) = -u + u^3$ in (2.9), the kink u_* is available in the exact form $u_*(x) = \tanh(x/\sqrt{2})$. The Schrödinger operator $\mathcal{L}_0$ is given by

$$\mathcal{L}_0 = -\partial_x^2 + 2 - 3\,\mathrm{sech}^2\left(\frac{x}{\sqrt{2}}\right),$$

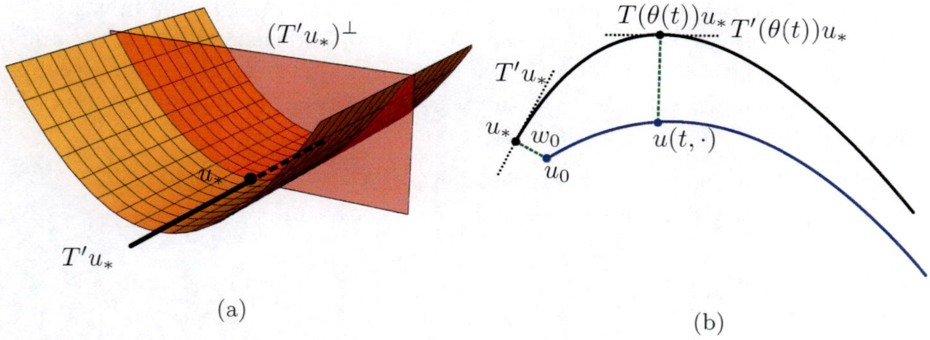

FIGURE 2.4. (a) The degeneracy of u_* due to the symmetry $T(\theta)$ can be compensated by restricting to an orthogonal subspace $(T'u_*)^\perp$. (b) A sketch of the orthogonal decomposition of the solution $u(t,\cdot)$ into $T(\theta(t))u_*$ and the orthogonal subspace $(T'(\theta(t))u_*)^\perp$ in the proof of Theorem 2.3.

and its spectrum is like the one in Figure 2.3 with $V''(u_+) = 2$ and $\lambda_+ = \frac{3}{2}$ due to the exact solutions

$$\mathcal{L}_0 \operatorname{sech}^2\left(\frac{x}{\sqrt{2}}\right) = 0, \quad \mathcal{L}_0 \tanh\left(\frac{x}{\sqrt{2}}\right) \operatorname{sech}\left(\frac{x}{\sqrt{2}}\right) = \frac{3}{2} \tanh\left(\frac{x}{\sqrt{2}}\right) \operatorname{sech}\left(\frac{x}{\sqrt{2}}\right),$$

where the eigenfunctions for the first and second eigenvalues have even and odd parity, respectively.

In order to extend the stability result of Theorem 2.2 to the case of general potentials $V(u)$ in the generalized NW equation (2.9), we need a stability theory which allows us to control the degeneracy of the critical point u_* due to a symmetry $T(\theta)$. We have to decompose the solution into an orbit of the critical point u_* translated along the symmetry $T(\theta)$ with $\theta \in \mathbb{R}$ and an orthogonal remainder term, see the sketch in Figure 2.4.

The following theorem gives the orbital stability of a local minimizer of energy in the presence of a symmetry $T(\theta)$, which is an extension of Theorem 2.1. The local minimizer is not strict, but the degeneracy is only due to the nontrivial kernel (2.23) induced by the symmetry $T(\theta)$.

THEOREM 2.3. *Let Assumption* 2.1 *be satisfied. Let* $T(\theta) : X \to X$ *with* $\theta \in \mathbb{R}$ *be the symmetry operator satisfying Definition* 2.4 *with* $X \subseteq \operatorname{Dom}(T')$. *Let* $u_* \in X$ *be a critical point of* $H(u) : X \to \mathbb{R}$ *such that* $H \in C^2(X, \mathbb{R})$ *and* $T'u_* \in X$. *Assume that there exists* $\epsilon_0 > 0$ *such that for every* $u \in X$ *such that* $\|u - u_*\|_X < \epsilon_0$ *there is* $\theta_0 \in \mathbb{R}$ *such that*

$$\|u - T(\theta_0)u_*\|_X = \inf_{\theta \in \mathbb{R}} \|u - T(\theta)u_*\|_X. \tag{2.30}$$

If there exists $C_0 > 0$ *such that the linear operator* $\mathcal{L} := \nabla^2 H(u_*) : X \to X^*$ *satisfies*

$$\langle \mathcal{L}w, w\rangle \geq C_0 \|w\|_X^2, \quad \forall w \in X \quad \text{with } \langle w, T'u_*\rangle = 0, \tag{2.31}$$

then u_* *is orbitally stable in the sense of Definition* 2.5.

PROOF. The proof of this theorem consists of two steps. In the first step, we project the solution $u(t,\cdot) \in X$ to the orbit $T(\theta(t))u_*$ and the orthogonal remainder term to the tangent vector $T'(\theta(t))u_*$, see Figure 2.4. In the second step, we control the orthogonal remainder term in the time evolution of the Hamiltonian system (2.1) using the energy function.

STEP 1 (Decomposition). For a solution $u \in X$ of (2.1) such that $\|u - u_*\|_X < \epsilon_0$, there exists $C_0 > 0$ unique $\theta \in \mathbb{R}$ and $w \in X$ satisfying

$$u = T(\theta)(u_* + w), \qquad \langle w, T'u_* \rangle = 0, \tag{2.32}$$

such that $\|w\|_X \leq C_0\epsilon$, where

$$\epsilon := \inf_{\theta \in \mathbb{R}} \|u - T(\theta)u_*\|_X < \epsilon_0. \tag{2.33}$$

PROOF OF STEP 1. Let $u \in X$ be a solution of (2.1) such that $\|u - u_*\|_X < \epsilon_0$. Then the infimum in (2.33) is attained for a finite value of $\theta_0 \in \mathbb{R}$ by assumption (2.30) and we have that

$$\epsilon = \|u - T(\theta_0)u_*\|_X = \inf_{\theta \in \mathbb{R}} \|u - T(\theta)u_*\|_X \leq \|u - u_*\| < \epsilon_0.$$

We consider the smooth function $f(\theta) : \mathbb{R} \to \mathbb{R}$ given by

$$f(\theta) := \langle u - T(\theta)u_*, T'(\theta)u_* \rangle.$$

Then $f(\theta) = 0$ if and only if u can be represented by (2.32) since $T'(\theta) = T(\theta)T'$ and $T(\theta)$ is an isometry of the Hilbert space $\mathcal{H}$. By the Cauchy–Schwarz inequality, we have

$$|f(\theta_0)| \leq \|u - T(\theta_0)u_*\| \|T'u_*\| = \|T'u_*\|\epsilon.$$

In order to differentiate with respect to θ, we rewrite

$$f(\theta) = \langle T(-\theta)(u - T(\theta)u_*), T'u_* \rangle$$

and obtain

$$f'(\theta) = -\|T'u_*\|^2 - \langle T'(u - T(\theta)u_*), T'(\theta)u_* \rangle,$$

since $T'(\theta) = T(\theta)T'$. By the Cauchy–Schwarz inequality, the skew-adjointness of T' and since $T'u_* \in X \subseteq \mathrm{Dom}(T')$, there exists $C > 0$ such that

$$|f'(\theta_0) + \|T'u_*\|^2| \leq \|u - T(\theta_0)u_*\| \|T'T'(\theta_0)u_*\| \leq C\epsilon.$$

By making ϵ_0 small enough for $\epsilon \in (0, \epsilon_0)$, it follows that $f'(\theta_0) \neq 0$. Consider $\tilde{f}(\theta) := f(\theta) - f(\theta_0)$ where $\tilde{f}(\theta_0) = 0$, $\tilde{f}'(\theta_0) \neq 0$ and $f(\theta_0)$ is small of order $\mathcal{O}(\epsilon)$. By the inverse function theorem, there exists $C_1 > 0$ such that $\tilde{f}(\theta) = -f(\theta_0)$, and hence $f(\theta) = 0$, has a unique solution $\theta \in \mathbb{R}$ satisfying $|\theta - \theta_0| \leq C_1\epsilon$. Thus, the decomposition (2.32) is justified and we obtain by the triangle inequality and the assumption $T'u_* \in X$ that

$$\begin{aligned}
\|w\|_X &= \|T(\theta)w\|_X \\
&= \|u - T(\theta)u_*\|_X \\
&\leq \|u - T(\theta_0)u_*\|_X + \|T(\theta)u_* - T(\theta_0)u_*\|_X \\
&\leq \epsilon + C_2|\theta - \theta_0| \\
&\leq (1 + C_1C_2)\epsilon,
\end{aligned}$$

with the ϵ-independent Lipschitz constant $C_2 > 0$. Thus, $\|w\|_X \leq C_0\epsilon$ with $C_0 = 1 + C_1 C_2$. □

STEP 2 (Control of the error terms). There exists $C > 0$ such that for every $\epsilon \in (0, \epsilon_0)$, there exists $\delta \in (0, \delta_0)$ with $\delta_0 := \min\{\epsilon_0, C^{-1}\epsilon\}$ such that for every $u_0 \in X$ satisfying

$$\|u_0 - u_*\|_X < \delta,$$

the unique weak solution $u \in C^0([0, \tau_0), X)$ of the Hamiltonian system (2.1) with any given $\tau_0 > 0$ can be written in the form

$$u(t,\cdot) = T(\theta(t))[u_* + w(t,\cdot)], \quad \langle w(t,\cdot), T'u_*\rangle = 0, \quad t \in [0, \tau_0) \tag{2.34}$$

with unique $\theta(t)$ and $w(t,\cdot) \in X$ satisfying $\|w(t,\cdot)\|_X < \epsilon$ for all $t \in [0, \tau_0)$.

PROOF OF STEP 2. By Assumption 2.1, the Hamiltonian system (2.1) is locally well-posed in X and so there exists a unique local solution $u \in C^0([0, \tau_0), X)$ for every $u_0 \in X$ with some $\tau_0 > 0$. Since the local solution is continuous in time, it satisfies $\|u(t,\cdot) - u_*\| < \epsilon_0$ at least for small $\tau_0 > 0$ since $\delta < \epsilon_0$. Therefore, by Step 1, there exist unique $\theta(t)$ and $w(t,\cdot) \in X$ such that $u(t,\cdot)$ can be written in the form (2.34) for $t \in [0, \tau_0)$. Expanding $H(u)$ about u_* and taking into account the criticality of u_* for the Hamiltonian functional $H(u)$ and its invariance with respect to the symmetry $T(\theta)$ we find that

$$\begin{aligned} H(u(t,\cdot)) &= H(T(\theta(t))(u_* + w(t,\cdot))) \\ &= H(u_* + w(t,\cdot)) \\ &= H(u_*) + \frac{1}{2}\langle \mathcal{L}w(t,\cdot), w(t,\cdot)\rangle + R_H(w(t,\cdot)). \end{aligned}$$

Since $H \in C^2(X, \mathbb{R})$ there exists a constant $C_+ > 0$ such that

$$H(u_0) - H(u_*) \leq C_+\|u_0 - u_*\|^2.$$

In view of the coercivity bound in (2.31), there exists $C_- > 0$ such that

$$H(u(t,\cdot)) - H(u_*) \geq C_-\|u(t,\cdot) - u_*\|_X^2.$$

Combining these estimates by using that $H(u(t,\cdot)) = H(u_0)$ we find that

$$\|w(t,\cdot)\|_X \leq C\|u_0 - u_*\|_X, \quad t \in [0, \tau_0), \tag{2.35}$$

where $C := \sqrt{C_+ C_-^{-1}}$. If $\delta < C^{-1}\epsilon$, then $\|w(t,\cdot)\|_X < \epsilon$. □

END OF PROOF OF THEOREM 2.3. Orbital stability of the critical point u_* of $H(u)$ follows by combining the two steps. For every $\epsilon \in (0, \epsilon_0)$, let $\delta > 0$ be defined as in Step 2, such that if $\|u_0 - u_*\|_X < \delta$ then $\|w(t,\cdot)\|_X < \epsilon$ for every $t \in [0, \tau_0)$. It follows from the decomposition (2.34) that

$$\inf_{\theta\in\mathbb{R}} \|u(t,\cdot) - T(\theta)u_*\|_X \leq \|u(t,\cdot) - T(\theta(t))u_*\|_X = \|w(t,\cdot)\|_X < \epsilon, \quad t \in [0, \tau_0).$$

Hence the bound (2.33) obtained in Step 1 continues to be satisfied for $t \in [0, \tau_0)$ and in the limit $t \to \tau_0$, so that the local solution $u \in C^0([0, \tau_0), X)$ and the decomposition (2.34) in Step 2 are extended beyond the time $t = \tau_0$. Since the bound (2.35) is independent of t, the solution $u \in C^0(\mathbb{R}_+, X)$ and the decomposition (2.34) can be extended to all $t \in \mathbb{R}_+$ so that u_* is orbitally stable in the sense of Definition 2.5 with $\omega = 0$. □

REMARK 2.13. In the proof of Theorem 2.3, we do not have to consider the equations of motion for $\theta(t)$ or to differentiate the solution $u \in C^0(\mathbb{R}_+, X)$ in time t. This information can be added to characterize the dynamics of the perturbations if it is needed.

REMARK 2.14. The same method can be extended to relative equilibria $T(\omega t)u_\omega$ by replacing the energy $H(u)$ with the augmented energy $\Lambda_\omega(u) := H(u) - \omega Q(u)$, where $Q(u)$ is the conserved quantity associated to the symmetry $T(\theta)$ satisfying (2.5). However, one needs to make sure that the critical point u_ω is a minimizer of Λ_ω so that the bound (2.31) remains valid for $\omega \neq 0$.

REMARK 2.15. The assumption of Theorem 2.3 that the infimum in (2.30) is attained at a finite value of θ has to be checked for each particular symmetry operator $T(\theta)$. It is satisfied trivially if $T(\theta)$ is periodic in θ. If $T(\theta)$ is not periodic in θ, the assumption is satisfied if

$$\liminf_{|\theta|\to\infty} \|u_* - T(\theta)u_*\|_X > 0$$

and $\epsilon_0 > 0$ is sufficiently small.

The result of Theorem 2.3 can be used to establish the orbital stability of kinks in the generalized NW equation (2.9) with a general potential $V(u)$. Moreover, one can deal with both stationary and traveling kinks. These results are formulated and proven as the following two corollaries.

COROLLARY 2.1. *Assume that $V(u) \in C^2(\mathbb{R})$ and there exists $u_- < u_+$ such that*

- $V(u_\pm) = 0$, $V'(u_\pm) = 0$, $V''(u_\pm) > 0$,
- $V(u) > 0$ *for* $u \in (u_-, u_+)$.

Then there exists a kink $u_* \in C^3(\mathbb{R})$ satisfying $u_*(x) \to u_\pm$ as $x \to \pm\infty$ such that $(u, v) = (u_*, 0)$ is the critical point of $H(u, v)$ defined in (2.10). The kink is orbitally stable in the sense of Definition 2.5 with respect to perturbations in $X := H^1(\mathbb{R}) \times L^2(\mathbb{R})$.

PROOF. We have the same linear operator $\mathcal{L}$ as in (2.29) with $\sigma(\mathcal{L}) \geq 0$. As in the proof of Theorem 2.2, $\partial_x u_*$ spans $\mathrm{Ker}(\mathcal{L}_0)$ and $\sigma_c(\mathcal{L}_0) = [V''(u_+), \infty)$ with $V''(u_+) > 0$. To obtain the coercivity estimate (2.31) it suffices to notice that the eigenvectors of $\mathcal{L}_0$ for nonzero eigenvalues are all orthogonal to $\partial_x u_*$. Therefore, there exists $\lambda_+ > 0$ such that $\sigma(\mathcal{L}_0|_{\{\partial_x u_*\}^\perp}) \geq \lambda_+$ in $L^2(\mathbb{R})$, where $\mathcal{L}_0|_{\{\partial_x u_*\}^\perp} = \Pi_0 \mathcal{L}_0 \Pi_0$ is defined by the orthogonal projection to the complement of $\partial_x u_*$ in $L^2(\mathbb{R})$:

$$\Pi_0 f = f - \frac{\langle f, \partial_x u_* \rangle}{\|\partial_x u_*\|_{L^2}^2} \partial_x u_*, \quad f \in L^2(\mathbb{R}).$$

The coercivity bound (2.31) in $X = H^1(\mathbb{R}) \times L^2(\mathbb{R})$ follows from the same argument as at the end of the proof of Theorem 2.2.

The generator of the translational symmetry operator $T(\theta)$ defined in (2.11) is given by $T' = \partial_x$ with $\mathrm{Dom}(T') = H^1(\mathbb{R}) \times H^1(\mathbb{R})$. We have $T'u_* \in X$ since $u_* \in C^3(\mathbb{R})$ and $u_*(x) \to u_\pm$ as $x \to \pm\infty$ exponentially fast. Furthermore, the translational symmetry operator T satisfies

$$\liminf_{|\theta|\to\infty} \|u_* - T(\theta)u_*\|_{H^1} > 0,$$

and so the assumption (2.30) of Theorem 2.3 hold for sufficiently small $\epsilon_0 > 0$. Therefore, one can use the decomposition

$$\begin{bmatrix} u \\ v \end{bmatrix} = T(\theta) \begin{bmatrix} u_* + w \\ v \end{bmatrix}$$

with $(w, v) \in X = H^1(\mathbb{R}) \times L^2(\mathbb{R})$ satisfying $\langle w, \partial_x u_* \rangle = 0$. By Theorem 2.3, the kink is orbitally stable in the sense of Definition 2.5. □

REMARK 2.16. The stability arguments for the kinks in the generalized NW equation (2.9) are the same as for the kinks in the nonlinear diffusion (ND) equation

$$u_t = u_{xx} - V'(u), \tag{2.36}$$

which share the same stationary solutions. The energy function for the ND equation (2.36) is given by

$$H(u) = \int_{\mathbb{R}} \left[\frac{1}{2} u_x^2 + V(u) \right] dx.$$

Since $H(u(t, \cdot))$ is monotonically decreasing in t for every weak solution

$$u \in C^0([0, \tau_0), X)$$

in the energy space X, one can prove similarly to Theorem 2.2 and Corollary 2.1 that the kink is orbitally stable in the time evolution of the ND equation (2.36) with respect to perturbations in $H^1(\mathbb{R})$.

COROLLARY 2.2. *Under the same assumptions as in Corollary 2.1, there exists the traveling kink $u(t, x) = u_c(x + ct)$ for every $c \in (-1, 1)$. The traveling kink with the profile $(u_c, c\partial_x u_c)$ is orbitally stable in the sense of Definition 2.5 with respect to perturbations in $X := H^1(\mathbb{R}) \times L^2(\mathbb{R})$.*

PROOF. The traveling wave solution is given by $(u, v) = (u_c, v_c)$, where $v_c = c\partial_x u_c$ and u_c satisfies

$$-(1 - c^2) u_c'' + V'(u_c) = 0.$$

Let u_* be the kink satisfying $-u_*'' + V'(u) = 0$. Then, u_c can be written in the explicit form

$$u_c(x) = u_* \left(\frac{x}{\sqrt{1 - c^2}} \right), \tag{2.37}$$

for every $c \in (-1, 1)$. Expanding the augmented energy $\Lambda_c(u, v)$ given by (2.13) with the decomposition $(u, v) = (u_c, c\partial_x u_c) + (w, v)$ yields the linearized operator $\mathcal{L}_c$ in the form

$$\mathcal{L}_c := \begin{bmatrix} \mathcal{L}_0 & c\partial_x \\ -c\partial_x & 1 \end{bmatrix}, \quad \mathcal{L}_0 := -\partial_x^2 + V''(u_c).$$

The same reasoning as in the proof of Theorem 2.2 and Corollary 2.1 can be extended to the case of traveling kinks if one can show that the zero eigenvalue of $\mathcal{L}_c$ is simple and the rest of its spectrum is strictly positive and bounded away from zero. Since $\mathcal{L}_c$ is a matrix operator with c-dependent off-diagonal elements this does not follow simply from the theory of Schrödinger operators as in Theorem 2.2. To show this property, we differentiate the equation (2.14) in x and verify that

$$\mathcal{L}_c u_{\text{sym}} = 0, \qquad u_{\text{sym}} := \begin{bmatrix} \partial_x u_c \\ c\partial_x^2 u_c \end{bmatrix}, \tag{2.38}$$

hence $0 \in \sigma(\mathcal{L}_c)$. Then, we consider solutions of the homogeneous equation

$$\mathcal{L}_c \begin{bmatrix} w \\ v \end{bmatrix} = \begin{bmatrix} \mathcal{L}_0 & c\partial_x \\ -c\partial_x & 1 \end{bmatrix} \begin{bmatrix} w \\ v \end{bmatrix} = \begin{bmatrix} 0 \\ 0 \end{bmatrix},$$

which yields the second-order equation

$$-(1-c^2)w'' + V''(u_c)w = 0.$$

This equation admits two linearly independent solutions. One solution is given by $w = \partial_x u_c$ such that $w(x) \to 0$ as $|x| \to \infty$ exponentially fast. Since the Wronskian determinant of the two solutions is constant in x, the second, linearly independent solution grows as $|x| \to \infty$. Hence, 0 is a simple eigenvalue of $\mathcal{L}_c$ for every $c \in (-1, 1)$.

For $c = 0$ it follows from the proof of Theorem 2.2 that $\sigma(\mathcal{L}_{c=0}) \geq 0$. Since $\lambda = 0$ is an isolated simple eigenvalue of $\mathcal{L}_c$ for every $c \in (-1,1)$, no eigenvalues of $\mathcal{L}_c$ may cross zero and become negative for $c \neq 0$ by continuity in c. Hence $\sigma(\mathcal{L}_c) \geq 0$ for every $c \in (-1,1)$ with an isolated simple zero eigenvalue. As before, $\mathrm{Ker}(\mathcal{L}_c) = \mathrm{Span}\{u_{\mathrm{sym}}\}$ in (2.38) and since eigenfunctions of the self-adjoint operator $\mathcal{L}_c$ are orthogonal, we obtain the coercivity bound (2.31) in $X = H^1(\mathbb{R}) \times L^2(\mathbb{R})$. We still have $u_{\mathrm{sym}} \in X$ since $u_* \in C^3(\mathbb{R})$ and $u_*(x) \to u_\pm$ as $x \to \pm\infty$ exponentially fast. The assumptions of Theorem 2.3 hold so that one can use the decomposition

$$\begin{bmatrix} u \\ v \end{bmatrix} = T(\theta) \begin{bmatrix} u_c + \tilde{u} \\ v_c + \tilde{v} \end{bmatrix}$$

with $(\tilde{u}, \tilde{v}) \in X = H^1(\mathbb{R}) \times L^2(\mathbb{R})$ satisfying

$$\langle (\tilde{u}, \tilde{v}), u_{\mathrm{sym}} \rangle = \langle \tilde{u}, \partial_x u_c \rangle + c\langle \tilde{v}, \partial_x^2 u_c \rangle = 0.$$

By Theorem 2.3, the traveling kink is orbitally stable in the sense of Definition 2.5. □

REMARK 2.17. Orbital stability of kinks in the NW equation was proven for the first time in [**128**]. The recent studies deal with the asymptotic stability of kinks due to dispersive decay of perturbations. Asymptotic stability of kinks under odd and general perturbations in weighted energy spaces was considered in [**83, 101, 164, 165**]. Inverse scattering was used to consider the asymptotic stability of kinks and multi-soliton solutions of the sine–Gordon equation in [**49**] for the particular case $V(u) = 1 - \cos(u)$.

2.3. Saddle points of energy

If a critical point $u_* \in X$ of energy $H \in C^2(X, \mathbb{R})$ is a saddle point, then the coercivity bounds (2.24) or (2.31) are no longer true for the linear operator $\mathcal{L} := \nabla^2 H(u_*)$. Consequently, there exist $w_\pm \in X$ such that

$$\langle \mathcal{L}w_+, w_+ \rangle > 0 \quad \text{and} \quad \langle \mathcal{L}w_-, w_- \rangle < 0.$$

The situation is similar to the saddle points of finite-dimensional Hamiltonian systems considered in Section 1.4. Recall that Lyapunov's instability theory (Theorem 1.5) is applicable to isolated saddle points of gradient systems but not Hamiltonian systems, see Examples 1.8 and 1.9. The Hamilton–Krein theory (Theorem 1.7) implies instability of isolated saddle points of Hamiltonian systems in certain cases, but Examples 1.9 and 1.12 show that the theory is inconclusive in other cases. However, in the cases when the saddle points of Hamiltonian systems are linearly unstable, their nonlinear instability can be shown similarly to Theorem 1.3. These concepts will now be extended to the infinite-dimensional Hamiltonian system (2.1) in the energy space $X \subset \mathcal{H}$.

In the following definition we introduce the concepts of linear and spectral stability and instability in the nonlinear Hamiltonian system (2.1) linearized at a critical point.

DEFINITION 2.6. Let $u_* \in X$ be a critical point of the energy $H \in C^2(X, \mathbb{R})$. We say that u_* is **spectrally unstable** if there exists an eigenvalue $\lambda_0 \in \mathbb{C}$ with $\text{Im}(\lambda_0) > 0$ and an eigenvector $w_0 \in \mathcal{D} \subset \mathcal{H}$ of the spectral stability problem

$$J\mathcal{L}w_0 = \lambda_0 w_0, \tag{2.39}$$

with J defined in (2.1) and $\mathcal{L} := \nabla^2 H(u_*)$ with dense domain $\mathcal{D} := \{w \in \mathcal{H} : \mathcal{L}w \in \mathcal{H}\}$. If no such (unstable) eigenvalue λ_0 exists, we say that u_* is spectrally stable.

DEFINITION 2.7. We say that u_* is **linearly stable** if the linearized initial-value problem

$$\begin{cases} \dfrac{dw}{dt} = J\mathcal{L}w, & t > 0, \\ w|_{t=0} = w_0, \end{cases} \tag{2.40}$$

admits a unique weak solution $w \in C^0(\mathbb{R}_+, X) \cap C^1(\mathbb{R}_+, X^*)$ satisfying the global bound

$$\|w(t, \cdot)\|_X \leq C\|w_0\|_X, \qquad t \in \mathbb{R}_+,$$

for some $C > 0$. Otherwise, we say that u_* is linearly unstable.

REMARK 2.18. Since the operators J and $\mathcal{L}$ are t-independent, linear stability implies spectral stability and spectral instability implies linear instability.

The following theorem specifies sufficient conditions for the instability of a saddle point u_* of the energy $H(u)$. The proof of the theorem is obtained by repeating the estimates in the proof of Theorem 1.3.

THEOREM 2.4. *Let Assumption* 2.1 *be satisfied and let* $w \in C^0(\mathbb{R}_+, X) \cap C^1(\mathbb{R}_+, X^*)$ *be a unique weak solution to the linearized initial-value problem* (2.40). *Let* $u_* \in X$ *be a critical point of* $H(u)$ *such that*

(a) u_* *is spectrally unstable in the sense of Definition* 2.6.
(b) $\nabla H(u) : X \mapsto X^*$ *is a* C^1 *operator satisfying the bound*

$$\|\nabla H(u_* + w) - \mathcal{L}w\|_X \leq k\|w\|_X^{1+\nu}, \quad \|w\|_X \leq \epsilon, \tag{2.41}$$

for some $\nu > 0$, $k > 0$, *and small* $\epsilon > 0$.

Then u_* *is unstable in the sense of Definition* 2.3.

REMARK 2.19. The zero eigenvalue of $\mathcal{L}$ does not play any role in the proof of instability of a critical point u_* of $H(u)$ as long as u_* is spectrally unstable and the vector field in (2.1) satisfies the bound (2.41).

Theorem 2.4 can be used to prove instability of the standing and traveling pulses in the generalized NW equation (Example 2.3) as we will see in the following two theorems.

THEOREM 2.5. *Assume that* $V(u) \in C^{2+\nu}(\mathbb{R})$ *with* $\nu > 0$ *and that there exists* $u_+ > 0$ *such that*

- $V(0) = 0$, $V'(0) = 0$, $V''(0) > 0$,
- $V(u_+) = 0$, $V'(u_+) < 0$,
- $V(u) > 0$ *for* $u \in (0, u_+)$.

Then there exists a pulse $u_* \in C^{3+\nu}(\mathbb{R})$ *such that* $(u_*, 0)$ *is a saddle point of the Hamiltonian* $H(u, v)$ *defined in* (2.10). *The pulse is unstable with respect to perturbations in* $X := H^1(\mathbb{R}) \times L^2(\mathbb{R})$.

PROOF. The existence of u_* follows from Lemma 2.3 applied to

$$-u_*'' + V'(u_*) = 0,$$

as in Example 2.3. Expanding $H(u, v)$ given by (2.10) by using the decomposition $(u, v) = (u_*, 0) + (w, v)$ with $(w, v) \in X = H^1(\mathbb{R}) \times L^2(\mathbb{R})$ gives (2.20) with

$$\mathcal{L} := \begin{bmatrix} \mathcal{L}_0 & 0 \\ 0 & 1 \end{bmatrix}, \quad \mathcal{L}_0 := -\partial_x^2 + V''(u_*).$$

By the translational symmetry (2.11), it follows from (2.23) that

$$\mathcal{L} u_{\text{sym}} = 0, \qquad u_{\text{sym}} := \begin{bmatrix} \partial_x u_* \\ 0 \end{bmatrix}.$$

The unbounded Schrödinger operator $\mathcal{L}_0$ with dense domain $H^2(\mathbb{R})$ is self-adjoint in $L^2(\mathbb{R})$. By Weyl's theorem, $\sigma_c(\mathcal{L}_0) \geq V''(0) > 0$, hence the zero eigenvalue is isolated from the rest of $\sigma(\mathcal{L}_0)$ in $L^2(\mathbb{R})$. By Sturm's oscillation theorem and the fact that $\partial_x u_*$ has only one zero on $\mathbb{R}$, it follows that $\sigma(\mathcal{L}_0)$ includes a simple negative eigenvalue $-\lambda_0^2 < 0$, a simple zero eigenvalue, and the rest of its spectrum is strictly positive. Hence, $(u_*, 0)$ is a saddle point of $H(u, v)$.

We will now verify the assumptions of Theorem 2.4. Assumption 2.1 is satisfied by the local well-posedness theory for the generalized NW equation (2.9) in the energy space, see [**181**]. Similarly, the linearized evolution (2.40) is globally well-posed in $X = H^1(\mathbb{R}) \times L^2(\mathbb{R})$. The spectral stability problem (2.39) is rewritten for the perturbation (w, v) in the form:

$$\begin{bmatrix} 0 & 1 \\ -1 & 0 \end{bmatrix} \begin{bmatrix} \mathcal{L}_0 w \\ v \end{bmatrix} = \lambda \begin{bmatrix} w \\ v \end{bmatrix}, \tag{2.42}$$

which is equivalent to $\mathcal{L}_0 w = -\lambda^2 w$. Hence, the simple negative eigenvalue $-\lambda_0^2 < 0$ of $\mathcal{L}_0$ gives the positive real eigenvalue $\lambda_0 > 0$ of the spectral stability problem (2.42), which ensures assumption (a). Assumption (b) with the property (2.41) is satisfied due to the assumption that $V(u) \in C^{2+\nu}(\mathbb{R})$ since

$$\nabla H(u_* + w, v) - \mathcal{L} \begin{bmatrix} w \\ v \end{bmatrix} = \begin{bmatrix} V'(u_* + w) - V'(u_*) - V''(u_*) w \\ v \end{bmatrix}.$$

By Theorem 2.4, $(u_*, 0)$ is unstable with respect to perturbations in $X = H^1(\mathbb{R}) \times L^2(\mathbb{R})$. □

REMARK 2.20. If $V(u) = \frac{1}{2}u^2 - \frac{1}{4}u^4$ with $V'(u) = u - u^3$ in (2.9), the pulse with the profile $(u_*, 0)$ is available in the exact form $u_*(x) = \sqrt{2}\,\text{sech}(x)$. The Schrödinger operator $\mathcal{L}_0$ is given by

$$\mathcal{L}_0 = -\partial_x^2 + 1 - 6\,\text{sech}^2(x).$$

Similarly to Remark 2.12, $\mathcal{L}_0$ admits two simple eigenvalues isolated away from $\sigma_c(\mathcal{L}_0) = [1, \infty)$. One eigenvalue is negative at -3 and the other eigenvalue is at 0 due to the following exact solutions

$$\mathcal{L}_0 \operatorname{sech}^2(x) = -3 \operatorname{sech}^2(x), \qquad \mathcal{L}_0 \tanh(x) \operatorname{sech}(x) = 0.$$

REMARK 2.21. The instability arguments for the pulses in the NW equation (2.9) are the same as for the pulses of the ND equation (2.36). The linearized operator at the pulse u_* is $\mathcal{L} = -\mathcal{L}_0$ and since $\mathcal{L}_0$ has a negative eigenvalue $-\lambda_0^2 < 0$, this yields the positive real eigenvalue $\lambda_0^2 > 0$ of the spectral stability problem, which ensures assumption (a) of Theorem 2.4. Other assumptions are checked verbatim. Thus, the pulse with the profile u_* is unstable in the time evolution of the ND equation (2.36) with respect to perturbations in $H^1(\mathbb{R})$.

THEOREM 2.6. *Under the same assumptions as in Theorem 2.5, there exists the traveling pulse $u(t,x) = u_c(x+ct)$ for every $c \in (-1,1)$. The traveling pulse with the profile $(u_c, \partial_x u_c)$ is unstable with respect to perturbations in $X := H^1(\mathbb{R}) \times L^2(\mathbb{R})$.*

PROOF. The traveling pulse with the profile $(u_c, \partial_x u_c)$ is related to the standing pulse with the profile $(u_*, 0)$ by the formula (2.37), which holds for every $c \in (-1,1)$. The linearized operator $\mathcal{L}_c$ is given by

$$\mathcal{L}_c := \begin{bmatrix} \mathcal{L}_0 & c\partial_x \\ -c\partial_x & 1 \end{bmatrix}, \quad \mathcal{L}_0 := -\partial_x^2 + V''(u_c).$$

The same reasoning as in the proof of Corollary 2.2 ensures that the zero eigenvalue of $\mathcal{L}_c$ remains simple for every $c \in (-1,1)$. Hence, the simple negative eigenvalue of $\mathcal{L}_c$ for $c = 0$ remains a simple negative eigenvalue for every $c \in (-1,1)$ by continuity although its position may depend on c. Hence, (u_c, v_c) with $v_c := c\partial_x u_c$ is a saddle point of the augmented energy $\Lambda_c(u,v) := H(u,v) - cP(u,v)$, where H and P are given by (2.10) and (2.12). To be able to apply Theorem 2.4 one only needs to check assumption (a) since the other assumptions are identical to those in the proof of Theorem 2.5. In order to establish the existence of an unstable eigenvalue of $J\mathcal{L}_c$ in the spectral stability problem (2.39) we write it in the explicit form

$$\begin{bmatrix} 0 & 1 \\ -1 & 0 \end{bmatrix} \begin{bmatrix} \mathcal{L}_0 & c\partial_x \\ -c\partial_x & 1 \end{bmatrix} \begin{bmatrix} w \\ v \end{bmatrix} = \lambda \begin{bmatrix} w \\ v \end{bmatrix}. \tag{2.43}$$

Since the spectrum of $\mathcal{L}_c$ includes a simple negative eigenvalue, it follows from the extensions of the Hamiltonian–Krein theorem (Theorem 1.7), see [**55, 151, 180**], that the spectral problem (2.43) admits a positive real eigenvalue $\lambda_0 > 0$ which depends on c. Hence, the traveling pulse $(u_c, c\partial_x u_c)$ is spectrally unstable for every $c \in (-1,1)$ and by Theorem 2.4, it is nonlinearly unstable with respect to perturbations in X. □

REMARK 2.22. Characterization of spectral stability of solitary waves from eigenvalues of the spectral problem (2.39) started with two papers of M. Grillakis [**114, 115**] and two papers of C. Jones [**143, 144**]. More recently, unstable eigenvalues of the spectral problem (2.39) were counted similarly to the Hamiltonian–Krein theorem (Theorem 1.7) in [**55, 151, 180**]. The spectral problem (2.43) is equivalent to the quadratic eigenvalue problem

$$\mathcal{L}_0 w + c^2 \partial_x^2 w + 2c\lambda \partial_x w + \lambda^2 w = 0.$$

Extensions of the Hamiltonian–Krein theorem to the quadratic eigenvalue problem were developed in [**33, 231**].

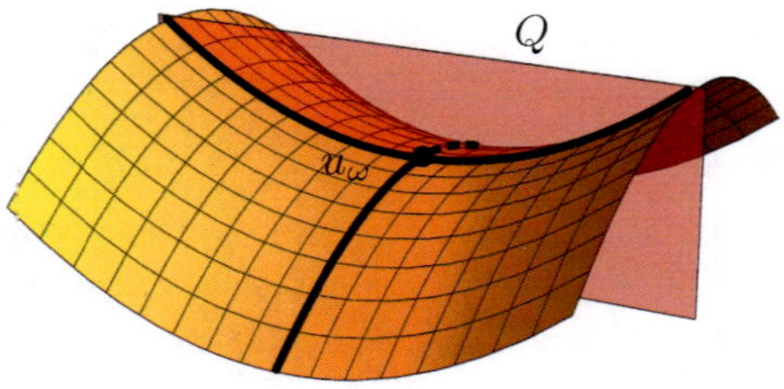

FIGURE 2.5. A sketch of a saddle point u_ω of the augmented energy functional $\Lambda_\omega(u) = H(u) - \omega Q(u)$ which is a minimum of H under the constraint of fixed Q.

REMARK 2.23. The case when the saddle points of energy are spectrally stable is more special. In this case, similar to the finite-dimensional systems considered in Section 1.4, the solitary waves could be either stable or unstable nonlinearly and each case needs special investigation. Examples when the saddle points of energy are unstable can be found in [**72, 74**]. Examples when saddle points of energy are orbitally stable can be constructed for many integrable nonlinear equations such as the cubic NLS equation and the massive Thirring model (Chapter 4).

2.4. Constrained minimizers of energy

In many Hamiltonian systems, e.g., in the generalized NLS equation (2.15), solitary waves are saddle points rather than minimizers of energy. We would not normally expect saddle points of energy to be stable as was discussed in Section 2.3. However, in the presence of symmetries, there exist additional conserved quantities such as momentum and mass of the NLS equation (Example 2.4). If the saddle points of energy are constrained minimizers of energy when momentum or mass or both are fixed, see the sketch in Figure 2.5, the energy function methods from Section 2.2 can be adapted to prove orbital stability of such constrained minimizers.

2.4.1. Setup. To develop a stability theory for constrained minimizers we assume again the local well-posedness of weak solutions of the Hamiltonian system (2.1) as stated in Assumption 2.1. For clarity of presentation, we consider only one symmetry group. A more general case of several symmetries will be considered in Section 3.3.

ASSUMPTION 2.2. Let $T(\theta) : \mathcal{H} \to \mathcal{H}$ with $\theta \in \mathbb{R}$ be the symmetry operator in Definition 2.4 and let $Q(u) : X \to \mathbb{R}$ be the associated conserved quantity given by (2.4). Let $u_\omega \in X$ be a critical point of the augmented energy $\Lambda_\omega(u) := H(u) - \omega Q(u)$ for some $\omega \in \mathcal{I} \subset \mathbb{R}$. Assume that $H(u), Q(u) \in C^2(X, \mathbb{R})$ for u near u_ω.

We want to prove that if the critical point u_ω of the augmented energy Λ_ω is a minimizer of H subject to fixed Q, then it is orbitally stable in the sense of Definition 2.5. Similarly as before, we control perturbations of u_ω by analysing the second variation of $\Lambda_\omega(u)$ at u_ω. Since u_ω is a critical point of $\Lambda_\omega(u) = H(u) -$

$\omega Q(u)$ and $H(u), Q(u) \in C^2(X,\mathbb{R})$ by Assumption 2.2, we can expand $\Lambda_\omega(u)$ near u_ω by using the perturbation $w := u - u_\omega$ as

$$\Lambda_\omega(u_\omega + w) = \Lambda_\omega(u_\omega) + \underbrace{\langle \nabla \Lambda_\omega(u_\omega), w\rangle}_{=0} + \frac{1}{2}\langle \mathcal{L}_\omega w, w\rangle + R_\Lambda(w), \quad w \in X, \tag{2.44}$$

where $\mathcal{L}_\omega := \nabla^2 \Lambda_\omega(u_\omega) : X \to X^*$ is a self-adjoint operator in the Hilbert space $\mathcal{H}$ with $X \subset \mathcal{H} \subset X^*$ and $R_\Lambda \in C^2(X,\mathbb{R})$ is the superquadratic remainder term satisfying

$$\lim_{\|w\|_X \to 0} \frac{|R_\Lambda(w)|}{\|w\|_X^2} = 0. \tag{2.45}$$

The quadratic form of $\mathcal{L}_\omega$ is bounded in X with the bound

$$|\langle \mathcal{L}_\omega w, w\rangle| \leq C_\omega \|w\|_X^2, \quad \forall w \in X, \tag{2.46}$$

with some constant $C_\omega > 0$. The domain of $\mathcal{L}_\omega$ is denoted by $\mathcal{D} \subset \mathcal{H}$.

REMARK 2.24. If u_ω is a constrained minimizer of $H(u)$ under fixed $Q(u)$, then $\omega \in \mathcal{I} \subset \mathbb{R}$ is a Lagrange multiplier of the constrained minimization problem. In addition to the parameter $\omega \in \mathcal{I}$, there is always a parameter $\theta \in \mathbb{R}$ along the symmetry group $T(\theta)T(\omega t)u_\omega = T(\omega t + \theta)u_\omega$. Both the parameters ω and θ satisfy modulation equations in the time flow of the Hamiltonian system (2.1), however, the actual dynamics of the parameter θ is irrelevant in the definition of the orbital stability. On the other hand, the Lagrange multiplier ω must be controlled in the time flow together with the perturbations.

2.4.2. Constrained minimization. The perturbation $w := u - u_\omega$ in the expansion of the augmented energy $\Lambda_\omega(u)$ near the critical point u_ω has to satisfy two constraints: one coming from the symmetry $T(\theta)$ and one because the associated conserved quantity Q should be fixed. Let us clarify what this means: first, due to the symmetry operator $T(\theta)$, the linearized operator $\mathcal{L}_\omega$ admits a nontrivial kernel $u_{\text{sym}} := T' u_\omega$, which by (2.5) implies that $J\nabla Q(u_\omega) \in \text{Ker}(\mathcal{L}_\omega)$ and has to be avoided. Second, to also leave Q invariant the perturbation should lie inside the admissible space

$$\mathcal{A}_\omega := \{w \in X : \quad \langle w, J\nabla Q(u_\omega)\rangle = 0, \quad Q(u_\omega + w) = Q(u_\omega)\}.$$

This space is nonlinear, but since we consider only small perturbations it is sufficient to develop the coercivity estimate in the tangent plane to the admissible space $\mathcal{A}_\omega$ at $w = 0$. If we look at the Taylor expansion of Q at u_ω we find that

$$Q(u_\omega + w) - Q(u_\omega) = \langle \nabla Q(u_\omega), w\rangle + \mathcal{O}(\|w\|_X^2).$$

Therefore, we can redefine the **admissible space** for the perturbation $w := u - u_\omega$ as

$$\mathcal{B}_\omega := \{w \in X : \quad \langle w, J\nabla Q(u_\omega)\rangle = 0, \quad \langle w, \nabla Q(u_\omega)\rangle = 0\},$$

which is the tangent plane to the nonlinear admissible space $\mathcal{A}_\omega$.

In the definition below we make precise the notion of constrained minimizers and saddle points of $H(u)$ under the constraint of fixed $Q(u)$, based on the admissible space $\mathcal{B}_\omega$. To shorten the exposition, we denote the orthogonal complement of a fixed element $w_0 \in \mathcal{H}$ by

$$\mathcal{H}|_{\{w_0\}^\perp} := \{f \in \mathcal{H} : \quad \langle f, w_0\rangle = 0\}$$

and use the notation $X|_{\{w_0\}^\perp} = X \cap \mathcal{H}|_{\{w_0\}^\perp}$.

DEFINITION 2.8. Let u_ω be a critical point of $\Lambda_\omega(u)$. We say that u_ω is a strict local minimizer of $H(u)$ under fixed $Q(u)$ up to the symmetry operator $T(\theta)$ if there exists $C_\omega > 0$ such that

$$\langle \mathcal{L}_\omega w, w\rangle \geq C_\omega \|w\|_X^2, \quad \forall w \in X|_{\{J\nabla Q(u_\omega), \nabla Q(u_\omega)\}^\perp}. \tag{2.47}$$

We say that u_ω is a saddle point of $H(u)$ under fixed $Q(u)$ if there exist $w_\pm \in X|_{\{J\nabla Q(u_\omega), \nabla Q(u_\omega)\}^\perp}$ such that

$$\langle \mathcal{L}_\omega w_+, w_+\rangle > 0 \quad \text{and} \quad \langle \mathcal{L}_\omega w_-, w_-\rangle < 0. \tag{2.48}$$

The following preliminary result will be needed when we establish a criterion for the critical point u_ω to be a strict minimizer of $H(u)$ with fixed $Q(u)$ in Theorem 2.8. Formal differentiation of $\nabla H(u_\omega) - \omega \nabla Q(u_\omega) = 0$ in $\omega \in \mathcal{I} \subset \mathbb{R}$ yields the linear nonhomogeneous equation

$$\mathcal{L}_\omega \partial_\omega u_\omega = \nabla Q(u_\omega). \tag{2.49}$$

The following lemma justifies the formal equation (2.49).

LEMMA 2.4. *Let Assumption 2.2 be satisfied and assume that* $\operatorname{Ker}(\mathcal{L}_\omega) = \operatorname{Span}\{J\nabla Q(u_\omega)\}$ *and that the following spectral gap condition is satisfied:*

$$\exists \delta > 0: \quad (-\delta, \delta) \cap \sigma(\mathcal{L}_\omega) = \{0\}. \tag{2.50}$$

For every $\omega_0 \in \mathcal{I} \subset \mathbb{R}$*, there exists a unique* C^1 *mapping* $\mathcal{I} \ni \omega \mapsto u_\omega \in X|_{\{J\nabla Q(u_{\omega_0})\}^\perp}$ *near* $\omega = \omega_0$ *with a unique solution* $\partial_\omega u_\omega|_{\omega=\omega_0} \in X|_{\{J\nabla Q(u_{\omega_0})\}^\perp}$ *satisfying* (2.49).

PROOF. Let $\omega_0 \in \mathcal{I}$. By Assumption 2.2 for $\omega = \omega_0$, we have that $\nabla\Lambda_{\omega_0}(u_{\omega_0}) = 0$. Define $F(u,\omega) : X \times \mathcal{I} \mapsto X^*$, $F(u,\omega) := \nabla\Lambda_\omega(u) = \nabla H(u) - \omega\nabla Q(u)$ and note that F is C^1 in both arguments. Expanding $F(u,\omega) = 0$ near $(u,\omega) = (u_{\omega_0}, \omega_0)$ yields

$$\mathcal{L}_{\omega_0}(u - u_{\omega_0}) = (\omega - \omega_0)\nabla Q(u_{\omega_0}) + R(u,\omega), \tag{2.51}$$

since $D_u F(u_{\omega_0}, \omega_0) = \mathcal{L}_{\omega_0}$, $\partial_\omega F(u_{\omega_0}, \omega_0) = -\nabla Q(u_{\omega_0})$, and $R(u,\omega) : X \times \mathcal{I} \mapsto X^*$ is superlinear with respect to both $(u - u_{\omega_0})$ and $(\omega - \omega_0)$.

We will show that $\mathcal{L}_{\omega_0}$ is invertible on $\mathcal{H}|_{\{J\nabla Q(u_{\omega_0})\}^\perp}$, the orthogonal complement of the kernel, to prove that the implicit equation (2.51) has a unique solution. To this end, we rewrite the implicit equation (2.51) in a form for which the implicit function theorem can be used. Since $\mathcal{L}_{\omega_0}$ is self-adjoint and its continuous spectrum is bounded away from zero, $\mathcal{L}_{\omega_0}$ is a Fredholm operator of index zero with one-dimensional kernel $\operatorname{Ker}(\mathcal{L}_{\omega_0}) = \operatorname{Span}\{J\nabla Q(u_{\omega_0})\}$ (Lemma B.1). Therefore, it follows from Fredholm's theorem (Theorem B.2) that a solution of the inhomogeneous equation $\mathcal{L}_{\omega_0} g = f$ exists if and only if $\langle f, J\nabla Q(u_{\omega_0})\rangle = 0$ and it is uniquely defined under the constraint $\langle g, J\nabla Q(u_{\omega_0})\rangle = 0$. Since J is skew-symmetric, we have

$$\langle \nabla Q(u_{\omega_0}), J\nabla Q(u_{\omega_0})\rangle = 0. \tag{2.52}$$

In addition, since $\mathcal{L}_{u_{\omega_0}}$ is self-adjoint and $\mathcal{L}_{u_{\omega_0}} J\nabla Q(u_{\omega_0}) = 0$ we also have

$$\langle R(u,\omega), J\nabla Q(u_{\omega_0})\rangle = 0. \tag{2.53}$$

Therefore, $\mathcal{L}_{\omega_0}$ is invertible on $\mathcal{H}|_{\{J\nabla Q(u_{\omega_0})\}^\perp}$ with a bounded inverse.

Let $\Pi_0 : \mathcal{H} \mapsto \mathcal{H}|_{\{J\nabla Q(u_{\omega_0})\}^\perp}$ be the orthogonal projection operator. Since $\mathcal{L}_{\omega_0}$ is self-adjoint in $\mathcal{H}$ and the conditions (2.52) and (2.53) hold, the implicit equation (2.51) can be equivalently written in the form

$$u - u_{\omega_0} = (\omega - \omega_0)\Pi_0 \mathcal{L}_{\omega_0}^{-1} \Pi_0 \nabla Q(u_{\omega_0}) + \Pi_0 \mathcal{L}_{\omega_0}^{-1} \Pi_0 R(u, \omega). \tag{2.54}$$

By the implicit function theorem, there exists $\varepsilon > 0$ such that for every $\omega \in \mathcal{I}$ satisfying $|\omega - \omega_0| < \varepsilon$, there exists a unique solution $u \in X|_{\{J\nabla Q(u_{\omega_0})\}^\perp}$ of the implicit equation (2.54), which we denote by u_ω, such that the mapping $\omega \mapsto u_\omega$ is C^1 and $u_\omega = u_{\omega_0}$ at $\omega = \omega_0$. Dividing (2.54) with $u = u_\omega$ by $\omega - \omega_0$ we find that

$$\frac{u_\omega - u_{\omega_0}}{\omega - \omega_0} = \Pi_0 \mathcal{L}_{\omega_0}^{-1} \Pi_0 \nabla Q(u_{\omega_0}) + \Pi_0 \mathcal{L}_{\omega_0}^{-1} \Pi_0 \frac{R(u_\omega, \omega)}{\omega - \omega_0}.$$

Taking the limit $\omega \to \omega_0$ we obtain that

$$\partial_\omega u_\omega|_{\omega=\omega_0} = \Pi_0 \mathcal{L}_{\omega_0}^{-1} \Pi_0 \nabla Q(u_{\omega_0}),$$

where we have used that $R(u_\omega, \omega)$ is superlinear with respect to $(u_\omega - u_{\omega_0})$ and $(\omega - \omega_0)$, and that $\mathcal{L}_{\omega_0}^{-1}$ is bounded on $\mathcal{H}|_{\{J\nabla Q(u_{\omega_0})\}^\perp}$ and hence continuous. Therefore, the unique $\partial_\omega u_\omega|_{\omega=\omega_0} \in X|_{\{J\nabla Q(u_{\omega_0})\}^\perp}$ satisfies the linear equation (2.49). □

The following theorem represents the criterion for the critical point u_ω of $\Lambda_\omega(u)$ to be the strict constrained minimizer of $H(u)$ under fixed $Q(u)$. The result was obtained in [**239**] for the NLS equation and then re-discovered for other systems in [**27, 228, 244**].

THEOREM 2.7. *Assume that the spectrum of $\mathcal{L}_\omega$ in $(-\infty, \delta)$ for some $\delta > 0$ consists of finitely many eigenvalues and* $\mathrm{Ker}(\mathcal{L}_\omega) = \mathrm{Span}\{J\nabla Q(u_\omega)\}$. *Then, the constrained minimizer condition* (2.47) *is satisfied if*

- *either $\sigma(\mathcal{L}_\omega) \geq 0$*
- *or $\mathcal{L}_\omega$ has only one simple negative eigenvalue and $\langle \partial_\omega u_\omega, \nabla Q(u_\omega) \rangle < 0$.*

The saddle point condition (2.48) *is satisfied if*

- *either $\mathcal{L}_\omega$ has only one simple negative eigenvalue and $\langle \partial_\omega u_\omega, \nabla Q(u_\omega) \rangle > 0$*
- *or $\mathcal{L}_\omega$ has two or more negative eigenvalues.*

PROOF. If $\sigma(\mathcal{L}_\omega) \geq 0$, then $\mathcal{L}_\omega$ is positive semi-definite with a simple zero eigenvalue in view of the assumption on the kernel. Hence, u_ω is a strict local minimizer of $\Lambda_\omega(u)$ up to the symmetry transformation $T(\theta)$ and the coercivity condition (2.31) is satisfied. Imposing the additional constraint $\langle w, \nabla Q(u_\omega) \rangle = 0$ in the minimizer conidition (2.47) does not affect the minimization property of u_ω with respect to $H(u)$, since restricting the operator $\mathcal{L}_\omega$ only implies that $\sigma(\mathcal{L}_\omega|_{\{\nabla Q(u_\omega)\}^\perp}) \geq \sigma(\mathcal{L}_\omega)$ (Theorem B.5). Moreover, $\mathcal{L}_\omega|_{\{\nabla Q(u_\omega)\}^\perp}$ has a simple zero eigenvalue whose eigenspace is orthogonal to $\nabla Q(u_\omega)$ in view of the orthogonality condition (2.52). Therefore, the coercivity bound (2.47) is satisfied if $\sigma(\mathcal{L}_\omega) \geq 0$.

If $\mathcal{L}_\omega$ admits finitely many negative eigenvalues, then u_ω is a saddle point of $\Lambda_\omega(u)$ and cannot satisfy the coercivity condition (2.31) without constraining the operator. Restricting the operator to the admissible space $\mathcal{B}_\omega$ we can study the constrained minimization problem by analysing $\sigma(\mathcal{L}_\omega|_{\{J\nabla Q(u_\omega), \nabla Q(u_\omega)\}^\perp})$. The smallest eigenvalue in this set, denoted by λ_*, is found by taking the infimum of the Rayleigh quotient [**110**] in $\mathcal{B}_\omega$, that is,

$$\lambda_* = \inf_{w \neq 0,\, w \in \mathcal{B}_\omega} \frac{\langle \mathcal{L}_\omega w, w \rangle}{\|w\|^2}.$$

This constrained minimization problem can be rewritten as

$$\lambda_* = \inf_{\substack{w \in X,\ w \neq 0 \\ (\nu,\mu)\in\mathbb{R}^2}} \frac{\langle \mathcal{L}_\omega w, w\rangle - 2\nu\langle w, \nabla Q(u_\omega)\rangle - 2\mu\langle w, J\nabla Q(u_\omega)\rangle}{\|w\|^2}, \tag{2.55}$$

where ν and μ are Lagrange multipliers [**189**]. The corresponding Euler–Lagrange equation is given by

$$(\mathcal{L}_\omega - \lambda I)w = \nu\nabla Q(u_\omega) + \mu J\nabla Q(u_\omega), \quad w \in \mathcal{B}_\omega, \quad w \neq 0 \tag{2.56}$$

and λ_* is the smallest eigenvalue of the constrained spectral problem (2.56).

Let us show that $\mu = 0$ in (2.56) due to the constraint $\langle w, J\nabla Q(u_\omega)\rangle = 0$. Indeed, note that

$$\langle \mathcal{L}_\omega w, J\nabla Q(u_\omega)\rangle = \langle w, \mathcal{L}_\omega J\nabla Q(u_\omega)\rangle = 0$$

in view of the assumption on the kernel and self-adjointness of $\mathcal{L}_\omega$. Together with (2.52) this implies that if $w \in \mathcal{B}_\omega$ is an eigenvector of the constrained spectral problem (2.56), then

$$\mu\|J\nabla Q(u_\omega)\|^2 = 0,$$

from which we deduce that $\mu = 0$. Therefore, we can rewrite the constrained spectral problem (2.56) in the form

$$(\mathcal{L}_\omega - \lambda I)w = \nu\nabla Q(u_\omega), \quad w \in \mathcal{B}_\omega, \tag{2.57}$$

where $\langle J\nabla Q(u_\omega), w\rangle = 0$ is satisfied automatically for every $\lambda \neq 0$. To find the eigenvector $w \in \mathcal{B}_\omega$, we define the function $F(\lambda) : \mathbb{R}\backslash\sigma(\mathcal{L}_\omega) \to \mathbb{R}$,

$$F(\lambda) := \langle(\mathcal{L}_\omega - \lambda I)^{-1}\nabla Q(u_\omega), \nabla Q(u_\omega)\rangle.$$

We have the following dichotomy for solutions to (2.57):

(A) Either $\lambda_* \in \sigma(\mathcal{L}_\omega)$, then $\nu = 0$

(B) Or $\lambda_* \notin \sigma(\mathcal{L}_\omega)$ and $F(\lambda_*) = 0$, then $\nu \neq 0$.

Both cases can be well understood from the properties of $F(\lambda)$, which are studied next.

(1) The function $F(\lambda)$ is analytic for $\lambda \notin \sigma(\mathcal{L}_\omega)$ and may have infinite jump discontinuities at $\lambda \in \sigma(\mathcal{L}_\omega)$.

(2) Let $\lambda_- < 0$ be the lowest eigenvalue of $\mathcal{L}_\omega$. Then, for every $\lambda \in (-\infty, \lambda_-)$, we have (Theorem B.3)

$$\|(\mathcal{L}_\omega - \lambda I)^{-1} f\| \leq |\lambda - \lambda_-|^{-1}\|f\|, \quad \text{for every } f \in \mathcal{H},$$

so that

$$F(\lambda) \to 0^+ \quad \text{as} \quad \lambda \to -\infty.$$

(3) The function $F(\lambda)$ is monotonically increasing for every $\lambda \notin \sigma(\mathcal{L}_\omega)$ since

$$F'(\lambda) = \langle(\mathcal{L}_\omega - \lambda I)^{-2}\nabla Q(u_\omega), \nabla Q(u_\omega)\rangle = \|(\mathcal{L}_\omega - \lambda I)^{-1}\nabla Q(u_\omega)\|^2 > 0.$$

(4) Let $\lambda_0 \in \sigma(\mathcal{L}_\omega)$ be a simple eigenvalue with eigenvector $w_0 \in \mathrm{Ker}(\mathcal{L}_\omega - \lambda_0 I)$. Then, F is continuous at $\lambda = \lambda_0$ if and only if $\langle \nabla Q(u_\omega), w_0 \rangle = 0$. Otherwise, $F(\lambda) \to \pm\infty$ as $\lambda \to \lambda_0^{\mp}$.

Indeed, let us use an orthogonal decomposition of the space $\mathcal{H} = \{w_0\} \oplus \mathcal{H}|_{\{w_0\}^\perp}$, where both $\{w_0\}$ and $\mathcal{H}|_{\{w_0\}^\perp}$ are invariant under $\mathcal{L}_\omega$. Therefore, we can write

$$\nabla Q(u_\omega) = \alpha_0 w_0 + w^\perp, \quad f := (\mathcal{L}_\omega - \lambda I)^{-1} \nabla Q(u_\omega) = \beta_0 w_0 + f^\perp,$$

where

$$\alpha_0 := \frac{\langle \nabla Q(u_\omega), w_0 \rangle}{\|w_0\|^2} \quad \text{and} \quad w^\perp \in \mathcal{H}|_{\{w_0\}^\perp},$$

whereas $\beta_0 \in \mathbb{R}$ and $f^\perp \in \mathcal{H}|_{\{w_0\}^\perp}$ are found from

$$(\mathcal{L}_\omega - \lambda I) f = \beta_0 (\lambda_0 - \lambda) w_0 + (\mathcal{L}_\omega - \lambda I) f^\perp = \alpha_0 w_0 + w^\perp$$

in the form

$$\beta_0 = \frac{\alpha_0}{\lambda_0 - \lambda} \text{ and } f^\perp = (\mathcal{L}_\omega - \lambda I)^{-1} w^\perp.$$

Using the orthogonal decomposition, we find that

$$\begin{aligned} F(\lambda) &= \langle f, \nabla Q(u_\omega) \rangle \\ &= \alpha_0 \beta_0 \|w_0\|^2 + \langle f^\perp, w^\perp \rangle \\ &= \frac{\alpha_0^2}{\lambda_0 - \lambda} \|w_0\|^2 + \langle (\mathcal{L}_\omega - \lambda I)^{-1} w^\perp, w^\perp \rangle \\ &= \frac{|\langle \nabla Q(u_\omega), w_0 \rangle|^2}{(\lambda_0 - \lambda)\|w_0\|^2} + \hat{F}(\lambda), \end{aligned}$$

where $\hat{F}(\lambda) = \langle (\mathcal{L}_\omega - \lambda I)^{-1} w^\perp, w^\perp \rangle$ is continuous at λ_0. The statement now follows from the decomposition.

(5) F is continuous at $\lambda = 0$ and

$$F(0) = \lim_{\lambda \uparrow 0} F(\lambda) = \langle \Pi_0 \mathcal{L}_\omega^{-1} \Pi_0 \nabla Q(u_\omega), \nabla Q(u_\omega) \rangle = \langle \partial_\omega u_\omega, \nabla Q(u_\omega) \rangle.$$

Indeed, in the notation of (4) for $\lambda_0 = 0$ we have $w_0 = J \nabla Q(u_\omega)$, and hence $\langle \nabla Q(u_\omega), w_0 \rangle = 0$ in view of the orthogonality condition (2.52). The expression for $F(0)$ then follows from (2.49).

Now we are ready to deal with the different cases mentioned in the statement of the theorem, which are illustrated in Figures 2.6–2.8.

- Let $\mathcal{L}_\omega$ admit one simple negative eigenvalue, say $\lambda_0 < 0$, with the eigenvector $w_0 \in \mathrm{Ker}(\mathcal{L}_\omega - \lambda_0 I)$, and $F(0) = \langle \partial_\omega u_\omega, \nabla Q(u_\omega) \rangle < 0$. Assume that F is continuous at λ_0, i.e., $\langle \nabla Q(u_\omega), w_0 \rangle = 0$, then monotonicity and continuity of $F(\lambda)$ imply that $F(0) > 0$ in contradiction with $\langle \partial_\omega u_\omega, \nabla Q(u_\omega) \rangle < 0$. Hence $\langle \nabla Q(u_\omega), w_0 \rangle \neq 0$ so that $F(\lambda) \to \pm\infty$ as $\lambda \to \lambda_0^{\mp}$. Since $F(0) < 0$, monotonicity and continuity of $F(\lambda)$ imply that $F(\lambda) \neq 0$ for $\lambda \in (-\infty, 0] \backslash \{\lambda_0\}$, see Figure 2.6. Hence, $\sigma(\mathcal{L}_\omega|_{\{\nabla Q(u_\omega)\}^\perp}) \geq 0$ and $\mathrm{Ker}(\mathcal{L}_\omega|_{\{\nabla Q(u_\omega)\}^\perp}) = \mathrm{Ker}(\mathcal{L}_\omega) = \mathrm{Span}\{J \nabla Q(u_\omega)\}$. Thus, $\lambda_* > 0$ in (2.55) implying that the minimizer condition (2.47) is satisfied.

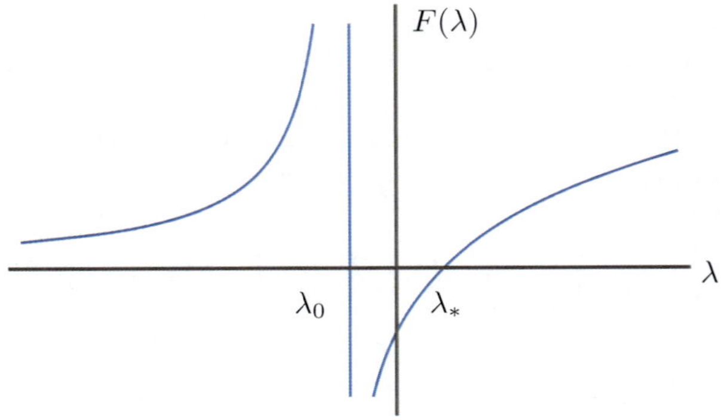

FIGURE 2.6. Sketch of $F(\lambda)$ implying that $\sigma(\mathcal{L}_\omega|_{\{J\nabla Q(u_\omega),\nabla Q(u_\omega)\}^\perp}) > 0$.

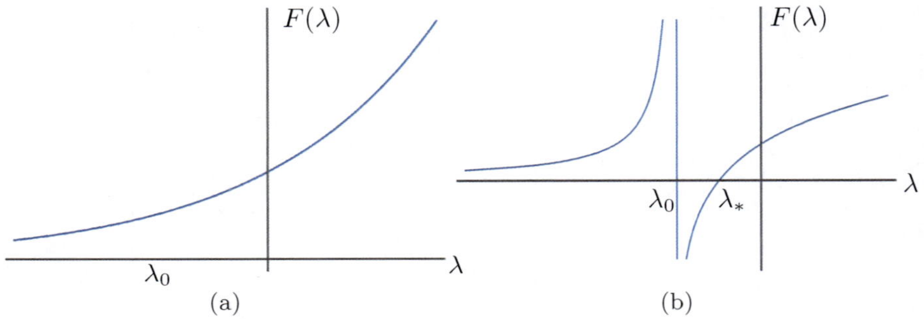

FIGURE 2.7. Sketches of $F(\lambda)$ implying that $\sigma(\mathcal{L}_\omega|_{\{J\nabla Q(u_\omega),\nabla Q(u_\omega)\}^\perp})$ contains a simple negative eigenvalue.

- Let $\mathcal{L}_\omega$ admit one simple negative eigenvalue λ_0 with eigenvector $w_0 \in \mathrm{Ker}(\mathcal{L}_\omega - \lambda_0 I)$ and $F(0) = \langle \partial_\omega u_\omega, \nabla Q(u_\omega)\rangle > 0$. Then either w_0 is orthogonal to $\nabla Q(u_\omega)$ and $F(\lambda)$ is continuous across λ_0 so that $F(0) > 0$ and $\lambda_* = \lambda_0$, see Figure 2.7 (a), or $F(\lambda)$ has an infinite jump discontinuity across λ_0 and admits a simple zero $\lambda_* \in (\lambda_0, 0)$, see Figure 2.7 (b). In either case, $\mathcal{L}_\omega|_{\{J\nabla Q(u_\omega),\nabla Q(u_\omega)\}^\perp}$ admits a simple negative eigenvalue and the saddle point condition (2.48) is satisfied.

- Let $\mathcal{L}_\omega$ admit two or more simple negative eigenvalues. Then monotonicity of $F(\lambda)$ implies that either some eigenvectors of $\mathcal{L}_\omega$ for negative eigenvalues are orthogonal to $\nabla Q(u_\omega)$ and F is continuous there or F has jump discontinuities at those values and admits negative zeros between the smallest negative eigenvalue and the zero eigenvalue, see Figure 2.8. In either case, $\mathcal{L}_\omega|_{\{J\nabla Q(u_\omega),\nabla Q(u_\omega)\}^\perp}$ admits negative eigenvalues and the condition (2.48) is satisfied.

The cases considered above complete the proof of the theorem. □

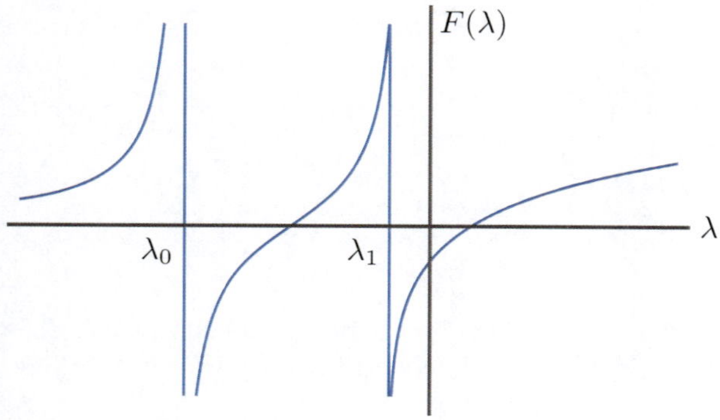

FIGURE 2.8. Sketch of $F(\lambda)$ implying that $\sigma(\mathcal{L}_\omega|_{\{J\nabla Q(u_\omega), \nabla Q(u_\omega)\}^\perp})$ contains a simple negative eigenvalue.

REMARK 2.25. The condition $\langle \partial_\omega u_\omega, \nabla Q(u_\omega)\rangle < 0$ can be expressed as

$$\frac{d}{d\omega} Q(u_\omega) < 0, \tag{2.58}$$

which is referred to as the Vakhitov–Kolokolov criterion or the slope condition [**239**].

REMARK 2.26. If $\mathcal{L}_\omega$ admits only one negative eigenvalue and

$$\langle \partial_\omega u_\omega, \nabla Q(u_\omega)\rangle = 0,$$

then $\sigma(\mathcal{L}_\omega|_{\{\nabla Q(u_\omega)\}^\perp}) \geq 0$ but $\mathrm{Ker}(\mathcal{L}_\omega|_{\{\nabla Q(u_\omega)\}^\perp}) \neq \mathrm{Ker}(\mathcal{L}_\omega)$ and $\lambda_* = 0$. Therefore, the second derivative test fails to clarify if u_ω is a constrained minimizer of $H(u)$ subject to fixed $Q(u)$ due to the additional degeneracy of $\mathrm{Ker}(\mathcal{L}_\omega|_{\{\nabla Q(u_\omega)\}^\perp})$. Another degenerate case occurs if $\mathrm{Ker}(\mathcal{L}_\omega) \neq \mathrm{Span}\{J\nabla Q(u_\omega)\}$. These degenerate cases will be studied in Section 2.5.

REMARK 2.27. If $\mathcal{L}_\omega$ admits infinitely many negative eigenvalues or a negative continuous spectrum, then the saddle point condition (2.48) is satisfied since one can construct v_- in (2.48) by a linear combination of infinitely many functions from the negative subspace of $\mathcal{L}_\omega$ which are orthogonal to $\mathrm{Ker}(\mathcal{L}_\omega) = \mathrm{Span}\{J\nabla Q(u_\omega)\}$ and satisfy a single constraint of orthogonality to $\nabla Q(u_\omega)$.

REMARK 2.28. Since the condition $\sigma(\mathcal{L}_\omega) \geq 0$ corresponds to u_ω being an unconstrained local minimizer of $\Lambda_\omega(u)$, the stability of such minimizers was studied in Theorem 2.3 (Section 2.2). Hence we will only focus on the second case of Theorem 2.7, when u_ω is a saddle point of $\Lambda_\omega(u)$ and a strict minimizer of $H(u)$ under fixed $Q(u)$. The other cases of Theorem 2.7, when u_ω is a saddle point of $H(u)$ under fixed $Q(u)$, can be studied as in Section 2.3.

EXAMPLE 2.5. In the proof of Theorem 2.6, we have shown for the generalized NW equation (2.9) that the critical point (u_c, v_c) of the augmented energy $\Lambda_c(u,v) = H(u,v) - cP(u,v)$ in (2.13) is a saddle point of Λ_c. We shall now verify that the critical point (u_c, v_c) is a saddle point of energy $H(u,v)$ under the constraint of fixed momentum $P(u,v)$. We use the slope condition in Theorem 2.7

with $Q \equiv P$ and $\omega \equiv c$. It follows from (2.12) and (2.37) with $v_c = c\partial_x u_c$ that

$$P(u_c, v_c) = \int_{\mathbb{R}} c(\partial_x u_c)^2 dx = \frac{c}{\sqrt{1-c^2}} \int_{\mathbb{R}} (\partial_x u_*)^2 dx.$$

Since the mapping $(-1,1) \ni c \mapsto P(u_c, v_c)$ is strictly monotonically increasing, we have

$$\frac{d}{dc} P(u_c, v_c) > 0.$$

The saddle point condition (2.48) is satisfied by Theorem 2.7. As a result, the spectrum of $\mathcal{L}_c$ in $\mathcal{H} = L^2(\mathbb{R}) \times L^2(\mathbb{R})$ constrained by the two orthogonality conditions

$$\langle (\tilde{u}, \tilde{v}), J\nabla P(u_c, v_c) \rangle = \langle \tilde{u}, \partial_x u_c \rangle + c\langle \tilde{v}, \partial_x^2 u_c \rangle = 0$$

and

$$\langle (\tilde{u}, \tilde{v}), \nabla P(u_c, v_c) \rangle = -c\langle \tilde{u}, \partial_x^2 u_c \rangle + \langle \tilde{v}, \partial_x u_c \rangle = 0$$

admits a simple negative eigenvalue $\lambda_* < 0$. Therefore, the spectral stability problem (2.43) admits an unstable (positive real) eigenvalue, see [**55, 151, 180**]. This was used in the proof of instability of the traveling pulse in energy space $X = H^1(\mathbb{R}) \times L^2(\mathbb{R})$ in Theorem 2.6.

2.4.3. Orbital stability of a constrained minimizer. We are now ready to explain how to control the Lagrange multiplier ω and the perturbations near an orbit of the constrained minimizer u_ω by using the energy methods. The following theorem gives the result which was obtained for the first time in [**116**]. The generalization to the case of several symmetries was made in [**117**].

THEOREM 2.8. *Let Assumption* 2.1 *be satisfied. Let $T(\theta) : X \to X$ with $\theta \in \mathbb{R}$ be the symmetry operator satisfying Definition* 2.4 *with $X \subseteq \mathrm{Dom}(T')$. Let $u_\omega \in X$ be a critical point of $\Lambda_\omega(u) : X \to \mathbb{R}$ such that $\Lambda_\omega \in C^2(X, \mathbb{R})$ and $T' u_\omega \in X$. Assume that there exists $\epsilon_0 > 0$ such that for every $u \in X$ satisfying $\|u - u_\omega\|_X < \epsilon_0$ there exists $\theta_0 \in \mathbb{R}$ such that*

$$\|u - T(\theta_0) u_{\omega_0}\|_X = \inf_{\theta \in \mathbb{R}} \|u - T(\theta) u_{\omega_0}\|_X. \tag{2.59}$$

Assume the spectral gap condition (2.50) *and $\mathrm{Ker}(\mathcal{L}_\omega) = \mathrm{Span}\{J\nabla Q(u_\omega)\}$ with $J\nabla Q(u_\omega) \in X$. If $\mathcal{L}_\omega$ admits only one simple negative eigenvalue and*

$$\langle \partial_\omega u_\omega, \nabla Q(u_\omega) \rangle < 0. \tag{2.60}$$

then u_ω is orbitally stable in the sense of Definition 2.5.

PROOF. By Lemma 2.4, the mapping $\omega \mapsto u_\omega \in X|_{\{J\nabla Q(u_\omega)\}^\perp}$ is C^1 in $\omega \in \mathcal{I}$. By Theorem 2.7, u_ω is a strict minimizer of $H(u)$ under fixed $Q(u)$ up to the symmetry $T(\theta)$ in the sense of Definition 2.8, so that the coercivity bound (2.47) holds.

The proof of the theorem follows along the lines of the proof of Theorem 2.3 but extends it to two constraints in the admissible space $\mathcal{B}_\omega$. In order to incorporate the two constraints, we have to continue both θ and ω with respect to time t. In order to distinguish $\omega(t)$ from ω in the formulation of the theorem, we fix $\omega_0 \in \mathcal{I}$, set $\omega(0) = \omega_0$, and prove the orbital stability of the stationary state $T(\omega_0 t) u_{\omega_0}$. The proof consists again of two steps.

STEP 1 (Decomposition). For a solution $u \in X$ of (2.1) such that $\|u - u_\omega\|_X < \epsilon_0$, there exists $C_0 > 0$ and unique $\theta \in \mathbb{R}$, $\omega \in \mathbb{R}$, and $w \in X$ such that

$$u = T(\theta)(u_\omega + w), \quad \langle w, T' u_\omega \rangle = 0, \quad \langle w, \nabla Q(u_\omega) \rangle = 0, \tag{2.61}$$

such that $|\omega - \omega_0| + \|w\|_X \leq C_0 \epsilon$, where

$$\epsilon := \inf_{\theta \in \mathbb{R}} \|u - T(\theta) u_{\omega_0}\|_X < \epsilon_0. \tag{2.62}$$

PROOF OF STEP 1. Let $u \in X$ be a solution of (2.1) such that $\|u - u_\omega\|_X < \epsilon_0$. Then the infimum in (2.62) is attained for a finite value of $\theta_0 \in \mathbb{R}$ by assumption (2.59) and we have that

$$\epsilon = \|u - T(\theta_0) u_\omega\|_X = \inf_{\theta \in \mathbb{R}} \|u - T(\theta) u_\omega\|_X \leq \|u - u_\omega\| < \epsilon_0.$$

We consider the smooth function $f : \mathbb{R}^2 \to \mathbb{R}^2$,

$$f(\theta, \omega) := \begin{bmatrix} \langle u - T(\theta) u_\omega, T'(\theta) u_\omega \rangle \\ \langle u - T(\theta) u_\omega, T(\theta) \nabla Q(u_\omega) \rangle \end{bmatrix}, \quad \theta \in \mathbb{R}, \quad \omega \in \mathcal{I} \subset \mathbb{R}.$$

Then $f(\theta, \omega) = 0$ if and only if u can be represented by (2.61) since $T'(\theta) = T(\theta) T'$ and $T(\theta)$ is an isometry of the Hilbert space $\mathcal{H}$. By the Cauchy–Schwarz inequality, we have

$$|f(\theta_0, \omega_0)| \leq C_1 \epsilon$$

for some ϵ-independent constant $C_1 > 0$. From the properties of the symmetry operator $T(\theta)$ we find that the Jacobian of f is given by

$$\begin{aligned} Df(\theta, \omega) = &- \begin{bmatrix} \|T' u_\omega\|^2 & \langle \partial_\omega u_\omega, T' u_\omega \rangle \\ 0 & \langle \partial_\omega u_\omega, \nabla Q(u_\omega) \rangle \end{bmatrix} \\ &+ \begin{bmatrix} -\langle T'(u - T(\theta) u_\omega), T'(\theta) u_\omega \rangle & \langle u - T(\theta) u_\omega, T'(\theta) \partial_\omega u_\omega \rangle \\ \langle u - T(\theta) u_\omega, T'(\theta) \nabla Q(u_\omega) \rangle & \langle u - T(\theta) u_\omega, T(\theta) \partial_\omega \nabla Q(u_\omega) \rangle \end{bmatrix}. \end{aligned}$$

Note that in the first term on the right of the above expression, the $(1,1)$ entry is the same as in the proof of Theorem 2.3 and the $(2,1)$ entry is zero because

$$\langle T' u_\omega, \nabla Q(u_\omega) \rangle = \langle J \nabla Q(u_\omega), \nabla Q(u_\omega) \rangle = 0.$$

The second term in the expansion of $Df(\theta, \omega)$ is bounded by $C_2 \epsilon$ for another ϵ-independent constant $C_2 > 0$ and the first term is independent of ϵ. By choosing ϵ_0 small enough for $\epsilon \in (0, \epsilon_0)$, the Jacobian $Df(\theta_0, \omega_0)$ is invertible as it is ϵ-close to an upper triangular matrix with nonzero diagonal entries in view of (2.60). By the local inverse mapping theorem, for any $u \in X$ satisfying (2.62) there exists a unique solution $(\theta, \omega) \in \mathbb{R}^2$ of $f(\theta, \omega) = 0$ satisfying

$$|\theta - \theta_0| + |\omega - \omega_0| \leq C_3 \epsilon$$

for another ϵ-independent constant $C_3 > 0$. Thus, the decomposition (2.61) is justified and we obtain by the triangle inequality, the assumption $T' u_\omega \in X$, and smoothness of the map $\mathcal{I} \ni \omega \mapsto u_\omega \in X$ that

$$\begin{aligned} \|w\|_X &= \|T(\theta) w\|_X \\ &= \|u - T(\theta) u_\omega\|_X \\ &\leq \|u - T(\theta_0) u_{\omega_0}\|_X + \|T(\theta_0) u_{\omega_0} - T(\theta_0) u_\omega\|_X + \|T(\theta_0) u_\omega - T(\theta) u_\omega\|_X \\ &\leq \epsilon + C_4 |\omega - \omega_0| + C_5 |\theta - \theta_0| \\ &\leq (1 + 2C_3(C_4 + C_5))\epsilon \end{aligned}$$

for some ϵ-independent constants $C_4, C_5 > 0$. Hence we have $|\omega-\omega_0|+\|w\|_X \leq C_0\epsilon$ with $C_0 = \max\{C_3, 1+2C_3(C_4+C_5)\}$. □

STEP 2 (Control of the error terms). There exists $C > 0$ such that for every $\epsilon \in (0, \epsilon_0)$ there exists $\delta \in (0, \delta_0)$ with $\delta_0 := \min\{\epsilon_0, C^{-1}\epsilon\}$ such that for every $u_0 \in X$ satisfying

$$\|u_0 - u_{\omega_0}\|_X < \delta,$$

the unique weak solution $u \in C^0([0, \tau_0), X)$ of the Hamiltonian system (2.1) with any given $\tau_0 > 0$ can be written in the form

$$u = T(\theta(t))[u_{\omega(t)} + w(t, \cdot)] \quad \text{such that} \tag{2.63}$$
$$\langle w(t,\cdot), T' u_{\omega(t)}\rangle = 0, \quad \langle w(t,\cdot), \nabla Q(u_{\omega(t)})\rangle = 0, \quad t \in [0, \tau_0)$$

for unique $\theta(t)$, $\omega(t)$, and $w(t, \cdot) \in X$ satisfying $|\omega(t) - \omega_0| + \|w(t, \cdot)\|_X < \epsilon$ for all $t \in [0, \tau_0)$.

PROOF OF STEP 2. By Assumption 2.1 the Hamiltonian system (2.1) is locally well-posed in X and so there exists a unique local weak solution $u \in C^0([0, \tau_0), X)$ for every $u_0 \in X$. Since u depends continuously on inital data, it satisfies $\|u - u_\omega\|_X < \epsilon_0$ at least for small $\tau_0 > 0$ if $\delta \in (0, \epsilon_0)$. Therefore, by Step 1, there exist unique $\theta(t)$, $\omega(t)$, and $w(t, \cdot) \in X$ such that u can be written in the form of the orthogonal decomposition (2.63) for $t \in [0, \tau_0)$.

To expand the conserved quantities, we define

$$V(t) := H(u(t,\cdot)) - H(u_{\omega_0}) - \omega(t)\left[Q(u(t,\cdot)) - Q(u_{\omega_0})\right]$$

and

$$\Delta(\omega(t)) := H(u_{\omega(t)}) - H(u_{\omega_0}) - \omega(t)\left[Q(u_{\omega(t)}) - Q(u_{\omega_0})\right]. \tag{2.64}$$

Expanding $\Lambda_\omega(u)$ about u_ω and subtracting $H(u_{\omega_0}) - \omega(t)Q(u_{\omega_0})$ from both sides we obtain in view of the invariance of H and Q with respect to $T(\theta)$ that

$$V(t) = \Delta(\omega(t)) + \frac{1}{2}\langle \mathcal{L}_{\omega(t)} w(t,\cdot), w(t,\cdot)\rangle + R_\Lambda(w(t,\cdot)), \tag{2.65}$$

where the superquadratic remainder term R_Λ satisfies (2.45). Since the mapping $\omega \mapsto \Delta(\omega)$ is C^2 in $\omega \in \mathcal{I}$, we differentiate it in ω and obtain

$$\Delta'(\omega) = Q(u_{\omega_0}) - Q(u_\omega).$$

Hence, $\Delta(\omega_0) = \Delta'(\omega_0) = 0$. Moreover, it follows from (2.58) that

$$\Delta''(\omega_0) = -\partial_\omega Q(u_\omega)|_{\omega=\omega_0} > 0.$$

It follows from (2.45), (2.47), and (2.64) that the expansion (2.65) implies that there exist constants $C_1, C_2 > 0$ such that

$$V(t) \geq C_1|\omega(t) - \omega_0|^2 + C_2\|w(t,\cdot)\|_X^2$$

as long as $|\omega(t) - \omega_0|$ and $\|w(t,\cdot)\|_X$ remain small. On the other hand, due to the conservation of $H(u)$ and $Q(u)$, we have

$$\begin{aligned} V(t) &= H(u_0) - H(u_{\omega_0}) - \omega(t)\left[Q(u_0) - Q(u_{\omega_0})\right] \\ &= V(0) - [\omega(t) - \omega_0]\left[Q(u_0) - Q(u_{\omega_0})\right], \end{aligned}$$

where

$$|V(0)| \leq C_3\delta^2, \quad |Q(u_0) - Q(u_{\omega_0})| \leq 2C_4\delta,$$

for some δ-independent constants $C_3, C_4 > 0$ due to bounds (2.45) and (2.46) with $\omega(0) = \omega_0$. Adding all bounds together yields

$$C_1|\omega(t) - \omega_0|^2 + C_2\|w(t,\cdot)\|_X^2 \leq C_3\delta^2 + 2C_4\delta|\omega(t) - \omega_0|.$$

By using Young's inequality $2|a||b| \leq \nu a^2 + \nu^{-1}b^2$ for every $\nu > 0$ and $a, b \in \mathbb{R}$, we pick ν sufficiently large so that the previous bound can be written in the form

$$|\omega(t) - \omega_0| + \|w(t,\cdot)\|_X \leq C_5\delta, \quad t \in [0, \tau_0), \tag{2.66}$$

for another δ-independent constant $C_5 > 0$. Choosing $\delta < C_5^{-1}\epsilon$, proves Step 2. □

END OF THE PROOF OF THEOREM 2.8. Orbital stability of the constrained minimizer u_{ω_0} in the sense of Definition 2.5 now follows by combining the two steps. For every $\epsilon \in (0, \epsilon_0)$, let $\delta > 0$ be defined as in Step 2, such that if $\|u_0 - u_\omega\|_X < \delta$ then $|\omega(t) - \omega_0| + \|w(t,\cdot)\|_X < \epsilon$ for every $t \in [0, \tau_0)$. By using the triangle inequality and smoothness of the map $\mathcal{I} \ni \omega \mapsto u_\omega \in X$, it follows from the decomposition (2.63) that

$$\begin{aligned}
&\inf_{\theta\in\mathbb{R}} \|u(t,\cdot) - T(\theta)u_{\omega_0}\|_X \\
&\quad\leq \|u(t,\cdot) - T(\theta(t))u_{\omega_0}\|_X \\
&\quad\leq \|u(t,\cdot) - T(\theta(t))u_{\omega(t)}\|_X + \|T(\theta(t))u_{\omega(t)} - T(\theta(t))u_{\omega_0}\|_X \\
&\quad\leq \|w(t,\cdot)\|_X + C_6|\omega(t) - \omega_0| \\
&\quad\leq C\delta
\end{aligned}$$

where $C := C_5(1 + C_6)$, $\theta(t)$ and $\omega(t)$ are defined in the orthogonal decomposition (2.63) for $t \in [0, \tau_0)$ and $C_6 > 0$ is another δ-independent constant. Assuming that $\delta \in (0, C^{-1}\epsilon)$, we conclude that the bound (2.62) obtained in Step 1 continues to be satisfied for $t \in [0, \tau_0)$ and in the limit $t \to \tau_0$, so that the local solution $u \in C^0([0, \tau_0), X)$ and the decomposition (2.63) in Step 2 are extended beyond the time $t = \tau_0$. Since the bound (2.66) is independent of t, the solution $u \in C^0(\mathbb{R}_+, X)$ and the decomposition (2.63) can be extended to all $t \in \mathbb{R}_+$ so that u_ω is orbitally stable in the sense of Definition 2.5. □

REMARK 2.29. Theorem 2.8 gives stability of local constrained minimizers of $H(u)$ for fixed $Q(u)$ which are degenerate only due to the symmetry operator $T(\theta)$. The ground state of the Hamiltonian system (2.1) with the associated conserved quantity $Q(u)$ is defined as the global minimizer of the variational problem

$$\mathcal{H}_\mu := \inf_{u\in X} \{H(u) : \quad Q(u) = \mu\}. \tag{2.67}$$

If it is attained at a function $u_\mu \in X$, then $H(u_\mu) = \mathcal{H}_\mu$ and $Q(u_\mu) = \mu$. The concentration compactness method was introduced in [**183, 184**] in order to prove existence of such global minimizers for some $H(u)$ and $Q(u)$. If nondegeneracy and the spectral gap condition can be proven for these global minimizers, then Theorem 2.8 guarantees their orbital stability in the time evolution of the Hamiltonian system (2.1).

We shall now use Theorem 2.8 to prove stability of bright solitons in the generalized NLS equation (2.15) (Example 2.4) in the case when the bright solitons are constrained minimizers of energy for fixed mass.

THEOREM 2.9. *Assume that* $W \in C^2(\mathbb{R}_+)$ *and there exists* $\psi_+ > 0$ *such that*

- $W(0) = 0$, $W'(0) > 0$,

- $W(\psi_+^2) = 0$, $W'(\psi_+^2) < 0$,
- $W(\psi^2) > 0$ *for* $\psi \in (0, \psi_+)$.

Then there exists $\mathcal{I} \subset \mathbb{R}^2$ *that contains* $(0,0)$ *such that for every* $(c,\omega) \in \mathcal{I}$ *there exists a bright soliton* $\psi_{c,\omega} \in C^3(\mathbb{R})$ *satisfying* $\psi_{c,\omega}(x) \to 0$ *as* $x \to \pm\infty$ *such that* $(\psi_{c,\omega}, \bar{\psi}_{c,\omega})$ *is a critical point of the augmented energy* $\Lambda_{c,\omega}(\psi, \bar{\psi})$ *defined in* (2.18). *Moreover, the bright soliton can be written in the form*

$$\psi_{c,\omega}(x) = u_{\omega + \frac{1}{4}c^2}(x) e^{-\frac{i}{2}cx}, \quad u_\omega(x) : \mathbb{R} \mapsto \mathbb{R}. \tag{2.68}$$

If

$$\frac{d}{d\gamma}\|u_\gamma\|^2 < 0, \quad \gamma := \omega + \frac{1}{4}c^2, \tag{2.69}$$

then the bright soliton $\psi_{c,\omega}$ *is orbitally stable in the sense of Definition* 2.5 *with respect to the perturbations in* $X := H^1(\mathbb{R}, \mathbb{C})$.

PROOF. Assumption 2.1 is verified by the local well-posedness theory for the generalized NLS equation (2.15) with $W \in C^2(\mathbb{R}_+)$ in $X = H^1(\mathbb{R}, \mathbb{C})$ [**40**]. Critical points of $\Lambda_{c,\omega}$ satisfy the second-order differential equation (2.19). After the transformation (2.68), this differential equation is written in the form

$$u_\gamma'' + \gamma u_\gamma - W'(|u_\gamma|^2) u_\gamma = 0, \tag{2.70}$$

where $\gamma := \omega + \frac{1}{4}c^2$.

By Lemma 2.3 with $P(u) = -W(u^2)$, there exists a positive solution $u_{\gamma=0} \in C^3(\mathbb{R})$ such that $u_{\gamma=0}(x) \to 0$ as $|x| \to \infty$ due to assumptions on W. The solution is continued as a positive solution $u_\gamma \in H^1(\mathbb{R})$ of the differential equation (2.70) for every small γ since assumptions of Lemma 2.3 are satisfied for $P(u) = \frac{1}{2}\gamma u^2 - W(u^2)$ for small γ. Hence the bright soliton exists in a region $\mathcal{I} \subset \mathbb{R}^2$ that contains $(0,0)$.

It follows by multiplying (2.70) by $\bar{u}_\gamma$ and taking complex conjugates that

$$\bar{u}_\gamma(x) u_\gamma'(x) - u_\gamma(x) \bar{u}_\gamma'(x) = 0.$$

Since the bright soliton satisfies $u_\gamma(x) \to 0$ as $|x| \to \infty$, it follows by integration of this equality that there exists $\theta \in \mathbb{R}$ such that $u_\gamma(x) = |u_\gamma(x)| e^{-i\theta}$. Hence, u_γ is real-valued and positive up to the symmetry transformation $T_2(\theta) = e^{-i\theta}$.

The differential equation (2.70) is the Euler–Lagrange equation for the augmented energy $\Lambda_\gamma(u, \bar{u}) := H(u, \bar{u}) - \gamma Q(u, \bar{u})$. Hence, for the orbital stability argument, we can consider local minimizers of H under only one constraint on fixed mass Q. On the other hand, there are two symmetry operators (2.16) and (2.17) which should be used to define the orbit of the bright soliton u_γ. Since the generalized NLS equation (2.15) is invariant under the Galilean transformation,

$$\psi(t,x) \mapsto \psi(t, x+ct) e^{-\frac{i}{2}cx - \frac{i}{4}c^2 t},$$

the orbital stability of the bright solitons $\psi_{c,\omega}$ follows from that of u_γ with $\gamma = \omega + \frac{1}{4}c^2$ if the infimum in Definition 2.5 is extended to both symmetry groups as follows:

$$\inf_{(\theta,\xi) \in \mathbb{R}^2} \|u(t,\cdot) - e^{-i\theta} u_\gamma(\cdot + \xi)\|_X. \tag{2.71}$$

Note that the translation symmetry operator T_1 satisfies

$$\liminf_{|\theta| \to \infty} \|u_\omega - T_1(\theta) u_\omega\|_{H^1} > 0,$$

and so the assumption (2.59) of Theorem 2.8 holds for sufficiently small $\epsilon_0 > 0$. The phase rotation symmetry operator T_2 satisfies the assumption (2.59) trivially since $T_2(\theta)$ is periodic in θ.

It remains to prove that $\mathcal{L}_\gamma := \nabla^2 \Lambda_\gamma(u_\gamma)$ admits only one simple negative eigenvalue and a double zero eigenvalue. To prove this, we consider the perturbation $u + iv \in H^1(\mathbb{R}, \mathbb{C})$ to the bright soliton $u_\gamma \in H^1(\mathbb{R}, \mathbb{R})$. Expanding $\Lambda_\gamma(u, \bar{u})$ about u_γ gives

$$\Lambda_\gamma(u_\gamma + u + iv, u_\gamma + u - iv) = \Lambda_\gamma(u_\gamma, u_\gamma) + \frac{1}{2}\langle \mathcal{L}_+ u, u\rangle + \frac{1}{2}\langle \mathcal{L}_- v, v\rangle + R_\Lambda(w), \tag{2.72}$$

where the second variation is defined by the linear operators

$$\begin{cases} \mathcal{L}_+ := -\partial_x^2 - \gamma + W'(u_\gamma^2) + 2u_\gamma^2 W''(u_\gamma^2), \\ \mathcal{L}_- := -\partial_x^2 - \gamma + W'(u_\gamma^2). \end{cases} \tag{2.73}$$

By the translational symmetry (2.16) it follows that $\mathcal{L}_+ \partial_x u_\gamma = 0$, where $\partial_x u_\gamma$ has exactly one zero on $\mathbb{R}$. The unbounded Sturm–Liouville operator $\mathcal{L}_+$ with a dense domain $H^2(\mathbb{R}) \subset L^2(\mathbb{R})$ is self-adjoint in $L^2(\mathbb{R})$. By Sturm's oscillation theorem (Theorem B.9) and single-zero property of $\partial_x u_\gamma$, it follows that $\mathcal{L}_+$ admits exactly one simple negative eigenvalue. By Weyl's theorem (Theorem B.7), $\sigma_c(\mathcal{L}_+) \geq -\gamma + W'(0) > 0$ since $W'(0) > 0$ and γ is defined near $\gamma = 0$. Furthermore, all eigenvalues below $\sigma_c(\mathcal{L}_+)$ are simple because a second-order differential equation with constant Wronskian cannot have two linearly independent solutions vanishing at infinity. We conclude that the simple zero eigenvalue is isolated from the rest of $\sigma(\mathcal{L}_+)$ in $L^2(\mathbb{R})$ so that the assumptions of Theorem 2.8 on nondegeneracy of the kernel of $\mathcal{L}_+$ and the spectral gap condition are satisfied.

By the phase rotation symmetry (2.17) it follows that $\mathcal{L}_- u_\gamma = 0$, where $u_\gamma > 0$. The unbounded Sturm–Liouville operator $\mathcal{L}_-$ with a dense domain $H^2(\mathbb{R}) \subset L^2(\mathbb{R})$ is self-adjoint in $L^2(\mathbb{R})$. By Sturm's oscillation theorem and positivity of u_γ it follows that $\sigma(\mathcal{L}_-) \geq 0$. By Weyl's theorem, $\sigma_c(\mathcal{L}_-) \geq -\gamma + W'(0) > 0$. Again, the simple zero eigenvalue is isolated from the rest of $\sigma(\mathcal{L}_-)$ in $L^2(\mathbb{R})$ so that the assumptions of Theorem 2.8 on nondegeneracy of the kernel of $\mathcal{L}_-$ and the spectral gap condition are satisfied.

The slope condition (2.60) holds since

$$\langle \partial_\gamma u_\gamma, \nabla Q(u_\gamma)\rangle = \langle \partial_\gamma u_\gamma, u_\gamma\rangle = \frac{1}{2}\frac{d}{d\gamma}\|u_\gamma\|^2 < 0$$

by the assumption (2.69). The orthogonality conditions in (2.47) are written explicitly for the perturbation $u + iv \in H^1(\mathbb{R}, \mathbb{C})$ due to the two symmetries and the constraint of fixed $Q(u, \bar{u})$:

$$\begin{aligned} &\langle \mathcal{L}_+ u, u\rangle + \langle \mathcal{L}_- v, v\rangle \geq C\|u + iv\|_{H^1}^2, \\ &\text{for all } u, v \text{ such that } \langle u, \partial_x u_\gamma\rangle = 0, \;\; \langle u, u_\gamma\rangle = 0, \;\; \langle v, u_\gamma\rangle = 0, \end{aligned} \tag{2.74}$$

where the coercivity bound on $\mathcal{L}_\gamma := \mathrm{diag}(\mathcal{L}_+, \mathcal{L}_-)$ follows from the spectral properties of operators $\mathcal{L}_\pm$ in $L^2(\mathbb{R})$. Hence, by Theorem 2.8 with the distance defined in (2.71) and using the unitarity of $T_{1,2}(\theta)$, the bright soliton u_γ (and hence $\psi_{c,\omega}$) is orbitally stable with respect to perturbations in $X = H^1(\mathbb{R}, \mathbb{C})$. □

REMARK 2.30. Under the slope condition (2.69), the bright soliton $\psi_{c,\omega}$ of the NLS equation (2.15) is a constrained minimizer of energy H for fixed mass Q. If we add the additional condition of fixed momentum P, then the fourth orthogonality condition $\langle v, \partial_x u_\gamma \rangle = 0$ needs to be added to the three orthogonality conditions in (2.74). However, coercivity of $\mathcal{L}_\omega$ in $H^1(\mathbb{R}, \mathbb{C})$ is already obtained under the three conditions and hence the fourth constraint is excessive. If we add this additional constraint, then we have to add parameter c in the decomposition near $\psi_{c,\omega}$ to satisfy the constraint. However, it is not possible to control the parameter c in the time evolution by using the method of the proof of Theorem 2.8 since only one parameter $\gamma = \omega + \frac{c^2}{4}$ appears in the expansion (2.72).

REMARK 2.31. In the proof of Theorems 2.8 and 2.9, we do not follow the actual dynamics of $\omega(t)$, $\theta(t)$, and $\xi(t)$. If we are to study these dynamics, it may be more useful to replace the orthogonality conditions $\langle u, \partial_x u_\gamma \rangle = 0$ and $\langle v, u_\gamma \rangle = 0$ in (2.74) by the symplectic orthogonality conditions $\langle u, x u_\gamma \rangle = 0$ and $\langle v, \partial_\gamma u_\gamma \rangle = 0$ which are preserved in the time evolution of the NLS equation [**71**, **201**].

REMARK 2.32. A similar proof of orbital stability can be developed for the black solitons of the NLS equation (Example 2.4). However, due to the nonzero boundary conditions, special considerations have to be made for the choice of the energy space X where the perturbations to the black solitons are considered. This problem will be studied in Section 7.3.

EXAMPLE 2.6. Consider the NLS equation with the power nonlinearity

$$i\psi_t + \psi_{xx} + |\psi|^{2m}\psi = 0, \tag{2.75}$$

where $m > 0$. The bright soliton u_γ, with $\gamma := \omega + \frac{1}{4}c^2$, satisfies the differential equation

$$u_\gamma'' + \gamma u_\gamma + u_\gamma^{1+2m} = 0, \tag{2.76}$$

where u_γ is real and positive. The bright soliton exists if $\gamma < 0$, in which case it is available in the exact form

$$u_\gamma(x) = [(m+1)|\gamma|]^{\frac{1}{2m}} \operatorname{sech}^{\frac{1}{m}}(m\sqrt{|\gamma|}x),$$

up to spatial translations (2.16). Since the mass is given by

$$\|u_\gamma\|_{L^2}^2 = \frac{(m+1)^{\frac{1}{m}}}{m} |\gamma|^{\frac{1}{m}-\frac{1}{2}} \int_{\mathbb{R}} \operatorname{sech}^{\frac{2}{m}}(z) dz,$$

the slope condition (2.69) is satisfied for $\gamma < 0$ if $m < 2$. By Theorem 2.9, the bright solitons of the NLS equation (2.75) are orbitally stable in $H^1(\mathbb{R})$ for $m \in (0, 2)$.

REMARK 2.33. By Theorem 2.7, since the slope condition (2.69) is satisfied, the bright soliton u_γ of the NLS equation (2.75) (Example 2.6) is a constrained minimizer of energy in $X = H^1(\mathbb{R}, \mathbb{C})$ if $m < 2$, a degenerate critical point of energy if $m = 2$, and a saddle point of energy subject to fixed mass if $m > 2$. It is well-known [**1**] that the variational problem (2.67) does not have minimizers for every $\mu > 0$ if $m > 2$ and has a minimizer for the unique value of

$$\mu = \frac{\sqrt{3}}{2} \int_{\mathbb{R}} \operatorname{sech}(z) dz$$

if $m = 2$. An alternative variational characterization of the bright soliton of the NLS equation (2.75) with every $m > 0$ was developed by Weinstein [**243**], who considered the following variational problem

$$\mathcal{J}_m := \inf_{u \in H^1(\mathbb{R})} \left\{ \frac{\|u'\|_{L^2}^m \|u\|_{L^2}^{2+m}}{\|u\|_{L^{2m+2}}^{2m+2}} \right\}. \tag{2.77}$$

The Euler–Lagrange equation for the variational problem (2.77) can be scaled to the differential equation

$$u_*'' - u_* + u_*^{1+2m} = 0,$$

which coincides with (2.76) for $\gamma = -1$. A minimizer of the variational problem (2.77) exists for every $m > 0$ and gives the best constant $C = \mathcal{J}_m^{-1}$ in the Gagliardo–Nirenberg inequality:

$$\|u\|_{L^{2m+2}}^{2m+2} \leq C\|u'\|_{L^2}^{m}\|u\|_{L^2}^{2+m}, \quad u \in H^1(\mathbb{R}).$$

Minimizers of the variational problem (2.77) coincide with minimizers of the constrained variational problem (2.67) for $m \leq 2$ (up to the scaling transformation) [**244**]. See also [**209**].

EXAMPLE 2.7. The Camassa–Holm equation

$$u_t - u_{xxt} + 3uu_x + 2ku_x = 2u_x u_{xx} + uu_{xxx},$$

is a model equation for shallow water waves, where $u = u(t, x)$ represents the horizontal component of fluid velocity at a certain depth beneath the water surface and $k > 0$ is fixed. The corresponding model equation for the free surface of the water is given by

$$\begin{aligned} u_t + u_x + 6uu_x - 6u^2 u_x + 12u^3 u_x \\ + u_{xxx} - u_{xxt} + 14uu_{xxx} + 28u_x u_{xx} = 0. \end{aligned}$$

The traveling solitary waves $u(t, x) = u_c(x - ct)$ with $u_c(x) \to 0$ as $|x| \to \infty$ are orbitally stable in the energy space of both shallow water models [**65, 89**]. An extended family of Camassa–Holm equations is available (Example 6.1) and orbital stability of the smooth traveling solitary waves in the extended family has also been proven [**168, 177, 187**].

2.5. Degenerate critical points of energy

When the assumptions of Theorem 2.8 are satisfied, the critical point u_ω of $\Lambda_\omega(u) = H(u) - \omega Q(u)$ is a local minimizer of energy $H(u)$ for fixed $Q(u)$ in the linear admissible space $\mathcal{B}_\omega$. Although u_ω is not a strict constrained minimizer because of the symmetry mode $u_{\text{sym}} = J\nabla Q(u_\omega)$, the two orthogonality constraints in $\mathcal{B}_\omega$ remove the degeneracy of the kernel due to symmetry and fix the value of $Q(u)$ up to the linear approximation.

Here we investigate what goes wrong if the second derivative test on the constrained minimization fails due to degeneracy of the kernels of $\mathcal{L}_\omega$ and $\mathcal{L}_\omega|_{\{\nabla Q(u_\omega)\}^\perp}$. In other words, we assume that $\sigma(\mathcal{L}_\omega|_{\{\nabla Q(u_\omega)\}^\perp}) \geq 0$ and

$$\text{(2.78)} \qquad \begin{cases} \text{either } \operatorname{Ker}(\mathcal{L}_\omega) \neq \operatorname{Span}\{J\nabla Q(u_\omega)\} \\ \text{or } \operatorname{Ker}(\mathcal{L}_\omega|_{\{\nabla Q(u_\omega)\}^\perp}) \neq \operatorname{Ker}(\mathcal{L}_\omega) = \operatorname{Span}\{J\nabla Q(u_\omega)\}. \end{cases}$$

We deal with the two cases in (2.78) separately.

2.5.1. Case $\mathrm{Ker}(\mathcal{L}_\omega) \neq \mathrm{Span}\{J\nabla Q(u_\omega)\}$. In order to get some general predictions, we assume the spectral gap condition (2.50) and the C^1 continuation of the mapping $\omega \mapsto u_\omega$. The latter property is no longer taken for granted since the first assumption of Lemma 2.4 is violated. Nevertheless, we assume that $\partial_\omega u_\omega \in X$ exists and satisfies (2.49). We also assume that $\mathrm{Ker}(\mathcal{L}_\omega)$ is two-dimensional with the symmetry mode $u_{\mathrm{sym}} \in X$ given by $u_{\mathrm{sym}} = J\nabla Q(u_\omega)$ and another linearly independent degeneracy mode $u_{\mathrm{deg}} \in X$. The setup is summarized in the following assumption.

ASSUMPTION 2.3. Assume $\mathrm{Ker}(\mathcal{L}_\omega) = \mathrm{Span}\{u_{\mathrm{sym}}, u_{\mathrm{deg}}\}$ and the spectral gap condition (2.50). Assume that the mapping $\omega \mapsto u_\omega$ is C^1 for $\omega \in \mathcal{I} \subset \mathbb{R}$ such that $\partial_\omega u_\omega \in X$ satisfies (2.49). Assume that $\mathcal{L}_\omega$ admits only one simple negative eigenvalue and the slope condition $\langle \partial_\omega u_\omega, \nabla Q(u_\omega)\rangle < 0$ holds.

The argument to study the stability of the critical points in the case of degeneracy is based on the **normal form reduction**, which is constructed in the following theorem.

THEOREM 2.10. *Let Assumptions 2.2 and 2.3 be satisfied. There exists $\delta > 0$ such that for every $c \in \mathbb{R}$ satisfying $|c| \leq \delta$, there exists a unique minimizer $U_\omega(c)$ of the variational problem*

$$M_\omega(c) := \inf_{U \in X|_{\{u_{\mathrm{sym}}, u_{\mathrm{deg}}, \nabla Q(u_\omega)\}^\perp}} \{\Lambda_\omega(u_\omega + c u_{\mathrm{deg}} + U) - \Lambda_\omega(u_\omega)\} \tag{2.79}$$

such that $\|U_\omega(c)\|_X$ is superlinear in c and $M(c)$ is superquadratic in c as $c \to 0$. Consequently, u_ω is a saddle point of $H(u)$ in $\mathcal{B}_\omega$ if $M_\omega(c)$ is either negative or sign-indefinite for $c \neq 0$ and a constrained minimizer of $H(u)$ in $\mathcal{B}_\omega$ if $M_\omega(c)$ is strictly positive for $c \neq 0$.

PROOF. We set up the Euler–Lagrange equation for the constrained minimization problem in (2.79). Let $\Pi_0 : \mathcal{H} \mapsto \mathcal{H}|_{\{u_{\mathrm{sym}}, u_{\mathrm{deg}}, \nabla Q(u_\omega)\}^\perp}$ be the projection operator with the adjoint operator $\Pi_0^* : \mathcal{H} \mapsto \mathcal{H}|_{\{u_{\mathrm{sym}}, u_{\mathrm{deg}}, \partial_\omega u_\omega\}^\perp}$. As before, we assume that the projection operators are extended to X and X^*.

It follows from (2.49) and Assumption 2.3 that there exists a unique solution $v \in X|_{\{u_{\mathrm{sym}}, u_{\mathrm{deg}}, \nabla Q(u_\omega)\}^\perp}$ of the nonhomogeneous equation $\mathcal{L}_\omega v = \Pi_0^* f$ for any $f \in X^*$. Indeed, by Fredholm Theorem (Theorem B.2), a solution exists since $\Pi_0^* f \perp \{u_{\mathrm{sym}}, u_{\mathrm{deg}}\}$ and the solution is unique if $w \perp \{u_{\mathrm{sym}}, u_{\mathrm{deg}}\}$. Furthermore,

$$\langle w, \nabla Q(u_\omega)\rangle = \langle w, \mathcal{L}_\omega \partial_\omega u_\omega\rangle = \langle \mathcal{L}_\omega w, \partial_\omega u_\omega\rangle = \langle \Pi^* f, \partial_\omega u_\omega\rangle = 0,$$

so that if $\Pi^* f \perp \partial_\omega u_\omega$, then $w \perp \nabla Q(u_\omega)$. Under Assumption 2.3, there exists $C_\omega > 0$ such that

$$\langle \mathcal{L}_\omega w, w\rangle \geq C_\omega \|w\|_X^2, \quad w \in X|_{\{u_{\mathrm{sym}}, u_{\mathrm{deg}}, \nabla Q(u_\omega)\}^\perp}. \tag{2.80}$$

We define the function

$$F(U,c) : X|_{\{u_{\mathrm{sym}}, u_{\mathrm{deg}}, \nabla Q(u_\omega)\}^\perp} \times \mathbb{R} \mapsto X^*|_{\{u_{\mathrm{sym}}, u_{\mathrm{deg}}, \partial_\omega u_\omega\}^\perp},$$

which is given by

$$F(U,c) := \Pi_0^* \mathcal{L}_\omega \Pi_0 U + \Pi_0^* R'_\Lambda(c u_{\mathrm{deg}} + U),$$

where we have used the Taylor expansion (2.44) and R'_Λ is superlinear according to (2.45). The function F satisfies the conditions of the implicit function theorem:

- F is C^1 in (U,c) since $H(u), Q(u) : X \to \mathbb{R}$ are C^2 due to Assumption 2.2.

- $F(0,0) = 0$.
- $D_U F(0,0) = \Pi_0^* \mathcal{L}_\omega \Pi_0 : X|_{\{u_{\rm sym}, u_{\rm deg}, \nabla Q(u_\omega)\}^\perp} \mapsto X^*|_{\{u_{\rm sym}, u_{\rm deg}, \partial_\omega u_\omega\}^\perp}$ is invertible with bounded inverse from X^* to X due to Assumption 2.3.

By the implicit function theorem, there are $r > 0$ and $\delta > 0$ such that for each $c \in \mathbb{R}$ with $|c| \leq \delta$, there exists a unique solution $U \in X|_{\{u_{\rm sym}, u_{\rm deg}, \nabla Q(u_\omega)\}^\perp}$ of the implicit equation $F(U,c) = 0$ with $\|U\|_X \leq r$, which we can denote by $U_\omega(c)$. The mapping $c \mapsto U_\omega(c)$ is C^1 near $c = 0$ and $U_\omega(0) = 0$. Since $F(0,c) = \Pi_0^* R'_\Lambda(cu_{\rm deg})$ is superlinear in c as $c \to 0$, we have $D_c F(0,c) = 0$ so that $D_c U_\omega(0) = 0$ and $\|U_\omega(c)\|_X$ is superlinear in c as $c \to 0$. Since $\langle \mathcal{L}_\omega u_{\rm deg}, u_{\rm deg}\rangle = 0$, the mapping $c \mapsto M(c)$ is superquadratic as $c \to 0$.

Due to the coercivity bound (2.80), the critical point

$$U = U_\omega(c) \in X|_{\{u_{\rm sym}, u_{\rm deg}, \nabla Q(u_\omega)\}^\perp}$$

of $\Lambda_\omega(u_\omega + cu_{\rm deg} + U)$ obtained with the implicit function theorem is the unique infimum of the constrained variational problem (2.79) for a fixed small c. Using this $U_\omega(c)$ yields

$$\Lambda_\omega(u_\omega + cu_{\rm deg} + U_\omega(c)) = \Lambda_\omega(u_\omega) + M_\omega(c), \quad c \in \mathbb{R}.$$

If $M_\omega(c)$ is either negative or sign-indefinite, there is c_0 such that

$$\Lambda_\omega(u_\omega + c_0 u_{\rm deg} + U_\omega(c_0)) < \Lambda_\omega(u_\omega),$$

hence, u_ω is the saddle point of $H(u)$ in $\mathcal{B}_\omega$. On the other hand, if $M_\omega(c)$ is strictly positive for all $c \neq 0$, then for all $c \in \mathbb{R}$ satisfying $0 \neq |c| \leq \delta$ we have

$$\Lambda_\omega(u_\omega + c_0 u_{\rm deg} + U_\omega(c_0)) > \Lambda_\omega(u_\omega),$$

hence u_ω is a constrained minimizer of $H(u)$ in $\mathcal{B}_\omega$. □

REMARK 2.34. If $H(u), Q(u) : X \to \mathbb{R}$ are C^k with $k \geq 3$, the mapping $c \mapsto U_\omega(c) \in X|_{\{u_{\rm sym}, u_{\rm deg}, \nabla Q(u_\omega)\}^\perp}$ is C^{k-1}. Since M_ω is computed from quadratic and superquadratic terms in $cu_{\rm deg} + U_\omega(c)$, the mapping $c \mapsto M_\omega(c)$ is C^k. Since $M_\omega(c)$ is superquadratic in c, the C^k function M_ω can be expanded as

$$M_\omega(c) = m_3 c^3 + m_4 c^4 + \cdots + m_k c^k + \widetilde{M}_\omega(c), \tag{2.81}$$

where $\widetilde{M}_\omega(c)$ is smaller than $|c|^k$ as $c \to 0$. Expansion (2.81) implies for small $|c| \leq \delta$ that

- If $m_3 \neq 0$, then $M_\omega(c)$ is sign-indefinite.
- If $m_3 = 0$ and $m_4 < 0$, then $M_\omega(c)$ is negative.
- If $m_3 = 0$ and $m_4 > 0$, then $M_\omega(c)$ is positive.

In the first two cases, u_ω is a saddle point of $H(u)$ in $\mathcal{B}_\omega$. In the last case, u_ω is a constrained minimizer of $H(u)$ in $\mathcal{B}_\omega$.

REMARK 2.35. If $m_3 \neq 0$ in the expansion (2.81), no information on $U_\omega(c)$ is needed for the normal form argument because $U_\omega(c)$ is quadratic in c and

$$\langle \mathcal{L}_\omega(cu_{\rm deg} + U_\omega(c)), (cu_{\rm deg} + U_\omega(c))\rangle = \langle \mathcal{L}_\omega U_\omega(c), U_\omega(c)\rangle = \mathcal{O}(c^4).$$

However, if $m_3 = 0$, then the quadratic term of $U_\omega(c)$ in c is needed for computations of m_4 in the expansion (2.81).

REMARK 2.36. It is natural to expect in the setup of Assumptions 2.1, 2.2 and 2.3 that if u_ω is a constrained minimizer of $H(u)$ in $\mathcal{B}_\omega$, then it is orbitally stable in the sense of Definition 2.5, whereas if it is a saddle point of $H(u)$ in $\mathcal{B}_\omega$, then it is orbitally unstable. However, it is hard to prove such a general result. Instead, we give an example where u_ω corresponds to the saddle point with $m_3 \neq 0$ in the expansion (2.81) and where it has been shown with the normal form $M_\omega(c)$ that u_ω is orbitally unstable.

2.5.2. NLS equation on a star graph. The degenerate critical points occur in the NLS equation posed on the star graph, which consists of N half-lines attached at a common vertex. The Hilbert space on the star graph is $\mathcal{H} := \oplus_{j=1}^N L^2(\mathbb{R}_+)$ and the energy space is

$$X := \left\{\oplus_{j=1}^N H^1(\mathbb{R}_+) : \quad \psi_1(0) = \psi_2(0) = \dots \psi_N(0)\right\}.$$

We use capital letters for the vector of components on N half-lines:

$$\Psi := (\psi_1, \psi_2, \dots, \psi_N).$$

For simplicity, let us consider the cubic NLS equation

$$i\Psi_t + \Psi_{xx} + |\Psi|^2\Psi = 0, \tag{2.82}$$

where each component of Ψ satisfies (2.75) with $m = 1$ on $\mathbb{R}_+$. Strong solutions of the cubic NLS equation (2.82) on the star graph are defined in the domain $\mathcal{D}$ given by

$$\mathcal{D} := \left\{\oplus_{j=1}^N H^2(\mathbb{R}_+) : \quad \psi_1(0) = \psi_2(0) = \dots \psi_N(0), \quad \sum_{j=1}^N \psi_j'(0) = 0\right\}.$$

The boundary conditions in $\mathcal{D}$ are called **Neumann–Kirchhoff conditions**.

The differential equation (2.76) with $m = 1$ and $\gamma = \omega$ admits the half-soliton

$$u_\omega(x) = \sqrt{2|\omega|}\,\mathrm{sech}(\sqrt{|\omega|}x)$$

for every $\omega < 0$. The solution $U_\omega := (u_\omega, u_\omega, \dots, u_\omega)$ belongs to X and satisfies the Neumann–Kirchhoff conditions in $\mathcal{D}$. The symmetry mode $U_{\rm sym} = iU_\omega$ is due to the phase rotation symmetry (2.17) whereas $\nabla Q(u_\omega) = U_\omega$ forms a constraint $\langle U, U_\omega\rangle = 0$ for the real-valued perturbation U.

Stability of the half-soliton with the profile U_ω for $N = 1$ and $N = 2$ follows from Theorem 2.9 because the slope condition (2.69) is satisfied due to

$$\|U_\omega\|^2 = 2N\sqrt{|\omega|} \quad \Rightarrow \quad \frac{d}{d\omega}\|U_\omega\|^2 < 0.$$

For $N = 1$, the NLS equation (2.82) is posed on the half line $\mathbb{R}_+$ subject to the Neumann condition on $\psi \equiv \psi_1$ at $x = 0$. For $N = 2$, the NLS equation (2.82) is posed on the line $\mathbb{R}$ after concatenating two half lines to $\mathbb{R}$ as

$$\psi(x) = \begin{cases} \psi_1(x), & x \in \mathbb{R}_+, \\ \psi_2(-x), & x \in \mathbb{R}_-, \end{cases} \tag{2.83}$$

such that $\psi(x)$ is C^1 at $x = 0$ without any specific boundary conditions at $x = 0$. The spectral properties of Theorem 2.9 are satisfied in both cases.

For $N \geq 3$, the half-soliton state U_ω is a degenerate critical point of Λ_ω in X. In order to show this, we recall the linearized operators $\mathcal{L}_\pm$ defined in (2.73) with $W(|u|^2) = -\frac{1}{2}|u|^4$ and $\gamma = \omega < 0$. These operators are well-defined in $\mathcal{H}$ with the dense domain $\mathcal{D} \subset \mathcal{H}$. Similarly to the proof of Theorem 2.9, it follows that $\sigma(\mathcal{L}_-) \geq 0$ because $\mathcal{L}_- U_\omega = 0$ and $U_\omega > 0$. The zero eigenvalue of $\mathcal{L}_-$ is simple and is bounded away from the rest of the spectrum of $\mathcal{L}_-$. On the other hand, the spectral problem for $\mathcal{L}_+ W = \lambda W$ with $W = (w_1, w_2, \dots, w_N)$ can be reduced to the Schrödinger equation on the half-line

$$-w'' + |\omega| w - 6u_\omega^2 w = \lambda w, \quad x \in \mathbb{R}_+ \tag{2.84}$$

by the transformation $w_j(x) = c_j w(x)$. The boundary conditions in $\mathcal{D}$ are satisfied if and only if $(c_1, c_2, \dots, c_N)$ satisfies

$$c_1 w(0) = c_2 w(0) = \cdots = c_N w(0), \quad \sum_{j=1}^N c_j w'(0) = 0.$$

The associated matrix

$$A = \begin{bmatrix} w(0) & -w(0) & 0 & \dots & 0 \\ w(0) & 0 & -w(0) & \dots & 0 \\ w(0) & 0 & 0 & \dots & 0 \\ \vdots & \vdots & \vdots & \ddots & \vdots \\ w'(0) & w'(0) & w'(0) & \dots & w'(0) \end{bmatrix}$$

has the following determinant

$$\det(A) = N[w(0)]^{N-1} w'(0)$$

which shows that λ is an eigenvalue of multiplicity $N-1$ if $w(0) = 0$ and is a simple eigenvalue if $w'(0) = 0$, where $w(x)$ is a solution of (2.84) decaying to zero as $x \to +\infty$. Since the exponentially decaying solutions of (2.84) for $\lambda \in (-\infty, |\omega|)$ exists in the closed analytical form (see [**146**]):

$$w(x) = e^{-\sqrt{|\omega|-\lambda}x} \frac{3|\omega| - \lambda + 3\sqrt{|\omega|}\sqrt{|\omega| - \lambda}\tanh(\sqrt{|\omega|}x) - 3|\omega| \operatorname{sech}^2(\sqrt{|\omega|}x)}{3|\omega| - \lambda + 3\sqrt{|\omega|}\sqrt{|\omega| - \lambda}},$$

it follows that $\sigma(\mathcal{L}_+)$ admits the simple negative eigenvalue $\lambda = -3|\omega|$ for which $w(x) = \operatorname{sech}^2(\sqrt{|\omega|}x)$ and the zero eigenvalue $\lambda = 0$ of multiplicity $(N-1)$ for which $w(x) = u_\omega'(x)$. Assumption 2.3 is satisfied with

$$u_{\deg} = \operatorname{Span}\{u_\omega' \mathbf{e}^{(1)}, u_\omega' \mathbf{e}^{(2)}, \dots, u_\omega' \mathbf{e}^{(N-1)}\},$$

where $\{\mathbf{e}^{(1)}, \mathbf{e}^{(2)}, \dots, \mathbf{e}^{(N-1)}\}$ is a set of orthogonal vectors in $\mathbb{R}^N$ satisfying the scalar constraint $\sum_{j=1}^N \mathbf{e}_j^{(k)} = 0$ for each $1 \leq k \leq N-1$.

EXAMPLE 2.8. For $N = 2$, only one eigenvector $u_{\deg} = \operatorname{Span}\{u_\omega' \mathbf{e}^{(1)}\}$ exists with $\mathbf{e}^{(1)} = (1, -1)$, and it corresponds to the translational mode of the NLS equation (2.82) with $N = 2$ posed on the line by the transformation (2.83). For $N = 3$, $u_{\deg}$ is spanned by two eigenvectors with $\mathbf{e}^{(1)} = (1, -1, 0)$ and $\mathbf{e}^{(2)} = (1, 1, -2)$. The set of orthogonal vectors $\{\mathbf{e}^{(1)}, \mathbf{e}^{(2)}, \dots, \mathbf{e}^{(N-1)}\}$ satisfying the scalar constraint $\sum_{j=1}^N \mathbf{e}_j^{(k)} = 0$ can be constructed algorithmically for every $N \geq 3$.

By using Theorem 2.10, we subsitute $\Psi = U_\omega + \sum_{j=1}^{N-1} c_j u_\omega' \mathbf{e}^{(j)} + \mathfrak{U}$ with real $\mathfrak{U}$ into

$$\Lambda_\omega(\Psi, \bar{\Psi}) = H(\Psi, \bar{\Psi}) - \omega Q(\Psi, \bar{\Psi}),$$

and obtain the function $M_\omega(\mathbf{c})$ defined by (2.79) with $\mathbf{c} = (c_1, c_2, \dots, c_{N-1}) \in \mathbb{R}^{N-1}$ under the orthogonality conditions

$$\langle \mathfrak{U}, U_\omega \rangle = 0, \quad \langle \mathfrak{U}, u_\omega' \mathbf{e}^{(1)} \rangle = \langle \mathfrak{U}, u_\omega' \mathbf{e}^{(2)} \rangle = \cdots = \langle \mathfrak{U}, u_\omega' \mathbf{e}^{(N-1)} \rangle = 0.$$

Since the constrainted minimizer $\mathfrak{U} = \mathfrak{U}_\omega(c)$ of $\Lambda_\omega(\Psi, \bar{\Psi})$ with

$$\Psi = U_\omega + \sum_{j=1}^{N-1} c_j u_\omega' \mathbf{e}^{(j)} + \mathfrak{U}$$

exists by Theorem 2.10 and is quadratic in $\mathbf{c}$, we expand $M_\omega(\mathbf{c})$ up to the cubic order and obtain

$$M_\omega(\mathbf{c}) = -4 \sum_{i=1}^{N-1} \sum_{j=1}^{N-1} \sum_{k=1}^{N-1} c_i c_j c_k \langle u_\omega, (u_\omega')^3 \rangle \langle \mathbf{e}^{(i)} \mathbf{e}^{(j)}, \mathbf{e}^{(k)} \rangle + \widetilde{M}_\omega(\mathbf{c}),$$

where the remainder term $\widetilde{M}_\omega$ is quartic in $\mathbf{c}$. The cubic structure of the leading order terms in $M_\omega(\mathbf{c})$ guarantees the saddle point geometry of $\Lambda_\omega(\Psi, \bar{\Psi})$ at $\Psi = U_\omega$ in the linear constrained space provided that at least one coefficient $\langle \mathbf{e}^{(i)} \mathbf{e}^{(j)}, \mathbf{e}^{(k)} \rangle$ is nonzero since $\langle u_\omega, (u_\omega')^3 \rangle \neq 0$ due to the explicit computation:

$$\langle u_\omega, (u_\omega')^3 \rangle = -|\omega|^3 \int_0^\infty \operatorname{sech}^4(x) \tanh^3(x) dx = -\frac{1}{12} |\omega|^3.$$

EXAMPLE 2.9. For $N = 2$, we have $\mathbf{e}^{(1)} = (1, -1)$ so that $\langle \mathbf{e}^{(1)} \mathbf{e}^{(1)}, \mathbf{e}^{(1)} \rangle = 0$. The saddle point geometry is not present for $N = 2$ and this agrees with the orbital stability of the bright soliton u_ω of the cubic NLS equation on the line $\mathbb{R}$ by Theorem 2.9 and Example 2.6. For $N = 3$, we have $\mathbf{e}^{(1)} = (1, -1, 0)$ and $\mathbf{e}^{(2)} = (1, 1, -2)$ so that

$$\langle \mathbf{e}^{(1)} \mathbf{e}^{(1)}, \mathbf{e}^{(1)} \rangle = 0, \quad \langle \mathbf{e}^{(1)} \mathbf{e}^{(1)}, \mathbf{e}^{(2)} \rangle = 2, \quad \langle \mathbf{e}^{(2)} \mathbf{e}^{(2)}, \mathbf{e}^{(1)} \rangle = 0, \quad \langle \mathbf{e}^{(2)} \mathbf{e}^{(2)}, \mathbf{e}^{(2)} \rangle = -6.$$

Hence, U_ω is a saddle point of energy H in the linear constrained space $\mathcal{B}_\omega$ by Theorem 2.10. Similar conclusion holds on the star graph for any $N \geq 3$ [**145**].

REMARK 2.37. Orbital instability of the saddle point U_ω of energy H in the time evolution of the NLS equation (2.82) on the star graph with $N \geq 3$ was proven in [**145**], where the two steps in the proof of Theorem 2.8 were combined with the secondary decomposition, symplectic orthogonality conditions, and the normal form $M_\omega(\mathbf{c})$ from Theorem 2.10.

REMARK 2.38. Compared to the setup of Assumption 2.3 and Theorem 2.10, the degeneracy mode $u_{\rm deg} = \operatorname{Span}\{u_\omega' \mathbf{e}^{(1)}, u_\omega' \mathbf{e}^{(2)}, \dots, u_\omega' \mathbf{e}^{(N-1)}\} \in \operatorname{Ker}(\mathcal{L}_+)$ is spanned by $(N-1)$ eigenvectors. Another difference is that the symmetry mode $u_{\rm sym} = iU_\omega$ is not present in the analysis since the perturbation U is restricted to be real-valued.

REMARK 2.39. An interesting modification of the cubic NLS equation (2.82) was considered in [**146, 147**], where the generalized Neumann–Kirchhoff conditions were applied to a star graph. Bright solitons on such star graphs can be translated away from the vertex points with a translational parameter. There exists parameter intervals for which these bright solitons are degenerate constrained minimizers of energy in $\mathcal{B}_\omega$, yet with the degenerate second variation of energy [**146**]. It has been shown in [**147**] that the degenerate constrained minimizers are orbitally unstable since the degeneracy leads to their irreversible drift towards the vertex point, where the degenerate minimizers transform to the saddle points of energy with the same instability as in Remark 2.37.

REMARK 2.40. The degenerate critical point u_ω may not satisfy the conditions in Assumption 2.3 if $\mathrm{Ker}(\mathcal{L}_\omega) = \mathrm{Span}\{u_{\mathrm{sym}}, u_{\mathrm{deg}}\}$ but the mapping $\omega \mapsto u_\omega$ is not C^1. In this case, the degenerate critical point u_ω is at bifurcation and its stability should be analyzed together with bifurcations of new critical points with respect to parameter ω. The standard bifurcations are saddle-node, pitchfork, and transcritical bifurcations. These bifurcations were studied in [**246**, **248**] in a rather general setting with examples involving the NLS equations with the x-dependent potentials.

REMARK 2.41. Another special situation when the degenerate critical point does not satisfy Assumption 2.3 occurs when the spectral gap condition (2.50) is violated because the continuous spectrum of $\mathcal{L}_\omega$ is positive but ends at zero. This situation is typical for peaked solitary waves (Section 6.4).

2.5.3. Case $\mathrm{Ker}(\mathcal{L}_\omega|_{\{\nabla Q(u_\omega)\}^\perp}) \neq \mathrm{Ker}(\mathcal{L}_\omega) = \mathrm{Span}\{J\nabla Q(u_\omega)\}$**.** If the spectral gap condition (2.50) is satisfied, the C^1 continuation of the mapping $\omega \mapsto u_\omega$ is provided by Lemma 2.4 since $\mathrm{Ker}(\mathcal{L}_\omega) = \mathrm{Span}\{J\nabla Q(u_\omega)\}$. On the other hand, since

$$\mathrm{Ker}(\mathcal{L}_\omega|_{\{\nabla Q(u_\omega)\}^\perp}) \neq \mathrm{Ker}(\mathcal{L}_\omega),$$

the slope condition is degenerate with $\langle \partial_\omega u_\omega, \nabla Q(u_\omega)\rangle = 0$ in Theorem 2.7.

Recall $\Delta(\omega)$ given by (2.64) in the proof of Theorem 2.8. This quantity is computed from the mapping $\omega \mapsto \Lambda_\omega(u_\omega)$. Since

$$\begin{aligned}
&\frac{d}{d\omega}\Lambda_\omega(u_\omega) = -Q(u_\omega),\\
&\frac{d^2}{d\omega^2}\Lambda_\omega(u_\omega) = -\frac{d}{d\omega}Q(u_\omega) = -\langle \partial_\omega u_\omega, \nabla Q(u_\omega)\rangle = 0,
\end{aligned}$$

it becomes relevant to compute the third-order derivative in ω:

$$\frac{d^3}{d\omega^3}\Lambda_\omega(u_\omega) = -\frac{d^2}{d\omega^2}Q(u_\omega) = -\langle \partial_\omega^2 u_\omega, \nabla Q(u_\omega)\rangle - \langle \partial_\omega u_\omega, \partial_\omega \nabla Q(u_\omega)\rangle.$$

Computing the derivative of the linear nonhomogeneous equation (2.49) in ω yields the linear nonhomogeneous equation

$$\mathcal{L}_\omega \partial_\omega^2 u_\omega = \partial_\omega \nabla Q(u_\omega) - \partial_\omega(\mathcal{L}_\omega)\partial_\omega u_\omega =: g_\omega, \tag{2.85}$$

where $g_\omega \perp \mathrm{Ker}(\mathcal{L}_\omega)$ if the mapping $\omega \mapsto u_\omega$ is C^2 for $\omega \in \mathcal{I} \subset \mathbb{R}$. The following assumption summarizes the setup.

ASSUMPTION 2.4. Assume $\mathrm{Ker}(\mathcal{L}_\omega) = \mathrm{Span}\{J\nabla Q(u_\omega)\}$ and the spectral gap condition (2.50). Assume that the mapping $\omega \mapsto u_\omega$ is C^2 for $\omega \in \mathcal{I} \subset \mathbb{R}$ such that $\partial_\omega u_\omega$ defined by (2.49) satisfies the degenerate slope condition $\langle \partial_\omega u_\omega, \nabla Q(u_\omega)\rangle = 0$ and $\partial_\omega^2 u_\omega$ is defined by (2.85). Assume that $\mathcal{L}_\omega$ admits only one simple negative eigenvalue.

The stability of the critical points in the case of degeneracy in the sense of Assumption 2.4 is studied again with the normal form reduction constructed in the following theorem.

THEOREM 2.11. *Let Assumptions 2.2 and 2.4 be satisfied. There exists $\delta > 0$ such that for every $c \in \mathbb{R}$ satisfying $|c| \le \delta$, there exists a unique minimizer $U_\omega(c)$ of the variational problem*

$$D_\omega(c) := \inf_{U \in X|_{\{u_{\rm sym}, \nabla Q(u_\omega), g_\omega\}^\perp}} \{\Lambda_\omega(u_\omega + c\partial_\omega u_\omega + U) - \Lambda_\omega(u_\omega)\} \tag{2.86}$$

such that $\|U_\omega(c)\|_X$ is superlinear in c and $D_\omega(c)$ is superquadratic in c as $c \to 0$. Consequently, u_ω is a saddle point of $H(u)$ in $\mathcal{B}_\omega$ if $D_\omega(c)$ is either negative or sign-indefinite and a constrained minimizer of $H(u)$ in $\mathcal{B}_\omega$ if $D_\omega(c)$ is strictly positive for $c \neq 0$.

PROOF. The proof repeats the steps of the proof of Theorem 2.10 but for the function

$$F(U,c) : X|_{\{u_{\rm sym}, \nabla Q(u_\omega), g_\omega\}^\perp} \times \mathbb{R} \mapsto X^*|_{\{u_{\rm sym}, \partial_\omega u_\omega, \partial_\omega^2 u_\omega\}^\perp},$$

where the linear nonhomogeneous equation (2.85) is used to define the projection operators $\Pi_0 : \mathcal{H} \mapsto \mathcal{H}|_{\{u_{\rm sym}, \nabla Q(u_\omega), g_\omega\}^\perp}$ and $\Pi_0^* : \mathcal{H} \mapsto \mathcal{H}|_{\{u_{\rm sym}, \partial_\omega u_\omega, \partial_\omega^2 u_\omega\}^\perp}$. Using the Taylor expansion (2.44) yields the Euler–Lagrange equation for the constrained minimization problem (2.86) in the form

$$F(U,c) := \Pi_0^* \mathcal{L}_\omega \Pi_0 U + \Pi_0^* R'_\Lambda(c\partial_\omega u_\omega + U) = 0.$$

By the same application of the implicit function theorem, there exists a unique solution $U \in X|_{\{u_{\rm sym}, \nabla Q(u_\omega), g_\omega\}^\perp}$ of the implicit equation $F(U,c) = 0$ denoted by $U_\omega(c)$ such that the mapping $c \mapsto U_\omega(c)$ is C^1 near $c = 0$ and superlinear in c as $c \to 0$. Since

$$\langle \mathcal{L}_\omega \partial_\omega u_\omega, \partial_\omega u_\omega \rangle = \langle \nabla Q(u_\omega), \partial_\omega u_\omega \rangle = 0,$$

the mapping $c \mapsto D_\omega(c)$ is superquadratic as $c \to 0$. The rest of the proof repeats the proof of Theorem 2.10 verbatim. □

REMARK 2.42. If $H(u), Q(u) : X \to \mathbb{R}$ are C^k with $k \ge 3$, the mapping $c \mapsto D_\omega$ is C^k and can be expanded as

$$D_\omega(c) = d_3 c^3 + d_4 c^4 + \cdots d_k c^k + \widetilde{D}_\omega(c), \tag{2.87}$$

where $\widetilde{D}_\omega(c)$ is smaller than $|c|^k$ as $c \to 0$. If $d_3 \neq 0$ in the expansion (2.87), then u_ω is a saddle point of $H(u)$ in $\mathcal{B}_\omega$. In this case, orbital instability of the saddle point u_ω can be proven [**57**] (see generalizations in [**196**] and [**212**]).

EXAMPLE 2.10. Consider the nonlinear Schrödinger equation with two power nonlinearities

$$i\psi_t + \psi_{xx} - (m+2)|\psi|^m \psi + (m+1)|\psi|^{2m}\psi = 0,$$

where $m > 0$. The bright soliton u_ω for $\omega < 0$ is available in the exact form

$$u_\omega(x) = \left(\frac{|\omega|}{\sqrt{1+|\omega|}\cosh(m\sqrt{|\omega|}x) - 1} \right)^{\frac{1}{m}}$$

up to the spatial translation in (2.16). The mass is given

$$Q(u_\omega) = \|u_\omega\|_{L^2}^2 = \frac{1}{m}|\omega|^{\frac{2}{m}-\frac{1}{2}} \int_{\mathbb{R}} \frac{dz}{[\sqrt{1+|\omega|}\cosh(z) - 1]^{\frac{2}{m}}}.$$

For $1 < m < 2$, the mapping $\omega \mapsto Q(u_\omega)$ exhibits a global minimum point for some $\omega_0 \in (-\infty, 0)$ such that $\frac{d^2}{d\omega^2} Q(u_\omega) > 0$, see Figure 2.9 for $m = 1.5$. Assumption 2.4

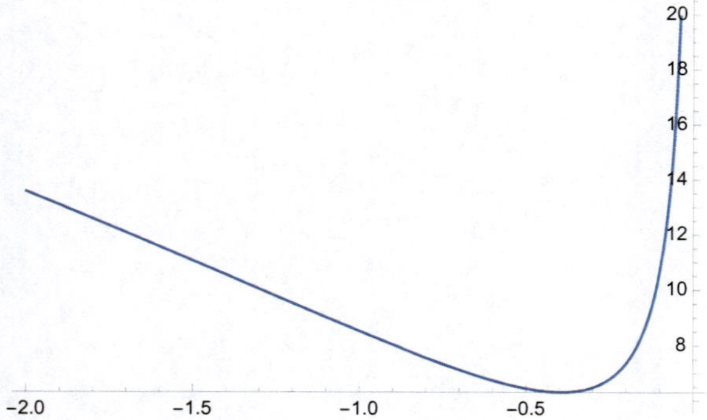

FIGURE 2.9. The dependence of $Q(u_\omega)$ versus ω in Example 2.10 with $m = 1.5$.

is satisfied at $\omega = \omega_0$. Since $\frac{d}{d\omega}Q(u_\omega) < 0$ for $\omega < \omega_0$, the corresponding bright solitons are constrained energy minimizers and they are orbitally stable in $X = H^1(\mathbb{R}, \mathbb{C})$ by Theorem 2.9. On the other hand, since $\frac{d}{d\omega}Q(u_\omega) > 0$ for $\omega \in (\omega_0, 0)$, the corresponding bright solitons are saddle points of the constrained energy. They are unstable in $X = H^1(\mathbb{R}, \mathbb{C})$ by Theorem 2.4 since they are linearly unstable (Section 3.5).

CHAPTER 3

Stability of periodic waves

Compared to the case of spatially decaying waves in Chapter 2, we will be dealing with spatially periodic waves of Hamiltonian systems in Chapter 3. Instead of working in the Hilbert space $\mathcal{H} = L^2(\mathbb{R})$, we will now consider functions in the Hilbert space $\mathcal{H} = L^2(0,T)$ with periodic boundary conditions at the end points of the interval $[0,T]$.

At first glance it may seem that the study of stability of periodic waves is easier than that of solitary waves because the real line $\mathbb{R}$ is now replaced by the compact interval $[0,T]$. However, the periodic waves have usually additional parameters compared to the solitary waves and it is harder to control minimization properties of the periodic waves as the critical points of conserved quantities. In addition, perturbations to the periodic waves do not have to be periodic with the same period but could be periodic with a multiple of the original period or be slowly decaying to zero at infinity. As a result, the stability problem might have to be formulated on a larger periodic domain $[0,NT]$ with $N \in \mathbb{N}$ or on the real line $\mathbb{R}$.

We will here extend the concepts of Chapter 2 to the stability of periodic waves on a periodic domain. We will give more advanced results on constrained minimization used in the proof of orbital stability of the periodic waves. We will also include the case where the skew-adjoint operator J is unbounded, in which conserved quantities arise both from symmetries and from the kernel of J (the so-called Casimir functionals). Stability of periodic waves with respect to perturbations of multiple periods and slowly decaying perturbations are studied in Chapter 4 for integrable Hamiltonian systems.

The basic setup and examples of periodic waves as critical points of the energy function are described in Section 3.1. A useful tool to study the stability of periodic solutions of second-order differential equations called the period function is introduced in Section 3.2. Conditions on the constrained minimization of energy due to several constraints are derived and used for applications in Section 3.3. Alternative variational characterizations of periodic waves are discussed in Section 3.4. Saddle points of energy under constraints are studied with the extension of the Hamiltonian–Krein theorem in Section 3.5.

3.1. Hamiltonian systems with periodic waves

As before, let $\mathcal{H}$ be a Hilbert space equipped with the inner product $\langle \cdot, \cdot \rangle$. Let X be a Banach space and X^* be its dual in $\mathcal{H}$ satisfying $X \subset \mathcal{H} \subset X^*$. The norm in X is denoted by $\| \cdot \|_X$ and the induced norm in $\mathcal{H}$ is denoted by $\| \cdot \|$.

We say that an evolution problem is cast in Hamiltonian form if there exists a C^1 energy function $H(u) : X \subset \mathcal{H} \to \mathbb{R}$ and a densely defined, skew-adjoint operator $J : \mathrm{Dom}(J) \subset \mathcal{H} \to \mathcal{H}$ such that

$$\begin{cases} \dfrac{du}{dt} = J\nabla H(u), & t > 0, \\ u(0) = u_0. \end{cases} \tag{3.1}$$

The skew-adjoint property of J can be expressed in the form

$$\langle Jf, g\rangle = -\langle f, Jg\rangle, \quad f, g \in \mathrm{Dom}(J) \subset \mathcal{H}.$$

The evolution problem in the form (3.1) is well-defined if a solution u exists in a subset of X for which $\nabla H(u) \in \mathrm{Dom}(J)$. As before, we will not be concerned with the local well-posedness of (3.1) in the energy space X and make the following assumption.

ASSUMPTION 3.1. Assume that the Hamiltonian system (3.1) admits a unique local weak solution $u \in C^0([0, \tau_0), X)$ such that $H(u(t, \cdot)) = H(u_0)$ for every $t \in [0, \tau_0)$. Moreover, the solution depends continuously on the initial data $u_0 \in X$.

We further assume the existence of the symmetry operator $T(\theta)$ with the properties (a)–(e) according to Definition 2.4. In addition, we assume that $J^{-1}T'$ is a well-defined operator on X so that the conserved quantity $Q(u) = \frac{1}{2}\langle J^{-1}T'u, u\rangle$, $u \in X$ exists, as in Lemma 2.1, and $Q(u(t, \cdot)) = Q(u_0)$ for every $t \in [0, \tau_0)$ if $u \in C^0([0, \tau_0), X)$ is the local weak solution in the energy space X in Assumption 3.1.

In addition to the conserved quantity $Q(u)$ coming from the symmetry operator $T(\theta)$, the kernel of the operator J can also induce conserved quantities as detailed in the following definition.

DEFINITION 3.1. We say that $M(u) : X \mapsto \mathbb{R}$ is a **Casimir functional** if $\nabla M(u) \in \mathrm{Dom}(J) \subset \mathcal{H}$ and $J\nabla M(u) = 0$ for every $u \in X$.

REMARK 3.1. The Casimir functional $M(u)$ is conserved in the time evolution of the Hamiltonian system (3.1) since

$$\frac{d}{dt}M(u) = \langle \nabla M(u), J\nabla H(u)\rangle = -\langle J\nabla M(u), \nabla H(u)\rangle = 0,$$

for every solution for which $\nabla M(u), \nabla H(u) \in \mathrm{Dom}(J)$. Hence, for every weak solution $u \in C^0([0, \tau_0), X)$ in Assumption 3.1, we have $M(u(t, \cdot)) = M(u_0)$ for every $t \in [0, \tau_0)$.

REMARK 3.2. In the setting of spatially decaying waves, Casimir functionals may exist but are generally not relevant for the stability analysis. However, in the setting of spatially periodic waves, Casimir functionals are always relevant as they are related to additional parameters which periodic waves depend on, as we will see in Examples 3.1 and 3.2.

The following two examples illustrate why periodic waves usually have additional parameters compared to solitary waves, some of which are related to the symmetry operator $T(\theta)$ while others are related to the kernel of J. The first is an example with bounded operator J, while in the second example J is unbounded.

EXAMPLE 3.1. We continue Example 2.4 on the generalized nonlinear Schrödinger (gNLS) equation

$$i\psi_t = -\psi_{xx} + W'(|\psi|^2)\psi \tag{3.2}$$

with $\psi(t,x) : \mathbb{R} \times [0,T] \to \mathbb{C}$ and $W \in C^2(\mathbb{R}_+, \mathbb{R})$. The energy space is $X = H^1_{\text{per}}$, where

$$H^1_{\text{per}} := \left\{ f \in H^1(0,T) : \quad f(0) = f(T) \right\}. \tag{3.3}$$

The energy space H^1_{per} is compactly embedded into $L^q(0,T)$ for any $q > 0$ thanks to the Sobolev embedding $\|f\|_{L^\infty} \le C\|f\|_{H^1}$ and the Hölder inequality $\|f\|_{L^q} \le C_{p,q}\|f\|_{L^p}$ for every $q \in (0,p)$ on the compact interval $[0,T]$. The Hamiltonian $H(u,\bar{u})$, the momentum $P(u,\bar{u})$, and the mass $Q(u,\bar{u})$ of the gNLS equation (3.2) are given in the periodic case by

$$H(u,\bar{u}) = \int_0^T \left(|u_x|^2 + W(|u|^2) \right) dx,$$

$$P(u,\bar{u}) = \frac{i}{2} \int_0^T (\bar{u}u_x - \bar{u}_x u) dx,$$

and

$$Q(u,\bar{u}) = \int_0^T |u|^2 dx.$$

The three conserved quantities are related to the time translation, space translation, and phase rotation symmetries of the gNLS equation (3.2), respectively, and they are well-defined in $X = H^1_{\text{per}}$. There exists another conserved quantity $M(u,\bar{u})$ given by

$$M(u,\bar{u}) = \frac{1}{2i} \int_0^T \left(\frac{u_x}{u} - \frac{\bar{u}_x}{\bar{u}} \right) dx,$$

provided $u \neq 0$ everywhere. Since $\nabla M(u,\bar{u}) = 0$, the additional conserved quantity $M(u,\bar{u})$ is a Casimir functional in the sense of Definition 3.1. It is well-defined in $X = H^1_{\text{per}}$ provided $u \neq 0$ everywhere. The traveling waves of the gNLS equation (3.2) are given by $\psi(t,x) = \psi_{\omega,c}(x+ct)e^{-i\omega t}$, where $\psi_{\omega,c}$ is a solution of the second-order differential equation

$$\omega\psi_{c,\omega} + ic\psi'_{c,\omega} + \psi''_{c,\omega} - W'(|\psi_{c,\omega}|^2)\psi_{c,\omega} = 0.$$

With the transformation $\psi_{c,\omega}(x) = u_{\omega+\frac{1}{4}c^2}(x)e^{-\frac{i}{2}cx}$, the equation can be written as

$$u''_\gamma + \gamma u_\gamma - W'(|u_\gamma|^2)u_\gamma = 0, \quad \gamma := \omega + \frac{1}{4}c^2. \tag{3.4}$$

Compared to the solitary waves in Theorem 2.9, u_γ is no longer real (up to the phase rotation). Separating the variables in polar form as $u_\gamma = R_\gamma e^{i\theta_\gamma}$ with $R_\gamma, \theta_\gamma \in \mathbb{R}$ yields two differential equations

$$R''_\gamma - R_\gamma(\theta'_\gamma)^2 + \gamma R_\gamma - W'(R^2_\gamma)R_\gamma = 0$$

and

$$2R'_\gamma\theta'_\gamma + R_\gamma\theta''_\gamma = 0.$$

The second equation can be integrated after multiplication by R_γ so that $\theta'_\gamma = a/R^2_\gamma$, where a is an integration constant. Hence the periodic wave depends on two parameters γ and a and the profile $R = R_{\gamma,a}$ satisfies Newton's equation

$$R'' - \frac{a^2}{R^3} + \gamma R - W'(R^2)R = 0. \tag{3.5}$$

The solitary wave with $R_{\gamma,a}(x) \to 0$ as $|x| \to \infty$ corresponds to $a = 0$, whereas the periodic waves with $R_{\gamma,a}(x+T) = R_{\gamma,a}(x)$ correspond to nonzero values of

a. If $a \neq 0$, then $R_{\gamma,a}(x) \neq 0$ everywhere because of the singular term $-a^2/R^3$ in (3.5). The additional integration constant a suggests a different characterization of periodic waves as critical points of the augmented energy compared to the solitary waves. Recall that (3.4) is the Euler–Lagrange equation for the augmented energy

$$\Lambda_\gamma(u,\bar{u}) := H(u,\bar{u}) - \gamma Q(u,\bar{u}). \tag{3.6}$$

Instead, (3.5) is the Euler–Lagrange equation for the augmented energy

$$\Lambda_{\gamma,a}(R) := H(R) - \gamma Q(R) - aM(R), \tag{3.7}$$

where

$$H(R) = \int_0^T \left[(R')^2 + W(R^2)\right] dx,$$
$$Q(R) = \int_0^T R^2 dx,$$
$$M(R) = \int_0^T \theta' dx = a\int_0^T R^{-2} dx.$$

Two constraints are needed to characterize periodic waves with the profile $R_{\gamma,a}$ as constrained minimizers of $\Lambda_{\gamma,a}$ defined in (3.7), whereas one constraint is needed to characterize the solitary waves with $a = 0$ as constrained minimizers of Λ_γ in (3.6).

REMARK 3.3. Although the space $L^2(0,T)$ is too weak to control the periodic boundary conditions compared to $H^1_{\rm per}$ in (3.3), it is customary to use the notation $L^2_{\rm per}$ instead of $L^2(0,T)$ to indicate periodicity of functions on the periodic domain $[0,T]$.

EXAMPLE 3.2. Another example with periodic and solitary waves is given by the generalized Korteweg–de Vries (gKdV) equation:

$$u_t + V'(u)u_x + u_{xxx} = 0, \tag{3.8}$$

where $u(t,x) : \mathbb{R} \times [0,T] \to \mathbb{R}$ and $V(u) \in C^2(\mathbb{R})$. The gKdV equation (3.8) can be written in the evolution form (3.1) associated with the Hamiltonian function

$$H(u) = \frac{1}{2}\int_0^T u_x^2 dx - \int_0^T \left(\int_0^u V(u')du'\right) dx$$

and $J = \partial_x$. The energy space is given again by $X := H^1_{\rm per}$ defined in (3.3). On the other hand, $J = \partial_x$ is an unbounded operator on $\mathcal{H} = L^2_{\rm per}$ with $\mathrm{Dom}(J) = H^1_{\rm per}$ and $\mathrm{Ker}(J) = \mathrm{Span}\{1\} \in L^2_{\rm per}$. The gKDV is invariant under spatial translations defined by the symmetry operator

$$u(t,x) \mapsto T(\theta)u(t,x) := u(t,x+\theta), \quad \theta \in \mathbb{R},$$

which induces the conservation of momentum.

$$Q(u) = \frac{1}{2}\int_0^T u^2 dx, \tag{3.9}$$

This follows formally from Lemma 2.1 with $J^{-1}T' = \mathrm{id}_{\mathcal{H}}$. To properly define $J^{-1} = \partial_x^{-1}$ one can use the Fourier series with zero mean and write

$$f(x) = \sum_{n\in\mathbb{Z}\backslash\{0\}} f_n e^{\frac{2\pi i n x}{T}}$$

for all $f \in L^2_{\rm per}$, which implies that the anti-derivative can be defined as

$$\partial_x^{-1} f(x) := \sum_{n \in \mathbb{Z}\backslash\{0\}} \frac{T f_n}{2\pi i n} e^{\frac{2\pi i n x}{T}}.$$

Since $T' = \partial_x$ we find that $J^{-1}T'$ is the identity on the subspace of periodic functions in $L^2_{\rm per}$ with zero mean. In addition, we have conservation of mass,

$$M(u) = \int_0^T u dx, \tag{3.10}$$

where $M(u)$ is a Casimir functional since $\nabla M(u) = 1$ and $J\nabla M(u) = 0$. Both $Q(u)$ and $M(u)$ are well defined in $H^1_{\rm per}$, where the local well-posedness theory of the gKdV equation (3.8) is established [**181**]. The traveling waves of the gKdV equation (3.8) are given by $u(t,x) = T(ct)\phi = \phi(x+ct)$, where ϕ is a solution of the third-order differential equation

$$\phi''' + V'(\phi)\phi' + c\phi' = 0.$$

Integration of this equation in the moving frame variable x yields the second-order differential equation

$$\phi'' + V(\phi) + c\phi + b = 0, \tag{3.11}$$

where b is an integration constant. Hence, the periodic wave with profile $\phi = \phi_{c,b}$ depends on the two parameters c and b, whereas the solitary wave depends on only one parameter c because $b = -V(0)$ is uniquely selected due to the decay $\phi_{c,b}(x) \to 0$ as $|x| \to \infty$. The differential equation (3.11) is the Euler–Lagrange equation for the augmented energy

$$\Lambda_{c,b}(u) := H(u) - cQ(u) - bM(u). \tag{3.12}$$

Two constraints are needed to characterize periodic waves with profile $\phi_{c,b}$ as constrained minimizers of $\Lambda_{c,b}$ in (3.12).

REMARK 3.4. The time evolution problem and the spectral stability of traveling solitary waves are challenging for the gKdV equation (3.8) in $L^2(\mathbb{R})$ compared to the Hamiltonian systems with bounded and invertible J considered in Chapter 2. The skew-adjoint operator $J = \partial_x$ is unbounded in $\mathcal{H} = L^2(\mathbb{R})$ with unbounded inverse in $L^2(\mathbb{R})$. Both problems become easier in the periodic setting, $\mathcal{H} = L^2_{\rm per}$, when perturbations to the traveling periodic waves are restricted to periodic functions with zero mean.

REMARK 3.5. Due to the mass conservation (3.10), the mean value of periodic solutions of the gKdV equation (3.8) is constant in time. Note that by Parseval's equality and orthogonality of Fourier modes, we can rewrite the conserved momentum (3.9) as

$$Q(u) = \frac{1}{2}\int_0^T u^2 dx = \frac{1}{2}\langle J^{-1}T'u, u\rangle + \frac{1}{2T}\left(\int_0^T u dx\right)^2,$$

where $J^{-1}T'$ is the identity on the subspace of periodic functions in $L^2_{\rm per}$ with zero mean. Only the variable part of u with the zero mean is relevant in the time evolution and stability problems since the constant mean value of u can be included as an arbitrary parameter of the family of periodic waves.

3.2. Period function

For many physically relevant infinite-dimensional Hamiltonian systems, including those considered in the examples in this book, the traveling waves are characterized by a second-order differential equation (which is the Euler–Lagrange equation for the augmented energy). The periodic traveling waves then correspond to periodic orbits around a center of the associated planar system and can be parametrized by the energy level sets of the first integral of this system. The period function assigns the minimal period to each periodic solution. It plays an important role in the stability analysis of periodic waves.

Let us consider Newton's equation,

$$\frac{d^2u}{dx^2} + P'(u) = 0, \tag{3.13}$$

where $P \in C^2(\mathbb{R})$ is the potential energy. The total energy

$$E(u,v) = \frac{1}{2}v^2 + P(u), \qquad v := \frac{du}{dx}, \tag{3.14}$$

is constant along every classical solution of Newton's equation (3.13). Periodic solutions of Newton's equation exist if P admits a local nondegenerate minimum, see the illustration in Figure 3.1. The following definition introduces the period function of the center of the planar system corresponding to Newton's equation (3.13).

Definition 3.2. Let $P \in C^2(\mathbb{R})$ and $u_0 \in \mathbb{R}$ be a local nondegenerate minimum of P such that $P'(u_0) = 0$, and $P''(u_0) > 0$. Then $(u_0, 0)$ is a center in the phase plane (u,v) corresponding to Newton's equation (3.13) and there exists $\mathcal{E}_0 > 0$ such that the periodic orbits surrounding $(u_0, 0)$ belong to the level curve $E(u,v) = \mathcal{E}$ for $\mathcal{E} \in (0, \mathcal{E}_0)$. The periodic solutions have the period

$$T := 2\int_{u_-}^{u_+} \frac{du}{\sqrt{2(\mathcal{E} - P(u))}},$$

where the turning points $u_\pm$ are given by the roots of $P(u) = \mathcal{E}$ closest to u_0 and satisfy the ordering $u_- < u_0 < u_+$. The mapping $T(\mathcal{E}) : (0, \mathcal{E}_0) \to \mathbb{R}_+$ is called **period function**.

Assume that the spatial profile of a periodic waves is represented by u in Newton's equation (3.13). We single out the periodic waves which only admit one oscillation over the periodic domain $[0, T]$.

Definition 3.3. We say that a periodic wave with profile u_* is **single-lobe** if u_* has only one maximum and minimum on the periodic domain $[0, T]$.

In the following four examples we show how the period function can be explicitly expressed in cases when the spatial profiles of the periodic waves are given in terms of Jacobi elliptic functions (Appendix A). The main parameter of the Jacobi elliptic functions is the elliptic modulus $k \in (0,1)$. The examples arise in the context of the periodic waves of the cubic NLS equation (Example 3.1 with $W(|\psi|^2) = \pm|\psi|^4$), the modified KdV equation (Example 3.2 with $V(u) = \pm u^3$), and the KdV equation (Example 3.2 with $V(u) = u^2$). We will use the results on the monotonicity of the period function in Section 3.3 to prove orbital stability of periodic waves in these equations.

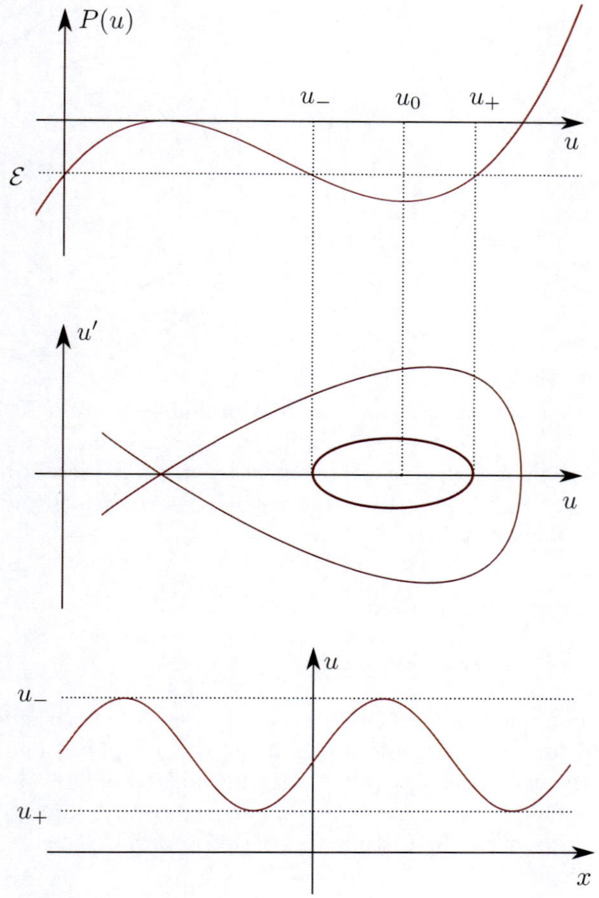

FIGURE 3.1. Top: a schematic illustration of the potential $P(u)$ with the local minimum at u_0 and two turning points $u_- < u_0 < u_+$ for the energy level $E(u,v) = \mathcal{E}$. Middle: the corresponding phase portrait of Newton's equation on the (u,v) plane. Bottom: the periodic solution for the energy level $\mathcal{E}$ oscillating between u_- and u_+.

EXAMPLE 3.3. Consider Duffing's equation

$$u'' - cu + 2u^3 = 0, \tag{3.15}$$

which corresponds to Newton's equation (3.13) with $P(u) = -\frac{1}{2}cu^2 + \frac{1}{2}u^4$. The Jacobi elliptic function $\psi_0(z) = \mathrm{dn}(z;k)$ satisfies the normalized equation

$$\psi_0'' + (k^2 - 2)\psi_0 + 2\psi_0^3 = 0$$

and the normalized first-order invariant

$$\frac{1}{2}(\psi_0')^2 + \frac{1}{2}(k^2 - 2)\psi_0^2 + \frac{1}{2}\psi_0^4 = \frac{1}{2}(k^2 - 1).$$

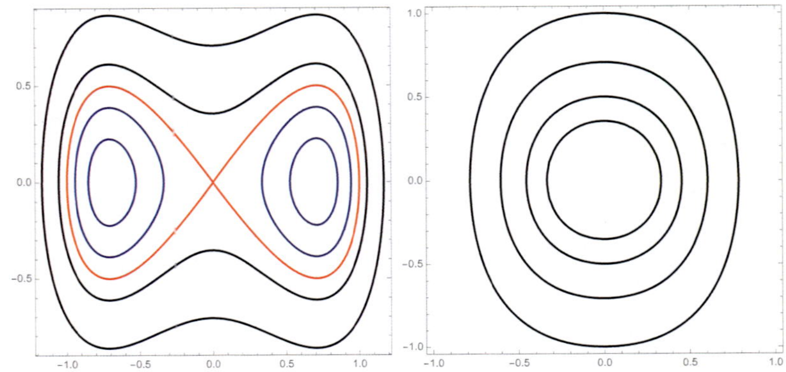

FIGURE 3.2. The phase portrait of Duffing's equation (3.15) on the (u, v) plane for $c > 0$ (left) and $c < 0$ (right). The sign indefinite periodic solutions correspond to the orbits depicted in black while the sign definite positive and negative periodic solutions correspond to the orbits depicted in blue.

By using the scaling transformation,

$$u(x) = \alpha\ \mathrm{dn}(\alpha x; k), \quad \alpha := \frac{\sqrt{c}}{\sqrt{2-k^2}}, \tag{3.16}$$

the positive periodic solutions of Duffing's equation (3.15) with fixed parameter $c > 0$ are obtained from the periodic solution $\psi_0(z) = \mathrm{dn}(z; k)$. The phase portrait for $c > 0$ is shown on Figure 3.2 (left) and the positive solutions parameterized by $k \in (0, 1)$ exist inside one of the two homoclinic orbits connected to the saddle point $(0, 0)$. It follows from (3.16) that the period function $T = T(\mathcal{E})$ is expressed in the parametric form:

$$T = 2\alpha^{-1}K(k), \quad \mathcal{E} = \frac{1}{2}\alpha^4(k^2 - 1),$$

where $K(k)$ is the complete elliptic integral of the first kind. For fixed $c > 0$, the parameter α in (3.16) depends on k and the chain rule gives

$$\begin{aligned} \frac{dT}{dk} &= \frac{2}{\sqrt{c(2-k^2)}}\left[(2-k^2)\frac{d}{dk}K(k) - kK(k)\right] \\ &= \frac{2}{\sqrt{c(2-k^2)}}\frac{(2-k^2)E(k) - 2(1-k^2)K(k)}{k(1-k^2)} \end{aligned}$$

and

$$\frac{d\mathcal{E}}{dk} = \frac{c^2k^3}{(2-k^2)^3},$$

where $E(k)$ is the complete elliptic integral of the second kind and we have used (A.7) for elliptic integrals. Positivity of

$$f(k) := (2-k^2)E(k) - 2(1-k^2)K(k), \quad k \in (0, 1)$$

follows from $f(0) = 0$ and $f'(k) = 3k[K(k) - E(k)] > 0$ due to properties (A.6), (A.7), and (A.8). Hence we have $\frac{dT}{dk} > 0$ for $k \in (0, 1)$. Since $\frac{d\mathcal{E}}{dk} > 0$ for $k \in (0, 1)$, the chain rule implies that the mapping $\mathcal{E} \mapsto T(\mathcal{E})$ is monotonically increasing.

EXAMPLE 3.4. Continuing Example 3.3, let us now consider the sign-indefinite solutions of the Duffing equation (3.15), which exist for every $c \in \mathbb{R}$. The Jacobi elliptic function $\psi_0(z) = k\ \mathrm{cn}(z;k)$ satisfies the normalized equation

$$\psi_0'' + (1-2k^2)\psi_0 + 2\psi_0^3 = 0$$

and the normalized first-order invariant

$$\frac{1}{2}(\psi_0')^2 + \frac{1}{2}(1-2k^2)\psi_0^2 + \frac{1}{2}\psi_0^4 = \frac{1}{2}k^2(1-k^2).$$

By using the scaling transformation,

$$\psi(x) = \alpha k\,\mathrm{cn}(\alpha x;k), \quad \alpha := \sqrt{\frac{c}{2k^2-1}}, \tag{3.17}$$

the sign-indefinite periodic solutions of the Duffing equation (3.15) are obtained from the periodic solution $\psi_0(z) = k\ \mathrm{cn}(z;k)$. It is understood in (3.17) that $k \in (0, \frac{1}{\sqrt{2}})$ if $c > 0$ and $k \in (\frac{1}{\sqrt{2}}, 1)$ if $c < 0$. If $c = 0$, then $k = \frac{1}{\sqrt{2}}$ and the parameter $\alpha > 0$ is arbitrary. The phase portraits of Duffing's equation (3.15) for $c > 0$ and $c < 0$ are shown on Figure 3.2 (left and right panels, respectively), where the sign-indefinite solutions parameterized by $k \in (0,1)$ exist either outside the two homoclinic orbits connected to the saddle point $(0,0)$ for $c > 0$ or surround the center point $(0,0)$ for $c < 0$. It follows from (3.17) that the period function $T = T(\mathcal{E})$ is expressed in the parametric form:

$$T = 4\alpha^{-1}K(k), \quad \mathcal{E} = \frac{1}{2}\alpha^4 k^2(1-k^2). \tag{3.18}$$

For fixed $c \neq 0$, parameter α in (3.17) depends on k and the chain rule gives

$$\begin{aligned}\frac{dT}{dk} &= \frac{4\ \mathrm{sgn}(c)}{\sqrt{c(2k^2-1)}}\left[(2k^2-1)\frac{d}{dk}K(k) + 2kK(k)\right] \\ &= \frac{4\ \mathrm{sgn}(c)}{\sqrt{c(2k^2-1)}k(1-k^2)}\left[(1-k^2)(K(k)-E(k)) + k^2E(k)\right]\end{aligned}$$

and

$$\frac{d\mathcal{E}}{dk} = \frac{kc^2}{(1-2k^2)^3}.$$

Since $E(k) < K(k)$ and $k \in (0,1)$, we have $\frac{dT}{dk} < 0$ and $\frac{d\mathcal{E}}{dk} > 0$ for $c < 0$ and $k \in (0, \frac{1}{\sqrt{2}})$, while $\frac{dT}{dk} > 0$ and $\frac{d\mathcal{E}}{dk} < 0$ for $c > 0$ and $k \in (\frac{1}{\sqrt{2}}, 1)$. Therefore, the mapping $\mathcal{E} \mapsto T(\mathcal{E})$ is monotonically decreasing in either case. The only difference is that $T(\mathcal{E}) \in (0, 2\pi|c|^{-1/2})$ for $c < 0$, and $T(\mathcal{E}) \in (0,\infty)$ for $c > 0$. For $c = 0$ and $k = \frac{1}{\sqrt{2}}$, parameter α in (3.17) is arbitrary and it follows from (3.18) that $\mathcal{E} = C/T^4$ for some $C > 0$ so that the mapping $\mathcal{E} \mapsto T(\mathcal{E})$ is also monotonically decreasing.

EXAMPLE 3.5. Consider Duffing's equation in the form

$$u'' + cu - 2u^3 = 0, \tag{3.19}$$

which corresponds to Newton's equation (3.13) with $P(u) = \frac{1}{2}cu^2 - \frac{1}{2}u^4$. The Jacobi elliptic function $\psi_0(z) = k\,\mathrm{sn}(z;k)$ satisfies the normalized equation

$$\psi_0'' + (1+k^2)\psi_0 - 2\psi_0^3 = 0$$

and the normalized first-order invariant

$$\frac{1}{2}(\psi_0')^2 + \frac{1}{2}(1+k^2)\psi_0^2 - \frac{1}{2}\psi_0^4 = \frac{1}{2}k^2.$$

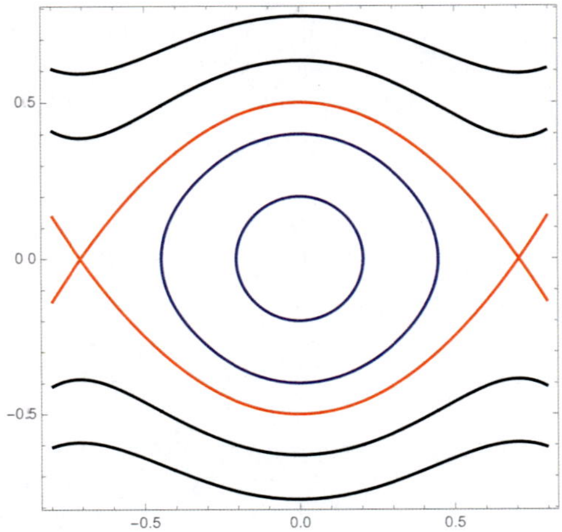

FIGURE 3.3. The phase portrait of Duffing's equation (3.19) on the (u, v) plane for $c > 0$. The sign indefinite periodic solution corresponds to the blue periodic orbit inside the heteroclinic orbits.

By using the scaling transformation,

$$u(x) = \alpha k \,\mathrm{sn}(\alpha x; k), \quad \alpha := \frac{\sqrt{c}}{\sqrt{1+k^2}}, \tag{3.20}$$

the sign-indefinite periodic solutions of the Duffing equation (3.19) with fixed parameter $c > 0$ are obtained from the periodic solution $\psi_0(z) = k\,\mathrm{sn}(z; k)$. The phase portrait for $c > 0$ is shown on Figure 3.3 and the sign-indefinite solutions parameterized by $k \in (0, 1)$ exist inside the heteroclinic orbits connected to the two symmetric saddle points. It follows from (3.20) that the period function $T = T(\mathcal{E})$ is expressed in the parametric form:

$$T = 4\alpha^{-1} K(k), \quad \mathcal{E} = \frac{1}{2}\alpha^4 k^2.$$

For fixed $c > 0$, parameter α in (3.20) depends on k and the chain rule gives

$$\frac{dT}{dk} = \frac{4}{\sqrt{c(1+k^2)}} \left[(1+k^2) \frac{d}{dk} K(k) + k K(k) \right]$$

and

$$\frac{d\mathcal{E}}{dk} = \frac{c^2 k (1-k^2)}{(1+k^2)^3}.$$

Since $\frac{d}{dk}K(k) > 0$, we have $\frac{dT}{dk} > 0$ and $\frac{d\mathcal{E}}{dk} > 0$ for $k \in (0, 1)$. The chain rule implies that the mapping $\mathcal{E} \mapsto T(\mathcal{E})$ is monotonically increasing.

EXAMPLE 3.6. Consider the differential equation

$$u'' - cu + u^2 = 0, \tag{3.21}$$

which corresponds to Newton's equation (3.13) with $P(u) = -\frac{1}{2}cu^2 + \frac{1}{3}u^3$. The Jacobi elliptic function $\psi_0(z) = k^2 \,\mathrm{cn}^2(z; k)$ satisfies the normalized equation

$$\psi_0'' + 4(1 - 2k^2)\psi_0 + 6\psi_0^2 = 2k^2(1 - k^2)$$

and the normalized first-order invariant

$$\frac{1}{4}(\psi_0')^2 = \psi_0(k^2 - \psi_0)(1 - k^2 + \psi_0).$$

By using the scaling transformation and the translation,

$$(3.22)\qquad \begin{aligned} u(x) &= 2\alpha^2 \left[1 - 2k^2 + \sqrt{1 - k^2 + k^4} + 3k^2 \operatorname{cn}^2(\alpha x; k)\right], \\ \alpha &:= \frac{\sqrt{c}}{2\sqrt[4]{1 - k^2 + k^4}}, \end{aligned}$$

the positive periodic solutions of the differenetial equation (3.21) with fixed parameter $c > 0$ are obtained from the periodic solution $\psi_0(z) = k^2 \operatorname{cn}^2(z; k)$. The phase portrait for $c > 0$ is shown on Figure 3.4 and the positive periodic solutions parameterized by $k \in (0, 1)$ exist inside the only homoclinic orbit connectd to the saddle point $(0, 0)$. It follows from (3.22) that the period function $T = T(\mathcal{E})$ is expressed in the parametric form:

$$T = 2\alpha^{-1}K(k), \quad \mathcal{E} = \frac{8}{3}\alpha^6(1 - 2k^2 + \sqrt{1 - k^2 + k^4})^2(1 - 2k^2 - 2\sqrt{1 - k^2 + k^4}).$$

For fixed $c > 0$, parameter α in (3.20) depends on k and the chain rule gives

$$\begin{aligned} \frac{dT}{dk} &= \frac{2}{\sqrt{c}(1 - k^2 + k^4)^{3/4}} \left[2(1 - k^2 + k^4)\frac{d}{dk}K(k) + k(2k^2 - 1)K(k)\right] \\ &= \frac{2}{\sqrt{c}(1 - k^2 + k^4)^{3/4}k(1 - k^2)} \left[2(1 - k^2 + k^4)E(k) - (1 - k^2)(2 - k^2)K(k)\right] \end{aligned}$$

and

$$\frac{d\mathcal{E}}{dk} = \frac{9c^3k^3(1 - k^2)}{8(1 - k^2 + k^4)^{5/2}}.$$

It follows from the inequalities used in Examples 3.3 and 3.4 that

$$\begin{aligned} &2(1 - k^2 + k^4)E(k) - (1 - k^2)(2 - k^2)K(k) \\ &= [(2 - k^2)E(k) - 2(1 - k^2)K(k)] + k^2(1 - k^2)[K(k) - E(k)] + k^4E(k) > 0 \end{aligned}$$

so that $\frac{dT}{dk} > 0$ and $\frac{d\mathcal{E}}{dk} > 0$ for $k \in (0, 1)$. Hence, the mapping $\mathcal{E} \mapsto T(\mathcal{E})$ is monotonically increasing.

REMARK 3.6. Monotonicity of the period function for the positive periodic solutions in Examples 3.3 and 3.6 is a classical result [**50, 70, 241, 247**]. Monotonicity of the period function for the sign-indefinite periodic solutions in Examples 3.4 and 3.5 is less known with some partial results obtained in [**54, 100**]. Stability of the periodic waves in the NLS, KdV, and modified KdV equations related to these examples have been studied in many details, see, e.g., [**15, 18, 79, 206, 207**].

REMARK 3.7. Quite often, the monotonicity of the period function cannot be obtained simply by explicit computations as in the above examples. Nevertheless, there exist tools to study its monotonicity as for instance the monotonicity criterion for planar systems [**51**], and criteria to bound the number of critical periods [**198**]. These tools have been used in the stability analysis of the reduced Ostrovsky equation [**104**], the Camassa–Holm equation [**108, 109**], and the Degasperis–Procesi equation [**107**].

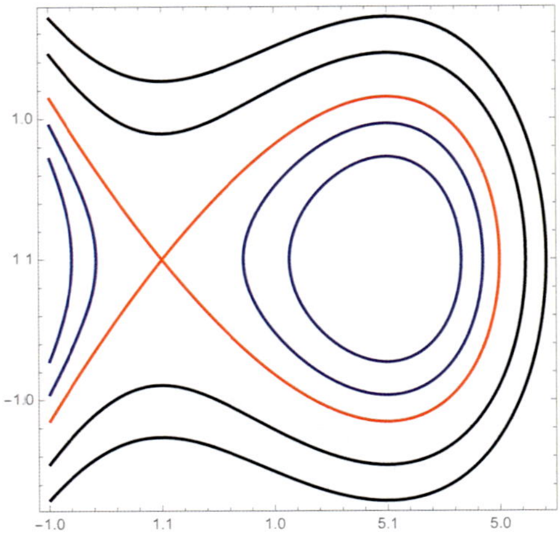

FIGURE 3.4. The phase portrait of the differential equation (3.21) on the (u, v) plane for $c > 0$. The sign definite periodic solution corresponds to the blue periodic orbit inside the homoclinic orbit.

In order to show how the period function is used in the stability analysis of periodic waves, we recall that the periodic wave with the profile u_* in the Hamiltonian system (3.1) arises as a critical point of the augmented energy Λ. If Newton's equation (3.13) represents the Euler–Lagrange equation for Λ, then the second variation of $\Lambda(u)$ at $u = u_*$ is determined by the self-adjoint Schrödinger operator

$$\mathcal{L} := -\partial_x^2 - P''(u_*),$$

which is considered in $L^2_{\rm per}$ with the domain $H^2_{\rm per}$ given by

$$H^2_{\rm per} := \left\{ f \in H^2(0,T) : \quad f(0) = f(T), \quad f'(0) = f'(T) \right\}.$$

Due to smoothness of $u_* \in C^3_{\rm per}$ if $P \in C^2(\mathbb{R})$ and the translation symmetry of Newton's equation (3.13), we have that $\partial_x u_* \in H^2_{\rm per}$ and $\mathcal{L}\partial_x u_* = 0$. If u_* is a single-lobe periodic wave as defined in Definition 3.3, then $\partial_x u_*$ admits exactly two zeros on the periodic domain $[0,T]$. However, contrary to Sturm's oscillation theorem on the unbounded domain (Theorem B.9), this property on the periodic domain does not imply that the zero eigenvalue is the third eigenvalue of $\mathcal{L}$ in $L^2_{\rm per}$ with exactly two simple negative eigenvalues because eigenvalues of the Schrödinger operator $\mathcal{L}$ in $L^2_{\rm per}$ could be double. The period function gives the tool to clarify if the zero eigenvalue is double or simple, and if there exist only one or two simple negative eigenvalues of $\mathcal{L}$ in $L^2_{\rm per}$.

By differentiating Newton's equation (3.13) with respect to the energy $\mathcal{E}$, as given by the first-order invariant (3.14), it follows that $\mathcal{L}\partial_{\mathcal{E}} u_* = 0$, where $\partial_{\mathcal{E}} u_* \in H^2(0,T)$ is evaluated at the energy level $\mathcal{E}$ corresponding to u_*. In general, $\partial_{\mathcal{E}} u_* \notin H^2_{\rm per}$ because it may not be periodic. Whether or not this is the case can be studied with the monotonicity of the period function $\mathcal{E} \mapsto T(\mathcal{E})$.

Without loss of generality, we may assume that u_* is even in x, which can be achieved by translation due to reversibility of the differential equation (3.13): if $u(x)$ is a solution, then $u(-x)$ is also a solution. If u_* is even in x, then $\partial_{\mathcal{E}} u_*$ is even and $\partial_x u_*$ is odd, so that the general solution of the second-order homogeneous equation $\mathcal{L}v = 0$ is given by

$$v(x) = c_1 \partial_{\mathcal{E}} u_*(x) + c_2 \partial_x u_*(x),$$

where c_1 and c_2 are x-independent constants. Note that $\partial_x u_*$ is always T-periodic in x. The following lemma determines the necessary and sufficient condition for $\partial_{\mathcal{E}} u_*$ being T-periodic in x as well.

LEMMA 3.1. $\mathrm{Ker}(\mathcal{L}) = \mathrm{Span}\{\partial_x u_*\}$ *in* L^2_{per} *if and only if* $T'(\mathcal{E}) \neq 0$.

PROOF. Assuming that the T-periodic function u_* is even in x, we have

$$\partial_x u_*(0) = \partial_x u_*(T) = 0$$

so that the two functions

$$y_1(x) := \frac{\partial_{\mathcal{E}} u_*(x)}{\partial_{\mathcal{E}} u_*(0)}, \quad y_2(x) := \frac{\partial_x u_*(x)}{\partial_x^2 u_*(0)}, \tag{3.23}$$

form the fundamental set of solutions associated to the equation $\mathcal{L}v = 0$ in $H^2(0, T)$ such that

$$\begin{cases} y_1(0) = 1, \\ y_1'(0) = 0, \end{cases} \qquad \begin{cases} y_2(0) = 0, \\ y_2'(0) = 1. \end{cases} \tag{3.24}$$

Note that y_1, y_2 are well-defined: $\partial_{\mathcal{E}} u_*(0) \neq 0$ and $\partial_x^2 u_*(0) \neq 0$ holds because $1 = P'(u_*(0))\partial_{\mathcal{E}} u_*(0)$ and $\partial_x^2 u_*(0) = -P'(u_*(0))$ where $u_*(0)$ is not a critical point of P by Definition 3.2.

Differentiating the boundary conditions $u_*(T) = u_*(0)$ and $u_*'(T) = 0$ in $\mathcal{E}$, where $T = T(\mathcal{E})$, yields $y_1(T) = y_1(0) = 1$ and

$$y_1'(T) = -\frac{T'(\mathcal{E})}{\partial_{\mathcal{E}} u_*(0)} \partial_x^2 u_*(0) = \frac{T'(\mathcal{E})}{[\partial_{\mathcal{E}} u_*(0)]^2}. \tag{3.25}$$

Hence, $y_1 \in H^2_{\mathrm{per}}$ if and only if $T'(\mathcal{E}) = 0$, which yields the assertion of the lemma. □

If the periodic wave is a saddle point of the augmented energy, the linearization operator $\mathcal{L}$ with the dense domain $H^2_{\mathrm{per}} \subset L^2_{\mathrm{per}}$ admits negative eigenvalues in L^2_{per}. For the single-lobe periodic waves in the sense of Definition 3.3, we know from Lemma 3.1 and from two zeros of $\partial_x u_*$ on the periodic domain $[0, T]$ that $\mathcal{L}$ has either one or two simple negative eigenvalues in L^2_{per} if $T'(\mathcal{E}) \neq 0$. The following theorem gives the precise count of the number of negative eigenvalues of $\mathcal{L}$ depending on the sign of $T'(\mathcal{E})$.

THEOREM 3.1. *Let* u_* *be the single-lobe periodic wave in Definition* 3.3 *satisfying* (3.13) *and* (3.14) *with* $T'(\mathcal{E}) \neq 0$. *The zero eigenvalue of* $\mathcal{L}$ *in* L^2_{per} *is the second eigenvalue if* $T'(\mathcal{E}) > 0$ *and the third eigenvalue of* $\mathcal{L}$ *if* $T'(\mathcal{E}) < 0$.

PROOF. Since H^2_{per} is compactly embedded into L^2_{per}, the spectrum of $\mathcal{L}$ in L^2_{per} consists of isolated eigenvalues of finite multiplicities. By Lemma 3.1, $\mathrm{Ker}(\mathcal{L}) = \mathrm{Span}\{\partial_x u_*\}$ since $T'(\mathcal{E}) \neq 0$ and the fundamental set of solutions in (3.23) satisfying (3.24) also satisfies

$$y_1(x + T) = y_1(x) + y_1'(T) y_2(x), \quad y_2(x + T) = y_2(x),$$

where $y_1'(T)$ is given by (3.25). Assume again that the T-periodic function u_* is even in x so that y_2 is odd in x.

Note that, since u_* is even and T periodic, it has two critical points on $[0, T)$, and therefore y_2 has two zeros on the periodic domain. It follows from Sturm's oscillation theory in the periodic setting that zero is either the second or third eigenvalue of $\mathcal{L}$ on L^2_{per} (see for instance [**197**, Theorem 2.14] or [**236**, Theorem 5.34]). The remainder of the proof is devoted to distinguishing these two cases based on the sign of $y_1'(T) \neq 0$, which by (3.25) is the same as the sign of $T'(\mathcal{E}) \neq 0$.

Consider the spectral problem $\mathcal{L}y = \lambda y$ on $H^2(0, T)$. We claim that the location of a nondegenerate extremal point of $y(\cdot, \lambda)$ on $(0, T)$ is a monotonically decreasing function of λ, where $y(x, \lambda)$ is a solution of

$$-y''(x, \lambda) - P''(u_*(x))y(x, \lambda) = \lambda y(x, \lambda), \quad x \in [0, T] \tag{3.26}$$

starting with $y(0, \lambda) = 1$ and $y'(0, \lambda) = 0$, where the prime denotes derivatives in x. Let $x(\lambda) \in (0, T)$ be the location of $y'(x(\lambda), \lambda) = 0$. Without loss of generality, we can consider the maximum point with $y(x(\lambda), \lambda) > 0$ and $y''(x(\lambda), \lambda) < 0$. By differentiating $y'(x(\lambda), \lambda) = 0$ we obtain

$$\partial_\lambda y'(x(\lambda), \lambda) + x'(\lambda) y''(x(\lambda), \lambda) = 0.$$

By differentiating (3.26) in λ, multiplying it by $y(x, \lambda)$, and integrating on $[0, x(\lambda)]$, we obtain using integration by parts and equation (3.26) that

$$\left[y'(x, \lambda)\partial_\lambda y(x, \lambda) - y(x, \lambda)\partial_\lambda y'(x, \lambda)\right]\Big|_{x=0}^{x=x(\lambda)} = \int_0^{x(\lambda)} y(x, \lambda)^2 dx.$$

Since $y'(0, \lambda) = 0$ and $y'(x(\lambda), \lambda) = 0$ for every λ, the two equations yield

$$x'(\lambda) = -\frac{\partial_\lambda y'(x(\lambda), \lambda)}{y''(x(\lambda), \lambda)} = \frac{\int_0^{x(\lambda)} y(x, \lambda)^2 dx}{y(x(\lambda), \lambda) y''(x(\lambda), \lambda)} < 0,$$

where the last inequality is due to $y(x(\lambda), \lambda) > 0$ and $y''(x(\lambda), \lambda) < 0$. The same is true for the minimum point with $y(x(\lambda), \lambda) < 0$ and $y''(x(\lambda), \lambda) > 0$, which proves the claim.

Consider now the even solution y_1 of $\mathcal{L}y = 0$ satisfying (3.24). If we denote by $\{\lambda_n\}_{n \in \mathbb{N}}$ the eigenvalues of $\mathcal{L}$ with respect to periodic boundary conditions on $[0, T]$ and by $\{\mu_n\}_{n \in \mathbb{N}}$ the eigenvalues of $\mathcal{L}$ with respect to antiperiodic boundary conditions on $[0, T]$, then the following interlacing formula (see [**236**, Theorem 5.31] or [**90**]) holds:

$$\lambda_1 < \mu_1 \leq \mu_2 < \lambda_2 \leq \lambda_3 < \mu_3 \leq \mu_4 < \dots,$$

where either $\lambda_2 = 0$ or $\lambda_3 = 0$. In either case, $\mu_1 \leq \mu_2 < 0$. The eigenfunction for either μ_1 or μ_2 satisfies the Neuman boundary conditions $y'(0) = 0$ and $y'(T) = 0$, with a maximum at $x = 0$ and a minimum at $x = T$. Since the location of a nondegenerate extremal point of $y(\cdot, \lambda)$ on $(0, T)$ is monotonically decreasing in λ, this implies that y_1 has a global minimum in $(0, T)$.

It remains to distinguish the two cases of the sign of $y_1'(T) \neq 0$.

- If $y_1'(T) > 0$, then y_1 has only one extremal point on $(0, T)$, see Figure 3.5 (left). By monotonicity of the extremal points versus λ, this means that for all $\lambda \leq 0$ the even solutions of $\mathcal{L}y = 0$ do not have a maximum inside $(0, T)$, and so the second eigenvalue of $\mathcal{L}$ in the subspace of even functions must be positive, that is, $\lambda_3 > 0 = \lambda_2$.

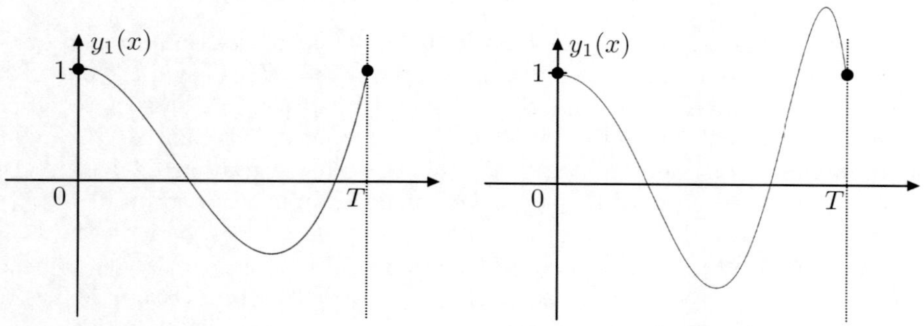

FIGURE 3.5. A schematic graph of $y_1(x)$ on $[0,T]$ for $y_1'(T) > 0$ (left) and $y_1'(T) < 0$ (right).

- If $y_1'(T) < 0$, then y_1 has two extremal points on $(0,T)$, see Figure 3.5 (right). Since the extremal points of y_1 are decreasing functions of λ, the maximum was obtained at $x = T$ for some negative eigenvalue, and so the second eigenvalue of $\mathcal{L}$ in the subspace of even functions is negative, that is, $\lambda_2 < 0 = \lambda_3$.

This dichotomy completes the proof since the sign of $y_1'(T)$ coincides with the sign of $T'(\mathcal{E})$ in view of (3.25). □

REMARK 3.8. Theorem 3.1 was proven in [**140**] and [**208**] for applications to the spectral stability of periodic waves. It was used in similar contexts in [**18**, **108**].

REMARK 3.9. If $T'(\mathcal{E}) = 0$, then Lemma 3.1 and Theorem 3.1 suggest that the zero eigenvalue of $\mathcal{L}$ in $L^2_{\rm per}$ is double, that is, $\lambda_2 = \lambda_3 = 0$. In this case, there exists only one simple negative eigenvalue λ_1 of $\mathcal{L}$ in $L^2_{\rm per}$ associated with the positive even eigenfunction.

EXAMPLE 3.7. Consider the harmonic oscillator equation

$$u'' + \omega_0^2 u = 0,$$

which corresponds to Newton's equation (3.13) with $P(u) = \frac{1}{2}\omega_0^2 u^2$. The period of the harmonic oscillator is fixed at $T = \frac{2\pi}{\omega_0}$ independently of its energy $\mathcal{E} > 0$. The exact solution is

$$u(x) = \sqrt{2\mathcal{E}}\omega_0^{-1}\cos(\omega_0 t + \theta_0)$$

with arbitrary phase $\theta_0 \in \mathbb{R}$. Thus, $T'(\mathcal{E}) = 0$ and the zero eigenvalue of $\mathcal{L} = -\partial_x^2 - \omega_0^2$ in $L^2_{\rm per}$ is the second (double) eigenvalue by Lemma 3.1 and Theorem 3.1. Indeed, the Fourier modes $e^{in\omega_0 t}$, $n \in \mathbb{Z}$, are eigenfunctions of $\mathcal{L}$ in $L^2_{\rm per}$ for the eigenvalues $\{\omega_0^2(n^2-1),\ n \in \mathbb{Z}\}$ so that

$$\lambda_1 = -\omega_0^2 < \lambda_2 = \lambda_3 = 0 < \lambda_4 = \lambda_5 = 3\omega_0^2 < \cdots,$$

which gives precisely the result we obtain from Lemma 3.1 and Theorem 3.1.

3.3. Constrained minimizers of energy

In Section 2.4, the stationary state with profile u_ω of the Hamiltonian system (2.1) is a critical point of the augmented energy $\Lambda_\omega(u) = H(u) - \omega Q(u)$, constructed from the Hamiltonian $H(u)$ and the conserved quantity $Q(u)$ associated to the symmetry operator $T(\theta)$. In that setting, fixing the conserved quantity $Q(u)$, we are able to eliminate one negative eigenvalue of the second variation $\mathcal{L}_\omega = \nabla^2 \Lambda_\omega(u_\omega)$ to obtain orbital stability of the strict constrained minimizer with profile u_ω.

In Section 3.1, it is shown that in contrast to solitary waves, periodic waves may have more parameters due to Casimir functionals, which are conserved quantities related to the operator J. For instance, in the gKdV equation (Example 3.2), the periodic wave with the single-lobe profile $u_{c,b}$ has the two parameters c and b related to the conserved quantities $Q(u)$ and $M(u)$, so that $u_{c,b}$ is a critical point of the augmented energy $\Lambda_{c,b}(u) = H(u) - cQ(u) - bM(u)$. In Section 3.2 we show that the linear operator $\mathcal{L}_{c,b} = \nabla^2 \Lambda_{c,b}(u_{c,b})$ may have up to two negative eigenvalues, depending on the monotonicity of the period function, see Theorem 3.1. Similar to the strategy used in Section 2.4, we can fix the two conserved quantities $Q(u)$ and $M(u)$ independently of each other and check if the periodic waves with the profile $u_{c,b}$ are constrained minimizers of the energy $H(u)$. One constraint can only eliminate one negative eigenvalue of $\mathcal{L}_{c,b}$ (Theorem 2.7) but two constraints may be sufficient to remove up to two negative eigenvalues of $\mathcal{L}_{c,b}$. If this is the case, we can obtain orbital stability of the strict constrained minimizer with profile $u_{c,b}$.

For clarity and generality of presentation, we consider the case of N conserved quantities and N parameters labeled by the same letter (with subscripts). The following assumption sets the scene, after which we will extend the results of Lemma 2.4, Theorem 2.7, and Theorem 2.8 to the case of N constraints.

ASSUMPTION 3.2. Let $Q_k(u) : X \mapsto \mathbb{R}$ be functionals which are conserved in the Hamiltonian system (3.1) for $1 \leq k \leq N$. Let u_ω be a critical point of the augmented energy

$$\Lambda_\omega(u) := H(u) - \sum_{k=1}^{N} \omega_k Q_k(u),$$

where $\omega := (\omega_1, \omega_2, \dots, \omega_N) \in \mathcal{I} \subset \mathbb{R}^N$. Assume that $H(u)$ and $\{Q_k(u)\}_{1 \leq k \leq N}$ are C^2 functions near u_ω and that they are all functionally independent.

3.3.1. Constrained minimization under N constraints. Similarly as in Section 2.4, we will characterize the critical point u_ω of the augmented energy $\Lambda_\omega(u)$ as a constrained minimizer of $H(u)$. To do so, we consider a perturbation $w := u - u_\omega$ and study the second variation of $\Lambda_\omega(u)$ at u_ω, that is, the linear operator $\mathcal{L}_\omega := \nabla^2 \Lambda_\omega(u_\omega) : X \to X^*$, which is self-adjoint on the Hilbert space $\mathcal{H}$.

The admissible set of perturbations w to the critical point u_ω for the nonlinear constrained minimization problem is given by

$$\mathcal{A}_\omega := \{w \in X : \quad Q_k(u_\omega + w) = Q_k(u_\omega), \quad 1 \leq k \leq N\}.$$

However, since we are only concerned with local constrained minimizers, it is sufficient to consider the tangent plane to the nonlinear admissible space $\mathcal{A}_\omega$ and consider perturbations w to the critical point u_ω in the linear admissible space

$$\mathcal{B}_\omega := \{w \in X : \quad \langle w, \nabla Q_k(u_\omega) \rangle = 0, \quad 1 \leq k \leq N\}, \tag{3.27}$$

where the set $\{\nabla Q_k(u_\omega)\}_{1 \leq k \leq N}$ is linearly independent by Assumption 3.2.

To simplify notation we denote $w_k := \nabla Q_k(u_\omega)$, $1 \leq k \leq N$, and $\mathcal{L}_\omega|_{\mathcal{B}_\omega} := \mathcal{L}_\omega|_{\{w_1,\ldots,w_N\}^\perp}$. The number of negative eigenvalues, counting their multiplicities, of the self-adjoint operator $\mathcal{L}_\omega$ in the Hilbert space $\mathcal{H}$ is denoted by $n(\mathcal{L}_\omega)$ and the multiplicity of the zero eigenvalue is denoted by $z(\mathcal{L}_\omega)$. Similarly, we denote by $n(\mathcal{L}_\omega|_{\mathcal{B}_\omega})$ and $z(\mathcal{L}_\omega|_{\mathcal{B}_\omega})$ the number of negative eigenvalues and the multiplicity of the zero eigenvalue of the constrained operator $\mathcal{L}_\omega|_{\mathcal{B}_\omega}$.

DEFINITION 3.4. Let u_ω be a critical point of the augmented energy $\Lambda_\omega(u)$. We say that u_ω is a strict local minimizer of $H(u)$ in $\mathcal{B}_\omega \subset X$ if the Hamiltonian system (3.1) admits Z-many symmetry operators related to the conserved quantities $\{Q_1(u), \cdots, Q_Z(u)\}$ with $Z \leq N$ and there exists C_ω such that

$$\langle \mathcal{L}_\omega w, w\rangle \geq C_\omega \|w\|_X^2, \quad w \in \mathcal{B}_\omega|_{\{J\nabla Q_1(u_\omega),\cdots,J\nabla Q_Z(u_\omega)\}^\perp}. \tag{3.28}$$

The following lemma clarifies how to identify the constrained minimizer of Definition 3.4 by counting the number of negative eigenvalues and the multiplicity of the zero eigenvalue of $\mathcal{L}_\omega$ and $\mathcal{L}_\omega|_{\mathcal{B}_\omega}$.

LEMMA 3.2. *Let Assumption* 3.2 *be satisfied. Assume that the Hamiltonian system* (3.1) *admits Z-many symmetry operators. The critical point u_ω of $\Lambda_\omega(u)$ is a strict local minimizer of $H(u)$ in $\mathcal{B}_\omega \subset X$ in the sense of Definition* 3.4 *if the spectrum of $\mathcal{L}_\omega$ in $(-\infty, \delta)$ for some $\delta > 0$ consists of finitely many eigenvalues and*

$$n(\mathcal{L}_\omega|_{\mathcal{B}_\omega}) = 0, \quad z(\mathcal{L}_\omega|_{\mathcal{B}_\omega}) = z(\mathcal{L}_\omega) = Z. \tag{3.29}$$

PROOF. If N functionally independent conserved quantities are fixed, then the energy space X is constrained by N linearly independent constraints in (3.27). If the conditions (3.29) are satisfied, then the operator $\mathcal{L}_\omega|_{\mathcal{B}_\omega}$ is positive semi-definite. Since the number of zero eigenvalues of $\mathcal{L}_\omega|_{\mathcal{B}_\omega}$ is equal to the number of symmetry operators and $\mathcal{L}_\omega$ admits the spectral gap in $(-\delta, \delta)$ by assumption, this implies the coercivity condition (3.28) in Definition 3.4. □

The following theorem gives a tool to count the number of negative eigenvalues and the multiplicity of the zero eigenvalue of a self-adjoint operator in a Hilbert space under N linearly independent constraints. It extends Theorem 2.7 to allow for several constraints. This theorem was first proven [**117**] and further generalized in [**76**].

THEOREM 3.2. *Assume that $\mathcal{L}_\omega : \mathrm{Dom}(\mathcal{L}_\omega) \subset \mathcal{H} \to \mathcal{H}$ is a self-adjoint operator and its spectrum in $(-\infty, \delta)$ for some $\delta > 0$ consists of finitely many eigenvalues. Assume that $\{w_1, \cdots, w_N\} \in \mathcal{H}$ are linearly independent. Let $A(\lambda)$ be the matrix-valued function defined by*

$$A_{ij}(\lambda) := \langle (\mathcal{L}_\omega - \lambda I)^{-1} w_i, w_j\rangle, \quad 1 \leq i, j \leq N, \quad \lambda \notin \sigma(\mathcal{L}_\omega). \tag{3.30}$$

Let n_0, z_0, and p_0 be the numbers of negative, zero, and positive eigenvalues of $A(\lambda)$ in the limit $\lambda \to 0^-$ (counting their multiplicities), and $z_\infty := N - n_0 - z_0 - p_0$ be the number of eigenvalues of $A(\lambda)$ diverging in the limit $\lambda \to 0^-$. Then we have

$$\begin{cases} n(\mathcal{L}_\omega|_{\mathcal{B}_\omega}) = n(\mathcal{L}_\omega) - n_0 - z_0, \\ z(\mathcal{L}_\omega|_{\mathcal{B}_\omega}) = z(\mathcal{L}_\omega) + z_0 - z_\infty. \end{cases} \tag{3.31}$$

PROOF. Since $\mathcal{L}_\omega$ is bounded from below, the Spectral Theorem (Theorem B.3) implies that $A_{ij}(\lambda) \to 0$ as $\lambda \to -\infty$ for $1 \leq i, j \leq N$. On the other hand, since

$$A'_{i,j}(\lambda) = \langle (\mathcal{L}_\omega - \lambda I)^{-2} w_i, w_j \rangle, \quad \lambda \notin \sigma(\mathcal{L}_\omega),$$

the matrix representation of $\langle (\mathcal{L}_\omega - \lambda I)^{-2} w, w \rangle$ on $w \in \text{Span}\{w_1, \dots, w_N\}$ is positive definite in the sense that

$$\langle (\mathcal{L}_\omega - \lambda I)^{-2} \sum_{j=1}^{N} x_j w_j, \sum_{j=1}^{N} x_j w_j \rangle > 0$$

for every nonzero $\mathbf{x} = (x_1, \dots, x_N) \in \mathbb{C}^N$. Therefore, the matrix $A'(\lambda)$ is positive for every $\lambda \notin \sigma(\mathcal{L}_\omega)$. Because $A(\lambda)$ is a symmetric matrix, we can define the set of its real eigenvalues $\{\alpha_i(\lambda)\}_{i=1}^N$ and the set of its orthonormal eigenvectors $\{\mathbf{v}_i(\lambda)\}_{i=1}^N$ in $\mathbb{C}^N$. Since A is symmetric, differentiation of $A(\lambda)\mathbf{v}_i(\lambda) = \alpha(\lambda)\mathbf{v}_i(\lambda)$ and projection onto $\mathbf{v}_i(\lambda)$ yield

$$\alpha_i'(\lambda) = \langle A'(\lambda)\mathbf{v}_i(\lambda), \mathbf{v}_j(\lambda) \rangle_{\mathbb{C}^N}.$$

Therefore, since $A'(\lambda)$ is positive, the mapping $\lambda \mapsto \alpha_i(\lambda)$ is monotonically increasing for $\lambda \notin \sigma(\mathcal{L}_\omega)$ and every $1 \leq i \leq N$.

Let λ_k be an isolated eigenvalue of the self-adjoint operator $\mathcal{L}_\omega$ of multiplicity m_k and $\{u_k^j\}_{j=1}^{m_k} \in L^2_{\text{per}}$ be an orthonormal set of eigenvectors. Denote the projection operator to the eigenspace of $\text{Ker}(\mathcal{L}_\omega - \lambda_k I)$ by $\Pi_k : L^2_{\text{per}} \to U_k \subset L^2_{\text{per}}$, where $U_k = \text{Span}\{u_k^1, \dots, u_k^{m_k}\}$. The matrix $A(\lambda)$ can then be written in the form

$$A(\lambda) = \frac{1}{\lambda_k - \lambda} A_k + B_k(\lambda),$$

where $[A_k]_{i,j} = \langle \Pi_k w_i, \Pi_k w_j \rangle$ and $B_k(\lambda)$ is holomorphic at $\lambda = \lambda_k$. Therefore, the eigenvalue problem for $A(\lambda)$ can be written as

$$[A_k + B_k(\lambda)(\lambda_k - \lambda)] \mathbf{v}_i(\lambda) = \alpha_i(\lambda)(\lambda_k - \lambda)\mathbf{v}_i(\lambda).$$

Eigenvalues of the holomorphic Hermitian matrix $A_k + B_k(\lambda)(\lambda_k - \lambda)$ are holomorphic at $\lambda = \lambda_k$, hence we obtain

$$\alpha_i(\lambda)(\lambda_k - \lambda) = \alpha_i^0 + (\lambda_k - \lambda)\beta_i(\lambda),$$

where α_i^0 are eigenvalues of A_k and $\beta_i(\lambda)$ are holomorphic at $\lambda = \lambda_k$. Therefore,

$$\alpha_i(\lambda) = \frac{\alpha_i^0}{\lambda_k - \lambda} + \beta_i(\lambda)$$

near $\lambda = \lambda_k$, which means that the behavior of $\alpha_i(\lambda)$ in the neighborhood of $\lambda = \lambda_k$ depends on the rank of the matrix A_k. We note that the matrix A_k is nonnegative for all $\mathbf{x} \in \mathbb{C}^N$ in the sense

$$\langle A_k \mathbf{x}, \mathbf{x} \rangle_{\mathbb{C}^N} = \left\| \Pi_k \sum_{i=1}^{N} x_i w_i \right\|^2 \geq 0.$$

Therefore, $\alpha_i^0 \geq 0$ for $1 \leq i \leq N$. Given $N_k = \text{rank}(A_k)$, where $0 \leq N_k \leq \min(m_k, N)$, there are precisely N_k linearly independent $\{\Pi_k w_i\}_{i=1}^{N_k}$. We can hence construct an orthonormal set of eigenvectors $\{u_k^j\}_{j=1}^{m_k}$ corresponding to λ_k, such that $u_k^j \notin \mathcal{B}_\omega$ for $1 \le j \le N_k$ and $u_k^j \in \mathcal{B}_\omega$ for $N_k + 1 \le j \le m_k$. In other words, $\alpha_i^0 > 0$ for $1 \le i \le N_k$ and these $\alpha_i(\lambda)$ diverge at λ_k as follows

$$\lim_{\lambda \uparrow \lambda_k} \alpha_i(\lambda) = +\infty, \qquad \lim_{\lambda \downarrow \lambda_k} \alpha_i(\lambda) = -\infty.$$

On the other hand, $\alpha_i^0 = 0$ for $N_k + 1 \le i \le N$ and these $\alpha_i(\lambda)$ are continuous at λ_k.

Assume now that the ith eigenvalue $\alpha_i(\lambda)$ has vertical asymptotes at

$$\lambda_{i_1} < \lambda_{i_2} < \cdots < \lambda_{i K_i} < 0.$$

On the interval $(\lambda_{i_l}, \lambda_{i_{l+1}})$ for $1 \le l \le K_i - 1$, the eigenvalue $\alpha_i(\lambda)$ has a simple zero at $\lambda_l^* \in (\lambda_{i_l}, \lambda_{i_{l+1}})$, for which there exists the unique solution $w = \sum_{j=1}^N x_j(\mathcal{L}_\omega - \lambda_l^* I)^{-1} w_j$ of the constrained eigenvalue problem

$$(\mathcal{L}_\omega - \lambda_l^* I)w = \sum_{j=1}^N x_j w_j, \quad w \in \mathcal{B}_\omega,$$

associated with the Lagrange multipliers $(x_1, x_2, \ldots, x_N)$, since $A(\lambda_l^*)\mathbf{x} = \mathbf{0}$. Therefore, $(K_i - 1)$ negative eigenvalues of operator $\mathcal{L}_\omega|_{\mathcal{B}_\omega}$ are located at $\lambda_l^* \in (\lambda_{i_l}, \lambda_{i_{l+1}})$ for $1 \le l \le K_i - 1$. Due to the monotonicity, $\alpha_i(\lambda) < 0$ on $\lambda \in (\lambda_{i K_i}, 0)$ if $\alpha_i(0) < 0$ or has precisely one zero at $\lambda_{K_i}^* \in (\lambda_{i K_i}, 0)$ if $\alpha_j(0) > 0$ or $\lim_{\lambda \to 0^-} \alpha_i(\lambda) = +\infty$.

We are now ready to count the numbers of negative and zero eigenvalues of operator $\mathcal{L}_\omega|_{\mathcal{B}_\omega}$. Sorting the K negative eigenvalues of $\mathcal{L}_\omega$ in increasing order

$$\lambda_1 < \lambda_2 < \cdots < \lambda_K < 0,$$

we have $n(\mathcal{L}_\omega) = \sum_{k=1}^K m_k$ and $z(\mathcal{L}_\omega) = m_{K+1}$.

Let $\Theta(x)$ be the Heaviside step function such that $\Theta(x) = 1$ for $x \ge 0$ and $\Theta(x) = 0$ for $x < 0$. Each eigenvalue $\alpha_i(\lambda)$ has K_i jump discontinuities and $K_i - \Theta(\alpha_i(0))$ zeros on $(-\infty, 0)$. On the other hand, for each eigenvalue $\lambda_k < 0$, only $N_k \le \min(m_k, N)$ eigenvalues of $\{\alpha_i(\lambda)\}_{i=1}^N$ diverge at λ_k and there exist $m_k - N_k$ eigenvectors $\{u_k^j\}_{j=1}^{m_k}$ that belong to $\mathcal{B}_\omega$. Summing all contributions together and noting that $\sum_{i=1}^N K_i = \sum_{k=1}^K N_k$, we obtain

$$n(\mathcal{L}_\omega|_{\mathcal{B}_\omega}) = \sum_{i=1}^N [K_i - \Theta(\alpha_i(0))] + \sum_{k=1}^K [m_k - N_k] = n(\mathcal{L}_\omega) - (n_0 + z_0),$$

where $n_0 + z_0$ is the total number of nonpositive eigenvalues of $\lim_{\lambda \to 0^-} A(\lambda)$. Counting what is left over at $\lambda = 0$, we obtain

$$z(\mathcal{L}_\omega|_{\mathcal{B}_\omega}) = z_0 + m_{K+1} - N_{K+1} = z(L) + z_0 - z_\infty,$$

where $z_\infty = N_{K+1}$ is the number of unbounded eigenvalues of $A(\lambda)$ as $\lambda \to 0^-$. □

REMARK 3.10. Let $\{w_j\}_{1 \le j \le N}$ be the set of orthogonal eigenvectors of the self-adjoint operator $\mathcal{L}_\omega$ for N negative eigenvalues $\{\lambda_j\}_{1 \le j \le N}$. Then, we have

$$A_{ij}(\lambda) = (\lambda_i - \lambda)^{-1} \langle w_i, w_j \rangle, \quad i \le i, j \le N,$$

so that $A(\lambda)$ is a diagonal matrix with $n_0 = N$, $z_0 = 0$, and $z_\infty = 0$. The count formula (3.31) yields the obvious relation:

$$n(\mathcal{L}_\omega|_{\mathcal{B}_\omega}) = n(\mathcal{L}_\omega) - N, \quad z(\mathcal{L}_\omega|_{\mathcal{B}_\omega}) = z(\mathcal{L}_\omega).$$

REMARK 3.11. If $z(\mathcal{L}_\omega) = 0$, the self-adjoint operator $\mathcal{L}_\omega$ is invertible so that $z_\infty = 0$. If $z(\mathcal{L}_\omega) \geq 1$, then z_∞ is nonzero if $\mathrm{Ker}(\mathcal{L}_\omega) \notin \mathcal{B}_\omega$. This follows from the fact that in the limit $\lambda \to 0^-$ some entries of (3.30) diverge to infinity. Indeed, since by Fredholm's Alternative $\mathcal{L}_\omega^{-1} w_i$ exists if and only if $w_i \perp \mathrm{Ker}(\mathcal{L}_\omega)$, there is a linear combination of w_i for which $\mathcal{L}_\omega^{-1} w_i$ diverges if $\mathrm{Ker}(\mathcal{L}_\omega) \notin \mathcal{B}_\omega$

3.3.2. Orbital stability of constrained minimizers under N constraints. It is natural to expect that the local constrained minimizers of Lemma 3.2 are orbitally stable in the time evolution of the Hamiltonian system (3.1) in X. We extend Definition 2.5 of orbital stability to critical points u_ω of the augmented energy Λ_ω with respect to Z-many symmetry operators $T(\theta)$.

Orbital stability of the periodic wave with the profile u_ω in X is definitely true when $n_0 = n(\mathcal{L}_\omega) = N$, $z_0 = z_\infty = 0$, and $z(\mathcal{L}_\omega) = Z$, i.e., if u_ω is a constrained minimizer by Lemma 3.2, because the proof of Theorem 2.8 based on the control of the remainder term in the decomposition of the solution and energy bounds extend trivially to the case of N constraints due to N conserved quantities $\{Q_k(u)\}_{k=1}^N$ and N parameters $\{\omega_k\}_{k=1}^N$ and to the case of Z symmetries.

However, the proof might be difficult if $n(\mathcal{L}_\omega) < N$ because some constraints are no longer needed and if they are used, they actually appear with wrong signs in the energy bounds in the proof of Theorem 2.8. An alternative energy method to prove the orbital stability theorem is developed in [**11**] for the particular case of $n(\mathcal{L}_\omega) = 1$ and $z(\mathcal{L}_\omega) = 1$. If $Z = 1 < N$, then there exists only one symmetry operator of the Hamiltonian system (3.1) and $(N-1)$ conserved quantities are Casimir functionals in the sense of Definition 3.1. The following theorem combines these ideas, together with the control of the remainder term in the decomposition and the energy bounds used in the proof of Theorem 2.8.

THEOREM 3.3. *Let Assumptions* 3.1 *and* 3.2 *be satisfied. Assume that the linearized operator $\mathcal{L}_\omega$ admits a simple negative and a simple zero eigenvalue, while the rest of its spectrum in $\mathcal{H}$ is strictly positive. Let* $\mathrm{Ker}(\mathcal{L}_\omega) = \mathrm{Span}\{u_{\mathrm{sym}}\}$, *where* $u_{\mathrm{sym}} = J\nabla Q_1(u_\omega)$. *If the conditions* (3.29) *are satisfied with* $Z = 1$, *then u_ω is orbitally stable in the time evolution of the Hamiltonian system* (3.1) *in* X.

PROOF. The proof consists of four steps, where the last two repeat the two steps of the proof of Theorem 2.8. In the first step, we show that the conditions in (3.29) can be reformulated in terms of a scalar conserved quantity $Q(u)$ computed from the N conserved quantities $\{Q_1(u), \cdots, Q_N(u)\}$. In the second step, we use $Q(u)$ to set up a reduced constraint space on which $\mathcal{L}_\omega$ is coercive in the sense of the bound (3.28). In the third step, we introduce the decomposition near the orbit of the periodic wave $T_1(\theta)u_\omega$, where $T_1(\theta)$ is the symmetry operator associated to the conserved quantity $Q_1(u)$ which generates the kernel of $\mathcal{L}_\omega$. In the fourth step, energy bounds are used to control the error terms.

STEP 1 (Scalar conserved quantity $Q(u)$). There exists a particular choice of $c := (c_1, c_2, \dots, c_N) \in \mathbb{R}^N$ such that

$$Q(u) = \sum_{k=1}^{N} c_k Q_k(u) \tag{3.32}$$

satisfies $\langle \nabla Q(u_\omega), u_{\mathrm{sym}} \rangle = 0$. Moreover, there exists $\Upsilon \in X$ such that $\langle \mathcal{L}_\omega \Upsilon, v \rangle = 0$ for every $v \in X$ such that $\langle \nabla Q(u_\omega), v \rangle = 0$ and $\langle \mathcal{L}_\omega \Upsilon, \Upsilon \rangle < 0$.

PROOF OF STEP 1. For $k = 1$ we have $u_{\rm sym} = J\nabla Q_1(u_\omega)$ and hence

$$\langle \nabla Q_1(u_\omega), u_{\rm sym}\rangle = \langle \nabla Q_1(u_\omega), J\nabla Q_1(u_\omega)\rangle = 0.$$

For $2 \le k \le N$, we have

$$\langle \nabla Q_k(u_\omega), u_{\rm sym}\rangle = -\langle J\nabla Q_k(u_\omega), \nabla Q_1(u_\omega)\rangle = 0,$$

since $\{Q_k(u)\}_{2\le k\le N}$ are Casimir functionals which are not related to the symmetry operator $T_1(\theta)$. Since $\nabla Q(u) = \sum_{k=1}^N c_k \nabla Q_k(u)$, we have $\langle \nabla Q(u_\omega), u_{\rm sym}\rangle = 0$. Similarly to the proof of Lemma 2.4, the nondegeneracy of $\mathrm{Ker}(\mathcal{L}_\omega) = \mathrm{Span}\{J\nabla Q_1(u_\omega)\}$ and the spectral gap assumption imply that the mapping $\omega \mapsto u_\omega$ is C^1 for $\omega \in \mathcal{I} \subset \mathbb{R}^N$ so that there exists the unique solution $w_k := \partial_{\omega_k} u_\omega \in X|_{\{J\nabla Q_k(u_\omega)\}^\perp}$ of the linear nonhomogeneous equation

$$\mathcal{L}_\omega w_k = \nabla Q_k(u_\omega), \quad 1 \le k \le N. \tag{3.33}$$

Solvability of the linear nonhomogeneous equation (3.33) implies that $z_\infty = 0$ in Theorem 3.2. The conditions (3.29) with $Z = 1$ applied to the count (3.31) imply that $z_0 = 0$ and $n_0 = n(\mathcal{L}_\omega) = 1$. We proceed by constructing $\Upsilon \in X$ as

$$\Upsilon := \sum_{k=1}^N c_k \partial_{\omega_k} u_\omega, \tag{3.34}$$

so that $\mathcal{L}_\omega \Upsilon = \sum_{k=1}^N c_k \nabla Q_k(u_\omega) = \nabla Q(u_\omega)$ and $\langle \mathcal{L}_\omega \Upsilon, v\rangle = 0$ for every $v \in X$ such that $\langle \nabla Q(u_\omega), v\rangle = 0$. A straightforward calculation gives us

$$\langle \mathcal{L}_\omega \Upsilon, \Upsilon\rangle = \sum_{k=1}^N \sum_{m=1}^N c_k c_m \langle \mathcal{L}_\omega^{-1} v_k, v_m\rangle, \tag{3.35}$$

where the quadratic form (3.35) is defined by the same symmetric matrix $\lim_{\lambda\to 0^-} A(\lambda)$ as in Theorem 3.2. If $n_0 = 1$, there exists a choice of $c \in \mathbb{R}^N$ such that $\langle \mathcal{L}_\omega \Upsilon, \Upsilon\rangle < 0$. □

STEP 2 (Reduced constrained space). There exists $C_\omega > 0$ such that the linearized operator $\mathcal{L}_\omega$ satisfies

$$\langle \mathcal{L}_\omega w, w\rangle \ge C_\omega \|w\|_X^2, \quad w \in X, \quad \langle u_{\rm sym}, w\rangle = 0, \quad \langle \nabla Q(u_\omega), w\rangle = 0, \tag{3.36}$$

where $Q(u)$ is given by (3.32) with the particular choice of $c \in \mathbb{R}^N$ obtained in Step 1.

PROOF OF STEP 2. Recall $\Upsilon \in X$ given by (3.34) and decompose it as $\Upsilon = a_0\chi + b_0 u_{\rm sym} + \eta_0$, where χ is the eigenfunction of $\mathcal{L}_\omega$ for the only negative eigenvalue $\lambda_0 < 0$ and $\eta_0 \perp \mathrm{Span}\{\chi, u_{\rm sym}\}$ belong to the positive subspace of $\mathcal{L}_\omega$. Then, we have

$$\begin{aligned}\langle \mathcal{L}_\omega \Upsilon, \Upsilon\rangle &= \langle \mathcal{L}_\omega(a_0\chi + b_0 u_{\rm sym} + \eta_0), (a_0\chi + b_0 u_{\rm sym} + \eta_0)\rangle \\ &= \lambda_0 a_0^2 \|\chi\|^2 + \langle \mathcal{L}_\omega \eta_0, \eta_0\rangle < 0,\end{aligned} \tag{3.37}$$

where the last inequality was obtained in Step 1.

Taking an arbitrary $w \perp \mathrm{Span}\{u_{\rm sym}, \nabla Q(u_\omega)\}$, we can write it as $w = a_1\chi + \eta_1$ with $\eta_1 \perp \mathrm{Span}\{\chi, u_{\rm sym}\}$ in the positive subspace of $\mathcal{L}_\omega$ and

$$\begin{aligned}\langle \mathcal{L}_\omega \Upsilon, w\rangle &= \langle \mathcal{L}_\omega(a_0\chi + b_0 u_{\rm sym} + \eta_0), (a_1\chi + \eta_1)\rangle \\ &= \lambda_0 a_0 a_1 \|\chi\|^2 + \langle \mathcal{L}_\omega \eta_0, \eta_1\rangle = 0.\end{aligned} \tag{3.38}$$

By the Cauchy–Schwarz inequality, we have

$$|\langle \mathcal{L}_\omega \eta_0, \eta_1\rangle|^2 \leq \langle \mathcal{L}_\omega \eta_0, \eta_0\rangle \langle \mathcal{L}_\omega \eta_1, \eta_1\rangle. \tag{3.39}$$

Using (3.37), (3.38), and (3.39) yields

$$\begin{aligned}\langle \mathcal{L}_\omega w, w\rangle &= \lambda_0 a_1^2 \|\chi\|^2 + \langle \mathcal{L}_\omega \eta_1, \eta_1\rangle \\ &\geq \lambda_0 a_1^2 \|\chi\|^2 + \frac{|\langle \mathcal{L}_\omega \eta_0, \eta_1\rangle|^2}{\langle \mathcal{L}_\omega \eta_0, \eta_0\rangle} \\ &= \lambda_0 a_1^2 \|\chi\|^2 + \frac{\lambda_0^2 a_0^2 a_1^2 \|\chi\|^4}{\langle \mathcal{L}_\omega \eta_0, \eta_0\rangle} \\ &= \frac{|\mu_0 \lambda_0|}{|\lambda_0| a_0^2 \|\chi\|^2 + \mu_0} a_1^2 \|\chi\|^2,\end{aligned}$$

where $\mu_0 := \langle \mathcal{L}_\omega \Upsilon, \Upsilon\rangle < 0$ and $|\lambda_0| a_0^2 \|\chi\|^2 + \mu_0 > 0$ since $\langle \mathcal{L}_\omega \eta_0, \eta_0\rangle > 0$. Since $w \perp \text{Span}\{u_{\text{sym}}, \nabla Q(u_\omega)\}$ is arbitrarily chosen, a_1 can be used as the scaling parameter such that $\|\eta_1\| \leq C_1 |a_1| \|\chi\|$ for some positive constant C_1. Therefore,

$$\|w\|^2 = a_1^2 \|\chi\|^2 + \|\eta_1\|^2 \leq (1 + C_1^2) a_1^2 \|\chi\|^2$$

and the previous bound can be rewritten in the form

$$\langle \mathcal{L}_\omega w, w\rangle \geq \frac{|\mu_0 \lambda_0|}{(1 + C_1^2)(|\lambda_0| a_0^2 \|\chi\|^2 + \mu_0)} \|w\|^2.$$

This yields the bound $\langle \mathcal{L}_\omega w, w\rangle \geq C_\omega \|w\|^2$ in (3.36). □

STEP 3 (Decomposition). There exists $\epsilon_0 > 0$ and $C_0 > 0$ such that for every $u \in X$ satisfying

$$\epsilon := \|u - T_1(\theta_0) u_\omega\|_X = \inf_{\theta \in \mathbb{R}} \|u - T_1(\theta) u_\omega\|_X < \epsilon_0, \tag{3.40}$$

there exists a unique choice of $\theta \in \mathbb{R}$, $\zeta \in \mathbb{R}$, and $w \in X$ such that

$$u = T_1(\theta)(u_{\omega+\zeta c} + w), \quad \langle w, u_{\text{sym}}\rangle = 0, \quad \langle w, \nabla Q(u_\omega)\rangle = 0 \tag{3.41}$$

satifying $|\zeta| + \|w\|_X \leq C_0 \epsilon$, where $Q(u)$ is given by (3.32) with the particular choice of $c \in \mathbb{R}^n$ in Step 1.

PROOF OF STEP 3. We consider the smooth function $f(\theta, \zeta) : \mathbb{R}^2 \mapsto \mathbb{R}^2$ given by

$$f(\theta, \zeta) := \begin{bmatrix} \langle u - T_1(\theta) u_{\omega+\zeta c}, T_1'(\theta) u_{\omega+\zeta c}\rangle \\ \langle u - T_1(\theta) u_{\omega+\zeta c}, T_1(\theta) \nabla Q(u_\omega)\rangle \end{bmatrix}, \quad \theta \in \mathbb{R}, \quad \omega \in \mathcal{I} \subset \mathbb{R}$$

so that $f(\theta, \zeta) = 0$ if and only if u can be represented by (3.41). Compared to the proof of Theorem 2.8, we have denoted $u_{\text{sym}} := T_1^{-1}(\theta) T_1'(\theta) u_\omega$ and have fixed the value of ω to which ζc is added with $\zeta \in \mathbb{R}$. Computing the Jacobian of $f : \mathbb{R}^2 \mapsto \mathbb{R}^2$ at $\zeta = 0$ gives

$$\begin{aligned} Df(\theta, 0) = &- \begin{bmatrix} \|u_{\text{sym}}\|^2 & \langle \Upsilon, u_{\text{sym}}\rangle \\ \langle u_{\text{sym}}, \nabla Q(u_\omega)\rangle & \langle \Upsilon, \nabla Q(u_\omega)\rangle \end{bmatrix} \\ &+ \begin{bmatrix} -\langle T_1'(u - T_1(\theta) u_\omega), T_1'(\theta) u_\omega\rangle & \langle u - T_1(\theta) u_\omega, T_1'(\theta) \Upsilon\rangle \\ \langle u - T_1(\theta) u_\omega, T_1'(\theta) \nabla Q(u_\omega)\rangle & \langle u - T_1(\theta) u_\omega, T_1(\theta) \partial_\omega \nabla Q(u_\omega)\rangle \end{bmatrix}, \end{aligned}$$

where $\Upsilon := \partial_\zeta u_{\omega+\zeta c}|_{\zeta=0}$ is given by (3.34). Since we have obtained in Step 1 that $\langle u_{\text{sym}}, \nabla Q(u_\omega)\rangle = 0$ and

$$\langle \Upsilon, \nabla Q(u_\omega)\rangle = \langle \mathcal{L}_\omega \Upsilon, \Upsilon\rangle < 0,$$

whereas the second term in $Df(\theta,0)$ is small due to the bound (3.40), $Df(\theta,0)$ is invertible. It follows by the local inverse function theorem that there exists a solution $(\theta,\zeta) \in \mathbb{R}^2$ of $f(\theta,\zeta) = 0$ such that

$$|\theta - \theta_0| + |\zeta| \leq C_0 \epsilon$$

for an ϵ-independent constant $C_0 > 0$. The bound $\|w\|_X \leq C_0\epsilon$ follows from the same argument as in the proof of Theorem 2.8. □

STEP 4 (Control of the error terms). There exists $C > 0$ such that for every $\epsilon \in (0, \epsilon_0)$ with ϵ_0 defined in (3.40) and for every $u_0 \in X$ satisfying

$$\|u_0 - u_\omega\|_X < \delta := \min\{\epsilon_0, C^{-1}\epsilon\}$$

the unique weak solution $u \in C^0([0,\tau_0), X)$ can be written in the form (3.41) for unique $\theta(t)$, $\zeta(t)$, and $p(t,\cdot) \in X$ satisfying $|\zeta(t)| + \|p(t,\cdot)\|_X < \epsilon$ for $t \in [0,\tau_0)$.

PROOF OF STEP 4. It repeats the proof of Theorem 2.8 with the scalar function

$$V(t) := H(u(t,\cdot)) - H(u_\omega) - \sum_{k=1}^{N} (\omega_k + c_k\zeta(t)) \left[Q_k(u(t,\cdot)) - Q_k(u_\omega)\right],$$

where the scalar function $\zeta(t)$ is defined for fixed $\omega \in \mathbb{R}^N$ and $c \in \mathbb{R}^N$. □

END OF THE PROOF OF THEOREM 3.3. Combining the third and fourth steps together and iterating in time like in the proof of Theorem 2.8, we obtain the orbital stability of the constrained minimizer with the profile u_ω in the time evolution of the Hamiltonian system (3.1) in X in the sense of Definition 2.5 with only one symmetry operator $T_1(\theta)$. □

The result of Theorem 3.3 can be generalized to any number of negative eigenvalues and any multiplicity of the zero eigenvalue of $\mathcal{L}_\omega$ with $1 \leq n(\mathcal{L}_\omega) \leq N$ and $1 \leq z(\mathcal{L}_\omega) \leq N$, as long as the conditions (3.29) are satisfied. This is given in the following corollary.

COROLLARY 3.1. *Let Assumptions* 3.1 *and* 3.2 *be satisfied. Assume that the Hamiltonian system* (3.1) *admits Z-many symmetry operators and that $\{Q_k(u)\}_{k=1}^N$ satisfy the following nondegeneracy condition:*

$$\langle \nabla Q_k(u_\omega), u_{\mathrm{sym}}\rangle = 0, \quad \forall u_{\mathrm{sym}} \in \mathrm{Ker}(\mathcal{L}_\omega), \quad 1 \leq k \leq N. \tag{3.42}$$

If the spectrum of $\mathcal{L}_\omega$ in $(-\infty,\delta)$ for some $\delta > 0$ consists of finitely many eigenvalues and the minimization conditions (3.29) *are satisfied, then u_ω is orbitally stable in the time evolution of the Hamiltonian system* (3.1) *in X.*

PROOF. The nondegeneracy condition (3.42) implies that $z_\infty = 0$ in Theorem 3.2, whereas the minimization conditions (3.29) imply that $z_0 = 0$, $n_0 = n(\mathcal{L}_\omega)$. and $Z = z(\mathcal{L}_\omega)$. The first two steps in the proof of Theorem 3.3 are extended for $n(\mathcal{L}_\omega) > 1$ and $z(\mathcal{L}_\omega) > 1$ with $n(\mathcal{L}_\omega)$ functions replacing $Q(u)$ in (3.32) obtained from the conserved quantities $\{Q_k(u)\}_{1\leq k\leq N}$. The last two steps in the proof of Theorem 3.3 are extended for $f(\theta,\zeta) : \mathbb{R}^{z(\mathcal{L}_\omega)+n(\mathcal{L}_\omega)} \mapsto \mathbb{R}^{z(\mathcal{L}_\omega)+n(\mathcal{L}_\omega)}$ related to $z(\mathcal{L}_\omega)$ symmetry operators with parameters $\theta \in \mathbb{R}^{z(\mathcal{L}_\omega)}$ and to $n(\mathcal{L}_\omega)$ choices of $c \in \mathbb{R}^N$ near $\omega \in \mathbb{R}^N$ with parameters $\zeta \in \mathbb{R}^{n(\mathcal{L}_\omega)}$. □

3.3.3. Applications to the gKdV equation. Theorem 3.3 and Corollary 3.1 can be used to prove orbital stability of periodic waves in the gKdV equation (Example 3.2). The following theorem specifies the sufficient condition for orbital stability, which is subsequently applied to the focusing and defocusing mKdV equations (Example 3.3).

THEOREM 3.4. *Let $\phi_{c,b} \in H^2_{\rm per}$ be the profile of the traveling T-periodic wave of the gKdV equation* (3.8) *satisfying the differential equation* (3.11) *for some* $(c,b) \in \mathbb{R}^2$. *Assume that* $\mathrm{Ker}(\mathcal{L}_{c,b}) = \mathrm{Span}\{\partial_x \phi_{c,b}\}$ *in* $L^2_{\rm per}$. *Then the mapping* $(c,b) \mapsto \phi_{c,b} \in H^2_{\rm per}$ *is* C^1 *in an open neighborhood* $\mathcal{I} \subset \mathbb{R}^2$. *The periodic wave is orbitally stable in* $H^1_{\rm per}$ *if the Jacobian of the mapping*

$$(3.43) \qquad (c,b) \mapsto (Q(\phi_{c,b}), M(\phi_{c,b})), \quad (c,b) \in \mathcal{I} \subset \mathbb{R}^2,$$

with $Q(u)$ *and* $M(u)$ *given by* (3.9) *and* (3.10), *respectively, has no zero eigenvalues and as many negative eigenvalues as* $n(\mathcal{L}_{c,b})$.

PROOF. Local well-posedness of the gKdV equation (3.8) with $V \in C^2(\mathbb{R})$ in $H^1_{\rm per}$ is established in [**182**]. The differential equation (3.11) is the Euler–Lagrange equation for the augmented energy (3.12). The corresponding linearized operator $\mathcal{L}_{c,b}$ in $L^2_{\rm per}$ with dense domain $H^2_{\rm per}$ is given by

$$\mathcal{L}_{c,b} = -\partial_x^2 - V'(\phi_{c,b}) - c : H^2_{\rm per} \subset L^2_{\rm per} \mapsto L^2_{\rm per}.$$

Without loss of generality, we assume $\phi_{c,b}$ to be even in x, then $\partial_x \phi_{c,b}$ is odd in x. Since $\mathrm{Ker}(\mathcal{L}_{c,b}) = \mathrm{Span}\{\partial_x \phi_{c,b}\}$ and the spectral gap condition is satisfied due to the fact that the spectrum of $\mathcal{L}_{c,b}$ in $L^2_{\rm per}$ consists of isolated eigenvalues, it follows from the implicit function theorem similarly to the proof of Lemma 2.4 that the mapping $(c,b) \mapsto \phi_{c,b} \in H^2_{\rm per}$ is C^1 in an open neighborhood $\mathcal{I} \subset \mathbb{R}^2$, where we have restricted $\phi_{c,b}$ to the subspace of even functions. The period T of the periodic domain $[0,T]$ in $H^2_{\rm per}$ is fixed in the parameter continuation of $\phi_{c,b}$ in $(c,b) \in \mathcal{I}$.

Differentiating (3.11) with respect to c and b yields

$$\mathcal{L}_{c,b} \partial_c \phi_{c,b} = \phi_{c,b}, \quad \mathcal{L}_{c,b} \partial_b \phi_{c,b} = 1.$$

This allows us to uniquely define

$$\partial_c \phi_{c,b} = \mathcal{L}_{c,b}^{-1} \phi_{c,b} \quad \text{and} \quad \partial_b \phi_{c,b} = \mathcal{L}_{c,b}^{-1} 1,$$

in the subspace of even functions in $H^2_{\rm per}$. Since $\nabla Q(\phi_{c,b}) = \phi_{c,b}$ and $\nabla M(\phi_{c,b}) = 1$ are orthogonal to $\mathrm{Ker}(\mathcal{L}_{c,b}) = \mathrm{Span}\{\partial_x \phi_{c,b}\}$, we obtain that $z_\infty = 0$ in Theorem 3.2. Therefore, the following limit is uniquely defined:

$$(3.44) \quad \lim_{\lambda \to 0^-} A(\lambda) = \begin{bmatrix} \langle \mathcal{L}_{c,b}^{-1} \phi_{c,b}, \phi_{c,b} \rangle & \langle \mathcal{L}_{c,b}^{-1} \phi_{c,b}, 1 \rangle \\ \langle \mathcal{L}_{c,b}^{-1} 1, \phi_{c,b} \rangle & \langle \mathcal{L}_{c,b}^{-1} 1, 1 \rangle \end{bmatrix} = \begin{bmatrix} \partial_c Q(\phi_{c,b}) & \partial_c M(\phi_{c,b}) \\ \partial_b Q(\phi_{c,b}) & \partial_b M(\phi_{c,b}) \end{bmatrix}.$$

Under the assumption on the Jacobian of the mapping (3.43), we have $n_0 = n(\mathcal{L}_{c,b})$ and $z_0 = 0$ so that the conditions (3.29) are satisfied from the equalities (3.31) with $z(\mathcal{L}_{c,b}) = 1$. Hence $\phi_{c,b}$ is a local nondegenerate constrained minimizer of the energy $H(u)$ in $H^1_{\rm per}|_{\{\nabla Q(\phi_{c,b}), \nabla M(\phi_{c,b})\}^\perp}$. Orbital stability of the traveling periodic wave with the profile $\phi_{c,b}$ in $H^1_{\rm per}$ follows by Theorem 3.3 (if $n(\mathcal{L}_{c,b}) = 1$) or Corollary 3.1 (if $n(\mathcal{L}_{c,b}) = 2$). □

REMARK 3.12. The translational symmetry $T_1(\theta)$ in the gKdV equation (3.8) is related to the conserved quantity $Q(u)$. The mass conservation of $M(u)$ is not related to any symmetry of the KdV equation, and therefore does not produce a nontrivial element in the kernel of the linear operator $\mathcal{L}_{c,b}$. Hence we have $Z = 1 = z(\mathcal{L}_{c,b})$ under the conditions of Theorem 3.4.

EXAMPLE 3.8. Continuing Example 3.3, we will prove orbital stability of the dnoidal waves in the focusing modified KdV equation

$$u_t + 6u^2 u_x + u_{xxx} = 0. \tag{3.45}$$

The traveling wave $u(t,x) = \phi_{c,b=0}(x - ct)$, for $c > 0$, is explicitly given by

$$\phi_{c,b=0}(x) = \frac{\sqrt{c}}{\sqrt{2-k^2}} \,\mathrm{dn}\left(\frac{\sqrt{c}}{\sqrt{2-k^2}}x; k\right), \tag{3.46}$$

due to the exact solution (3.16). The linearized operator $\mathcal{L}_{c,b=0} : H^2_{\text{per}} \subset L^2_{\text{per}} \to L^2_{\text{per}}$ is given by

$$\mathcal{L}_{c,b=0} = -\partial_x^2 + c - 6\phi^2_{c,b=0}.$$

Since the mapping $\mathcal{E} \mapsto T(\mathcal{E})$ is monotonically increasing (Example 3.3), Theorem 3.1 implies that $n(\mathcal{L}_{c,b=0}) = z(\mathcal{L}_{c,b=0}) = 1$. In order to apply Theorem 3.4, we compute

$$Q(\phi_{c,b=0}) = \frac{\sqrt{c}}{\sqrt{2-k^2}} \int_0^{K(k)} \mathrm{dn}^2(x;k)dx = \frac{\sqrt{c}}{\sqrt{2-k^2}} E(k).$$

There are two parameters in the family of dnoidal waves: c and k. However, in order for the mapping $c \mapsto \phi_{c,b=0} \in H^2_{\text{per}}$ to have a fixed period $T(\mathcal{E}) = T$, the elliptic modulus $k \in (0,1)$ must be uniquely found from the relation

$$\sqrt{c}T = 2\sqrt{2-k^2}K(k).$$

Since $\frac{dT}{dk} > 0$ for fixed $c > 0$ (Example 3.3), it follows that $\frac{dc}{dk} > 0$ for fixed $T > 0$. Hence, we obtain for fixed $T > 0$ that

$$\begin{aligned}
\frac{d}{dk} Q(\phi_{c,b=0}) &= \frac{2}{T}\frac{d}{dk}\left[E(k)K(k)\right] \\
&= \frac{2}{Tk(1-k^2)}\left[E^2(k) - (1-k^2)K^2(k)\right],
\end{aligned}$$

where properties (A.7) have been used. The function

$$f(k) := E^2(k) - (1-k^2)K^2(k), \quad k \in (0,1),$$

is positive since $f(0) = 0$ and $f'(k) = 2k^{-1}[E(k) - K(k)]^2 > 0$ for $k \in (0,1)$. Therefore, $\frac{d}{dk}Q(\phi_{c,b=0}) > 0$ for $k \in (0,1)$. Since $\frac{dc}{dk} > 0$, we have $\frac{d}{dc}Q(\phi_{c,b=0}) > 0$ for fixed $T > 0$. In contrast to first diagonal entry of the matrix in (3.44), here we have that

$$\mathcal{L}_{c,b=0}\partial_c \phi_{c,b=0} = -\phi_{c,b=0},$$

with a minus sign on the right-hand side. Since $\frac{d}{dc}Q(\phi_{c,b=0}) > 0$, this implies that the quadratic form $\langle \mathcal{L}^{-1}_{c,b=0}\phi_{c,b=0}, \phi_{c,b=0}\rangle$ is strictly negative. Hence, the number of negative eigenvalues of the matrix $A(\lambda)$ in (3.30) in the limit $\lambda \to 0^-$ is $n_0 \geq 1$, that is, at least one of the two eigenvalues of $\lim_{\lambda\to 0^-} A(\lambda)$ is negative. On the other hand, since $n(\mathcal{L}_{c,b=0}) = 1$ the count (3.31) implies that $z_0 + n_0 \leq 1$. Therefore, we must have $n_0 = 1 = n(\mathcal{L}_{c,b=0})$ and $z_0 = 0$ which implies by Theorem 3.4 that the dnoidal waves of the focusing modified KdV equation (3.45) are orbitally stable in H^1_{per}.

REMARK 3.13. In Example 3.8, we have avoided extending the periodic wave of the fixed period T with the profile $\phi_{c,b=0}$ with respect to parameter $b \neq 0$. It follows that only one constraint of fixed momentum $Q(u)$ is sufficient for showing that the periodic wave with the profile $\phi_{c,b=0}$ is a local nondegenerate constrained minimizer of the energy since $n(\mathcal{L}_{c,b=0}|_{\{\phi_{c,b=0}\}^\perp}) = 0$.

EXAMPLE 3.9. Continuing Example 3.5, we will prove the orbital stability of the snoidal waves in the defocusing modified KdV equation

$$u_t + 6u^2 u_x - u_{xxx} = 0. \tag{3.47}$$

The traveling wave $u(t,x) = \phi_{c,b=0}(x-ct)$, $c > 0$, is explicitly given by

$$\phi_{c,b=0}(x) = \frac{\sqrt{c}k}{\sqrt{1+k^2}} \,\text{sn}\left(\frac{\sqrt{c}}{\sqrt{1+k^2}}x; k\right), \tag{3.48}$$

due to the exact solution (3.20). The linearized operator $\mathcal{L}_{c,b=0} : H^2_{\text{per}} \subset L^2_{\text{per}} \to L^2_{\text{per}}$ is given by

$$\mathcal{L}_{c,b=0} = -\partial_x^2 - c + 6\phi^2_{c,b=0}.$$

Again, the mapping $\mathcal{E} \mapsto T(\mathcal{E})$ is monotonically increasing (Example 3.5) so that Theorem 3.1 implies that

$$n(\mathcal{L}_{c,b=0}) = z(\mathcal{L}_{c,b=0}) = 1.$$

Since we have $\mathcal{L}_{c,b=0}\partial_c\phi_{c,b=0} = \phi_{c,b=0}$, the entries in $\lim_{\lambda\to 0^-} A(\lambda)$ are the same as in (3.44). We first try the strategy of Example 3.8 and compute

$$Q(\phi_{c,b=0}) = \frac{2\sqrt{c}k^2}{\sqrt{1+k^2}} \int_0^{K(k)} \text{sn}^2(x;k)dx = \frac{2\sqrt{c}}{\sqrt{1+k^2}} \left[K(k) - E(k)\right],$$

in view of (A.5). The elliptic modulus $k \in (0,1)$ is uniquely defined by the relation

$$\sqrt{c}T = 4\sqrt{1+k^2}K(k),$$

for fixed $T(\mathcal{E}) = T$. Since $\frac{dT}{dk} > 0$ for fixed $c > 0$ (Example 3.5), it follows that $\frac{dc}{dk} > 0$ for fixed $T > 0$. Hence, we obtain for fixed $T > 0$ that

$$\begin{aligned}\frac{d}{dk}Q(\phi_{c,b=0}) &= \frac{8}{T}\frac{d}{dk}\left[K(k)\left[K(k) - E(k)\right]\right] \\ &= \frac{8}{Tk(1-k^2)}\left[E(k)\left[K(k)-E(k)\right] + K(k)\left[E(k) - (1-k^2)K(k)\right]\right],\end{aligned}$$

where properties (A.7) have been used. We have $\frac{d}{dk}Q(\phi_{c,b=0}) > 0$ for every $k \in (0,1)$ because of properties (A.8). Since $\frac{dc}{dk} > 0$, we have $\frac{d}{dc}Q(\phi_{c,b=0}) > 0$ for fixed T. It follows from the first diagonal entry in the matrix $A(\lambda)$ in (3.30) in the limit $\lambda \to 0^-$ that $n(\mathcal{L}_{c,b=0}|_{\{\phi_{c,b=0}\}^\perp}) = n(\mathcal{L}_{c,b=0}) = 1$. In order to show that $n(\mathcal{L}_{c,b=0}|_{\{\phi_{c,b=0},1\}^\perp}) = 0$, it is not sufficient to consider the profile $\phi_{c,b}$ under the restriction $b = 0$, and so the other entries of $\lim_{\lambda\to 0^-} A(\lambda)$ have to be inspected as well. To do so, we observe that $M(\phi_{c,b=0}) = 0$ because $\phi_{c,b=0}$ has zero mean which implies that $\lim_{\lambda\to 0^-} A(\lambda)$ is a lower triangular matrix. Since the first diagonal entry is positive because $\frac{d}{dc}Q(\phi_{c,b=0}) > 0$, then $n_0 = 1$, $z_0 = 0$ is true for $b = 0$ if and only if $\langle 1, \partial_b\phi_{c,t=0}\rangle < 0$, where $v_0 := \partial_b\phi_{c,b=0}$ is a solution of the linear inhomogeneous equation $\mathcal{L}_{c,b=0}v_0 = 1$. By using the scaling transformation

$$z = \frac{\sqrt{c}}{\sqrt{1+k^2}}x, \quad \hat{v}_0(z) = \frac{c}{1+k^2}v_0(x),$$

it follows that $\hat{v}_0$ is a solution of the linear inhomogeneous equation $\hat{\mathcal{L}}(k)\hat{v}_0 = 1$, where

$$\hat{\mathcal{L}}(k) := -\partial_z^2 - 1 + 5k^2 - 6k^2 \operatorname{cn}^2(z;k).$$

The first five eigenvalues and eigenfunctions of $\mathcal{L}(k)$ in the space of $4K(k)$-periodic functions are known explicitly [**15**, **79**]:

$$\begin{aligned}
&\hat{\lambda}_0 = 1 + k^2 - 2\sqrt{1-k^2+k^4}, \quad \varphi_0(z) = 1 + k^2 + \sqrt{1-k^2+k^4} - 3k^2\operatorname{sn}(z;k)^2,\\
&\hat{\lambda}_1 = 0, \quad \varphi_1(z) = \operatorname{cn}(z;k)\operatorname{dn}(z;k),\\
&\hat{\lambda}_2 = 3k^2, \quad \varphi_2(z) = \operatorname{sn}(z;k)\operatorname{dn}(z;k),\\
&\hat{\lambda}_3 = 3, \quad \varphi_3(z) = \operatorname{sn}(z;k)\operatorname{cn}(z;k),\\
&\hat{\lambda}_4 = 1 + k^2 + 2\sqrt{1-k^2+k^4}, \quad \varphi_4(z) = 1 + k^2 - \sqrt{1-k^2+k^4} - 3k^2\operatorname{sn}(z;k)^2.
\end{aligned}$$

As a result, the solution $\hat{v}_0$ of $\hat{\mathcal{L}}(k)\hat{v}_0 = 1$ can be written in the form

$$\hat{v}_0 = \frac{1}{2\sqrt{1-k^2+k^4}}\left(\frac{\hat{\lambda}_4\varphi_0 - \hat{\lambda}_0\varphi}{\hat{\lambda}_0\hat{\lambda}_4}\right).$$

Using

$$\int_0^{K(k)} k^2\operatorname{sn}^2(z;k)dz = \int_0^{K(k)} (1 - \operatorname{dn}^2(z;k))dz = K(k) - E(k),$$

we obtain

$$\begin{aligned}
\langle 1, \hat{v}_0\rangle &= \frac{1}{2\sqrt{1-k^2+k^4}\hat{\lambda}_0\hat{\lambda}_4}\left(\hat{\lambda}_4\langle 1,\varphi_0\rangle - \hat{\lambda}_0\langle 1,\varphi_4\rangle\right)\\
&= \frac{1}{2\sqrt{1-k^2+k^4}\hat{\lambda}_0\hat{\lambda}_4}\Big(K(k)\left[(k^2-2)(\hat{\lambda}_4-\hat{\lambda}_0) + \sqrt{1-k^2+k^4}(\hat{\lambda}_4+\hat{\lambda}_0)\right]\\
&\qquad\qquad + 3E(k)(\hat{\lambda}_4-\hat{\lambda}_0)\Big)\\
&= -\frac{12}{3+2k^2+3k^4}\left(2E(k) - (1-k^2)K(k)\right) < 0,
\end{aligned}$$

since $E(k) > (1-k^2)K(k)$ due to (A.8). It follows from the scaling transformation that $\langle 1, v_0\rangle < 0$ since $\langle 1, \hat{v}_0\rangle < 0$, which implies that $\lim_{\lambda\to 0^-} A(\lambda)$ in (3.44) has one positive and one negative eigenvalue. Therefore, $n_0 = 1 = n(\mathcal{L}_{c,b=0})$, $z_0 = 0$, and $z(\mathcal{L}_{c,b=0}) = 1$. By Theorem 3.4, the snoidal waves of the defocusing modified KdV equation (3.47) are orbitally stable in H^1_{per}.

REMARK 3.14. In Example 3.9, we have computed $\partial_b\phi_{c,b=0}$ which is the first step towards extending the solution family with respect to parameter b. It follows that the constraint of fixed mass $M(u)$ is sufficient for showing that the periodic wave with the profile $\phi_{c,b=0}$ is a constrained minimizer of the energy since $n(\mathcal{L}_{c,b=0}|_{\{1\}^\perp}) = 0$.

REMARK 3.15. The proof of orbital stability of the cnoidal waves in the focusing modified KdV equation (Example 3.4) and in the KdV equation (Example 3.6) will be given in Examples 3.11 and 3.12.

REMARK 3.16. Theorems 3.3 and Corollary 3.1 can be formulated in exactly the same way for solitary waves in $\mathcal{H} = L^2(\mathbb{R})$. The corresponding results have been used to study stability of multi-component solitary waves of coupled NLS systems

[**213**] and multi-soliton solutions of the KdV and NLS equations in [**193**] and [**149**], respectively.

3.4. Other variational characterizations

Constrained minimization of the energy $H(u)$ for the periodic waves of the gKdV equation (3.8) with profile $\phi_{c,b}$ is obtained from Theorem 3.4 if the profile $\phi_{c,b}$ with a fixed period $T > 0$ is C^1 with respect to parameters $(c, b) \in \mathcal{I} \subset \mathbb{R}^2$. This is true by the assumption that $\text{Ker}(\mathcal{L}_{c,b}) = \text{Span}\{\partial_x \phi_{c,b}\}$. If this assumption is not true and the C^1 property is not valid, alternative variational characterizations of periodic waves can be developed to study their stability in the time evolution of the gKdV equation (3.8) and similar Hamiltonian systems.

Here we describe a useful characterization of periodic waves based on minimization of the quadratic part of the augmented energy in the space of periodic functions with zero mean under the constraint of fixed nonquadratic part of the augmented energy. This presentation is based on two main examples of the fractional KdV equations with quadratic and cubic nonlinearities.

3.4.1. The fractional Korteweg–de Vries (fKdV) equation. The fKdV equation can be written in the normalized form:

$$u_t + 2uu_x = (-\Delta)^{\alpha/2} u_x, \tag{3.49}$$

where the fractional Laplacian $(-\Delta)^{\alpha/2}$ is defined by the Fourier series. Since the scaling transformation $u(t, x) \mapsto \beta^\alpha u(\beta^{1+\alpha} t, \beta x)$ leaves (3.49) invariant for every $\beta > 0$, the period T can be normalized to any number. Hence we normalize the periodic domain to $\mathbb{T} := \mathbb{R} \backslash \{2\pi\mathbb{Z}\}$ and define

$$f(x) = \sum_{j \in \mathbb{Z}} f_j e^{ijx}, \quad [(-\Delta)^{\alpha/2} f](x) = \sum_{j \in \mathbb{Z}} |j|^\alpha f_j e^{ijx}, \quad x \in \mathbb{T}.$$

Similar to Example 3.2, the fKdV equation (3.49) admits conservation of energy

$$H(u) = \frac{1}{2} \int_{\mathbb{T}} [(-\Delta)^{\alpha/4} u]^2 - \frac{1}{3} \int_{\mathbb{T}} u^3 dx,$$

in addition to the momentum and mass given by

$$Q(u) = \frac{1}{2} \int_{\mathbb{T}} u^2 dx \quad \text{and} \quad M(u) = \int_{\mathbb{T}} u dx.$$

The energy space is $H^{\alpha/2}_{\text{per}}$ if $\alpha > \frac{1}{3}$, since the cubic term of $H(u)$ is well-defined due to the compact embedding of $H^{\alpha/2}_{\text{per}}$ into $L^3(\mathbb{T})$ if $\alpha > \frac{1}{3}$. This follows from the Sobolev inequality [**13**]:

$$\|f\|_{L^{k+1}} \leq C \|f\|_{H^{\alpha/2}}, \quad \forall f \in H^{\alpha/2}_{\text{per}},$$

where $k + 1 < 2/(1 - \alpha)$ and $C > 0$ is independent of $H^{\alpha/2}_{\text{per}}$. For $k = 2$ this yields the restriction $\alpha > 1/3$. Stability of traveling periodic waves in the fKdV equation (3.49) was developed in [**136**] and [**206**], whereas stability of traveling solitary waves was considered in [**16**]. Further applications for stability of traveling periodic waves in a similar Benjamin–Bona–Mahony model can be found in [**12**].

Traveling wave $u(t, x) = \phi_{c,b}(x - ct)$ with the wave speed c is obtained from the stationary equation

$$(-\Delta)^{\alpha/2} \phi_{c,b} + c\phi_{c,b} - \phi_{c,b}^2 + b = 0, \tag{3.50}$$

where b is the integration constant and the sign convention for c and b is opposite to the one in Example 3.2. The stationary equation (3.50) is the Euler–Lagrange equation for critical points of the augmented energy

$$\Lambda_{c,b}(u) = H(u) + cQ(u) + bM(u).$$

Although the traveling periodic waves can be characterized as constrained minimizers of energy $H(u)$ subject to fixed $Q(u)$ and $M(u)$ for every $\alpha \in \left(\frac{1}{3}, 2\right]$ [**136**], the C^1 property of the constrained minimizers with respect to Lagrange multipliers c and b cannot be taken for granted [**206**].

An alternative variational characterization of the traveling periodic waves can be developed on the minimization of the quadratic form

$$\mathcal{F}_c(u) := \frac{1}{2}\int_{\mathbb{T}} \left[[(-\Delta)^{\alpha/4}u]^2 + cu^2\right] dx \tag{3.51}$$

in the constrained set

$$Y_c := \left\{ u \in H^{\alpha/2}_{\text{per}} : \quad \int_{\mathbb{T}} u^3 dx = 1, \quad \int_{\mathbb{T}} u dx = 0 \right\}.$$

This has been done in [**206**] based on the pioneering study of traveling periodic waves with zero mean profiles in [**17**].

If $\varphi_c \in Y_c$ is a minimizer of $\mathcal{F}_c(u)$, it satisfies the Euler–Lagrange equation

$$(-\Delta)^{\alpha/2}\varphi_c + c\varphi_c = C_1\varphi_c^2 + C_2, \tag{3.52}$$

for some Lagrange multiplier C_1 and C_2 given by

$$C_1 = 2\mathcal{F}_c(\varphi_c), \quad C_2 = -\frac{2}{\pi}Q(\varphi_c)\mathcal{F}_c(\varphi_c).$$

The scaling transformation $\phi_c = C_1\varphi_c$ maps the Euler–Lagrange equation (3.52) to the form (3.50) with $b = b(c)$ computed uniquely from the zero mean condition for ϕ_c:

$$b(c) := \frac{1}{2\pi}\int_{\mathbb{T}} \phi_c^2 dx = \frac{1}{\pi}Q(\phi_c). \tag{3.53}$$

Thus, the constrained minimizers of $\mathcal{F}_c(u)$ in Y_c define the zero-mean solutions of the stationary equation (3.50) for which only one parameter c is independent. In what follows, we write ϕ_c instead of $\phi_{c,b}$.

REMARK 3.17. If $\phi_c \in H^{\alpha/2}_{\text{per}}$ satisfies the stationary equation (3.50) with $b = b(c)$ given by (3.53), then it follows by direct substitution that

$$\psi_\omega := \phi_c - \frac{1}{2}\left(c - \sqrt{c^2 + 4b(c)}\right), \quad \omega := \sqrt{c^2 + 4b(c)}. \tag{3.54}$$

is a solution of the stationary equation

$$(-\Delta)^{\alpha/2}\psi_\omega + \omega\psi_\omega - \psi_\omega^2 = 0. \tag{3.55}$$

If ϕ_c has zero mean, then ψ_ω has nonzero mean. Traveling periodic waves of the fKdV equation (3.49) of the form $u(t,x) = \psi_\omega(x - \omega t)$ are obtained with the transformation (3.54) from the traveling periodic waves $u(t,x) = \phi_c(x - ct)$.

For each solution ϕ_c of the stationary equation (3.50) with $b = b(c)$, we define the linearization operator $\mathcal{L}_c$ in L^2_{per} with the dense domain H^α_{per} according to

$$\mathcal{L}_c := (-\Delta)^{\alpha/2} + c - 2\phi_c.$$

Two important relations follows immediately from (3.50):

$$\mathcal{L}_c \phi_c = -\phi_c^2 - b(c) \tag{3.56}$$

and

$$\mathcal{L}_c 1 = -2\phi_c + c. \tag{3.57}$$

By the translational symmetry, it is also true that $\mathcal{L}_c \partial_x \phi_c = 0$.

Assuming the nondegeneracy of minimizers of $\mathcal{F}_c(u)$ in Y_c, we will show that the mapping $c \mapsto b(c)$ allows us to compute indices $n(\mathcal{L}_c)$ and $z(\mathcal{L}_c)$ and to clarify the orbital stability of the periodic waves with profile ϕ_c in the fKdV equation (3.49). The following theorem gives the existence of constrained minimizers of $\mathcal{F}_c(u)$ in Y_c.

THEOREM 3.5. *For every $\alpha > \frac{1}{3}$ and $c > -1$, there exists the ground state of the constrained minimization problem*

$$\inf_{u \in H^{\frac{\alpha}{2}}_{\rm per}} \left\{ \mathcal{F}_c(u) : \quad \int_{\mathbb{T}} u^3 dx = 1, \quad \int_{\mathbb{T}} u dx = 0 \right\}, \tag{3.58}$$

where $\mathcal{F}_c(u)$ is given by (3.51). If $\alpha \leq 2$, the ground state has the single-lobe profile in the sense of Definition 3.3.

PROOF. It follows that $\mathcal{F}_c$ is a smooth functional bounded on $H^{\alpha/2}_{\rm per}$. Moreover, $\mathcal{F}_c$ is proportional to the quadratic form of the operator $c + (-\Delta)^{\alpha/2}$ with the spectrum in $L^2_{\rm per}$ given by $\{c + |m|^\alpha, \ m \in \mathbb{Z}\}$. It follows from the zero-mass constraint in (3.58) that for every $c > -1$,

$$\mathcal{F}_c(u) \geq \frac{1}{2}(c+1)\|u\|^2_{L^2}, \quad u \in Y_c,$$

which implies the existence of $C > 0$ such that

$$\mathcal{F}_c(u) \geq C\|u\|^2_{H^{\alpha/2}}, \quad u \in Y_c.$$

Hence $\mathcal{F}_c$ is equivalent to the squared norm in $H^{\alpha/2}_{\rm per}$ for functions in Y_c, yielding

$$q_c := \inf_{u \in Y_c} \mathcal{F}_c(u) \geq 0.$$

Let $\{u_n\}_{n\in\mathbb{N}}$ be a minimizing sequence for the constrained minimization problem (3.58), that is, a sequence in Y_c satisfying

$$\mathcal{F}_c(u_n) \to q_c \quad \text{as} \quad n \to \infty.$$

Since $\{u_n\}_{n\in\mathbb{N}}$ is bounded in $H^{\alpha/2}_{\rm per}$, there exists $\varphi_c \in H^{\alpha/2}_{\rm per}$ such that, up to a subsequence,

$$u_n \rightharpoonup \varphi_c \quad \text{in } H^{\alpha/2}_{\rm per}, \quad \text{as} \quad n \to \infty.$$

For every $\alpha > \frac{1}{3}$, the energy space $H^{\alpha/2}_{\rm per}$ is compactly embedded in $L^3(\mathbb{T})$. Thus,

$$u_n \to \varphi_c \quad \text{in } L^3(\mathbb{T}), \quad \text{as} \quad n \to \infty.$$

Using the estimate

$$\begin{aligned} \left| \int_{-\pi}^{\pi} (u_n^3 - \varphi_c^3) dx \right| &\leq \int_{-\pi}^{\pi} |u_n^3 - \varphi_c^3| dx \\ &\leq \left(\|\varphi_c\|^2_{L^3} + \|\varphi_c\|_{L^3}\|u_n\|_{L^3} + \|u_n\|^2_{L^3} \right) \|u_n - \varphi_c\|_{L^3}, \end{aligned}$$

it follows that $\int_{-\pi}^{\pi} \varphi_c^3 dx = 1$. By a similar argument, since $H^{\alpha/2}_{\rm per}$ is also compactly embedded in $L^1(\mathbb{T})$, it follows that $\int_{-\pi}^{\pi} \varphi_c dx = 0$. Hence, $\varphi_c \in Y_c$. It follows by the weak lower semi-continuity of $\mathcal{F}_c$ that $\mathcal{F}_c(\varphi_c) \leq \liminf_{n\to\infty} \mathcal{F}_c(u_n) = q_c$. Therefore, $\mathcal{F}_c(\varphi_c) = q_c$ and φ_c is the ground state of the constrained minimization problem (3.58).

If $\alpha \in (0,2]$, the symmetric decreasing rearrangements of u do not increase $\mathcal{F}_c(u)$ while leaving the constraints in Y_c invariant thanks to the fractional Polya–Szegö inequality (see Appendix A in [**56**]). As a result, the ground state with the profile $\varphi_c \in Y_c$ of $\mathcal{F}_c(u)$ must decrease away symmetrically from the maximum point. Hence, the ground state has the single-lobe profile in the sense of Definition 3.3. □

REMARK 3.18. By the translational invariance, the symmetric function φ_c can attain its maximum at $x = 0$, which yields the periodic wave with the even profile ϕ_c after the transformation of the Euler–Lagrange equation (3.52) to the form (3.50) with $b = b(c)$.

The following lemma gives a smooth continuation of the periodic wave with the profile ϕ_c with respect to the wave speed c.

LEMMA 3.3. *Fix $\alpha \in (\frac{1}{3}, 2]$ and let $\phi_0 \in H^{\alpha/2}_{\rm per}|_{\{1\}^\perp}$ be the even solution of the stationary equation* (3.50) *for a fixed $c = c_0 \in (-1,\infty)$ and $b = b(c_0)$ obtained from Theorem* 3.5. *Assume* $\mathrm{Ker}(\mathcal{L}_{c_0}|_{\{1\}^\perp}) = \mathrm{Span}\{\partial_x \phi_0\}$. *There exists a unique continuation of even solutions of the stationary equation* (3.50) *with zero mean in an open interval $\mathcal{I} \subset (-1,\infty)$ containing c_0 such that the mapping*

$$\mathcal{I} \ni c \mapsto \phi(\cdot, c) \in H^{\alpha/2}_{\rm per}|_{\{1\}^\perp} \tag{3.59}$$

is C^1 and $\phi(\cdot, c_0) = \phi_0$.

PROOF. Let Π_0 be the projection operator reducing the mean value of the 2π-periodic function to zero so that $\Pi_0 f := f - \frac{1}{2\pi}\int_{-\pi}^{\pi} f(x)dx$. The difference function $\tilde{\phi} := \phi - \phi_0 \in H^{\alpha/2}_{\rm per}|_{\{1\}^\perp}$ satisfies the following equation:

$$\mathcal{L}_{c_0}|_{\{1\}^\perp}\tilde{\phi} = -(c - c_0)(\phi_0 + \tilde{\phi}) + \Pi_0\tilde{\phi}^2, \tag{3.60}$$

where $\mathcal{L}_{c_0}|_{\{1\}^\perp} = \Pi_0 \mathcal{L}_{c_0} \Pi_0$ can be written explicitly by

$$\mathcal{L}_{c_0}|_{\{1\}^\perp} f = \mathcal{L}_{c_0} f + \frac{1}{\pi}\int_{\mathbb{T}} f(x)\phi_0(x)dx \tag{3.61}$$

since it follows from (3.57) that $\langle \mathcal{L}_{c_0} f, 1\rangle = -2\langle f, \phi_0\rangle$ if $\langle f, 1\rangle = 0$.

Assume $\mathrm{Ker}(\mathcal{L}_{c_0}|_{\{1\}^\perp}) = \mathrm{Span}\{\partial_x\phi_0\}$ and consider the subspace of even functions with zero mean for which ϕ_0 belongs. Then, $\mathcal{L}_{c_0}|_{\{1\}^\perp}$ is invertible and the persistence equation (3.60) can be written as the fixed-point equation:

$$\tilde{\phi} = -(c - c_0)\left(\mathcal{L}_{c_0}|_{\{1\}^\perp}\right)^{-1}(\phi_0 + \tilde{\phi}) + \left(\mathcal{L}_{c_0}|_{\{1\}^\perp}\right)^{-1}\Pi_0\tilde{\phi}^2. \tag{3.62}$$

By the implicit function theorem, there exist an open interval containing c_0, an open ball $B_r \in H^{\alpha/2}_{\rm per}|_{\{1\}^\perp}$ of radius $r > 0$ centered at 0, and a unique C^1 mapping $\mathcal{I} \ni c \mapsto \tilde{\phi}(\cdot, c) \in B_r$ such that $\tilde{\phi}(\cdot, c)$ is a solution to the fixed-point equation (3.62) for every $c \in \mathcal{I}$ and $\tilde{\phi}(\cdot, c_0) = 0$. □

REMARK 3.19. The proof of Lemma 3.3 repeats a more abstract proof of Lemma 2.4 with precise computations suitable for the stationary equation (3.50).

REMARK 3.20. Although the solution ϕ_0 is obtained from a global minimizer of the variational problem (3.58) with $c = c_0$, the solution $\phi(\cdot, c)$ in Lemma 3.3 is continued from the Euler–Lagrange equation. Therefore, even if the solution $\phi(\cdot, c)$ is C^1 with respect to c in $\mathcal{I}$ as in Lemma 3.3, this solution may not coincide with the global minimizer of $\mathcal{F}_c$ in Y_c for $c \neq c_0$ if the minimizers in Theorem 3.5 are not unique. Uniqueness of minimizers and the nondegeneracy assumption $\text{Ker}(\mathcal{L}_{c_0}|_{\{1\}^\perp}) = \text{Span}\{\partial_x \phi_0\}$ are notoriously difficult problems which have only been solved for the fKdV equation in the context of solitary waves [**95**].

REMARK 3.21. We claim that the nondegeneracy condition $\text{Ker}(\mathcal{L}_{c_0}|_{\{1\}^\perp}) = \text{Span}\{\partial_x \phi_0\}$ in Lemma 3.3 coincides with the nondegeneracy of minimizers in Theorem 3.5 so that

$$\text{Ker}(\mathcal{L}_{c_0}|_{\{1\}^\perp}) = \text{Ker}(\mathcal{L}_{c_0}|_{\{1,\phi_0^2\}^\perp}).$$

If $f \in \text{Ker}(\mathcal{L}_{c_0}|_{\{1\}^\perp})$, then it follows from (3.56) and (3.61) that $\langle f, \phi_0^2 \rangle = 0$ so that $f \in \text{Ker}(\mathcal{L}_{c_0}|_{\{1,\phi_0^2\}^\perp})$. In the opposite direction, one can write

$$\mathcal{L}_{c_0}|_{\{1,\phi_0^2\}^\perp} f = \mathcal{L}_{c_0} f + \frac{1}{\pi}\langle f, \phi_0 \rangle - \frac{\langle \mathcal{L}_{c_0} f, \Pi_0 \phi_0^2 \rangle}{\langle \phi_0^2, \Pi_0 \phi_0^2 \rangle} \Pi_0 \phi_0^2,$$

where $\langle \phi_0^2, \Pi_0 \phi_0^2 \rangle = \|\phi_0\|_{L^4}^4 - \frac{1}{2\pi}\|\phi_0\|_{L^2}^2$. Since the single-lobe profile ϕ_0 cannot be constant, we have $\langle \phi_0^2, \Pi_0 \phi_0^2 \rangle > 0$. If $f \in \text{Ker}(\mathcal{L}_{c_0}|_{\{1,\phi_0^2\}^\perp})$, then it follows from (3.56) that

$$0 = \langle f, \mathcal{L}_{c_0} \phi_0 \rangle = \langle \mathcal{L}_{c_0} f, \phi_0 \rangle = \frac{\langle \mathcal{L}_{c_0} f, \Pi_0 \phi_0^2 \rangle}{\langle \phi_0^2, \Pi_0 \phi_0^2 \rangle} \langle \Pi_0 \phi_0^2, \phi_0 \rangle$$

and since $\langle \Pi_0 \phi_0^2, \phi_0 \rangle = \int_{-\pi}^{\pi} \phi_0^3 dx > 0$ by the variational construction in Theorem 3.5, then $\langle \mathcal{L}_{c_0} f, \Pi_0 \phi_0^2 \rangle = 0$ so that $f \in \text{Ker}(\mathcal{L}_{c_0}|_{\{1\}^\perp})$.

One important consequence of Lemma 3.3 is that the mapping (3.59) can be differentiated in c if the assumption is satisfied. Differentiating of (3.50) with $b = b(c)$ in c gives

$$\mathcal{L}_c \partial_c \phi_c = -\phi_c - b'(c). \tag{3.63}$$

It follows from (3.57) and (3.63) that if $c + 2b'(c) = 0$, then $1 - 2\partial_c \phi_c \in \text{Ker}(\mathcal{L}_c)$ in addition to $\partial_x \phi_c \in \text{Ker}(\mathcal{L}_c)$. This characterization of $\text{Ker}(\mathcal{L}_c)$ is one advantage of having constructed the periodic waves with zero mean from the variational principle in Theorem 3.5 and continued the family smoothly with respect to wave speed in Lemma 3.3 provided that the minimizers are nondegenerate in $L^2_{\text{per}}|_{\{1\}^\perp}$.

The following theorem gives other advantages, such as the count of negative eigenvalues of the linearized operator $\mathcal{L}_c$ and the criterion for orbital stability of the periodic wave obtained from the mapping $c \mapsto b(c)$.

THEOREM 3.6. *Fix* $\alpha \in (\frac{1}{3}, 2]$ *and let* $\phi_c \in H^{\alpha/2}_{\text{per}}|_{\{1\}^\perp}$ *be the even solution of the stationary equation* (3.50) *for a fixed* $c \in (-1, \infty)$ *obtained from Theorem* 3.5. *Assume* $\text{Ker}(\mathcal{L}_c|_{\{1\}^\perp}) = \text{Span}\{\partial_x \phi_c\}$. *Then the spectrum of* $\mathcal{L}_c$ *in* L^2_{per} *includes*

- *a simple negative eigenvalue and a simple zero eigenvalue if* $c + 2b'(c) > 0$,
- *a simple negative eigenvalue and a double zero eigenvalue if* $c + 2b'(c) = 0$,
- *two simple negative eigenvalues and a simple zero eigenvalue if*

$$c + 2b'(c) < 0.$$

The periodic wave ϕ_c is orbitally stable in the fKdV equation (3.49) in $H^{\alpha/2}_{\rm per}$ if $b'(c) > 0$, under the assumption that the initial-value problem is locally well-posed in $H^{\alpha/2}_{\rm per}$.

PROOF. It follows from the variational characterization in Theorem 3.5 and the nondegeneracy assumption in Remark 3.21 that

$$n(\mathcal{L}_c|_{\{1,\phi_c^2\}^\perp}) = 0, \quad z(\mathcal{L}_c|_{\{1,\phi_c^2\}^\perp}) = 1.$$

To use the count (3.31) in Theorem 3.2, we construct the symmetric 2-by-2 matrix related to the two constraints:

$$A_1(\lambda) := \begin{bmatrix} \langle (\mathcal{L}_c - \lambda I)^{-1}\phi_c^2, \phi_c^2\rangle & \langle (\mathcal{L}_c - \lambda I)^{-1}\phi_c^2, 1\rangle \\ \langle (\mathcal{L}_c - \lambda I)^{-1}1, \phi_c^2\rangle & \langle (\mathcal{L}_c - \lambda I)^{-1}1, 1\rangle \end{bmatrix}, \quad \lambda \notin \sigma(\mathcal{L}_c).$$

Let us first assume that $c + 2b'(c) \neq 0$. By using (3.56), (3.57), and (3.63), we compute

$$\begin{aligned} \langle \mathcal{L}_c^{-1}1, 1\rangle &= \frac{2\pi}{c+2b'(c)}, \\ \langle \mathcal{L}_c^{-1}1, \phi_c^2\rangle &= \frac{2\pi}{c+2b'(c)}\left[b(c) - \frac{2}{3}\gamma'(c)\right], \\ \langle \mathcal{L}_c^{-1}\phi_c^2, 1\rangle &= -\frac{2\pi b(c)}{c+2b'(c)}, \\ \langle \mathcal{L}_c^{-1}\phi_c^2, \phi_c^2\rangle &= -2\pi\gamma(c) - \frac{2\pi b(c)}{c+2b'(c)}\left[b(c) - \frac{2}{3}\gamma'(c)\right], \end{aligned}$$

where $\gamma(c) := \frac{1}{2\pi}\int_{-\pi}^{\pi}\phi_c^3 dx$. It follows by symmetry $\langle \mathcal{L}_c^{-1}1, \phi_c^2\rangle = \langle \mathcal{L}_c^{-1}\phi_c^2, 1\rangle$ that $\gamma'(c) = 3b(c)$. Hence, we have

$$\lim_{\lambda\to 0^-} \det A_1(\lambda) = -\frac{4\pi^2\gamma(c)}{c+2b'(c)}. \tag{3.64}$$

Since $\gamma(c) > 0$ and $c + 2b'(c) \neq 0$, it follows from (3.31) in Theorem 3.2 and the computation (3.64) that $z_0 = 0$, $z_\infty = 0$, $n(\mathcal{L}_c) = n_0$, and $z(\mathcal{L}_c) = 1$, where $n_0 = 1$ if $c + 2b'(c) > 0$ and $n_0 = 2$ if $c + 2b'(c) < 0$.

If $c + 2b'(c) = 0$, then $z_\infty = 1$, $z_0 = 0$, and $n_0 = 1$, so that the count (3.31) implies that $n(\mathcal{L}_c) = 1$ and $z(\mathcal{L}_c) = 2$.

It remains to address the orbital stability of the periodic wave with the profile ϕ_c. It follows from Theorem 3.3 that we need to check the stability condition (3.29), namely

$$n\left(\mathcal{L}_c|_{\{1,\phi_c\}^\perp}\right) = 0, \quad z\left(\mathcal{L}_c|_{\{1,\phi_c\}^\perp}\right) = 1. \tag{3.65}$$

To use the count (3.31) in Theorem 3.2, we construct the symmetric 2-by-2 matrix related to the two constraints:

$$A_2(\lambda) := \begin{bmatrix} \langle (\mathcal{L}_c - \lambda I)^{-1}\phi_c, \phi_c\rangle & \langle (\mathcal{L}_c - \lambda I)^{-1}\phi_c, 1\rangle \\ \langle (\mathcal{L}_c - \lambda I)^{-1}1, \phi_1\rangle & \langle (\mathcal{L}_c - \lambda I)^{-1}1, 1\rangle \end{bmatrix}, \quad \lambda \notin \sigma(\mathcal{L}_c).$$

We again first assume that $c + 2b'(c) \neq 0$ and compute

$$\langle \mathcal{L}_c^{-1} 1, 1\rangle = \frac{2\pi}{c+2b'(c)},$$

$$\langle \mathcal{L}_c^{-1} 1, \phi_c\rangle = -\frac{2\pi b'(c)}{c+2b'(c)},$$

$$\langle \mathcal{L}_c^{-1} \phi_c, 1\rangle = -\frac{2\pi b'(c)}{c+2b'(c)},$$

$$\langle \mathcal{L}_c^{-1} \phi_c, \phi_c\rangle = -\pi b'(c) + \frac{2\pi [b'(c)]^2}{c+2b'(c)},$$

so that

$$\lim_{\lambda \to 0^-} \det A_2(0) = -\frac{2\pi^2 b'(c)}{c+2b'(c)}. \tag{3.66}$$

If $c+2b'(c) \neq 0$, then $z_\infty = 0$ and $z(\mathcal{L}_c) = 1$. It follows from the count of $n(\mathcal{L}_c)$ and the relation (3.66) that the stability condition (3.65) is satisfied if $b'(c) > 0$ both for $c + 2b'(c) > 0$ and $c + 2b'(c) < 0$. If $c + 2b'(c) = 0$, then $z_\infty = 1$ and $z(\mathcal{L}_c) = 2$. The condition (3.65) are again satisfied if $b'(c) > 0$. Therefore, independently of the value of $c+2b'(c)$, the periodic wave with the profile ϕ_c is a local nondegenerate constrained minimizer of energy $H(u)$ so that its orbital stability in $H^{\alpha/2}_{\rm per}$ follows by Theorem 3.3 if the initial-value problem is locally well-posed for the fKdV equation (3.49) in $H^{\alpha/2}_{\rm per}$. □

REMARK 3.22. If $b'(c) = 0$, we obtain

$$n\left(\mathcal{L}_c|_{\{1,\phi_c\}^\perp}\right) = 0, \quad z\left(\mathcal{L}_c|_{\{1,\phi_c\}^\perp}\right) = 2.$$

The second variation test fails and we are in the marginal situation where further analysis is needed (Section 2.5). If $b'(c) < 0$, we obtain

$$n\left(\mathcal{L}_c|_{\{1,\phi_c\}^\perp}\right) = 1, \quad z\left(\mathcal{L}_c|_{\{1,\phi_c\}^\perp}\right) = 1.$$

The periodic wave with the profile ϕ_c is a saddle point of the constrained energy $H(u)$ for fixed $Q(u)$ and $M(u)$. Based on the spectral stability theory (Section 3.5), we can prove the existence of an unstable (real positive) eigenvalue in the spectral stability problem if $b'(c) < 0$ (see Remark 3.42).

REMARK 3.23. The stability condition in Theorem 3.6 is obtained from the monotonicity of the mapping $c \mapsto b(c) = \pi^{-1} Q(\phi_c)$ with the momentum $Q(u)$. This coincides with the stability criterion for solitary waves in the same fKdV equation [**16**].

EXAMPLE 3.10. The case $\alpha = 1$ of the fKdV equation (3.49) corresponds to the Benjamin–Ono (BO) equation

$$u_t + 2uu_x + H(u_{xx}) = 0, \quad H(u) := \frac{1}{\pi}\text{p.v.} \int_{-\infty}^{\infty} \frac{u(y)}{x-y} dy,$$

where $H(u)$ is referred to as the Hilbert transform. The single-lobe periodic solution of the stationary equation (3.55) is known in the exact form:

$$\omega = \coth\theta, \quad \psi_\omega(x) = \frac{\sinh\theta}{\cosh\theta - \cos x},$$

where $\theta \in (0, \infty)$ is a free parameter. Since it follows by (3.54) that

$$\omega - c = \frac{1}{\pi} \int_{\mathbb{T}} \psi_\omega(x) dx \quad \text{and} \quad b = \frac{1}{4}(\omega^2 - c^2),$$

we obtain from $\int_{-\pi}^{\pi} \psi_\omega(x) dx = 2\pi$ that $c = \omega - 2$ and $b = \omega - 1$. Eliminating $\omega \in (1, \infty)$ yields $b(c) = c + 1$ so that $b'(c) = 1 > 0$ and $c + 2b'(c) = c + 2 > 0$ for every $c > -1$. Hence, $n(\mathcal{L}_c) = 1$ and $z(\mathcal{L}_c) = 1$. It follows from Theorem 3.6 that the periodic waves with the profile ψ_ω are orbitally stable in $H^{1/2}_{\rm per}$ since the local well-posedness of the BO equation in $H^{1/2}_{\rm per}$ has been shown in [**204**].

EXAMPLE 3.11. The case $\alpha = 2$ of the fKdV equation (3.49) corresponds to the KdV equation

$$u_t + 2uu_x + u_{xxx} = 0.$$

The periodic solution of the stationary equation (3.55) with the single-lobe profile is given by (3.22) with $\alpha = \frac{K(k)}{\pi}$ due to the normalized period 2π, or explicitly,

$$\omega = \frac{4K(k)^2}{\pi^2} \sqrt{1 - k^2 + k^4}$$

and

$$\psi_\omega(x) = \frac{2K(k)^2}{\pi^2} \left[\sqrt{1 - k^2 + k^4} + 1 - 2k^2 + 3k^2 \,\mathrm{cn}^2 \left(\frac{K(k)}{\pi} x; k \right) \right],$$

where the elliptic modulus $k \in (0, 1)$ is a free parameter. Since it follows by (3.54) that

$$\omega - c = \frac{1}{\pi} \int_{\mathbb{T}} \psi_\omega(x) dx \quad \text{and} \quad b = \frac{1}{4}(\omega^2 - c^2),$$

we obtain from

$$\int_{-\pi}^{\pi} \psi_\omega(x) dx = \frac{4K(k)^2}{\pi} \left[\sqrt{1 - k^2 + k^4} + 1 - 2k^2 \right] + \frac{12K(k)}{\pi} \left[E(k) + (k^2 - 1) K(k) \right]$$

that $b = b(c)$ is defined in the parametrized form:

$$c = \frac{4K(k)^2}{\pi^2} \left[2 - k^2 - \frac{3E(k)}{K(k)} \right]$$

and

$$b = \frac{4K(k)^4}{\pi^4} \left[-3(1 - k^2) + (2 - k^2) \frac{6E(k)}{K(k)} - \frac{9E(k)^2}{K(k)^2} \right],$$

which exist for $c > -1$. Plotting c and b versus k in Figure 3.6 verifies that $\frac{dc}{dk} > 0$ and $\frac{db}{dk} > 0$ so that $b'(c) > 0$ and $c + 2b'(c) > 0$. Hence, $n(\mathcal{L}_c) = 1$ and $z(\mathcal{L}_c) = 1$. It follows from Theorem 3.6 that the periodic waves with the profile ψ_ω are orbitally stable in $H^1_{\rm per}$ since the local well-posedness of the KdV equation in H^1 has been shown in [**157**].

REMARK 3.24. Another approach to study stability of the traveling periodic waves in the fKdV equation (3.49) is developed in [**120**], where periodic solutions are found from the minimization of energy $H(u)$ subject to the fixed momentum $Q(u)$. This corresponds to the periodic solutions with profile ψ_ω satisfying the stationary equation (3.55) for $\omega > 0$. Each minimizer is stable by the construction;

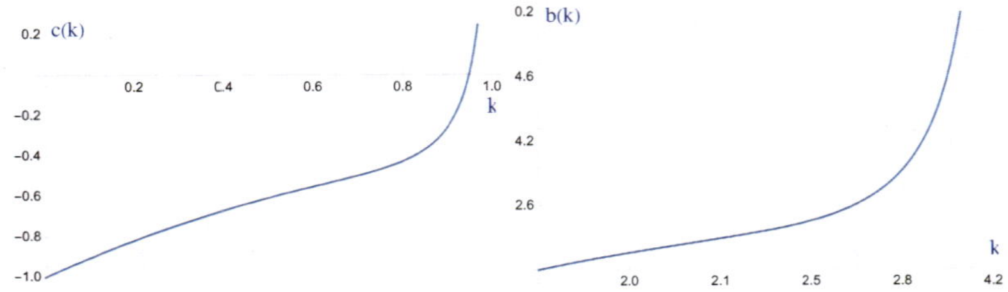

FIGURE 3.6. A plot of c (left) and b (right) versus $k \in (0, 1)$ for Example 3.11.

however, (stable) periodic waves with two negative eigenvalues of the Hessian of the augmented energy cannot be obtained by this variational method with a single constraint. Furthermore, the set of constrained minimizers in [**120**] includes the constant solution which is different from the periodic waves with the single-lobe profile.

The nondegeneracy assumption $\mathrm{Ker}(\mathcal{L}_c|_{\{1\}^\perp}) = \mathrm{Span}\{\partial_x\phi_c\}$ is the key to prove Lemma 3.3 and Theorem 3.6. The variational characterization of periodic waves in Theorem 3.5 is only useful if this nondegeneracy assumption is satisfied. The following lemma shows that this assumption is always satisfied if $n(\mathcal{L}_c) = 1$ and $z(\mathcal{L}_c) = 1$.

LEMMA 3.4. *The nondegeneracy assumption* $\mathrm{Ker}(\mathcal{L}_c|_{\{1\}^\perp}) = \mathrm{Span}\{\partial_x\phi_c\}$ *is satisfied if* $n(\mathcal{L}_c) = 1$ *and* $z(\mathcal{L}_c) = 1$.

PROOF. Assume the existence of $f_c \in \mathrm{Ker}(\mathcal{L}_c|_{\{1\}^\perp})$ such that f_c is linearly independent from $\partial_x\phi_c$. Remark 3.21 implies that $\langle f_c, 1\rangle = 0$ and $\langle f_c, \phi_c^2\rangle = 0$. Since f_c satisfies $\mathcal{L}_c f_c = -\frac{1}{\pi}\langle f_c, \phi_c\rangle$ due to (3.61) and $f_c \notin \mathrm{Ker}(\mathcal{L}_c)$ since $z(\mathcal{L}_c) = 1$, it follows that $\langle f_c, \phi_c\rangle \neq 0$. Hence we can normalize f_c such that it satisfies $\mathcal{L}_c f_c = -1$. It follows then from equations (3.56) and (3.57) that

$$\mathcal{L}_c(\phi_c - b(c)f_c) = -\phi_c^2, \quad \mathcal{L}_c(1 + cf_c) = -2\phi_c$$

so that

$$\langle \mathcal{L}_c^{-1}1, 1\rangle = 0, \quad \langle \mathcal{L}_c^{-1}1, \phi_c^2\rangle = \langle \mathcal{L}^{-1}\phi_c^2, 1\rangle = 0, \quad \langle \mathcal{L}_c^{-1}\phi_c^2, \phi_c^2\rangle = -2\pi\gamma(c),$$

where $\gamma(c) > 0$. The count formula in the proof of Theorem 3.6 implies that $n_0 = z_0 = 1$ and $z_\infty = 0$ so that $n(\mathcal{L}_c) = 2$, in a contradiction with $n(\mathcal{L}_c) = 1$. □

REMARK 3.25. It was found in [**136**] that

$$\text{if } \{1, \phi_c, \phi_c^2\} \in \mathrm{Ran}(\mathcal{L}_c), \quad \text{then } \mathrm{Ker}(\mathcal{L}_c) = \mathrm{Span}\{\partial_x\phi_c\}.$$

It follows from this property that if there exists $f_c \in \mathrm{Ker}(\mathcal{L}_c|_{\{1\}^\perp})$ such that f_c is linearly independent from $\partial_x\phi_c$, then $\mathrm{Ker}(\mathcal{L}_c) = \mathrm{Span}\{\partial_x\phi_c\}$ and $z(\mathcal{L}_c) = 1$. Therefore, the assumption $z(\mathcal{L}_c) = 1$ in Lemma 3.4 is redundant and can be removed. Indeed, if $f_c \in \mathrm{Ker}(\mathcal{L}_c)$ in addition to $f_c \in \mathrm{Ker}(\mathcal{L}_c|_{\{1\}^\perp})$, then $\langle f_c, 1\rangle = 0$, $\langle f_c, \phi_c\rangle = 0$, and $\langle f_c, \phi_c^2\rangle = 0$ so that $\{1, \phi_c, \phi_c^2\} \perp \mathrm{Ker}(\mathcal{L}_c)$ but since $\mathcal{L}_c$ is self-adjoint, then $\{1, \phi_c, \phi_c^2\} \in \mathrm{Ran}(\mathcal{L}_c)$ which contradicts the assumption $f_c \in \mathrm{Ker}(\mathcal{L}_c)$.

In case if the nondegeneracy assumption $\mathrm{Ker}(\mathcal{L}_c|_{\{1\}^\perp}) = \mathrm{Span}\{\partial_x \phi_c\}$ fails so that $n(\mathcal{L}_c) = 2$ follows from Lemma 3.4, the following lemma shows that the corresponding periodic waves do not correspond to minimizers of energy $H(u)$ subject to fixed $Q(u)$ and $M(u)$. Constrained minimizers of energy can only be obtained from nondegenerate minimizers of Theorem 3.5.

LEMMA 3.5. *If $n(\mathcal{L}_c) = 2$ and $\mathrm{Ker}(\mathcal{L}_c|_{\{1\}^\perp}) \neq \mathrm{Span}\{\partial_x \phi_c\}$, then the periodic wave with profile ϕ_c is a saddle point of energy $H(u)$ subject to fixed $Q(u)$ and $M(u)$.*

PROOF. Assume existence of $f_c \in \mathrm{Ker}(\mathcal{L}_c|_{\{1\}^\perp})$ such that $\mathcal{L}_c f_c = -1$, then

$$\tilde{f}_c := f_c - \frac{\phi_c}{2b(c)}$$

satisfies $\langle \tilde{f}_c, 1\rangle = 0$, $\langle \tilde{f}_c, \phi_c\rangle = 0$ and

$$\langle \mathcal{L}_c \tilde{f}_c, \tilde{f}_c\rangle = \frac{\langle \mathcal{L}_c \phi_c, \phi_c\rangle}{4b(c)^2} = -\frac{\gamma(c)}{4b(c)^2} < 0$$

where we have used (3.56). Hence $n\left(\mathcal{L}_c\big|_{\{1,\phi_c\}^\perp}\right) = 1$ and the periodic wave with profile ϕ_c is a saddle point of the constrained energy by Remark 3.22. □

3.4.2. The modified fractional Korteweg–de Vries (mfKdV) equation. The mfKdV equation can be written in the normalized form:

$$u_t + 6u^2 u_x = (-\Delta)^{\alpha/2} u_x, \qquad \mathbb{T} = \mathbb{R}\backslash\{2\pi\mathbb{Z}\}. \tag{3.67}$$

Compared to the fKdV equation (3.49), the energy is given by

$$H(u) = \frac{1}{2}\int_{\mathbb{T}} [(-\Delta)^{\alpha/4} u]^2 dx - \frac{1}{2}\int_{\mathbb{T}} u^4 dx,$$

with the same expressions for the momentum $Q(u)$ and mass $M(u)$. The energy space $X := H^{\alpha/2}_{\mathrm{per}}$ is compactly embeedded into $L^4(\mathbb{T})$ if $\alpha > \frac{1}{2}$. Traveling wave $u(t,x) = \phi_{c,b}(x-ct)$ with the wave speed c is obtained from the stationary equation

$$(-\Delta)^{\alpha/2}\phi_{c,b} + c\phi_{c,b} - 2\phi_{c,b}^3 + b = 0, \tag{3.68}$$

where b is the integration constant. Since this is the focusing version of the modified KdV equation, two families of solutions exist in the local case $\alpha = 2$ for the choice $b = 0$ (Examples 3.3 and 3.4) and these two families are continued for $b \neq 0$ in a rather nontrivial way [**42**]. For the scope of this presentation, we will avoid formulating variational results for $b \neq 0$ and characterize variationally the two families of periodic waves of the stationary equation (3.68) for $b = 0$ based on [**207**]. The case of $\alpha = 2$ and $b \neq 0$ was considered in [**172**].

The reason why the two families can be distinguished is that the cubic nonlinearity of the stationary equation (3.68) preserves parity of functions, which means that it maps even functions to even functions and odd functions to odd functions. For notational convience, we denote subspaces of H^s_{per} for even and odd functions by $H^s_{\mathrm{per,even}}$ and $H^s_{\mathrm{per,odd}}$.

The first family of periodic waves with $b = 0$ is characterized variationally similarly to Theorem 3.5, except that we replace the zero-mean constraint by the stronger condition of odd functions. It is clear that every odd function satisfies the zero-mean constraint. The proof of the following theorem is analogous to the proof of Theorem 3.5, see [**207**].

THEOREM 3.7. *For every $\alpha > \frac{1}{2}$ and $c > -1$, there exists the ground state of the constrained minimization problem*

$$\inf_{u \in H^{\alpha/2}_{\rm per,odd}} \left\{ \mathcal{F}_c(u) : \int_{-\pi}^{\pi} u^4 dx = 1, \right\}. \tag{3.69}$$

where $\mathcal{F}_c(u)$ is given by (3.51). If $\alpha \leq 2$, the ground state has the single-lobe profile in the sense of Definition 3.3, which is even with respect to the quarter period points at $x = \pm\frac{\pi}{2}$.

REMARK 3.26. The quarter-period symmetry of the minimizer in Theorem 3.7 follows from Polya–Szegö inequality for $\alpha \leq 2$ (see [**56, 172**]). Since the minimizer decreases symmetrically away from the maximum point and must be zero at three points on $\mathbb{T}$ ($x = 0$ and $x = \pm\pi$), the maximum and minimum points of the minimizers are located at $x = \pm\frac{\pi}{2}$.

REMARK 3.27. By using Euler–Lagrange equation

$$(-\Delta)^{\alpha/2}\varphi_c + c\varphi_c = \mu\varphi_c^3,$$

for the constrained minimization problem (3.69) with $\mu = 2\mathcal{F}_c(\varphi_c) > 0$ and the scaling transformation $\phi_c = C\varphi_c$ with

$$C := \sqrt{\mathcal{F}_c(\varphi_c)},$$

every minimizer $\varphi_c \in H^{\alpha/2}_{\rm per,odd}$ of Theorem 3.7 determines an odd solution $\phi_c \in H^{\alpha/2}_{\rm per,odd}$ of the stationary equation (3.68) with $b = 0$.

The quarter-period symmetry of the periodic waves obtained from Theorem 3.7 has nice consequences for the smooth continuation of the periodic waves with respect to parameter $c \in (-1, \infty)$ and for the count of negative and zero eigenvalues of the linearized operator $\mathcal{L}_c$ in $L^2_{\rm per}$ with the dense domain $H^{\alpha}_{\rm per}$ defined by

$$\mathcal{L}_c := (-\Delta)^{\alpha/2} + c - 6\phi_c^2.$$

The following lemma gives the corresponding properties of the periodic waves obtained from Theorem 3.7.

LEMMA 3.6. *Let $\alpha \in (\frac{1}{2}, 2]$ and $\phi_c \in H^{\alpha/2}_{\rm per,odd}$ be the periodic wave profile obtained from Theorem 3.7. Then, the mapping $(-1, \infty) \ni c \mapsto \phi_c(\cdot; c)$ is C^1, $n(\mathcal{L}_c) = 2$, and*

$$z(\mathcal{L}_c) = \begin{cases} 1, & \text{if } 1 \in \mathrm{Ran}(\mathcal{L}_c), \\ 2, & \text{if } 1 \notin \mathrm{Ran}(\mathcal{L}_c). \end{cases}$$

PROOF. We compute separately $n(\mathcal{L}_c)$ and $z(\mathcal{L}_c)$. As a by-product of the computations of $z(\mathcal{L}_c)$, we also prove the C^1 smoothness of the mapping $(-1, \infty) \ni c \mapsto \phi_c(\cdot; c)$.

Computation of $n(\mathcal{L}_c)$. Since $\phi_c \in H^{\alpha/2}_{\rm per,odd}$ is related to a minimizer of the constrained variational problem (3.69) with only one constraint, we have either $n(\mathcal{L}_c|_{L^2_{\rm odd}}) = 1$ or $n(\mathcal{L}_c|_{L^2_{\rm odd}}) = 0$. Since

$$\langle \mathcal{L}_c\phi_c, \phi_c\rangle_{L^2} = -4\|\phi_c\|^4_{L^4} < 0,$$

with odd ϕ_c, $n(\mathcal{L}_c|_{L^2_{\rm odd}})$ cannot be zero, so that $n(\mathcal{L}_c|_{L^2_{\rm odd}}) = 1$. Since $\partial_x\phi_c \in \operatorname{Ker}(\mathcal{L}_c)$ and $\partial_x\phi_c$ is even with two nodes on $\mathbb{T}$, then it follows from Sturm's oscillation theory (Remark 3.28) that 0 is not the first eigenvalue of $\mathcal{L}_c|_{L^2_{\rm even}}$. However, an even eigenfunction for the second eigenvalue of $\mathcal{L}_c|_{L^2_{\rm even}}$ must have two nodes and the nodes are located at the symmetry points $x = \pm\pi/2$. Since these properties coincide with properties of the even eigenfunction $\partial_x\phi_c \in \operatorname{Ker}(\mathcal{L}_c)$, we conclude that 0 is the second eigenvalue of $\mathcal{L}_c|_{L^2_{\rm even}}$, so that $n(\mathcal{L}_c|_{L^2_{\rm even}}) = 1$ and

$$n(\mathcal{L}_c) = n(\mathcal{L}_c|_{L^2_{\rm odd}}) + n(\mathcal{L}_c|_{L^2_{\rm even}}) = 2.$$

Computation of $z(\mathcal{L}_c)$. By the symmetry of ϕ_c and ϕ_c^2, the operator $\mathcal{L}_c$ has a π-periodic potential, which is even with respect to both $x = 0$ and $x = \pi/2$. The negative eigenvalue of $\mathcal{L}_c$ in $L^2_{\rm even}$, which is the lowest eigenvalue of $\mathcal{L}_c$ in $L^2(\mathbb{T})$, corresponds to the sign-definite π-periodic function, which is even with respect to both $x = 0$ and $x = \pi/2$. The negative eigenvalue in $L^2_{\rm odd}$, which is the second eigenvalue of $\mathcal{L}_c$ in $L^2(\mathbb{T})$, corresponds to the eigenfunction with two nodes on $\mathbb{T}$, which is even with respect to $x = \pi/2$. The eigenfunction $\partial_x\phi_c$ for the zero eigenvalue in $L^2_{\rm even}$, which is the third eigenvalue of $\mathcal{L}_c$ in $L^2(\mathbb{T})$, has two nodes and is odd with respect to $x = \pi/2$. The zero eigenvalue is the lowest eigenvalue for the eigenfunctions that are odd with respect to $x = \pi/2$. Eigenfunctions with four nodes on $\mathbb{T}$ have the same parity with respect to $x = 0$ and $x = \pi/2$, hence odd functions in $L^2_{\rm odd}$ are also odd with respect to $x = \pi/2$. Since the zero eigenvalue is the lowest eigenvalue for the eigenfunctions that are odd with respect to $x = \pi/2$ and the lowest eigenvalue is simple, the odd eigenfunction of $\mathcal{L}_c$ in $L^2_{\rm odd}$ with four nodes corresponds to the positive eigenvalue of $\mathcal{L}_c$. Therefore, $z(\mathcal{L}_c|_{L^2_{\rm odd}}) = 0$ and by the same argument as in the proof of Lemma 3.3, the mapping

$$c \mapsto \phi_c(\cdot; c) \in H^{\alpha/2}_{\rm per,odd}$$

is C^1 in c. Hence, we can use

$$\mathcal{L}_c\partial_c\phi_c = -\phi_c,$$

in addition to the other two equations

$$\mathcal{L}_c 1 = c - 6\phi_c^2, \quad \mathcal{L}_c\phi_c = -4\phi_c^3,$$

which follow from the stationary equation (3.68) with $b = 0$ and the explicit expression for $\mathcal{L}_c$. It remains to check if the even eigenfunction of $\mathcal{L}_c$ in $L^2_{\rm even}$ with four nodes (call it f_c) may belong to $\operatorname{Ker}(\mathcal{L}_c)$. Assume it does, so that $\operatorname{Ker}(\mathcal{L}_c) = \operatorname{Span}(\partial_x\phi_c, f_c)$. Either $\langle 1, f_c\rangle_{L^2} = 0$ or $\langle 1, f_c\rangle_{L^2} \neq 0$. If $\langle 1, f_c\rangle_{L^2} = 0$, then $1 \in \operatorname{Ran}(\mathcal{L}_c)$ and it follows from the previous equations that $\phi_c, \phi_c^2 \in \operatorname{Ran}(\mathcal{L}_c)$. Therefore, $\{1, \phi_c, \phi_c^2\} \in \operatorname{Ran}(\mathcal{L}_c)$ and by Remark 3.25, we have $\operatorname{Ker}(\mathcal{L}_c) = \operatorname{Span}(\partial_x\phi_c)$, so that the existence of $f_c \in \operatorname{Ker}\{\mathcal{L}_c\}$ leads to a contradiction. Hence, $z(\mathcal{L}_c) = 1$ if $1 \in \operatorname{Ran}(\mathcal{L}_c)$. If $\langle 1, f_c\rangle_{L^2} \neq 0$, then $1 \notin \operatorname{Ran}(\mathcal{L}_c)$ because $\operatorname{Ran}(\mathcal{L}_c)$ is orthogonal to $\operatorname{Ker}(\mathcal{L}_c)$. Hence, $z(\mathcal{L}_c) = 2$ if $1 \notin \operatorname{Ran}(\mathcal{L}_c)$. □

REMARK 3.28. Sturm's oscillation theory for the Schrödinger operators on the periodic domain used in the proof of Theorem 3.1 cannot be used in the proof of Lemma 3.6 for the fractional operator $\mathcal{L}_c$. However, it follows from the generalization of Sturm's oscillation theory in **[136]** that an eigenfunction of $\mathcal{L}_c$ which corresponds to the nth eigenvalue of $\mathcal{L}_c$ changes its sign at most $2(n-1)$ times on $\mathbb{T}$. Moreover, due to the quarter period symmetry of the potentials in $\mathcal{L}_c$, an

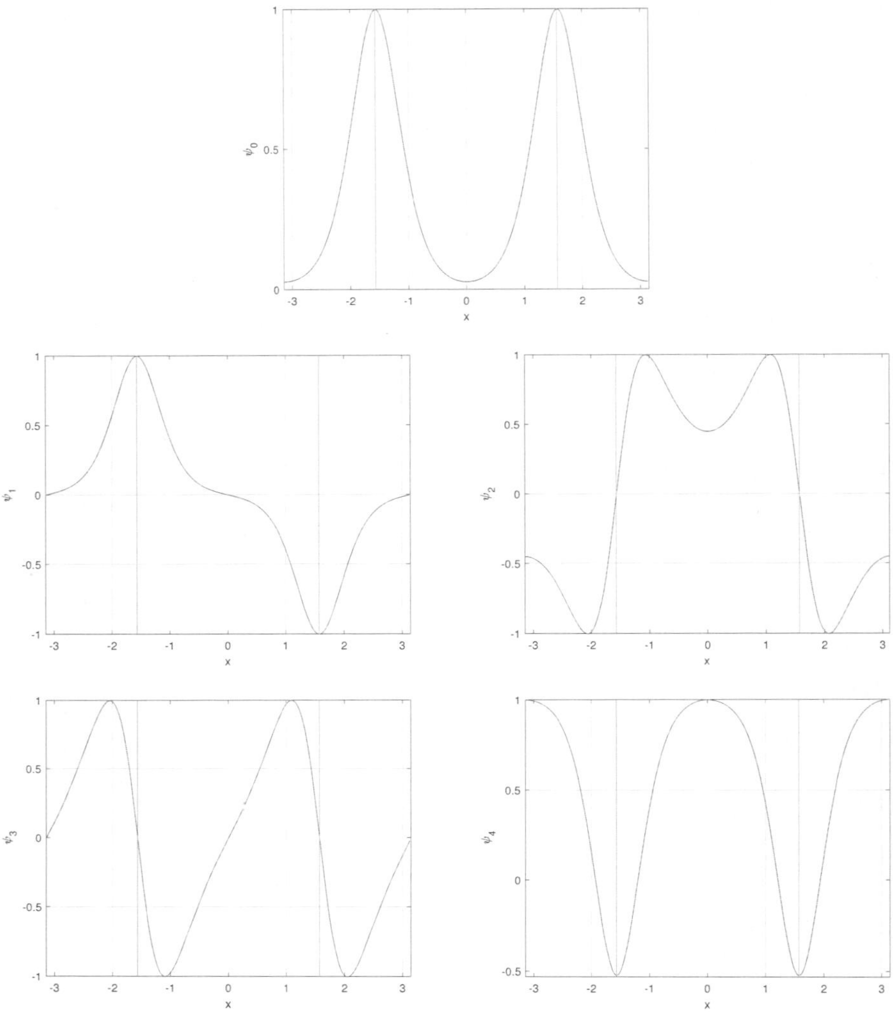

FIGURE 3.7. The first five eigenfunctions of $\mathcal{L}_c$ on $\mathbb{T}$ with their centers of symmetries for Example 3.12 with $\alpha = 2$.

eigenfunction of $\mathcal{L}_c|_{L^2_{\text{even}}}$ for the nth eigenvalue of $\mathcal{L}_c|_{L^2_{\text{even}}}$ changes its sign at most $2(n-1)$ times on $\mathbb{T}$.

REMARK 3.29. For illustration purposes, we plot in Figure 3.7 the first five eigenfunctions of the operator $\mathcal{L}_c$ in $L^2(\mathbb{T})$ for $\alpha = 2$ (Example 3.12). The first function is strictly positive and it is even with respect to both $x = 0$ and $x = \pm\frac{\pi}{2}$. The second (third) functions are even (odd) with respect to $x = \pm\frac{\pi}{2}$ and odd (even) with respect to $x = 0$. The fourth (fifth) functions are odd (even) both with respect to $x = 0$ and $x = \pm\frac{\pi}{2}$. The first two eigenfunctions correspond to negative eigenvalues, the third eigenfunction corresponds to the zero eigenvalue, and the fourth and fifth eigenfunctions correspond to positive eigenvalues of $\mathcal{L}_c$ for $\alpha = 2$.

By using the same computations as in the proof of Theorem 3.6, the stability criterion for the periodic waves obtained from Theorem 3.7 can be obtained. The following lemma gives the result under the simplification that $1 \in \mathrm{Ran}(\mathcal{L}_c)$.

LEMMA 3.7. *Let* $\alpha \in (\frac{1}{2}, 2]$ *and* $\phi_c \in H^{\alpha/2}_{\mathrm{per,odd}}$ *be the periodic wave profile obtained from Theorem* 3.7. *Assume that* $1 \in \mathrm{Ran}(\mathcal{L}_c)$. *Then, the periodic wave is spectrally stable if*

$$\langle \mathcal{L}_c^{-1} 1, 1 \rangle \leq 0, \quad \frac{d}{dc} \|\phi_c\|^2_{L^2} \geq 0, \tag{3.70}$$

and is spectrally unstable with exactly one real positive eigenvalue of $\partial_x \mathcal{L}_c$ *if*

$$\begin{cases} \text{either} & \langle \mathcal{L}_c^{-1} 1, 1 \rangle > 0, \;\; \frac{d}{dc}\|\phi_c\|^2_{L^2} > 0 \\ \text{or} & \langle \mathcal{L}_c^{-1} 1, 1 \rangle < 0, \;\; \frac{d}{dc}\|\phi_c\|^2_{L^2} < 0 \\ \text{or} & \langle \mathcal{L}_c^{-1} 1, 1 \rangle = 0, \;\; \frac{d}{dc}\|\phi_c\|^2_{L^2} < 0, \\ \text{or} & \langle \mathcal{L}_c^{-1} 1, 1 \rangle > 0, \;\; \frac{d}{dc}\|\phi_c\|^2_{L^2} = 0. \end{cases}$$

PROOF. If $1 \in \mathrm{Ran}(\mathcal{L}_c)$, it follows from $\mathcal{L}_c \partial_c \phi_c = -\phi_c$ that

$$\langle \mathcal{L}^{-1} 1, \phi_c \rangle = \langle \mathcal{L}^{-1} \phi_c, 1 \rangle = 0, \quad \langle \mathcal{L}^{-1} \phi_c, \phi_c \rangle = -\frac{1}{2} \frac{d}{dc} \|\phi_c\|^2_{L^2}.$$

By using the same matrix $A_2(\lambda)$ in the proof of Theorem 3.6 and the counting formula (3.31) from Theorem 3.2, we obtain in the case (3.70) that

$$n(\mathcal{L}_c|_{\{1,\phi_c\}^\perp}) = 0, \quad 1 \leq z(\mathcal{L}_c|_{\{1,\phi_c\}^\perp}) \leq 2,$$

which is sufficient for the spectral stability of the periodic wave with the profile ϕ_c. In the other cases, we obtain

$$n(\mathcal{L}_c|_{\{1,\phi_c\}^\perp}) = 1, \quad 1 \leq z(\mathcal{L}_c|_{\{1,\phi_c\}^\perp}) \leq 2,$$

which is sufficient for the spectral instability. □

REMARK 3.30. The case $\langle \mathcal{L}_c^{-1} 1, 1 \rangle > 0$ and $\frac{d}{dc}\|\phi_c\|^2_{L^2} < 0$ is inconclusive for the spectral stability or instability because

$$n(\mathcal{L}_c|_{\{1,\phi_c\}^\perp}) = 2, \quad z(\mathcal{L}_c|_{\{1,\phi_c\}^\perp}) = 1.$$

Numerical examples in [**207**] show that this case does not occur for every $\alpha \in (\frac{1}{2}, 2]$. Similarly, numerical examples in [**207**] show that the assumption $1 \in \mathrm{Ran}(\mathcal{L}_c)$ is satisfied for every $\alpha \in (\frac{1}{2}, 2]$.

The second family of periodic waves with $b = 0$ is characterized variationally in the space of even functions. This characterization can be used for every nonlinearity but, unfortunately, it does not distinguish between minimizers given by constant solutions and minimizers given by the periodic solutions with the single-lobe profile.

THEOREM 3.8. *For every* $\alpha > \frac{1}{2}$ *and* $c > 0$, *there exists the ground state of the constrained minimization problem:*

$$\inf_{u \in H^{\alpha/2}_{\mathrm{per,even}}} \left\{ \mathcal{F}_c(u) : \int_{-\pi}^{\pi} u^4 dx = 1 \right\}, \tag{3.71}$$

with the same $\mathcal{F}_c(u)$ *as in* (3.51). *If* $\alpha \leq 2$, *the ground state is the constant solution for* $c \in (0, \frac{1}{2}]$ *and has the single-lobe profile for* $c \in (\frac{1}{2}, \infty)$.

PROOF. Since $\mathcal{F}_c$ is a smooth, bounded, and coercive functional on $H^{\alpha/2}_{\rm per,even}$ if $c > 0$, hence $q_{c,\rm even} > 0$ and the existence of the minimizer of the constrained minimization problem (3.71) is proven exactly like in the proof of Theorem 3.5. Moreover, for $\alpha \in (0, 2]$, the symmetric rearrangements and Polya–Szygö inequality (see Appendix A in [**56**]) suggest that the minimizer is either constant or it must decrease symmetrically away from the maximum point.

In order to ensure that the minimizer has the single-lobe profile, we need to eliminate the constant solution in $H^{\alpha/2}_{\rm per,even}$. By Lagrange's Multiplier Theorem, the constrained minimizer $\chi \in H^{\frac{\alpha}{2}}_{\rm per,even}$ of the constrained variational problem (3.71) satisfies the Euler–Lagrange equation

$$(-\Delta)^{\alpha/2}\varphi_c + c\varphi_c = \mu\varphi_c^3,$$

where $\mu = 2\mathcal{F}_c(\varphi_c)$ due to the normalization in (3.71). Since $\mathcal{F}_c(\varphi_c) > 0$, the scaling transformation $\phi_c = C\varphi_c$ with $C := \sqrt{\mathcal{F}_c(\varphi_c)}$ maps the Euler–Lagrange equation to the form (3.68) with $b = 0$. The constant nonzero solution to the stationary equation (3.68) with $b = 0$ is given by $\phi_c(x) = \sqrt{c/2}$ up to a sign choice. The linearization operator $\mathcal{L}_c$ evaluated at the constant solution is given by

$$\mathcal{L}_c := (-\Delta)^{\alpha/2} + c - 6\phi_c^2 = (-\Delta)^{\alpha/2} - 2c.$$

Since $n(\mathcal{L}_c) = 1$ if and only if $c \in \left(0, \frac{1}{2}\right]$, the constant wave corresponds to the constrained minimizer of (3.71) for $c \in \left(0, \frac{1}{2}\right]$ and to a saddle point for $c \in \left(\frac{1}{2}, \infty\right)$. For $c > \frac{1}{2}$, the global minimizer can not be the constant solution so that it is a nonconstant solution with the single-lobe profile. □

REMARK 3.31. The minimizer constructed in Theorem 3.7 is even after translation $x \mapsto x - \pi/2$. However, since $n(\mathcal{L}_c) = 2$ by Lemma 3.6, the odd periodic wave translated into an even function cannot be a solution of the constrained minimization problem (3.71) with a single constraint. Hence, it differs from the minimizer constructed in Theorem 3.8.

REMARK 3.32. For $\alpha = 2$, the minimizers of Theorem 3.7 are equivalent to the cnoidal periodic wave (Example 3.4) and the minimizers of Theorem 3.8 are equivalent to the dnoidal periodic wave (Example 3.3) [**207**]. Stability of the dnoidal periodic wave in the modified KdV equation (3.45) was considered in Example 3.8. Stability of the cnoidal periodic wave is considered in Example 3.12.

In the next examples, we only consider the periodic wave solutions of the first family, related to the minimizers of Theorem 3.7. Examples of the periodic wave solutions of the second family, related to the minimizers of Theorem 3.8, can be found in [**207**].

EXAMPLE 3.12. Minimizers of Theorem 3.7 for $c > -1$ and $\alpha = 2$ define the cnoidal wave solutions (Example 3.4). Since the mapping $\mathcal{E} \mapsto T$ is monotonically decreasing, Theorem 3.1 implies that $n(\mathcal{L}_c) = 2$ and $z(\mathcal{L}_c) = 1$, where the linearized operator $\mathcal{L}_c : H^2_{\rm per} \subset L^2_{\rm per} \to L^2_{\rm per}$ is given by

$$\mathcal{L}_c = -\partial_x^2 + c - 6\phi_c^2.$$

The cnoidal wave is given by (3.17) with $\alpha = \frac{2K(k)}{\pi}$ due to the normalized period 2π, or explicitly,

$$c = \frac{4}{\pi^2}K(k)^2(2k^2 - 1).$$

and

$$\phi_c(x) = \frac{2}{\pi} k K(k) \operatorname{cn}\left(\frac{2}{\pi} K(k) x; k\right). \tag{3.72}$$

To consider stability of the cnoidal waves in the modified KdV equation (3.45) with an application of Theorem 3.4, we have to compute a continuation of the solution ϕ_c with respect to both parameters c and b in the stationary equation (3.68), similarly to Example 3.9. However, Lemma 3.7 formulates the stability criterion (3.70) for $b = 0$ and this criterion can be verified from the explicit profile ϕ_c. By using the scaling transformation $z := \frac{2}{\pi} K(k) x$, we can rewrite $\mathcal{L}_c$ in the form

$$\mathcal{L}_c = \frac{4}{\pi^2} K(k)^2 \mathcal{L}(k), \quad \mathcal{L}(k) = -\partial_z^2 + 2k^2 - 1 - 6k^2 \operatorname{cn}(z;k)^2.$$

The first five eigenvalues and eigenfunctions of $\mathcal{L}(k)$ in the space of $4K(k)$-periodic functions are known explicitly [**15, 79**]:

$$\lambda_0 = 1 - 2k^2 - 2\sqrt{1-k^2+k^4}, \quad \varphi_0(z) = 1 + k^2 + \sqrt{1-k^2+k^4} - 3k^2 \operatorname{sn}(z;k)^2,$$
$$\lambda_1 = -3k^2, \quad \varphi_1(z) = \operatorname{cn}(z;k)\operatorname{dn}(z;k),$$
$$\lambda_2 = 0, \quad \varphi_2(z) = \operatorname{sn}(z;k)\operatorname{dn}(z;k),$$
$$\lambda_3 = 3(1-k^2), \quad \varphi_3(z) = \operatorname{sn}(z;k)\operatorname{cn}(z;k),$$
$$\lambda_4 = 1 - 2k^2 + 2\sqrt{1-k^2+k^4}, \quad \varphi_4(z) = 1 + k^2 - \sqrt{1-k^2+k^4} - 3k^2 \operatorname{sn}(z;k)^2.$$

To compute $\langle \mathcal{L}_c^{-1} 1, 1\rangle$ in (3.70), we obtain

$$\frac{1}{2\sqrt{1-k^2+k^4}} \mathcal{L}(k) \left[\frac{\lambda_4 \varphi_0 - \lambda_0 \varphi_4}{\lambda_0 \lambda_4}\right] = 1,$$

from which it follows that

$$\langle \mathcal{L}(k)^{-1} 1, 1\rangle = \frac{\lambda_4 \langle \varphi_0, 1\rangle - \lambda_0 \langle \varphi_4, 1\rangle}{2\sqrt{1-k^2+k^4}\lambda_0\lambda_4} = -4\left[2E(k) - K(k)\right].$$

Since $2E(k) - K(k) > 0$ as $k \to 0$ and $2E(k) - K(k) < 0$ as $k \to 1$, there exists at least one value of k in $(0,1)$, where $\langle \mathcal{L}(k)^{-1} 1, 1\rangle = 0$. However, since

$$\frac{d}{dk} k E(k) = 2E(k) - K(k) \quad \text{and}$$
$$\frac{d^2}{dk^2} k E(k) = \frac{(1-k^2)[E(k) - K(k)] - k^2 E(k)}{k(1-k^2)} < 0,$$

there exists exactly one value of k labeled as k^* such that $\langle \mathcal{L}(k)^{-1} 1, 1\rangle < 0$ for $k \in (0, k^*)$ and $\langle \mathcal{L}(k)^{-1} 1, 1\rangle > 0$ for $k \in (k^*, 1)$. The sign of $\langle \mathcal{L}_c^{-1} 1, 1\rangle$ coincides with the sign of $\langle \mathcal{L}(k)^{-1} 1, 1\rangle$ due to the scaling transformation. To compute $\frac{d}{dc}\|\phi_c\|_{L^2}^2$ in (3.70), we obtain

$$\|\phi_c\|_{L^2}^2 = \frac{4}{\pi} K(k) \int_0^{2K(k)} k^2 \operatorname{cn}^2(z;k) dz = \frac{8}{\pi} K(k) \left[E(k) - (1-k^2) K(k)\right] > 0$$

and

$$\frac{d}{dk}\|\phi_c\|_{L^2}^2$$
$$= \frac{9}{\pi k (1-k^2)} \left[(1-k^2) K(k)[K(k) - E(k)] + E(k)[E(k) - (1-k^2)K(k)]\right] > 0,$$

for every $k \in (0, 1)$, where the properties (A.7) and (A.8) have been used. Similarly, we obtain

$$\frac{dc}{dk} = \frac{8K(k)}{\pi^2 k(1-k^2)} \left[(1-k^2)(K(k) - E(k)) + k^2 E(k)\right] > 0,$$

hence, $\frac{d}{dc}Q(\phi_c) > 0$. By Lemma 3.7, the periodic wave with the profile ϕ_c is spectrally stable if $k \in (0, k_*)$ and is spectrally unstable if $k \in (k_*, 1)$. By monotonicity of $\frac{dc}{dk} > 0$, the unique value of $k_* \in (0, 1)$ corresponds to the unique value of $c_* \in (-1, \infty)$, so that the cnoidal waves are spectrally stable for $c \in (-1, c_*)$ and spectrally unstable for $c \in (c_*, \infty)$.

REMARK 3.33. The stability and instability results for cnoidal waves in the focusing modified KdV equation (3.45) were obtained in [**18, 32, 79, 215**].

REMARK 3.34. The change of the sign of $\langle \mathcal{L}_c^{-1} 1, 1\rangle$ in Example 3.12 implies a change in the variational characterization of the cnoidal waves with the profile ϕ_c as constrained minimizers of the variational problem

$$\inf_{u \in H^{\alpha/2}_{\rm per}} \left\{ \mathcal{F}_c(u) : \int_{-\pi}^{\pi} u^4 dx = 1, \quad \int_{-\pi}^{\pi} u dx = 0 \right\}. \tag{3.73}$$

Since ϕ_c is odd after translation by a quarter-period (and so is ϕ_c^3) and $\mathcal{L}_c^{-1} 1$ is even after translation by a quarter-period, we have $\langle \mathcal{L}_c^{-1} 1, \phi_c^3\rangle = \langle \mathcal{L}_c^{-1} \phi_c^3, 1\rangle = 0$. On the other hand, we have

$$\langle \mathcal{L}_c^{-1} \phi_c^3, \phi_c^3 \rangle = -\frac{1}{4}\langle \phi_c, \phi_c^3\rangle = -\frac{1}{4}\int_{-\pi}^{\pi} \phi_c^4 dx < 0.$$

Therefore, $n(\mathcal{L}_c|_{\{1,\phi_c^3\}^\perp}) = 0$ for $c \in (-1, c_*]$ and $n(\mathcal{L}_c|_{\{1,\phi_c^3\}^\perp}) = 1$ for $c \in (c_*, \infty)$ so that the cnoidal waves obtained from the constrained minimizers in the variational problem (3.69) correspond to the constrained minimizers in (3.73) for $c \in (-1, c_*)$ and the constrained saddle points in (3.73) for $c \in (c_*, \infty)$.

EXAMPLE 3.13. Figure 3.8 illustrates the periodic wave solutions of Example 3.12. The top left panel shows the profiles ϕ_c satisfying the stationary equation (3.68) with $b = 0$ for three different values of c which are minimizers of the variational problem (3.69). The top right panel shows the profiles of ϕ_c satisfying the stationary equation (3.68) with $b \neq 0$ which are minimizers of the variational problem (3.73). The vertical lines show the symmetry points at $x = \pm\pi/2$. The family of solutions with $b = 0$ has odd symmetry with respect to these points, whereas the family of solutions with $b \neq 0$ does not have this symmetry; both families are even at $x = 0$ and $x = \pm\pi$. The bottom left panel of Figure 3.8 shows the dependence of b in the stationary equation (3.68) versus speed c. The pitchfork bifurcation point is located at $c_* \approx 1.425$ and corresponds to the stability threshold obtained in Example 3.12. The bottom right panel of Figure 3.8 shows the momentum $Q(\phi_c) := \frac{1}{2}\|\phi_c\|^2_{L^2}$ versus c for the two families of solutions. The momentum $Q(\phi_c)$ is an increasing function of c for both the families. However, $\langle \mathcal{L}_c^{-1} 1, 1\rangle$ for the solutions with $b = 0$ is negative for $c < c_*$ and positive for $c > c_*$. Consequently, the solutions with $b = 0$ are spectrally stable for $c < c_*$ and spectrally unstable for $c > c_*$. More details about bifurcation of the cnoidal waves at $c = c_*$ can be found in [**207**]. Generalizations of the families of the dnoidal and cnoidal solutions for $\alpha = 2$ in the context of the variational problem (3.73) with two constraints were studied in [**172**].

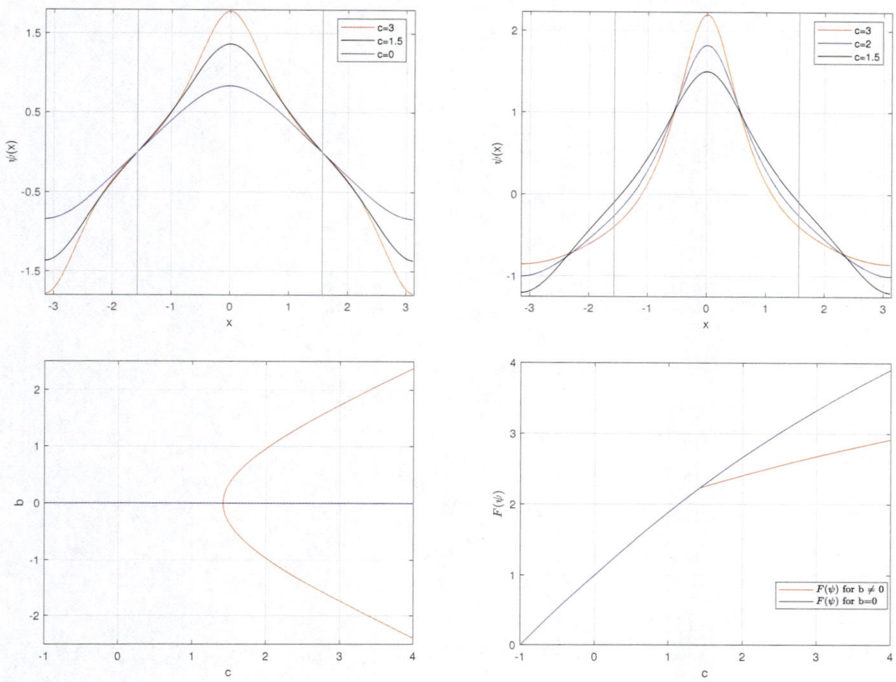

FIGURE 3.8. Periodic waves of the mfKdV equation (3.67) for $\alpha = 2$. Top left: Profiles of ϕ_c with $b = 0$ for three different values of c. Top right: Profiles of ϕ_c with $b \neq 0$ for three values of c. Bottom left: Dependence of b versus c showing the pitchfork bifurcation point c_*. Bottom right: Dependence of the momentum $Q(\phi_c)$ versus c for the two families of solutions.

EXAMPLE 3.14. Numerical results similar to those on Figure 3.8 can be obtained for the modified Benjamin–Ono equation

$$u_t + 6u^2 u_x + H(u_{xx}) = 0,$$

which corresponds to the case $\alpha = 1$ of the mfKdV equation (3.67). Figure 3.9 presents the same four panels as on Figure 3.8. The bifurcation point c_* moves to the left and becomes $c_* \approx -0.310$. The existence and stability of the periodic solutions of the stationary equation (3.68) with $b = 0$ is very similar with the only difference that the dependence of the momentum $Q(\phi_c)$ versus speed c approaches the horizontal asymptote as $c \to \infty$. Nevertheless, since $\langle \mathcal{L}_c 1, 1 \rangle$ is positive for $c > c_*$, the solutions with $b = 0$ are unstable for $c > c_*$. The pitchfork bifurcation point of c_* leads to the appearance of the periodic solutions of the stationary equation (3.68) with $b \neq 0$, for which $Q(\phi_c)$ is a decreasing function of c. A generalization of Lemma 3.7 holds for such solutions to guarantee their spectral instability for $c > c_*$. All periodic solutions for $c > c_*$ are spectrally unstable.

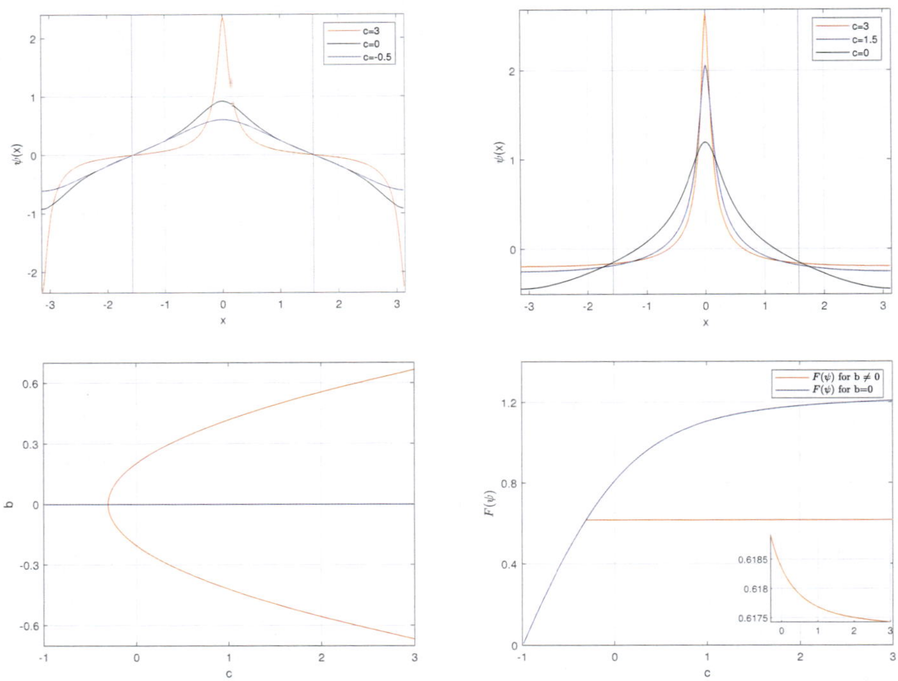

FIGURE 3.9. The same as Figure 3.8 but for $\alpha = 1$.

3.5. Spectral stability and instability

3.5.1. Spectral stability of constrained minimizers of energy. In the setting of Section 3.3, the stationary state with the profile u_ω is a saddle point of the augmented energy $\Lambda_\omega(u)$ and a constrained minimizer of the energy $H(u)$. In Theorem 3.3 and Corollary 3.1 we prove the orbital stability of nondegenerate constrained minimizers of the energy. In this section we will show that under the same conditions, the stationary state with the profile u_ω is spectrally and linearly stable in the linearized Hamiltonian system, according to Definitions 2.6 and 2.7.

By expanding the augmented energy Λ_ω about the perturbation $w = u - u_\omega$ in the Hamiltonian system (3.1) and neglecting higher order terms, one obtains the linearized Hamiltonian system at the stationary state with the profile u_ω in the form

$$\begin{cases} \dfrac{dw}{dt} = J\mathcal{L}_\omega w, \quad t > 0, \\ w(0) = w_0, \end{cases} \tag{3.74}$$

where $\mathcal{L}_\omega := \nabla^2 \Lambda_\omega(u_\omega) : X \to X^*$ and $w_0 \in X$. If the Hamiltonian system (3.1) is locally well-posed in the Banach space X as in Assumption 3.1, then the linearized Hamiltonian system (3.74) is globally well-posed in X with the unique global solution $w \in C^0(\mathbb{R}, X)$ for every $w_0 \in X$.

The following theorem ensures **linear orbital stability** of the stationary state with profile u_ω in the linearized Hamiltonian system (3.74) under the same assumptions as in Theorem 3.3. The linear orbital stability differs from the linear stability in Definition 2.7 only insofar as the solution $w(t,\cdot) \in X$ is allowed to have a growing component in the direction of the symmetry mode $u_{\text{sym}} := T' u_\omega$ induced by the symmetry operator $T(\theta)$. This result is similar to Theorem 1.11 for the periodic orbits in finite-dimensional systems.

THEOREM 3.9. *Let Assumptions* 3.1 *and* 3.2 *be satisfied and let the admissible space of perturbations* $\mathcal{B}_\omega$ *be given by* (3.27). *Assume that the spectrum of* $\mathcal{L}_\omega$ *in* $(-\infty, \delta)$ *for some* $\delta > 0$ *consists of finitely many eigenvalues and* $\operatorname{Ker} \mathcal{L}_\omega = \operatorname{Span}\{u_{\text{sym}}\}$, *where* $u_{\text{sym}} = J\nabla Q_1(u_\omega)$ *and assume that* $J u_{\text{sym}} \in X$. *If the conditions* (3.29) *are satisfied, then there exists a positive constant* $C > 0$ *such that for every* $w_0 \in \mathcal{B}_\omega \subset X$, *the unique solution* $w \in C^0(\mathbb{R}, \mathcal{B}_\omega)$ *of the linearized Hamiltonian system* (3.74) *satisfies*

$$\|w(t,\cdot) - a(t) u_{\text{sym}}\|_X \leq C \|w_0\|_X, \quad t > 0, \tag{3.75}$$

where the unique function $a(t)$ *satisfies* $|a'(t)| \leq C\|w_0\|_X$ *for all* $t > 0$.

PROOF. In view of the conditions (3.29), Lemma 3.2 implies that u_ω is a strict local minimizer of $H(u)$ satisfying the coercivity bound (3.28) for $w \in \mathcal{B}_\omega|_{\{u_{\text{sym}}\}^\perp}$. We decompose the unique global solution $v \in C^0(\mathbb{R}, X)$ of the linearized system (3.74) in the form

$$w(t,\cdot) = a(t) u_{\text{sym}} + p(t,\cdot), \quad t > 0, \tag{3.76}$$

where the decomposition is unique if $p(t,\cdot)$ satisfies the orthogonality condition

$$\langle u_{\text{sym}}, p(t,\cdot)\rangle = 0, \quad t > 0.$$

Recall that $\langle \nabla Q_k(u_\omega), J\nabla Q_j(u_\omega)\rangle = 0$ for $1 \leq k, j \leq N$, because $Q_1(u)$ is related to the symmetry mode $u_{\text{sym}} = J\nabla Q_1(u_\omega)$ and $\{Q_k(u)\}_{2\leq k\leq N}$ are Casimir functionals satisfying $J\nabla Q_k(u) = 0$ due to the assumptions. Since $\langle \nabla Q_k(u_\omega), u_{\text{sym}}\rangle = 0$, we have $u_{\text{sym}} \in \mathcal{B}_\omega$. Hence $w(t,\cdot) \in \mathcal{B}_\omega$ if and only if $p(t,\cdot) \in \mathcal{B}_\omega|_{\{u_{\text{sym}}\}^\perp}$. Moreover,

$$\frac{d}{dt}\langle \nabla Q_k(u_\omega), w(t,\cdot)\rangle = \langle \nabla Q_k(u_\omega), J\mathcal{L}_\omega w(t,\cdot)\rangle = 0, \quad 1 \leq k \leq N,$$

because $\mathcal{L}_\omega J\nabla Q_1(u_\omega) = 0$ and $J\nabla Q_k(u_\omega) = 0$ for $2 \leq k \leq N$. Therefore, if $w_0 \in \mathcal{B}_\omega$, then $w(t,\cdot) \in \mathcal{B}_\omega$ so that the unique solution of the linearized system (3.74) belongs to $w \in C^0(\mathbb{R}, \mathcal{B}_\omega)$ if $w_0 \in \mathcal{B}_\omega$. It follows from (3.74) that

$$\frac{1}{2}\frac{d}{dt}\langle \mathcal{L}_\omega w, w\rangle = \langle \mathcal{L}_\omega w, J\mathcal{L}_\omega w\rangle = 0,$$

and hence $\langle \mathcal{L}_\omega w(t,\cdot), w(t,\cdot)\rangle$ is constant in time for $w \in C^0(\mathbb{R}, \mathcal{B}_\omega)$. Since the quadratic form is bounded in X, the bound (3.75) follows from the coercivity bound (3.28) as follows:

$$\begin{aligned}
C_\omega^- \|p(t,\cdot)\|_X^2 &\leq \langle \mathcal{L}_\omega p(t,\cdot), p(t,\cdot)\rangle \\
&= \langle \mathcal{L}_\omega w(t,\cdot), w(t,\cdot)\rangle \\
&= \langle \mathcal{L}_\omega w_0, w_0\rangle \\
&\leq C_\omega^+ \|w_0\|_X^2,
\end{aligned}$$

for some positive constants $C_\omega^- \le C_\omega^+$, which implies (3.75). In order to control the evolution of $a(t)$, we substitute the decomposition (3.76) into the linearized Hamiltonian system (3.74) and obtain

$$a'(t)u_{\rm sym} + p'(t,\cdot) = J\mathcal{L}_\omega p(t,\cdot)$$

since $u_{\rm sym} \in \operatorname{Ker}\mathcal{L}_\omega$. Taking the inner product with $u_{\rm sym}$ and observing that $\langle p'(t,\cdot), u_{\rm sym}\rangle = 0$ due to the orthogonality condition on $p(t,\cdot)$, we obtain that

$$a'(t)\,\|u_{\rm sym}\|^2 = \langle J\mathcal{L}_\omega p(t,\cdot), u_{\rm sym}\rangle = -\langle p(t,\cdot), \mathcal{L}_\omega J u_{\rm sym}\rangle.$$

Using bound (3.75), together with the Cauchy–Schwarz inequality and the assumption that $Ju_{\rm sym} \in X$, we obtain the bound $|a'(t)| \le C\|w_0\|_X$ for a positive constant $C > 0$. □

REMARK 3.35. The statement and proof of Theorem 3.9 can be easily generalized to a more general case of Z-many symmetry operators after straightforward modifications.

REMARK 3.36. The linear orbital stability of the stationary state with the profile u_ω in Theorem 3.9 implies its spectral stability according to Definition 2.6. Indeed, since eigenvalues of $J\mathcal{L}_\omega$ are symmetric relative to $i\mathbb{R}$ (Lemma 1.1), if there is an eigenvalue of $J\mathcal{L}_\omega$ outside $i\mathbb{R}$, then there is an unstable eigenvalue $\lambda_0 \in \mathbb{C}$ with $\mathrm{Re}(\lambda_0) > 0$ and corresponding eigenvector $w_0 \in \mathcal{D}$ in the spectral problem Definition 2.6. Moreover, since $\lambda_0 \neq 0$, we have

$$\lambda_0\langle\nabla Q_k(u_\omega), w_0\rangle = \langle\nabla Q_k(u_\omega), J\mathcal{L}_\omega w_0\rangle = -\langle\mathcal{L}_\omega J\nabla Q_k(u_\omega), w_0\rangle = 0, \quad 1 \le k \le N,$$

so that the eigenvector $w_0 \in \mathcal{D}$ belongs to $\mathcal{B}_\omega$. If the eigenvalue λ_0 with $\mathrm{Re}(\lambda_0) > 0$ exists, then the exponentially growing solution $w(t,\cdot) = e^{\lambda_0 t}w_0 \in C^0(\mathbb{R}, \mathcal{B}_\omega)$ contradicts Theorem 3.9. Hence the linear orbital stability implies spectral stability and, in the opposite direction, the spectral instability implies the linear orbital instability.

REMARK 3.37. If the stationary state with the profile u_ω is a saddle point of the energy $H(u)$ under the constraints in $\mathcal{B}_\omega$ defined by (3.27), then the linear constrained operator $\mathcal{L}_\omega|_{\mathcal{B}_\omega}$ is no longer sign-definite. Saddle points of energy are orbitally unstable by Theorem 2.4 if they are spectrally unstable and if the nonlinear vector field enjoys the estimate (2.41). Therefore, to conclude on the orbital instability of the saddle points of energy, one needs to obtain their spectral instability.

3.5.2. Spectral instability of saddle points of energy when J is invertible. The existence of unstable eigenvalues in the spectral problem can be studied by generalizing of the Hamiltonian–Krein Theorem (Theorem 1.7) to the infinite-dimensional Hamiltonian system (3.1). We define the Krein matrix $K(\lambda_0)$ of an eigenvalue λ_0 of $J\mathcal{L}_\omega$ in the same way as in Definition 1.7. Krein matrices satisfy the same properties as in Lemma 1.2.

When J is invertible, as in Example 3.1, the degeneracies of the kernel of $J\mathcal{L}_\omega$, analyzed in Section 2.5, can be eliminated by using the following assumptions.

ASSUMPTION 3.3. We assume that the spectrum of $\mathcal{L}_\omega$ in $\mathcal{H}$ in $(-\infty, \delta)$ for some $\delta > 0$ consists of finitely many eigenvalues, whereas the eigenvectors for the zero eigenvalue of $\mathcal{L}_\omega$ satisfy the nondegeneracy conditions:

$$\operatorname{Ker}(\mathcal{L}_\omega) = \operatorname{Span}\{J\nabla Q_1(u_\omega), \dots, J\nabla Q_N(u_\omega)\} \tag{3.77}$$

and

$$\langle J\nabla Q_j(u_\omega), \nabla Q_k(u_\omega)\rangle = 0, \quad 1 \le j,k \le N. \tag{3.78}$$

In addition, we assume that $z(\mathcal{L}_\omega|_{\mathcal{B}_\omega}) = z(\mathcal{L}_\omega) = N$, where $\mathcal{B}_\omega \subset X$ is defined by equation (3.27).

The following theorem relates the finite number of negative eigenvalues of $\mathcal{L}_\omega$ to the number of unstable and neutrally stable eigenvalues of $J\mathcal{L}_\omega$.

THEOREM 3.10. *Let Assumptions* 3.2 *and* 3.3 *be satisfied. Assume that the spectrum of* $J\mathcal{L}_\omega$ *in* $\mathcal{H}$ *is purely discrete and denote*

- *the number of real positive eigenvalues of* $J\mathcal{L}_\omega$ *by* N_{real},
- *the (even) number of complex eigenvalues of* $J\mathcal{L}_\omega$ *with positive real part by* $2N_{\text{comp}}$,
- *the number of positive (resp. negative) eigenvalues of the Krein matrix* $K(\lambda_0)$ *for each eigenvalue* $\lambda_0 \in i\mathbb{R}_+$ *of* $J\mathcal{L}_\omega$ *by* p_{λ_0} *(resp.* n_{λ_0}*),*

where all numbers of eigenvalues are accounted for with their algebraic multiplicities. These eigenvalues satisfy the following completeness relation:

$$n(\mathcal{L}_\omega|_{\mathcal{B}_\omega}) = N_{\text{real}} + 2N_{\text{comp}} + 2\sum_{\lambda_0 \in i\mathbb{R}_+} n_{\lambda_0}. \tag{3.79}$$

In addition, the zero eigenvalue of $J\mathcal{L}_\omega$ *has geometric multiplicity* N *and algebraic multiplicity* $2N$.

PROOF. Since J is invertible, the zero eigenvalue of $J\mathcal{L}_\omega$ has geometric multiplicity N due to the condition (3.77). By extending the arguments in Lemma 2.4 to the Assumption 3.3, it follows that the mapping $\omega \mapsto u_\omega$ is smooth. It follows by differentiating $\nabla\Lambda_\omega(u_\omega) = 0$ with respect to ω_k that $\mathcal{L}_\omega\partial_{\omega_k}u_\omega = \nabla Q_k(u_\omega)$, for $1 \le k \le N$. This implies that $\partial_{\omega_k}u_\omega \in \mathcal{D} \subset \mathcal{H}$ by Fredholm's theorem (Theorem B.2) due to the condition (3.78). Moreover, it implies together with condition (3.77) that

$$J\mathcal{L}_\omega J\nabla Q_k(u_\omega) = 0, \quad J\mathcal{L}_\omega\partial_{\omega_k}u_\omega = J\nabla Q_k(u_\omega), \quad 1 \le k \le N.$$

Hence, the zero eigenvalue of $J\mathcal{L}_\omega$ has algebraic multiplicity of at least $2N$. Recalling the limiting matrix $\lim_{\lambda\to 0^-} A(\lambda)$ in Theorem 3.2 with $v_k = \nabla Q_k(u_\omega)$, we find that it does not have zero eigenvalue, so that $z_0 = 0$, since Assumption 3.3 implies that $z(\mathcal{L}_\omega|_{\mathcal{B}_\omega}) = z(\mathcal{L}_\omega) = N$ and $z_\infty = 0$. Hence, for every nonzero $(c_1, c_2, \dots, c_N) \in \mathbb{R}^N$, the linear combination

$$\sum_{k=1}^N c_k\partial_{\omega_k}u_\omega = \sum_{k=1}^N c_k\mathcal{L}_\omega^{-1}\nabla Q_k(u_\omega)$$

does not belong to the Span$\{\nabla Q_1(u_\omega), \cdots, \nabla Q_N(u_\omega)\}$, since otherwise $z_0 \ge 1$. This implies by Fredholm's theorem (Theorem B.2) that there exist no solution $v \in \mathcal{D} \subset \mathcal{H}$ of the linear inhomogeneous equation

$$J\mathcal{L}_\omega w = \sum_{k=1}^N c_k\partial_{\omega_k}u_\omega,$$

for nonzero $(c_1, c_2, \dots, c_N) \in \mathbb{R}^N$. Therefore, the algebraic multiplicity of the zero eigenvalue of $J\mathcal{L}_\omega$ is exactly $2N$.

Let us now consider the nonzero eigenvalues of $J\mathcal{L}_\omega$ with the eigenvector $w \in \mathcal{D}$ of the spectral stability problem $J\mathcal{L}_\omega w = \lambda w$. Since $\lambda \neq 0$, $w \in \mathcal{D}$ is a solution of $J\mathcal{L}_\omega w = \lambda w$ if and only if $w \in \mathcal{D} \cap \mathcal{B}_\omega$. Let $\Pi_0 : \mathcal{H} \mapsto \mathcal{H} \cap \mathcal{B}_\omega$ be the orthogonal projection operator so that

$$\mathcal{L}_\omega \Pi_0 w = \Pi_0^* \mathcal{L}_\omega \Pi_0 w + \sum_{k=1}^{N} c_k(w) \nabla Q_k(u_\omega),$$

where the coefficients $\{c_k(w)\}_{k=1}^N$ of the decomposition depend on w. Since $w \in \mathcal{B}_\omega$, $J\nabla Q_k(u_\omega) \in \mathrm{Ker}(\mathcal{L}_\omega)$ and $J\nabla Q_k(u_\omega) \in \mathcal{B}_\omega$, $1 \leq k \leq N$ due to conditions (3.77) and (3.78), the spectral problem $J\mathcal{L}_\omega \Pi_0 w = \lambda \Pi_0 w$ for $w \in \mathcal{D} \cap \mathcal{B}_\omega$ and $\lambda \neq 0$ can be rewritten in the form

$$(\Pi_0^* J \Pi_0)(\Pi_0^* \mathcal{L}_\omega \Pi_0) \tilde{w} = \lambda \tilde{w},$$

where $\tilde{w} \in \mathcal{D} \cap \mathcal{B}_\omega$ is defined by

$$\tilde{w} := \Pi_0 w - \lambda^{-1} \sum_{k=1}^{N} c_k(w) J \nabla Q(u_\omega).$$

The constrained operators $\mathcal{L}_\omega|_{\mathcal{B}_\omega} := \Pi_0^* \mathcal{L}_\omega \Pi_0$ and $J|_{\mathcal{B}_\omega} := \Pi_0^* J \Pi_0$ are self-adjoint and skew-adjoint operators in $\mathcal{H}$, respectively. By applying the proof of Theorem 1.7 to the negative $\mathcal{L}_\omega|_{\mathcal{B}_\omega}$-invariant subspace of $\mathcal{B}_\omega$ of dimension $n(\mathcal{L}_\omega|_{\mathcal{B}_\omega})$, we obtain the completeness relation (3.79). □

REMARK 3.38. The constrained operator $\mathcal{L}_\omega|_{\mathcal{B}_\omega} := \Pi_0^* \mathcal{L}_\omega \Pi_0$ has zero eigenvalue of algebraic and geometric multiplicity N due to the conditions (3.77) and (3.78), and the zero eigenvalue is isolated from the rest of the spectrum of $\mathcal{L}_\omega|_{\mathcal{B}_\omega}$. Strictly speaking, the semi-simple zero eigenvalue was not allowed by the assumptions of Theorem 1.7. However, it can be included without any changes in the completeness relation (3.79). Indeed, since J is invertible, one can define **symplectic projections** of $\mathcal{H}$ to the orthogonal complement of the $2N$-generalized kernel of $J\mathcal{L}_\omega$ by augmenting $\mathcal{B}_\omega$ to

$$\widetilde{\mathcal{B}}_\omega := \left\{ w \in X : \quad \langle w, \nabla Q_k(u_\omega) \rangle = \langle w, J^{-1} \partial_{\omega_k} u_\omega \rangle = 0, \quad 1 \leq k \leq N \right\}.$$

If $w \in \mathcal{D} \cap \mathcal{B}_\omega$ is a solution of $J\mathcal{L}_\omega w = \lambda w$ with $\lambda \neq 0$, then

$$\begin{aligned} \langle w, J^{-1} \partial_{\omega_k} u_\omega \rangle &= \lambda^{-1} \langle J\mathcal{L}_\omega w, J^{-1} \partial_{\omega_k} u_\omega \rangle \\ &= -\lambda^{-1} \langle w, \mathcal{L}_\omega \partial_{\omega_k} u_\omega \rangle \\ &= -\lambda^{-1} \langle w, \nabla Q_k(u_\omega) \rangle = 0, \end{aligned}$$

so that $w \in \mathcal{D} \cap \widetilde{\mathcal{B}}_\omega$. Hence the number of nonzero eigenvalues of $J\mathcal{L}_\omega$ in Theorem 3.10 is not affected. Furthermore, it follows from Assumption 3.3 as in the proof of Theorem 3.10 that

$$\sum_{k=1}^{N} c_k J^{-1} \partial_{\omega_k} u_\omega \notin \mathrm{Ker}(\mathcal{L}_\omega)$$

for every nonzero $(c_1, \cdots, c_N) \in \mathbb{R}^N$. This implies $z(\mathcal{L}_\omega|_{\widetilde{\mathcal{B}}_\omega}) = 0$ so that the proof of Theorem 1.7 can be applied verbatim for $\mathcal{L}_\omega|_{\widetilde{\mathcal{B}}_\omega}$. It results in the same relation (3.79) if $n(\mathcal{L}_\omega|_{\widetilde{\mathcal{B}}_\omega}) = n(\mathcal{L}_\omega|_{\mathcal{B}_\omega})$. To verify the latter relation, we can consider eigenvalues of the $2N \times 2N$ matrix associated with the $2N$ constraints in $\widetilde{\mathcal{B}}_\omega$ in Theorem 3.2:

$$\widetilde{A}(\lambda) := \begin{bmatrix} A(\lambda) & B(\lambda) \\ B(\lambda) & C(\lambda) \end{bmatrix},$$

where elements of the $N \times N$ matrices $A(\lambda)$, $B(\lambda)$, and $C(\lambda)$ are given by

$$\begin{aligned} A_{jk}(\lambda) &= \langle (\mathcal{L}_\omega - \lambda I)^{-1} \nabla Q_j(u_\omega), \nabla Q_k(u_\omega) \rangle, \\ B_{jk}(\lambda) &= \langle (\mathcal{L}_\omega - \lambda I)^{-1} \nabla Q_j(u_\omega), J^{-1} \partial_{\omega_k} u_\omega \rangle, \\ C_{jk}(\lambda) &= \langle (\mathcal{L}_\omega - \lambda I)^{-1} J^{-1} \partial_{\omega_j} u_\omega, J^{-1} \partial_{\omega_k} u_\omega \rangle. \end{aligned}$$

for $1 \leq j, k \leq N$. The matrix $A(\lambda)$ is the same as the $N \times N$ matrix associated with the N constraints in $\mathcal{B}_\omega$. It has a finite limit as $\lambda \to 0^-$ with $z_0 = 0$ and n_0 negative eigenvalues. The limit of $B(\lambda)$ as $\lambda \to 0^-$ also exists because $\mathcal{L}_\omega^{-1} \nabla Q_j(u_\omega) = \partial_{\omega_j} u_\omega$. However, all N eigenvalues of $C(\lambda)$ diverge as $\lambda \to 0^-$ because $z(\mathcal{L}_\omega|_{\widetilde{\mathcal{B}}_\omega}) = 0$ and $z_0 = 0$ in the second counting formula (3.31) implies $z_\infty = N$. The remaining N eigenvalues of the matrix $\widetilde{A}(\lambda)$ converge as $\lambda \to 0^-$ to the N eigenvalues of $\lim_{\lambda \to 0^-} A(\lambda)$. Hence, the first counting formula (3.31) yields $n(\mathcal{L}_\omega|_{\widetilde{\mathcal{B}}_\omega}) = n(\mathcal{L}_\omega) - n_0 = n(\mathcal{L}_\omega|_{\mathcal{B}_\omega})$.

EXAMPLE 3.15. Consider the standing waves $u(t, x) = \phi_c(x) e^{ict}$ of the focusing nonlinear Schrödinger (NLS) equation

$$iu_t + u_{xx} + 2|u|^2 u = 0. \tag{3.80}$$

If ϕ_c is real-valued, it solves the same differential equation (3.15) as the one arising for traveling waves of the focusing modified KdV equation (3.45). Consequently, the two families of periodic solutions are given by the dnoidal waves (3.46) and by the cnoidal waves (3.72) as in Examples 3.8 and 3.12. Since ϕ_c is real-valued, the linearized operator for the focusing NLS equation (3.80) at ϕ_c is diagonalized into two Schrödinger operators $\mathcal{L}_\pm : H^2_{\text{per}} \subset L^2_{\text{per}} \to L^2_{\text{per}}$ given by

$$\mathcal{L}_- = -\partial_x^2 + c - 2\phi_c^2, \qquad \mathcal{L}_+ = -\partial_x^2 + c - 6\phi_c^2.$$

We recall that $\mathcal{L}_- \phi_c = 0$ and $\mathcal{L}_+ \partial_x \phi_c = 0$ by the rotational and translational symmetries of the NLS equation (3.80) as in Example 2.4. The corresponding conserved quantities of mass Q and momentum P introduce two constraints on the complex-valued perturbation w to the real-valued profile ϕ_c:

$$\langle \phi_c, \text{Re}(w) \rangle = 0 \quad \text{and} \quad \langle \partial_x \phi_c, \text{Im}(w) \rangle = 0.$$

In order to compute the left-hand side of the completeness relation (3.79), we need to compute $n(\mathcal{L}_+|_{\{\phi_c\}^\perp})$ and $n(\mathcal{L}_-|_{\{\partial_x \phi_c\}^\perp})$ as well as to check Assumption 3.3.

- For the dnoidal wave (3.46), it follows from positivity of ϕ_c that $n(\mathcal{L}_-) = 0$ and $z(\mathcal{L}_-) = 1$. Hence we have $n(\mathcal{L}_-|_{\{\partial_x \phi_c\}^\perp}) = 0$ and $z(\mathcal{L}_-|_{\{\partial_x \phi_c\}^\perp}) = 1$. Example 3.8 suggests that $n(\mathcal{L}_+) = 1$ and $z(\mathcal{L}_+) = 1$ with $n(\mathcal{L}_+|_{\{\phi_c\}^\perp}) = 0$ and $z(\mathcal{L}_+|_{\{\phi_c\}^\perp}) = 1$. Assumption 3.3 is satisfied and the completeness relation (3.79) of Theorem 3.10 yields $N_{\text{real}} = N_{\text{comp}} = 0$ so that the dnoidal wave is spectrally stable in the NLS equation (3.80). By Theorem 3.3, the dnoidal wave is also orbitally stable in the energy space H^1_{per}.

- For the cnoidal wave (3.72), Example 3.12 suggests that $n(\mathcal{L}_+) = 2$ and $z(\mathcal{L}_+) = 1$ with $n(\mathcal{L}_+|_{\{\phi_c\}^\perp}) = 1$ and $z(\mathcal{L}_+|_{\{\phi_c\}^\perp}) = 1$. We use the scaling transformation
$$z := \frac{2}{\pi}K(k)x, \quad \mathcal{L}_- = \frac{4}{\pi^2}K(k)^2\hat{\mathcal{L}}_-(k),$$
to obtain a normalized Schrödinger operator
$$\hat{\mathcal{L}}_-(k) := -\partial_z^2 + 2k^2 - 1 - 2k^2 \operatorname{cn}(z;k)^2.$$
The first three eigenvalues and eigenfunctions of $\hat{\mathcal{L}}_-(k)$ in the space of $4K(k)$-periodic functions are given explicitly by
$$\begin{aligned} &\lambda_0 = -(1-k^2), \quad \varphi_0(z) = \operatorname{dn}(z;k), \\ &\lambda_1 = 0, \quad \varphi_1(z) = \operatorname{cn}(z;k), \\ &\lambda_2 = k^2, \quad \varphi_2(z) = \operatorname{sn}(z;k). \end{aligned}$$
Hence, we have $n(\mathcal{L}_-) = 1$ and $z(\mathcal{L}_-) = 1$ and since φ_0 and φ_1 are even and $\partial_x\phi_c$ is odd, we have $n(\mathcal{L}_-|_{\{\partial_x\phi_c\}^\perp}) = 1$ and $z(\mathcal{L}_-|_{\{\partial_x\phi_c\}^\perp}) = 1$. Assumption 3.3 is satisfied and the completeness relation (3.79) of Theorem 3.10 yields
$$N_{\text{real}} + N_{\text{comp}} + 2n_{\lambda_0} = 2.$$
Hence, the spectral stability of the cnoidal wave (3.72) is inconclusive in the NLS equation (3.80), since the linearized system could have either two purely imaginary eigenvalues (stable), two real eigenvalues (unstable) or one complex quadruplet (unstable).

REMARK 3.39. In order to derive a more definite spectral stability conclusion for the cnoidal wave (3.72) in Example 3.15, perturbations to the periodic wave can be restricted to the set of functions

$$Y_c := \left\{ w \in H^1_{\text{per}} : \quad w(x + T/2) = -w(x) \right\}. \tag{3.81}$$

This constraint Y_c eliminates the lowest negative eigenvalue λ_0 both for $\mathcal{L}_+$ and $\mathcal{L}_-$ while preserving the next two eigenvalues λ_1 and λ_2 of both $\mathcal{L}_+$ and $\mathcal{L}_-$. Therefore, $n(\mathcal{L}_+|_{Y_c}) = 1$, $z(\mathcal{L}_+|_{Y_c}) = 1$ and $n(\mathcal{L}_-|_{Y_c}) = 0$, $z(\mathcal{L}_-|_{Y_c}) = 1$. At the same time, we have $\phi_c \in Y_c$ so that it follows by the same computation as in Example 3.12 that $n(\mathcal{L}_+|_{Y_c\cap\{\phi_c\}^\perp}) = 0$. Since $n(\mathcal{L}_-|_{Y_c}) = 0$, we also have $n(\mathcal{L}_-|_{Y_c\cap\{\partial_x\phi_c\}^\perp}) = 0$. By the completeness relation (3.79) of Theorem 3.10, we have $N_{\text{real}} = N_{\text{comp}} = 0$, so that the cnoidal wave is spectrally stable in the NLS equation (3.80) under perturbations in Y_c. By Theorem 3.3, the cnoidal wave is orbitally stable in $Y_c \subset H^1_{\text{per}}$.

EXAMPLE 3.16. Consider the standing waves $u(t,x) = \phi_c(x)e^{-ict}$ of the defocusing NLS equation

$$iu_t + u_{xx} - 2|u|^2u = 0. \tag{3.82}$$

If ϕ_c is real-valued, it solves the same differential equation (3.19) as the one arising for traveling waves of the defocusing modified KdV equation (3.47). Consequently, the family of periodic solutions is given by the snoidal waves (3.48) as in Example 3.9. Since ϕ_c is real-valued, the linearized operator for the defocusing

NLS equation (3.82) at ϕ_c is again diagonalized into two Schrödinger operators $\mathcal{L}_\pm : H^2_{\rm per} \subset L^2_{\rm per} \to L^2_{\rm per}$ given by

$$\mathcal{L}_- = -\partial_x^2 - c + 2\phi_c^2, \qquad \mathcal{L}_+ = -\partial_x^2 - c + 6\phi_c^2.$$

For the snoidal wave (3.48), Example 3.9 suggests that $n(\mathcal{L}_+) = 1$, $z(\mathcal{L}_+) = 1$ with $n(\mathcal{L}_+|_{\{\phi_c\}^\perp}) = 1$, $z(\mathcal{L}_+|_{\{\phi_c\}^\perp}) = 1$. We use the scaling transformation

$$z := \frac{2}{\pi} K(k) x, \quad \mathcal{L}_- = \frac{4}{\pi^2} K(k)^2 \hat{\mathcal{L}}_-(k),$$

to obtain a normalized Schrödinger operator

$$\hat{\mathcal{L}}_-(k) := -\partial_z^2 + k^2 - 1 - 2k^2 \,\mathrm{cn}(z;k)^2.$$

The first three eigenvalues and eigenfunctions of $\hat{\mathcal{L}}_-(k)$ in the space of $4K(k)$-periodic functions are given explicitly by

$$\begin{aligned} \lambda_0 &= -1, \quad \varphi_0(z) = \mathrm{dn}(z;k), \\ \lambda_1 &= -k^2, \quad \varphi_1(z) = \mathrm{cn}(z;k), \\ \lambda_2 &= 0, \quad \varphi_2(z) = \mathrm{sn}(z;k). \end{aligned}$$

Hence, we have $n(\mathcal{L}_-) = 2$ and $z(\mathcal{L}_-) = 1$. We show next that $n(\mathcal{L}_-|_{\{\partial_x \phi_c\}^\perp}) = 1$ and $z(\mathcal{L}_-|_{\{\partial_x \phi_c\}^\perp}) = 1$ so that the completeness relation (3.79) of Theorem 3.10 yields

$$N_{\rm real} + N_{\rm comp} + 2n_{\lambda_0} = 2,$$

and the spectral stability for the snoidal wave is inconclusive in the defocusing NLS equation (3.82). The assertion $n(\mathcal{L}_-|_{\{\partial_x \phi_c\}^\perp}) = 1$ follows from $\langle \mathcal{L}_-^{-1} \partial_x \phi_c, \partial_x \phi_c \rangle < 0$ or, in the rescaled version, it follows from $\langle \hat{\mathcal{L}}_-(k)^{-1} v_0, v_0 \rangle < 0$, where $v_0(z) := \mathrm{sn}(z;k)\,\mathrm{dn}(z;k)$. We verify by direct computations that

$$\begin{aligned} \hat{\mathcal{L}}_-(k)\,\mathrm{cn}(z;k)\,\mathrm{dn}(z;k) &= -4k^2 \,\mathrm{cn}(z;k)\,\mathrm{dn}(z;k)\,\mathrm{sn}^2(z;k), \\ \hat{\mathcal{L}}_-(k)\,\mathrm{sn}(z;k) E(z;k) &= -2\,\mathrm{cn}(z;k)\,\mathrm{dn}(z;k)\left[1 - 2k^2 \,\mathrm{sn}^2(z;k)\right], \\ \hat{\mathcal{L}}_-(k) z\,\mathrm{sn}(z;k) &= -2\,\mathrm{cn}(z;k)\,\mathrm{dn}(z;k). \end{aligned}$$

Therefore, the function

$$V_0(z) := \mathrm{cn}(z;k)\,\mathrm{dn}(z;k) + \mathrm{sn}(z;k)\left[E(z;k) - \frac{E(k)}{K(k)} z\right], \tag{3.83}$$

is periodic with period $T = 4K(k)$ and satisfies the inhomogeneous equation

$$\hat{\mathcal{L}}_-(k) V_0 = -2\left[1 - \frac{E(k)}{K(k)}\right] \mathrm{cn}(z;k)\,\mathrm{dn}(z;k), \tag{3.84}$$

where $E(k) < K(k)$ for every $k \in (0,1)$. Hence, we compute

$$\begin{aligned} \langle \hat{\mathcal{L}}_-(k)^{-1} v_0, v_0 \rangle &= -\frac{K(k)}{2K(k) - 2E(k)} \langle V_0, v_0 \rangle \\ &= -\frac{1}{k^2 K(k) - k^2 E(k)} \left[(k^2 - 1) K(k)^2 + 2E(k)K(k) - E(k)^2\right], \end{aligned}$$

where the right-hand side is negative because

$$f(k) := (k^2 - 1) K(k)^2 + 2E(k)K(k) - E(k)^2$$

satisfies $f(0) = 0$ and $f'(k) = \frac{2kE(k)^2}{1-k^2)} > 0$ so that $f(k) > 0$ for every $k \in (0,1)$.

REMARK 3.40. When perturbations to the periodic wave are restricted to the set of functions Y_c given by (3.81), the constraint removes the lowest negative eigenvalue both for $\mathcal{L}_+$ and $\mathcal{L}_-$ so that $n(\mathcal{L}_+|_{Y_c}) = 0$, $z(\mathcal{L}_+|_{Y_c}) = 1$ and $n(\mathcal{L}_-|_{Y_c}) = 1$, $z(\mathcal{L}_-|_{Y_c}) = 1$. Since $\partial_x\phi_c \in Y_c$, it follows from the computations in Example 3.16 that $n(\mathcal{L}_-|_{Y_c\cap\{\partial_x\phi_c\}^\perp}) = 0$. Since $n(\mathcal{L}_+|_{Y_c}) = 0$, we also have $n(\mathcal{L}_+|_{Y_c\cap\{\phi_c\}^\perp}) = 0$. By the completeness relation (3.79) of Theorem 3.10, we have $N_{\text{real}} = N_{\text{comp}} = 0$, so that the snoidal wave is spectrally stable in the NLS equation (3.82) under the perturbations in Y_c. By Theorem 3.3, the snoidal wave is orbitally stable in $Y_c \subset H^1_{\text{per}}$.

REMARK 3.41. Orbital stability of dnoidal waves of the NLS equation in H^1_{per} was shown in [**15**]. Orbital stability of cnoidal and snoidal waves of the NLS equation with respect to periodic perturbations in the set $Y_c \subset H^1_{\text{per}}$ given by equation (3.81) was proven in [**96**].

3.5.3. Spectral instability of saddle points of energy when J is not invertible. Here we are concerned with applying the theory to the gKdV equation (Example 3.2), where $J = \partial_x$ and the admissible space $\mathcal{B}_\omega$ for perturbations is determined from the conservation of momentum Q and mass M, a Casimir functional satisfying $\nabla M(u) = 1$ and giving the zero mean constraint $\langle 1, w\rangle = 0$ on the perturbation w to the periodic wave with the profile u_ω. Hence, we can simplify the nondegeneracy assumptions for this particular case. Using these simplifications, the following theorem provides a count of unstable and neutrally stable eigenvalues of $\partial_x\mathcal{L}_\omega$ in terms of the finite number of negative eigenvalues of $\mathcal{L}_\omega$.

THEOREM 3.11. *In addition to Assumption 3.2 with $J = \partial_x$, $N = 1$, and $Q_1 \equiv Q$, assume that the spectrum of $\mathcal{L}_\omega$ in $\mathcal{H}$ in $(-\infty, \delta)$ for some $\delta > 0$ consists of finitely many eigenvalues, and that the nondegeneracy condition*

$$\operatorname{Ker}(\mathcal{L}_\omega) = \operatorname{Span}\{\partial_x\nabla Q(u_\omega)\}$$

holds. Assume that $z(\mathcal{L}_\omega|_{\{1,\nabla Q(u_\omega)\}^\perp}) = z(\mathcal{L}_\omega) = 1$. Then the eigenvalues of $\partial_x\mathcal{L}_\omega$ satisfy the following completeness relation

$$n(\mathcal{L}_\omega|_{\{1,\nabla Q(u_\omega)\}^\perp}) = N_{\text{real}} + 2N_{\text{comp}} + 2\sum_{\lambda_0\in i\mathbb{R}_+} n_{\lambda_0}, \tag{3.85}$$

where N_{real}, N_{comp}, and n_{λ_0} are defined as in Theorem 3.10. In addition, the zero eigenvalue of $\partial_x\mathcal{L}_\omega$ has geometric multiplicity 1 and algebraic multiplicity 2.

PROOF. Since $\langle\partial_x\nabla Q(u_\omega), \nabla Q(u_\omega)\rangle = 0$, there exists a unique $\partial_\omega u_\omega \in \mathcal{D} \subset \mathcal{H}$ such that $\mathcal{L}_\omega\partial_\omega u_\omega = \nabla Q(u_\omega)$. The double zero eigenvalue of $\partial_x\mathcal{L}_\omega$ is related to the Jordan chain:

$$\partial_x\mathcal{L}_\omega\partial_x\nabla Q(u_\omega) = 0, \quad \partial_x\mathcal{L}_\omega\partial_\omega u_\omega = \partial_x\nabla Q(u_\omega).$$

Since $z(\mathcal{L}_\omega|_{\{1,\nabla Q(u_\omega)\}^\perp}) = z(\mathcal{L}_\omega)$, there exists no solution $w \in \mathcal{D}$ of $\partial_x\mathcal{L}_\omega w = \partial_\omega u_\omega$ so that the zero eigenvalue of $\partial_x\mathcal{L}_\omega$ is algebraically double.

For nonzero eigenvalues λ of $\partial_x\mathcal{L}_\omega$ with the eigenvectors $w \in \mathcal{D}$, we realize that if $\lambda \neq 0$, then $w \perp \operatorname{Span}\{1, \nabla Q(u_\omega)\}$. Let $\Pi_0 : \mathcal{H} \mapsto \mathcal{H}|_{\{1,\nabla Q(u_\omega)\}^\perp}$ be the orthogonal projection operator so that

$$\mathcal{L}\Pi_0 w = \Pi_0^*\mathcal{L}_\omega\Pi_0 w + c_1\nabla Q(u_\omega) + c_2$$

for some coefficients (c_1, c_2) that depend on w. Since $\partial_x 1 = 0$ and $\lambda \neq 0$, we can define $\tilde{w} \in \mathcal{H}|_{\{1,\nabla Q(u_\omega)\}^\perp}$ by

$$\tilde{w} := w - \lambda^{-1} c_1 \partial_x \nabla Q(u_\omega)$$

and rewrite the spectral problem $\partial_x \mathcal{L}_\omega w = \lambda w$ in the form

$$(\Pi_0^* \partial_x \Pi_0)(\Pi_0^* \mathcal{L}_\omega \Pi_0)\tilde{w} = \lambda \tilde{w}.$$

The constrained operator $\mathcal{L}_\omega|_{\{1,\nabla Q(u_\omega)\}^\perp} := \Pi_0^* \mathcal{L}_\omega \Pi_0$ is self-adjoint in $\mathcal{H}$, whereas the constrained operator $\Pi_0^* \partial_x \Pi_0$ is skew-adjoint and invertible with a bounded inverse in $\mathcal{H}$. By applying the proof of Theorem 1.7 for the negative $\mathcal{L}_\omega|_{\{1,\nabla Q(u_\omega)\}^\perp}$-invariant subspace of $\mathcal{H}$, we obtain the completeness relation (3.85). $\square$

REMARK 3.42. Theorem 3.11 can be applied to the focusing and defocusing versions of the modified KdV equation as in Examples 3.8, 3.9, and 3.12 and to the KdV equation as in Example 3.11. For these applications, we have $\nabla Q(u_\omega) = u_\omega$ so that $n(\mathcal{L}_\omega|_{\{1,u_\omega\}^\perp})$ is computed from a 2-by-2 matrix as in Theorems 3.4 and 3.6, where we have used wave speed c instead of ω. Compared to these results which yield orbital stability of periodic waves with the profile u_ω in the case of $n(\mathcal{L}_\omega|_{\{1,u_\omega\}^\perp}) = 0$ and $z(\mathcal{L}_\omega|_{\{1,u_\omega\}^\perp}) = z(\mathcal{L}_\omega) = 1$, Theorem 3.11 can be used in the case $n(\mathcal{L}_\omega|_{\{1,u_\omega\}^\perp}) \geq 1$ under the same nondegeneracy assumption $z(\mathcal{L}_\omega|_{\{1,u_\omega\}^\perp}) = z(\mathcal{L}_\omega) = 1$. If $n(\mathcal{L}_\omega|_{\{1,u_\omega\}^\perp})$ is odd, Theorem 3.11 implies the spectral instability of periodic waves with the profile u_ω. This was used in Remark 3.22 and Example 3.12.

REMARK 3.43. The completeness relation (3.79) in the case of periodic waves was first derived in [**123**] as an extension of similar studies for solitary waves with invertible J [**76, 150, 213**]. For the periodic waves of the gKdV equation (Example 3.2), the completeness relation (3.85) was obtained by using momentum and mass conservation [**79**]. Extensions to the case of solitary waves in the gKdV equation were developed in [**153, 215**]. Further extensions of the spectral stability problem to the problems of the class $\mathcal{L}_\omega v = \lambda \partial_x v$ can be found in [**232**]. A rather general theory which covers many other cases of spectral stability for solitary waves with noninvertible J can be found in [**180**].

REMARK 3.44. Solitary waves of the gKdV equation (Example 3.2) correspond to $b = 0$. They are critical points of the augmented energy $\Lambda_\omega(u) = H(u) - \omega Q(u)$. The orbital stability problem is solved exactly like in Chapter 2 and depends on the monotonicity of the mapping $\omega \mapsto Q(u_\omega)$. Hence, the constraint $\langle 1, v \rangle = 0$ for the perturbation v to the solitary wave with the profile u_ω is irrelevant for the orbital stability problem in the gKdV equation. It is interesting that the same conclusion holds for the spectral stability problem since the completeness relation for eigenvalues of $J\mathcal{L}_\omega$ is computed from $n(\mathcal{L}_\omega|_{\{u_\omega\}^\perp})$, see [**153, 180, 215**].

REMARK 3.45. The numbers $n(\mathcal{L}_\omega|_{\mathcal{B}_\omega})$ in Theorem 3.10 or $n(\mathcal{L}_\omega|_{\{1,\nabla Q(u_\omega)\}^\perp})$ in Theorem 3.11 are typically small if the periodic perturbations are considered in L^2_{per}. However, if the periodic wave is perturbed by **subharmonic perturbations**, for which the period is a multiple of the wave period, then the numbers grow proportionally to the period multiplier. The energy method becomes inconclusive for such saddle points with many negative eigenvalues. In the limit when the multiple period goes to infinity and the perturbation in $L^2(\mathbb{R})$ are spatially decaying, the number of negative eigenvalues diverges since the continuous spectrum of the associated linearized operator is located both in the negative and positive subspace of the Hilbert space. This case will be treated in Section 5.1.

CHAPTER 4

Orbital stability in integrable Hamiltonian systems

Among the infinite-dimensional Hamiltonian systems, there exists a special class of systems with additional special properties. It is customary to refer to them as **integrable Hamiltonian systems**, partially because these systems admit infinitely many conserved quantities, not just the conserved energy, momentum, and mass.

Besides the existence of infinitely many conserved quantities, the integrable Hamiltonian systems arise as compatibility conditions of systems of linear equations with a spectral parameter. The relation between nonlinear and linear systems has rich consequences and enables many methods for analysis of such Hamiltonian systems.

Since the primary subject of this monograph is stability of nonlinear waves, we will only review the relevant properties of integrable Hamiltonian systems. These include the existence of a hierarchy of commuting flows, the Bäcklund–Darboux transformation between different solutions of the same system, and the relation between the spectral parameter of the linear system and the stability spectrum of the nonlinear waves.

There are many obstacles why the standard stability theory in Chapters 2 and 3 is only applicable to some but not to all cases. First, the nonlinear waves must arise as nondegenerate constrained minimizers of energy which is often impossible if the energy functionals are not sign-definite. Second, the function spaces where the conserved quantities are defined must be compatible with the weak solutions to the initial-value problems. Third, the spectral information about the linearized operators is often not available without asymptotic or numerical approximations.

Illustrated here for selected examples of integrable Hamiltonian systems (Section 4.1), a superpower of integrability allows us to overcome many obstacles in the standard theory of stability of nonlinear waves. Saddle points of energy may become constrained minimizers of the higher-order energy defined in Sobolev spaces of higher regularity (Section 4.2). Stability in Sobolev spaces of lower regularity can be shown without variational characterization of nonlinear waves by using the Bäcklund–Darboux transformation (Section 4.3). Spectral stability of nonlinear waves can be determined from the spectral properties of an associated linear system and this will be the subject of Chapter 5.

4.1. Integrable Hamiltonian systems

As before, let us consider an abstract Hamiltonian system in the form

$$\frac{du}{dt} = J\nabla H(u), \tag{4.1}$$

where J is a skew-symmetric operator and $H(u) : X \mapsto \mathbb{R}$ is a Hamiltonian function defined in a Banach space X equipped with the norm $\|\cdot\|_X$. We assume that $\nabla H(u) \in X^*$, where $X \subset \mathcal{H} \subset X^*$ are defined relative to a Hilbert space $\mathcal{H}$ equipped with the inner product $\langle \cdot, \cdot \rangle$ and the induced norm $\|\cdot\|$. In order to declare that the Hamiltonian system (4.1) is integrable, we assume that it exhibits the following two properties.

($\mathcal{P}_1$): Solutions of the Hamiltonian system (4.1) are supposed to be related to solutions of the following system of linear equations

$$\varphi_x = U(\lambda, u)\varphi, \quad \varphi_t = V(\lambda, u)\varphi, \tag{4.2}$$

where $U(\lambda, u)$ and $V(\lambda, u)$ are some multiplicative potentials associated with the spectral parameter $\lambda \in \mathbb{C}$, which is independent of (t, x). The relation between (4.1) and (4.2) is expressed in the following sense: the nonlinear equation (4.1) arises from the compatibility condition $\varphi_{xt} = \varphi_{tx}$ of the linear equations (4.2), which reads explicitly as

$$\frac{\partial}{\partial t}U(\lambda, u) - \frac{\partial}{\partial x}V(\lambda, u) + U(\lambda, u)V(\lambda, u) - V(\lambda, u)U(\lambda, u) = 0. \tag{4.3}$$

An arbitrary Hamiltonian system can not be related to the linear system (4.2) and its compatibility condition (4.3). Only integrable systems admit this property with specific U and V depending on λ and u, such that the compatibility condition eliminates λ and expresses a constraint on u which is equivalent to the integrable system written in Hamiltonian form (4.1).

REMARK 4.1. The linear systems (4.2) with the compatibility condition (4.3) are referred to as **Lax systems** due to the pioneering contribution of P. D. Lax [**170**] in the context of the Korteweg–de Vries (KdV) equation (Example 3.11).

($\mathcal{P}_2$): The Hamiltonian system (4.1) is a member of a hierarchy of Hamiltonian systems

$$\frac{du}{dt_k} = J\nabla H_k(u), \quad k \in \mathbb{N}_0 := \{0, 1, 2, \dots\}, \tag{4.4}$$

associated with the same $J : \mathcal{H} \to \mathcal{H}$ and a sequence of time coordinates $\{t_k\}_{k\in\mathbb{N}_0}$ and Hamiltonians $\{H_k(u)\}_{k\in\mathbb{N}_0}$ with $H_k(u) : X_k \to \mathbb{R}$ defined on a suitable function space $X_k \subset \mathcal{H}$. The Hamiltonians $\{H_k(u)\}_{k\in\mathbb{N}_0}$ are generated recursively by using a recursion operator R such that

$$\nabla H_{k+1}(u) = R\nabla H_k(u), \quad k \in \mathbb{N}_0, \tag{4.5}$$

where the recursion operator R satisfies the relation $JR = R^*J$, where R^* is the adjoint of R relative to $\mathcal{H}$.

REMARK 4.2. In (4.4), it is convenient to enumerate the hierarchy starting with $k = 0$. The Hamiltonian system (4.1) is just one member of the hierarchy of such systems (4.4) for some fixed k.

We shall now elaborate the relation of the hierarchy of Hamiltonians $\{H_k(u)\}_{k\in\mathbb{N}_0}$ to the time evolution of a particular Hamiltonian system (4.1). We start with the following definition.

DEFINITION 4.1. We say that two Hamiltonians H_k, H_m are in **involution** if

$$\langle J\nabla H_k(u), \nabla H_m(u)\rangle = 0, \tag{4.6}$$

for $u \in X_k \cap X_m$. This property implies that the gradient of H_m is orthogonal to the flow associated to the Hamiltonian system corresponding to H_k and vice versa.

REMARK 4.3. If property (4.6) holds, then all Hamiltonians $\{H_m(u)\}_{m\in\mathbb{N}_0}$ are conserved quantities of the integrable Hamiltonian system (4.1) with $H(u) = H_k(u)$ for some k. Since J is skew-adjoint, each $H_m(u)$ is a conserved quantity of the kth Hamiltonian system (4.4):

$$\frac{d}{dt_k}H_m(u) = \langle \frac{du}{dt_k}, \nabla H_m(u)\rangle = \langle J\nabla H_k(u), \nabla H_m(u)\rangle = 0, \quad k, m \in \mathbb{N}_0\,.$$

If a Hamiltonian system with N degrees of freedom has N independent conserved quantities which are in involution, then it is said to be **completely integrable**. By analogy, the infinite dimensional Hamiltonian system (4.1) with infinitely many conserved quantities $\{H_m(u)\}_{m\in\mathbb{N}_0}$ is said to be completely integrable.

The following lemma guarantees that the recursion relation (4.5) is sufficient for the Hamiltonians $\{H_k(u)\}_{k\in\mathbb{N}_0}$ to be in involution. This result was derived in [**238**].

LEMMA 4.1. *Let R be the recursion operator satisfying* (4.5) *and $JR = R^*J$. Then, Hamiltonians in the set $\{H_k(u)\}_{k\in\mathbb{N}_0}$ are in involution.*

PROOF. In order to show the property (4.6), we use the method of mathematical induction. It follows from $J = -J^*$ that $\langle J\nabla H_0(u), \nabla H_0(u)\rangle = 0$. For every $k \in \mathbb{N}_0$ and every $0 \le m \le k$, we obtain from $J = -J^*$ and $JR = R^*J$ that

$$\begin{aligned}\langle J\nabla H_{k+1}(u), \nabla H_m(u)\rangle &= \langle JR\nabla H_k(u), \nabla H_m(u)\rangle = \langle R^*J\nabla H_k(u), \nabla H_m(u)\rangle \\ &= -\langle \nabla H_k(u), JR\nabla H_m(u)\rangle = -\langle \nabla H_k(u), J\nabla H_{m+1}(u)\rangle.\end{aligned}$$

Iterating this further yields for even $(k-m)$ that

$$\begin{aligned}\langle J\nabla H_{k+1}(u), \nabla H_m(u)\rangle &= (-1)^{k+1-m}\langle \nabla H_m(u), J\nabla H_{k+1}(u)\rangle \\ &= -\overline{\langle J\nabla H_{k+1}(u), \nabla H_m(u)\rangle},\end{aligned}$$

so that $\langle J\nabla H_{k+1}(u), \nabla H_m(u)\rangle$ is purely imaginary and hence zero. For odd $(k-m)$, we obtain

$$\langle J\nabla H_{k+1}(u), \nabla H_m(u)\rangle = (-1)^{\frac{k+1-m}{2}}\langle J\nabla H_{\frac{k+1-m}{2}}(u), \nabla H_{\frac{k+1-m}{2}}(u)\rangle = 0,$$

since J is skew-adjoint. By induction, this yields (4.6) for every $k, m \in \mathbb{N}_0$. ☐

REMARK 4.4. Since $J^* = -J$ and $JR = R^*J$, we have

$$(JR)^* = R^*J^* = -R^*J = -JR,$$

so that JR is another skew-symmetric operator, which defines another Hamiltonian structure of the same integrable system (4.1) written as

$$\frac{du}{dt} = \tilde{J}\nabla\tilde{H}(u) \tag{4.7}$$

where $\tilde{J} := JR$ and $\tilde{H}$ is obtained from H by using $\nabla H(u) = R\nabla\tilde{H}(u)$. Solutions of the Hamiltonian system (4.1) are also solutions of the Hamiltonian system (4.7). Consequently, the Hamiltonian system (4.1) is referred to as a **bi-Hamiltonian** system.

Traveling nonlinear waves arise as solutions of a hierarchy of stationary equations called **the Lax–Novikov equations** named after the works of P. D. Lax [**171**] and S. P. Novikov [**210**]. The stationary (Lax–Novikov) equations are formed by a linear combination of the commuting flows of the hierarchy (4.4). We say that the kth order Lax–Novikov equation is given by

$$\nabla H_k(u) + \sum_{j=0}^{k-1} c_j \nabla H_j(u) = 0, \tag{4.8}$$

where $\{c_j\}_{j=0}^{k-1}$ are some real valued coefficients. The traveling nonlinear waves in the main examples considered next arise at the order $k = 2$ with two arbitrary parameters.

Since iterations of the recursion operator R in (4.5) give $\nabla H_k(u) = R^k \nabla H_0(u)$, it follows formally that solutions of the lower-order Lax–Novikov equations satisfy the higher-order Lax–Novikov equations when the linear combinations of the conserved quantities in (4.8) are carefully selected. The following lemma states this property for the particular case of $k = 2$.

LEMMA 4.2. *Let R be the recursion operator satisfying* (4.5). *If u is a solution of the second-order Lax–Novikov equation*

$$\nabla H_2(u) + \gamma_1 \nabla H_1(u) + \gamma_0 \nabla H_0(u) = 0, \tag{4.9}$$

for given γ_1 and γ_0, then it also satisfies for every $k \geq 3$,

$$\nabla H_k(u) + \gamma_1 \nabla H_{k-1}(u) + \gamma_0 \nabla H_{k-2}(u) = 0. \tag{4.10}$$

PROOF. By using $\nabla H_k(u) = R^k \nabla H_0(u)$, we obtain

$$\begin{aligned} \nabla H_k(u) &+ \gamma_1 \nabla H_{k-1}(u) + \gamma_0 \nabla H_{k-2}(u) \\ &= R^{k-2} \left[\nabla H_2(u) + \gamma_1 \nabla H_1(u) + \gamma_0 \nabla H_0(u)\right], \end{aligned}$$

so that (4.9) implies (4.10) for every $k \geq 3$. □

REMARK 4.5. If the recursion operator R is nonlocal, the relation (4.5) and hence Lemma 4.2 are usually satisfied for solitary waves only, for which $u(x) \to 0$ as $|x| \to \infty$. For the periodic waves, additional constants arise in the recursion relation (4.5) and in the linear combination of the higher-order Lax–Novikov equations.

The following examples illustrate properties of integrable Hamiltonian systems and will be used as toy models for explaning the superpower of integrability in the stability theory of nonlinear waves. We will also explain how the linear combinations of the conserved quantities are selected in the higher-order Lax–Novikov equations in order to contain all solutions of the lower-order Lax–Novikov equations (Remark 4.5).

EXAMPLE 4.1. The nonlinear Schrödinger (NLS) equation

$$iu_t + \frac{1}{2} u_{xx} + \sigma |u|^2 u = 0, \tag{4.11}$$

which is used either for $\sigma = +1$ (focusing case) or $\sigma = -1$ (defocusing case), is obtained as the compatibility condition of the Lax system (4.2) with

$$U(\lambda, u) = \begin{bmatrix} \lambda & u \\ -\sigma \bar{u} & -\lambda \end{bmatrix}, \qquad V(\lambda, u) = i \begin{bmatrix} \lambda^2 + \frac{1}{2}\sigma |u|^2 & \frac{1}{2} u_x + \lambda u \\ \frac{1}{2}\sigma \bar{u}_x - \lambda \sigma \bar{u} & -\lambda^2 - \frac{1}{2}\sigma |u|^2 \end{bmatrix}. \tag{4.12}$$

The standard Hamiltonian structure (4.1) of the NLS equation (4.11) is defined in the complex-conjugate variables $(u, \bar{u})$ by using

$$J = \begin{bmatrix} 0 & -i \\ i & 0 \end{bmatrix}, \qquad H(u, \bar{u}) = \frac{1}{2} \int_{\mathbb{R}} \left(|u_x|^2 - \sigma |u|^4 \right) dx,$$

so that

$$\frac{d}{dt} \begin{bmatrix} u \\ \bar{u} \end{bmatrix} = \begin{bmatrix} 0 & -i \\ i & 0 \end{bmatrix} \begin{bmatrix} \nabla_u H(u, \bar{u}) \\ \nabla_{\bar{u}} H(u, \bar{u}) \end{bmatrix} = \begin{bmatrix} \frac{i}{2} u_{xx} + i\sigma |u|^2 u \\ -\frac{i}{2} \bar{u}_{xx} - i\sigma |u|^2 \bar{u} \end{bmatrix}.$$

The Hilbert space $\mathcal{H} := L^2(\mathbb{R}, \mathbb{C})$ used in Lemma 4.1 is associated with the inner product

$$\langle f, g \rangle := \int_{\mathbb{R}} (f\bar{g} + \bar{f}g) dx$$

and the recursion operator R is defined by

$$R = i \begin{bmatrix} \partial_x + 2\sigma \bar{u} \partial_x^{-1} u & -2\sigma \bar{u} \partial_x^{-1} \bar{u} \\ 2\sigma u \partial_x^{-1} u & -\partial_x - 2\sigma u \partial_x^{-1} \bar{u} \end{bmatrix},$$

where ∂_x^{-1} is a formal anti-derivative operator. For solitary waves, the anti-derivative operator can be defined by

$$\partial_x^{-1} u(x) := \int_{-\infty}^{x} u(x') dx'.$$

For periodic waves, an arbitrary constant of integration arise in the definition of the anti-derivative operator. To show how the recursion operator R generates the conserved quantities $\{H_k(u, \bar{u})\}_{k \in \mathbb{N}_0}$, we recall the first three conserved quantities of the NLS equation (4.11) (Example 2.4):

$$\begin{aligned} H_0(u, \bar{u}) &= \int_{\mathbb{R}} |u|^2 dx, \\ H_1(u, \bar{u}) &= \frac{i}{2} \int_{\mathbb{R}} (u\bar{u}_x - u_x \bar{u})\, dx, \\ H_2(u, \bar{u}) &= \int_{\mathbb{R}} \left(|u_x|^2 - \sigma |u|^4 \right) dx. \end{aligned}$$

Assuming the fast decay of $u(x) \to 0$ as $|x| \to \infty$, the recursion relation (4.5) recovers $H_1(u, \bar{u})$ and $H_2(u, \bar{u})$ correctly from the relations

$$\nabla H_0 = \begin{bmatrix} \bar{u} \\ u \end{bmatrix}, \quad \nabla H_1 = \begin{bmatrix} i\bar{u}_x \\ -iu_x \end{bmatrix}, \quad \nabla H_2 = \begin{bmatrix} -\bar{u}_{xx} - 2\sigma |u|^2 \bar{u} \\ -u_{xx} - 2\sigma |u|^2 u \end{bmatrix},$$

so that

$$R \nabla H_0 = \nabla H_1 \quad \text{and} \quad R \nabla H_1 = \nabla H_2.$$

Continuing iterations, we can compute the higher-order Hamiltonians

$$\begin{aligned} H_3(u, \bar{u}) &= \frac{i}{2} \int_{\mathbb{R}} \left[u_x \bar{u}_{xx} - u_{xx} \bar{u}_x - 3\sigma |u|^2 (u\bar{u}_x - u_x \bar{u}) \right] dx, \\ H_4(u, \bar{u}) &= \int_{\mathbb{R}} \left[|u_{xx}|^2 - 6\sigma |u|^2 |u_x|^2 - \sigma (\bar{u} u_x + u\bar{u}_x)^2 + 2|u|^6 \right] dx. \end{aligned}$$

The NLS equation (4.11) corresponds to $k = 2$ with $H = \frac{1}{2} H_2$ and $t = t_2$. The stationary NLS equation with parameters c and b,

$$u'' + 2\sigma |u|^2 u + 2icu' + 4bu = 0, \tag{4.13}$$

is obtained from the NLS equation (4.11) for the class of traveling waves:

$$u(t,x) \mapsto u(x+ct)e^{-2ibt}.$$

The second-order equation (4.13) is integrable with two conserved quantities

$$i(\bar{u}u' - \bar{u}'u) - 2c|u|^2 = 4a, \tag{4.14}$$

$$|u'|^2 + \sigma|u|^4 + 4b|u|^2 = 8d, \tag{4.15}$$

where a and d are additional parameters. The stationary NLS equation (4.13) appears as the Lax–Novikov equation (4.8) for $k = 2$, since it is obtained by the variational derivative of $\Lambda_2 := H_2 + 2cH_1 - 4bH_0$. Similarly,

$$\Lambda_3 := H_3 + 2cH_2 - 4bH_1 - 8a\sigma H_0$$

and

$$\Lambda_4 := H_4 + 2cH_3 - 4bH_2 - 8a\sigma H_1 + 16d\sigma H_0$$

generate higher-order Lax–Novikov equations

$$u''' + 6\sigma|u|^2u' + 2ic(u'' + 2\sigma|u|^2u) + 4bu' + 8i\sigma au = 0, \tag{4.16}$$

and

$$\begin{aligned} &u'''' + 8\sigma|u|^2u'' + 2\sigma u^2\bar{u}'' + 4\sigma u|u'|^2 + 6\sigma(u')^2\bar{u} + 6|u|^4u \\ &+ 2ic(u''' + 6\sigma|u|^2u') + 4b(u'' + 2\sigma|u|^2u) + 8ia\sigma u' + 16d\sigma u = 0, \end{aligned} \tag{4.17}$$

where the coefficients of the lower-order conserved quantities are carefully selected to preserve solutions of the second-order equation (4.13) with the first-order invariants (4.14) and (4.15) as solutions of the third-order and fourth-order equations (4.16) and (4.17). If $a = d = 0$ for the class of solitary waves satisfying $u(x) \to 0$ as $|x| \to \infty$, it is clear that Λ_3 and Λ_4 agree with (4.10) for $k = 3$ and $k = 4$. However, a and d are no longer zero for the class of periodic waves, for which Λ_3 and Λ_4 do not agree with (4.10) for $k = 3$ and $k = 4$ because they include a linear combination of H_0, H_1 in addition to H_2, H_3, and H_4.

EXAMPLE 4.2. The derivative nonlinear Schrödinger (DNLS) equation

$$iu_t + u_{xx} + i(|u|^2u)_x = 0, \tag{4.18}$$

is obtained as the compatibility condition of the Lax system (4.2) with

$$U(\lambda, u) = \begin{bmatrix} -i\lambda^2 & \lambda u \\ -\lambda\bar{u} & i\lambda^2 \end{bmatrix} \tag{4.19}$$

and

$$V(\lambda, u) = \begin{bmatrix} -2i\lambda^4 + i\lambda^2|u|^2 & 2\lambda^3u + \lambda(iu_x - |u|^2u) \\ -2\lambda^3\bar{u} + \lambda(i\bar{u}_x + |u|^2\bar{u}) & 2i\lambda^4 - i\lambda^2|u|^2 \end{bmatrix}. \tag{4.20}$$

The list of the first five conserved quantities is given by

$$
\begin{aligned}
H_0(u,\bar{u}) &= \int_{\mathbb{R}} |u|^2 dx,\\
H_1(u,\bar{u}) &= \frac{1}{2}\int_{\mathbb{R}} \left[i(u\bar{u}_x - u_x\bar{u}) + |u|^4\right] dx,\\
H_2(u,\bar{u}) &= \int_{\mathbb{R}} \left[|u_x|^2 - \frac{3i}{4}|u|^2(\bar{u}u_x - \bar{u}_x u) + \frac{1}{2}|u|^6\right] dx,\\
H_3(u,\bar{u}) &= \frac{1}{2}\int_{\mathbb{R}} \left[i(u_x\bar{u}_{xx} - u_{xx}\bar{u}_x) + \frac{5}{4}(|u|^4 + i(\bar{u}_x u - \bar{u}u_x))^2\right.\\
&\qquad \left. + \frac{9}{4}(\bar{u}u_x + \bar{u}_x u)^2 + |u|^2|u_x|^2\right] dx,\\
H_4(u,\bar{u}) &= \int_{\mathbb{R}} \left[|u_{xx}|^2 + \frac{5i}{2}|u|^2(\bar{u}_{xx}u_x - \bar{u}_x u_{xx}) + \frac{5}{2}|u|^2(\bar{u}^2u_x^2 + \bar{u}_x^2 u^2)\right.\\
&\qquad \left. + \frac{25}{2}|u|^4|u_x|^2 - \frac{35i}{16}|u|^6(\bar{u}u_x - \bar{u}_x u) + \frac{7}{8}|u|^{10}\right] dx.
\end{aligned}
$$

The stationary DNLS equation with parameters c and b,

$$
u'' + i(|u|^2u)' + 2icu' + 4bu = 0. \tag{4.21}
$$

is obtained from the DNLS equation (4.18) for the class of traveling waves:

$$
u(t,x) \mapsto u(x+2ct)e^{-4ibt}.
$$

It is integrable again with two conserved quantities

$$
2i(\bar{u}u' - \bar{u}'u) - 3|u|^4 - 4c|u|^2 = 8a, \tag{4.22}
$$

$$
2|u'|^2 - |u|^6 - 2c|u|^4 + 8(b-a)|u|^2 = 16d, \tag{4.23}
$$

where a and d are additional parameters. The stationary DNLS equation (4.21) is obtained from the variational derivative of $\Lambda_2 := H_2 + 2cH_1 + 4(a-b)H_0$ given by

$$
u'' + 3i|u|^2u' - \frac{3}{2}|u|^4u + 2c(iu' - |u|^2u) + 4(b-a)u = 0,
$$

after equation (4.22) is used to eliminate the parameter a. Similarly to Example 4.1, if u is a solution of the second-order equation (4.21) with the first-order invariants (4.22) and (4.23), then u is simultaneously a solution of the higher-order Lax–Novikov equation with carefully selected coefficients. The third-order equation

$$
\begin{aligned}
&iu''' - 4|u|^2u'' - u^2\bar{u}'' - 2|u'|^2u - 3(u')^2\bar{u} - \frac{15i}{2}|u|^4u' + \frac{5}{2}|u|^6u\\
&\quad - 2c\left[u'' + 3i|u|^2u' - \frac{3}{2}|u|^4u\right] + 4(b-a)\left[iu' - |u|^2u\right] - 8du = 0
\end{aligned}
$$

is generated from the linear combination

$$
\Lambda_3 := H_3 + 2cH_2 + 4(a-b)H_1 - 8dH_0.
$$

The third-order equation is satisfied identically after a and d are eliminated from (4.22) and (4.23) and the second-order equation (4.21) is differentiated in x. Similarly, one can show that the fourth-order equation is satisfied identically if it is obtained from the linear combination

$$
\Lambda_4 := H_4 + 2cH_3 + 4(a-b)H_2 - 8dH_1 + 8(2ab + 2cd - a^2)H_0.
$$

Constants a and d are again zero for the class of solitary waves and nonzero for the class of periodic waves. It is only in the former case, Λ_3 and Λ_4 can be obtained by using equations (4.9) and (4.10) in Lemma 4.2.

EXAMPLE 4.3. The massive Thirring model (MTM)

$$\left\{\begin{array}{l} i(u_t + u_x) + v = |v|^2 u, \\ i(v_t - v_x) + u = |u|^2 v, \end{array}\right. \tag{4.24}$$

is obtained as the compatibility condition of the Lax system (4.2) with

$$U(\lambda, u) = -\frac{i}{4}(|u|^2 - |v|^2)\sigma_3 - \frac{i\lambda}{2}\begin{bmatrix} 0 & \overline{v} \\ v & 0 \end{bmatrix} - \frac{i}{2\lambda}\begin{bmatrix} 0 & \overline{u} \\ u & 0 \end{bmatrix} - \frac{i}{4}\left(\lambda^2 - \frac{1}{\lambda^2}\right)\sigma_3 \tag{4.25}$$

and

$$V(\lambda, u) = +\frac{i}{4}(|u|^2 + |v|^2)\sigma_3 - \frac{i\lambda}{2}\begin{bmatrix} 0 & \overline{v} \\ v & 0 \end{bmatrix} + \frac{i}{2\lambda}\begin{bmatrix} 0 & \overline{u} \\ u & 0 \end{bmatrix} - \frac{i}{4}\left(\lambda^2 + \frac{1}{\lambda^2}\right)\sigma_3, \tag{4.26}$$

where $\sigma_3 = \mathrm{diag}(1, -1)$. Mass, momentum, and energy of the MTM are given by

$$Q = \int_{\mathbb{R}} (|u|^2 + |v|^2) dx, \tag{4.27}$$

$$P = \frac{i}{2}\int_{\mathbb{R}} (u\bar{u}_x - u_x\bar{u} + v\bar{v}_x - v_x\bar{v})\, dx, \tag{4.28}$$

and

$$E = \int_{\mathbb{R}} \left[\frac{i}{2}(u\bar{u}_x - u_x\bar{u} - v\bar{v}_x + v_x\bar{v}) - v\bar{u} - u\bar{v} + |u|^2|v|^2\right] dx \tag{4.29}$$

However, the MTM admits higher-order conserved quantities due to integrability and, in particular, the following higher-order energy found in [**221**]:

$$\begin{aligned} H = \int_{\mathbb{R}} \Big[& |u_x|^2 + |v_x|^2 - \frac{i}{4}(u_x\overline{u} - \overline{u}_x u)(|u|^2 + 2|v|^2) + \frac{i}{4}(v_x\overline{v} - \overline{v}_x v)(2|u|^2 + |v|^2) \\ & -\frac{1}{2}(u\overline{v} + \overline{u}v)(|u|^2 + |v|^2) + \frac{1}{2}|u|^2|v|^2(|u|^2 + |v|^2)\Big]\, dx. \end{aligned} \tag{4.30}$$

The standing waves of the form $u(t,x) \to u(x)e^{i\omega t}$ and $v(t,x) \to v(x)e^{i\omega t}$, where $\omega \in \mathbb{R}$, satisfy the stationary MTM system

$$\left\{\begin{array}{l} +iu' + v = |v|^2 u + \omega u, \\ -iv' + u = |u|^2 v + \omega v, \end{array}\right. \tag{4.31}$$

which is obtained from the variational derivative of $\Lambda_2 := E + \omega Q$. The two equations of system (4.31) can be reduced with $v = \bar{u}$ to the first-order equation

$$iu' + \bar{u} = |u|^2 u + \omega u. \tag{4.32}$$

Let us now consider the following system of second-order equations:

$$\left\{\begin{array}{l} u'' + i(|u|^2 + |v|^2)u' + iuv\overline{v}' + \frac{1}{2}\left[(2|u|^2 + |v|^2)v + u^2\overline{v}\right] \\ \qquad -\frac{1}{2}(2|u|^2 + |v|^2)|v|^2 u = \Omega u, \\ v'' - i(|u|^2 + |v|^2)v' - iuv\overline{u}' + \frac{1}{2}\left[(|u|^2 + 2|v|^2)u + v^2\overline{u}\right] \\ \qquad -\frac{1}{2}(|u|^2 + 2|v|^2)|u|^2 v = \Omega v, \end{array}\right. \tag{4.33}$$

which is obtained from the variational derivative of $\Lambda_4 := H + \Omega Q$. The two equations of system (4.33) can be reduced with $v = \bar{u}$ to the second-order equation

$$u'' + 3i|u|^2 u' + \frac{3}{2}|u|^2\bar{u} + \frac{1}{2}u^3 - \frac{3}{2}|u|^4 u = \Omega u. \tag{4.34}$$

Substituting the first-order equation (4.32) to the second-order equation (4.34) yields the constraint

$$(1-\omega^2)u + \frac{1}{2}\left(|u|^4 + 2\omega|u|^2 - u^2 - \bar{u}^2\right)u = \Omega u,$$

which is satisfied if $\Omega := 1 - \omega^2$ and $u(x) \to 0$ as $|x| \to \infty$ because $|u|^4 + 2\omega|u|^2 - u^2 - \bar{u}^2 = 0$ follows from the first-order equation (4.32) for solitary wave solutions.

REMARK 4.6. The MTM (Example 4.3) is a particular case of the nonlinear Dirac equations in one spatial dimension [**31**]. The quadratic part of the energy function E is not sign-definite because the spectrum of the associated Dirac operator

$$D := \begin{bmatrix} i\partial_x & 1 \\ 1 & -i\partial_x \end{bmatrix} \tag{4.35}$$

is located in $(-\infty, -1] \cup [1, \infty)$. As a result, the standing solitary waves of the nonlinear Dirac equations are saddle points of the constrained energy E subject to fixed mass Q and momentum P in the energy space $H^{1/2}(\mathbb{R}) \cap L^4(\mathbb{R})$. The corresponding Hessian operator has infinite-dimensional positive and negative subspaces in $L^2(\mathbb{R})$. The standard theory of orbital stability of solitary waves from Chapter 2 is not applicable to the MTM. Unexpectively, the solitary waves of the MTM are constrained minimizers of the higher-order energy H subject to fixed mass Q and momentum P so that the standard theory of orbital stability can be developed in $H^1(\mathbb{R})$ (Section 4.2).

REMARK 4.7. The periodic waves of the defocusing NLS equation (Example 4.1) are not minimizers of the constrained energy H_2 subject to fixed mass H_0 with respect to subharmonic (N-periodic, with $N \geq 2$) perturbations in $H^1_{\rm Nper}$. Nevertheless, they are minimizers of the constrained higher-order energy H_4 subject to fixed mass H_0 and energy H_2 so that the standard theory of orbital stability can be developed in $H^2_{\rm Nper}$ (Section 4.2).

REMARK 4.8. The solitary waves of the focusing NLS equation (Example 4.1) are orbitally stable in $H^1(\mathbb{R})$ (Theorem 2.9). When rougher solutions are considered in $L^2(\mathbb{R})$, only the mass conservation can be used, which alone is not sufficient for the proof of their orbital stability. Nevertheless, the orbital stability of the solitary waves in $L^2(\mathbb{R})$ follows from stability of the zero solution in $L^2(\mathbb{R})$ by using the Lax system (4.2) and the associated Bäcklund–Darboux transformation (Section 4.3).

4.2. Stability in Sobolev spaces of higher regularity

If nonlinear waves are saddle points of energy for fixed momentum and mass, the presence of higher-order conserved quantities allows us to develop an alternative variational characterization of the same nonlinear waves. Although there are no general results, it is a common belief confirmed by many examples that if the nonlinear waves are stable in the integrable Hamiltonian system, then there exists a suitably chosen linear combination of higher-order conserved quanitites for which the nonlinear wave is a constrained minimizer in a Sobolev space of higher regularity.

Note, however, that integrability of the Hamiltonian system does not exclude the possibility that the nonlinear waves are unstable in the corresponding time evolution (Section 5.1).

We will illustrate the stability theory with two examples: solitary waves in the MTM (Example 4.3) and periodic waves in the defocusing NLS equation (Example 4.1) based on [**221**] and [**29, 97**], respectively. Further examples including multi-solitons and breathers can be treated with similar ideas (Section 5.3).

4.2.1. Solitary waves in the MTM. Standing solitary waves of the systems (4.31) and (4.33) with $\Omega = 1 - \omega^2$ are expressed in the exact form

$$\begin{cases} u = U_\omega(x+\xi_0)e^{i\alpha_0}, \\ v = \bar{U}_\omega(x+\xi_0)e^{i\alpha_0}, \end{cases} \tag{4.36}$$

where $\xi_0, \alpha_0 \in \mathbb{R}$ are arbitrary parameters, $\omega \in (-1,1)$ belongs to the gap in the spectrum $(-\infty,-1]\cup[1,\infty)$ of the Dirac operator D given by (4.35), and the profile U_ω satisfying (4.32) and (4.34) with $\Omega = 1-\omega^2$ is given by

$$U_\omega(x) = \frac{\sqrt{2(1-\omega^2)}}{\sqrt{1+\omega}\cosh\left(\sqrt{1-\omega^2}x\right) + i\sqrt{1-\omega}\sinh\left(\sqrt{1-\omega^2}x\right)}. \tag{4.37}$$

As is shown in Example 4.3, the solitary wave with the profile $(u,v) = (U_\omega, \bar{U}_\omega)$ is simultaneously a critical point of $E + \omega Q$ and $H + (1-\omega^2)Q$, where Q, E, and H are given by (4.27), (4.29) and (4.30), respectively. In what follows, we only use the augmented higher-order energy $\Lambda_\omega := H + (1-\omega^2)Q$.

Two arbitrary parameters ξ_0 and α_0 in (4.36) are related to the translational and rotational symmetries of the system (4.24):

$$\begin{cases} u(t,x) \mapsto u(t,x+\xi_0)e^{i\alpha_0}, \\ v(t,x) \mapsto v(t,x+\xi_0)e^{i\alpha_0}, \end{cases} \qquad \xi_0, \alpha_0 \in \mathbb{R}. \tag{4.38}$$

The additional (Lorentz) transformation

$$\begin{cases} u(t,x) \mapsto \left(\frac{1+c}{1-c}\right)^{1/4} u\left(\frac{t-cx}{\sqrt{1-c^2}}, \frac{x-ct}{\sqrt{1-c^2}}\right), \\ v(t,x) \mapsto \left(\frac{1-c}{1+c}\right)^{1/4} v\left(\frac{t-cx}{\sqrt{1-c^2}}, \frac{x-ct}{\sqrt{1-c^2}}\right), \end{cases} \qquad c \in (-1,1), \tag{4.39}$$

translates the standing solitary waves to the traveling solitary waves. Although we formulate the main result for $c = 0$, it can be extended with straightforward modifications to every $c \in (-1,1)$. The following theorem gives the orbital stability of the standing solitary waves of the MTM in $H^1(\mathbb{R},\mathbb{C}^2)$.

THEOREM 4.1. *There exists $\omega_0 \in (0,1]$ such that for any fixed $\omega \in (-\omega_0,\omega_0)$, the solitary wave with the profile $(u,v) = (U_\omega, \bar{U}_\omega)$ is orbitally stable in $H^1(\mathbb{R},\mathbb{C}^2)$ in the time evolution of the MTM. In other words, for every $\epsilon > 0$, there is $\delta > 0$ such that if the initial datum satisfies*

$$\|u_0 - U_\omega\|_{H^1} + \|v_0 - \bar{U}_\omega\|_{H^1} < \delta,$$

then the unique solution $(u,v)\in C^0(\mathbb{R},H^1(\mathbb{R},\mathbb{C}^2))$ of the system (4.24) with $u(0,\cdot) = u_0$ and $v(0,\cdot) = v_0$ satisfies

$$\inf_{\xi,\alpha\in\mathbb{R}} \|u(t,\cdot) - e^{i\alpha}U_\omega(\cdot+\xi)\|_{H^1} + \|v(t,\cdot) - e^{i\alpha}\bar{U}_\omega(\cdot+\xi)\|_{H^1} < \epsilon,$$

for every $t > 0$.

PROOF. We will show that the standing solitary wave is a local minimizer of the higher-order energy H in $H^1(\mathbb{R},\mathbb{C}^2)$ subject to fixed mass Q and momentum P, where Q, P, H are given by (4.27), (4.28), and (4.30), respectively. The local minimizer is only degenerate by the two symmetries (4.38), hence the coercivity of the linearized operator holds under four orthogonality constraints due to two lower-order conserved quantities and two symmetry modes. The MTM is locally and globally well-posed in $H^1(\mathbb{R},\mathbb{C}^2)$ [**233**]. Since all assumptions of Theorem 2.8 are satisfied, the standing solitary wave is orbitally stable in $H^1(\mathbb{R},\mathbb{C}^2)$.

The proof of local minimization and nondegeneracy of the solitary wave is based on four steps, where the following precise computations are performed.

STEP 1. The Hessian operator for $\Lambda_\omega := H + (1-\omega^2)Q$ at the solitary wave with the profile $(u,v) = (U_\omega, \bar{U}_\omega)$ is defined by two self-adjoint operators $\mathcal{L}_\pm$ in $L^2(\mathbb{R},\mathbb{C})$ with dense domain $H^2(\mathbb{R},\mathbb{C})$, which are given by

$$\mathcal{L}_+ = \begin{bmatrix} L_+ & -3\omega U_\omega^2 \\ -3\omega \bar{U}_\omega^2 & \bar{L}_+ \end{bmatrix}, \quad \mathcal{L}_- = \begin{bmatrix} L_- & \omega U_\omega^2 \\ \omega \bar{U}_\omega^2 & \bar{L}_- \end{bmatrix}, \tag{4.40}$$

where

$$L_+ := -\frac{d^2}{dx^2} - 3i|U_\omega|^2\frac{d}{dx} + \frac{3}{2}|U_\omega|^4 - \frac{3}{2}U_\omega^2 + \frac{3}{2}\bar{U}_\omega^2 - 3\omega|U_\omega|^2 + 1 - \omega^2,$$
$$L_- := -\frac{d^2}{dx^2} - i|U_\omega|^2\frac{d}{dx} - \frac{1}{2}|U_\omega|^4 - \frac{1}{2}U_\omega^2 + \frac{1}{2}\bar{U}_\omega^2 - \omega|U_\omega|^2 + 1 - \omega^2.$$

PROOF OF STEP 1. Adding a perturbation (u,v) to the solitary wave with the profile $(U_\omega, \bar{U}_\omega)$ and expanding the augmented energy $\Lambda_\omega := H + (1-\omega^2)Q$ in variables $(u, v, \bar{u}, \bar{v})$ yields the Hessian operator in the form

$$\mathcal{L}_\omega := \begin{bmatrix} L_1 & 2L_2 & L_2 & L_3 \\ 2\bar{L}_2 & \bar{L}_1 & \bar{L}_3 & \bar{L}_2 \\ \bar{L}_2 & \bar{L}_3 & \bar{L}_1 & 2\bar{L}_2 \\ L_3 & L_2 & 2L_2 & L_1 \end{bmatrix}, \tag{4.41}$$

where

$$L_1 = -\frac{d^2}{dx^2} - 2i|U_\omega|^2\frac{d}{dx} - 2i\bar{U}_\omega U_\omega' + \frac{5}{2}|U_\omega|^4 - U_\omega^2 - \bar{U}_\omega^2 + 1 - \omega^2,$$
$$L_2 = -iU_\omega U_\omega' + U_\omega^2|U_\omega|^2 - |U_\omega|^2,$$
$$L_3 = -i|U_\omega|^2\frac{d}{dx} - i\bar{U}_\omega U_\omega' + 2|U_\omega|^4 - \frac{1}{2}U_\omega^2 - \frac{1}{2}\bar{U}_\omega^2.$$

The 4×4 matrix operator $\mathcal{L}_\omega$ can be diagonalized into two 2×2 matrix operators $\mathcal{L}_\pm$ by means of the orthogonal similarity transformation

$$S^T\mathcal{L}_\omega S = \begin{bmatrix} \mathcal{L}_+ & 0 \\ 0 & \mathcal{L}_- \end{bmatrix}, \quad S = \frac{1}{\sqrt{2}}\begin{bmatrix} 1 & 0 & -1 & 0 \\ 0 & 1 & 0 & 1 \\ 0 & 1 & 0 & -1 \\ 1 & 0 & 1 & 0 \end{bmatrix}, \tag{4.42}$$

with $\mathcal{L}_\pm$ defined in (4.40). The first-order equation (4.32) has been used to eliminate U_ω'. The matrix operators $\mathcal{L}_\pm$ act on functions of the form $(f, \bar{f})$ in $L^2(\mathbb{R},\mathbb{C})$. The scalar operators $L_\pm$ are self-adjoint operators in $L^2(\mathbb{R})$ since the relation $L_\pm^* = L_\pm$ is checked by using the same equation (4.32). $\square$

STEP 2. For every $\omega \in (-1,1)$, the spectrum of the self-adjoint operator $\mathcal{L}_-$ in $L^2(\mathbb{R},\mathbb{C}^2)$ consists of two simple eigenvalues $\lambda_-(\omega)$ and 0 and the continuous spectrum on $[1-\omega^2,\infty)$, where $\mathrm{sgn}(\lambda_-(\omega)) = \mathrm{sgn}(\omega)$.

PROOF OF STEP 2. The operator $\mathcal{L}_-$ is a 2-by-2 matrix Sturm–Liouville operator with the exponentially decaying potentials. By Weyl's theorem (Theorem B.7), the continuous spectrum of $\mathcal{L}_-$ is given by $[1-\omega^2,\infty)$, where $1-\omega^2>0$. The zero eigenvalue is present due to the explicit solution

$$\mathcal{L}_- \begin{bmatrix} U_\omega \\ -\bar{U}_\omega \end{bmatrix} = \begin{bmatrix} 0 \\ 0 \end{bmatrix}, \tag{4.43}$$

which is related to the rotational symmetry in (4.38). It remains to show that besides the continuous spectrum on $[1-\omega^2,\infty)$ and the simple zero eigenvalue, $\sigma(\mathcal{L}_-)$ includes exactly one simple eigenvalue $\lambda_-(\omega)$ such that $\mathrm{sgn}(\lambda_-(\omega)) = \mathrm{sgn}(\omega)$.

Let us consider the eigenvalue problem $\mathcal{L}_-\mathbf{f} = \lambda\mathbf{f}$, where $\mathbf{f} = (f,\bar{f})$ and $\lambda \in \mathbb{R}$. Using the transformation

$$f(x) = \varphi(x) e^{-i\int_0^x |U_\omega(x')|^2 dx'}$$

and the relation

$$U_\omega^2 e^{i\int_0^x |U_\omega(x')|^2 dx'} = \frac{2(1-\omega^2)}{\omega+\cosh(2\sqrt{1-\omega^2}x)} = |U_\omega|^2,$$

we obtain the equivalent spectral problem

$$\begin{bmatrix} -\partial_x^2 + 1 - \omega^2 - \omega|U_\omega|^2 - \frac{3}{4}|U_\omega|^4 & \omega|U_\omega|^2 \\ \omega|U_\omega|^2 & -\partial_x^2 + 1 - \omega^2 - \omega|U_\omega|^2 - \frac{3}{4}|U_\omega|^4 \end{bmatrix} \begin{bmatrix} \varphi \\ \bar{\varphi} \end{bmatrix} = \lambda \begin{bmatrix} \varphi \\ \bar{\varphi} \end{bmatrix},$$

with $(\varphi,\bar{\varphi})$ as a new eigenfunction. Because the off-diagonal entries are real, we set

$$\psi_\pm(z) := \varphi(x) \pm \bar{\varphi}(x), \quad z := \sqrt{1-\omega^2}x, \quad \lambda := (1-\omega^2)\mu$$

to diagonalize the spectral problem into two following spectral problems:

$$-\frac{d^2\psi_+}{dz^2} + \left[1 - \frac{3(1-\omega^2)}{(\omega+\cosh(2z))^2}\right]\psi_+ = \mu\psi_+ \tag{4.44}$$

and

$$-\frac{d^2\psi_-}{dz^2} + \left[1 - \frac{3(1-\omega^2)}{(\omega+\cosh(2z))^2} - \frac{4\omega}{\omega+\cosh(2z)}\right]\psi_- = \mu\psi_-. \tag{4.45}$$

The spectral problem (4.45) for $\mu = 0$ admits the eigenfunction

$$\psi_0(z) = \frac{1}{(\omega+\cosh(2z))^{1/2}},$$

due to (4.43). Because the eigenfunction ψ_0 is positive definite, Sturm's oscillation theorem (Theorem B.9) states that the simple zero eigenvalue of the spectral problem (4.45) is at the bottom of the spectrum of the associated Schrödinger operator for any $\omega \in (-1,1)$. Furthermore, the function

$$\psi_1(z) = \frac{\sinh(2z)}{\omega+\cosh(2z)}$$

satisfies the spectral problem

$$-\frac{d^2\psi}{dz^2} + \left[1 - \frac{8(1-\omega^2)}{(\omega+\cosh(2z))^2} - \frac{4\omega}{\omega+\cosh(2z)}\right]\psi = \mu\psi, \tag{4.46}$$

for $\mu = 1$. Since ψ_1 is bounded at infinity and admits exactly one zero on the line, there is only one isolated eigenvalue below $\mu = 1$ for the spectral problem (4.46). Now the difference between the potentials of the spectral problems (4.45) and (4.46) is

$$\Delta V(z) = \frac{5(1-\omega^2)}{(\omega + \cosh(2z))^2} > 0, \quad z \in \mathbb{R}$$

for every $\omega \in (-1, 1)$. By Sturm's comparison theorem (Theorem B.10), the spectral problem (4.45) has exactly one simple isolated eigenvalue and this is the zero eigenvalue with the eigenfunction ψ_0.

The difference between the potentials of the spectral problems (4.44) and (4.45) is given by

$$\Delta V(z) = \frac{4\omega}{\omega + \cosh(2z)}$$

so that $\operatorname{sgn}(\Delta V(z)) = \operatorname{sgn}(\omega)$ for every $z \in \mathbb{R}$. By Sturm's comparison theorem (Theorem B.10), there exists a simple eigenvalue $\mu_-(\omega) \in (-\infty, 1)$ of the spectral problem (4.44) such that $\operatorname{sgn}(\mu_-(\omega)) = \operatorname{sgn}(\omega)$. This simple isolated eigenvalue gives $\lambda_-(\omega) = (1-\omega^2)\mu_-(\omega)$ after the scaling transformation.

It remains to show that this eigenvalue $\mu_-(\omega)$ is the only isolated eigenvalue of the spectral problem (4.44). To do so, we note that

$$\omega + \cosh(2z) \geq \omega + 1 + 2z^2, \quad z \in \mathbb{R}$$

and consider the spectral problem

$$-\frac{d^2\psi}{dz^2} + \left[1 - \frac{3(1-\omega^2)}{(\omega + 1 + 2z^2)^2}\right]\psi = \mu\psi. \tag{4.47}$$

Rescaling the independent variable $z := \frac{\sqrt{1+\omega}}{\sqrt{2}} y$ and denoting $\psi(z) := \tilde{\psi}(y)$, we rewrite (4.47) in the equivalent form

$$-\frac{d^2\tilde{\psi}}{dy^2} - \frac{3}{(1+y^2)^2}\left(1 - \frac{1+\omega}{2}\right)\tilde{\psi} = \frac{(\mu-1)(1+\omega)}{2}\tilde{\psi}. \tag{4.48}$$

It follows that the function

$$\tilde{\psi}_c(y) = \frac{y}{\sqrt{1+y^2}}$$

satisfies the spectral problem

$$-\frac{d^2\tilde{\psi}}{dy^2} - \frac{3}{(1+y^2)^2}\tilde{\psi} = \frac{(\mu-1)(1+\omega)}{2}\tilde{\psi} \tag{4.49}$$

for $\mu = 1$. Since $\tilde{\psi}_c$ is bounded at infinity and admits exactly one zero on the line, there is only one isolated eigenvalue below $\mu = 1$ for the spectral problem (4.49). Because the difference between potentials of the spectral problems (4.48) and (4.49) as well as those of the spectral problems (4.44) and (4.47) is strictly positive for every $\omega \in (-1, 1)$, it follows from Sturm's comparison theorem (Theorem B.10) that the spectral problem (4.44) has exactly one simple isolated eigenvalue in $(-\infty, 1)$ for every $\omega \in (-1, 1)$ and this eigenvalue is $\mu_-(\omega)$. □

STEP 3. There is $\omega_0 \in (0, 1]$ such that for every $\omega \in (-\omega_0, \omega_0)$, the spectrum of the self-adjoint operator $\mathcal{L}_+$ in $L^2(\mathbb{R}, \mathbb{C}^2)$ consists of two simple eigenvalues $\lambda_+(\omega)$ and 0 and the continuous spectrum on $[1-\omega^2, \infty)$, where $\operatorname{sgn}(\lambda_-(\omega)) = -\operatorname{sgn}(\omega)$.

PROOF OF STEP 3. The operator $\mathcal{L}_+$ is a 2-by-2 matrix Sturm–Liouville operator with the exponentially decaying potentials. By Weyl's theorem (Theorem B.7), the continuous spectrum of $\mathcal{L}_+$ is given by $[1-\omega^2,\infty)$, where $1-\omega^2>0$. The zero eigenvalue is present due to the explicit solution

$$\mathcal{L}_+ \begin{bmatrix} U_\omega' \\ \bar{U}_\omega' \end{bmatrix} = \begin{bmatrix} 0 \\ 0 \end{bmatrix}, \tag{4.50}$$

which is related to the translational symmetry in (4.38). It remains to show that besides the continuous spectrum on $[1-\omega^2,\infty)$ and the simple zero eigenvalue, $\sigma(\mathcal{L}_+)$ includes exactly one simple eigenvalue $\lambda_+(\omega)$ for $\omega \in (-\omega_0,\omega_0)$ such that $\operatorname{sgn}(\lambda_+(\omega)) = -\operatorname{sgn}(\omega)$.

Let us consider the eigenvalue problem $\mathcal{L}_+\mathbf{f} = \lambda\mathbf{f}$, where $\mathbf{f} = (f,\bar{f})$ and $\lambda \in \mathbb{R}$. Using the transformation

$$f(x) = \varphi(x) e^{-3i\int_0^x |U_\omega(x')|^2 dx'}$$

we obtain an equivalent spectral problem:

$$\begin{bmatrix} -\partial_x^2 + 1 - \omega^2 - 3\omega|U_\omega|^2 - \frac{3}{4}|U_\omega|^4 & -3\omega W_\omega \\ -3\omega\bar{W}_\omega & -\partial_x^2 + 1 - \omega^2 - 3\omega|U_\omega|^2 - \frac{3}{4}|U_\omega|^4 \end{bmatrix} \begin{bmatrix} \varphi \\ \bar{\varphi} \end{bmatrix} = \lambda \begin{bmatrix} \varphi \\ \bar{\varphi} \end{bmatrix},$$

with $(\varphi,\bar{\varphi})$ being a new eigenfunction and where

$$\begin{aligned} W_\omega &:= U_\omega^2 e^{3i\int_0^x |U_\omega(x')|^2 dx'} \\ &= 2(1-\omega^2)\frac{\left(1+\omega\cosh\left(2\sqrt{1-\omega^2}x\right) + i\sqrt{1-\omega^2}\sinh\left(2\sqrt{1-\omega^2}x\right)\right)^2}{\left(\omega+\cosh\left(2\sqrt{1-\omega^2}x\right)\right)^3}. \end{aligned}$$

Setting now $z := \sqrt{1-\omega^2}x$ and $\lambda := (1-\omega^2)\mu$, we rewrite the spectral problem in the form

$$\begin{bmatrix} -\partial_z^2 + 1 + V_1(z) & V_2(z) \\ \bar{V}_2(z) & -\partial_z^2 + 1 + V_1(z) \end{bmatrix} \begin{bmatrix} \varphi \\ \bar{\varphi} \end{bmatrix} = \mu \begin{bmatrix} \varphi \\ \bar{\varphi} \end{bmatrix}, \tag{4.51}$$

where

$$\begin{aligned} V_1(z) &:= -\frac{3(1-\omega^2)}{(\omega+\cosh(2z))^2} - \frac{6\omega}{\omega+\cosh(2z)}, \\ V_2(z) &:= -6\omega\frac{\left(1+\omega\cosh(2z) + i\sqrt{1-\omega^2}\sinh(2z)\right)^2}{\left(\omega+\cosh(2z)\right)^3}. \end{aligned}$$

The spectral problem (4.51) for $\mu = 0$ admits the eigenvector $(\varphi_0,\bar{\varphi}_0)$ with

$$\varphi_0(z) = \frac{\omega\sinh(2z) + i\sqrt{1-\omega^2}\cosh(2z)}{(\omega+\cosh(2z))^{3/2}},$$

thanks to the existence of the eigenvector of L_+ in (4.50). If $\omega = 0$, then $\mu = 0$ is a double zero eigenvalue of the spectral problem (4.51), with the second linearly independent eigenvector $(\varphi_0,-\bar{\varphi}_0)$. The end point $\mu = 1$ of the continuous spectrum of the spectral problem (4.51) with $\omega = 0$ does not admit a bounded eigenfunction due to the same comparison arguments as in Step 2. No other eigenvalues μ exist for $\omega = 0$.

If $\omega \neq 0$, the double zero eigenvalue splits and no additional eigenvalues may bifurcate below $\mu = 1$. To study the splitting, we compute the quadratic form of the operator on the left-hand side of the spectral problem (4.51) at the vector $(\varphi_0, -\bar{\varphi}_0)$ and obtain

$$-2\int_{\mathbb{R}} (V_2\bar{\varphi}_0^2 + \bar{V}_2\varphi_0^2)\, dz = -12\omega \int_{\mathbb{R}} \frac{3 - 2\omega^2 - \cosh(4z)}{(\omega + \cosh(2z))^4} dz.$$

Since

$$\int_{\mathbb{R}} \frac{3 - \cosh(4z)}{\cosh(2z)^4} dz = \frac{2}{3}$$

the perturbation theory for small $\omega \neq 0$ implies that the small nonzero eigenvalue $\mu_+(\omega)$ with $\mu_+(0) = 0$ satisfies $\operatorname{sgn}(\mu_+(\omega)) = -\operatorname{sgn}(\omega)$ if $\omega \in (-\omega_0, \omega_0)\backslash\{0\}$ with small $\omega_0 > 0$. This simple isolated eigenvalue gives $\lambda_+(\omega) = (1 - \omega^2)\mu_+(\omega)$ after the scaling transformation. □

STEP 4. For every $\omega \in (-\omega_0, \omega_0)$, we have $\sigma(\mathcal{L}_+|_{X_+}) \geq 0$ and $\operatorname{Ker}(\mathcal{L}_+|_{X_+}) = \operatorname{Span}\{\mathbf{U}_+\}$, where $\mathbf{U}_+ := (U_\omega', \bar{U}_\omega')$ and

$$X_+ := \left\{u \in L^2(\mathbb{R}, \mathbb{C}^2) : \quad \int_{\mathbb{R}} (\bar{U}_\omega u + U_\omega \bar{u})\, dx = 0\right\}. \tag{4.52}$$

For every $\omega \in (-1, 1)$, we have $\sigma(\mathcal{L}_-|_{X_-}) \geq 0$ and $\operatorname{Ker}(\mathcal{L}_-|_{X_-}) = \operatorname{Span}\{\mathbf{U}_-\}$, where $\mathbf{U}_- := (U_\omega, -\bar{U}_\omega)$ and

$$X_- := \left\{u \in L^2(\mathbb{R}, \mathbb{C}^2) : \quad \int_{\mathbb{R}} (\bar{U}_\omega' u - U_\omega' \bar{u})\, dx = 0\right\}. \tag{4.53}$$

PROOF OF STEP 4. For operator $\mathcal{L}_+$, we differentiate the second-order equation (4.34) with respect to ω and obtain

$$\mathcal{L}_+ \begin{bmatrix} \partial_\omega U_\omega \\ \partial_\omega \bar{U}_\omega \end{bmatrix} = 2\omega \begin{bmatrix} U_\omega \\ \bar{U}_\omega \end{bmatrix} =: 2\omega \mathbf{V}_+.$$

By using Theorem 2.7, we compute

$$\begin{aligned} \langle \mathcal{L}_+^{-1}\mathbf{V}_+, \mathbf{V}_+\rangle &= \frac{1}{2\omega} \int_{\mathbb{R}} \left(\bar{U}_\omega \frac{\partial U_\omega}{\partial \omega} + U_\omega \frac{\partial \bar{U}_\omega}{\partial \omega}\right) dx \\ &= \frac{1}{2\omega}\frac{d}{d\omega} \int_{\mathbb{R}} |U_\omega|^2 dx \\ &= -\frac{1}{\omega\sqrt{1-\omega^2}}, \end{aligned}$$

where we have used the exact expression $\|U_\omega\|^2 = 2\arccos(\omega)$. Since $\lambda_+(\omega) < 0$ and $\langle \mathcal{L}_+^{-1}\mathbf{V}_+, \mathbf{V}_+\rangle < 0$ for $\omega > 0$, whereas $\lambda_+(\omega) > 0$ for $\omega < 0$, we have $\sigma(\mathcal{L}_+|_{X_+}) \geq 0$ for every $\omega \in (-\omega_0, \omega_0)$. For $\omega = 0$, the eigenvector $\mathbf{U}_+' := (U_0', -\bar{U}_0')$ of the zero eigenvalue of $\mathcal{L}_+$ does not belong to X_+ because

$$\int_{\mathbb{R}} (\bar{U}_0 U_0' - U_0 \bar{U}_0')\, dx = -i \int_{\mathbb{R}} (4|U_0|^4 - U_0^2 - \bar{U}_0^2)\, dx = -2i \neq 0. \tag{4.54}$$

On the other hand, for every $\omega \in (-1, 1)$, the eigenvector $\mathbf{U}_+ := (U_\omega', \bar{U}_\omega')$ for the zero eigenvalue of $\mathcal{L}_+$ belongs to X_+. Thus, we have $\operatorname{Ker}(\mathcal{L}_+|_{X_+}) = \operatorname{Span}\{\mathbf{U}_+\}$ for every $\omega \in (-\omega_0, \omega_0)$.

For operator $\mathcal{L}_-$, it follows from both (4.32) and (4.34) that

$$\mathcal{L}_- \left(-\frac{1}{2}\begin{bmatrix} xU_\omega \\ -x\bar{U}_\omega \end{bmatrix} + \frac{1}{4i\omega}\begin{bmatrix} U_\omega \\ \bar{U}_\omega \end{bmatrix}\right) = \begin{bmatrix} U_\omega' \\ -\bar{U}_\omega' \end{bmatrix} =: \mathbf{V}_-,$$

where the expression in the left-hand side follows from the derivative of the traveling solitary wave (4.39) with respect to the wave speed c at $c = 0$. By using Theorem 2.7, we integrate by parts and compute

$$\begin{aligned}\langle \mathcal{L}_-^{-1}\mathbf{V}_-, \mathbf{V}_-\rangle &= \int_{\mathbb{R}} \left(\frac{1}{2}|U_\omega|^2 - \frac{1}{4i\omega}\left(\bar{U}_\omega U_\omega' - U_\omega \bar{U}_\omega'\right)\right) dx \\ &= \frac{1}{4\omega}\int_{\mathbb{R}} \left(2|U_\omega|^4 - U_\omega^2 - \bar{U}_\omega^2 + 4\omega|U_\omega|^2\right) dx \\ &= \frac{1-\omega^2}{\omega}\int_{\mathbb{R}} \frac{1+\omega\cosh(2\sqrt{1-\omega^2}x)}{(\omega+\cosh(2\sqrt{1-\omega^2}x))^2} dx \\ &= \frac{\sqrt{1-\omega^2}}{\omega}.\end{aligned}$$

Since $\lambda_-(\omega) < 0$ and $\langle \mathcal{L}_-^{-1}\mathbf{V}_-, \mathbf{V}_-\rangle < 0$ for $\omega < 0$, whereas $\lambda_-(\omega) > 0$ for $\omega > 0$, we have $\sigma(\mathcal{L}_-|_{X_-}) \geq 0$ for every $\omega \in (-1,1)$. For $\omega = 0$, the eigenvector $\mathbf{U}_-' := (U_0, \bar{U}_0)$ for the zero eigenvalue of $\mathcal{L}_-$ does not belong to X_- because of the same computation (4.54). On the other hand, for every $\omega \in (-1,1)$, the eigenvector $\mathbf{U}_- := (U_\omega, -\bar{U}_\omega)$ of the zero eigenvalue of $\mathcal{L}_-$ belongs to X_-. Thus, we have $\operatorname{Ker}(\mathcal{L}_-|_{X_-}) = \operatorname{Span}\{\mathbf{U}_-\}$ for every $\omega \in (-1,1)$. □

End of the proof of Theorem 4.1. The perturbation $(u,v) \in \mathbb{C}^2$ to the solitary wave with the profile $(U_\omega, \bar{U}_\omega)$ must satisfy four constraints which can be formulated as two complex-valued constraints:

$$\int_{\mathbb{R}} \left(\bar{U}_\omega u + U_\omega v\right) dx = 0, \quad \int_{\mathbb{R}} \left(\bar{U}_\omega' u + U_\omega' v\right) dx = 0. \tag{4.55}$$

The real part of the first constraint and the imaginary part of the second constraint in (4.55) are due to fixed Q and P in (4.27) and (4.28). After the transformation (4.42), these two constraints define the constraints in X_+ and X_- given in (4.52) and (4.53). The imaginary part of the first constraint and the real part of the second constraint in (4.55) are due to the orthogonality of the perturbation to the symmetry modes associated to the symmetries (4.38). After the transformation (4.42), these two constraints remove the simple zero eigenvalue of the operators $\mathcal{L}_-|_{X_-}$ and $\mathcal{L}_+|_{X_+}$ in (4.43) and (4.50). In a subspace of $L^2(\mathbb{R},\mathbb{C}^2)$ with four orthogonality constraints, the Hessian operator $\mathcal{L}_\omega$ in (4.41) is coercive and the orbital stability of the standing solitary wave with the profile $(U_\omega, \bar{U}_\omega)$ with respect to the perturbation (u,v) in $H^1(\mathbb{R},\mathbb{C}^2)$ follows by Theorem 2.8. □

Remark 4.9. The result of Theorem 4.1 is restricted to $(-\omega_0, \omega_0) \subset (-1,1)$ due to the perturbation theory used in Step 3. It was numerically verified in [**221**] that the spectral problem (4.51) has exactly two isolated eigenvalues for every $\omega \in (-1,1)$ with $\operatorname{sgn}(\lambda_+(\omega)) = -\operatorname{sgn}(\omega)$. The orbital stability of solitary waves of the MTM for every $\omega \in (-1,1)$ was proven in [**68**] in a subspace of $L^2(\mathbb{R})$ by using the Bäcklund–Darboux transformations (Section 4.3).

4.2.2. Periodic waves in the defocusing NLS equation. Periodic wave solutions of the defocusing NLS equation (4.11) with $\sigma = -1$ satisfying (4.13) with $c = 0$ and $b = \frac{1}{2}$ can be expressed in the exact form (Example 3.5) as $u(t,x) =$

$u_0(x)e^{-it}$ with

$$u_0(x) = \frac{\sqrt{2}k}{\sqrt{1+k^2}} \, \mathrm{sn} \left(\frac{\sqrt{2}x}{\sqrt{1+k^2}}; k \right),$$

where $k \in (0,1)$ is an arbitrary parameter. The spatial period is given by $T = 2\sqrt{2(1+k^2)}K(k)$, where $K(k)$ is the complete elliptic integral of the first kind, see Appendix A. The periodic wave with the profile u_0 is a critical point of the energy

$$E(u,\bar{u}) = \frac{1}{2}\int_0^T \left(|u_x|^2 + (1-|u|^2)^2\right) dx,$$

which is obtained from $H_2 - 2H_0$ (Example 4.1). The energy E is defined in H^1_{per} on the periodic domain $\mathbb{R}\backslash(T\,\mathbb{Z})$ and the Euler–Lagrange equation is written in the form

$$u_0'' + 2(1-u_0^2)u_0 = 0. \tag{4.56}$$

The first-order invariant for (4.56) can be written in the form:

$$(u_0')^2 - u_0^4 + 2u_0^2 = 1 - \mathcal{E}^2, \qquad \mathcal{E} := \frac{1-k^2}{1+k^2} \in (0,1). \tag{4.57}$$

The critical point of E is a saddle point of E subject to fixed mass H_0 and momentum H_1 in H^1_{per} (Example 3.16). The second variation is expressed by two Schrödinger operators $\mathcal{L}_+$ and $\mathcal{L}_-$ such that $n(\mathcal{L}_+|_{\{u_0\}^\perp}) = 1$, $z(\mathcal{L}_+|_{\{u_0\}^\perp}) = 1$, $n(\mathcal{L}_-|_{\{\partial_x u_0\}^\perp}) = 1$, and $z(\mathcal{L}_-|_{\{\partial_x u_0\}^\perp}) = 1$, where the constraints are due to fixed H_0 and H_1 and the zero eigenvalues are due to translational and rotational symmetry:

$$u(t,x) \mapsto u(t,x+\xi_0)e^{i\alpha_0}, \quad \xi_0, \alpha_0 \in \mathbb{R}. \tag{4.58}$$

Theorem 3.3 and Corollary 3.1 are not applicable to claim the orbital stability of the periodic wave with the profile u_0 in H^1_{per} (Example 3.16). Only if the symmetry constraint $u(x+T/2) = -u(x)$ is added for the perturbation $u \in H^1_{\mathrm{per}}$ of u_0, orbital stability of the periodic wave can be proven in such a restrictive subspace, see [**96**].

The situation is getting worse if we consider subharmonic perturbations to the periodic wave, which are periodic with the multiple period NT for some integer $N \geq 2$. Then, the critical point with the profile u_0 becomes a saddle point of E in H^1_{Nper} for fixed mass H_0 and momentum H_1 with $2N$ negative eigenvalues. By Theorem 3.10, such critical points may, in principle, be spectrally and orbitally unstable in H^1_{Nper}. To rule out such instability, we use the higher-order energy to prove orbital stability of the periodic waves with respect to perturbations with the multiple period NT in a Sobolev space of higher regularity.

By using the hierarchy of conserved quantities (Example 4.1), we know that the periodic wave with the profile u_0 satisfying (4.13), (4.14), and (4.15) with $\sigma = -1$, $a = c = 0$, $b = \frac{1}{2}$, and $d = \frac{1}{8}(1-\mathcal{E}^2)$, see (4.56) and (4.57), is also a critical point of the higher-order energy $H_4 - 2H_2 - 2(1-\mathcal{E}^2)H_0$ in H^2_{per}. Since the combination

$H_2 - 2H_0$ is redundant, it follows that u_0 is a critical point in $H^2_{\rm per}$ of the higher-order energy

$$H(u,\bar{u}) = \frac{1}{4}\int_0^T \left(|u_{xx}|^2 + 6|u|^2|u_x|^2 + (\bar{u}u_x + u\bar{u}_x)^2 + 2|u|^6\right) dx$$
$$- \frac{1}{2}(3-\mathcal{E}^2)\int_0^T |u|^2 dx.$$

Both variational formulations for $E(u,\bar{u})$ and $H(u,\bar{u})$ can be combined together with an arbitrary coefficient $\gamma \in \mathbb{R}$. Thus, for every $\gamma \in \mathbb{R}$, the periodic wave with the profile u_0 is a critical point of the augmented energy functional

$$\Lambda_\gamma := H - \gamma E$$

in $H^2_{\rm per}$. The main idea of [**29, 97**] is to find the admissible range of values of γ, for which the periodic wave is a local (unconstrained) minimizer of Λ_γ, which is only degenerate due to the translational and rotational symmetry in (4.58). This construction yields the following theorem on the orbital stability of the periodic wave with the profile u_0 in $H^2_{N\rm per}$ for every fixed $N \in \mathbb{N}$.

THEOREM 4.2. *For any $k \in (0,1)$ and $N \in \mathbb{N}$, the periodic wave with the profile u_0 is orbitally stable in $H^2_{N\rm per}$ in the time evolution of the defocusing NLS equation. In other words, for any $\epsilon > 0$, there exist $\delta > 0$ such that if the initial datum satisfies*

$$\|u(0,\cdot) - u_0\|_{H^2_{N\rm per}} < \delta$$

then the unique global solution $u \in C^0(\mathbb{R}_+, H^2_{N\rm per})$ of the defocusing NLS equation satisfies

$$\inf_{\xi,\alpha \in \mathbb{R}^2} \|u(t,\cdot) - e^{i\alpha}u_0(\cdot+\xi)\|_{H^2_{N\rm per}} < \epsilon,$$

for every $t > 0$.

PROOF. We will establish that the periodic wave with the profile u_0 is a local minimizer of Λ_γ in $H^2_{N\rm per}$ for $\gamma = 2$. The local minimizer is only degenerate by the two symmetries in (4.58), hence the coercivity of the linearized operator holds under two orthogonality constraints due to two symmetry modes. The defocusing NLS equation is locally and globally well-posed in $H^s_{N\rm per}$ for any integer $s \geq 0$ [**30, 253**]. Since all assumptions of Theorem 2.3 are satisfied, the periodic wave is orbitally stable in $H^2_{N\rm per}$.

The proof of local minimization and nondegeneracy of the periodic wave is based on three steps, where the following precise computations are performed.

STEP 1. The Hessian operator for Λ_γ at the periodic wave with the profile u_0 is defined by two self-adjoint operators $\mathcal{M}_\pm - \gamma\mathcal{L}_\pm$ with dense domain $H^4_{N\rm per}$ in $L^2_{N\rm per}$, where

$$(4.59)\qquad \begin{cases} \mathcal{L}_+ &= -\frac{1}{2}\partial_x^2 + 3u_0^2 - 1, \\ \mathcal{L}_- &= -\frac{1}{2}\partial_x^2 + u_0^2 - 1, \end{cases} \qquad \begin{cases} \mathcal{M}_+ &= \frac{1}{4}\partial_x^4 - \frac{5}{2}\partial_x u_0^2 \partial_x - 5u_0^4 + 15u_0^2 - 4 + 3\mathcal{E}^2, \\ \mathcal{M}_- &= \frac{1}{4}\partial_x^4 - \frac{3}{2}\partial_x u_0^2 \partial_x + u_0^2 - 1. \end{cases}$$

PROOF OF STEP 1. Adding a perturbation $w + iv$ to u_0 and expanding the functional $\Lambda_\gamma = H - \gamma E$, we obtain the second variations of E and H in the following form:

$$\int_0^{NT} \left[\frac{1}{2}w_x^2 + (3u_0^2 - 1)w^2\right] dx + \int_0^{NT} \left[\frac{1}{2}v_x^2 + (u_0^2 - 1)v^2\right] dx$$

and

$$\int_0^{NT} \left[\frac{1}{4}w_{xx}^2 + \frac{5}{2}u_0^2 w_x^2 + (-5u_0^4 + 15u_0^2 - 4 + 3\mathcal{E}^2)w^2\right] dx$$
$$+ \int_0^{NT} \left[\frac{1}{4}v_{xx}^2 + \frac{3}{2}u_0^2 v_x^2 + (u_0^2 - 1)v^2\right] dx.$$

Hence the second variations of E and H are block-diagonal in the sense that the contributions of u and v do not mix if u_0 is real-valued. We thus express them in terms of the self-adjoint Hessian operators $\mathcal{L}_\pm$ and $\mathcal{M}_\pm$ given by (4.59). □

STEP 2. For every $k \in (0, 1)$, the self-adjoint operator $\mathcal{K}_- := \mathcal{M}_- - 2\mathcal{L}_-$ is positive in $L^2_{\rm Nper}$ and $\operatorname{Ker}(\mathcal{K}_-) = \operatorname{Span}\{u_0\}$.

PROOF OF STEP 2. It is clear that $u_0 \in \operatorname{Ker}(\mathcal{K}_-)$ due to the rotational symmetry in (4.58):

$$\mathcal{M}_- u_0 = 0, \quad \mathcal{L}_- u_0 = 0.$$

Next we show that the quadratic form $\langle \mathcal{K}_- v, v\rangle$ can be decomposed into a sum of squared quantities. We obtain by using the definition (4.59) and integrating by parts that

$$\begin{aligned} \|\mathcal{L}_- v\|_{L^2}^2 &= \int_0^{NT} \left[\frac{1}{4}v_{xx}^2 + (1 - u_0^2)v v_{xx} + (1 - u_0^2)^2 v^2\right] dx \\ &= \int_0^{NT} \left[\frac{1}{4}v_{xx}^2 - (1 - u_0^2)v_x^2 - (u_0 u_0')' v^2 + (1 - u_0^2)^2 v^2\right] dx. \end{aligned}$$

Similarly, we obtain

$$\|u_0 v_x - u_0' v\|_{L^2}^2 = \int_0^{NT} [u_0^2 v_x^2 + (u_0 u_0')' v^2 + (u_0')^2 v^2]\, dx.$$

As a consequence, we have

$$\begin{aligned} &\|\mathcal{L}_- v\|_{L^2}^2 + \frac{1}{2}\|u_0 v_x - u_0' v\|_{L^2}^2 \\ &\quad = \int_{\mathbb{R}} \left[\frac{1}{4}v_{xx}^2 + \frac{1}{2}(3u_0^2 - 2)v_x^2 + [(1 - u_0^2)^2 - \frac{1}{2}u_0 u_0'']v^2\right] dx. \end{aligned}$$

Since $(1 - u_0^2)^2 - \frac{1}{2}u_0 u_0'' = 1 - u_0^2$, this yields

$$\langle \mathcal{K}_- v, v\rangle = \|\mathcal{L}_- v\|_{L^2}^2 + \frac{1}{2}\|u_0 v_x - u_0' v\|_{L^2}^2,$$

where the right-hand side vanishes if and only if $v = C u_0$ for some constant C. □

STEP 3. For every $k \in (0, 1)$, the self-adjoint operator $\mathcal{K}_+ := \mathcal{M}_+ - 2\mathcal{L}_+$ is positive in $L^2_{\rm Nper}$ and $\operatorname{Ker}(\mathcal{K}_+) = \operatorname{Span}\{\partial_x u_0\}$.

PROOF OF STEP 3. It is clear that $\partial_x u_0 \in \text{Ker}(\mathcal{K}_+)$ due to the translational symmetry in (4.58):

$$\mathcal{M}_+ \partial_x u_0 = 0, \quad \mathcal{L}_+ \partial_x u_0 = 0.$$

The quadratic form $\langle \mathcal{K}_+ u, u\rangle$ does not admit such a nice representation as the one for $\langle \mathcal{K}_- v, v\rangle$ in Step 2. Therefore, we have to use a different method and to adopt a continuation argument from the limit $u_0 = 0$ $(k = 0)$ combined with analysis of the second-order Schrödinger operators with T-periodic coefficients.

For $u_0 = 0$ $(k = 0)$, we have

$$\langle \mathcal{K}_+ w, w\rangle = \int_0^{NT} \left[\frac{1}{4} w_{xx}^2 - w_x^2 + w^2\right] dx = \int_0^{NT} \left(\frac{1}{2} w_{xx} + w\right)^2 dx \geq 0. \tag{4.60}$$

The zero is attained at the double zero eigenvalue. The standard perturbation theory [**97**] shows that for (small) $k > 0$, the double zero eigenvalue of $\mathcal{K}_+$ in L^2_{Nper} splits into a simple zero and a simple positive eigenvalue. Hence, the claim is true for small $k \in (0, 1)$.

Next we develop the continuation argument for every $k \in (0, 1)$. It follows from the hierarchy of commuting Hamiltonian flows associated with H_0, H_2, and H_4 in (4.4) that the linearized flows at the periodic wave with the profile u_0 written as

$$\frac{\partial}{\partial t_1} \begin{bmatrix} w \\ v \end{bmatrix} = \begin{bmatrix} 0 & \mathcal{L}_- \\ -\mathcal{L}_+ & 0 \end{bmatrix} \begin{bmatrix} w \\ v \end{bmatrix}, \qquad \frac{\partial}{\partial t_2} \begin{bmatrix} w \\ v \end{bmatrix} = \begin{bmatrix} 0 & \mathcal{M}_- \\ -\mathcal{M}_+ & 0 \end{bmatrix} \begin{bmatrix} w \\ v \end{bmatrix}$$

also commute so that $\partial_{t_1} \partial_{t_2} = \partial_{t_2} \partial_{t_1}$. As a result, the self-adjoint operators $\mathcal{L}_\pm$ and $\mathcal{M}_\pm$ satisfy the following intertwining relations

$$\mathcal{L}_- \mathcal{M}_+ = \mathcal{M}_- \mathcal{L}_+, \qquad \mathcal{L}_+ \mathcal{M}_- = \mathcal{M}_+ \mathcal{L}_-, \tag{4.61}$$

which can also be verified by a direct calculation using the differential equations (4.56) and (4.57). If we can show that for every $k \in (0, 1)$, we have

$$\begin{aligned} &\text{Ker}(\mathcal{M}_+) = \text{Ker}(\mathcal{L}_+) = \text{Span}\{\partial_x u_0\} \\ \text{and} \quad &\text{Ker}(\mathcal{M}_-) = \text{Ker}(\mathcal{L}_-) = \text{Span}\{u_0\}, \end{aligned} \tag{4.62}$$

then the proof of Step 3 follows from positivity of $\mathcal{K}_+$ in (4.60) as $k \to 0$, simplicity of the zero eigenvalue for (small) $k \in (0, 1)$, and the continuation argument for every $k \in (0, 1)$. Indeed, all eigenvalues of $\mathcal{K}_+$ in L^2_{Nper} are strictly positive for (small) $k \in (0, 1)$ with a simple zero eigenvalue. Since it follows from (4.62) that eigenvalues of $\mathcal{K}_+$ do not cross the zero level in parameter continuations in k, the spectrum of $\mathcal{K}_+$ remains positive in L^2_{Nper} for every $k \in (0, 1)$.

Next we show the relations (4.62) for every $k \in (0, 1)$. Simplicity of $\text{Ker}(\mathcal{L}_-)$ in L^2_{Nper} follows from the general solution of $\mathcal{L}_- v = 0$ given by $v(x) = c_1 u_0(x) + c_2 v_0(x)$ with

$$v_0(x) = u_0'(x) - u_0(x) \int_0^x u_0(y)^2 dy,$$

which is nonperiodic in x. Simplicity of $\text{Ker}(\mathcal{M}_-)$ in L^2_{Nper} follows from simplicity of $\text{Ker}(\mathcal{K}_-)$ and $\text{Ker}(\mathcal{L}_-)$.

Simplicity of $\text{Ker}(\mathcal{L}_+)$ in L^2_{Nper} follows from the general solution of $\mathcal{L}_+ w = 0$ given by $w = c_1 \partial_x u_0 + c_2 \partial_{\mathcal{E}} u_0$, where the second solution is nonperiodic in x since the period function $T(\mathcal{E})$ does not have a critical point for $\mathcal{E} \in (0, 1)$ (Example 3.5).

Simplicity of $\text{Ker}(\mathcal{M}_+)$ in L^2_{Nper} follows from simplicity of $\text{Ker}(\mathcal{K}_+)$, which can be shown from the intertwining relation

$$\mathcal{L}_+ \mathcal{K}_- = \mathcal{K}_+ \mathcal{L}_-.$$

Consider all smooth $v \in H^{\infty}_{\text{Nper}}$ such that $\mathcal{K}_+\mathcal{L}_-v = 0$. Then, $\mathcal{L}_+\mathcal{K}_-v = 0$ so that $\mathcal{K}_-v = C\partial_x u_0$ for some constant C. As is shown in Example 3.16, there is another constant $B \neq 0$ such that $\tilde{\mathcal{L}}_-V_0 = B\partial_x u_0$, where V_0 is given in (3.83) and $\tilde{\mathcal{L}}_-$ is obtained after a scaling transformation $\mathcal{L}_- = (1+k^2)\tilde{\mathcal{L}}_-$. Using a similar scaling transformation

$$\mathcal{M}_- = (1+k^2)^2\tilde{\mathcal{M}}_-, \quad \tilde{\mathcal{M}}_- := \partial_z^4 - 6k^2\,\text{sn}(z;k)^2\partial_z + 2k^2(1+k^2)\,\text{sn}(z;k)^2 - (1+k^2)^2,$$

we obtain directly from (3.83) that

$$\tilde{\mathcal{M}}_-V_0 = 4\left[k^2 - \left(1 - \frac{E(k)}{K(k)}\right)(1+k^2)\right]\text{cn}(z;k)\,\text{dn}(z;k).$$

By using (3.84) for $\tilde{\mathcal{K}}_- := \tilde{\mathcal{M}}_- - 2(1+k^2)\tilde{\mathcal{L}}_-$, we finally obtain

$$\tilde{\mathcal{K}}_-V_0 = 4k^2\,\text{cn}(z;k)\,\text{dn}(z;k),$$

so that the equation $\mathcal{K}_-v = C\partial_x u_0$ has the general solution $v = C_1u_0 + C_2v_0$ for some constants C_1 and C_2, where v_0 is obtained from V_0 after the inverse scaling transformation. Hence, $\mathcal{L}_-v = C\partial_x u_0$ for some constant C and $\mathcal{K}_+w = 0$ with $w := \mathcal{L}_-v$ is satisfied if and only if $w = C\partial_x u_0$. □

END OF THE PROOF OF THEOREM 4.2. To summarize, we have shown for every $k \in (0,1)$ that the second variation of the functional $\Lambda_{\gamma=2}$ at the periodic wave with the profile u_0 is positive except for two neutral modes due to symmetries in (4.58). Perturbation $w + iv \in H^2_{\text{Nper}}$ must satisfy the following two constraints:

$$\langle \partial_x u_0, w(t,\cdot)\rangle = 0, \qquad \langle u_0, v(t,\cdot)\rangle = 0. \tag{4.63}$$

In a subspace of L^2_{Nper} with the two orthogonality conditions (4.63), the Hessian operator for $\Lambda_{\gamma=2}$ at u_0 is coercive and the orbital stability of the periodic wave with the profile u_0 in H^2_{Nper} follows by Theorem 2.3. □

REMARK 4.10. The proof of Theorem 4.2 shows that for $\epsilon \in (0,1]$, one can take $\delta = \epsilon/C_N$ for some constant $C_N \geq 1$ that depends on $N \in \mathbb{N}$. However, since 0 is attained at the periodic eigenfunctions, there is no coercivity for the perturbations in $L^2(\mathbb{R})$ in the limit of $N \to \infty$, since $C_N \to \infty$ as $N \to \infty$. This indicates that, although a given periodic wave is orbitally stable with respect to perturbations with arbitrary large period NT, the size of the stability basin becomes very small when N is large.

REMARK 4.11. The exact range (γ_-, γ_+) of values of γ for which the second variation of the functional Λ_γ at u_0 is positive except for two neutral modes in (4.62) is known from [**29**] and [**97**]. It is given by

$$\gamma_\pm = 2 \pm \frac{2k}{1+k^2}$$

with $\gamma = 2$ being the midpoint of the interval $[\gamma_-, \gamma_+]$.

The following lemma clarifies modulation equations for translational parameters along the orbit of the periodic wave.

LEMMA 4.3. *There exists* $\xi, \alpha \in C^1(\mathbb{R})$ *in the orthogonal decomposition*

$$e^{it+i\alpha(t)}u(t, x+\xi(t)) = u_0(x) + w(t,x) + iv(t,x), \quad x \in \mathbb{R}, \tag{4.64}$$

where $w(t,\cdot)+iv(t,\cdot)\in H^2_{\rm Nper}$ *satisfy* (4.63) *and* $\xi(t)$, $\alpha(t)$ *satisfy*

$$|\dot\xi(t)|+|\dot\alpha(t)|\le C\epsilon,\quad t\in\mathbb{R}, \tag{4.65}$$

for some positive constant C.

PROOF. Validity of the decomposition (4.64) can be shown similarly to what is done in Step 1 in the proof of Theorem 2.3. In order to control the time evolution of $\xi,\alpha\in C^1(\mathbb{R})$, we obtain the time evolution of (w,v) from the system

$$\begin{cases} w_t = \mathcal{L}_- v + \dot\xi(u_0'+w_x) - \dot\alpha v + (2u_0 w + w^2+v^2)v, \\ -v_t = \mathcal{L}_+ w - \dot\xi v_x - \dot\alpha(u_0+w) + (3u_0w+w^2+v^2)w + u_0v^2. \end{cases}$$

Using the orthogonality conditions (4.63), we eliminate the time derivatives w_t, v_t by taking the scalar product of the first equation with $\partial_x u_0$ and of the second equation with u_0. This gives the following linear system for the derivatives $\dot\xi$ and $\dot\alpha$:

$$\begin{aligned}\begin{bmatrix} -\|\partial_x u_0\|^2 - \langle \partial_x u_0, w_x\rangle & \langle \partial_x u_0, v\rangle \\ \langle u_0, v_x\rangle & \|u_0\|^2+\langle u_0,w\rangle\end{bmatrix}\begin{bmatrix}\dot\xi\\ \dot\alpha\end{bmatrix} \\ = \begin{bmatrix} \langle \partial_x u_0, \mathcal{L}_- v\rangle + \langle \partial_x u_0, (2u_0w+w^2+v^2)v\rangle \\ \langle u_0, \mathcal{L}_+ w\rangle + \langle u_0, (3u_0w+w^2+v^2)w + u_0v^2\rangle\end{bmatrix}.\end{aligned}$$

Since $\|w(\cdot,t)+iv(\cdot,t)\|_{H^2_{\rm Nper}}<\epsilon$ for all $t\in\mathbb{R}$ with small $\epsilon>0$, the matrix in the left-hand side is invertible, whereas the right-hand side is bounded by $C\epsilon$, so that the bound (4.65) holds for every $t\in\mathbb{R}$. □

REMARK 4.12. In the case when the profile u_0 is a saddle point of energy E, we would also need to use additional two constraints on the perturbation $w+iv$:

$$\langle u_0, w(t,\cdot)\rangle = 0,\quad \langle \partial_x u_0, v(t,\cdot)\rangle = 0,$$

which are related to the conditions of fixed mass H_0 and momentum H_1. These constraints are preserved in the time evolution of the linearized system but we would need to add two parameters to the decomposition (4.64) in order to preserve these constraints in the time evolution of the NLS equation. All of this is not needed if the profile u_0 is a minimizer of $\Lambda_{\gamma=2}:=H-2E$.

REMARK 4.13. Recall the operators

$$\mathcal{J} := \begin{bmatrix} 0 & 1\\ -1 & 0\end{bmatrix},\quad \mathcal{L} := \begin{bmatrix} \mathcal{L}_+ & 0 \\ 0 & \mathcal{L}_-\end{bmatrix},\quad \mathcal{M} := \begin{bmatrix} \mathcal{M}_+ & 0\\ 0 & \mathcal{M}_-\end{bmatrix},$$

where $\mathcal{L}_\pm$ and $\mathcal{M}_\pm$ are defined in (4.59). Denote

$$\mathcal{K} := \begin{bmatrix} \mathcal{K}_+ & 0 \\ 0 & \mathcal{K}_-\end{bmatrix},$$

where $\mathcal{K}_\pm := \mathcal{M}_\pm - 2\mathcal{L}_\pm$. Then

$$\mathcal{J}\mathcal{L}\mathcal{J}\mathcal{K} = \mathcal{J}\mathcal{K}\mathcal{J}\mathcal{L} \tag{4.66}$$

because the operators $\mathcal{J}\mathcal{L}$ and $\mathcal{J}\mathcal{K}$ commute due to the intertwining relations (4.61).

The following lemma shows that (4.66) and the positivity of the operators $\mathcal{K}_\pm$ imply the spectral stability of the periodic wave with the profile u_0. The result extends the statement of Theorem 3.10 valid for positive operators $\mathcal{L}_\pm$ to positive operators $\mathcal{K}_\pm$ commuting with nonpositive operators $\mathcal{L}_\pm$.

LEMMA 4.4. *The periodic wave with the profile u_0 is spectrally stable with respect to perturbations with multiple period NL. In other words, $\sigma(\mathcal{J}\mathcal{L}) \subset i\,\mathbb{R}$ in $L^2_{\rm Nper}$.*

PROOF. Let $\lambda \in \sigma(\mathcal{J}\mathcal{L})$ be an eigenvalue of the spectral problem $J\mathcal{L}\psi = \lambda\psi$ with the eigenfunction $\psi \in \mathrm{Dom}(\mathcal{J}\mathcal{L})$. If $\psi \in H^\infty_{\rm Nper}$, then (4.66) suggest that

$$\mathcal{J}\mathcal{L}\mathcal{J}\mathcal{K}\psi = \lambda\mathcal{J}\mathcal{K}\psi.$$

As $\mathcal{J}$ is invertible, we thus have $\mathcal{L}\mathcal{J}\mathcal{K}\psi = \lambda\mathcal{K}\psi$. If we now take the scalar product of both sides with ψ in $L^2_{\rm Nper}$, we obtain

$$\lambda\langle\psi, \mathcal{K}\psi\rangle_{L^2} = \langle\psi, \mathcal{L}\mathcal{J}\mathcal{K}\psi\rangle_{L^2} = -\langle\mathcal{J}\mathcal{L}\psi, \mathcal{K}\psi\rangle_{L^2} = -\bar{\lambda}\langle\psi, \mathcal{K}\psi\rangle_{L^2},$$

where we have used the fact that $\mathcal{L}$ is self-adjoint and $\mathcal{J}$ is skew-adjoint. If $\lambda \neq 0$, then ψ is not a linear combination of the two neutral eigenfunctions $(u_0', 0)$ and $(0, u_0)$. In that case, we have $\langle\psi, \mathcal{K}\psi\rangle_{L^2} > 0$ so that $\lambda = -\bar{\lambda}$, that is, $\lambda \in i\mathbb{R}$. This yields $\sigma(\mathcal{J}\mathcal{L}) \subset i\,\mathbb{R}$. □

REMARK 4.14. Spectral stability of periodic waves in the Korteweg–de Vries equation was shown in [**28**] by using explicit expressions for the eigenfunctions of the operator $\mathcal{J}\mathcal{L}$. Spectral stability of smooth periodic waves in the reduced Ostrovsky equation was shown in [**141**] by direct computations of a positive operator $\mathcal{K}$ from a linear combination of nonpositive operators $\mathcal{L}$ and $\mathcal{M}$ from two commuting flows. Spectral stability via positive commuting operators was developed in [**124**] in connection to the transverse stability of periodic waves in the Kadomtsev–Petviashvill-II equation.

4.3. Stability in Sobolev spaces of lower regularity

Well-posedness of integrable Hamiltonian systems in Sobolev spaces of lower regularity has recently become an active area of research, where many sharp results were obtained, e.g., for the KdV equation (Example 3.11) [**35, 158**], the cubic NLS equation (Example 4.1) [**161, 163**], and the derivative NLS equation (Example 4.2) [**126**]. Generalized conserved quantities were constructed in Sobolev spaces of lower regularity and were used to obtain orbital stability of solitary and periodic waves [**159, 162**].

We will focus this discussion on how stability of solitary waves can be obtained in $L^2(\mathbb{R})$ where only mass conservation can be used. To control deviations of the time-dependent solution from the orbit of solitary wave solutions we will use an algebraic tool called **Bäcklund–Darboux (BD) transformation**. The fact that such a transformation exists is an intrinsic property of any integrable Hamiltonian system.

The BD transformation generates a new solution of an integrable Hamiltonian system from another solution of the same system. For instance, a small neighborhood of the zero solution can be mapped to a neighborhood of the solitary wave and vice verse. The transformation involves solutions of the Lax system (4.2) defined for a specific choice of the spectral parameter λ. Since several integrable Hamiltonian systems may share the same spectral problem $\varphi_x = U(\lambda, u)\varphi$ with the differences appearing only in the time-evolution problem $\varphi_t = V(\lambda, u)\varphi$, the same analytical expression of the BD transformation is applicable to all systems of the same hierarchy.

Ideas of using the BD transformation for stability analysis of solitary waves go back to the works [**200**], [**202**], and [**130**] where the asymptotic stability of solitary waves was established for the KdV equation, the Toda lattice, and the related integrable equations. The following section relies on the works [**203**] and [**68**] where the orbital stability of solitary waves in $L^2(\mathbb{R})$ was obtained for the cubic NLS equation (Example 4.1) and the MTM system (Example 4.3), respectively. Asymptotic stability of solitary waves in a weighted $L^2(\mathbb{R})$ space can be obtained by combining the BD transformation with the steepest descent method [**73, 75**].

The BD transformation was also used in other contexts involving analysis of integrable Hamiltonian systems. For example, global existence of large-norm solutions of the derivative NLS equation has long been an open problem of analysis, which was solved only recently with the use of the inverse scattering transform methods in [**139, 186, 222**]. Global existence results were extended to the case of finitely many solitons by using the BD transformation in [**126, 220**].

4.3.1. Orbital stability of solitary waves in the NLS equation. Let us rewrite the cubic NLS equation (4.11) in the focusing case $\sigma = 1$ as

$$iu_t + \frac{1}{2}u_{xx} + |u|^2 u = 0. \tag{4.67}$$

The Lax equations (4.2) with (4.12) can be rewritten as

$$\partial_x \begin{bmatrix} \varphi_1 \\ \varphi_2 \end{bmatrix} = \begin{bmatrix} \lambda & u \\ -\bar{u} & -\lambda \end{bmatrix} \begin{bmatrix} \varphi_1 \\ \varphi_2 \end{bmatrix} \tag{4.68}$$

and

$$\partial_t \begin{bmatrix} \varphi_1 \\ \varphi_2 \end{bmatrix} = i \begin{bmatrix} \lambda^2 + \frac{1}{2}|u|^2 & \frac{1}{2}u_x + \lambda u \\ \frac{1}{2}\bar{u}_x - \lambda \bar{u} & -\lambda^2 - \frac{1}{2}|u|^2 \end{bmatrix} \begin{bmatrix} \varphi_1 \\ \varphi_2 \end{bmatrix}. \tag{4.69}$$

The spatial part (4.68) of the linear system is referred to as the **Zakharov–Shabat spectral problem** [**250**].

The family of solitary waves of the NLS equation (4.67) has generally four arbitrary parameters:

$$u(t,x) = U_{k,c}(t - t_0, x - x_0), \quad U_{k,c}(t,x) := Q_k(x - ct)\, e^{icx + \frac{i}{2}(k^2 - c^2)t}, \tag{4.70}$$

where $Q_k(x) := k\,\mathrm{sech}(kx)$, $k > 0$, $c \in \mathbb{R}$, $x_0 \in \mathbb{R}$, and $t_0 \in \mathbb{R}$. The family of solitary waves (4.70) can be obtained from the zero solution with the BD transformation.

The following lemma defines the BD transformation for NLS equation (4.67).

LEMMA 4.5. *Let u be a C^1 solution of the NLS equation* (4.67) *and $\varphi = (p,q)^T$ be a nonzero C^2 solution of the linear system* (4.68) *and* (4.69) *with the specific value $\lambda = \eta$. Then, the function*

$$\hat{u} := u + \frac{2(\eta + \bar{\eta})p\bar{q}}{|p|^2 + |q|^2} \tag{4.71}$$

is a new solution of the same NLS equation (4.67) *and the vector*

$$\begin{bmatrix} \hat{p} \\ \hat{q} \end{bmatrix} := \frac{(\eta + \bar{\eta})}{|p|^2 + |q|^2} \begin{bmatrix} -\bar{q} \\ \bar{p} \end{bmatrix} \tag{4.72}$$

is a solution of the linear system (4.68) *and* (4.69) *with $\lambda = \eta$ and the new solution $\hat{u}$.*

PROOF. The proof is based on substitutions and can be easily verified with the use of symbolic software packages. □

EXAMPLE 4.4. If we take the zero solution ($u = 0$) of the NLS equation (4.67) and set $\eta = \frac{1}{2}(k + ic)$ with $k > 0$ and $c \in \mathbb{R}$, then the solution $\varphi = (p, q)^T$ of the Lax equations (4.68) and (4.69) can be written in the form

$$p(t,x) = e^{\frac{1}{2}(k+ic)x+\frac{i}{4}(k+ic)^2t}, \quad q(t,x) = -e^{-\frac{1}{2}(k+ic)x-\frac{i}{4}(k+ic)^2t}.$$

The BD transformation (4.71) yields $\hat{u}$ as the solitary wave solution in the form (4.70) with $t_0 = x_0 = 0$. The corresponding solution $\varphi = (\hat{p}, \hat{q})^T$ of the Lax equations with $u = \hat{u}$ and $\lambda = \eta$ obtained from (4.72) can be written in the form:

$$\hat{p}(t,x) = \frac{1}{2}Q_k(x - ct)e^{-\frac{1}{2}(k-ic)x+\frac{i}{4}(k-iv)^2t},$$

$$\hat{q}(t,x) = \frac{1}{2}Q_k(x - ct)e^{\frac{1}{2}(k-ic)x-\frac{i}{4}(k-ic)^2t}.$$

Since $\varphi = (\hat{p}, \hat{q})^T$ is exponentially decaying as $|x| \to \infty$ for every $t \in \mathbb{R}$, $\lambda = \eta$ is an eigenvalue of the spectral problem (4.68) with $u = \hat{u}$. If the BD transformation is iterated with $\varphi = (\hat{p}, \hat{q})^T$ and $u = \hat{u}$ as

$$\hat{\hat{u}} = \hat{u} + \frac{2(\eta + \bar{\eta})\hat{p}\bar{\hat{q}}}{|\hat{p}|^2 + |\hat{q}|^2},$$

then we return back to the zero solution $\hat{\hat{u}} = 0$.

By using Lemma 4.5 and the well-posedness theory for the NLS equation in $L^2(\mathbb{R})$ [**181**], we prove the following theorem, which tells us that solutions of the NLS equation which are close initially to the solitary wave in the $L^2(\mathbb{R})$-norm remain close to a nearby solitary wave in the family (4.70) for all times and the four parameters of the family are almost the same as those of the original solitary wave. Without loss of generality, we can give the result for $k = 1$ and $c = 0$ due to symmetries of the NLS equation (4.67).

THEOREM 4.3. *Let $u(t,x)$ be a unique solution of the NLS equation* (4.67) *in the class*

$$u \in C^0(\mathbb{R}_+; L^2(\mathbb{R})) \cap L^8_{loc}(\mathbb{R}_+; L^4(\mathbb{R})).$$

There exist positive constants C and ε such that if

$$\|u(0,\cdot) - Q_1\|_{L^2} \le \varepsilon, \tag{4.73}$$

then there exist real constants k_0 and c_0 such that

$$\inf_{\xi,\alpha\in\mathbb{R}} \|e^{i\alpha}u(t,\cdot+\xi) - U_{k_0,c_0}\|_{L^2} + |k_0 - 1| + |c_0| \le C\|u(0,\cdot) - Q_1\|_{L^2}, \tag{4.74}$$

for every $t > 0$.

PROOF. First, we fix an initial datum $u(0,\cdot) \in H^3(\mathbb{R})$ such that $u(0,\cdot)$ is sufficiently close to Q_1 in L^2-norm, according to the bound (4.73). We perform three steps, see Figure 4.1, in which we map the initial data near Q_1 to a small ball, use the well-posedness theory of the NLS equation in $H^3(\mathbb{R})$ and in $L^2(\mathbb{R})$, and map the time-dependent solution in a small ball in $L^2(\mathbb{R})$ to the time-dependent solution $u \in C^0(\mathbb{R}_+, H^3(\mathbb{R}))$ near U_{k_0,c_0} with appropriately defined (k_0, c_0) near $(1,0)$. Finally, we use an approximating sequence of solutions in $H^3(\mathbb{R})$ in order to prove the bound (4.74) for the limiting solution in $L^2(\mathbb{R})$.

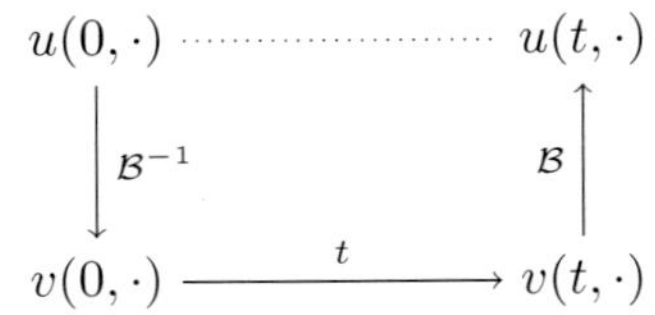

FIGURE 4.1. A diagram explaining the proof of Theorem 4.3 using the BD transformation $\mathcal{B}(u) = \hat{u}$ given in (4.71).

STEP 1. There exist positive constants C, ε and real constants k, c such that if the bound (4.73) is satisfied for $u_0 := u(0, \cdot) \in H^3(\mathbb{R})$, then there exist a solution $\varphi = (\varphi_1, \varphi_2)^T \in H^1(\mathbb{R})$ of the spectral problem (4.68) with $\lambda = \frac{1}{2}(k_0 + ic_0)$ such that

$$|k_0 - 1| + |c_0| \leq C\|u_0 - Q_1\|_{L^2}\,. \tag{4.75}$$

Moreover,

$$v_0 := u_0 + \frac{2k_0\varphi_1\overline{\varphi}_2}{|\varphi_1|^2 + |\varphi_2|^2} \tag{4.76}$$

satisfies $v_0 \in H^3(\mathbb{R})$ and $\|v_0\|_{L^2} \leq C\|u_0 - Q_1\|_{L^2}$.

PROOF OF STEP 1. This step is proven with the Lyapunov–Schmidt reduction method for a simple eigenvalue. The spectral problem (4.68) is equivalent to an eigenvalue problem

$$L\varphi - M(u_0 - Q_1)\varphi = \left(\lambda - \frac{1}{2}\right)\varphi, \tag{4.77}$$

where $L : H^1(\mathbb{R}, \mathbb{C}^2) \subset L^2(\mathbb{R}, \mathbb{C}^2) \to L^2(\mathbb{R}, \mathbb{C}^2)$ and $M : L^2(\mathbb{R}, \mathbb{C}^2) \to L^2(\mathbb{R}, \mathbb{C}^2)$ are given by

$$L := \begin{bmatrix} \partial_x - \frac{1}{2} & -Q_1 \\ -Q_1 & -\partial_x - \frac{1}{2} \end{bmatrix}, \quad M(u) := \begin{bmatrix} 0 & u \\ \bar{u} & 0 \end{bmatrix}.$$

The fundamental solution $\{\Psi_1(x), \Psi_2(x)\}$ of $L\varphi = 0$ is given by

$$\Psi_1(x) = \begin{bmatrix} e^{-x/2} \\ e^{x/2} \end{bmatrix} \operatorname{sech}(x), \quad \Psi_2(x) = \begin{bmatrix} (e^x + 2(1+x)e^{-x})e^{x/2} \\ -(e^{-x} - 2xe^x)e^{-x/2} \end{bmatrix} \operatorname{sech}(x)$$

We have $\mathrm{Ker}(L) = \mathrm{Span}\{\Psi_1\}$ and $\mathrm{Ker}(L^*) = \mathrm{Span}\{\mathbf{\Theta}_1\}$ with

$$\mathbf{\Theta}_1(x) = \begin{bmatrix} e^{x/2} \\ e^{-x/2} \end{bmatrix} \operatorname{sech}(x).$$

Writing the decomposition as

$$u_0 = Q_1 + S, \quad \lambda = \frac{1}{2} + \Lambda, \quad \phi = \Psi_1 + \mathbf{\Phi}, \quad \langle \Psi_1, \mathbf{\Phi}\rangle_{L^2} = 0,$$

we can decompose the eigenvalue problem (4.77) in the block-diagonal form

$$L\mathbf{\Phi} = \Pi_1\left[(\Lambda I + M(S))(\Psi_1 + \mathbf{\Phi})\right], \tag{4.78}$$

and

$$\langle \mathbf{\Theta}_1, (\Lambda I + M(S))(\Psi_1 + \mathbf{\Phi}))\rangle_{L^2} = 0, \tag{4.79}$$

where Π_1 be the spectral projection operator given by

$$\Pi_1 \mathbf{\Phi} := \mathbf{\Phi} - \frac{1}{4}\langle \mathbf{\Phi}, \mathbf{\Theta}_1 \rangle_{L^2} \Psi_1.$$

Since $L_0 := \mathrm{diag}(\partial_x - \frac{1}{2}, -\partial_x - \frac{1}{2})$ is a closed operator on $L^2(\mathbb{R};\mathbb{C}^2)$ with $\mathrm{Ran}(L_0) = L^2(\mathbb{R};\mathbb{C}^2)$ and $M(Q_1)L_0^{-1}$ is a compact operator on $L^2(\mathbb{R};\mathbb{C}^2)$, it follows that L is a Fredholm operator with

$$\mathrm{Ran}(L) = \{\mathbf{\Phi} \in L^2(\mathbb{R},\mathbb{C}^2) : \quad \langle \mathbf{\Phi}, \mathbf{\Theta}_1 \rangle_{L^2} = 0\}.$$

Thus we can define L^{-1} as a bounded operator

$$L^{-1} : L^2(\mathbb{R},\mathbb{C}^2)|_{\{\mathrm{Ker}(L^*)\}^\perp} \to H^1(\mathbb{R},\mathbb{C}^2)|_{\{\mathrm{Ker}(L)\}^\perp}.$$

If $S \in L^2(\mathbb{R})$ and $\Lambda \in \mathbb{C}$ are small, there exists a unique solution $\mathbf{\Phi} \in H^1(\mathbb{R}^2,\mathbb{C}^2)$ of the linear inhomogeneous equation (4.78) such that

$$\|\mathbf{\Phi}\|_{H^1(\mathbb{R},\mathbb{C}^2)} \leq C(\|S\|_{L^2} + |\Lambda|)\,, \tag{4.80}$$

where C is a constant that does not depend on S and Λ. On the other hand, the scalar equation (4.79) can be written in the form

$$\langle \Theta_1, \Psi \rangle_{L^2} = 2\langle Q_1, \mathrm{Re}(S)\rangle_{L^2} - 2i\langle \partial_x Q_1, \mathrm{Im}(S)\rangle_{L^2} \langle \Theta_1, M(S)\Psi\rangle_{L^2}.$$

In view of the bound (4.80), the latter equation gives

$$\left|\Lambda - \frac{1}{2}\langle Q_1, \mathrm{Re}(S)\rangle_{L^2} + \frac{i}{2}\langle \partial_x Q_1, \mathrm{Im}(S)\rangle_{L^2}\right| \leq C\|S\|_{L^2}^2\,,$$

which concludes the proof of the bound (4.75) for $\Lambda = \frac{1}{2}(k_0 - 1) + \frac{i}{2}c_0$. To prove the second statement, we use $\phi = \psi_1 + \Psi$ and write

$$\begin{aligned}
&\frac{2k_0\varphi_1\overline{\varphi}_2}{|\varphi_1|^2 + |\varphi_2|^2} \\
&\qquad = 2k_0 e^{i(v_0 x(x-\xi)+\alpha)} \frac{1 + \varepsilon_1(x) + e^{k_0(x-\xi)}\varepsilon_2(x) + e^{-k_0(x-\xi)}\varepsilon_3(x)}{e^{k_0(x-\xi)}(1+\varepsilon_4(x)) + \varepsilon_5(x) + e^{-k_0(x-\xi)}(1+\varepsilon_6(x))}\,,
\end{aligned}$$

in terms of some functions $\varepsilon_1, \ldots, \varepsilon_6$ found in [**202**, Lemma 3.3] satisfying

$$\begin{aligned}
\|\varepsilon_1\|_{L^\infty} + \|\varepsilon_2\|_{L^2\cap L^\infty} + \|\varepsilon_3\|_{L^2\cap L^\infty} + \|\varepsilon_4\|_{L^\infty} + \|\varepsilon_5\|_{L^2\cap L^\infty} + \|\varepsilon_6\|_{L^\infty} + |\alpha| + |\xi| \\
\leq C\|u_0 - Q_1\|_{L^2},
\end{aligned}$$

as a generalization of the bound (4.80) with a u-independent constant $C > 0$. Combining this bound with the expansion,

$$\begin{aligned}
&\frac{1 + \varepsilon_1(x) + e^{k_0(x-\xi)}\varepsilon_2(x) + e^{-k_0(x-\xi)}\varepsilon_3(x)}{e^{k_0(x-\xi)}(1+\varepsilon_4(x)) + \varepsilon_5(x) + e^{-k_0(x-\xi)}(1+\varepsilon_6(x))} \\
&= \frac{1 + \varepsilon_1(x)}{e^{k_0(x-\xi)}(1+\varepsilon_4(x)) + \varepsilon_5(x) + e^{-k_0(x-\xi)}(1+\varepsilon_6(x))} + \mathcal{O}(|\varepsilon_2(x)| + |\varepsilon_3(x)|) \\
&= \frac{1}{2}\,\mathrm{sech}\,(k_0(x-\xi))\,(1 + \mathcal{O}(|\varepsilon_1(x)| + |\varepsilon_4(x)| + |\varepsilon_5(x)| + |\varepsilon_6(x)|)) \\
&\quad + \mathcal{O}(|\varepsilon_2(x)| + |\varepsilon_3(x)|)\,,
\end{aligned}$$

we obtain with the triangle inequality that

$$\|v_0\|_{L^2} \leq \|u_0 - Q_1\|_{L^2} + \left\|Q_1 - \frac{2k_0\varphi_1\overline{\varphi}_2}{|\varphi_1|^2 + |\varphi_2|^2}\right\|_{L^2} \leq C\|u_0 - Q_1\|_{L^2},$$

for another u-independent constant $C > 0$. Finally, if $u_0 \in H^3(\mathbb{R})$, then the first three derivatives of $\varepsilon_1, \ldots, \varepsilon_6$ are bounded in their relevant norms so that $v_0 \in H^3(\mathbb{R})$. □

STEP 2. For every $v_0 \in L^2(\mathbb{R})$ in Step 1, there exists a unique solution

$$v \in C^0(\mathbb{R}; L^2(\mathbb{R})) \cap L^8_{loc}(\mathbb{R}; L^4(\mathbb{R}))$$

of the NLS equation (4.67) such that $v(0, \cdot) = v_0$ and

$$\|v(t, \cdot)\|_{L^2} = \|v(0, \cdot)\|_{L^2}.$$

Moreover, if $v_0 \in H^3(\mathbb{R})$, then there exists a unique solution $v \in C^0(\mathbb{R}, H^3(\mathbb{R}))$ of the NLS equation (4.67) for which the L^2 norm is preserved.

PROOF OF STEP 2. The global well-posedness result in $L^2(\mathbb{R})$ follows by [**181**, Theorem 5.2]. Similarly, the global well-posedness theory holds in $H^k(\mathbb{R})$ for every $k \in \mathbb{N}$ [**26**]. Mass is conserved both in $L^2(\mathbb{R})$ and in $H^k(\mathbb{R})$, $k \in \mathbb{N}$. □

STEP 3. Let $v \in C^0(\mathbb{R}, H^3(\mathbb{R}))$ be a solution of the NLS equation (4.67) in Step 2 such that $\|v(0, \cdot)\|_{L^2} < \varepsilon$. The Lax system (4.68) and (4.69) with $\lambda = \frac{1}{2}$ admits a general solution in the form

$$\varphi(t, x) = c_1 e^{x/2} \phi(t, x) + c_2 e^{-x/2} \chi(t, x),$$

where $c_1, c_2 \in \mathbb{C}$ and the two components of ϕ and χ satisfy the bounds

$$\|\phi_1(t, \cdot) - e^{it/2}\|_{L^\infty} + \|\phi_2(t, \cdot)\|_{L^2 \cap L^\infty} \leq C\|v(0, \cdot)\|_{L^2}, \tag{4.81}$$

$$\|\chi_1(t, \cdot)\|_{L^2 \cap L^\infty} + \|\chi_2(t, \cdot) + e^{-it/2}\|_{L^\infty} \leq C\|v(0, \cdot)\|_{L^2}. \tag{4.82}$$

The complex parameters c_1 and c_2 may depend on time t but are independent of x. Let $c_1 = ae^{(\xi + i\alpha)/2}$, $c_2 = ae^{-(\xi + i\alpha)/2}$ with $a \neq 0$, $\xi \in \mathbb{R}$, $\alpha \in \mathbb{R}$ and define

$$u(t, x) = v(t, x) + \frac{2\varphi_1(t, x)\overline{\varphi}_2(t, x)}{|\varphi_1(t, x)|^2 + |\varphi_2(t, x)|^2}. \tag{4.83}$$

Then $u \in C^0(\mathbb{R}, H^3(\mathbb{R}))$ is a solution of the NLS equation (4.67) and there is an ε-independent constant $C > 0$ such that

$$\|u(t, \cdot) - e^{i(t+\alpha)} Q_1(\cdot - \xi)\|_{L^2} \leq C\|v(0, \cdot)\|_{L^2} \tag{4.84}$$

for every $t > 0$.

PROOF OF STEP 3. Since the proof of (4.81) and (4.82) is identical for every $t \in \mathbb{R}$, we give it for $t = 0$ and drop the time variable from the list of arguments for now. The fundamental system of solutions of the spectral problem (4.68) with $u = 0$ and $\lambda = \frac{1}{2}$ is given by

$$\begin{bmatrix} e^{x/2} \\ 0 \end{bmatrix}, \quad \begin{bmatrix} 0 \\ e^{-x/2} \end{bmatrix}.$$

When v is small in the $L^2(\mathbb{R})$ norm, a fundamental system of the spectral problem (4.68) with $u = v$ and $\lambda = \frac{1}{2}$ is found as a perturbation of the two linearly independent solutions. Separating variables as $e^{x/2}\phi(x)$ and $e^{-x/2}\chi(x)$ yields the following systems

$$\begin{cases} \phi_1' = v\phi_2, \\ \phi_2' = -\bar{v}\phi_1 - \phi_2, \end{cases} \qquad \begin{cases} \chi_1' = \chi_1 + v\chi_2, \\ \chi_2' = -\bar{v}\chi_1, \end{cases}$$

subject to the boundary conditions

$$\begin{cases} \lim_{x\to+\infty} \phi_1(x) = 1, \\ \lim_{x\to-\infty} e^x \phi_2(x) = 0, \end{cases} \qquad \begin{cases} \lim_{x\to+\infty} e^{-x}\chi_1(x) = 0, \\ \lim_{x\to-\infty} \chi_2(x) = 1\,. \end{cases}$$

Existence and uniqueness of solutions of the two boundary-value problems is studied in the class of functions

$$(\phi_1\,,\phi_2) \in L^\infty \times (L^2 \cap L^\infty)\,, \quad (\chi_1\,,\chi_2) \in (L^2 \cap L^\infty) \times L^\infty.$$

satisfying the bounds (4.81) and (4.82) for $t = 0$. To do so, we translate the boundary-value problem for ϕ into a system of integral equations

$$\text{(4.85)} \qquad \begin{cases} \phi_1 = T_1(\phi_1, \phi_2), & T_1(\phi_1,\phi_2)(x) := 1 - \int_x^\infty v(y)\phi_2(y)dy, \\ \phi_2 = T_2(\phi_1, \phi_2), & T_2(\phi_1,\phi_2)(x) := -\int_{-\infty}^x e^{-(x-y)}\overline{v(y)}\phi_1(y)dy. \end{cases}$$

We claim that $T = (T_1, T_2) : Z \to Z$ is a contraction mapping on the Banach space $Z := L^\infty \times (L^\infty \cap L^2)$ equipped with the norm

$$\|(u_1, u_2)\|_Z = \|u_1\|_{L^\infty} + \|u_2\|_{L^\infty\cap L^2}\,.$$

Using Schwarz's inequality and Young's inequality, we obtain

$$\begin{aligned} \|T_1(\phi_1,\phi_2) - T_1(\tilde\phi_1,\tilde\phi_2)\|_{L^\infty} &= \sup_{x\in\mathbb{R}} \left| \int_x^\infty v(y)(\phi_2(y) - \tilde\phi_2(y))dy \right| \\ &\le \|v\|_{L^2}\|\phi_2 - \tilde\phi_2\|_{L^2} \end{aligned}$$

and

$$\begin{aligned} \|T_2(\phi_1,\phi_2) - T_2(\tilde\phi_1,\tilde\phi_2)\|_{L^2\cap L^\infty} &= \left\| \int_{-\infty}^x e^{-(x-y)}\overline{v}(y)(\phi_1(y) - \tilde\phi_1(y))dy \right\|_{L^2\cap L^\infty} \\ &\le \|e^{-\cdot}\|_{L^1(\mathbb{R}_+)\cap L^2(\mathbb{R}_+)}\|v\|_{L^2}\|\phi_1 - \tilde\phi_1\|_{L^\infty} \\ &\le \|v\|_{L^2}\|\phi_1 - \tilde\phi_1\|_{L^\infty} \end{aligned}$$

If $\|v\|_{L^2} \le \varepsilon$ is sufficiently small, then $T = (T_1, T_2)$ is a contraction mapping on Z. Therefore $T = (T_1\,, T_2)$ has a unique fixed point $(\phi_1, \phi_2) \in Z$ and

$$\begin{aligned} \|\phi_1 - 1\|_{L^\infty} + \|\phi_2\|_{L^\infty\cap L^2} &\le \|v\|_{L^2}\|(\phi_1,\phi_2)\|_Z \\ &\le \|v\|_{L^2}(1 + \|\phi_1 - 1\|_{L^\infty} + \|\phi_2\|_{L^\infty\cap L^2})\,, \end{aligned}$$

which gives the bound (4.81) for $t = 0$. The boundary condition $\lim_{x\to\infty}\phi_1(x) = 1$ is obtained from (4.85) and the fact that $v \in L^2(\mathbb{R})$ and $\phi_2 \in L^2(\mathbb{R})$. The boundary condition $\lim_{x\to-\infty} e^x\phi_2(x) = 0$ is obtained since ϕ_2 is bounded.

The boundary-value problem for $\chi(x)$ is solved similarly, which yields the bound (4.82) for $t = 0$. The result for every $t \in \mathbb{R}$ is obtained in exactly the same way because $\|v(t,\cdot)\|_{L^2} = \|v(0,\cdot)\|_{L^2}$ remains small for every $t \in \mathbb{R}$.

Let us now control the linear time evolution of the two solutions $\varphi(t,x) = e^{x/2}\phi(t,x)$, $\varphi(t,x) = e^{-x/2}\chi(t,x)$ of the Lax equation (4.69), which can be rewritten in the matrix form

$$\text{(4.86)} \qquad \partial_t\varphi(t,x) = A(t,x)\varphi(t,x)\,,$$

where $A(t,x)$ is obtained from (4.69) at $\lambda = \frac{1}{2}$. The two solutions satisfy the following system for every $t \in \mathbb{R}$:

$$\begin{cases} \partial_x \phi_1(t,x) = v(t,x)\phi_2(t,x), \\ \partial_x \phi_2(t,x) = -\bar{v}(t,x)\phi_1(t,x) - \phi_2(t,x), \end{cases} \tag{4.87}$$
$$\begin{cases} \partial_x \chi_1(t,x) = \chi_1(t,x) + v(t,x)\chi_2(t,x)\,, \\ \partial_x \chi_2(t,x) = -\bar{v}(t,x)\chi_1(t,x), \end{cases}$$

subject to the boundary conditions

$$\begin{cases} \lim\limits_{x\to+\infty} \phi_1(t,x) = e^{\frac{i}{4}t}, \\ \lim\limits_{x\to-\infty} e^x\phi_2(t,x) = 0, \end{cases} \qquad \begin{cases} \lim\limits_{x\to+\infty} e^{-x}\chi_1(t,x) = 0, \\ \lim\limits_{x\to-\infty} \chi_2(t,x) = e^{-\frac{i}{4}t}. \end{cases}$$

Again, we only deal with the boundary-value problem for $\phi(t,x)$ as the boundary-value problem for $\chi(t,x)$ is similar.

The coefficient matrix $A(t,x)$ of the system (4.86) is continuous in (t,x) and C^1 in x since $v(t,x) \in C^0(\mathbb{R}, H^3(\mathbb{R}))$ so that $\phi(0,x)$ is C^1 in x. Solving the Cauchy problem for the linear evolution equation (4.86), we find that $\phi_{1,2}(t,x)$ are in $C^1(\mathbb{R}\times\mathbb{R})$. By a bootstrapping argument, we conclude that $\partial_x\partial_t\phi(t,x)$ and $\partial_t\partial_x\phi(t,x)$ are in $C^0(\mathbb{R}\times\mathbb{R})$ and thus $\partial_x\partial_t\phi(t,x) = \partial_t\partial_x\phi(t,x)$. Let

$$B(t,x) = \begin{bmatrix} 0 & v(t,x) \\ -\overline{v(t,x)} & -1 \end{bmatrix}, \quad \mathbf{F}(t,x) = \partial_x\phi(t,x) - B(t,x)\phi(t,x)\,.$$

Since v is a solution of the NLS equation (4.67), the matrices A and B satisfy the compatibility condition $\partial_x A - \partial_t B + [A\,,B] = 0$. As a result, we obtain

$$\begin{aligned} \partial_t \mathbf{F} =& \partial_t\partial_x\phi - (\partial_t B)\phi - B\partial_t\phi \\ =& \partial_x(A\phi) - (\partial_t B)\phi - BA\phi \\ =& (\partial_x A + [A\,,\,B] - \partial_t B)\phi + A\mathbf{F} \\ =& A\mathbf{F}\,. \end{aligned}$$

Applying Gronwall's equality, we see that for any $T > 0$, there exists a constant $C(T)$ such that

$$|\mathbf{F}(t)| \le C(T)|\mathbf{F}(0)|, \quad t \in [-T,T].$$

Since $\mathbf{F}(0) = \mathbf{0}$ by the assumption, it follows that $\mathbf{F}(t) = \mathbf{0}$ for every $t \in \mathbb{R}$. Thus we have proven that ϕ satisfies the first system in (4.87) for every $t \in \mathbb{R}$.

Next we will prove $\phi_1(t,\cdot) \in L^\infty(\mathbb{R})$ and $\phi_2(t,\cdot) \in L^2(\mathbb{R}) \cap L^\infty(\mathbb{R})$ for every $t \in \mathbb{R}$. By the linear evolution (4.86), we have

$$\begin{aligned} \left|\partial_t(|\phi_1(t,x)|^2 + |\phi_2(t,x)|^2)\right| &= 4\left|\operatorname{Im} v(t,x)\overline{\phi_1(t,x)}\phi_2(t,x)\right| \\ &\le 2\|v(t,\cdot)\|_{L^\infty}(|\phi_1(t,x)|^2 + |\phi_2(t,x)|^2). \end{aligned}$$

Applying Gronwall's inequality again, we have

$$|\phi_1(t,x)|^2 + |\phi_2(t,x)|^2 \le e^{\alpha|t|}(|\phi_1(0,x)|^2 + |\phi_2(0,x)|^2), \quad t \in \mathbb{R}$$

where $\alpha = 2\|v\|_{L^\infty(\mathbb{R}\times\mathbb{R})}$. Since $\phi(0,\cdot) \in L^\infty(\mathbb{R})$, it follows that $\phi(t,\cdot) \in L^\infty(\mathbb{R})$ for every $t \in \mathbb{R}$. Using the linear system (4.86) again, we have

$$\begin{aligned} \partial_t|\phi_2(t,x)|^2 &\le 2|\partial_x v(t,x) - v(t,x)|\phi_1(t,x)||\phi_2(t,x)| \\ &\le |\phi_2(t,x)|^2 + |\phi_1(t,x)|^2|\partial_x v(t,x) - v(t,x)|^2. \end{aligned}$$

By Gronwall's inequality, for any $T > 0$ there exists a $C(T) > 0$ such that

$$|\phi_2(t,x)|^2 \leq |\phi_2(s,x)|^2 + C(T)\int_s^t |\partial_x v(\tau,x) - v(\tau,x)|^2 d\tau, \quad 0 \leq s \leq t \leq T, \quad x \in \mathbb{R}.$$

Therefore, we have

$$\|\phi_2(t,\cdot)\|_{L^2}^2 \leq \|\phi_2(s,\cdot)\|_{L^2}^2 + C(T)\int_s^t \|\partial_x v(\tau,\cdot) - v(\tau,\cdot)\|_{L^2}^2 d\tau.$$

Since $\phi_2(0,\cdot) \in L^2(\mathbb{R})$, it follows that $\phi_2(t,\cdot) \in L^2(\mathbb{R})$ for every $t \in \mathbb{R}$ and $\|\phi_2(t,\cdot)\|_{L^2}$ is continuous in t. Since $A(t,\cdot) \in C^0(\mathbb{R}, H^2(\mathbb{R}))$ and $\|\phi_1(t,\cdot)\|_{L^\infty}$ and $\|\phi_2(t,\cdot)\|_{L^2 \cap L^\infty}$ are bounded locally in time, the linear system (4.86) implies that $\phi_1(t,\cdot)$ and $\phi_2(t,\cdot)$ are continuous in $L^\infty(\mathbb{R})$ and thus $\phi_2(t,\cdot)$ is continuous in $L^2(\mathbb{R})$.

It remains to prove the boundary conditions for $\phi(t,x)$. Since $\phi_2(t,x)$ is bounded in x for every $t \in \mathbb{R}$, we have $\lim_{x\to-\infty} e^x \phi_2(t,x) = 0$. By the variation of constants formula, we have

$$\phi(t,x) = e^{\frac{i}{4}\sigma_3 t}\phi(0,x) + \int_0^t e^{\frac{i}{4}\sigma_3(t-s)} A_1(s,x)\phi(s,x)ds,$$

where $A_1(t,x) := A(t,x) - \frac{i}{4}\sigma_3$, where $\sigma_3 = \mathrm{diag}(1,-1)$. By the assumption that $v \in C^0(\mathbb{R}, H^3(\mathbb{R}))$, we have

$$\sup_{x\in\mathbb{R}} \sup_{0\leq s\leq t} |A_1(s,x)| < \infty \quad \text{and} \quad \lim_{x\to\pm\infty} A_1(s,x) = 0.$$

Applying Lebesgue's dominated convergence theorem to the integral equation yields

$$\lim_{x\to\infty} \left|\phi(t,x) - e^{\frac{i}{4}\sigma_3 t}\phi(0,x)\right| = 0\,.$$

Combining this with the boundary condition

$$\lim_{x\to+\infty} \phi_1(0,x) = 1 \text{ yields } \lim_{x\to+\infty} \phi_1(t,x) = e^{\frac{i}{4}t}$$

for every $t \in \mathbb{R}$.

Finally, we use the BD transformation (4.83) and specify properties of the new solution $u \in C^0(\mathbb{R}, H^3(\mathbb{R}))$ of the NLS equation (4.67) satisfying the bound (4.84). Since φ solves the Lax system (4.68) and (4.69), Lemma 4.5 implies that if $v(t,x)$ is a solution of (4.67), so is $u(t,x)$. Let

$$\begin{aligned} R(t,x) :=& u(t,x) - v(t,x) \\ =& \frac{2(c_1\phi_1(t,x) + c_2 e^{-x}\chi_1(t,x))(\overline{c_1 e^x \phi_2(t,x)} + \overline{c_2\chi_2(t,x)})}{|c_1\phi_1(t,x) + c_2 e^{-x}\chi_1(t,x)|^2 e^x + |c_1 e^x\phi_2(t,x) + c_2\chi_2(t,x)|^2 e^{-x}}. \end{aligned}$$

For $x \geq -\xi$, we have

$$R(t,x) = \frac{2e^{-x-\xi+i\alpha}\phi_1(t,x)\overline{\chi_2(t,x)}}{|\phi_1(t,x)|^2 + e^{-2(x+\xi)}|\chi_2(t,x)|^2} + \mathcal{O}(|\phi_2(t,x)| + e^{-x-\xi}|\chi_1(t,x)|)$$

since $|\phi_1|$, $|\chi_2| \sim 1$ and ϕ_2, $\chi_1 \sim 0$. Similarly, for $x \leq -\xi$, we have

$$R(t,x) = \frac{2e^{x+\xi+i\alpha}\phi_1(t,x)\overline{\chi_2(t,x)}}{|\chi_2(t,x)|^2 + e^{2(x+\xi)}|\phi_1(t,x)|^2} + \mathcal{O}(|\chi_1(t,x)| + e^{x+\xi}|\phi_2(t,x)|)\,.$$

Combining these two estimates yields

$$\begin{aligned} |R(t,x) - e^{\frac{i}{2}(t+\alpha)}\,\mathrm{sech}(x+\xi)| \leq& \, Ce^{-|x+\xi|}(\|\phi_1 - e^{\frac{i}{4}t}\|_{L^\infty} + \|\chi_2 - e^{-\frac{i}{4}t}\|_{L^\infty}) \\ &+ C(|\phi_2(t,x)| + |\chi_1(t,x)|)\,, \end{aligned}$$

where C is a constant depending only on $\|v(0,\cdot)\|_{L^2}$. Thus, there is $C > 0$ such that

$$\|R(t,\cdot) - e^{\frac{i}{2}(t+\alpha)}\operatorname{sech}(\cdot + \xi)\|_{L^2} \leq C\|v(0,\cdot)\|_{L^2}\,, \quad t \in \mathbb{R}\,.$$

Combining the above with the L^2-conservation law $\|v(t,\cdot)\|_{L^2} = \|v(0,\cdot)\|_{L^2}$ and with the triangle inequality, we conclude that

$$\begin{aligned}\|u(t,\cdot) - e^{\frac{i}{2}(t+\alpha)}\operatorname{sech}(\cdot + \xi)\|_{L^2} &\leq \|R(t,\cdot) - e^{i(t+\alpha)}\operatorname{sech}(\cdot + \xi)\|_{L^2} + \|v(t,\cdot)\|_{L^2} \\ &\leq C\|v(0,\cdot)\|_{L^2},\end{aligned}$$

which completes the proof of the bound (4.84). □

END OF THE PROOF OF THEOREM 4.3. Assuming that $u_0 := u(0,\cdot) \in H^3(\mathbb{R})$, Step 1 implies that if the bound (4.73) is satisfied, then there exists an eigenvalue $\lambda = \frac{1}{2}(k_0 + ic_0)$ of the spectral problem (4.68) satisfying the bound (4.75). The BD transformation (4.76) with the eigenvector $\varphi = (\varphi_1, \varphi_2)^T$ gives new initial data $v_0 \in H^3(\mathbb{R})$ such that $\|v_0\|_{L^2} \leq C\|u_0 - Q_1\|_{L^2}$.

Step 2 generates a new solution of the NLS equation (4.67) from the initial data $v(0,\cdot) = v_0 \in H^3(\mathbb{R})$ such that $v \in C^0(\mathbb{R}, H^3(\mathbb{R}))$ and $\|v(t,\cdot)\|_{L^2} = \|v(0,\cdot)\|_{L^2}$ for every $t \in \mathbb{R}$.

Step 3 is obtained for the particular case $\lambda = \frac{1}{2}$ rather than for the correct eigenvalue $\lambda = \frac{1}{2}(k_0 + ic_0)$ obtained in Step 1. However, the change of variables

$$\begin{cases}\tilde{\varphi}_1(y) = e^{-\frac{i}{2}(c_0x+\alpha)}\tilde{\varphi}_1(x+\xi), \\ \tilde{\varphi}_2(y) = e^{\frac{i}{2}(c_0x+\alpha)}\tilde{\varphi}_2(x+\xi)\,,\end{cases}$$

and

$$\tilde{u}(y) = k_0^{-1}e^{-i(c_0x+\alpha)}u(x+\xi)\,,$$

with $y = k_0x$ transforms the spectral problem (4.68) with $\lambda = \frac{1}{2}(k_0 + ic_0)$ into the one with $\lambda = \frac{1}{2}$, whereas the NLS equation (4.67) is invariant under the scaling transformation

$$k_0\tilde{u}(k_0^2t, k_0x) = e^{\frac{i}{2}c_0x - \frac{i}{4}c_0t}u(t, x - c_0t),$$

where $k_0 > 0$ and $c_0 \in \mathbb{R}$ are arbitrary constants. As a consequence, the results of Step 3 can be applied to the correct eigenvalue $\lambda = \frac{1}{2}(k_0 + ic_0)$ after the symmetry transformations.

Given the corresponding solution $v(t,x)$ of the NLS equaton (4.67), we use the linear superposition principle and introduce complex parameters c_1 and c_2 in the solution $\varphi(t,x)$ to the Lax system (4.68) and (4.69). The complex parameters

$$c_1 = ae^{\frac{1}{2}(\xi+i\theta)}, \quad c_2 = ae^{-\frac{1}{2}(\xi+i\theta)}, \quad a \in \mathbb{R},$$

may depend on time t but are independent of x. The BD transformation (4.83) defines the new solution $u \in C^0(\mathbb{R}, H^3(\mathbb{R}))$ of the NLS equation (4.67) satisfying the bound (4.84). With the bounds in Step 1 and the scaling transformation from $\lambda = \frac{1}{2}$ to $\lambda = \frac{1}{2}(k_0 + ic_0)$, this gives the bound (4.74).

It remains to extend the result from $u_0 \in H^3(\mathbb{R})$ to $u_0 \in L^2(\mathbb{R})$. To do so, we use the approximating sequence $\{u_{0,n}\}_{n\in\mathbb{N}} \in H^3(\mathbb{R})$ that converges to u_0 in the $L^2(\mathbb{R})$ norm. By triangle inequality, there exists $\varepsilon > 0$ (small) such

$$\|u_{0,n} - Q_1\|_{L^2} \leq \varepsilon.$$

Let $\{u_n(t,x)\}_{n\in\mathbb{N}}$ be the corresponding sequence of solutions of NLS equation (4.67) with $u_n(0,\cdot) = u_{n,0}$ such that $u_n \in C^0(\mathbb{R}, H^3(\mathbb{R}))$. Since all bounds in Steps 1,2,3 are obtained in the $L^2(\mathbb{R})$ norm with the constants independent on the solution, it follows that there exist a positive constant C and real numbers k_n, c_n for $n \in \mathbb{N}$ such that

$$\inf_{\xi_n,\alpha_n\in\mathbb{R}} \|e^{i\alpha_n} u_n(t,\cdot+\xi_n) - U_{k_n,c_n}\|_{L^2} + |k_n - 1| + |c_n| \le C\|u_{0,n} - Q_1\|_{L^2},$$

for every $t \in \mathbb{R}$. Since the upper bound is bounded, there exists a subsequence $\{u_{n_j}(t,x)\}_{j\in\mathbb{N}}$ and the limiting solution $u \in C^0(\mathbb{R}, L^2(\mathbb{R}))$ of NLS equation (4.67) with the initial data $u(0,\cdot) = u_0 \in L^2(\mathbb{R})$ such that

$$\lim_{j\to\infty} \|u_{n_j}(t,\cdot) - u(t,\cdot)\|_{L^2} = 0,$$

for every $t \in \mathbb{R}$. Similarly, there exist k_0 and c_0 such that $\lim_{j\to\infty} k_{n_j} = k_0$ and $\lim_{j\to\infty} c_{n_j} = c_0$. The same bound (4.74) holds for the limiting solution $u \in C^0(\mathbb{R}, L^2(\mathbb{R}))$ satisfying (4.73) due to the convergence. □

REMARK 4.15. The H^3-regularity of a solution to the NLS equation (4.67) is only used to prove commutativity properties of solutions to the Lax system (4.68) and (4.69) and has not been used to prove the bounds (4.81) and (4.82). This is the reason why we can prove Theorem 4.3 for any initial data $u(0,\cdot) \in L^2(\mathbb{R})$ satisfying $\|u(0,\cdot) - Q_1\|_{L^2} < \varepsilon$ by using an approximation argument.

REMARK 4.16. Translational parameters ξ and α in the bound (4.74) may depend on time t. Modulation equations for these parameters are not defined in the proof of Theorem 4.3. This can be viewed as an advantage of the BD transformation, which does not rely on the global control of the dynamics of parameters $c_1(t)$ and $c_2(t)$ which determine translational parameters $\xi(t)$ and $\alpha(t)$. On the other hand, parameters k_0 and c_0 in the eigenvalue $\lambda = \frac{1}{2}(k_0 + ic_0)$ near the value $\lambda = \frac{1}{2}$ determines the solitary wave U_{k_0,c_0} in (4.74) near $U_{1,0} = Q_1$. The proof of Theorem 4.3 uses the fact that the eigenvalue $\lambda = \frac{1}{2}(k_0 + ic_0)$ is preserved in the time evolution of the NLS equation (4.67). The stability bound

$$\inf_{\xi,\alpha\in\mathbb{R}} \|e^{i\alpha} u(t,\cdot+\xi) - Q_1\|_{L^2} \le C\|u(0,\cdot) - Q_1\|_{L^2}$$

follows from the bound (4.74) by the triangle inequality.

REMARK 4.17. The solution $u \in C^0(\mathbb{R}, L^2(\mathbb{R}))$ to the NLS equation (4.67) in an L^2-neighborhood of the zero solution could contain some additional solitons with small $L^2(\mathbb{R})$ norm, which are related to the discrete spectrum of the spectral problem (4.68). If the small solitons do appear, the asymptotic decay of $u(t,\cdot)$ to $U_{k_0,c_0}(t,\cdot)$ can not be proved because the transformed solution $v(t,\cdot)$ after the BD transformation does not decay to the zero solution as $t \to \pm\infty$. A more restrictive hypothesis on the initial data $u(t,\cdot)$ in the weighted $L^2(\mathbb{R})$ space is needed to establish the asymptotic stability of the solitary waves in the cubic NLS equation (4.67) [**75**].

4.3.2. Orbital stability of solitary waves in MTM system. Compared to the NLS equation (4.67), the proof of orbital stability of solitary waves in the MTM system (4.24) in the energy space cannot be achieved from the conserved quantities Q, P, and E in (4.27), (4.28), and (4.29) since the augmented energy is not bounded from neither above nor below. Although Theorem 4.1 corrects this issue by using the higher-order Hamiltonian H in (4.30), the result is only limited to the solitary waves (4.37) with $\omega \in (-\omega_0, \omega_0)$ for small $\omega_0 \in (0,1)$.

The method based on the BD transformation is free of this limitation and can be used to get orbital stability of solitary waves both in space of higher regularity $H^1(\mathbb{R})$ as in Theorem 4.1 and in space of lower regularity $L^2(\mathbb{R})$ compared to the regularity required in the energy space $H^{1/2}(\mathbb{R}) \cap L^4(\mathbb{R})$. The method is based on the availability of the linear system (4.2) with (4.25) and (4.26). The following lemma explaining the BD transformation for the MTM system (4.24) was shown in [**68**] with symbolic computations.

LEMMA 4.6. *Let (u,v) be a C^1 solution of the MTM system (4.24) and $\varphi = (p,q)^T$ be a C^2 nonzero solution of the linear system (4.2) with (4.25) and (4.26) for a specific value $\lambda = \delta e^{i\gamma/2}$. Then,*

$$\hat{u} = -u\frac{e^{-i\gamma/2}|p|^2 + e^{i\gamma/2}|q|^2}{e^{i\gamma/2}|p|^2 + e^{-i\gamma/2}|q|^2} + \frac{2i\delta^{-1}\sin\gamma\bar{p}q}{e^{i\gamma/2}|p|^2 + e^{-i\gamma/2}|q|^2}$$

and

$$\hat{v} = -v\frac{e^{i\gamma/2}|p|^2 + e^{-i\gamma/2}|q|^2}{e^{-i\gamma/2}|p|^2 + e^{i\gamma/2}|q|^2} - \frac{2i\delta\sin\gamma\bar{p}q}{e^{-i\gamma/2}|p|^2 + e^{i\gamma/2}|q|^2}$$

generates a new C^1 solution $(\hat{u},\hat{v})$ of the MTM system (4.24). Moreover,

$$\hat{p} = \frac{\bar{q}}{|e^{i\gamma/2}|p|^2 + e^{-i\gamma/2}|q|^2|}, \quad \hat{q} = \frac{\bar{p}}{|e^{i\gamma/2}|p|^2 + e^{-i\gamma/2}|q|^2|} \tag{4.88}$$

yields a C^2 nonzero solution $\varphi = (\hat{p},\hat{q})^T$ of the linear system (4.2) with (4.25) and (4.26) for $\lambda = \delta e^{i\gamma/2}$ and the new potential $(\hat{u},\hat{v})$.

For the zero solution $(u,v) = (0,0)$ of the MTM system (4.24), we obtain the following solution of the linear system (4.2) with (4.25) and (4.26) for $\lambda = \delta e^{i\gamma/2}$,

$$\begin{cases} p(t,x) = e^{-\frac{i}{4}(\lambda^2-\lambda^{-2})x - \frac{i}{4}(\lambda^2+\lambda^{-2})t}, \\ q(t,x) = e^{\frac{i}{4}(\lambda^2-\lambda^{-2})x + \frac{i}{4}(\lambda^2+\lambda^{-2})t}. \end{cases}$$

The BD transformation of Lemma 4.6 generates the solitary wave solution of the MTM system (4.24) in the form

$$\begin{cases} u_\lambda(t,x) = \delta^{-1}\sin\gamma\,\mathrm{sech}\left[\alpha(x+\nu t) + i\frac{\gamma}{2}\right] e^{i\beta(t+\nu x)}, \\ v_\lambda(t,x) = \delta\sin\gamma\,\mathrm{sech}\left[\alpha(x+\nu t) - i\frac{\gamma}{2}\right] e^{i\beta(t+\nu x)}, \end{cases}$$

with

$$\nu = \frac{\delta^2 - \delta^{-2}}{\delta^2 + \delta^{-2}}, \quad \alpha = \frac{\delta^2 + \delta^{-2}}{2}\sin\gamma, \quad \beta = \frac{\delta^2 + \delta^{-2}}{2}\cos\gamma.$$

Figure 4.2 shows the existence region for these solutions in the (β,ν) plane.

The corresponding solution (4.88) of the linear system (4.2) with (4.25) and (4.26) for $\lambda = \delta e^{i\gamma/2}$ and the new potential $(u,v) = (u_\lambda, v_\lambda)$ is given by

$$\begin{cases} p_\lambda(t,x) = e^{-\frac{1}{2}x\sin\gamma - \frac{i}{2}t\cos\gamma}\left|\mathrm{sech}\left(x\sin\gamma + i\frac{\gamma}{2}\right)\right|, \\ q_\lambda(t,x) = e^{\frac{1}{2}x\sin\gamma + \frac{i}{2}t\cos\gamma}\left|\mathrm{sech}\left(x\sin\gamma + i\frac{\gamma}{2}\right)\right|, \end{cases}$$

and it decays exponentially as $|x| \to \infty$ for every $t \in \mathbb{R}$. The BD transformation iterated with $(u,v) = (u_\lambda, v_\lambda)$ and $\varphi = (p_\lambda, q_\lambda)$ returns back the zero solution $(u,v) = (0,0)$. This property allows us to map a small neighborhood of the zero solution to a small neighborhood of the solitary wave solution and vice versa.

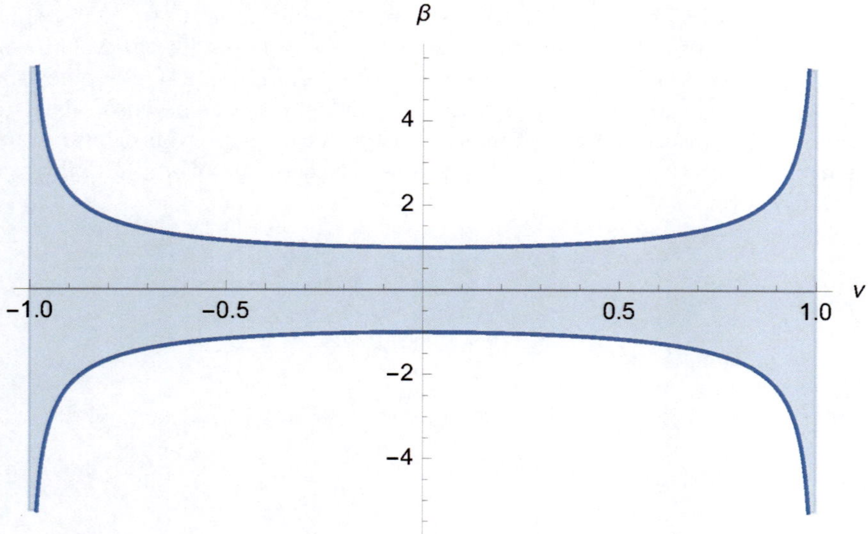

FIGURE 4.2. A plot of the boundaries of the existence region for the traveling Dirac solitons (u_λ, v_λ) in the (ν, β) plane.

Traveling solitary waves with $\nu \neq 0$ can be recovered from the stationary solitary waves with $\nu = 0$ by the Lorentz transformation:

$$\begin{cases} \tilde{u}(t,x) := \delta^{-1} u(k_1 t + k_2 x, k_1 x + k_2 t), \\ \tilde{v}(t,x) := \delta v(k_1 t + k_2 x, k_1 x + k_2 t), \end{cases} \qquad k_1 := \frac{\delta^2 + \delta^{-2}}{2}, \quad k_2 := \frac{\delta^2 - \delta^{-2}}{2},$$

where (u, v) and $(\tilde{u}, \tilde{v})$ are solutions of the MTM system (4.24). The stationary solitary wave corresponds to $\delta = 1$, $\nu = 0$, and $\omega = \cos\gamma$. The corresponding transformation of the Lax system takes the form

$$\tilde{\varphi}(t,x) := \varphi(k_1 t + k_2 x, k_1 x + k_2 t),$$

which does not change the spectral parameter λ. Hence, without loss of generality, one can choose $\lambda_0 = e^{i\gamma_0/2}$ for a fixed $\gamma_0 \in (0, \pi)$ to consider the orbital stability of the solitary waves of the MTM system (4.24).

The proof of Theorem 4.3 can be extended very similarly to the MTM system (4.24). Global well-posedness of the MTM system (4.24) in a subset of $L^2(\mathbb{R})$ was proved in [**38, 132**]. Based on these ingredients, the following theorem was proven in [**68**] similarly to the proof of Theorem 4.3.

THEOREM 4.4. *Let $(u, v) \in C^0(\mathbb{R}; L^2(\mathbb{R}))$ be a unique solution of the MTM system (4.24). Fix $\lambda_0 = e^{i\gamma_0/2}$. There exists a real positive ε such that if the initial value $(u_0, v_0) \in L^2(\mathbb{R})$ satisfies*

$$\|u_0 - u_{\lambda_0}(0,\cdot)\|_{L^2} + \|v_0 - v_{\lambda_0}(0,\cdot)\|_{L^2} \leq \varepsilon,$$

then there exists $\lambda \in \mathbb{C}$ such that $|\lambda - \lambda_0| \leq C\varepsilon$ and

$$\inf_{\xi,\alpha\in\mathbb{R}} \left(\|u(t,\cdot+\xi) - e^{-i\alpha} u_\lambda(t,\cdot)\|_{L^2} + \|v(t,\cdot+\xi) - e^{-i\alpha} v_\lambda(t,\cdot)\|_{L^2} \right) \leq C\varepsilon,$$

for every $t \in \mathbb{R}$, where the positive constant C is independent of ε and t.

REMARK 4.18. The approximating sequence of initial data $\{(u_{n,0}, v_{n,0})\}_{n\in\mathbb{N}}$ is defined in $H^2(\mathbb{R})$ to obtain commutability of the Lax system for the MTM system (4.24). Local and global well-posedness of the MTM system (4.24) in Sobolev spaces $H^k(\mathbb{R})$ with $k \in \mathbb{N}$ can be obtained with the regular methods based on fixed-point iterations and a priori energy estimates [**214**]. As a result, the orbital stability result in $L^2(\mathbb{R})$ given by Theorem 4.4 is generalized to the orbital stability result in $H^k(\mathbb{R})$ for every $k \in \mathbb{N}$.

CHAPTER 5

Spectral stability in integrable Hamiltonian systems

Spectral stability of nonlinear waves is determined by the spectrum of a linearized operator which is obtained by expanding the nonlinear evolution equation near the nonlinear wave and dropping the quadratic and higher-order terms. The precise information about the spectrum of the linearized operator is hardly available if the nonlinear waves are spatially dependent, e.g., in the class of solitary or periodic waves. As a result, most studies are focused on properties of the self-adjoint Hessian operators in relation to the number of unstable eigenvalues of the non-self-adjoint linearized operators, e.g., the Hamiltonian–Krein theory (Sections 1.4 and 3.5).

For the class of integrable Hamiltonian systems, spectral stability theory benifits from the relation between the nonlinear and linear (Lax) systems. We will show here that the spectral information of the Lax system fully determines the spectrum of the linearized equations at the nonlinear waves via the squared eigenfunction relation.

In the case of periodic waves, perturbations are typically restricted to the same period so that the Hilbert space $\mathcal{H}$ is the space of periodic functions, e.g., $\mathcal{H} = L^2_{\rm per}$. If perturbations to the periodic waves are defined on the infinite line and belong to the space of localized functions, e.g., $\mathcal{H} = L^2(\mathbb{R})$, then the stability analysis relies on the Floquet–Bloch theory. One intermediate case between the two limits is the space of periodic functions with N-multiple periods, e.g., $\mathcal{H} = L^2_{\rm Nper}$. The intermediate case recovers $L^2_{\rm per}$ if $N = 1$ and approximates $L^2(\mathbb{R})$ if $N \to \infty$, for which we can introduce the concept of modulational instability of periodic waves (Section 5.1).

Nonlinear waves in many integrable Hamiltonian systems can be characterized algebraically to include multi-solitons and multi-periodic solutions (Section 5.2). These solutions satisfy the higher-order stationary Lax–Novikov equations. An additional tool of the multi-fold Bäcklund–Darboux transformation is used to generate multi-solitons from the exponential solutions of the Lax equations at the zero solution. By using the methods based on the higher-order energy and the linearization via the Lax system, the orbital and spectral stability theory can be extended to double-solitons, time-periodic breathers, and double-periodic solutions of the integrable Hamiltonian systems (Section 5.3).

5.1. Modulational stability of periodic waves

We consider the spectral stability problem in the form

$$J\mathcal{L}v = \lambda v, \tag{5.1}$$

where $\mathcal{L}$ is a self-adjoint operator in a Hilbert space $\mathcal{H}$ with a dense domain $\text{Dom}(\mathcal{L})$ and $J : \text{Dom}(J) \subset \mathcal{H} \to \mathcal{H}$ is a skew-adjoint operator. These operators are obtained from the Hamiltonian formulation of the nonlinear evolution equation. The eigenvector $v \neq 0$ is defined in $\text{Dom}(J\mathcal{L}) \subset \mathcal{H}$ for an eigenvalue $\lambda \in \mathbb{C}$. The following definition distinguishes the stable spectrum of the linear operator $J\mathcal{L}$ in the case when $\mathcal{L}$ is obtained from the linearization at a periodic wave with the spatial period $L > 0$.

DEFINITION 5.1. We say that the periodic wave is spectrally stable with respect to N-periodic perturbations in $\mathcal{H} = L^2_{\text{Nper}}$ if the spectral stability problem (5.1) admits no eigenvalues $\lambda \in \mathbb{C}$ with $\text{Re}(\lambda) > 0$ for the eigenvectors $v \in \text{Dom}(J\mathcal{L}) \subset L^2_{\text{Nper}}$. We say that the periodic wave is spectrally stable with respect to spatially decaying perturbations in $\mathcal{H} = L^2(\mathbb{R})$ if he spectral stability problem (5.1) admits no values of $\lambda \in \mathbb{C}$ with $\text{Re}(\lambda) > 0$ such that $v \in L^\infty(\mathbb{R})$. If the wave is not spectrally stable, then it is spectrally unstable.

REMARK 5.1. In most cases related to the periodic waves in integrable Hamiltonian systems, $\text{Dom}(J\mathcal{L})$ is compactly embedded into L^2_{Nper} so that the spectrum of $J\mathcal{L}$ in L^2_{Nper} consists of isolated eigenvalues. On the other hand, it follows from the Floquet theorem (Theorem 1.9) for linear differential equations with periodic coefficients that the spectrum of $J\mathcal{L}$ is purely continuous in $L^2(\mathbb{R})$.

REMARK 5.2. The space $L^2(\mathbb{R})$ can be approximated by the space L^2_{Nper} in the limit $N \to \infty$. Isolated eigenvalues λ of the point spectrum of $J\mathcal{L}$ in L^2_{Nper} fill densely the continuous spectrum of $J\mathcal{L}$ in $L^2(\mathbb{R})$ as $N \to \infty$.

If the periodic wave is spectrally unstable in $L^2(\mathbb{R})$, then parts of the continuous spectrum of $J\mathcal{L}$ in $L^2(\mathbb{R})$ are located outside $i\mathbb{R}$ with $\text{Re}(\lambda) > 0$. By the dense approximation of the discrete spectrum of $J\mathcal{L}$ in L^2_{Nper}, there exist unstable eigenvalues for every $N \in \mathbb{N}$ sufficiently large. This allows us to define **the modulational instability** of the periodic waves in addition to their spectral instability.

DEFINITION 5.2. If no unstable eigenvalues λ with $\text{Re}(\lambda) > 0$ of $J\mathcal{L}$ in L^2_{Nper} exist in a local neighborhood of $\lambda = 0$ for $N \in \mathbb{N}$ sufficiently large, then the periodic wave is said to be modulationally stable. Otherwise, it is said to be modulationally unstable.

REMARK 5.3. A modulationally stable periodic wave may still be spectrally unstable due to other eigenvalues of $J\mathcal{L}$ in L^2_{Nper} outside the local neighborhood of $\lambda = 0$.

REMARK 5.4. The concept of modulational instability in Definition 5.2 is important for applications in physics. The spatial perturbations with longer periods break the periodic symmetry of the nonlinear wave and may result in the disintegration of the periodic wave into spatially and temporally decaying **rogue waves** (Remark 5.58).

For nonintegrable Hamiltonian systems, the modulational stability problem can be treated analytically by using perturbation theory applied to the spectral stability problem (5.1) [**34**], and this is the extent of what can be achieved analytically. All other investigations of spectral stability typically rely on numerical approximations.

For integrable Hamiltonian systems (4.1), the spectral stability problem (5.1) can be solved by using the linear Lax system (4.2). Eigenfunctions of the two linear problems are related to each other via the squared eigenfunctions. The following two examples give details of the relevant analysis for the focusing NLS equation (Example 4.1) and the derivative NLS equation (Example 4.2). The presentation is based on [**46, 80**] and [**44, 82**], respectively. More examples can be found for the KdV equation [**28**], the defocusing NLS equation [**29**], and the sine–Gordon equation [**81**].

5.1.1. Modulational stability in the focusing NLS equation. We first identify some objects used in the modulational stability analysis.

Let $u = u(t,x)$ be a solution of the focusing NLS equation (4.67). The perturbation $v = v(t,x)$ satisfies the linearized NLS equation

$$iv_t + \frac{1}{2}v_{xx} + 2|u|^2 v + u^2 \bar{v} = 0 \tag{5.2}$$

where the quadratic and cubic terms in v were dropped. Let $\varphi = \varphi(t,x)$ be a solution of the linear system (4.68) and (4.69) with the potential $u = u(t,x)$. Bounded solutions $\varphi(t,x)$ in x are referred to as **eigenfunctions** and the corresponding values λ are referred to as **eigenvalues**. The set of eigenvalues λ form the **Lax spectrum** of the linear system (4.68) and (4.69).

REMARK 5.5. The Lax spectrum is defined in the space $C_b^0(\mathbb{R})$ of x-dependent functions which are bounded and continuous on $\mathbb{R}$. However, depending on the properties of the solution to the NLS equation, other function spaces may be considered. If $u = u(t,x)$ is L-periodic in x, the space of L-periodic functions $L^2_{\rm per}$, N-multiple L-periodic functions $L^2_{\rm Nper}$, and spatially decaying functions $L^2(\mathbb{R})$ can be considered for $\varphi = \varphi(t,x)$ in x. The choice of the function space affects the nature of the Lax spectrum. The Lax spectrum of the linear system (4.68) and (4.69) is purely discrete for $L^2_{\rm per}$ and $L^2_{\rm Nper}$ and purely continuous for $L^2(\mathbb{R})$.

The following lemma connects solutions of the linearized NLS equation (5.2) and solutions of the linear system (4.68) and (4.69).

LEMMA 5.1. *Assume u is a solution to the NLS equation* (4.67) *and $\varphi = (\varphi_1,\varphi_2)^T$ is a solution to the linear system* (4.68) *and* (4.69) *for some λ. Then,*

$$v = \varphi_1^2 - \bar{\varphi}_2^2, \quad \bar{v} = -\varphi_2^2 + \bar{\varphi}_1^2 \tag{5.3}$$

and

$$v = i(\varphi_1^2 + \bar{\varphi}_2^2), \quad \bar{v} = -i(\varphi_2^2 + \bar{\varphi}_1^2) \tag{5.4}$$

are solutions to the linearized NLS equation (5.2).

PROOF. Solutions of the linear system (4.68) and (4.69) satisfy the following symmetry: If $\varphi = (\varphi_1,\varphi_2)^T$ is a solution for λ, then $\varphi = (-\bar{\varphi}_2,\bar{\varphi}_1)^T$ is a solution for $-\bar{\lambda}$. Due to this symmetry and the linear superposition principle, it is sufficient to confirm that $v = \varphi_1^2$ and $\bar{v} = -\varphi_2^2$ satisfy the linearized NLS equation (5.2). This is obtained directly:

$$\begin{aligned} iv_t + \frac{1}{2}v_{xx} + 2|u|^2 v + u^2\bar{v} &= \varphi_1(2i\varphi_{1t} + \varphi_{1xx}) + (\varphi_{1x})^2 + 2|u|^2\varphi_1^2 - u^2\varphi_2^2 \\ &= \varphi_1(-|u|^2\varphi_1 - 2\lambda^2\varphi_1 - 2\lambda u\varphi_2 - u_x\varphi_2 \\ &\quad + \lambda(\lambda\varphi_1 + u\varphi_2) + u_x\varphi_2 + u(-\bar{u}\varphi_1 - \lambda\varphi_2)) \\ &\quad + (\lambda\varphi_1 + u\varphi_2)^2 + 2|u|^2\varphi_1^2 - u^2\varphi_2^2 = 0. \end{aligned}$$

TABLE 1. Table of possible solutions of the linearized NLS equation (5.2) generated from two solutions $\varphi = (\varphi_1, \varphi_2)^T$ and $\phi = (\phi_1, \phi_2)^T$ to the linear system (4.68) and (4.69) for the same value of λ.

Pair I	Pair II	Pair III
$v = \varphi_1^2 - \bar{\varphi}_2^2$	$v = \varphi_1\phi_1 - \bar{\varphi}_2\bar{\phi}_2$	$v = \phi_1^2 - \bar{\phi}_2^2$
$v = i\varphi_1^2 + i\bar{\varphi}_2^2$	$v = i\varphi_1\phi_1 + i\bar{\varphi}_2\bar{\phi}_2$	$v = i\phi_1^2 + i\bar{\phi}_2^2$

Extending this solution with the symmetric solution $v = \bar{\varphi}_2^2$ and $\bar{v} = -\bar{\varphi}_1^2$ in the linear combinations (5.3) and (5.4) ensures that $\bar{v}$ is a complex conjugate of v. $\square$

REMARK 5.6. The result in Lemma 5.1 applies to two linearly independent solutions $\varphi = (\varphi_1, \varphi_2)^T$ and $\phi = (\phi_1, \phi_2)^T$ that exist for the same value of λ in the linear system (4.68) and (4.69). These solutions generate three pairs of solutions of the linearized NLS equation (5.2) in Table 1.

REMARK 5.7. Periodic in x eigenfunctions $\varphi = \varphi(t,x)$ generate periodic in x solutions $v = v(t,x)$. The symmetry of the linear system (4.68) and (4.69) implies that the eigenvalues λ are symmetric with respect to reflection about $i\mathbb{R}$.

REMARK 5.8. We extend the notion of a periodic wave with L-periodic profile to a quasi-periodic wave with profile $u(x) = R(x)e^{i\Theta(x)}$, where R and Θ' are L-periodic in x. In this case, we can consider quasi-periodic eigenfunctions $\varphi = (\varphi_1, \varphi_2)^T$ with $\varphi_1(x) = \hat{\varphi}_1(x)e^{\frac{i}{2}\Theta(x)}$ and $\varphi_2(x) = \hat{\varphi}_2(x)e^{-\frac{i}{2}\Theta(x)}$, where $\hat{\varphi}_1$ and $\hat{\varphi}_2$ are L-periodic.

Traveling waves of the NLS equation (4.67) are solutions of the form

$$u(t,x) \mapsto u(x+ct)e^{-2ibt}$$

with parameters b and c (Example 4.1), where the profile u satisfies the stationary NLS equation

$$u'' + 2|u|^2u + 2icu' + 4bu = 0. \tag{5.5}$$

This equation is integrable with the two conserved quantities

$$\begin{cases} i(\bar{u}u' - \bar{u}'u) - 2c|u|^2 = 4a, \\ |u'|^2 + |u|^4 + 4b|u|^2 = 8d, \end{cases} \tag{5.6}$$

where a and d are additional parameters. Separation of variables in the linear system (4.68) and (4.69) leads to the representation

$$\varphi_1(t,x) = \chi_1(x+ct)e^{-ibt+\Omega t}, \quad \varphi_2(t,x) = \chi_2(x+ct)e^{+ibt+\Omega t},$$

where $\Omega \in \mathbb{C}$ is another spectral parameter and $\chi = (\chi_1, \chi_2)^T$ satisfies the spectral problems

$$\chi' = \begin{bmatrix} \lambda & u \\ -\bar{u} & -\lambda \end{bmatrix} \chi, \tag{5.7}$$

and

$$\Omega\chi + c\begin{bmatrix} \lambda & u \\ -\bar{u} & -\lambda \end{bmatrix}\chi = i\begin{bmatrix} \lambda^2 + \frac{1}{2}|u|^2 + b & \frac{1}{2}u' + \lambda u \\ \frac{1}{2}\bar{u}' - \lambda\bar{u} & -\lambda^2 - \frac{1}{2}|u|^2 - b \end{bmatrix}\chi, \tag{5.8}$$

where derivatives are now applied to functions of a single variable in the moving frame, which we call x again.

Similarly, the linearized NLS equation (5.2) for solutions of the form

$$v(t,x) \mapsto v(x+ct)e^{-2ibt}$$

can be written as follows:

$$iv_t + icv_x + 2bv + \frac{1}{2}v_{xx} + 2|u|^2 v + u^2\bar{v} = 0.$$

Separation of variables according to

$$v(t,x) = w_1(x)e^{t\Lambda}, \quad \bar{v}(t,x) = w_2(x)e^{t\Lambda}$$

transforms the linearized NLS equation to the spectral stability problem

$$(5.9)\quad i\Lambda \begin{bmatrix} 1 & 0 \\ 0 & -1 \end{bmatrix} w + \begin{bmatrix} \frac{1}{2}\partial_x^2 + 2|u|^2 + 2b + ic\partial_x & u^2 \\ \bar{u}^2 & \frac{1}{2}\partial_x^2 + 2|u|^2 + 2b - ic\partial_x \end{bmatrix} w = 0,$$

where Λ is an eigenvalue and $w = (w_1, w_2)^T$ is an eigenfunction.

REMARK 5.9. Components w_1 and w_2 are no longer complex-conjugate to each other if $\Lambda \notin \mathbb{R}$.

The following lemma connects bounded solutions of (5.7) and (5.8) with bounded solutions of (5.9).

LEMMA 5.2. *Let λ be in the Lax spectrum of the spectral problem* (5.7) *with the bounded eigenfunction $\chi(x)$. Eigenvalues and eigenfunctions of the spectral stability problem* (5.9) *are given by*

$$(5.10)\qquad w_1 = \chi_1^2, \quad w_2 = -\chi_2^2, \quad \Lambda = 2\Omega,$$

where $\Omega = \pm i\sqrt{P(\lambda)}$ with

$$(5.11)\qquad P(\lambda) = \lambda^4 + 2ic\lambda^3 + (2b - c^2)\lambda^2 + 2i(a+bc)\lambda + b^2 - 2ac + 2d.$$

PROOF. Since the spectral problem (5.8) is a linear algebraic system, it admits a nonzero solution if and only if the determinant of the coefficient matrix is zero. The latter condition yields the x-independent relation between Ω and λ in the form

$$\Omega^2 + P(\lambda) = 0$$

with $P(\lambda)$ given by

$$\begin{aligned} P(\lambda) &= (\lambda^2 + i\lambda c + \frac{1}{2}|u|^2 + b)^2 - (\lambda + ic)^2|u|^2 - \frac{1}{2}(\lambda + ic)(\bar{u}u' - \bar{u}'u) + \frac{1}{4}|u'|^2 \\ &= \lambda^4 + 2ic\lambda^3 + (2b - c^2)\lambda^2 + \frac{i}{2}\lambda(i(\bar{u}u' - \bar{u}'u) - 2c|u|^2 + 4bc) \\ &\quad + \frac{1}{4}(|u'|^2 + |u|^4 + 4(c^2+b)|u|^2 - 2ic(\bar{u}u' - \bar{u}'u) + 4b^2) \\ &= \lambda^4 + 2ic\lambda^3 + (2b - c^2)\lambda^2 + 2i(a+bc)\lambda + b^2 - 2ac + 2d, \end{aligned}$$

where parameters a and d appear in the two conserved quantities (5.6). By Lemma 5.1, we obtain the relations (5.10) with $\Omega = \pm i\sqrt{P(\lambda)}$. □

REMARK 5.10. If $u(x+L) = u(x)$ with $L > 0$, then the Lax spectrum for λ in the spectral problem (5.7) is found by using the Floquet theorem (Theorem 1.9). Bounded solutions are represented in the form

$$(5.12)\quad \chi(x) = \hat{\chi}(x)e^{i\nu x}, \qquad \text{where } \hat{\chi}(x+L) = \hat{\chi}(x) \quad \text{and} \quad \nu \in \mathbb{T}_L := \left[-\frac{\pi}{L}, \frac{\pi}{L}\right].$$

When $\nu = 0$ and $\nu = \pm\frac{\pi}{L}$, the bounded solutions are L-periodic and L-anti-periodic, respectively. The Lax spectrum is obtained by a collection of the continuous spectral bands $\{\lambda_n(\nu),\ \nu \in \mathbb{T}_L\}_{n\in\mathbb{N}}$ computed from the discrete spectrum of the periodic eigenvalue problem

$$\begin{bmatrix} \frac{d}{dx} + i\nu & -u \\ -\bar{u} & -\frac{d}{dx} - i\nu \end{bmatrix} \hat{\chi} = \lambda \hat{\chi}.$$

The main result (Theorem 5.1) shows that every traveling periodic wave of the focusing NLS equation is modulationally unstable. The proof of this result was elaborated in [**80, 82**] and in [**46**]. In order to simplify the presentation, we will first show that the parameter c among the four parameters of $P(\lambda)$ in (5.11) is redundant due to the Lorentz transformation and may be chosen arbitrarily.

LEMMA 5.3. *In the family of periodic waves satisfying* (5.5) *and* (5.6), *one can set* $c = 0$ *without loss of generality.*

PROOF. The system (5.5) and (5.6) is invariant under the Lorentz transformation

$$u(x) = \tilde{u}(x)e^{-icx}, \quad b = \tilde{b} - \frac{1}{4}c^2, \quad d = \tilde{d} + \frac{1}{2}ac, \quad a = \tilde{a},$$

and so c may be set to 0. Similarly, $P(\lambda)$ in (5.11) can be written as

$$P(\lambda) = \left(\lambda + \frac{i}{2}c\right)^4 - 2\tilde{b}\left(\lambda + \frac{i}{2}c\right)^2 + 2i\tilde{a}\left(\lambda + \frac{i}{2}c\right) + \tilde{b}^2 + 2\tilde{d},$$

which shows that the dependence of $P(\lambda)$ on c can be removed by the translation $-ic/2$ in the spectral plane of λ. □

The following theorem proves the modulational instability of every standing perodic wave of the focusing NLS equation (4.67) in the particular case $c = 0$. Due to Lemma 5.3, the result extends to every $c \in \mathbb{R}$ so that it is extended to every traveling periodic wave solutions of the focusing NLS equation (4.67).

THEOREM 5.1. *Every standing periodic wave of the NLS equation* (4.67) *is modulationally unstable in the spectral stability problem* (5.9).

PROOF. The proof of this theorem consists of three steps. In the first step, we obtain all possible standing periodic waves in the particular case $c = 0$ by integrating the differential equations (5.5) and (5.6) in the polar form. In the second step, we show that the Lax spectrum of the spectral problem (5.7) at such a solution includes two spectral bands outside $i\mathbb{R}$. In the third step, we prove that the two spectral bands define the unstable spectral bands of the spectral stability problem (5.9) intersecting the origin $\Lambda = 0$. This implies the modulational instability of the periodic wave according to Definition 5.2.

STEP 1. If $c = 0$, there exists a family of standing periodic waves in the following parameter region:

$$\mathcal{R} := \{(\alpha, \beta, k) \in \mathbb{R}^2 : \quad \alpha^2 k^2 \le \beta \le \alpha^2, \quad \alpha > 0, \quad 0 < k < 1\}.$$

PROOF OF STEP 1. Substituting the decomposition $u(x) = R(x)e^{i\Theta(x)}$ with real R and Θ into (5.5) and (5.6) with $c = 0$ yields the system of first-order equations

$$\begin{cases} (R')^2 + R^2(\Theta')^2 + R^4 + 4bR^2 = 8d, \\ R^2\Theta' = -2a. \end{cases}$$

Eliminating Θ' yields the following first-order invariant:

$$(R')^2 + 4a^2R^{-2} + R^4 + 4bR^2 = 8d.$$

The singularity $R = 0$ is unfolded with the transformation $\rho = R^2$ which yields

$$\frac{1}{4}(\rho')^2 + Q(\rho) = 0, \quad Q(\rho) := \rho^3 - 4b\rho^2 - 8d\rho + 4a^2.$$

Since $Q(0) = 4a^2 \geq 0$, one of the roots of the cubic polynomial Q is negative. Positive periodic solutions of the first-order invariant with $\rho = R^2 \geq 0$ exist if and only if three real roots of Q exist. We denote the roots by $\{\rho_1, \rho_2, \rho_3\}$ and order them as follows:

$$\rho_3 \leq 0 \leq \rho_2 \leq \rho_1.$$

This yields the first-order invariant:

$$(\rho')^2 + 4(\rho - \rho_1)(\rho - \rho_2)(\rho - \rho_3) = 0.$$

The periodic solutions with $\rho \geq 0$ are located in the interval $[\rho_2, \rho_1]$ and are given by the explicit formula:

$$\rho(x) = \rho_1 - (\rho_1 - \rho_2)\,\mathrm{sn}^2(\alpha x; k),$$

where α and k are related to (ρ_1, ρ_2, ρ_3) by

$$\alpha^2 = \rho_1 - \rho_3, \quad k^2 = \frac{\rho_1 - \rho_2}{\rho_1 - \rho_3}.$$

Setting $\beta = \rho_1$ yields ρ in the form $\rho(x) = \beta - \alpha^2k^2\,\mathrm{sn}^2(\alpha x; k)$. The periodic solutions are defined in the triangular region $\mathcal{R}$ for $\alpha^2k^2 \leq \beta \leq \alpha^2$. The upper boundary $\beta = \alpha^2$ corresponds to the dnoidal solution $\rho(x) = \alpha^2\,\mathrm{dn}^2(\alpha k; k)$ and the lower boundary $\beta = \alpha^2k^2$ corresponds to the cnoidal solution $\rho(x) = \alpha^2k^2\,\mathrm{cn}^2(\alpha x; k)$ (Example 3.15). □

REMARK 5.11. Roots ρ_1, ρ_2, and ρ_3 are related to parameters a, b, and d by

$$\begin{cases} 4b = -\rho_1 - \rho_2 - \rho_3, \\ 8d = -\rho_1\rho_2 - \rho_1\rho_3 - \rho_2\rho_3, \\ 4a^2 = -\rho_1\rho_2\rho_3. \end{cases}$$

Without loss of generality, we use the negative square root for $2a = -\sqrt{\rho_1\rho_2}\sqrt{-\rho_3}$. Solutions with the positive square root are obtained by the transformation: $u \mapsto \bar{u}$.

STEP 2. There exist at least two spectral bands of the Lax spectrum in the spectral problem (5.7) outside $i\mathbb{R}$.

PROOF OF STEP 2. The polynomial $P(\lambda)$ given by (5.11) with $c = 0$ can be written in terms of the roots $\{\rho_1, \rho_2, \rho_3\}$ as follows:

$$P(\lambda) = \lambda^4 - \frac{1}{2}(\rho_1 + \rho_2 + \rho_3)\lambda^2 - i\sqrt{\rho_1\rho_2}\sqrt{-\rho_3}\lambda$$
$$+ \frac{1}{16}(\rho_1^2 + \rho_2^2 + \rho_3^2 - 2\rho_1\rho_2 - 2\rho_1\rho_3 - 2\rho_2\rho_3).$$

Two pairs of roots of $P(\lambda)$ exist in the explicit form $\{\lambda_1, \lambda_2, -\bar{\lambda}_1, -\bar{\lambda}_2\}$, where

$$\lambda_1 = \frac{1}{2}(\sqrt{\rho_1} + \sqrt{\rho_2}) + \frac{i}{2}\sqrt{-\rho_3}, \qquad \lambda_2 = \frac{1}{2}(\sqrt{\rho_1} - \sqrt{\rho_2}) - \frac{i}{2}\sqrt{-\rho_3}. \tag{5.13}$$

Since $\rho_1 = \beta$, $\rho_2 = \beta - \alpha^2 k^2$, and $\rho_3 = \beta - \alpha^2$, it follows that $\mathrm{Re}(\lambda_{1,2}) \neq 0$ for every $(\alpha, \beta, k) \in \mathcal{R}$.

We will show that the four roots of $P(\lambda)$ are related to L-periodic squared eigenfunctions χ of the spectral problem (5.7). These eigenfunctions are either L-periodic or L-antiperiodic so that they correspond to either $\nu = 0$ or $\nu = \pm\frac{\pi}{L}$ in the representation (5.12). By the Floquet theorem (Theorem 1.9), there exist spectral bands of the Lax spectrum which connect the roots of $P(\lambda)$. Since there exist four distinct roots of $P(\lambda)$ and all four roots are located outside $i\mathbb{R}$, at least two spectral bands exist outside $i\mathbb{R}$.

To show that the four roots of $P(\lambda)$ are related to the L-periodic squared eigenfunctions χ, we prove that the eigenfunction $\chi = (p_1, q_1)^T$ for a particular root $\lambda = \lambda_1$ of $P(\lambda)$ satisfying $\lambda_1 + \bar{\lambda}_1 \neq 0$ (that is, $\lambda_1 \notin i\,\mathbb{R}$) can be related to the profile u in the family of standing periodic waves according to the algebraic relation

$$u = p_1^2 + \bar{q}_1^2. \tag{5.14}$$

Assuming (5.14), the spectral problem (5.7) becomes the following nonlinear complex Hamiltonian system

$$\frac{dp_1}{dx} = \frac{\partial H}{\partial q_1}, \quad \frac{dq_1}{dx} = -\frac{\partial H}{\partial p_1}, \tag{5.15}$$

where the Hamiltonian function is given by

$$H = \lambda_1 p_1 q_1 + \bar{\lambda}_1 \bar{p}_1 \bar{q}_1 + \frac{1}{2}(p_1^2 + \bar{q}_1^2)(\bar{p}_1^2 + q_1^2). \tag{5.16}$$

The value of H is a constant of motion for the Hamiltonian system (5.15) together with another constant of motion

$$F = i(p_1 q_1 - \bar{p}_1 \bar{q}_1). \tag{5.17}$$

The complex Hamiltonian system is completely integrable with two real-valued invariants in involution. Differentiating of (5.14) in x and using (5.15) and (5.17) yields

$$\frac{du}{dx} + 2iFu = 2(\lambda_1 p_1^2 - \bar{\lambda}_1 \bar{q}_1^2).$$

Second differentiating in x and using (5.15), (5.16), and (5.17) yield

$$\frac{d^2u}{dx^2} + 2|u|^2 u + 2iF\frac{du}{dx} - 4Hu = 4(\lambda_1^2 p_1^2 + \bar{\lambda}_1^2 \bar{q}_1^2).$$

Eliminating the squared eigenfunctions p_1^2 and $\bar{q}_1^2$ from (5.14) yields a closed second-order equation in u, which is equivalent to the second-order equation (5.5) with

$$\begin{cases} c = F + i(\lambda_1 - \bar{\lambda}_1), \\ -b = H + iF(\lambda_1 - \bar{\lambda}_1) + |\lambda_1|^2. \end{cases}$$

Hence, the constraint (5.14) can only be imposed to the family of standing periodic waves $u(t, x) \mapsto u(x+ct)e^{-2ibt}$ of the NLS equation (4.67). If the profile u is periodic in x, then p_1^2, p_1q_1, and $\bar{q}_1^2$ are periodic in x according to the inverse relations:

$$p_1^2 = \frac{u' + 2icu + 2\lambda_1 u}{2(\lambda_1 + \bar{\lambda}_1)},$$
$$\bar{q}_1^2 = \frac{-u' - 2icu + 2\bar{\lambda}_1 u}{2(\lambda_1 + \bar{\lambda}_1)},$$
$$p_1 q_1 = -\frac{|u|^2 + 2b + 2ic\lambda_1 + 2\lambda_1^2}{2(\lambda_1 + \bar{\lambda}_1)}.$$

The proof of Step 2 is complete if we will prove that λ_1 is a root of $P(\lambda)$. To do so, we represent the Hamiltonian system (5.15) in the form:

$$\frac{d}{dx} W(\lambda, u) = U(\lambda, u) W(\lambda, u) - W(\lambda, u) U(\lambda, u), \quad U(\lambda, u) = \begin{bmatrix} \lambda & u \\ -\bar{u} & -\lambda \end{bmatrix},$$

where u is given by (5.14) and $W(\lambda, u)$ is represented in the form

$$W(\lambda, u) = \begin{bmatrix} W_{11}(\lambda, u) & W_{12}(\lambda, u) \\ \overline{W}_{12}(-\lambda, u) & -\overline{W}_{11}(-\lambda, u) \end{bmatrix},$$

with the entries

$$W_{11}(\lambda, u) = 1 - \left(\frac{p_1 q_1}{\lambda - \lambda_1} - \frac{\bar{p}_1 \bar{q}_1}{\lambda + \bar{\lambda}_1} \right), \qquad W_{12}(\lambda, u) = \frac{p_1^2}{\lambda - \lambda_1} + \frac{\bar{q}_1^2}{\lambda + \bar{\lambda}_1}.$$

Substituting p_1^2, $\bar{q}_1^2$, and $p_1 q_1$ yields

$$W_{11}(\lambda, u) = \frac{\lambda^2 + ic\lambda + \frac{1}{2}|u|^2 + b}{(\lambda - \lambda_1)(\lambda + \bar{\lambda}_1)}, \qquad W_{12}(\lambda, u) = \frac{u\lambda + \frac{1}{2}u' + icu}{(\lambda - \lambda_1)(\lambda + \bar{\lambda}_1)},$$

so that $\det W(\lambda, u)$ can be written in the equivalent forms

$$\begin{aligned} \det W(\lambda, u) &= -\left[W_{11}(\lambda, u)\right]^2 - W_{12}(\lambda, u)\bar{W}_{12}(-\lambda) \\ &= -1 + \frac{2\left[H - iF(\lambda - \lambda_1 + \bar{\lambda}_1)\right]}{(\lambda - \lambda_1)(\lambda + \bar{\lambda}_1)} \\ &= -\frac{P(\lambda)}{(\lambda - \lambda_1)^2 (\lambda + \bar{\lambda}_1)^2}, \end{aligned}$$

where $P(\lambda)$ is given by (5.11). Since $\det W(\lambda, u)$ contains only simple poles at λ_1 and $-\bar{\lambda}_1$, it follows that λ_1 and $-\bar{\lambda}_1$ are roots of the polynomial $P(\lambda)$. Due to the symmetry of the four roots in (5.13), if λ_1 is a root of $P(\lambda)$, then so is $-\bar{\lambda}_1$. □

STEP 3. The two spectral bands of the Lax spectrum obtained in Step 2 generates spectral bands of the spectral stability problem (5.9) which intersect the origin $\Lambda = 0$ and contain parts outside $i\mathbb{R}$.

PROOF OF STEP 3. By Lemma 5.2, we have $\Lambda = 2\Omega = \pm 2i\sqrt{P(\lambda)}$ so that each root of $P(\lambda)$ is mapped to the origin $\Lambda = 0$. The corresponding squared eigenfunctions in (5.10) are L-periodic in x if u is L-periodic in x by Step 2. Thus $\Lambda = 0$ belongs to the spectral bands of the spectral stability problem (5.9).

It remains to show that the Lax spectrum near the roots of $P(\lambda)$ is mapped to the spectral bands of the spectral stability problem (5.9) outside $i\mathbb{R}$. To do so, we compute

$$P'(\lambda) = 4\lambda^3 - (\rho_1 + \rho_2 + \rho_3)\lambda - i\sqrt{\rho_1 \rho_2 |\rho_3|}$$

and find

$$P'(\lambda_{1,2}) = (\rho_3 \pm \sqrt{\rho_1\rho_2})(\sqrt{\rho_1} \pm \sqrt{\rho_2}) \pm i\sqrt{|\rho_3|}(\sqrt{\rho_1} \pm \sqrt{\rho_2})^2.$$

Since $P(\lambda_{1,2}) = 0$ and $\Lambda = \pm 2i\sqrt{P(\lambda_{1,2})} = 0$, the idea is to consider the Lax spectrum in a neighborhood of the points $\lambda = \lambda_{1,2}$ and to show that the corresponding spectral band near $\Lambda = 0$ is outside $i\mathbb{R}$.

It follows from algebraic computations in [**80**] that the Lax spectrum of the spectral problem (5.7) belongs to the set $\operatorname{Re} S(\lambda) = 0$, where $S(\lambda) : \mathbb{C} \mapsto \mathbb{C}$ is analytic with the derivative given by

$$-2i\Omega S'(\lambda) = 2(\rho_1 - \rho_3)E(k) - (\rho_1 + \rho_2 - \rho_3 - 4\lambda^2)K(k),$$

where $K(k)$ and $E(k)$ are complete elliptic integrals for the standing periodic waves in Step 1. Thus, we compute that

$$-2i\Omega S'(\lambda_{1,2}) = 2(\rho_1 - \rho_3)E(k) + 2(\rho_3 \pm \sqrt{\rho_1\rho_2} \pm i\sqrt{|\rho_3|}(\sqrt{\rho_1} \pm \sqrt{\rho_2}))K(k).$$

Recall that if $f(z) = u(x,y) + iv(x,y)$ is analytic in $z = x + iy$, then

$$f'(z) = u_x - iu_y,$$

where partial derivatives of $u(x,y)$ and $v(x,y)$ are related by the Cauchy–Riemann equations: $u_x = v_y$ and $u_y = -v_x$. For simplicity of notations, we can write $\Omega = \Omega_R + i\Omega_I$, $S = S_R + iS_I$, and $\lambda = \lambda_R + i\lambda_I$. Since the Lax spectrum consists of spectral bands along the level curve $S_R(\lambda_R, \lambda_I) = 0$, the tangent vector along the spectral band at $\lambda = \lambda_{1,2}$ in the λ plane is given by

$$\frac{\partial S_R}{\partial \lambda_R} d\lambda_R + \frac{\partial S_R}{\partial \lambda_I} d\lambda_I = 0.$$

By using the chain rule for $\Omega = \Omega(\lambda)$, we obtain that the image of the Lax spectrum in the Ω plane is tangential to the line with

$$\frac{d\Omega_I}{d\Omega_R} = \frac{\dfrac{\partial \Omega_I}{\partial \lambda_I}\dfrac{\partial S_R}{\partial \lambda_R} - \dfrac{\partial \Omega_I}{\partial \lambda_R}\dfrac{\partial S_R}{\partial \lambda_I}}{\dfrac{\partial \Omega_R}{\partial \lambda_I}\dfrac{\partial S_R}{\partial \lambda_R} - \dfrac{\partial \Omega_R}{\partial \lambda_R}\dfrac{\partial S_R}{\partial \lambda_I}}.$$

In view of the Cauchy–Riemann equations, it can be written as

$$\frac{d\Omega_I}{d\Omega_R} = \frac{2(\Omega_R + i\Omega_I)^2\left(\dfrac{\partial \Omega_R}{\partial \lambda_R}\dfrac{\partial S_R}{\partial \lambda_R} + \dfrac{\partial \Omega_R}{\partial \lambda_I}\dfrac{\partial S_R}{\partial \lambda_I}\right)}{2(\Omega_R + i\Omega_I)^2\left(\dfrac{\partial \Omega_R}{\partial \lambda_I}\dfrac{\partial S_R}{\partial \lambda_R} - \dfrac{\partial \Omega_R}{\partial \lambda_R}\dfrac{\partial S_R}{\partial \lambda_I}\right)}.$$

By using previous computations, we obtain at $\lambda = \lambda_{1,2}$:

$$2(\Omega_R + i\Omega_I)\frac{\partial \Omega_R}{\partial \lambda_R} = -(\rho_3 \pm \sqrt{\rho_1\rho_2})(\sqrt{\rho_1} \pm \sqrt{\rho_2}),$$

$$2(\Omega_R + i\Omega_I)\frac{\partial \Omega_R}{\partial \lambda_I} = \pm\sqrt{|\rho_3|}(\sqrt{\rho_1} \pm \sqrt{\rho_2})^2,$$

$$(\Omega_R + i\Omega_I)\frac{\partial S_R}{\partial \lambda_R} = \mp\sqrt{|\rho_3|}(\sqrt{\rho_1} \pm \sqrt{\rho_2})K(k),$$

$$(\Omega_R + i\Omega_I)\frac{\partial S_I}{\partial \lambda_I} = -(\rho_1 - \rho_3)E(k) - (\rho_3 \pm \sqrt{\rho_1\rho_2})K(k).$$

After elementary computations, we obtain the slopes of the spectral bands at $\Omega(\lambda_{1,2}) = 0$ given by

$$(5.18)\quad \frac{d\Omega_I}{d\Omega_R} = \pm \frac{\sqrt{|\rho_3|}(\sqrt{\rho_1} \pm \sqrt{\rho_2})(\rho_1 - \rho_3)E(k)}{(\rho_3 \pm \sqrt{\rho_1\rho_2})(\rho_1 - \rho_3)E(k) + (\rho_3^2 + \rho_1\rho_2 - \rho_3(\rho_1 + \rho_2))K(k)}.$$

Recall that $\rho_1 = \beta$, $\rho_2 = \beta - \alpha^2 k^2$, and $\rho_3 = \beta - \alpha^2$ so that the denominators in equation (5.18) after canceling $\alpha^2 = \rho_1 - \rho_3$ can be simplified to the form

$$(\beta - \alpha^2 \pm \sqrt{\beta(\beta - \alpha^2 k^2)}E(k) + \alpha^2(1 - k^2)K(k).$$

We claim that the expression for the upper sign is strictly positive for every $k \in (0,1)$ and $\alpha^2 k^2 \leq \beta \leq \alpha^2$. Indeed, for $\beta = \alpha^2 k^2$, it is positive since $K(k) > E(k)$. Computing the derivative with respect to β yields

$$\frac{2\sqrt{\beta(\beta - \alpha^2 k^2)} + \beta + \beta - \alpha^2 k^2}{2\sqrt{\beta(\beta - \alpha^2 k^2)}} E(k) > 0,$$

hence it is monotonically increasing in β from a positive value. Since the numerator in (5.18) is always positive, we have proven that $\frac{d\Omega_R}{d\Omega_I} \neq 0$ at $\lambda = \lambda_1$. Since $\Lambda = 2\Omega$, this implies that at least one spectral band in the spectral stability problem (5.9) near $\Lambda = 0$ is outside $i\,\mathbb{R}$. $\square$

REMARK 5.12. The parameter α in Step 1 can be set to one due to the scaling transformation of the NLS equation (4.67): if $u(t,x)$ is a solution, so is $\alpha u(\alpha^2 t, \alpha x)$, $\alpha > 0$.

REMARK 5.13. Since the traveling wave with the profile $u(x) = R(x)e^{i\Theta(x)}$ is quasi-periodic, the Floquet theorem in Step 2 can be applied after the transformation

$$p_1(x) = P_1(x)e^{\frac{i}{2}\Theta(x)}, \quad q_1(x) = Q_1(x)e^{-\frac{i}{2}\Theta(x)},$$

which reduces the spectral problem (5.7) with the quasi-periodic potential u to the one with a periodic potential R. A similar transformation

$$w_1(x) = W_1(x)e^{i\Theta(x)}, \quad w_2(x) = W_2(x)e^{-i\Theta(x)}$$

applies to the spectral stability problem (5.9).

REMARK 5.14. Numerical results show that there exist exactly two spectral bands of the Lax spectrum outside $i\mathbb{R}$ in Step 2, see Examples 5.1, 5.2, and 5.3.

REMARK 5.15. The spectral instability of the standing periodic waves is related to the spectral bands of the Lax spectrum outside $i\,\mathbb{R}$. Indeed, the polynomial $P(\lambda)$ can be written in the factorized form

$$P(\lambda) = (\lambda - \lambda_1)(\lambda - \lambda_2)(\lambda + \bar{\lambda}_1)(\lambda + \bar{\lambda}_2),$$

where $\mathrm{Re}(\lambda_{1,2}) \neq 0$. If $\lambda \in i\,\mathbb{R}$ belongs to the Lax spectrum, then $P(\lambda) : i\mathbb{R} \to \mathbb{R}$ with $P(\lambda) > 0$ since $P(\lambda)$ does not vanish for every $\lambda \in i\,\mathbb{R}$ and $P(\lambda) = \lambda^4 + \mathcal{O}(\lambda^3)$ as $|\lambda| \to \infty$. Since $\Lambda = \pm 2i\sqrt{P(\lambda)}$, it follows that $\Lambda \in i\,\mathbb{R}$ if $\lambda \in i\,\mathbb{R}$.

REMARK 5.16. The stability band intersecting $\Lambda = 0$ for $\lambda = \lambda_1$ is always transversal to $i\,\mathbb{R}$ as in Step 3. On the other hand, the stability band intersecting $\Lambda = 0$ for $\lambda = \lambda_2$ can be tangential to $i\,\mathbb{R}$ at the point where

$$(\beta - \alpha^2 - \sqrt{\beta(\beta - \alpha^2 k^2)}E(k) + \alpha^4(1 - k^2)K(k) = 0.$$

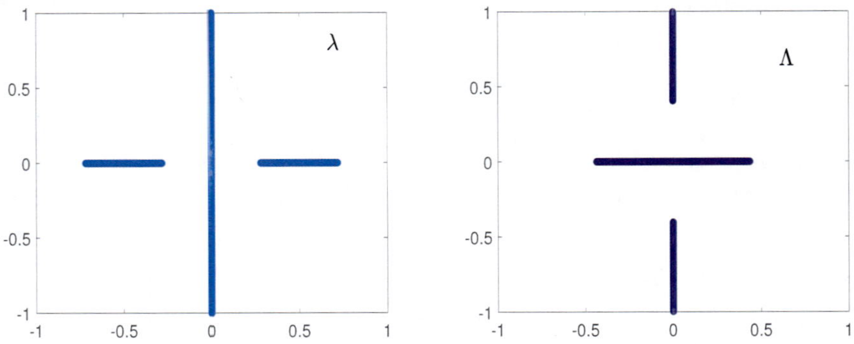

FIGURE 5.1. Lax spectrum on the λ-plane (left) and stability spectrum on the Λ-plane (right) for the dnoidal periodic wave with $\alpha = 1$ and $k = 0.9$.

It was shown in [**46**] for $\alpha = 1$ that this equation is satisfied for a unique value of $\beta \in (k^2, 1)$ and every value of $k \in (0, 1)$. Numerical results show that if the stability band is tangential to $i\,\mathbb{R}$, it nevertheless has parts outside $i\,\mathbb{R}$, see Example 5.3.

REMARK 5.17. The modulational stability analysis can be performed as the perturbation expansion of the spectral bands of the stability problem (5.9) for small characteristic exponent ν. The analysis performed in [**34**] recovers the slopes of the spectral band in the Λ plane intersecting $\Lambda = 0$.

EXAMPLE 5.1. Dnoidal periodic waves correspond to the boundary $\beta = \alpha^2$ in the region $\mathcal{R}$ (Step 1) and can be written as $u(x) = \alpha\,\mathrm{dn}(\alpha x; k)$. The four roots of $P(\lambda)$ (Step 2) are real and symmetric since $\rho_3 = 0$. They are located at $\{\pm\lambda_1, \pm\lambda_2\}$ given by

$$\lambda_1 = \frac{1}{2}\alpha(1 + \sqrt{1-k^2}), \quad \lambda_2 = \frac{1}{2}\alpha(1 - \sqrt{1-k^2}).$$

The Lax spectrum is shown in Figure 5.1 (left) with two symmetric spectral bands $(-\lambda_1, -\lambda_2)$ and (λ_2, λ_1) outside $i\mathbb{R}$. Since $P(\lambda) < 0$ for these two bands, the corresponding values of Λ are real. The stability spectrum of spectral problem (5.9) on the Λ plane is shown in Figure 5.1 (right). The dnoidal periodic wave is modulationally unstable (Step 3). The maximal growth rate is attained at

$$\max_{\lambda\in[\lambda_2,\lambda_1]} \Lambda = \max_{\lambda\in[\lambda_2,\lambda_1]} 2\sqrt{(\lambda_1^2 - \lambda^2)(\lambda^2 - \lambda_2^2)} = (\lambda_1^2 - \lambda_2^2) = \alpha\sqrt{1-k^2},$$

which is largest in the limit $k \to 0$ and vanishes in the limit $k \to 1$.

EXAMPLE 5.2. Cnoidal periodic waves correspond to the boundary $\beta = \alpha^2 k^2$ in the region $\mathcal{R}$ (Step 1) and can be written as $u(x) = \alpha k\,\mathrm{cn}(\alpha x; k)$. The four roots of $P(\lambda)$ (Step 2) are complex-conjugate since $\rho_2 = 0$. They are located at the quadruple $\{\lambda_1, \bar{\lambda}_1, -\lambda_1, -\bar{\lambda}_1\}$ since $\lambda_2 = \bar{\lambda}_1$, where

$$\lambda_1 = \frac{1}{2}\alpha(1 + i\sqrt{1-k^2}).$$

The Lax spectrum of the spectral problem (5.7) is shown on left panels of Figure 5.2 for $k = 0.85$ (top) and $k = 0.95$ (bottom). The unstable spectrum on the Λ-plane

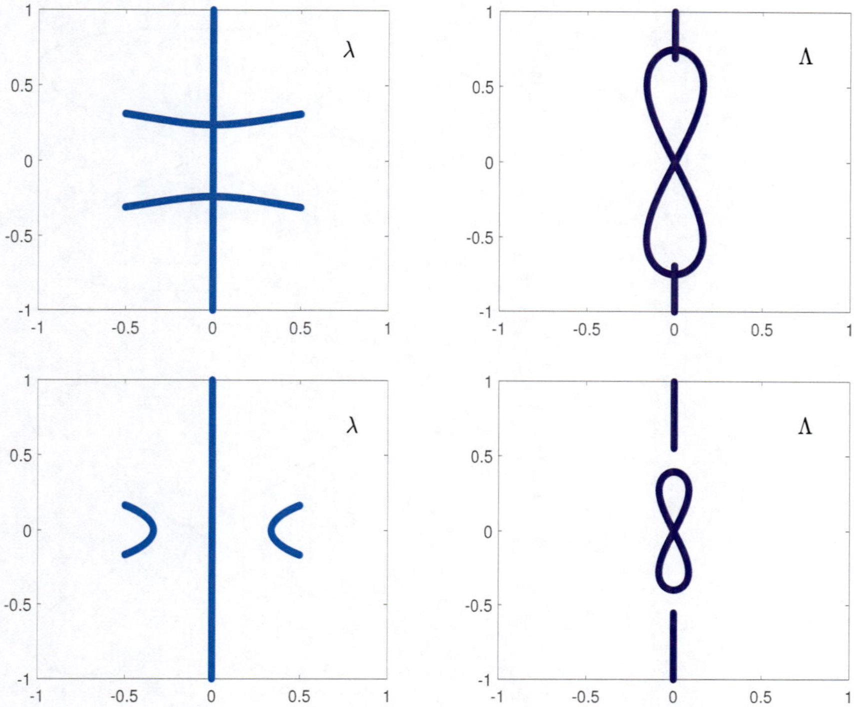

FIGURE 5.2. The same as Figure 5.1 but for the cnoidal periodic wave with $\alpha = 1$ and $k = 0.85$ (top) or $k = 0.95$ (bottom).

is obtained from the same expressions $\Lambda = \pm 2i\sqrt{P(\lambda)}$ when λ traverses along the bands of the Lax spectrum outside $i\mathbb{R}$. The unstable spectrum resembles the figure-eight as is shown on the right panels of Figure 5.2. The spectral bands for the figure-eight start and end at $\Lambda = 0$. The stability spectra for $k = 0.85$ and $k = 0.95$ are similar in spite of the differences in the Lax spectrum. The only difference is that the spectral bands for the figure-eight and the spectral bands on $i\mathbb{R}$ intersect for $k = 0.85$ (top) and do not intersect for $k = 0.95$ (bottom). The cnoidal periodic wave is modulationally unstable (Step 3).

EXAMPLE 5.3. Two examples of standing periodic waves with nontrivial phase $\Theta(x)$ in the profile $u(x) = R(x)e^{i\Theta(x)}$ are represented in Figure 5.3. The Lax spectrum of the spectral problem (5.7) is shown on the left and the stability spectrum of the spectral problem (5.9) is shown on the right. The symmetry of the Lax spectrum is broken compared to Figures 5.1 and 5.2. On the top panels, one unstable band becomes tangential to $i\mathbb{R}$ at $\Lambda = 0$ but still resembles the figure-eight and the other unstable band resembles a large figure-eight. On the bottom panels, the spectral bands of the Lax spectrum do not intersect $i\,\mathbb{R}$ and the unstable bands of the stability problem (5.9) represents a butterfly, also known as a propeller.

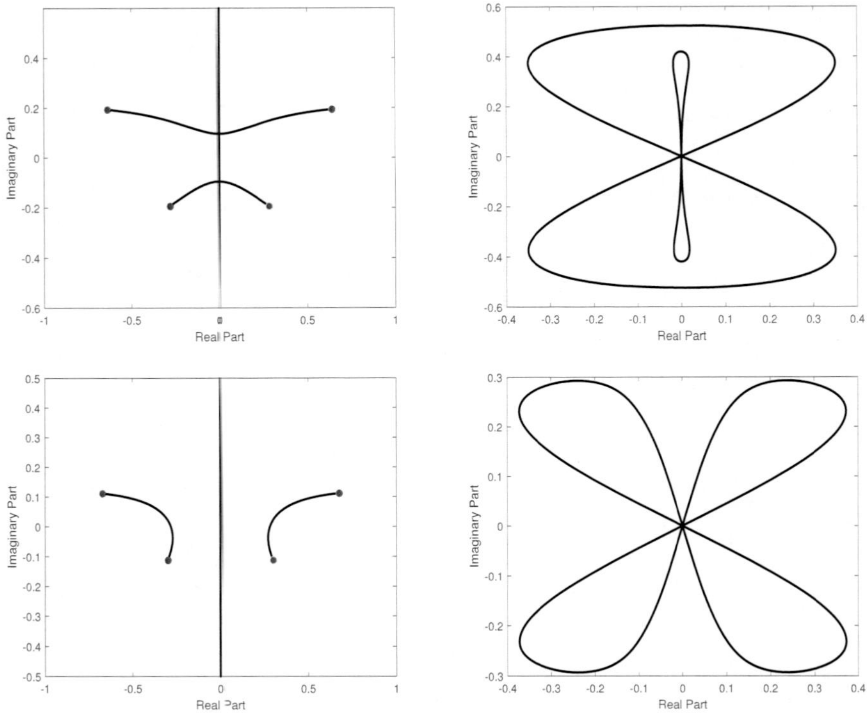

FIGURE 5.3. The same as Figure 5.1 but for the periodic wave with nontrivial phase for $(\alpha, \beta, k) = (1, 0.85, 0.85)$ (top) and $(\alpha, \beta, k) = (1, 0.95, 0.9)$ (bottom). Spectral bands on $i\mathbb{R}$ are not shown on the right panel.

5.1.2. Modulational stability in the derivative NLS equation. We follow the same algorithm as in the focusing NLS equation. This illustrates universality and broad applicability of this method for standing periodic waves in other integrable equations.

Let $u = u(t, x)$ be a solution of the DNLS equation (4.18). The perturbation $v = v(t, x)$ satisfies the linearized DNLS equation after the quadratic and cubic nonlinear terms are crossed out:

$$iv_t + v_{xx} + i(2|u|^2 v + u^2 \bar{v})_x = 0. \tag{5.19}$$

Let $\varphi = \varphi(t, x)$ be a solution of the linear system (4.2) with (4.19) and (4.20). Spatially bounded solutions $\varphi(t, x)$ are referred to as eigenfunctions and the corresponding values λ are referred to as eigenvalues. The set of eigenvalues λ form the Lax spectrum.

The following lemma connects solutions of the linearized DNLS equation (5.19) and solutions of the linear system (4.2) with (4.19) and (4.20).

LEMMA 5.4. *Assume u is a solution to the DNLS equation* (4.18) *and $\varphi = (\varphi_1, \varphi_2)^T$ is a solution to the linear system* (4.2) *with* (4.19) *and* (4.20) *for some*

λ. *Then,*

$$v = \partial_x \varphi_1^2 + \partial_x \bar{\varphi}_2^2, \quad \bar{v} = \partial_x \varphi_2^2 + \partial_x \bar{\varphi}_1^2 \tag{5.20}$$

and

$$v = i(\partial_x \varphi_1^2 - \partial_x \bar{\varphi}_2^2), \quad \bar{v} = i(\partial_x \varphi_2^2 - \partial_x \bar{\varphi}_1^2) \tag{5.21}$$

are solutions to the linearized DNLS equation (5.19).

PROOF. Solutions of the linear system (4.2) with (4.19) and (4.20) satisfy the following quadruple symmetry: If $\varphi = (\varphi_1, \varphi_2)^T$ is a solution for λ, then $\varphi = (-\bar{\varphi}_2, \bar{\varphi}_1)^T$ is a solution for $\bar{\lambda}$, $\varphi = (\bar{\varphi}_2, \bar{\varphi}_1)^T$ is a solution for $-\bar{\lambda}$, and $\varphi = (\varphi_1, -\varphi_2)^T$ is a solution for $-\lambda$. Due to this symmetry and the linear superposition principle, it is sufficient to show that $v = \partial_x \varphi_1^2$ and $\bar{v} = \partial_x \varphi_2^2$ satisfy (5.19). Since

$$\begin{aligned} \partial_x \varphi_1^2 &= 2\lambda(-i\lambda\varphi_1^2 + u\varphi_1\varphi_2) \\ \partial_x \varphi_2^2 &= 2\lambda(i\lambda\varphi_2^2 - \bar{u}\varphi_1\varphi_2), \end{aligned}$$

it suffices to show that

$$v = -i\lambda\varphi_1^2 + u\varphi_1\varphi_2, \quad \bar{v} = i\lambda\varphi_2^2 - \bar{u}\varphi_1\varphi_2,$$

satisfies the linearized DNLS equation (5.19). Plugging the two terms of v into equation (5.19) separately gives

$$\begin{aligned} &i\partial_t(\varphi_1^2) + \partial_x^2(\varphi_1^2) + i[2|u|^2\partial_x(\varphi_1^2) - u^2\partial_x(\varphi_2^2) + 2(u\bar{u}_x + u_x\bar{u})\varphi_1^2 - 2uu_x\varphi_2^2] \\ &\quad = 4u^2\varphi_2^2\lambda^2 + 4i|u|^2u\varphi_1\varphi_2\lambda + 2i(u\bar{u}_x + u_x\bar{u})\varphi_1^2 - 2iuu_x\varphi_2^2, \end{aligned}$$

and

$$\begin{aligned} &i\partial_t(u\varphi_1\varphi_2) + \partial_x^2(u\varphi_1\varphi_2) \\ &\qquad + i[2|u|^2\partial_x(u\varphi_1\varphi_2) - u^2\partial_x(\bar{u}\varphi_1\varphi_2) + 2(u\bar{u}_x + u_x\bar{u})u\varphi_1\varphi_2 - 2uu_x\bar{u}\varphi_1\varphi_2] \\ &\quad = 4iu^2\varphi_2^2\lambda^3 - 4|u|^2u\varphi_1\varphi_2\lambda^2 - 2(u\bar{u}_x + u_x\bar{u})\varphi_1^2\lambda + 2uu_x\varphi_2^2\lambda. \end{aligned}$$

Summing the first equality multiplied by $(-i\lambda)$ and the second equality yields zero. Hence $v = \partial_x \varphi_1^2$ and $\bar{v} = \partial_x \varphi_2^2$ satisfy the linearized DNLS equation (5.19). Extending this solution with the symmetric solution $v = \partial_x \bar{\varphi}_2^2$ and $\bar{v} = \partial_x \bar{\varphi}_1^2$ in the linear combinations (5.20) and (5.21) ensures that $\bar{v}$ is a complex conjugate of v. ☐

Traveling waves of the DNLS equation (4.18) are solutions of the form $u(t,x) \mapsto u(x+2ct)e^{-4ibt}$ with parameters b and c (Example 4.2), where the profile u satisfies the stationary DNLS equation (4.21). Separation of variables yields the representation

$$\varphi_1(t,x) = \chi_1(x+2ct)e^{-2ibt+2t\Omega}, \quad \varphi_2(t,x) = \chi_2(x+2ct)e^{+2ibt+2t\Omega},$$

where $\Omega \in \mathbb{C}$ is another spectral parameter and $\chi = (\chi_1, \chi_2)^T$ satisfies the spectral problems

$$\chi' = \begin{bmatrix} -i\lambda^2 & \lambda u \\ -\lambda\bar{u} & i\lambda^2 \end{bmatrix} \chi, \tag{5.22}$$

and

$$(5.23)\quad 2\Omega\chi + 2c\begin{bmatrix} -i\lambda^2 & \lambda u \\ -\lambda\bar{u} & i\lambda^2 \end{bmatrix}\chi = \begin{bmatrix} -2i\lambda^4 + i\lambda^2|u|^2 + 2ib & 2\lambda^3 u + \lambda(iu' - |u|^2 u) \\ -2\lambda^3\bar{u} + \lambda(i\bar{u}' + |u|^2\bar{u}) & 2i\lambda^4 - i\lambda^2|u|^2 - 2ib \end{bmatrix}\chi,$$

where the derivative is applied to the function χ of a single variable in the moving frame, which we call x again.

Similarly, the linearized DNLS equation (5.19) for solutions of the form

$$v(t,x) \mapsto v(x+2ct)e^{-4ibt}$$

can be written as follows:

$$iv_t - 2icv_x + 4bv + v_{xx} + i(2|u|^2 v + u^2\bar{v})_x = 0.$$

Separation of variables according to

$$v(t,x) = w_1(x)e^{t\Lambda}, \quad \bar{v}(t,x) = w_2(x)e^{t\Lambda}$$

transforms the linearized DNLS equation to the spectral stability problem

$$(5.24)\quad i\Lambda\begin{bmatrix} 1 & 0 \\ 0 & -1 \end{bmatrix} w + \begin{bmatrix} \partial_x^2 + 2i\partial_x|u|^2 + 4b + 2ic\partial_x & i\partial_x u^2 \\ -i\partial_x\bar{u}^2 & \partial_x^2 - 2i\partial_x|u|^2 + 4b - 2ic\partial_x \end{bmatrix} w = 0,$$

where Λ is an eigenvalue and $w = (w_1, w_2)^T$ is an eigenfunction. The following lemma connects bounded solutions of (5.22) and (5.23) with bounded solutions of (5.24).

LEMMA 5.5. *Let λ be in the Lax spectrum of the spectral problem* (5.22) *with the bounded eigenfunction $\chi(x)$. Eigenvalues and eigenfunctions of the spectral stability problem* (5.24) *are given by*

$$(5.25)\qquad w_1 = \partial_x\chi_1^2, \quad w_2 = \partial_x\chi_2^2, \quad \Lambda = 4\Omega.$$

where $\Omega = \pm i\sqrt{P(\lambda)}$ with

$$(5.26)\qquad P(\lambda) = \lambda^8 - 2c\lambda^6 + (2a - 2b + c^2)\lambda^4 + 2(d - ca + cb)\lambda^2 + b^2,$$

with a and d being parameters of the two conserved quantities (4.22) *and* (4.23).

PROOF. Since the spectral problem (5.23) is a linear algebraic system, it admits a nonzero solution if and only if the determinant of the coefficient matrix is zero. The latter condition yields the x-independent relation between Ω and λ in the form $\Omega^2 + P(\lambda) = 0$ with $P(\lambda)$ given by

$$P(\lambda) = \left(\lambda^4 - \frac{1}{2}\lambda^2(2c + |u|^2) - b\right)^2 + \frac{1}{4}\lambda^2\left(2\lambda^2 u + iu' - |u|^2 u - 2cu\right)\left(2\lambda^2\bar{u} - i\bar{u}' - |u|^2\bar{u} - 2c\bar{u}\right),$$

Using (4.22) and (4.23) in $P(\lambda)$ yields (5.26). By Lemma 5.4, if λ is in the Lax spectrum of the spectral problem (5.22) with the bounded eigenfunction $\chi \in L^\infty(\mathbb{R})$, then eigenvalues and eigenfunctions of the spectral stability problem (5.24) are given by (5.25). □

The following lemma explores the relation $\Lambda = \pm 4i\sqrt{P(\lambda)}$ in Lemma 5.5 and finds the conditions for which the Lax spectrum on $\mathbb{R} \cup i\mathbb{R}$ for λ is mapped to the stable line $i\mathbb{R}$ for Λ.

LEMMA 5.6. *Assume that $P(\lambda)$ is given by* (5.26) *with four pairs of roots*

$$(\pm\lambda_1, \pm\lambda_2, \pm\lambda_3, \pm\lambda_4) \in \mathbb{C}\backslash\mathbb{R}.$$

If $\lambda \in \mathbb{R}$, then $\Lambda \in i\mathbb{R}$. If $\lambda \in i\mathbb{R}$, then $\Lambda \in i\mathbb{R}$, provided one of the following conditions are true:

- *The roots form two complex quadruplets.*
- *Two pairs of roots are purely imaginary, e.g., $\lambda_{3,4} = i\beta_{3,4}$ with $0 < \beta_4 < \beta_3$, and $\lambda \in i\mathbb{R}\backslash\mathcal{S}$ with*

$$\mathcal{S} := \{\lambda \in i\mathbb{R} : \quad \text{Im}(\lambda) \in (-\beta_3, -\beta_4) \cup (\beta_4, \beta_3)\}. \tag{5.27}$$

- *Four pairs of roots are purely imaginary, e.g., $\lambda_{1,2,3,4} = i\beta_{1,2,3,4}$ with $0 < \beta_4 < \beta_3 < \beta_2 < \beta_1$, and $\lambda \in i\mathbb{R}\backslash\mathcal{S}$ with*

$$\mathcal{S} := \{\lambda \in i\mathbb{R} : \quad \text{Im}(\lambda) \in (-\beta_1, -\beta_2) \cup (-\beta_3, -\beta_4) \cup (\beta_4, \beta_3) \cup (\beta_2, \beta_1)\}. \tag{5.28}$$

PROOF. If $\lambda \in \mathbb{R}$, the polynomial (5.26) can be factorized by its roots as

$$P(\lambda) = (\lambda^2 - \lambda_1^2)(\lambda^2 - \lambda_2^2)(\lambda^2 - \lambda_3^2)(\lambda^2 - \lambda_4^2).$$

If $(\pm\lambda_1, \pm\lambda_2, \pm\lambda_3, \pm\lambda_4) \in \mathbb{C}\backslash\mathbb{R}$, then $P(\lambda) > 0$ for every $\lambda \in \mathbb{R}$ since $P(\lambda) \sim \lambda^8$ as $|\lambda| \to \infty$ and $P(\lambda)$ has no real roots. It follows from $\Lambda = \pm 4i\sqrt{P(\lambda)}$ and $P(\lambda) > 0$ that $\Lambda \in i\mathbb{R}$.

For $\lambda \in i\mathbb{R}$, we can rewrite the polynomial $P(\lambda)$ in the form

$$\tilde{P}(z) := P(-\lambda^2) = (z + \lambda_1^2)(z + \lambda_2^2)(z + \lambda_3^2)(z + \lambda_4^2),$$

where $z = -\lambda^2 \geq 0$. If the roots of $P(\lambda)$ form complex quadruplets, then $\tilde{P}(z) > 0$ for every $z \geq 0$. Indeed, $\tilde{P}(z) \sim z^4$ as $|z| \to \infty$ and $\tilde{P}(z)$ has no real roots for $z > 0$. Since

$$\Lambda = \pm 4i\sqrt{\tilde{P}(z)}$$

and $\tilde{P}(z) > 0$, we have $\Lambda \in i\mathbb{R}$.

If two or four pairs of purely imaginary eigenvalues occur, then $\tilde{P}(z) > 0$ for either $z \in (0, \beta_4^2) \cup (\beta_3^2, \infty)$ or $z \in (0, \beta_4^2) \cup (\beta_3^2, \beta_2^2) \cup (\beta_1^2, \infty)$, respectively. These intervals define the sets (5.27) and (5.28) which are excluded from the admissible values of $\lambda \in i\mathbb{R}$, for which $\Lambda \in i\mathbb{R}$. □

REMARK 5.18. It was also proven in [**82**] that if $\Lambda \in i\mathbb{R}$ for a given $\lambda \in \mathbb{R} \cup i\mathbb{R}$, then $\lambda \in \mathbb{R} \cup i\mathbb{R}$ belongs to the Lax spectrum of the spectral problem (5.22). By Lemma 5.6, this implies that $\mathbb{R} \cup i\mathbb{R}\backslash\mathcal{S}$ belongs to the Lax spectrum. Numerical results in [**44**] suggest that the set $\mathcal{S}$ does not belong to the Lax spectrum for every family of traveling waves in the DNLS equation (4.18).

Due to Lemma 5.6, modulational instability of the periodic waves in the DNLS equation (4.18) is expected if and only if there exists Lax spectrum of the spectral problem (5.22) outside $\mathbb{R} \cup i\mathbb{R}$ or inside the spectral gaps $\mathcal{S}$. The following theorem presents this result based on [**82**] and [**44**].

THEOREM 5.2. *Every traveling periodic wave of the DNLS equation* (4.18) *is modulationally unstable if there exists a Lax spectrum of the spectral problem* (5.22) *outside $\mathbb{R} \cup i\mathbb{R}$ or inside the spectral gaps $\mathcal{S}$ given by* (5.27) *and* (5.28).

PROOF. The proof of this theorem follows the three steps explained in the proof of Theorem 5.1 with additional complexity due to a higher degree of the corresponding polynomial $P(\lambda)$.

STEP 1. Two distinct families of traveling periodic waves exist in the stationary DNLS equation (4.21) given by (5.31) and (5.32).

PROOF OF STEP 1. Substituting the polar form $u(x) = R(x)e^{i\Theta(x)}$ with real R and Θ into (4.22) and (4.23) yields the following system of first-order equations:

$$\begin{cases} 2(R')^2 + 2R^2(\Theta')^2 - R^6 - 2cR^4 + 8(b-a)R^2 = 16d, \\ 4R^2\Theta' + 3R^4 + 4cR^2 = -8a. \end{cases}$$

Elimination of Θ' yields the first-order invariant

$$(R')^2 + \frac{4a^2}{R^2} + \frac{1}{16}R^6 + \frac{c}{2}R^4 + R^2\left(c^2 + 4b - a\right) + 4ac - 8d = 0. \tag{5.29}$$

The singularity $R = 0$ is unfolded with the transformation $\rho = \frac{1}{2}R^2$ which yields

$$(\rho')^2 + Q(\rho) = 0, \quad Q(\rho) := \rho^4 + 4c\rho^3 + 4(c^2 - a + 4b)\rho^2 + 8(ac - 2d)\rho + 4a^2. \tag{5.30}$$

The polynomial $Q(\rho)$ can be factorized by its roots (u_1, u_2, u_3, u_4) which can be related to the parameters (a, b, c, d) as follows:

$$\begin{cases} u_1 + u_2 + u_3 + u_4 = -4c, \\ u_1u_2 + u_1u_3 + u_1u_4 + u_2u_3 + u_2u_4 + u_3u_4 = 4(c^2 - a + 4b), \\ u_1u_2u_3 + u_1u_2u_4 + u_1u_3u_4 + u_2u_3u_4 = 8(2d - ac), \\ u_1u_2u_3u_4 = 4a^2. \end{cases}$$

The two distinct families of periodic waves correspond to two different cases: (i) four roots of Q are real and (ii) two roots of Q are real and one pair of roots is complex-conjugate. Because Q is real and the first-order invariant does not have real solutions if the two pairs of roots of Q are complex-conjugate, no other traveling periodic waves exist.

CASE (i) (Four roots of Q are real). Let us order the four real roots of Q as

$$u_4 \leq u_3 \leq u_2 \leq u_1.$$

Periodic solutions to (5.30) can be written explicitly with the help of Jacobi elliptic functions [**42**]:

$$\rho(x) = u_4 + \frac{(u_1 - u_4)(u_2 - u_4)}{(u_2 - u_4) + (u_1 - u_2)\,\mathrm{sn}^2(\nu x; k)}, \tag{5.31}$$

where positive parameters ν and k are uniquely expressed by

$$\nu = \frac{1}{2}\sqrt{(u_1 - u_3)(u_2 - u_4)}, \quad k = \frac{\sqrt{(u_1 - u_2)(u_3 - u_4)}}{\sqrt{(u_1 - u_3)(u_2 - u_4)}}.$$

The periodic solution ρ in (5.31) is located in the interval $[u_2, u_1]$ and has period $L = 2K(k)\nu^{-1}$. The solution is meaningful for $\rho = \frac{1}{2}R^2 \geq 0$ if and only if $u_2 \geq 0$.

CASE (ii) (Two roots of Q are real and one pair of roots is complex-conjugate). We take $u_{1,2}$ as real roots ordered as $u_2 \leq u_1$ and take $u_{3,4} = \gamma \pm i\eta$ as complex-conjugate roots. Periodic solutions to (5.30) can be written explicitly with the help of Jacobi elliptic functions [**42**]:

$$\rho(x) = u_1 + \frac{(u_2 - u_1)(1 - \mathrm{cn}(\mu x; k))}{1 + \delta + (\delta - 1)\,\mathrm{cn}(\mu x; k)}, \tag{5.32}$$

where positive parameters δ, μ, and k are uniquely expressed by

$$\delta = \frac{\sqrt{(u_2-\gamma)^2 + \eta^2}}{\sqrt{(u_1-\gamma)^2+\eta^2}}, \qquad \mu = \sqrt[4]{[(u_1-\gamma)^2+\eta^2]\,[(u_2-\gamma)^2+\eta^2]},$$

and

$$2k^2 = 1 - \frac{(u_1-\gamma)(u_2-\gamma)+\eta^2}{\sqrt{[(u_1-\gamma)^2+\eta^2]\,[(u_2-\gamma)^2+\eta^2]}}.$$

The periodic solution ρ in (5.32) is located in the interval $[u_2, u_1]$ and has period $L = 4K(k)\mu^{-1}$. The solution is meaningful for $\rho = \frac{1}{2}R^2 \geq 0$ if and only if $u_2 \geq 0$. $\square$

STEP 2. The family (5.31) is related to either at least four spectral bands of the Lax spectrum in the spectral problem (5.22) outside $\mathbb{R} \cup i\mathbb{R}$ or at least four spectral bands on $i\mathbb{R}$. The family (5.32) is related to at least two spectral bands outside $\mathbb{R} \cup i\mathbb{R}$ and at least two spectral bands on $i\mathbb{R}$.

PROOF OF STEP 2. Let $(\pm\lambda_1, \pm\lambda_2, \pm\lambda_3, \pm\lambda_4)$ be the roots of the polynomial P given by (5.26). The roots (u_1, u_2, u_3, u_4) of the polynomial Q given by (5.30) are related to the roots of P by the following explicit relations (see [**148**] and [**44**]):

$$\begin{cases} u_1 = -\frac{1}{2}(\lambda_1 - \lambda_2 + \lambda_3 - \lambda_4)^2, \\ u_2 = -\frac{1}{2}(\lambda_1 - \lambda_2 - \lambda_3 + \lambda_4)^2, \\ u_3 = -\frac{1}{2}(\lambda_1 + \lambda_2 - \lambda_3 - \lambda_4)^2, \\ u_4 = -\frac{1}{2}(\lambda_1 + \lambda_2 + \lambda_3 + \lambda_4)^2. \end{cases} \tag{5.33}$$

Let us consider separately the two families of the traveling periodic waves in Step 1.

CASE (i) (Four roots of Q are real). The four pairs of eigenvalues $\{\pm\lambda_1, \pm\lambda_2, \pm\lambda_3, \pm\lambda_4\}$ generate real roots $\{u_1, u_2, u_3, u_4\}$ of Q by the transformation formula (5.33) if and only if they satisfy the following three configurations: (a) they form two complex quadruplets; (b) they form four pairs of purely imaginary eigenvalues; or (c) they form four pairs of real eigenvalues.

For the two complex quadruplets in (a), we asume that the four pairs of eigenvalues form two complex quadruplets with

$$\lambda_1 = \bar{\lambda}_2 = \alpha_1 + i\beta_1, \qquad \lambda_3 = \bar{\lambda}_4 = \alpha_2 + i\beta_2.$$

Then the four real roots $\{u_1, u_2, u_3, u_4\}$ of Q satisfy the more precise ordering

$$u_4 \leq u_3 \leq 0 \leq u_2 \leq u_1.$$

For the four pairs of purely imaginary eigenvalues in (b), we assume that

$$\lambda_1 = i\beta_1, \qquad \lambda_2 = i\beta_2, \qquad \lambda_3 = i\beta_3, \qquad \lambda_4 = i\beta_4.$$

Then the four real roots $\{u_1, u_2, u_3, u_4\}$ of Q satisfy the more precise ordering

$$0 \leq u_4 \leq u_3 \leq u_2 \leq u_1.$$

For the four pairs of real eigenvalues in (c), the four real roots $\{u_1, u_2, u_3, u_4\}$ of Q satisfy the more precise ordering

$$u_4 \leq u_3 \leq u_2 \leq u_1 \leq 0.$$

No solutions correspond to (c) since $\rho \leq 0$ in (5.31).

CASE (ii) (Two roots of Q are real and one pair of roots is complex-conjugate). The four pairs of eigenvalues $\{\pm\lambda_1, \pm\lambda_2, \pm\lambda_3, \pm\lambda_4\}$ generate the two real roots $\{u_1, u_2\}$ and one pair of complex-conjugate roots $u_{3,4} = \gamma \pm i\eta$ of Q by the transformation formula (5.33) if and only if they satisfy the following two configurations: (a) they form one complex quadruplet and two pairs of purely imaginary eigenvalues or (b) they form one complex quadruplet and two pairs of real eigenvalues.

For (a), we assume that the four pairs of eigenvalues form one quadruplet $\{\pm\lambda_1, \pm\bar{\lambda}_1\}$ of complex eigenvalues and two pairs $\{\pm\lambda_3, \pm\lambda_4\}$ of purely imaginary eigenvalues. This yields $0 \leq u_2 \leq u_1$ and $u_3 = \bar{u}_4$ as required.

For (b), we assume that the four pairs of eigenvalues form one quadruplet $\{\pm\lambda_1, \pm\bar{\lambda}_1\}$ of complex eigenvalues and two pairs $\{\pm\lambda_3, \pm\lambda_4\}$ of real eigenvalues. Then $u_1 = \bar{u}_2$ and $u_4 \leq u_3 \leq 0$. No solutions correspond to (b) since $\rho \leq 0$ in (5.32).

The proof of Step 2 is complete if we show that the eight eigenvalues of $P(\lambda)$ are related to the L-periodic squared eigenfunctions χ of the spectral problem (5.22). In this case, by the Floquet theorem, there exist spectral bands of the Lax spectrum which connect the roots of $P(\lambda)$ either outside $\mathbb{R} \cup i\,\mathbb{R}$ or inside $i\,\mathbb{R}$. To prove periodicity of the squared eigenfunctions, we use the algebraic method that connects the traveling periodic wave with the profile u and the squared eigenfunctions of the spectral problem (5.22). This relation was elaborated in [**44**] in the form

$$\begin{cases} u = \lambda_1 p_1^2 + \bar{\lambda}_1 \bar{q}_1^2, \\ \bar{u} = \lambda_1 q_1^2 + \bar{\lambda}_1 \bar{p}_1^2, \end{cases} \tag{5.34}$$

if $\lambda_1 \in \mathbb{C} \backslash (\mathbb{R} \cup i\,\mathbb{R})$ or in the form

$$\begin{cases} u = i\beta_1 p_1^2 + i\beta_2 p_2^2, \\ \bar{u} = -i\beta_1 \bar{p}_1^2 - i\beta_2 \bar{p}_2^2, \end{cases} \tag{5.35}$$

if $\lambda_{1,2} = i\beta_{1,2} \in i\,\mathbb{R}$. We refer to [**44**] for further details of the algebraic method based on the relations (5.34) and (5.35). ☐

STEP 3. The spectral bands in Step 2 outside $\mathbb{R} \cup i\,\mathbb{R}$ or inside the spectral gaps $\mathcal{S}$ given by (5.27) and (5.28) contain the spectral bands of the spectral stability problem (5.24) outside $i\,\mathbb{R}$ and intersecting the origin $\Lambda = 0$.

PROOF OF STEP 3. Since $\Lambda = \pm 4i\sqrt{P(\lambda)}$, each root of $P(\lambda)$ is mapped to the origin $\Lambda = 0$. The corresponding squared eigenfunctions in (5.25) are L-periodic in x if u is L-periodic in x from Step 2 so that $\Lambda = 0$ belongs to the spectral bands of the spectral stability problem (5.24).

If there exists the Lax spectrum inside $\mathcal{S}$ given by either (5.27) or (5.28), then it follows from Lemma 5.6 that $\Lambda \in \mathbb{R}$. The end points of $\mathcal{S}$ correspond to $\Lambda = 0$ so that the periodic wave is modulationally unstable.

If there exists the Lax spectrum outside $\mathbb{R} \cup i\,\mathbb{R}$, then the spectral bands in the Λ plane intersect $\Lambda = 0$ transversally to $i\,\mathbb{R}$ so that the periodic wave is also

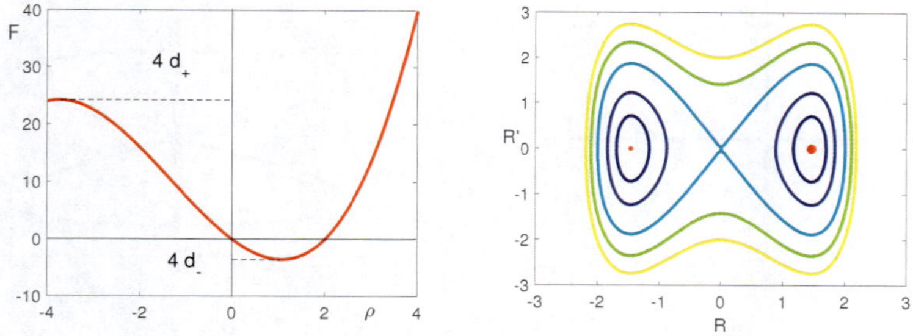

FIGURE 5.4. The graph of F versus ρ (left) and the phase portrait (R, R') (right) for $c^2 + 4b < 0$.

modulationally unstable. The transverse intersection can be shown by similar computations as in the proof of Theorem 5.1. □

REMARK 5.19. If $u_4 \geq 0$ in Step 2 [case (i), subcase (b)], there exists another periodic solution to (5.30) in the form

$$\rho(x) = u_2 - \frac{(u_2 - u_3)(u_2 - u_4)}{(u_2 - u_4) - (u_3 - u_4)\,\mathrm{sn}^2(\nu x; k)}, \tag{5.36}$$

with the same values of parameters ν and k as in (5.31). The periodic solution ρ is located in the interval $[u_4, u_3]$ and has the same period $L = 2K(k)\nu^{-1}$ as the periodic solution ρ in (5.31).

REMARK 5.20. The algebraic relation (5.34) involves two eigenvalues in the spectral problem (5.22). This follows from the symmetry of spectral problem (5.22): if $\lambda_1 \in \mathbb{C}\backslash\mathbb{R}$ is an eigenvalue with the eigenvector $\varphi = (p_1, q_1)^T$, then $\lambda_2 = \bar{\lambda}_1$ is also an eigenvalue with the eigenvector $\varphi = (\bar{q}_1, -\bar{p}_1)^T$.

REMARK 5.21. It follows from the spectral problem (5.22) that if $\lambda \in i\mathbb{R}\backslash\{0\}$ is a simple eigenvalue with the eigenvector $\varphi = (\varphi_1, \varphi_2)^T$, then there is $c \in \mathbb{C}$ with $|c| = 1$ such that $\varphi_1 = c\bar{\varphi}_2$. This property explains why the constraint (5.34) cannot be used with $\bar{\lambda}_1 = -i\beta_1$ if $\lambda_1 = i\beta_1 \in i\,\mathbb{R}$ as otherwise, it would only involve one eigenvalue. The two eigenvalues $\lambda_{1,2} = i\beta_{1,2} \in i\,\mathbb{R}$ are used in the constraint (5.35) and the symmetry of the eigenvectors $\varphi = (p_1, q_1)^T$ and $\varphi = (p_2, q_2)^T$ is taken into account with $q_{1,2} = -ip_{1,2}$.

EXAMPLE 5.4. Fix $a = 0$. The first-order invariant (5.29) can be rewritten in the nonsingular form

$$(R')^2 + F(R) = 8d, \quad F(R) = \frac{1}{16}R^6 + \frac{c}{2}R^4 + (c^2 + 4b)R^2,$$

which does not need to be unfolded into (5.30). With $\rho := \frac{1}{2}R^2$, $F(R)$ becomes a cubic polynomial in ρ with one root at zero. For $c^2 + 4b < 0$, Figure 5.4 shows F versus ρ (left) and the corresponding phase portrait in the (R, R') plane (right).

For $d \in (d_-, 0)$, Figure 5.5 shows four typical configurations for the Lax spectrum (left) and the corresponding stability spectrum (right). The four typical configurations of the Lax spectrum are: (a) with four spectral bands intersecting the real axis, (b) with four spectral bands intersecting the imaginary axis, (c) with two spectral bands intersecting the real axis and two spectral bands intersecting the

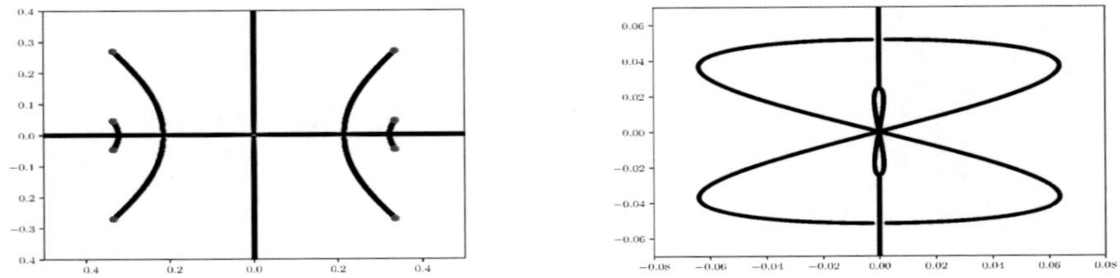

(a) $u_1 = 0.2$, $u_2 = 0.1$, $u_3 = 0$, $u_4 = -0.9$.

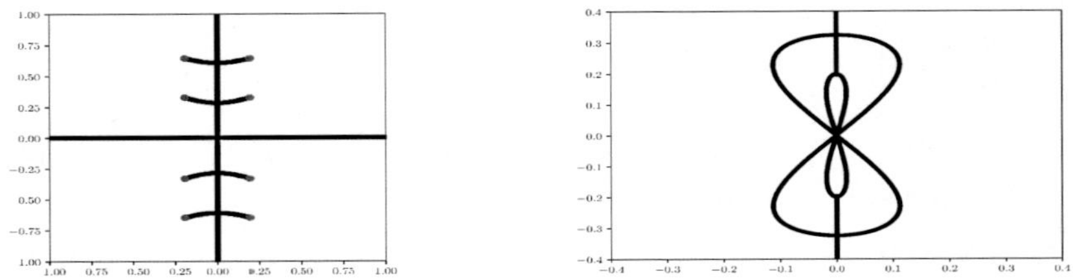

(b) $u_1 = 1.9$, $u_2 = 0.2$, $u_3 = 0$, $u_4 = -0.3$.

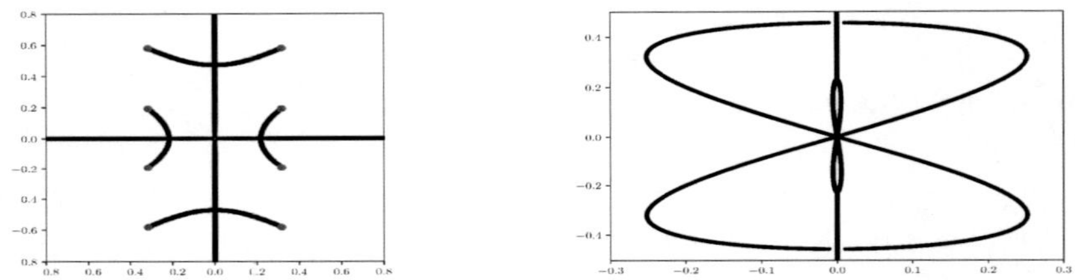

(c) $u_1 = 1.2$, $u_2 = 0.3$, $u_3 = 0$, $u_4 = -0.8$.

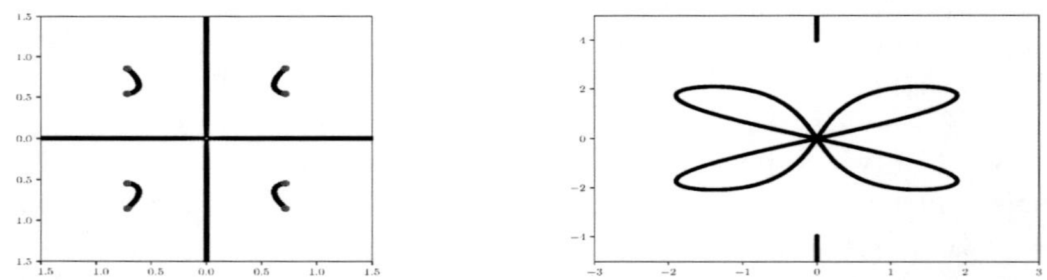

(d) $u_1 = 3.9$, $u_2 = 0.193012$, $u_3 = 0$, $u_4 = -4.090301$.

FIGURE 5.5. Lax spectrum (left) and stability spectrum (right) on the complex λ and Λ planes (Re. *vs.* Im.) for the four representative cases found for $c^2 + 4b < 0$ and $d \in (d_-, 0)$. Eight red dots on the left panel mark the roots of the polynomial $P(\lambda)$.

imaginary axis, and (d) with no intersections with the real or imaginary axes. The first three cases lead to a double figure-eight of the stability spectrum and the last case leads to a butterfly figure, which were seen for the traveling periodic waves in

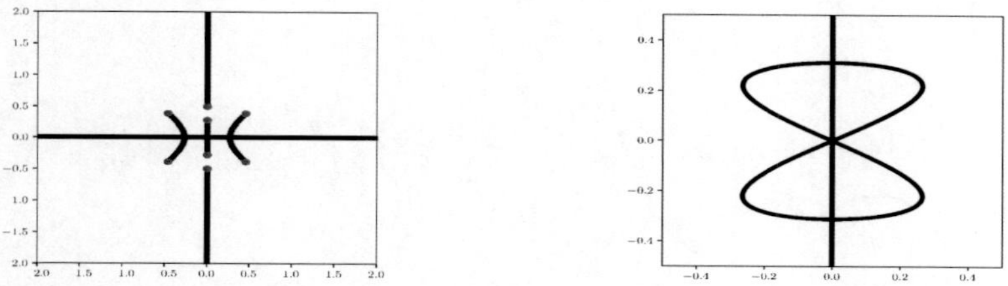

(a) $u_1 = 1.2$, $u_2 = 0$, $u_3 = -0.4 - 0.2i$, $u_4 = -0.4 + 0.2i$.

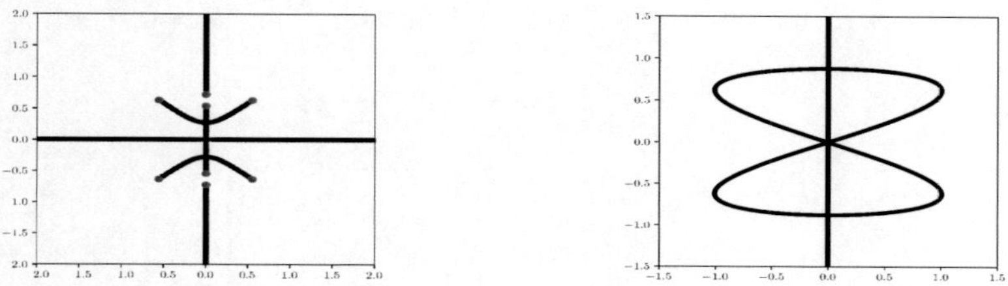

(b) $u_1 = 3.2$, $u_2 = 0$, $u_3 = -0.6 + 0.2i$, $u_4 = -0.6 - 0.2i$.

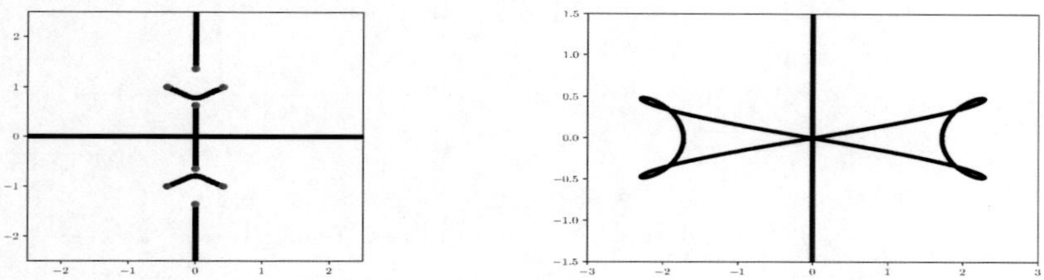

(c) $u_1 = 8$, $u_2 = 0$, $u_3 = -0.1 + 0.6i$, $u_4 = -0.1 - 0.6i$.

FIGURE 5.6. The same as Figure 5.5 but for $c^2 + 4b < 0$ and $d \in (d_+, \infty)$.

the NLS equation. These cases illustrate the two complex quadruplets in the roots of $P(\lambda)$ in Theorem 5.2, Step 2, case (i), subcase (a).

Similar patterns appear in the case $d \in (0, d_+)$, for which the only difference is that the periodic waves are sign-indefinite in R so that $\rho = \frac{1}{2}R^2$ attains zero on each period.

For $d \in (d_+, \infty)$, Figure 5.6 shows three typical configurations for the Lax spectrum (left) and the corresponding stability spectrum (right). This is the case of one complex quadruplet and two pairs of purely imaginary roots of $P(\lambda)$ in Theorem 5.2, Step 2, case (ii), subcase (a). The set $\mathcal{S}$ in (5.27) corresponds to two spectral gaps on the imaginary axis. The difference between the three configurations is as follows: (a) two spectral bands intersect the real axis, (b) two spectral bands intersect the imaginary axis in the inner spectral band on the imaginary axis, and (c) two spectral bands intersect the imaginary axis in the spectral gaps. The stability spectrum in cases (a) and (b) represent a single figure-eight whereas that in case (c)

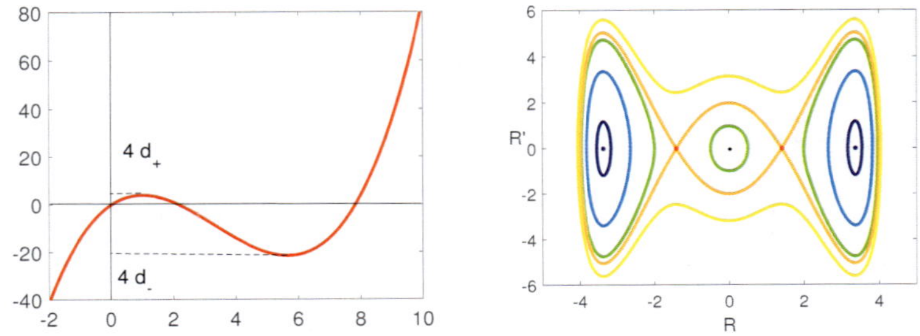

FIGURE 5.7. The same as Figure 5.4 but for $c^2 + 4b > 0$, $c < 0$, and $b < 0$.

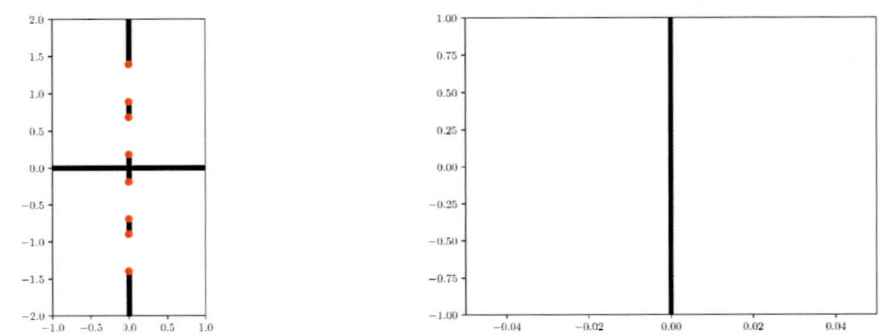

FIGURE 5.8. The same as Figure 5.5 but for $c^2 + 4b > 0$, $c < 0$, $b < 0$, and $d \in (0, d_+)$. The parameters are: $u_1 = 5$, $u_2 = 1$, $u_3 = 0.5$, $u_4 = 0$.

is a novel shape, which was not seen for the traveling periodic waves in the NLS equation.

EXAMPLE 5.5. Fix $a = 0$ and consider $c^2 + 4b > 0$, $c < 0$, and $b < 0$. Figure 5.7 shows F versus ρ (left) and the corresponding phase portrait in the (R, R') plane (right).

The cases $d \in (d_-, 0)$ and $d \in (d_+, \infty)$ are very similar to Figures 5.5 and 5.6, respectively. However, the case $d \in (0, d_+)$ is very different. Figure 5.8 shows only one typical configuration for the Lax spectrum (left) and the corresponding stability spectrum (right). This corresponds to four pairs of purely imaginary roots of $P(\lambda)$ in Theorem 5.2, Step 2, case (i), subcase (b). The set $\mathcal{S}$ in (5.28) corresponds to four spectral gaps on the imaginary axis. Since there is no Lax spectrum in $\mathcal{S}$ and outside of $\mathbb{R} \cup i\,\mathbb{R}$, the traveling periodic waves are spectrally and modulationally stable in agreement with the stability spectrum on Figure 5.8 (right). It is interesting that the stability holds for both periodic solutions (5.31) and (5.36) in $[u_2, u_1]$ and $[u_4, u_3]$, respectively, just because both periodic solutions correspond to the same configuration of the Lax spectrum with four pairs of purely imaginary eigenvalues.

REMARK 5.22. Traveling periodic waves were studied similarly to the analysis of Theorems 5.1 and 5.2 for the focusing modified KdV equation [**41, 42**]. Dnoidal and cnoidal solutions were shown to be related to one eigenvalue of the algebraic method, whereas more general periodic solutions expressed by a rational function of Jacobi elliptic functions were shown to be related to two eigenvalues of the algebraic method.

5.2. Algebraic tools for nonlinear waves

Here and in the rest of this chapter, we will work with only one example of an integrable Hamiltonian system given by the cubic NLS equation (Example 4.1). We will consider the focusing NLS equation (4.67) with the corresponding linear system given by (4.68) and (4.69). Similar techniques can be developed for the defocusing NLS equation (Example 4.1), the focusing and defocusing modified KdV equations (Examples 3.8 and 3.9), the KdV equation (Example 3.11), the derivative NLS equation (Example 4.2), the MTM system (Example 4.3), and many other integrable Hamiltonian systems.

5.2.1. Algebraic characterization of stationary (Lax–Novikov) equations. The stationary (Lax–Novikov) equations are related to Euler–Lagrange equations for the conserved quantities of the focusing NLS equation (4.67). The solitary and periodic waves arise in the second-order Lax–Novikov equation (5.5), which has parameters b and c. More complicated wave patterns, such as double-solitons, breathers, and double periodic waves (Section 5.3) are related to the third-order and fourth-order Lax–Novikov equations with additional parameters.

The following theorem clarifies the algebraic characterization of the fourth-order Lax–Novikov equation. It is generated by the algebraic method of nonlinearization of the linear Lax equations, which was adopted for the NLS equation in [**254, 255**] (see [**45**]). An application of this method with a single eigenvalue was used to characterize the second-order Lax–Novikov equation in Step 2 of the proof of Theorem 5.1. Here this method is developed with two distinct simple eigenvalues.

THEOREM 5.3. *Let $\varphi = (p_1, q_1)^T$ and $\varphi = (p_2, q_2)^T$ be two nonzero solutions of the linear equations* (4.68)–(4.69) *with fixed $\lambda = \lambda_1$ and $\lambda = \lambda_2$ such that $\lambda_1 + \bar{\lambda}_1 \neq 0$, $\lambda_2 + \bar{\lambda}_2 \neq 0$, $\lambda_1 + \bar{\lambda}_2 \neq 0$, and $\lambda_1 \neq \lambda_2$. Assume that the solution to the NLS equation* (4.67) *with the profile u is related to the squared eigenfunctions by*

$$u = p_1^2 + p_2^2 + \bar{q}_1^2 + \bar{q}_2^2. \tag{5.37}$$

Then, u is a solution of the fourth-order Lax–Novikov equation

$$\begin{aligned} & u'''' + 8|u|^2 u'' + 2u^2 \bar{u}'' + 4u|u'|^2 + 6(u')^2 \bar{u} + 6|u|^4 u \\ & + 2ic(u''' + 6|u|^2 u') + 4b(u'' + 2|u|^2 u) + 8iau' + 16du = 0, \end{aligned} \tag{5.38}$$

with parameters (a, b, c, d). The fourth-order equation (5.38) *is integrable with two complex-valued constants of motion.*

PROOF. We will use the following notations to make writing shorter:

$$\mathsf{p} := (p_1, p_2, \bar{q}_1, \bar{q}_2)^T, \quad \mathsf{q} := (q_1, q_2, -\bar{p}_1, -\bar{p}_2)^T, \quad \mathsf{L} := \mathrm{diag}(\lambda_1, \lambda_2, -\bar{\lambda}_1, -\bar{\lambda}_2).$$

Using the standard inner product in $\mathbb{C}^4$ as

$$\langle \mathsf{f}, \mathsf{g} \rangle := \bar{f}_1 g_1 + \bar{f}_2 g_2 + \bar{f}_3 g_3 + \bar{f}_4 g_4,$$

we can write (5.37) as $u = \langle \mathsf{p}, \mathsf{p} \rangle$. With this constraint, the spectral problem (4.68) becomes the following nonlinear complex Hamiltonian system

$$\frac{dp_{1,2}}{dx} = \frac{\partial H}{\partial q_{1,2}}, \quad \frac{dq_{1,2}}{dx} = -\frac{\partial H}{\partial p_{1,2}}, \tag{5.39}$$

where

$$H = \langle \mathsf{Lp}, \mathsf{q} \rangle + \frac{1}{2}\langle \mathsf{p}, \mathsf{p} \rangle \langle \mathsf{q}, \mathsf{q} \rangle.$$

With the constraint (5.37), the time-evolution problem (4.69) becomes another nonlinear complex Hamiltonian system given by

$$\frac{dp_{1,2}}{dt} = \frac{\partial K}{\partial q_{1,2}}, \quad \frac{dq_{1,2}}{dt} = -\frac{\partial K}{\partial p_{1,2}}, \tag{5.40}$$

where

$$K = i\left[\langle \mathsf{L}^2\mathsf{p}, \mathsf{q} \rangle + \frac{1}{2}\langle \mathsf{Lp}, \mathsf{p} \rangle \langle \mathsf{q}, \mathsf{q} \rangle + \frac{1}{2}\langle \mathsf{p}, \mathsf{p} \rangle \langle \mathsf{Lq}, \mathsf{q} \rangle + \frac{1}{2}\langle \mathsf{p}, \mathsf{q} \rangle \langle \mathsf{p}, \mathsf{p} \rangle \langle \mathsf{q}, \mathsf{q} \rangle\right].$$

There exist four real-valued constants of motion for the Hamiltonian systems (5.39) and (5.40), hence each Hamiltonian system is Liouville integrable. The four constants of motion are found in [**254, 255**] as follows:

$$\begin{aligned}
F_0 &= i\langle \mathsf{p}, \mathsf{q} \rangle, \\
F_1 &= \langle \mathsf{Lp}, \mathsf{q} \rangle + \frac{1}{2}\langle \mathsf{p}, \mathsf{p} \rangle \langle \mathsf{q}, \mathsf{q} \rangle - \frac{1}{2}\langle \mathsf{p}, \mathsf{q} \rangle^2, \\
F_2 &= i\left[\langle \mathsf{L}^2\mathsf{p}, \mathsf{q} \rangle + \frac{1}{2}\langle \mathsf{Lp}, \mathsf{p} \rangle \langle \mathsf{q}, \mathsf{q} \rangle + \frac{1}{2}\langle \mathsf{p}, \mathsf{p} \rangle \langle \mathsf{Lq}, \mathsf{q} \rangle - \langle \mathsf{p}, \mathsf{q} \rangle \langle \mathsf{Lp}, \mathsf{q} \rangle\right], \\
F_3 &= \langle \mathsf{L}^3\mathsf{p}, \mathsf{q} \rangle + \frac{1}{2}\langle \mathsf{L}^2\mathsf{p}, \mathsf{p} \rangle \langle \mathsf{q}, \mathsf{q} \rangle + \frac{1}{2}\langle \mathsf{Lp}, \mathsf{p} \rangle \langle \mathsf{Lq}, \mathsf{q} \rangle + \frac{1}{2}\langle \mathsf{p}, \mathsf{p} \rangle \langle \mathsf{L}^2\mathsf{q}, \mathsf{q} \rangle \\
&\quad - \frac{1}{2}\langle \mathsf{Lp}, \mathsf{q} \rangle^2 - \langle \mathsf{p}, \mathsf{q} \rangle \langle \mathsf{L}^2\mathsf{p}, \mathsf{q} \rangle,
\end{aligned}$$

where we note that $H = F_1 - \frac{1}{2}F_0^2$ and $K = F_2 + F_1F_0 - \frac{1}{2}F_0^3$. The Hamiltonian system (5.39) arises as a compatibility condition for the Lax equation

$$\frac{d}{dx}W(\lambda, u) = U(\lambda, u)W(\lambda, u) - W(\lambda, u)U(\lambda, u), \qquad \lambda \in \mathbb{C}, \tag{5.41}$$

where

$$U(\lambda, u) = \begin{bmatrix} \lambda & u \\ -\bar{u} & -\lambda \end{bmatrix}, \quad W(\lambda, u) = \begin{bmatrix} W_{11}(\lambda, u) & W_{12}(\lambda, u) \\ \overline{W}_{12}(-\lambda, u) & -\overline{W}_{11}(-\lambda, u) \end{bmatrix},$$

and the entries of W are

$$W_{11}(\lambda, u) = 1 - \sum_{j=1}^{2}\left(\frac{p_j q_j}{\lambda - \lambda_j} - \frac{\bar{p}_j \bar{q}_j}{\lambda + \bar{\lambda}_j}\right), \quad W_{12}(\lambda, u) = \sum_{j=1}^{2}\left(\frac{p_j^2}{\lambda - \lambda_j} + \frac{\bar{q}_j^2}{\lambda + \bar{\lambda}_j}\right).$$

The diagonal entry $W_{11}(\lambda, u)$ can be rewritten in the equivalent form

$$W_{11}(\lambda, u) = \frac{\lambda^4 + iT_1\lambda^3 + T_2\lambda^2 + iT_3\lambda + T_4}{(\lambda - \lambda_1)(\lambda - \lambda_2)(\lambda + \bar{\lambda}_1)(\lambda + \bar{\lambda}_2)}, \tag{5.42}$$

where

$$\begin{cases} T_1 = A_1 + i\langle \mathsf{p}, \mathsf{q}\rangle, \\ T_2 = A_2 - iA_1\langle \mathsf{p}, \mathsf{q}\rangle - \langle \Lambda\mathsf{p}, \mathsf{q}\rangle, \\ T_3 = A_3 + iA_2\langle \mathsf{p}, \mathsf{q}\rangle - A_1\langle \Lambda\mathsf{p}, \mathsf{q}\rangle + i\langle \Lambda^2\mathsf{p}, \mathsf{q}\rangle, \\ T_4 = A_4 - iA_3\langle \mathsf{p}, \mathsf{q}\rangle - A_2\langle \Lambda\mathsf{p}, \mathsf{q}\rangle - iA_1\langle \Lambda^2\mathsf{p}, \mathsf{q}\rangle - \langle \Lambda^3\mathsf{p}, \mathsf{q}\rangle \end{cases}$$

and

$$\begin{cases} A_1 = i(\lambda_1 + \lambda_2 - \bar{\lambda}_1 - \bar{\lambda}_2), \\ A_2 = (\lambda_1 - \bar{\lambda}_1)(\lambda_2 - \bar{\lambda}_2) - |\lambda_1|^2 - |\lambda_2|^2, \\ A_3 = i(\bar{\lambda}_1 - \lambda_1)|\lambda_2|^2 + i(\bar{\lambda}_2 - \lambda_2)|\lambda_1|^2, \\ A_4 = |\lambda_1|^2|\lambda_2|^2. \end{cases}$$

Substituting the four constants for $F_{0,1,2,3}$ into $T_{1,2,3,4}$ yield the more compact expressions

$$\begin{cases} T_1 = c, \\ T_2 = b + \frac{1}{2}|u|^2, \\ T_3 = a + \frac{1}{2}c|u|^2 - \frac{i}{4}(u'\bar{u} - u\bar{u}'), \\ T_4 = d + \frac{1}{2}b|u|^2 + \frac{i}{4}c(u'\bar{u} - u\bar{u}') + \frac{1}{8}(u\bar{u}'' + u''\bar{u} - |u'|^2 + 3|u|^4), \end{cases} \tag{5.43}$$

where

$$\begin{cases} c = A_1 + F_0, \\ b = A_2 - F_0A_1 - H, \\ a = A_3 + F_0A_2 - HA_1 + K, \\ d = A_4 - F_0A_3 - HA_2 - KA_1 - E. \end{cases}$$

and we have denoted $E := F_3 - F_0K + \frac{1}{2}H^2 = F_3 - F_0F_2 + \frac{1}{2}F_1^2 - \frac{3}{2}F_1F_0^2 + \frac{5}{8}F_0^4$.

The anti-diagonal entry $W_{12}(\lambda, u)$ can be rewritten in the equivalent form

$$W_{12}(\lambda, u) = \frac{S_0\lambda^3 + S_1\lambda^2 + S_2\lambda + S_3}{(\lambda - \lambda_1)(\lambda - \lambda_2)(\lambda + \bar{\lambda}_1)(\lambda + \bar{\lambda}_2)}, \tag{5.44}$$

where

$$\begin{cases} S_0 = \langle \mathsf{p}, \mathsf{p}\rangle, \\ S_1 = \langle \Lambda\mathsf{p}, \mathsf{p}\rangle + iA_1\langle \mathsf{p}, \mathsf{p}\rangle, \\ S_2 = \langle \Lambda^2\mathsf{p}, \mathsf{p}\rangle + iA_1\langle \Lambda\mathsf{p}, \mathsf{p}\rangle + A_2\langle \mathsf{p}, \mathsf{p}\rangle, \\ S_3 = \langle \Lambda^3\mathsf{p}, \mathsf{p}\rangle + iA_1\langle \Lambda^2\mathsf{p}, \mathsf{p}\rangle + A_2\langle \Lambda\mathsf{p}, \mathsf{p}\rangle + iA_3\langle \mathsf{p}, \mathsf{p}\rangle. \end{cases}$$

Differentiation of equation (5.37) in x and using the Hamiltonian system (5.39) together with F_0 yield

$$u' + 2iF_0u = 2\langle \Lambda\mathsf{p}, \mathsf{p}\rangle. \tag{5.45}$$

Differentiation of equation (5.45) in x and using F_0 and H yield

$$u'' + 2|u|^2u + 2iF_0u' - 4Hu = 4\langle \Lambda^2\mathsf{p}, \mathsf{p}\rangle. \tag{5.46}$$

Differentiation of equation (5.46) in x and using F_0, H, and K yield

$$u'' + 6|u|^2u' + 2iF_0\left(u'' + 2|u|^2u\right) - 4Hu' + 8iKu = 8\langle \Lambda^3\mathsf{p}, \mathsf{p}\rangle. \tag{5.47}$$

Substituting (5.37), (5.45), (5.46), and (5.47) into $S_{1,2,3,4}$ yield the more compact expressions

$$
\begin{cases}
S_0 = u, \\
S_1 = \frac{1}{2}u' + icu, \\
S_2 = \frac{1}{4}(u'' + 2|u|^2u) + \frac{i}{2}cu' + bu, \\
S_3 = \frac{1}{8}(u''' + 6|u|^2u') + \frac{i}{4}c(u'' + 2|u|^2u) + \frac{1}{2}bu' + iau.
\end{cases}
\tag{5.48}
$$

The $(1,2)$-component of the Lax equation (5.41) is written explicitly in the form

$$
\frac{d}{dx}W_{12}(\lambda,u) = 2\lambda W_{12}(\lambda,u) - 2\langle \mathsf{p},\mathsf{p}\rangle W_{11}(\lambda,u),
\tag{5.49}
$$

Substituting (5.42) and (5.44) with (5.43) and (5.48) into (5.49) yields constraints at different powers of λ. However, all constraints are satisfied identically at powers of λ^3, λ^2, and λ, whereas the constraint at λ^0 is equivalent to the fourth-order Lax–Novikov equation (5.38).

Integrability of the fourth-order Lax–Novikov equation (5.38) follows from properties of $\det W(\lambda,u)$. On the one hand, $\det W(\lambda,u)$ has only simple poles at λ_1, λ_2, $-\bar{\lambda}_1$, and $-\bar{\lambda}_2$ [**254, 255**]. On the other hand, it follows from (5.42) and (5.44) that

$$
\begin{aligned}
\det W(\lambda,u) &= -\left[W_{11}(\lambda,u)\right]^2 - W_{12}(\lambda,u)\bar{W}_{12}(-\lambda,u) \\
&= -\frac{P(\lambda)}{(\lambda-\lambda_1)^2(\lambda-\lambda_2)^2(\lambda+\bar{\lambda}_1)^2(\lambda+\bar{\lambda}_2)^2},
\end{aligned}
$$

where

$$
\begin{aligned}
P(\lambda) := {} & (\lambda^4 + iT_1\lambda^3 + T_2\lambda^2 + iT_3\lambda + T_4)^2 \\
& - (S_0\lambda^3 + S_1\lambda^2 + S_2\lambda + S_3)(\bar{S}_0\lambda^3 - \bar{S}_1\lambda^2 + \bar{S}_2\lambda - \bar{S}_3).
\end{aligned}
$$

Since λ_1, λ_2, $-\bar{\lambda}_1$, and $-\bar{\lambda}_2$ are roots of $P(\lambda)$, substituting (5.43) and (5.48) into $P(\lambda)$ and evaluating at λ_1 and λ_2 yield two complex-valued constants of motion for the fourth-order Lax–Novikov equation (5.38). □

REMARK 5.23. Roots of $P(\lambda)$ have the complex-conjugate symmetry due to the symmetry of the coefficients in $P(\lambda)$: If λ_0 is a root of $P(\lambda)$, so is $-\bar{\lambda}_0$. This ensures that if λ_1 and λ_2 are simple roots of $P(\lambda)$, then $\det W(\lambda,u)$ has simple poles at λ_1, λ_2, $-\bar{\lambda}_1$, $-\bar{\lambda}_2$.

REMARK 5.24. The third-order Lax–Novikov equation

$$
u''' + 6|u|^2u' + 2ic(u'' + 2|u|^2u) + 4bu' + 8iau = 0
\tag{5.50}
$$

and its constant of motion

$$
d + \frac{1}{2}b|u|^2 + \frac{i}{4}c(u'\bar{u} - u\bar{u}') + \frac{1}{8}(u\bar{u}'' + u''\bar{u} - |u'|^2 + 3|u|^4) = 0
\tag{5.51}
$$

are obtained from (5.43) and (5.48) by degeneration with $S_3 = 0$ and $T_4 = 0$. Since $P(\lambda)$ is independent on x, two additional constants of motion are obtained at powers λ^1 and λ^0:

$$
2f - a|u|^2 - \frac{1}{4}c(|u'|^2 + |u|^4) + \frac{i}{8}(u''\bar{u}' - u'\bar{u}'') = 0,
\tag{5.52}
$$

and

$$g - \frac{i}{2}a(u'\bar{u} - u\bar{u}') + \frac{1}{4}b(|u'|^2 + |u|^4) + \frac{1}{16}(|u'' + 2|u|^2u|^2 - (u'\bar{u} - u\bar{u}')^2) = 0, \tag{5.53}$$

where f and g are additional real-valued constants. The polynomial $P(\lambda)$ can be factorized as $P(\lambda) = \lambda^2 \tilde{P}(\lambda)$, where

$$\begin{aligned} \tilde{P}(\lambda) := \lambda^6 + 2ic\lambda^5 + (2b - c^2)\lambda^4 + 2i(a + bc)\lambda^3 + (b^2 - 2ac + 2d)\lambda^2 \\ + 2i(f + ab + cd)\lambda + g + 2bd - 2cf - a^2. \end{aligned} \tag{5.54}$$

REMARK 5.25. The second-order Lax–Novikov equation (5.5) and its two constants of motion (5.6) are obtained from (5.43) and (5.48) by degeneration with $S_2 = S_3 = 0$ and $T_3 = T_4 = 0$. Alternatively, the second-order Lax–Novikov equation (5.5) is derived from the algebraic method with one eigenvalue, as is done in Step 2 of Theorem 5.1, where (5.14) can be compared with (5.37).

REMARK 5.26. Higher-order Lax–Novikov equations can be generated similarly by using more eigenvalues in p, q, and L. This leads to polynomials $W_{11}(\lambda, u)$, $W_{12}(\lambda, u)$, and $P(\lambda)$ of higher degree in λ. With N distinct eigenvalues, this procedure gives the stationary Lax–Novikov equations of the order $2N$, which can be reduced to the stationary equations of every lower order.

5.2.2. N-fold Bäcklund–Darboux transformation. For the stability of N-solitons for any $N \in \mathbb{N}$, it is important to construct these N-soliton solutions explicitly. This task is possible by using the method of the N-fold Bäcklund–Darboux transformation, which extends the one-fold Bäcklund–Darboux transformation of Lemma 4.5.

The N-fold transformation can be used to generate N-solitons from the zero solution. In connection to the stationary Lax–Novikov equations, the N-soliton solutions satisfy the Lax–Novikov equations of the order $2N$.

The following theorem specifies the N-fold transformation. Computational details can be found in [**66**] for the cubic NLS equation and in [**41**] for the modified KdV equation. The pioneering results for the N-fold transformation using the dressing method were developed in [**251, 252**].

THEOREM 5.4. *Assume that $u = u(t,x)$ is a classical solution to the NLS equation (4.67) and let $\{s_k\}_{1\le k\le N}$ be nonzero solutions of the system (4.68) with λ in the set $\{-\bar{\lambda}_k\}_{1\le k\le N}$ with $\mathrm{Re}(\lambda_k) \neq 0$ for $1 \le k \le N$. Assume that the following matrix with entries*

$$M_{k,j} = \frac{\langle s_j, s_k\rangle}{\bar{\lambda}_k + \lambda_j}, \quad 1 \le k, j \le N \tag{5.55}$$

is invertible, where $\langle f, g\rangle := \bar{f}_1 g_1 + \bar{f}_2 g_2$ is the dot product between vectors in $\mathbb{C}^2$. Let $\{r_k\}_{1\le k\le N}$ be defined by the unique solution of the linear system

$$s_k = \sum_{j=1}^{N} \frac{\langle s_j, s_k\rangle}{\bar{\lambda}_k + \lambda_j} r_j, \quad 1 \le k \le N, \tag{5.56}$$

with the inverse

$$r_k = \sum_{j=1}^{N} \frac{\langle r_j, r_k\rangle}{\bar{\lambda}_j + \lambda_k} s_j, \quad 1 \le k \le N. \tag{5.57}$$

Then, $\{r_k\}_{1\le k\le N}$ are nonzero solutions of the system (4.68) with λ in the set $\{\lambda_k\}_{1\le k\le N}$ and with $u = \hat{u}(t,x)$ satisfying the NLS equation (4.67). The two solutions of the NLS equation (4.67) are related by the transformation formula:

$$Q(\hat{u}) = Q(u) + \sum_{k=1}^{N} r_k \otimes \bar{s}_k \sigma_3 - \sigma_3 s_k \otimes \bar{r}_k, \tag{5.58}$$

with

$$\sigma_3 := \begin{bmatrix} 1 & 0 \\ 0 & -1 \end{bmatrix}, \quad Q(u) := \begin{bmatrix} 0 & u \\ -\bar{u} & 0 \end{bmatrix}.$$

Moreover, if $\Phi(\lambda, u)$ is the fundamental matrix solution of the system (4.68), then $\hat{\Phi}(\lambda.\hat{u})$ is the fundamental matrix solution of the same system with the relation

$$\hat{\Phi}(\lambda, \hat{u}) := D(\lambda)\Phi(\lambda, u), \quad D(\lambda) := I + \sum_{k=1}^{N} \frac{r_k \otimes \bar{s}_k}{\lambda - \lambda_k}, \tag{5.59}$$

and $\otimes$ denotes an outer product of vectors in $\mathbb{C}^2$ (without complex conjugation).

Proof. Let us first ensure that the transformation (5.59) is invertible with the inverse

$$\hat{\Phi}^{-1}(\lambda, u) = \Phi^{-1}(\lambda, u) D^{+}(\lambda), \quad D^{+}(\lambda) = I - \sum_{k=1}^{N} \frac{s_k \otimes \bar{r}_k}{\lambda + \bar{\lambda}_k}.$$

We need to check that $D^{+}(\lambda)D(\lambda) = D(\lambda)D^{+}(\lambda) = I$. To do so, we use the partial fraction decomposition

$$\frac{1}{(\lambda + \bar{\lambda}_j)(\lambda - \lambda_k)} = \frac{1}{\bar{\lambda}_j + \lambda_k} \left[\frac{1}{\lambda - \lambda_k} - \frac{1}{\lambda + \bar{\lambda}_j} \right]$$

and observe that $D(\lambda)D^{+}(\lambda) = I$ is equivalent to the system

$$r_k \otimes \bar{s}_k = \sum_{j=1}^{N} \frac{(r_k \otimes \bar{s}_k)\,(s_j \otimes \bar{r}_j)}{\bar{\lambda}_j + \lambda_k},$$

which yields the system (5.56) after projection to r_k from the left. Since $D(\lambda)$ is a square $n \times n$ matrix and $D(\lambda)D^{+}(\lambda) = I$, then $|\det D(\lambda)| = 1$ and therefore, $D(\lambda)$ is invertible with $D^{-1}(\lambda) = D^{+}(\lambda)$. The equation $D^{+}(\lambda)D(\lambda) = I$ is equivalent to the system

$$r_k \otimes \bar{s}_k = \sum_{j=1}^{N} \frac{(s_j \otimes \bar{r}_j)\,(r_k \otimes \bar{s}_k)}{\bar{\lambda}_j + \lambda_k},$$

which yields the system (5.57) after projection to s_k from the right. Therefore, the system (5.57) is inverse to the system (5.56). All vectors in the sets $\{s_k\}_{1\le k\le N}$ and $\{r_k\}_{1\le k\le N}$ are nonzero because the matrix M in (5.55) is invertible.

Next we confirm that the set $\{r_k\}_{1\le k\le N}$ satisfies the system (4.68) with λ in $\{\lambda_k\}_{1\le k\le N}$ and with the potential $\hat{u}$ if the set $\{s_k\}_{1\le k\le N}$ satisfies the same system (4.68) with λ in $\{-\bar{\lambda}_k\}_{1\le k\le N}$ and with the potential u and if u and $\hat{u}$ are related by the transformation formula (5.58). Differentiating the linear system (5.56) in x and substituting (4.68) for s_k with $-\bar{\lambda}_k$, we obtain

$$-\bar{\lambda}_k \sigma_3 s_k + Q(u) s_k = -\sum_{j=1}^{N} \langle \sigma_3 s_j, s_k \rangle r_j + \sum_{j=1}^{N} \frac{\langle s_j, s_k \rangle}{\bar{\lambda}_k + \lambda_j} \partial_x r_j.$$

Using the transformation formula (5.58) and the inverse linear system (5.57), we rewrite this equation as follows:

$$\sum_{j=1}^{N} \frac{\langle s_j, s_k \rangle}{\bar{\lambda}_k + \lambda_j} \partial_x r_j = -\bar{\lambda}_k \sigma_3 s_k + \sum_{j=1}^{N} \frac{\langle s_j, s_k \rangle}{\bar{\lambda}_k + \lambda_j} Q(\hat{u}) r_j + \sigma_3 \sum_{j=1}^{N} \langle r_j, s_k \rangle s_j,$$

where

$$\begin{aligned} \sum_{m=1}^{N} \langle r_m, s_k \rangle s_m &= \sum_{m=1}^{N} \sum_{j=1}^{N} \frac{\langle r_m, r_j \rangle \langle s_j, s_k \rangle}{\bar{\lambda}_m + \lambda_j} s_m \\ &= \sum_{j=1}^{N} \langle s_j, s_k \rangle \sum_{m=1}^{N} \frac{\langle r_m, r_j \rangle}{\bar{\lambda}_m + \lambda_j} s_m \\ &= \sum_{j=1}^{N} \langle s_j, s_k \rangle r_j. \end{aligned}$$

Using the linear system (5.56) again, we obtain

$$\sum_{j=1}^{N} \frac{\langle s_j, s_k \rangle}{\bar{\lambda}_k + \lambda_j} \left[\partial_x r_j - \lambda_j \sigma_3 r_j - Q(\hat{u}) r_j \right] = 0,$$

which yields (4.68) for r_j with λ_j and $\hat{u}$ because the matrix M in (5.55) is invertible.

Finally, we verify that $\hat{\Phi}(\lambda, \hat{u})$ is a solution of the system

$$\partial_x \hat{\Phi}(\lambda, \hat{u}) = U(\lambda, \hat{u}) \hat{\Phi}(\lambda, \hat{u})$$

if $\Phi(\lambda, u)$ is a solution of same system

$$\partial_x \Phi(\lambda, u) = U(\lambda, u) \Phi(\lambda, u).$$

This is derived from the Darboux equation

$$U(\lambda, \hat{u}) = \partial_x \hat{\Phi}(\lambda, u) \hat{\Phi}(\lambda, \hat{u})^{-1} = \left[\partial_x D(\lambda) + D(\lambda) U(\lambda, u) \right] D^{+}(\lambda). \tag{5.60}$$

By using the partial fraction decompositions, we remove the residue terms at simple poles of equation (5.60). The residue terms at $\mathcal{O}\left(\lambda - \lambda_k\right)^{-1}$ are removed if

$$\left[\partial_x \left(r_k \otimes \bar{s}_k \right) + \lambda_k r_k \otimes \bar{s}_k \sigma_3 + r_k \otimes \bar{s}_k Q(u) \right] \left[I - \sum_{j=1}^{N} \frac{s_j \otimes \bar{r}_j}{\lambda_k + \bar{\lambda}_j} \right] = 0.$$

Because of the linear system (5.56), this equation simplifies to the form

$$r_k \otimes \left[\partial_x \bar{s}_k + \lambda_k \bar{s}_k \sigma_3 + \bar{s}_k Q(u) \right] \left[I - \sum_{j=1}^{N} \frac{s_j \otimes \bar{r}_j}{\lambda_k + \bar{\lambda}_j} \right] = 0.$$

Projection to r_k from the left and Hermite conjugation with the help of equation $[Q(u)]^{+} = -Q(u)$ yields

$$\left[I - \sum_{j=1}^{N} \frac{r_j \otimes \bar{s}_j}{\bar{\lambda}_k + \lambda_j} \right] \left[-\partial_x s_k - \bar{\lambda}_k \sigma_3 s_k + Q(u) s_k \right] = 0,$$

which is satisfied due to equation for s_k. The residue terms at $\mathcal{O}\left(\lambda+\bar{\lambda}_j\right)^{-1}$ are removed if

$$\left[\partial_x\left(\sum_{k=1}^N \frac{r_k\otimes\bar{s}_k}{\bar{\lambda}_j+\lambda_k}\right)+\bar{\lambda}_j\sigma_3-Q(u)\right.$$
$$\left.-\bar{\lambda}_j\sum_{k=1}^N \frac{r_k\otimes\bar{s}_k}{\bar{\lambda}_j+\lambda_k}\sigma_3+\sum_{k=1}^N \frac{r_k\otimes\bar{s}_k}{\bar{\lambda}_j+\lambda_k}Q(u)\right]s_j\otimes\bar{r}_j=0.$$

Projection to r_j from the right yields the new equation

$$\left[\partial_x\left(\sum_{k=1}^N \frac{r_k\otimes\bar{s}_k}{\bar{\lambda}_j+\lambda_k}\right)+\bar{\lambda}_j\sigma_3-Q(u)-\bar{\lambda}_j\sum_{k=1}^N \frac{r_k\otimes\bar{s}_k}{\bar{\lambda}_j+\lambda_k}\sigma_3+\sum_{k=1}^N \frac{r_k\otimes\bar{s}_k}{\bar{\lambda}_j+\lambda_k}Q(u)\right]s_j=0.$$

Substituting the linear system for s_k reduces this equation to the form

$$\left[\sum_{k=1}^N \frac{(\partial_x r_k)\otimes\bar{s}_k}{\bar{\lambda}_j+\lambda_k}+\bar{\lambda}_j\sigma_3-Q(u)-\sum_{k=1}^N r_k\otimes\bar{s}_k\sigma_3\right]s_j=0.$$

This equation is rewritten with the help of the reconstruction formula (5.58) in the form

$$\sum_{k=1}^N \frac{\langle s_k,s_j\rangle}{\bar{\lambda}_j+\lambda_k}\partial_x r_k+\bar{\lambda}_j\sigma_3 s_j-Q(\hat{u})s_j-\sigma_3\sum_{m=1}^N\langle r_m,s_j\rangle s_m=0,$$

which reduces with the help of the linear system (5.56) to the form

$$\sum_{k=1}^N \frac{\langle s_k,s_j\rangle}{\bar{\lambda}_j+\lambda_k}\left[\partial_x r_k-Q(\hat{u})r_k-\lambda_k\sigma_3 r_k\right]+\bar{\lambda}_j\sigma_3 s_j-\bar{\lambda}_j\sigma_3\sum_{k=1}^N \frac{\langle s_k,s_j\rangle}{\bar{\lambda}_j+\lambda_k}r_k=0.$$

The last two terms cancel out thanks again to the linear system (5.56), whereas the first term is zero due to the equation for r_k. Since all the residue terms are removed from equation (5.60), this equation reduces a single equation at $\mathcal{O}(1)$, which is nothing but the reconstruction formula (5.58). □

REMARK 5.27. Due to the symmetry of the Lax system (4.68)–(4.69), if $\varphi=(p_j,q_j)^T$ is a solution for $\lambda=\lambda_j$, so is $\varphi=(-\bar{q}_j,\bar{p}_j)^T$ for $\lambda=-\bar{\lambda}_j$. This symmetry can be used to relate the solutions $\{s_j\}_{1\le j\le N}$ of the Lax system (4.68)–(4.69) for $\lambda\in\{-\bar{\lambda}_j\}_{1\le j\le N}$ to the solutions of the Lax system (4.68)–(4.69) for $\lambda\in\{\lambda_j\}_{1\le j\le N}$.

REMARK 5.28. If $N=1$ and $u=0$, the outcome of Theorem 5.4 reproduces the Bäcklund–Darboux transformation of Lemma 4.5. Let $s_1=(\mathfrak{b}_1,\mathfrak{b}_2)^T$ be a solution of the Lax system (4.68)–(4.69) with $u=0$ and $\lambda=-\bar{\eta}$. Then, $r_1=(\mathfrak{a}_1,\mathfrak{a}_2)^T$ is given by the linear system (5.56), or explicitly

$$\begin{bmatrix}\mathfrak{a}_1\\ \mathfrak{a}_2\end{bmatrix}=\frac{\eta+\bar{\eta}}{|\mathfrak{b}_1|^2+|\mathfrak{b}_2|^2}\begin{bmatrix}\mathfrak{b}_1\\ \mathfrak{b}_2\end{bmatrix},$$

which coincides with (4.72) up to notations and normalization. The transformation formula (5.58) yields 1-soliton in the form

$$\hat{u}=-\mathfrak{a}_1\bar{\mathfrak{b}}_2-\bar{\mathfrak{a}}_2\mathfrak{b}_1=\frac{-2(\eta+\bar{\eta})\mathfrak{b}_1\bar{\mathfrak{b}}_2}{|\mathfrak{b}_1|^2+|\mathfrak{b}_2|^2},$$

which coincides with (4.71) for $u=0$.

REMARK 5.29. The N-fold Darboux transformation of Theorem 5.4 can be simplified to yield the explicit relation between u and $\hat{u}$. Let $D := \det(M) \neq 0$ and $D_{k,j}$ be the co-factor of the element $M_{k,j}$. A unique solution of the linear system (5.56) can be expressed in the explicit form

$$r_k = \sum_{j=1}^{n} \frac{D_{j,k}}{D} s_j. \tag{5.61}$$

Since the matrix M is Hermitian, we have $D_{j,k} = \bar{D}_{k,j}$ so that the transformation formula (5.58) yields

$$Q(\hat{u}) - Q(u) = \sum_{k=1}^{N} \begin{bmatrix} r_{k,1}\bar{s}_{k,1} - \bar{r}_{k,1}s_{k,1} & -r_{k,1}\bar{s}_{k,2} - \bar{r}_{k,2}s_{k,1} \\ r_{k,2}\bar{s}_{k,1} + \bar{r}_{k,1}s_{k,2} & -r_{k,2}\bar{s}_{k,2} + \bar{r}_{k,2}s_{k,2} \end{bmatrix}.$$

The diagonal entries are zeros and the off-diagonal entries yield the explicit transformation formula

$$\hat{u} - u = -\frac{2}{D} \sum_{k=1}^{N} \sum_{j=1}^{n} D_{j,k} s_{j,1} \bar{s}_{k,2}. \tag{5.62}$$

The proof of Theorem 5.4 did not include verification of validity of the time-dependent part (4.69) of the Lax system for solutions $\{s_j\}_{1 \leq j \leq N}$ and solutions $\{r_j\}_{1 \leq j \leq N}$. The following lemma gives the relevant computations, see also [**66**].

LEMMA 5.7. *Under the conditions of Theorem* 5.4, *the time-dependent equations* (4.69) *for solutions* $\{s_k\}_{1 \leq k \leq N}$ *with* λ *in the set* $\{-\bar{\lambda}_k\}_{1 \leq k \leq N}$ *and* u, *and solutions* $\{r_k\}_{1 \leq k \leq N}$ *with* λ *in the set* $\{\lambda_k\}_{1 \leq k \leq N}$ *and* $\hat{u}$, *are equivalent under* (5.58) *and*

$$|\hat{u}|^2 = |u|^2 + \partial_x^2 \log \det M, \tag{5.63}$$

where M *is given by* (5.55).

PROOF. We prove that the linear system (5.56) relate solutions $\{s_k\}_{1 \leq k \leq N}$ of the time-dependent equations (4.69) for λ in the set $\{-\bar{\lambda}_k\}_{1 \leq k \leq N}$ with the potential u to solutions $\{r_k\}_{1 \leq k \leq N}$ of the time-dependent equations (4.69) for λ in the set $\{\lambda_k\}_{1 \leq k \leq N}$ with the potential $\hat{u}$. To do so, we differentiate the linear system (5.56) in t and substitute (4.69) for s_k to obtain

$$\begin{aligned} & i(\bar{\lambda}_k^2 + \frac{1}{2}|u|^2)\sigma_3 s_k - i\bar{\lambda}_k Q(u) s_k - \frac{i}{2} Q(\partial_x u)\sigma_3 s_k \\ & = i \sum_{j=1}^{N} (\bar{\lambda}_k - \lambda_j) \langle \sigma_3 s_j, s_k \rangle r_j + i \sum_{j=1}^{N} \langle Q(u) s_j, s_k \rangle r_j + \sum_{j=1}^{N} \frac{\langle s_j, s_k \rangle}{\bar{\lambda}_k + \lambda_j} \partial_t r_j. \end{aligned}$$

This can be written as

$$\sum_{j=1}^{N} \frac{\langle s_j, s_k \rangle}{\bar{\lambda}_k + \lambda_j} \left[\partial_t r_j - i(|\lambda_k^2 + \frac{1}{2}|u|^2)\sigma_3 r_k - i\lambda_k Q(\hat{u}) r_k + \frac{i}{2} Q(\partial_x \hat{u})\sigma_3 r_k \right] = G_k,$$

where

$$
\begin{aligned}
G_k := & -i(\bar{\lambda}_k^2 + \frac{1}{2}|u|^2)\sigma_3 s_k + i\bar{\lambda}_k Q(u) s_k + \frac{i}{2} Q(\partial_x u)\sigma_3 s_k \\
& + i\sum_{j=1}^N (\bar{\lambda}_k - \lambda_j)\langle \sigma_3 s_j, s_k\rangle r_j \\
& + i\sum_{j=1}^N \langle Q(u) s_j, s_k\rangle r_j - i\sum_{j=1}^N \frac{\langle s_j, s_k\rangle}{\bar{\lambda}_k + \lambda_j}(\lambda_j^2 + \frac{1}{2}|\hat{u}|^2)\sigma_3 r_j \\
& - \sum_{j=1}^N \frac{\langle s_j, s_k\rangle}{\bar{\lambda}_k + \lambda_j}\lambda_j Q(\hat{u}) r_j + \frac{i}{2}\sum_{j=1}^N \frac{\langle s_j, s_k\rangle}{\bar{\lambda}_k + \lambda_j} Q(\partial_x \hat{u})\sigma_3 r_j.
\end{aligned}
$$

If $G_k = 0$, then invertibility of the matrix M in (5.55) implies the time-dependent equations (4.69) for $\{r_k\}_{1\leq k\leq N}$ with the potential $\hat{u}$. To show that $G_k = 0$, we use the linear system (5.56) and rewrite

$$
\begin{aligned}
G_k = & \frac{i}{2}(|\hat{u}|^2 - |u|^2)\sigma_3 s_k + i\bar{\lambda}_k \left[Q(u) - Q(\hat{u})\right] s_k + \frac{i}{2}\left[Q(\partial_x u) - Q(\partial_x \hat{u})\right]\sigma_3 s_k \\
& + i\sum_{j=1}^N (\bar{\lambda}_k - \lambda_j)\langle \sigma_3 s_j, s_k\rangle r_j + i\sum_{j=1}^N \langle Q(u) s_j, s_k\rangle r_j \\
& - i\sum_{j=1}^N (\bar{\lambda}_k - \lambda_j)\langle s_j, s_k\rangle \sigma_3 r_j - \sum_{j=1}^N \langle s_j, s_k\rangle Q(\hat{u}) r_j.
\end{aligned}
$$

We can now use the derivative of the transformation formula (5.58) in the following form:

$$
Q(\partial_x \hat{u}) - Q(\partial_x u) = 2\sum_{j=1}^N (\partial_x r_j)\otimes \bar{s}_j \sigma_3 + r_j \otimes (\partial_x \bar{s}_j)\sigma_3 - \partial_x R,
$$

where

$$
R := \sum_{j=1}^N r_j \otimes \bar{s}_j \sigma_3 + \sigma_3 s_j \otimes \bar{r}_j = \sum_{j=1}^N \begin{bmatrix} r_{j,1}\bar{s}_{j,1} + \bar{r}_{j,1} s_{j,1} & -r_{j,1}\bar{s}_{j,2} + \bar{r}_{j,2} s_{j,1} \\ r_{j,2}\bar{s}_{j,1} - \bar{r}_{j,1} s_{j,2} & -r_{j,2}\bar{s}_{j,2} - \bar{r}_{j,2} s_{j,2} \end{bmatrix}.
$$

Substituting this expression together with the linear systems for $\{s_k\}_{1\leq k\leq N}$ and $\{r_k\}_{1\leq k\leq N}$ to the expression for G_k, we obtain

$$
G_k = \frac{i}{2}\left[(|\hat{u}|^2 - |u|^2)I + \partial_x R\right]\sigma_3 s_k, \tag{5.64}
$$

where I is a 2×2 identity matrix. It follows from the explicit expression (5.61) that R is a diagonal matrix. Moreover, the difference between the two diagonal entries of the matrix R is constant in x because

$$\begin{aligned}\sum_{k=1}^{N}(r_{k,1}\bar{s}_{k,1}+\bar{r}_{k,1}s_{k,1}+r_{k,2}\bar{s}_{k,2}+\bar{r}_{k,2}s_{k,2}) &= \frac{2}{D}\sum_{k=1}^{N}\sum_{j=1}^{N}D_{j,k}\langle s_k, s_1\rangle \\ &= \frac{2}{D}\sum_{k=1}^{N}\sum_{j=1}^{N}(\bar{\lambda}_j+\lambda_k)D_{j,k}M_{j,k} \\ &= 2\sum_{j=1}^{N}(\bar{\lambda}_j+\lambda_k).\end{aligned}$$

Consequently, we have

$$\partial_x \sum_{k=1}^{n}(r_{k,1}\bar{s}_{k,1}+\bar{r}_{k,1}s_{k,1}+r_{k,2}\bar{s}_{k,2}+\bar{r}_{k,2}s_{k,2}) = 0,$$

so that

$$\begin{aligned}\partial_x R &= \frac{1}{2}\partial_x \sum_{k=1}^{N}(r_{k,1}\bar{s}_{k,1}+\bar{r}_{k,1}s_{k,1}-r_{k,2}\bar{s}_{k,2}-\bar{r}_{k,2}s_{k,2})\ I \\ &= \partial_x \frac{1}{D}\sum_{k=1}^{N}\sum_{j=1}^{N}D_{j,k}\langle\sigma_3 s_k, s_j\rangle\ I \\ &= -\partial_x \frac{1}{D}\sum_{k=1}^{N}\sum_{j=1}^{N}D_{j,k}\partial_x M_{j,k}\ I \\ &= -\partial_x^2 \log \det M\ I.\end{aligned}$$

Combining this expression with (5.64) shows $G_k = 0$ if and only if $|\hat{u}|^2$ and $|u|^2$ are related by the transformation formula (5.63). □

REMARK 5.30. The validity of the time-dependent Darboux eqution

$$V(\lambda,\hat{u}) = \partial_t\hat{\Phi}(\lambda,\hat{u})[\hat{\Phi}(\lambda,\hat{u})]^{-1} = [\partial_t D(\lambda)+D(\lambda)V(\lambda,u)]\,D^{+}(\lambda),$$

is verified similarly to the proof of Theorem 5.4.

REMARK 5.31. It is a rather nontrivial algebraic computation to prove that the constraint (5.63) is automatically satisfied if u and $\hat{u}$ are related by the transformation formula (5.62). This was shown for the modified KdV equation in [**41**].

EXAMPLE 5.6. The double-soliton solutions of the NLS equation (4.67) are obtained from the transformation formula (5.62) with $u = 0$ and $N = 2$. Solving the Lax system (4.68) and (4.69) with $u = 0$ and $\lambda = -\bar{\lambda}_{1,2}$ in the form

$$s_{1,2} = e^{-\bar{\lambda}_{1,2}(x-\xi_{1,2})\sigma_3 + i\bar{\lambda}_{1,2}^2 t\sigma_3 + \frac{i}{2}\alpha_{1,2}\sigma_3}\begin{bmatrix}1\\-1\end{bmatrix}$$

where $\lambda_{1,2} = \frac{1}{2}(k_{1,2}+ic_{1,2}) \in \mathbb{C}$ and $(\xi_{1,2},\alpha_{1,2}) \in \mathbb{R}^2$ are arbitrary parameters, yields the explicit expression $\hat{u} = 2\Sigma/D$, where

$$\begin{aligned}\Sigma := &\frac{e^{-k_2\varphi_2+i\psi_1}+e^{k_2\varphi_2+i\psi_1}}{k_2} + \frac{e^{-k_1\varphi_1+i\psi_2}+e^{k_1\varphi_1+i\psi_2}}{k_1} \\ &- 2\frac{e^{-k_2\varphi_2+i\psi_1}+e^{k_1\varphi_1+i\psi_2}}{k_1+k_2-i(c_1-c_2)} - 2\frac{e^{-k_1\varphi_1+i\psi_2}+e^{k_2\varphi_2+i\psi_1}}{k_1+k_2+i(c_1-c_2)}\end{aligned}$$

and

$$D := \frac{(e^{-k_1\varphi_1} + e^{k_1\varphi_1})(e^{-k_2\varphi_2} + e^{k_2\varphi_2})}{k_1 k_2} - \frac{4D_1}{(k_1+k_2)^2 + (c_1-c_2)^2}$$

with

$$\begin{aligned} D_1 &= \left(e^{-\frac{1}{2}(k_1\varphi_1 + k_2\varphi_2 - i\psi_1 + i\psi_2)} + e^{\frac{1}{2}(k_1\varphi_1 + k_2\varphi_2 - i\psi_1 + i\psi_2)}\right) \\ &\quad \times \left(e^{-\frac{1}{2}(k_1\varphi_1 + k_2\varphi_2 + i\psi_1 - i\psi_2)} + e^{\frac{1}{2}(k_1\varphi_1 + k_2\varphi_2 + i\psi_1 - i\psi_2)}\right) \end{aligned}$$

defined with the following phase parameters

$$\begin{cases} \varphi_1 = x - c_1 t - \xi_1, \\ \varphi_2 = x - c_2 t - \xi_2, \\ \psi_1 = \alpha_1 + c_1(x - \xi_1) + \frac{1}{2}(k_1^2 - c_1^2)t, \\ \psi_2 = \alpha_2 + c_2(x - \xi_2) + \frac{1}{2}(k_2^2 - c_2^2)t. \end{cases}$$

Figure 5.9 shows two particular types of the dynamics of the double-solitons: scattering of two solitons with nonequal speeds $c_1 \neq c_2$ (top) and oscillations of a bound state of two solitons with equal speeds $c_1 = c_2$ (bottom). In both cases, the constraint $k_1 \neq k_2$ was used to avoid degenerate solutions.

5.3. Stability of double-solitons, breathers, and double-periodic waves

For a nonlinear Hamiltonian system, we can consider more complicated solutions than the traveling solitary and periodic waves. A **double-soliton** is a nonlinear superposition of two solitary waves traveling with different speeds. A **breather** is a spatially localized and time-periodic wave. A **double-periodic** wave is periodic in both space and time.

Stability of double-solitons, breathers, and double-periodic waves is a challeging question. A major difficulty comes from the nontrivial dependence of these solutions on the time variable. The separation of variables for the traveling waves does not apply to these solutions and they are no longer critical points of the augmented energy. Only in the case of spatially localized and time-periodic breathers, Floquet theory in time can be used in the linearized equations as was done in [**256**] by using tools from the theory of semigroups and Fredholm operators. However, the double-solitons, the double-periodic, and other nonlinear waves have more complicated dependence on space and time variables, which makes determining their stability from linearization methods out of reach.

The situation is again different for integrable Hamiltonian systems, since integrability can help to overcome these difficulties. A rich variety of exact solutions which are more complicated than the traveling wave solutions and which include double-solitons, breathers, and double-periodic waves exist in explicit form as solutions of the stationary (Lax–Novikov) equations (Section 5.2). The linear and orbital stability of these solutions can be clarified by using the higher-order conserved quantities, the Bäcklund–Darboux transformation, and the Lax spectrum related to the linearized equations. These methods generalize the ideas used for stability of traveling wave solutions in Sections 4.2, 4.3, and 5.1, respectively.

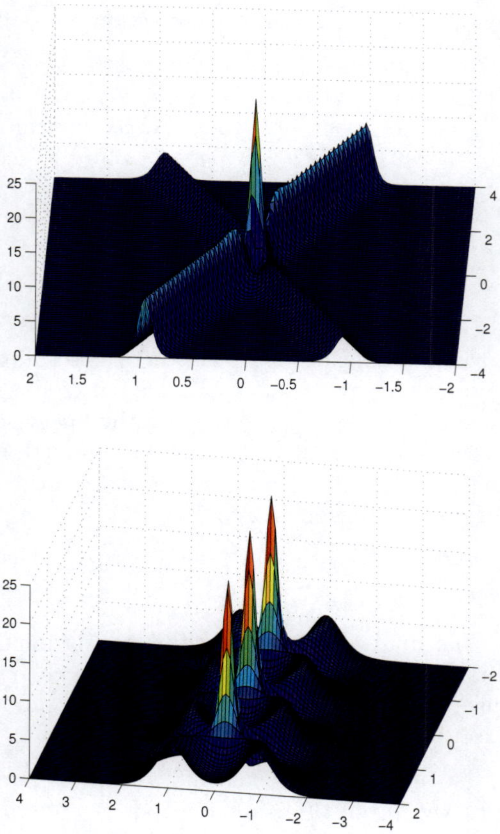

FIGURE 5.9. Surface plots of $|u|^2$ for the double-soliton solutions in Example 5.6 with $k_1 = 1$, $k_2 = 1.5$, and either $c_1 = -c_2 = -1$ (top) or $c_1 = c_2 = 0$ (bottom). The other translation parameters are set to zero.

5.3.1. Stability of double-solitons. A single soliton is the traveling wave solution of the NLS equation (4.67), hence its profile is a solution of the second-order Lax–Novikov equation (5.5) with two parameters b and c. The exact solution is given by

$$u(t,x) = U_{k,c}(t,x;\alpha_0,\xi_0) := k\,\mathrm{sech}(k(x+ct-\xi_0)e^{icx+\frac{i}{2}(k^2-c^2)t+i\alpha_0},$$

where $\alpha_0 \in \mathbb{R}$ and $\xi_0 \in \mathbb{R}$ are arbitrary parameters for the phase rotations and spatial translations, $c \in \mathbb{R}$ is an arbitrary parameter for the wave speed, and $k = \sqrt{-2b} > 0$ is the wave amplitude defined for $b < 0$. The derivatives in the second-order Lax–Novikov equation (5.5) for $u = u(t,x)$ apply only to the spatial coordinate x whereas the time coordinate t is arbitrarily fixed.

A double-soliton is a nonlinear superposition of two solitons, hence it has four arbitrary parameters. The four parameters for the wave speeds and amplitudes are related to the four parameters of the fourth-order Lax–Novikov equation (5.38). The following lemma clarifies the generic construction of double solitons.

LEMMA 5.8. *Let a solution with the profile u satisfying* (5.38) *be a nonlinear superposition of two solitons with parameters* (b_1, c_1) *and* (b_2, c_2). *Then the four parameters* (a, b, c, d) *of the fourth-order Lax–Novikov equation* (5.38) *are found uniquely by*

$$\begin{cases} a = c_1 b_2 + c_2 b_1, \\ b = b_1 + b_2 - c_1 c_2, \\ c = c_1 + c_2, \\ d = b_1 b_2. \end{cases} \tag{5.65}$$

PROOF. Since u is a nonlinear superposition of two solitons $U_{k_j, c_j}(t, x; \alpha_j, \xi_j)$ for $1 \leq j \leq 2$, it converges to $U_{k_1, c_1}(t, x; \alpha_1, \xi_1)$ if $|\alpha_2| + |\xi_2| \to \infty$ and vice versa. As a result, the tails of $U_{k_{1,2}, c_{1,2}}$ satisfying the linear second-order equation $u'' + 2ic_{1,2}u' + 4b_{1,2}u = 0$ must also satisfy the linear fourth-order equation $u'''' + 2icu''' + 4bu'' + 8iau' + 16du = 0$. These two linear equations are linearization of (5.5) and (5.38) at the zero solution. Factorizing

$$(\partial_x^2 + 2ic_1\partial_x + 4b_1)(\partial_x^2 + 2ic_2\partial_x + 4b_2) = \partial_x^4 + 2ic\partial_x^3 + 4b\partial_x^2 + 8ia\partial_x + 16d$$

yields the desired relations (5.65). □

REMARK 5.32. The double-soliton has four translational parameters (α_1, ξ_1) and (α_2, ξ_2) in addition to four parameters (b_1, c_1) and (b_2, c_2). The translational parameters arise naturally in the limits $|\alpha_1| + |\xi_1| \to \infty$ and $|\alpha_2| + |\xi_2| \to \infty$ for the double-soliton solution.

REMARK 5.33. The factorization relation of the linear parts of stationary Lax–Novikov equations in the proof of Lemma 5.8 does not imply that the fourth-order equation (5.38) with nonlinear terms can be factorized by two second-order equations (5.5).

Because of the relation between the stationary Lax–Novikov equation and the higher-order energies of the NLS equation (Example 4.1), one can place the double-solitons (and generally, the N-solitons) in a variational context and characterize them as constrained minimizers of the higher-order energy subject to the fixed conserved energies of all lower orders. This gives the orbital stability of multi-solitons in suitable Sobolev spaces of higher regularity (Section 4.2).

This variational characterization of multi-solitons was pioneered in [**193**] for the KdV equation and was developed in [**149**] for the NLS equation. More results were obtained recently with precise count of eigenvalues in the constrained minimization problem and in the linear stability problem for the modified KdV equation [**173**] and again for the KdV equation [**242**]. In the context of the NLS equation (4.67) and the double solitons, the orbital stability result of [**149**] can be formulated as the following theorem.

THEOREM 5.5. *Let* $u = U_{\vec{k}, \vec{c}}(t, x; \vec{\alpha}, \vec{\xi})$ *be the double-soliton solution of Lemma* 5.8 *with translational parameters* $\vec{\alpha}, \vec{\xi} \in \mathbb{R}^2$ *and with parameters* $\vec{k}, \vec{c} \in \mathbb{R}^2$. *For every* $\epsilon > 0$ *there exists* $\delta > 0$ *such that for every* $u_0 \in H^2(\mathbb{R})$ *satisfying*

$$\inf_{\vec{\alpha}, \vec{\xi} \in \mathbb{R}^2} \|u_0 - U_{\vec{k}, \vec{c}}(0, \cdot; \vec{\alpha}, \vec{\xi})\|_{H^2} < \delta,$$

the unique solution $u \in C^0(\mathbb{R}, H^2(\mathbb{R})) \cap C^1(\mathbb{R}, L^2(\mathbb{R}))$ *to the NLS equation* (4.67) *with* $u(0,\cdot) = u_0$ *satisfies*

$$\inf_{\vec{\alpha},\vec{\xi}\in\mathbb{R}^2} \|u(t,\cdot) - U_{\vec{k},\vec{c}}(t,\cdot;\vec{\alpha},\vec{\xi})\|_{H^2} < \epsilon,$$

for every $t \in \mathbb{R}$.

REMARK 5.34. Theorem 5.5 extends to every N-soliton solution of the cubic NLS equation (4.67). The corresponding solution satisfies the stationary Lax–Novikov equation of the $(2N)$th order and is defined as a constrained minimizer of the higher-order energy $H_{2N}(u,\bar{u})$ in Sobolev space $H^N(\mathbb{R})$ subject to fixed energies $H_0, H_1, \dots, H_{2N-1}$.

With the N-fold Darboux transformations (Section 5.2), one can extend the ideas of Section 4.3 to prove orbital stability of N-solitons in spaces of lower regularity. This analysis was pioneered in [**66**] and was later extended in fractional Sobolev spaces of lower regularity in [**161, 162**] both for the cubic NLS and the modified KdV equations.

The following theorem formulates the orbital stability result for the double solitons.

THEOREM 5.6. *Let* $u = U_{\vec{k},\vec{c}}(t,x;\vec{\alpha},\vec{\xi})$ *be the double soliton solution of Lemma 5.8 with* $\vec{\lambda} = \frac{1}{2}(\vec{k} + i\vec{c})$ *for* $\vec{\alpha},\vec{\xi} \in \mathbb{R}^2$ *and* $\vec{k},\vec{c} \in \mathbb{R}^2$ *such that* $\lambda_1 \neq \lambda_2$. *There exist positive constants* ϵ_0 *and* C *such that if* $u_0 \in L^{2,s}(\mathbb{R})$ *for any* $s > \frac{1}{2}$ *satisfies*

$$\epsilon := \|u_0 - U_{\vec{k},\vec{c}}(0,\cdot;\vec{\alpha},\vec{\xi})\|_{L^{2,s}} < \epsilon_0, \qquad \text{where } \|u\|_{L^{2,s}} := \|(1+x^2)^{s/2}u\|_{L^2},$$

then there exists a solution $u \in C^0(\mathbb{R}, L^{2,s}(\mathbb{R}))$ *to the cubic NLS equation* (4.67) *with* $u(0,\cdot) = u_0$ *and the double soliton solution*

$$u = U_{\vec{k}',\vec{c}'}(t,x;\vec{\alpha}',\vec{\xi}')$$

with $\vec{\alpha}',\vec{\xi}' \in \mathbb{R}^2$ *and* $\vec{k}',\vec{c}' \in \mathbb{R}^2$ *satisfying*

$$|\vec{\alpha}' - \vec{\alpha}| + |\vec{\xi}' - \vec{\xi}| + |\vec{k}' - \vec{k}| + |\vec{c}' - \vec{c}| \leq C\epsilon$$

and

$$\|u(t,\cdot) - U_{\vec{k}',\vec{c}'}(t,\cdot;\vec{\alpha}',\vec{\xi}')\|_{L^2} \leq C\epsilon,$$

for every $t \in \mathbb{R}$.

REMARK 5.35. The orbital stability bound of Theorem 5.6 is not expected to be true in the weighted L^2 norms because the variance $V(t) := \int_{\mathbb{R}} x^2|u(t,x)|^2 dx$ grows linearly in time t in the time evolution of the cubic NLS equation (4.67).

REMARK 5.36. The assumption that $\lambda_1 \neq \lambda_2$ is used as a technical simplification in the analysis of the Darboux transformation with simple eigenvalues, for which the Darboux transformation of Theorem 5.4 is invertible by construction. However, since the Darboux transformation is not onto, an arbitrary point in a small L^2-neighborhood of the double soliton is not generally mapped back to the small L^2-neighborhood of the zero solution. To avoid this technical difficulty, the restriction to the small distance in $L^{2,s}$ is used in the initial datum because a small perturbation of the double soliton in the $L^1(\mathbb{R})$ norm (and hence in the $L^{2,s}(\mathbb{R})$ norm with $s > \frac{1}{2}$) does not change the number of isolated eigenvalues of the Lax spectrum but only modifies their location.

REMARK 5.37. Multi-solitons in the case of their large spatial separations can be studied by direct functional analytic methods which do not rely on integrability of the Hamiltonian equations [**199, 223**]. Interactions of two solitons in the NLS equation (4.67) were studied in [**131**]. Because the Lax equations or higher-order conserved quantities are not used, the results obtained with functional analytic methods are usually weaker than those obtained with the use of integrability.

5.3.2. Stability of breathers. A breather solution with the profile $u = U(t,x)$ can be defined as T-periodic in time t and spatially decaying in x so that

$$U(t+T,x) = U(t,x), \qquad \lim_{x\to\pm\infty} U(t,x) = 0.$$

Stability of such solutions can be understood in the orbital sense, which combines Definition 1.8 for periodic orbits and Definition 2.5 for traveling waves.

DEFINITION 5.3. We say that the breather with the profile U is orbitally stable if for every $\epsilon > 0$ there exists $\delta > 0$ such that for every u_0 from Banach space X such that

$$\|u_0 - U(0,\cdot)\|_X < \delta,$$

the unique solution $u \in C^0(\mathbb{R}, X)$ satisfies

$$\inf_{(t_0,\xi_0)\in\mathbb{R}^2} \|u(t,\cdot) - U(t-t_0, \cdot - \xi_0)\|_X < \epsilon,$$

for every $t \in \mathbb{R}$. If the breather is not orbitally stable, then we say that it is orbitally unstable.

REMARK 5.38. In a more generalized sense, breathers are defined as solutions with the profile $u = U(\tau,\xi)$, where $\tau = x - at$ and $\xi = t - bx$ for some a and b such that $ab \neq 1$, where the profile U is periodic with respect to τ and decaying with respect to ξ. The condition $ab \neq 1$ is required since otherwise τ and ξ would be linearly dependent instead of being transversal. The same Definition 5.3 can be used for such generalizations of breathers due to translations $(t_0,\xi_0) \in \mathbb{R}^2$ used in the infimum.

Breathers do not exist generally in nonlinear Hamiltonian systems because of resonances of the nonlinearity-induced modes with multiple frequencies $n\omega$, $n \in \mathbb{N}$, where $\omega := \frac{2\pi}{T}$ is the breather frequency, with the linear dispersive modes at the same frequencies [**25, 78**]. However, these resonances are decoupled in the integrable Hamiltonian systems. For example, breathers exist in the focusing versions of the modified KdV equation and the cubic NLS equation.

REMARK 5.39. Breather solutions of the modified KdV equation are related to the double-soliton solutions but the parameters of the double-solitons have to be extended to the complex plain as the complex-conjugate pairs. Consequently, these solutions also satisfy the fourth-order Lax–Novikov equation and this enables a variational characterization of breathers as constrained minimizers of the higher-order energy subject to fixed lower-order energies in $H^2(\mathbb{R})$. This variational characterization was used in [**5, 9, 10**] to obtain the orbital stability of breathers in the sense of Definition 5.3 for the modified KdV equation, similarly to the orbital stability of double solitons in Theorem 5.5.

REMARK 5.40. Compared to the modified KdV equation, breather solutions of the NLS equation (4.67) corresponds to the double-soliton solutions with $c_1 = c_2$

(see the right panel on Figure 5.9) and the eigenvalues $\lambda_1 = \frac{1}{2}(k_1 + ic_1)$ and $\lambda_2 = \frac{1}{2}(k_2 + ic_2)$ are already complex. By Theorem 5.6, the double-solitons are orbitally stable but the parameters of the double-solitons change to $\lambda_1' = \frac{1}{2}(k_1' + ic_1')$ and $\lambda_2' = \frac{1}{2}(k_2' + ic_2')$. If $c_1' \neq c_2'$, the double-soliton solution is no longer a breather solution and the two solitons move away from each other with different velocities. As a result, the breather solutions of the NLS equation are orbitally unstable in the sense of Definition 5.3. This instability can not be clarified using the linearization methods.

In what follows, we consider breather solutions of the NLS equation (4.67) on the background of the constant-amplitude wave. Such breather solutions can be obtained with the Bäcklund–Darboux (BD) transformation of Lemma 4.5. In order to simplify the presentation, we use the transformation $u(t,x) \mapsto u(t,x)e^{it}$ and rewrite the NLS equation (4.67) in the equivalent form:

$$iu_t + \frac{1}{2}u_{xx} + (|u|^2 - 1)u = 0. \tag{5.66}$$

The normalized constant-amplitude wave $u(t,x) = e^{it}$ of (4.63) becomes the constant solution $u(t,x) = 1$ of (5.66).

EXAMPLE 5.7. Let $\lambda_0 \in (1,\infty)$ and define a particular solution $\varphi = (p,q)^T$ of the linear system (4.68)–(4.69) with $u = 1$ and $\lambda = \lambda_0$ in the following form:

$$\begin{cases} p(t,x) = \sqrt{\lambda_0 + \frac{1}{2}\beta_0}\, e^{\frac{1}{2}(\beta_0 x + i\alpha_0 t)} - \sqrt{\lambda_0 - \frac{1}{2}\beta_0}\, e^{-\frac{1}{2}(\beta_0 x + i\alpha_0 t)}, \\ q(t,x) = -\sqrt{\lambda_0 - \frac{1}{2}\beta_0}\, e^{\frac{1}{2}(\beta_0 x + i\alpha_0 t)} + \sqrt{\lambda_0 + \frac{1}{2}\beta_0}\, e^{-\frac{1}{2}(\beta_0 x + i\alpha_0 t)}, \end{cases} \tag{5.67}$$

where $\beta_0 = 2\sqrt{\lambda_0^2 - 1}$ and $\alpha_0 = \lambda_0\beta_0$. Elementary computations give

$$|p_0|^2 + |q_0|^2 = 4\left[\lambda_0 \cosh(\beta_0 x) - \cos(\alpha_0 t)\right],$$
$$p_0\bar{q}_0 = -2\cosh(\beta_0 x) + 2\lambda_0 \cos(\alpha_0 t) + i\beta_0 \sin(\alpha_0 t),$$

so that the BD transformation (4.71) yields a breather solution in the form:

$$\hat{u}(t,x) = -1 + \frac{2(\lambda_0^2 - 1)\cos(\alpha_0 t) + i\alpha_0 \sin(\alpha_0 t)}{\lambda_0 \cosh(\beta_0 x) - \cos(\alpha_0 t)}. \tag{5.68}$$

This solution is known as the **Kuznetsov–Ma (KM) breather** [**166, 191**]. The profile $\hat{u}$ is periodic in t with period $T = 2\pi/\alpha_0$ and decaying to -1 in x. The parameter $\lambda_0 \in (1,\infty)$ is arbitrary and the two translational parameters (t_0, ξ_0) can be added to (5.68) by using a more general linear combination of two solutions in (5.67).

EXAMPLE 5.8. Let $\lambda_0 \in (0,1)$ and define a particular solution $\varphi = (p,q)^T$ of the linear system (4.68)–(4.69) with $u = 1$ and $\lambda = \lambda_0$ in the following form:

$$\begin{cases} p(t,x) = \sqrt{\lambda_0 - \frac{i}{2}k_0}\, e^{\frac{1}{2}(-ik_0 x + \sigma_0 t)} - \sqrt{\lambda_0 + \frac{i}{2}k_0}\, e^{\frac{1}{2}(ik_0 x - \sigma_0 t)}, \\ q(t,x) = -\sqrt{\lambda_0 + \frac{i}{2}k_0}\, e^{\frac{1}{2}(-ik_0 x + \sigma_0 t)} + \sqrt{\lambda_0 - \frac{i}{2}k_0}\, e^{\frac{1}{2}(ik_0 x - \sigma_0 t)}, \end{cases}$$

where $k_0 = 2\sqrt{1-\lambda_0^2} \in (0,2)$ and $\sigma_0 = \lambda_0 k_0$. Elementary computations give

$$|p_0|^2 + |q_0|^2 = 4\left[\cosh(\sigma_0 t) - \lambda_0 \cos(k_0 x)\right],$$
$$p_0 \bar{q}_0 = 2\cos(k_0 x) - 2\lambda_0 \cosh(\sigma_0 t) + i k_0 \sinh(\sigma_0 t),$$

so that the BD transformation (4.71) yields a breather solution in the form:

$$\hat{u}(t,x) = -1 + \frac{2(1-\lambda_0^2)\cosh(\sigma_0 t) + i\sigma_0 \sinh(\sigma_0 t)}{\cosh(\sigma_0 t) - \lambda_0 \cos(k_0 x)}. \tag{5.69}$$

This solution is known as the **Akhmediev breather** [**3**]. The profile $\hat{u}$ is L-periodic in x with period $L = 2\pi/k_0 \in (\pi, \infty)$ and

$$\lim_{t\to\pm\infty} \hat{u}(t,x) = 1 - 2\lambda_0^2 \pm i k_0 \lambda_0 = \left(\sqrt{1-\lambda_0^2} \pm i\lambda_0\right)^2,$$

such that $\lim_{t\to\pm\infty} |\hat{u}(t,x)| = 1$.

REMARK 5.41. Other breather solutions of (5.66) on the constant solution $u = 1$ are possible if $\lambda_0 \in \mathbb{C}\backslash(\mathbb{R} \cup i\,\mathbb{R})$ and they are generally referred as the Tajiri–Watanabe breathers [**235**]. These solutions are periodic along one line in the (t,x) plane and decaying to $|u| = 1$ along a transverse line in the (t,x) plane.

REMARK 5.42. The two breathers of Examples 5.7 and 5.8 are related by the transformation $k_0 = i\beta_0$. In the limit $\lambda_0 \to 1$, each breather degenerates into the fully localized solution of (5.66) in the (t,x) plane given by the rational function

$$u(t,x) = -1 + \frac{4(1+2it)}{1+4(x^2+t^2)}.$$

This solution is known as the Peregrine's rogue wave [**224**].

REMARK 5.43. Breathers of Examples 5.7 and 5.8 with the profile $\hat{u}(t,x)$ satisfy the third-order Lax–Novikov equation (5.50) for arbitrarily fixed $t \in \mathbb{R}$. Consequently, the profile $\hat{u}$ also satisfies the fourth-order Lax–Novikov equaiton (5.38). This has been used in [**7, 8**] to explore the variational characterization of breathers in the energy space and to obtain their energetic instability as saddle points of the constrained variational problems. The KM breather (5.68) was found to be unstable with respect to decaying perturbations in $H^s(\mathbb{R})$ for $s > 1/2$, whereas the Akhmediev breather (5.69) was found to be unstable with respect to periodic perturbations in $H^s_{\rm per}$ for $s > 1/2$.

We show next that the linear instability of breathers can be concluded from the linear system (4.68)–(4.69) by using Lemma 5.1. This approach has been elaborated in [**37, 119**] for the Akhmediev breather in $L^2_{\rm per}$ and was extended in [**125**] to the KM breather in $L^2(\mathbb{R})$. Before formulating results for the breathers of Examples 5.7 and 5.8, we will deal with the modulational instability of the constant solution $u = 1$ in (5.66). This provides a simple case of study where all details can be made precise. The following theorem specifies the modulational instability of the constant solution.

THEOREM 5.7. *For every $v_0 \in L^2(\mathbb{R})$, there exists a unique solution $v \in C^0(\mathbb{R}, L^2(\mathbb{R}))$ to the linearized NLS equation at the constant solution $u = 1$ given by*

$$i v_t + \frac{1}{2} v_{xx} + v + \bar{v} = 0 \tag{5.70}$$

satisfying $v(0,\cdot) = v_0$. *The solution is given in the form of a linear superposition*

$$(5.71)\quad \begin{aligned} v(t,x) &= \int_0^\infty \left[c_k^+ \widetilde{v}_k^+(t) + c_k^- \widetilde{v}_k^-(t)\right] \cos(kx) dk \\ &\quad + \int_0^\infty \left[d_k^+ \widetilde{v}_k^+(t) + d_k^- \widetilde{v}_k^-(t)\right] \sin(kx) dk, \end{aligned}$$

where the coefficients $c_k^\pm$ *and* $d_k^\pm$ *are uniquely found from* $v_0 \in L^2(\mathbb{R})$, *and the functions* $\widetilde{v}_k^\pm(t)$ *for* $k > 0$ *are given by*

$$(5.72)\quad k \in (0,2): \quad \begin{cases} \widetilde{v}_k^+(t) = (2i\lambda + k)e^{k\lambda t}, \\ \widetilde{v}_k^-(t) = (2i\lambda - k)e^{-k\lambda t}, \end{cases} \qquad \lambda = \lambda(k) = \frac{1}{2}\sqrt{4-k^2}$$

and

$$(5.73)\quad k \in (2,\infty): \quad \begin{cases} \widetilde{v}_k^+(t) = k\cos(k\gamma t) - 2i\gamma \sin(k\gamma t), \\ \widetilde{v}_k^-(t) = 2i\gamma \cos(k\gamma t) + k\sin(k\gamma t), \end{cases} \qquad \gamma = \gamma(k) = \frac{1}{2}\sqrt{k^2-4}.$$

REMARK 5.44. Since the linearized NLS equation (5.70) has constant coefficients, the Fourier transform provides a basis of bounded solutions in x which can be used to represent a general solution in $L^2(\mathbb{R})$ (Remark 5.46). However, we give a longer proof of Theorem 5.7, which connects the Lax spectrum and the stability spectrum of the constant solution.

PROOF. The proof consists of two steps. In the first step, we solve the linear system (4.68)–(4.69) with $u = 1$ explicitly and obtain the Lax spectrum. In the second step, we use Lemma 5.1 and obtain the stability spectrum of the linearized NLS equation (5.70).

STEP 1. The Lax spectrum of the spectral problem (4.68) with $u = 1$ in the space $C_b^0(\mathbb{R})$ of bounded continuous functions is the set $\Sigma_0 = i\mathbb{R} \cup [-1,1]$ with the following properties:

(i) $\lambda = \pm 1$ are algebraically simple eigenvalues;
(ii) each $\lambda \in \Sigma_0 \backslash \{0, \pm 1\}$ is a geometrically and algebraically double eigenvalue;
(iii) $\lambda = 0$ is an eigenvalue with geometric multiplicity two and algebraic multiplicity four.

PROOF OF STEP 1. The linear system (4.68)–(4.69) with $u = 1$ can be solved explicitly. Two linearly independent solutions exist for every λ since the linear system is of the second order. We only consider real and purely imaginary values of λ because the solutions found for the other complex values λ are unbounded.

For $\lambda \in \mathbb{R}_+$, two solutions to the linear system (4.68)–(4.69) with $u = 1$ are given by

$$(5.74)\quad \begin{aligned} \varphi &= \begin{bmatrix} \sqrt{\lambda - \frac{i}{2}k(\lambda)} \\ -\sqrt{\lambda + \frac{i}{2}k(\lambda)} \end{bmatrix} e^{-\frac{1}{2}ik(\lambda)(x+i\lambda t)}, \\ \phi &= \begin{bmatrix} \sqrt{\lambda + \frac{i}{2}k(\lambda)} \\ -\sqrt{\lambda - \frac{i}{2}k(\lambda)} \end{bmatrix} e^{+\frac{1}{2}ik(\lambda)(x+i\lambda t)}, \end{aligned}$$

where $k(\lambda) := 2\sqrt{1-\lambda^2}$. These solutions are bounded and linearly independent for $\lambda \in (0,1)$ which corresponds to $k(\lambda) \in (0,2)$. Solutions for $\lambda \in \mathbb{R}_-$, and in particular, for $\lambda \in [-1,0)$, are found from the symmetry property of the linear system (4.68)–(4.69):

If $\varphi = (\varphi_1, \varphi_2)^T$ is a solution for λ, then $\varphi = (-\bar{\varphi}_2, \bar{\varphi}_1)^T$ is a solution for $-\bar{\lambda}$. This symmetry implies that any $\lambda \in (-1,0) \cup (0,1)$ is a geometrically double eigenvalue.

For $\lambda = i\gamma$ with $\gamma \in \mathbb{R}$, two solutions to the linear system (4.68)–(4.69) are given by

$$\varphi = \begin{bmatrix} \sqrt{\frac{1}{2}k(\gamma) - \gamma} \\ -i\sqrt{\frac{1}{2}k(\gamma) + \gamma} \end{bmatrix} e^{-\frac{1}{2}ik(\gamma)(x-\gamma t)}, \tag{5.75}$$
$$\phi = \begin{bmatrix} \sqrt{\frac{1}{2}k(\gamma) + \gamma} \\ i\sqrt{\frac{1}{2}k(\gamma) - \gamma} \end{bmatrix} e^{+\frac{1}{2}ik(\gamma)(x-\gamma t)},$$

where $k(\gamma) := 2\sqrt{1+\gamma^2}$. These solutions are bounded and linearly independent for every $\gamma > 0$ which corresponds to $k(\gamma) \in (2,\infty)$. Solutions for $\gamma \in \mathbb{R}_-$ are found from the same symmetry. Consequently, any $\lambda = i\gamma$ with $\gamma \in \mathbb{R}\backslash\{0\}$ is a geometrically double eigenvalue.

For $\lambda = 1$, two solutions of the linear system (4.68)–(4.69) are given by

$$\varphi = \begin{bmatrix} 1 \\ -1 \end{bmatrix}, \quad \phi = \begin{bmatrix} x + it + 1 \\ -x - it \end{bmatrix}. \tag{5.76}$$

Hence, the eigenvalues $\lambda = \pm 1$ are algebraically simple.

For $\lambda = 0$ ($\gamma = 0$), two solutions of the linear system (4.68)–(4.69) are given by

$$\varphi = \begin{bmatrix} 1 \\ -i \end{bmatrix} e^{-ix}, \quad \phi = \begin{bmatrix} 1 \\ i \end{bmatrix} e^{+ix}, \tag{5.77}$$

implying that $\lambda = 0$ is a geometrically double eigenvalue. However, eigenvalue $\lambda = 0$ is algebraically quadruple because the bounded solutions of $\mathcal{L}^2\varphi = 0$ are spanned by (5.77) and the two additional solutions

$$\varphi_{\rm g} = \begin{bmatrix} t \\ -1 - it \end{bmatrix} e^{-ix}, \quad \phi_{\rm g} = \begin{bmatrix} -t \\ -1 - it \end{bmatrix} e^{+ix}. \tag{5.78}$$

It remains to check the algebraic multiplicity of eigenvalues. For each eigenvalue λ, the bounded eigenfunctions φ and ϕ in $C_b^0(\mathbb{R})$ are periodic in x with some spatial period L. For the algebraic multiplicity of λ, we need to solve

$$(\mathcal{L} - \lambda I)\varphi_g = \varphi, \quad (\mathcal{L} - \lambda I)\phi_g = \phi$$

in the space of periodic functions with period L, where

$$\mathcal{L} := \begin{bmatrix} \partial_x & -1 \\ -1 & -\partial_x \end{bmatrix}.$$

Consequently, we can check the Fredholm condition for φ and ϕ in $L^2(0,L)$.

Let $\varphi = (\varphi_1, \varphi_2)^T$ be the bounded eigenfunction of the eigenvalue problem $(\mathcal{L} - \lambda I)\varphi = 0$. By the symmetry of the linear system (4.68)–(4.69), the adjoint problem $\left(\mathcal{L}^* - \bar{\lambda} I\right)\varphi^* = 0$ admits the eigenfunction $\varphi^* = (\bar{\varphi}_2, \bar{\varphi}_1)^T$. If $\lambda \in \Sigma_0\backslash\{+1,-1\}$, there exists another linearly independent eigenfunction $\phi = (\phi_1, \phi_2)^T$, for which we have similarly $\phi^* = (\bar{\phi}_2, \bar{\phi}_1)^T$. Since $\langle \psi^*, \varphi \rangle = \langle \varphi^*, \phi \rangle = 0$, the generalized eigenfunctions φ_g and ϕ_g exist if and only if $\langle \varphi^*, \varphi \rangle = 0$ and $\langle \phi^*, \phi \rangle = 0$.

For $\lambda \in (0,1)$, we obtain

$$\langle \varphi^*, \varphi \rangle = -2\lambda L e^{\lambda k(\lambda) t}, \quad \langle \phi^*, \phi \rangle = -2\lambda L e^{-\lambda k(\lambda) t},$$

which are both nonzero for $\lambda \neq 0$. For $\lambda = 1$, only one linearly independent eigenfunction φ in (5.76) exists and we check that $\langle \varphi^*, \varphi \rangle = -2L \neq 0$. For $\lambda = i\gamma$ with $\gamma \in \mathbb{R}$, we obtain

$$\langle \varphi^*, \varphi \rangle = -2i\gamma L e^{i\gamma k(\gamma) t}, \quad \langle \phi^*, \phi \rangle = -2i\gamma L e^{-i\gamma k(\gamma) t},$$

which are both nonzero for $\gamma \neq 0$. Hence, the algebraic multiplicity of all nonzero eigenvalues is equal to their geometric multiplicity.

For the eigenvalue $\lambda = 0$ with the eigenfunctions (5.77), we obtain

$$\langle \varphi^*, \varphi \rangle = \langle \phi^*, \phi \rangle = 0,$$

in agreement with the existence of the generalized eigenfunctions (5.78). On the other hand, we also have

$$\langle \varphi^*, \varphi_g \rangle = -L, \quad \langle \phi^*, \phi \rangle = -L,$$

which implies that no new solutions satisfying $\mathcal{L}^3 \varphi = 0$ exist. Hence, the algebraic multiplicity of $\lambda = 0$ is equal to four. □

STEP 2. The stability spectrum of the linearized NLS equation (5.70) in the space $C_b^0(\mathbb{R})$ is the set $i\mathbb{R} \cup [-1,1]$. For $\lambda \in [-1,1]$, the eigenfunctions are given by

$$v_\lambda^+(t,x) = -(2i\lambda + k)e^{\lambda k t} \sin(kx), \quad v_{-\lambda}^+(t,x) = (2i\lambda + k)e^{\lambda k t} \cos(kx), \tag{5.79}$$

and

$$v_\lambda^-(t,x) = (2i\lambda - k)e^{-\lambda k t} \sin(kx), \quad v_{-\lambda}^-(t,x) = (2i\lambda - k)e^{-\lambda k t} \cos(kx), \tag{5.80}$$

for which $k = k(\lambda) \in (0,2]$. For $\lambda = i\gamma$ with $\gamma \in \mathbb{R}_+$ the eigenfunctions are given by

$$\begin{cases} v_\lambda^+(t,x) = k\cos(kx - k\gamma t) + 2i\gamma \sin(kx - k\gamma t), \\ v_{-\lambda}^+(t,x) = -2i\gamma \cos(kx - k\gamma t) + k\sin(kx - k\gamma t), \end{cases} \tag{5.81}$$

while for γ replaced by $-\gamma$ the eigenfunctions are

$$\begin{cases} v_\lambda^-(t,x) = k\cos(kx + k\gamma t) - 2i\gamma \sin(kx + k\gamma t), \\ v_{-\lambda}^-(t,x) = 2i\gamma \cos(kx + k\gamma t) + k\sin(kx + k\gamma t). \end{cases} \tag{5.82}$$

for which $k = k(\gamma) \in (2,\infty)$.

PROOF OF STEP 2. By Lemma 5.1, solutions of the linearized NLS equation (5.70) can be constructed from solutions of the linear system (4.68)–(4.69) with $u = 1$ and $\lambda \in \mathbb{C}$. We use the Pairs I and III of solutions given in Table 1. Using φ and ϕ in (5.74) for $\lambda \in (0,1]$, we find that the four linearly independent bounded solutions are given by (5.79) and (5.80). These are, up to their sign, equal to the four solutions in the Fourier decomposition (5.71) given by (5.72) so that we have a one-to-one correspondence between the solutions provided by the linear system (4.68)–(4.69) with $\lambda \in (0,1]$ and the solutions in (5.71) with $k \in (0,2]$ through the equalities $k = k(\lambda)$ and $\lambda = \lambda(k)$. By the symmetry of the linear system (4.68)–(4.69), the solutions obtained for $\lambda = [-1, 0)$ are, up to their sign, the same. The stability spectrum for $\lambda \in [-1, 1]$ is located on

$$\{\pm k\lambda(k), \quad k \in [0, 2]\} = [-1, 1].$$

Using φ and ϕ in (5.75) for $\lambda = i\gamma$ with $\gamma \in \mathbb{R}_+$, we find that the two linearly independent solutions (5.81) and (5.82) are linear combinations of the four solutions in the Fourier decomposition (5.71) with $k \in (2, \infty)$ given by (5.73), and we have a one-to-one correspondence between these solutions through the equalities $k = k(\gamma)$ and $\gamma = \gamma(k)$. The stability spectrum for $\lambda = i\gamma \in i\,\mathbb{R}$ is located on

$$\{\pm ik\gamma(k), \quad k \in (2, \infty)\} = i\,\mathbb{R}. \qquad \square$$

End of the proof. Replacing the space $C_b^0(\mathbb{R})$ by $L^2(\mathbb{R})$ does not change the Lax and stability spectra. Summarizing, solutions $v \in C^0(\mathbb{R}, L^2(\mathbb{R}))$ to the linearized NLS equation (5.70) are expressed in the form:

(5.83)
$$v(t,x) = \int_0^\infty \left[c_k^+ v_{\lambda(k)}^+(t,x) + c_k^- v_{\lambda(k)}^-(t,x) + c_{-k}^+ v_{-\lambda(k)}^+(t,x) + c_{-k}^- v_{-\lambda(k)}^-(t,x) \right] dk,$$

where coefficients $c_{\pm k}^{\pm}$ are uniquely defined from the initial condition $v(\cdot, 0) = v_0 \in L^2(\mathbb{R})$, and $v_{\pm\lambda(k)}^{\pm}(t,x)$ are given by (5.79)–(5.80) if $k \in (0, 2]$ and by (5.81)–(5.82) if $k \in (2, \infty)$. This integral representation recovers the decomposition (5.71). Completeness of the basis of bounded functions in $L^2(\mathbb{R})$ follows from the Fourier theory. $\square$

Remark 5.45. For the particular values of $k = 0$ and $k = 2$, we have two solutions of the linearized NLS equation (5.70) in the form:

$$k = 0: \quad \begin{cases} \widetilde{v}_0^+(t) = 2i, \\ \widetilde{v}_0^-(t) = 1 + 2it, \end{cases} \tag{5.84}$$

and

$$k = 2: \quad \begin{cases} \widetilde{v}_2^+(t) = 2, \\ \widetilde{v}_2^-(t) = i + 2t. \end{cases} \tag{5.85}$$

The two solutions for $k = 0$ are recovered from Pair II of Table 1 by using φ and ϕ in either (5.74) or (5.75) with $\lambda = 1$,

$$v = 0, \quad \widetilde{v} = 2i, \tag{5.86}$$

and by using φ and ϕ from (5.76):

$$v = 1 + 2it, \quad \tilde{v} = i(2x + 1). \tag{5.87}$$

The zero solution $v = 0$ in (5.86) or the unbounded solution $\tilde{v}$ in (5.87) are neglected. The two solutions for $k = 2$ are recovered from Pair II of Table 1 by using φ and ϕ in (5.77) for $\lambda = 0$,

$$v_0^+(t,x) = -2\sin(2x), \quad v_{-0}^+(t,x) = 2\cos(2x) \tag{5.88}$$

and by using the eigenfunctions (5.77) and the generalized eigenfunctions (5.78),

$$\begin{aligned} v_0^-(t,x) &= 2(i + 2t)\cos(2x) - 2\sin(2x), \\ v_{-0}^-(t,x) &= 2(i + 2t)\sin(2x) + 2\cos(2x). \end{aligned} \tag{5.89}$$

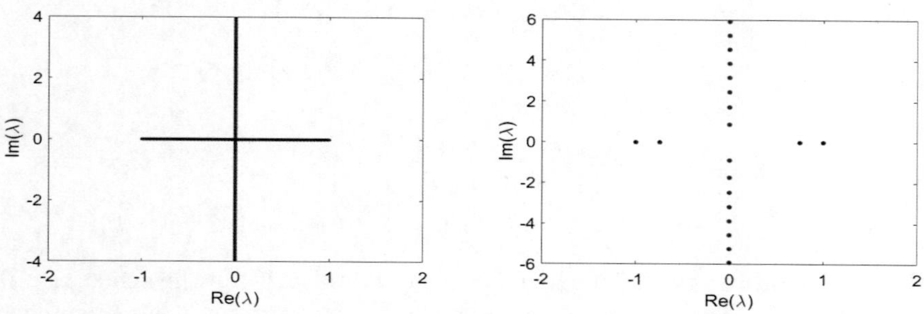

FIGURE 5.10. Left: The Lax spectrum Σ_0 in $L^2(\mathbb{R})$. Right: The union $\Sigma_0^{(P)} \cup \Sigma_0^{(A)}$ of the Lax spectra in $L^2_{\rm per}(0,L)$ and $L^2_{\rm antiper}(0,L)$ for $\pi < L < 2\pi$.

Since the points $k = 0$ and $k = 2$ are of measure zero in the integral (5.83), these solutions play no role in the dynamics of localized perturbations.

REMARK 5.46. The proof of Theorem 5.7 can also be obtained by using the separation of variables. Substituting $v(t,x) = \widetilde{v}_k(t)e^{ikx}$ into (5.70) yields the linear differential equation

$$i\frac{d}{dt}\widetilde{v}_k + \left(1 - \frac{k^2}{2}\right)\widetilde{v}_k + \overline{\widetilde{v}}_k = 0,$$

with two linearly independent solutions $\widetilde{v}_k^+(t)$ and $\widetilde{v}_k^-(t)$ given by (5.72) and (5.73) for $k \in (0,2)$ and $k \in (2,\infty)$. It also recovers (5.84) and (5.85) for $k = 0$ and $k = 2$.

REMARK 5.47. The stability spectrum in Step 2 of Theorem 5.7 implies the modulational instability of the constant solution $u = 1$ in the NLS equation (5.66).

REMARK 5.48. In the space $L^2_{\rm per}(0,L)$ of L-periodic functions, the Lax spectrum only contains the eigenvalues $\lambda \in \Sigma_0$ with L-periodic associated eigenfunctions, hence the pure point spectrum is located at

$$\Sigma_0^{(P)} = \{\pm\lambda_m^{(P)}, \ m \in \{0, \mathbb{N}_{\rm even}\}\}, \quad \lambda_m^{(P)} := \sqrt{1 - \frac{\pi^2}{L^2}m^2}. \tag{5.90}$$

Similarly, in the space $L^2_{\rm antiper}(0,L)$ of L-antiperiodic functions, the Lax spectrum only contains the eigenvalues $\lambda \in \Sigma_0$ with L-antiperiodic associated eigenfunctions, hence the pure point spectrum is located at

$$\Sigma_0^{(A)} = \{\pm\lambda_m^{(A)}, \ m \in \mathbb{N}_{\rm odd}\}, \quad \lambda_m^{(A)} := \sqrt{1 - \frac{\pi^2}{L^2}m^2}. \tag{5.91}$$

The algebraic and geometric multiplicities of these eigenvalues remain the same. Figure 5.10 illustrates the purely continuous Lax spectrum of Σ_0 in $L^2(\mathbb{R})$ (left) and the purely point Lax spectrum $\Sigma_0^{(P)}$ ($\Sigma_0^{(A)}$) in $L^2_{\rm per}(0,L)$ ($L^2_{\rm antiper}(0,L)$) (right).

The following corollary replaces Theorem 5.7 in the space of periodic functions with the period $L > 0$.

COROLLARY 5.1. *For every $v_0 \in L^2_{\rm per}(0,L)$, there exists a unique solution $v \in C^0(\mathbb{R}, L^2_{\rm per}(0,L))$ to the linearized NLS equation (5.70) satisfying $v(0,\cdot) = v_0$.*

The solution is given in the form of a linear superposition

$$(5.92)\qquad \begin{aligned} v(t,x) &= \sum_{m\in\mathbb{N}_0} \left[c^+_{k_m}\widetilde{v}^+_{k_m}(t) + c^-_{k_m}\widetilde{v}^-_{k_m}(t)\right]\cos(k_m x) \\ &\quad + \sum_{m\in\mathbb{N}} \left[d^+_{k_m}\widetilde{v}^+_{k_m}(t) + d^-_{k_m}\widetilde{v}^-_{k_m}(t)\right]\sin(k_m x), \end{aligned}$$

where coefficients $c^\pm_{k_m}$ *and* $d^\pm_{k_m}$ *are uniquely found from* $v_0 \in L^2_{\rm per}(0,L)$.

PROOF. Solutions of the linearized NLS equation (5.70) in the space $L^2_{\rm per}(0,L)$ are found by restricting the continuous Fourier decomposition (5.71) to the discrete values

$$k_m := \frac{2\pi m}{L}, \quad m \in \mathbb{N}_0 := \{0, \mathbb{N}\}.$$

The decomposition (5.92) is obtained by using eigenfunctions of the linear system (4.68)–(4.69) with $u = 1$ for both L-periodic and L-antiperiodic solutions, because the solutions of the linearized NLS equation (5.70) are constructed using squared eigenfunctions.

The Lax spectra $\Sigma_0^{(P)}$ in $L^2_{\rm per}(0,L)$ and $\Sigma_0^{(A)}$ in $L^2_{\rm antiper}(0,L)$ are given in (5.90) and (5.91), respectively. All functions in the Fourier series (5.92) are recovered from the eigenfunctions associated to the eigenvalues

$$\lambda \in \Sigma_0^{(P)} \cup \Sigma_0^{(A)}.$$

For $m = 0$ we have the eigenvalues $\pm\lambda(0) = \pm 1 \in \Sigma_0^{(P)}$ leading to the solutions $\widetilde{v}_0^+$ and $\widetilde{v}_0^-$ given by (5.84), which are constant in x. If $0 < \pi m < L$, then $\lambda(k_m) \in (0,1)$, and we have the four linearly independent solutions in (5.79)–(5.80) with $\lambda = \lambda(k_m)$. If $\pi m > L$, then $\lambda(k_m) = i\gamma(k_m)$ is purely imaginary, and we have the four linearly independent solutions in (5.81)–(5.82) with $\gamma = \gamma(k_m)$. In the particular case $L = \pi m$, for some $m \in \mathbb{N}$, we have $\lambda(k_m) = 0$ and four linearly independent solutions are given in (5.88) and (5.89). □

REMARK 5.49. The solution $v(t,x)$ in (5.92) grows in time t due to the linearly growing $\widetilde{v}_0^-(t)$, the exponentially growing $\widetilde{v}^\pm_{k_m}(t)$ with $\pi m < L$ (if $L > \pi$), and possibly linearly growing $\widetilde{v}^-_{k_m}(t)$ for $\pi m = L$ (if $L \in \pi\mathbb{N}$). Hence, the constant solution $u = 1$ is linearly unstable in the linearized NLS equation (5.70) in $L^2_{\rm per}(0,L)$.

By using the BD transformation (Examples 5.7 and 5.8), breathers with the profile $\hat{u}(t,x)$ can be obtained from the constant solution $u = 1$. The BD transformation also transforms the fundamental solution $\Phi(\lambda,u)$ of the linear system (4.68)–(4.69) to the fundamental solution $\hat{\Phi}(\lambda,\hat{u})$ of the same linear system but with new potential,

$$\hat{\Phi}(\lambda,\hat{u}) = D(\lambda)\Phi(\lambda,u),$$

where the Darboux matrix $D(\lambda)$ is given by

$$(5.93)\qquad D(\lambda) := I + \frac{1}{\lambda-\lambda_0}\begin{bmatrix}\hat{p}_0\\ \hat{q}_0\end{bmatrix}\begin{bmatrix}-q_0 & p_0\end{bmatrix}$$

for $\lambda \neq \{\lambda_0, -\bar{\lambda}_0\}$. Since

$$\det D(\lambda) = \frac{\lambda+\bar{\lambda}_0}{\lambda-\lambda_0},$$

the matrix $D(\lambda)$ is invertible for every $(t,x) \in \mathbb{R} \times \mathbb{R}$, and the correspondence between the 2×2 matrix solutions $\Phi(\lambda, u)$ and $\hat{\Phi}(\lambda, \hat{u})$ is one-to-one if $\lambda \neq \{\lambda_0, -\bar{\lambda}_0\}$. Based on the BD transformation, it is now possible to compute the Lax spectrum associated with the breather solutions and the corresponding bounded solutions of the linearized NLS equations either in the space $L^2(\mathbb{R})$ for the Kuznetsov–Ma breathers (Example 5.7) or in the space $L^2_{\rm per}(0,L)$ for the Akhmediev breathers (Example 5.8). The following two theorems were proven in [**125**].

THEOREM 5.8. *The Lax spectrum of the spectral problem* (4.68) *for the Kuznetsov–Ma breather with the profile* (5.68) *in* $L^2(\mathbb{R})$ *is the set*

$$\Sigma_{\rm KMB} = i\mathbb{R} \cup [-1,1] \cup \{\lambda_0, -\lambda_0\},$$

with the following properties:

(i) *For each* $\lambda \in i\mathbb{R} \cup (-1,1)$, *there exist two linearly independent bounded solutions.*
(ii) *For* $\lambda = 1$ *and* $\lambda = -1$, *there exists only one bounded solution.*
(iii) *The eigenvalues* $\lambda = \lambda_0$ *and* $\lambda = -\lambda_0$ *are algebraically simple.*

PROOF. The Darboux matrix $D(\lambda)$ given by (5.93) is invertible if $\lambda \neq \pm\lambda_0$. Moreover, both $D(\lambda)$ and $D(\lambda)^{-1}$ are bounded in x for every $t \in \mathbb{R}$ if $\lambda \neq \pm\lambda_0$. As a result, we have a one-to-one correspondence between the bounded solutions of the linear system (4.68)–(4.69) with $u = 1$ and with u being the breather profile (5.68) if $\lambda \neq \pm\lambda_0$. This implies that $\Sigma_0 \subset \Sigma_{\rm KMB}$ and that (i) and (ii) hold as in Step 1 of Theorem 5.7.

It remains to prove (iii). Due to the symmetry property of the linear system (4.68)–(4.69), it is enough to show the result for λ_0. The vector $\varphi = (p,q)^T$ given by (5.67) is a solution of the linear system (4.68)–(4.69) with $u = 1$ and $\lambda = \lambda_0$. The BD transformation (4.72) gives the solution $\varphi = (\hat{p}, \hat{q})^T$ of the linear system (4.68)–(4.69) with u being the breather profile (5.68) and $\lambda = \lambda_0$ so that $\hat{p}(t,x), \hat{q}(t,x) \to 0$ as $|x| \to \infty$ exponentially fast, hence $\varphi = (\hat{p}, \hat{q})^T$ is an eigenfunction in $L^2(\mathbb{R})$ associated with the eigenvalue λ_0. Furthermore, the linear system (4.68)–(4.69) having zero-trace, the Wronskian of any two solutions is constant both in x and t. Since one solution is decaying to zero as $|x| \to \infty$, another linearly independent solution is necessarily growing at infinity. Consequently, λ_0 is geometrically simple.

For the algebraic multiplicity of λ_0, we use the form $(\mathcal{L} - \lambda_0 I)\,\varphi_0 = 0$ with the eigenfunction $\varphi_0 = (\hat{p}, \hat{q})^T \in L^2(\mathbb{R})$ and show that the linear nonhomogeneous equation

$$(\mathcal{L} - \lambda_0 I)\,\psi_0 = \varphi_0,$$

does not have the generalized eigenfunction $\psi_0 \in L^2(\mathbb{R})$. The solvability condition for this equation is given by the Fredholm condition $\langle \varphi_0^*, \varphi_0 \rangle = 0$, where φ_0^* is an eigenfunction of the adjoint problem $(\mathcal{L}^* - \lambda_0 I)\,\varphi_0^* = 0$. A direct calculation shows that $\varphi_0^* = (\bar{q}, \bar{p})^T$, and then we compute

$$\begin{aligned}
\langle \varphi_0^*, \varphi_0 \rangle &= -\int_{\mathbb{R}} \frac{8\lambda_0^2 pq}{(|p|^2+|q|^2)^2} dx \\
&= \lambda_0^2 \int_{\mathbb{R}} \frac{\cosh(\beta_0 x + i\alpha_0 t) - \lambda_0}{(\lambda_0 \cosh(\beta_0 x) - \cos(\alpha_0 t))^2} dx \\
&= -\lambda_0 \beta_0^{-1} \frac{\lambda_0 \sinh(\beta_0 x) + i \sin(\alpha_0 t)}{\lambda_0 \cosh(\beta_0 x) - \cos(\alpha_0 t)} \Bigg|_{x \to -\infty}^{x \to +\infty} \\
&= -2\lambda_0 \beta_0^{-1}.
\end{aligned}$$

Consequently, $\langle \varphi_0^*, \varphi_C \rangle \neq 0$ so that λ_0 is a simple eigenvalue and property (iii) holds. □

THEOREM 5.9. *Assume $L \notin \pi\mathbb{N}$. The Lax spectrum of the spectral problem* (4.68) *for the Akhmediev breather with the profile* (5.69) *in $L^2_{\text{antiper}}(0, L)$ consists of isolated eigenvalues*

$$\Sigma_{\text{AB}}^{(A)} = \{\pm\lambda_m^{(A)}, \ m \in \mathbb{N}_{\text{odd}}\}, \quad \lambda_m^{(A)} := \sqrt{1 - \frac{\pi^2}{L^2} m^2},$$

with the following properties:

(i) *For each $m \in \mathbb{N}_{\text{odd}}$, $m \neq 1$, the eigenvalues $\pm\lambda_m^{(A)}$ are geometrically and algebraically double.*
(ii) *The eigenvalues $\lambda_1^{(A)} = \lambda_0$ and $-\lambda_1^{(A)} = -\lambda_0$ are geometrically simple and algebraically double.*

The Lax spectrum in $L^2_{\text{per}}(0, L)$ consists of isolated eigenvalues

$$\Sigma_{\text{AB}}^{(P)} = \{\pm\lambda_m^{(P)}, \ m \in \{0, \mathbb{N}_{\text{even}}\}\}, \quad \lambda_m^{(P)} := \sqrt{1 - \frac{\pi^2}{L^2} m^2},$$

with the following properties:

(i) *For each $m \in \mathbb{N}_{\text{even}}$, the eigenvalues $\pm\lambda_m^{(P)}$ are geometrically and algebraically double.*
(ii) *The eigenvalues $\lambda_0^{(P)} = 1$ and $-\lambda_0^{(P)} = -1$ are algebraically simple.*

PROOF. It is based on computations similar to the proof of Theorem 5.8; see [**125**]. □

Figure 5.11 shows the Lax spectrum Σ_{KMB} in $L^2(\mathbb{R})$ (left) and the union of the Lax spectra $\Sigma_{AB}^{(P)} \cup \Sigma_{AB}^{(A)}$ in $L^2_{\text{per}}(0, L)$ and $L^2_{\text{antiper}}(0, L)$ (right).

REMARK 5.50. Based on Theorem 5.8, bounded solutions of the linearized NLS equation (5.2) with u being the Kuznetsov–Ma breather with the profile (5.68) can be constructed from bounded solutions of the linear equations (4.68)–(4.69) for $\lambda \in \Sigma_{\text{KMB}}$, whereas the unbounded solutions correspond to $\lambda \notin \Sigma_{\text{KMB}}$. It was shown in [**125**] that the two eigenvalues at $\lambda = \pm\lambda_0$ give three bounded and three decaying solutions of the linearized NLS equation (5.2), whereas the rest of the spectrum $\Sigma_0 = \Sigma_{\text{KMB}} \backslash \{\lambda_0, -\lambda_0\}$ gives similar solutions to those for the linearized NLS equation (5.70). As a result, the Kuznetsov–Ma breathers with the profile (5.68) are linearly unstable with exactly the same growth rate as the constant solution $u = 1$.

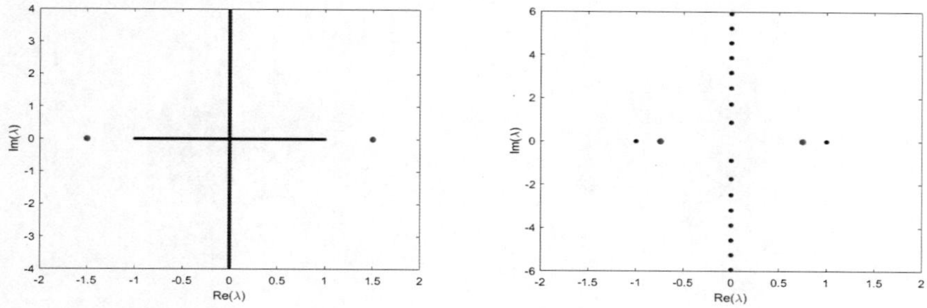

FIGURE 5.11. Left: The Lax spectrum $\Sigma_{\rm KMB}$ in $L^2(\mathbb{R})$. Right: The union $\Sigma_{AB}^{(P)} \cup \Sigma_{AB}^{(A)}$ of the Lax spectra in $L^2_{\rm per}(0, L)$ and $L^2_{\rm antiper}(0, L)$

REMARK 5.51. Based on Theorem 5.9, it was also shown in [**119**, **125**] that the eigenvalues at

$$\lambda = \pm\lambda_1^{(A)} = \pm\lambda_0 \text{ and } \lambda = \pm\lambda_0^{(P)} = \pm 1$$

generate six linearly independent x-periodic solutions of the linearized NLS equation (5.2) with u being the Akhmediev breather with the profile (5.69). Two solutions from these six solutions grow exponentially in time as $t \to \pm\infty$ similarly to the two modes in (5.92) which correspond to $m = 1$. Additionally, other eigenvalues in $\Sigma_{AB}^{(P)} \cup \Sigma_{AB}^{(A)}$ with $1 < m < L/\pi$ generate two exponentially growing unstable modes according to the transformation (5.93). As a result, the Akhmediev breathers with the profile (5.69) are linearly unstable with exactly the same growth rate and the same number of unstable modes as the constant solution $u = 1$.

REMARK 5.52. A given set $\mathcal{S} \subset L^2$ is called **complete** if every element of L^2 can be obtained as a unique linear combination of elements of $\mathcal{S}$. The completeness of solutions to the linearized NLS equation (5.2) is a particularly challenging question. Completeness of spatially localized squared eigenfunctions of the linear system (4.68)–(4.69) as solutions of the linearized integrable equations has been shown in [**156**] for zero boundary conditions and in [**24**] for nonzero boundary conditions, although the representation formulas in [**24**] lack precise details. In the case of periodic perturbations, completeness of periodic solutions was discussed in [**119**] also without precise details.

5.3.3. Stability of double-periodic waves. The double-periodic wave of the NLS equation (4.67) is a solution of the form $u(t, x) = U(t, x)e^{-2ibt}$, where the profile $U(t, x)$ is periodic both in temporal and spatial coordinates so that

$$U(t + T, x) = U(t, x + L) = U(t, x), \qquad (t, x) \in \mathbb{R} \times \mathbb{R}.$$

Although $u(t, x)$ is generally quasi-periodic in t (Remark 5.8), it is common to refer to these solutions as the double-periodic solutions. Such solutions of the NLS equation (4.67) were obtained by the separation of variables in [**3**] in the following form:

$$u(t, x) = [Q(t, x) + i\delta(t)]\, e^{i\theta(t)}, \tag{5.94}$$

where real δ and θ are x-independent and where $Q(t, x)$, $\delta(t)$, and $\theta'(t)$ are periodic in t and x with periods T and L, respectively.

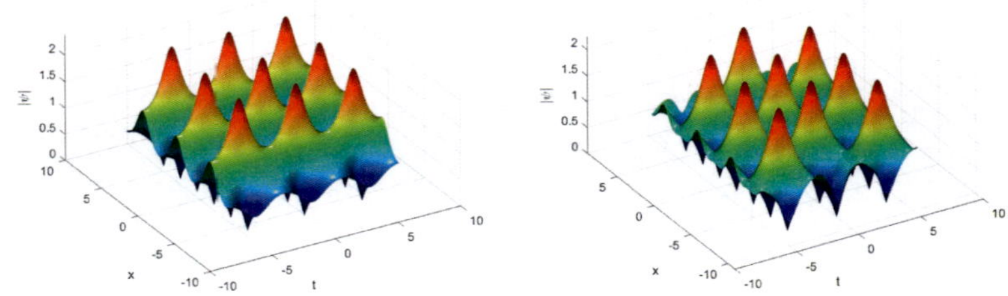

FIGURE 5.12. Solution surface for the double-periodic waves with the profile (5.95) for $k = 0.9$ (left) and (5.96) for $k = 0.8$ (right).

EXAMPLE 5.9. The following families of double-periodic waves were constructed in [**3**] for the NLS equation (4.67):

$$u(t,x) = k\frac{\mathrm{cn}(t;k)\,\mathrm{cn}(\sqrt{1+k}x;\kappa) + i\sqrt{1+k}\,\mathrm{sn}(t;k)\,\mathrm{dn}(\sqrt{1+k}x;\kappa)}{\sqrt{1+k}\,\mathrm{dn}(\sqrt{1+k}x;\kappa) - \mathrm{dn}(t;k)\,\mathrm{cn}(\sqrt{1+k}x;\kappa)}e^{it}, \quad \kappa := \sqrt{\frac{1-k}{1+k}} \tag{5.95}$$

and

$$u(t,x) = \frac{\mathrm{dn}(t;k)\,\mathrm{cn}(\sqrt{2}x;\kappa) + i\sqrt{k(1+k)}\,\mathrm{sn}(t;k)}{\sqrt{1+k} - \sqrt{k}\,\mathrm{cn}(t;k)\,\mathrm{cn}(\sqrt{2}x;\kappa)}e^{ikt}, \quad \kappa := \frac{\sqrt{1-k}}{\sqrt{2}}, \tag{5.96}$$

where the arbitrary parameter $k \in (0,1)$ is the elliptic modulus. It follows from (5.95) and (5.96) that the fundamental periods are $L = 4K(\kappa)/\sqrt{1+k}$ and $T = 4K(k)$ for (5.95) and $L = 2\sqrt{2}K(\kappa)$ and $T = 4K(k)$ for (5.96). In the limit $k \to 1$, both solutions degenerate to a particular version of the Akhmediev breather (Example 5.8):

$$u(t,x) = \frac{\cos(\sqrt{2}x) + i\sqrt{2}\sinh(t)}{\sqrt{2}\cosh(t) - \cos(\sqrt{2}x)}e^{it}. \tag{5.97}$$

Figure 5.12 shows the double-periodic waves with the profile (5.95) for $k = 0.9$ (left) and (5.96) for $k = 0.8$ (right). The former has periodically repeated maxima at the same positions, whereas the latter has periodically alternated maxima in a chess pattern.

The following theorem gives the most general construction of the double-periodic waves for the third-order Lax–Novikov equation (5.50) in the particular case $a = c = f = 0$.

THEOREM 5.10. *There exist two distinct families of double-periodic waves of the third-order Lax–Novikov equation* (5.50) *with $a = c = f = 0$ parameterized by three arbitrary parameters. The Lax spectrum for the double-periodic waves has at least three spectral bands outside $i\mathbb{R}$.*

PROOF. It follows from (5.52) with $a = c = f = 0$ that

$$\frac{d}{dx}\log\left(\frac{u'}{\bar{u}'}\right) = 0 \quad \Rightarrow \quad \frac{u'}{\bar{u}'} = e^{2i\theta},$$

with real, x-independent θ. Hence, $u'(x) = e^{i\theta}Q'(x)$ with real Q and integrating it again yields $u = (Q + i\delta)e^{i\theta}$ with x-independent δ. Substituting this expression into the third-order Lax–Novikov equation (4.16) with $a = c = 0$ yields

$$Q_{xxx} + 6(Q^2 + \delta^2)Q_x + 4bQ_x = 0,$$

which is integrated to the form

$$Q_{xx} + 2Q^3 + 6\delta^2 Q + 4bQ = A, \tag{5.98}$$

where A is x-independent. Substituting $u = (Q + i\delta)e^{i\theta}$ into (5.51) with $c = 0$ and using (5.98) yields the first-order invariant:

$$\left(\frac{dQ}{dx}\right)^2 + F(Q) = 0, \quad F(Q) = Q^4 + (6\delta^2 + 4b)Q^2 - 2AQ - 3\delta^4 - 4b\delta^2 - 8d. \tag{5.99}$$

Substituting $u = (Q + i\delta)e^{i\theta}$ with (5.98) and (5.99) into (5.53) with $a = c = 0$ yields

$$A^2 + 16\left[(\delta^2 + b)(\delta^4 + b\delta^2 + 2d) + g\right] = 0. \tag{5.100}$$

It remains to relate A with δ, for which we use the time evolution of the NLS equation (4.67). Substituting (5.94) into (4.67) and separating the variables yield the following system:

$$\begin{cases} Q_{xx} + 2(Q^2 + \delta^2 - \theta_t)Q - 2\delta_t = 0, \\ Q_t + (Q^2 + \delta^2 - \theta_t)\delta = 0. \end{cases} \tag{5.101}$$

Comparing (5.98) with the first equation of system (5.101) yields

$$A = 2\frac{d\delta}{dt}, \qquad \frac{d\theta}{dt} = -2(\delta^2 + b). \tag{5.102}$$

Setting $z := \delta^2$ reduces (5.100) to the first-order invariant

$$\left(\frac{dz}{dt}\right)^2 + G(z) = 0, \quad G(z) = 16z(z^3 + 2bz^2 + (b^2 + 2d)z + g + 2bd). \tag{5.103}$$

Let us parameterize the constant b, d, and g as follows:

$$\begin{cases} -2b = z_1 + z_2 + z_3, \\ 2d + b^2 = z_1 z_2 + z_1 z_3 + z_2 z_3, \\ g + 2bd = -z_1 z_2 z_3. \end{cases} \tag{5.104}$$

As a result, the invariants in (5.99) and (5.103) are given by

$$\begin{aligned} F(Q) = Q^4 &+ 2(3z - z_1 - z_2 - z_3)Q^2 - 4\delta_t Q - 3z^2 + 2z(z_1 + z_2 + z_3) \\ &- 2(z_1 z_2 + z_1 z_3 + z_2 z_3) + z_1^2 + z_2^2 + z_3^2 \end{aligned} \tag{5.105}$$

and

$$G(z) = 16z(z - z_1)(z - z_2)(z - z_3). \tag{5.106}$$

The polynomial $\tilde{P}(\lambda)$ in (5.54) is transformed under the parametrization (5.104) with $a = c = f = 0$ to the form

$$\tilde{P}(\lambda) = \lambda^6 - (z_1 + z_2 + z_3)\lambda^4 + (z_1 z_2 + z_1 z_3 + z_2 z_3)\lambda^2 - z_1 z_2 z_3,$$

which shows that $\pm\sqrt{z_1}$, $\pm\sqrt{z_2}$, $\pm\sqrt{z_3}$ are roots of $\tilde{P}(\lambda)$. These three parameters parametrize the two families of the double-periodic waves which we derive next.

For now, we can see that the three roots of $\tilde{P}(\lambda)$ are outside of $i\,\mathbb{R}$, therefore, at least three spectral bands exist outside of $i\,\mathbb{R}$.

Since $z = \delta^2 \geq 0$, one of the roots $z_{1,2,3}$ must be positive, whereas the other two roots are either real or complex-conjugate.

CASE (i) (All three roots $z_{1,2,3}$ are real). When the three roots $z_{1,2,3}$ are real, we order them by $z_1 \leq z_2 \leq z_3$. From positivity of the solutions $z(t)$ of the first-order invariant (5.103) with (5.106) it follows that $z_3 > 0$ and $z(t) \in [0, z_3]$. If we formally replace $\delta'(t) = 2\sqrt{(z_1 - z)(z_2 - z)(z_3 - z)}$ and factorize the quartic polynomial $F(Q)$ in (5.105) for $z \in [0, z_3]$ by

$$F(Q) = \left[Q^2 + 2\sqrt{z_3 - z}Q + z_3 + z - z_1 - z_2 + 2\sqrt{(z_1 - z)(z_2 - z)}\right] \times \left[Q^2 - 2\sqrt{z_3 - z}Q + z_3 + z - z_1 - z_2 - 2\sqrt{(z_1 - z)(z_2 - z)}\right],$$

then the discriminants of the two quadratic equations are given by

$$D_\pm = z_1 + z_2 - 2z \pm 2\sqrt{(z_1 - z)(z_2 - z)}.$$

It follows from the discriminants that all four roots of $F(Q)$ are complex-valued unless $z \leq z_1$. Hence $z_1 > 0$ and $z(t) \in [0, z_1]$ so that the three real roots satisfy the following ordering $0 \leq z_1 \leq z_2 \leq z_3$. The exact solution of the invariant (5.103) for $z(t) \in [0, z_1]$ under the ordering $0 \leq z_1 \leq z_2 \leq z_3$ is given by the explicit expression

$$z(t) = \frac{z_1 z_3 \operatorname{sn}^2(\mu t; k)}{z_3 - z_1 \operatorname{cn}^2(\mu t; k)},$$

where μ and k are related to (z_1, z_2, z_3) by

$$\mu^2 = 4z_2(z_3 - z_1), \quad k^2 = \frac{z_1(z_3 - z_2)}{z_2(z_3 - z_1)}.$$

Extracting the square root from $z = \delta^2$ in such a way that $\delta(t)$ remains smooth in t yields the exact solution

$$\delta(t) = \frac{\sqrt{z_1 z_3} \operatorname{sn}(\mu t; k)}{\sqrt{z_3 - z_1 \operatorname{cn}^2(\mu t; k)}}, \tag{5.107}$$

which is a smooth periodic function of t with period $T = 4K(k)/\mu$.

When $z(t) \in [0, z_1]$ under the ordering $0 \leq z_1 \leq z_2 \leq z_3$, the four roots of $F(Q)$ are real and can be ordered as $Q_4 \leq Q_3 \leq Q_2 \leq Q_1$. The following explicit formulae for the roots $Q_{1,2,3,4}$ were used in [**3**]:

$$\begin{cases} Q_1 = \sqrt{z_1 - z} + \sqrt{z_2 - z} + \sqrt{z_3 - z}, \\ Q_2 = -\sqrt{z_1 - z} - \sqrt{z_2 - z} + \sqrt{z_3 - z}, \\ Q_3 = -\sqrt{z_1 - z} + \sqrt{z_2 - z} - \sqrt{z_3 - z}, \\ Q_4 = \sqrt{z_1 - z} - \sqrt{z_2 - z} - \sqrt{z_3 - z}. \end{cases}$$

However, since $\sqrt{z_1 - z(t)}$ is nonsmooth at $t = K(k)/\mu$, this parametrization introduces singularities in the definition of $Q(t, x)$. Similarly, one can replace $\sqrt{z_1 - z}$ by $-\sqrt{z_1 - z}$ but this choice also introduces singularities in the definition of $Q(t, x)$.

In order to avoid the branch point singularities, we shall parameterize the roots $Q_{1,2,3,4}$ by using

$$\begin{cases} Q_1 = \sqrt{z_3 - z} + \sqrt{z_1 + z_2 - 2z + \frac{\delta'(t)}{\sqrt{z_3-z}}}, \\ Q_2 = \sqrt{z_3 - z} - \sqrt{z_1 + z_2 - 2z + \frac{\delta'(t)}{\sqrt{z_3-z}}}, \\ Q_3 = -\sqrt{z_3 - z} + \sqrt{z_1 + z_2 - 2z - \frac{\delta'(t)}{\sqrt{z_3-z}}}, \\ Q_4 = -\sqrt{z_3 - z} - \sqrt{z_1 + z_2 - 2z - \frac{\delta'(t)}{\sqrt{z_3-z}}}, \end{cases}$$

where all square roots stay away from zero, so that the definition of $Q(t,x)$ is smooth. The exact solution of the invariant (5.99) with

$$F(Q) = (Q - Q_1)(Q - Q_2)(Q - Q_3)(Q - Q_4)$$

for $Q \in [Q_2, Q_1]$ is given by

$$Q(t,x) = Q_4 + \frac{(Q_1 - Q_4)(Q_2 - Q_4)}{(Q_2 - Q_4) + (Q_1 - Q_2)\,\mathrm{sn}^2(\nu x;\kappa)}, \tag{5.108}$$

where

$$\begin{cases} 4\nu^2 = (Q_1 - Q_3)(Q_2 - Q_4), \\ 4\nu^2\kappa^2 = (Q_1 - Q_2)(Q_3 - Q_4), \end{cases} \quad \text{implies} \quad \begin{cases} \nu^2 = z_3 - z_1, \\ \nu^2\kappa^2 = z_2 - z_1. \end{cases}$$

The function Q is periodic in x with period $L = 2K(\kappa)/\nu$ and periodic in t with period $T = 4K(k)/\mu$ thanks to the T-periodicity of $z(t)$ and $\delta'(t)$ in $Q_{1,2,3,4}$.

CASE (ii) (One root is real and two are complex-conjugate). When one of the roots $z_{1,2,3}$ is real and positive while the other two roots are complex-conjugate, we can introduce the parametrization

$$z_1 \geq 0, \quad z_2 = \xi + i\eta, \quad z_3 = \xi - i\eta$$

with real-valued ξ and η. The exact solution of the invariant (5.103) for $z(t) \in [0, z_1]$ under this parametrization is given by

$$z(t) = \frac{z_1(1 - \mathrm{cn}(\mu t;k))}{(1+\zeta) + (\zeta - 1)\,\mathrm{cn}(\mu t;k)},$$

where $\zeta > 0$, $\mu > 0$, and $k \in (0,1)$ are parameters given by

$$\zeta = \frac{\sqrt{(z_1 - \xi)^2 + \eta^2}}{\sqrt{\xi^2 + \eta^2}},$$

$$\mu^2 = 16\sqrt{(z_1 - \xi)^2 + \eta^2}\sqrt{\xi^2 + \eta^2},$$

$$2k^2 = 1 + \frac{\xi(z_1 - \xi) - \eta^2}{\sqrt{(z_1 - \xi)^2 + \eta^2}\sqrt{\xi^2 + \eta^2}}.$$

We need to extract the square root from $\delta^2 = z$ so that $\delta(t)$ remain smooth in t. To do so, we use the half-argument formula

$$\mathrm{sn}^2\left(\frac{1}{2}\mu t;k\right) = \frac{1 - \mathrm{cn}(\mu t;k)}{1 + \mathrm{dn}(\mu t;k)},$$

which yields the exact solution

$$\delta(t) = \frac{\sqrt{z_1}\sqrt{1 + \mathrm{dn}(\mu t;k)}}{\sqrt{(1+\zeta) + (\zeta - 1)\,\mathrm{cn}(\mu t;k)}}\,\mathrm{sn}\left(\frac{1}{2}\mu t;k\right), \tag{5.109}$$

which is a smooth periodic function of t with period $T = 8K(k)/\mu$.

Since $z_1 \geq 0$ is now the largest positive root of $G(z)$, we can rewrite the factorization of $F(Q)$ in the equivalent form

$$\begin{aligned} F(Q) = &\left[Q^2 + 2\sqrt{z_1 - z}Q + z_1 + z - 2\xi + 2\sqrt{(\xi - z)^2 + \eta^2}\right] \\ \times &\left[Q^2 - 2\sqrt{z_1 - z}Q + z_1 + z - 2\xi - 2\sqrt{(\xi - z)^2 + \eta^2}\right], \end{aligned}$$

so that the discriminants of the two quadratic equations are given by

$$D_\pm = 2\xi - 2z \pm 2\sqrt{(\xi - z)^2 + \eta^2}.$$

Note that the factorization formula for $F(Q)$ is formal since the sign of $\sqrt{z_1 - z}$ should be defined from smooth continuations of the roots of $F(Q)$. It follows from the discriminants that two roots of $F(Q)$ are real-valued, ordered as $Q_2 \leq Q_1$, and two roots are complex-conjugate, $Q_3 = \bar{Q}_4$.

The following explicit formulae for the roots $Q_{1,2,3,4}$ were used in [**3**]:

$$\begin{cases} Q_1 = \sqrt{z_1 - z} + \sqrt{2}\sqrt{\sqrt{(\xi - z)^2 + \eta^2} + (\xi - z)}, \\ Q_2 = \sqrt{z_1 - z} - \sqrt{2}\sqrt{\sqrt{(\xi - z)^2 + \eta^2} + (\xi - z)}, \\ Q_3 = -\sqrt{z_1 - z} + i\sqrt{2}\sqrt{\sqrt{(\xi - z)^2 + \eta^2} - (\xi - z)}, \\ Q_4 = -\sqrt{z_1 - z} - i\sqrt{2}\sqrt{\sqrt{(\xi - z)^2 + \eta^2} - (\xi - z)}, \end{cases}$$

however, $\sqrt{z_1 - z(t)}$ is nonsmooth at $t = 2K(k)/\mu$, which introduces singularities in the definition of $Q(t, x)$. In order to avoid the branch point singularities, we parameterize the roots $Q_{1,2,3,4}$ by using

$$\begin{cases} Q_1 = \frac{\delta'(t)}{2\sqrt{(\xi - z)^2 + \eta^2}} + \sqrt{2}\sqrt{\sqrt{(\xi - z)^2 + \eta^2} + (\xi - z)}, \\ Q_2 = \frac{\delta'(t)}{2\sqrt{(\xi - z)^2 + \eta^2}} - \sqrt{2}\sqrt{\sqrt{(\xi - z)^2 + \eta^2} + (\xi - z)}, \\ Q_3 = -\frac{\delta'(t)}{2\sqrt{(\xi - z)^2 + \eta^2}} + i\sqrt{2}\sqrt{\sqrt{(\xi - z)^2 + \eta^2} - (\xi - z)}, \\ Q_4 = -\frac{\delta'(t)}{2\sqrt{(\xi - z)^2 + \eta^2}} - i\sqrt{2}\sqrt{\sqrt{(\xi - z)^2 + \eta^2} - (\xi - z)}, \end{cases}$$

which is smooth in t. Let us also denote $Q_3 = \alpha + i\beta$ and $Q_4 = \alpha - i\beta$ with real-valued α and β. The exact solution of the invariant (5.99) with

$$F(Q) = (Q - Q_1)(Q - Q_2)\left[(Q - \alpha)^2 + \beta^2\right]$$

for $Q \in [Q_2, Q_1]$ is given by

$$Q(t, x) = Q_1 + \frac{(Q_2 - Q_1)(1 - \mathrm{cn}(\nu x; \kappa))}{1 + \gamma + (\gamma - 1)\,\mathrm{cn}(\nu x; \kappa)}, \tag{5.110}$$

where $\gamma > 0$, $\nu > 0$, and $\kappa \in (0, 1)$ are parameters given by

$$\gamma = \frac{\sqrt{(Q_2 - \alpha)^2 + \beta^2}}{\sqrt{(Q_1 - \alpha)^2 + \beta^2}},$$

and

$$\begin{cases} \nu^2 = \sqrt{[(Q_1 - \alpha)^2 + \beta^2]\,[(Q_2 - \alpha)^2 + \beta^2]}, \\ 2\kappa^2 = 1 - \frac{(Q_1 - \alpha)(Q_2 - \alpha) + \beta^2}{\sqrt{[(Q_1 - \alpha)^2 + \beta^2][(Q_2 - \alpha)^2 + \beta^2]}}, \end{cases} \quad \text{implies} \quad \begin{cases} \nu^2 = 4\sqrt{(z_1 - \xi)^2 + \eta^2}, \\ 2\kappa^2 = 1 - \frac{z_1 - \xi}{\sqrt{(z_1 - \xi)^2 + \eta^2}}. \end{cases}$$

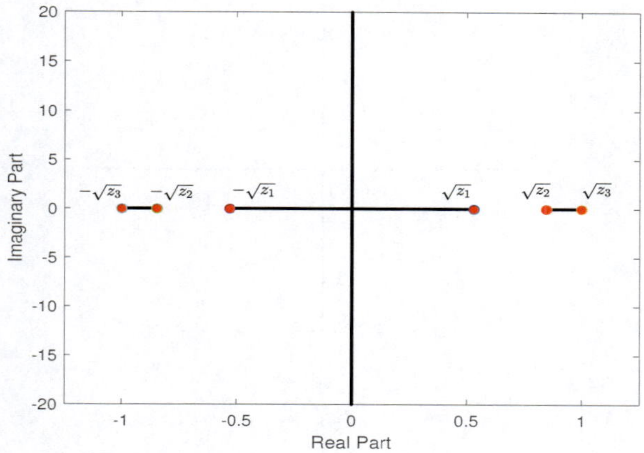

FIGURE 5.13. Lax spectrum for the double-periodic wave with the profile (5.95) for $k = 0.9$.

The function Q is periodic in x with the period $L = 4K(\kappa)/\nu$ and periodic in t with the period $T = 8K(k)/\mu$ thanks to the T-periodicity of $z(t)$ and $\delta'(t)$ in $Q_{1,2,3,4}$. □

REMARK 5.53. Figure 5.13 illustrates Theorem 5.10 by representing the Lax spectrum for the spectral problem (4.68) with u being the double-periodic wave with the profile (5.95) for $k = 0.9$. Figure 5.14 does the same for the profile (5.96) with $k = 0.8$ (top) and $k = 0.2$ (bottom). The black curves represent the purely continuous spectrum whereas the red dots represent roots $\pm\sqrt{z_1}$, $\pm\sqrt{z_2}$, and $\pm\sqrt{z_3}$ of the polynomial $\tilde{P}(\lambda)$. There exist exactly three spectral bands of the Lax spectrum outside $i\mathbb{R}$ if the band along the real axis passing through $\lambda = 0$ is counted as one spectral band.

REMARK 5.54. It is shown in [**45**] that the solution (5.95) is obtained from the double-periodic wave with the profiles (5.107) and (5.108) for

$$z_1 = \frac{1 - \sqrt{1-k^2}}{2}, \quad z_2 = \frac{1 + \sqrt{1-k^2}}{2}, \quad z_3 = 1,$$

whereas the solution (5.96) is obtained from the double-periodic wave with the profiles (5.109) and (5.110) for

$$z_1 = 2k, \quad z_{2,3} = \xi \pm i\eta, \quad \xi^2 + \eta^2 = \frac{1}{4}.$$

REMARK 5.55. The two families of explicit solutions given in terms of Jacobi elliptic functions in (5.108) and (5.110) are also relevant for other integrable equations such as the modified KdV equation [**42, 172**], where they represent traveling periodic waves.

Since the Akhmediev breather (5.97) is the limiting case of the double-periodic waves with the profiles (5.95) and (5.96), it is natural to see that both the Kuznetsov–Ma and Akhmediev breathers (Examples 5.7 and 5.8) are limiting cases in the family of solutions of the third-order Lax–Novikov equation (5.50). The following lemma shows that this is the case and gives the explicit characterization of the two breathers.

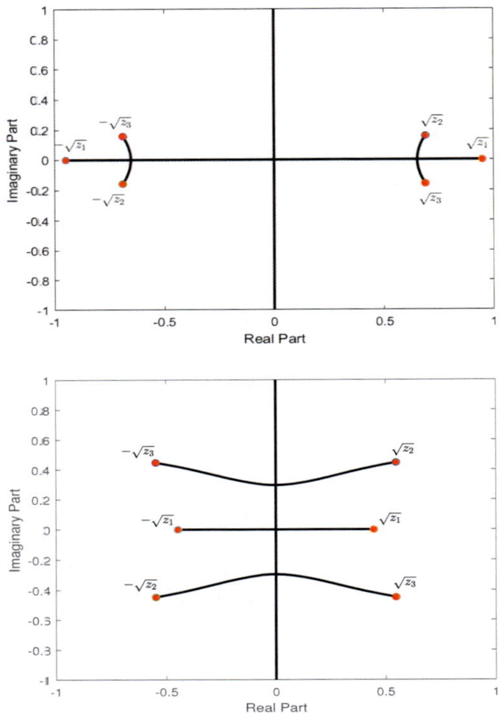

FIGURE 5.14. Lax spectrum for the double-periodic wave with the profile (5.96) for $k = 0.8$ (top) and $k = 0.2$ (bottom).

LEMMA 5.9. *The Kuznetsov–Ma breathers with the profile* (5.68) *are given by solutions of the system* (5.50), (5.51), (5.52), *and* (5.53) *with* $a = c = f = 0$ *for*

$$z_1 = 1, \qquad z_2 = z_3 = \lambda_0^2 \in (1, \infty),$$

whereas the Akhmediev breathers with the profile (5.69) *are obtained from solutions of the same equations with*

$$z_1 = z_2 = \lambda_0^2 \in (0, 1), \qquad z_3 = 1,$$

PROOF. We will prove the statement for the Kuznetsov–Ma breathers. The computations for the Akhmediev breathers are analogous.

If $z_1 = 1$ and $z_2 = z_3 = \lambda_0^2$, then parameters b, d, and g from (5.104) are expressed in terms of λ by

$$2b = -1 - 2\lambda_0^2, \quad 8d = 4\lambda_0^2 - 1, \quad 8g = 2\lambda_0^2 - 1.$$

The polynomial $\tilde{P}$ in (5.54) has a pair of double roots at $\pm\lambda_0$ and a pair of simple roots at ± 1. Parameters of (5.107) satisfy $\mu = \alpha_0 := 2\lambda_0\sqrt{\lambda_0^2 - 1}$ and $k = 0$ to yield

$$\delta(t) = \frac{\lambda_0 \sin(\alpha_0 t)}{\sqrt{\lambda_0^2 - \cos^2(\alpha_0 t)}}.$$

Solving the differential equation (5.102) gives

$$\theta(t) = t + \arctan\varphi(t), \quad \varphi(t) = \frac{\lambda_0}{\sqrt{\lambda_0^2 - 1}} \tan(\alpha_0 t).$$

The exact solution (5.94) can be written in the form

$$u(t,x) = \frac{1}{\sqrt{1+\varphi^2}} \left[Q - \delta\varphi + i\left(Q\varphi + \delta\right)\right] e^{it}.$$

It remains to compute $Q(t,x)$ explicitly. The turning points $Q_{1,2,3,4}$ are given by the following smooth functions of t:

$$\begin{cases} Q_1 = \dfrac{\sqrt{\lambda_0^2 - 1}}{\sqrt{\lambda_0^2 - \cos^2(\alpha_0 t)}} \left[\cos(\alpha_0 t) + 2\lambda_0\right], \\ Q_2 = Q_3 = -\dfrac{\sqrt{\lambda_0^2 - 1}}{\sqrt{\lambda_0^2 - \cos^2(\alpha_0 t)}} \cos(\alpha_0 t), \\ Q_4 = \dfrac{\sqrt{\lambda_0^2 - 1}}{\sqrt{\lambda_0^2 - \cos^2(\alpha_0 t)}} \left[\cos(\alpha_0 t) - 2\lambda_0\right]. \end{cases}$$

Parameters of (5.108) satisfy $\nu = \sqrt{\lambda_0^2 - 1}$ and $\kappa = 1$ to yield

$$Q(t,x) = \frac{\sqrt{\lambda_0^2 - 1}}{\sqrt{\lambda_0^2 - \cos^2(\alpha_0 t)}} \frac{2\lambda_0^2 - \lambda_0 \cos(\alpha_0 t)\cosh(\beta_0 x) - \cos^2(\alpha_0 t)}{\lambda_0 \cosh(\beta_0 x) - \cos(\alpha_0 t)},$$

where $\beta_0 := 2\sqrt{\lambda_0^2 - 1}$. Substituting $\delta(t)$, $\theta(t)$, and $Q(t,x)$ into $u(t,x)$ yields the exact expression (5.68). □

REMARK 5.56. Breathers with the profiles (5.68) and (5.69) can be obtained by degeneration of the double-periodic waves. Referring to Figures 5.13 and 5.14 (left), two spectral bands (either the real ones for the first family or the complex ones for the second family) shrink into a double eigenvalue, whereas the third spectral band is preserved on the real axis.

The linear instability of the double-periodic waves can be studied by using the linear system (4.68)–(4.69) and the Lax spectrum. Compared to the study of the modulational instability of the traveling periodic waves (Section 5.1), we should use Floquet theory in both variables x and t. The following lemma gives the relation between the Lax spectrum and the stability spectrum for the double-periodic waves.

LEMMA 5.10. *Let $u = U(t,x)e^{-2ibt}$ be the double-periodic solution of the NLS equation* (4.67) *with parameter $b \in \mathbb{R}$, where the profile $U(t,x)$ is periodic in t and x with periods T and L. The solution $\varphi = \varphi(t,x)$ to the linear system* (4.68)–(4.69) *can be represented in the form*

$$\varphi_1(t,x) = \chi_1(t,x) e^{-ibt + i\theta x + t\Omega}, \quad \varphi_2(t,x) = \chi_2(t,x) e^{ibt + i\theta x + t\Omega},$$

where $\chi_{1,2}(t,x)$ are periodic in t and x with periods T and L, $\theta \in \left[-\frac{\pi}{L}, \frac{\pi}{L}\right]$, and $\Omega = \Omega(\lambda) \in \mathbb{C}$ is the spectral parameter uniquely defined in $\mathrm{Im}(\Omega) \in \left[-\frac{\pi}{T}, \frac{\pi}{T}\right]$ *for every λ in the Lax spectrum. The corresponding solution $\varphi = \varphi(t,x)$ generates the solution $v = v(t,x)$ of the linearized NLS equation* (5.2) *by means of* (5.3) *and* (5.4).

PROOF. It follows from the linear system (4.68)–(4.69) that the vector $\chi = (\chi_1, \chi_2)^T$ satisfies the linear system

$$\chi_x + i\theta\chi = \begin{bmatrix} \lambda & U \\ -\bar{U} & -\lambda \end{bmatrix} \chi, \tag{5.111}$$

and

$$\chi_t + \Omega\chi = i \begin{bmatrix} \lambda^2 + \frac{1}{2}|U|^2 + b & \frac{1}{2}U_x + \lambda U \\ \frac{1}{2}\bar{U}_x - \lambda\bar{U} & -\lambda^2 - \frac{1}{2}|U|^2 - b \end{bmatrix} \chi. \tag{5.112}$$

Since $u = U(t,x)e^{-2ibt}$ satisfies the NLS equation (4.67), the system (5.111)–(5.112) is compatible in the sense that parameters $\theta \in \left[-\frac{\pi}{L}, \frac{\pi}{L}\right]$ and $\lambda, \Omega \in \mathbb{C}$ are independent of (t,x).

By Floquet theory, the admissible values of the spectral parameter $\lambda \in \mathbb{C}$ is determined from the spectral problem (5.111) subject to the periodicity conditions $\chi(t, x+L) = \chi(t,x)$ for every $t \in \mathbb{R}$ and for every given $\theta \in \left[-\frac{\pi}{L}, \frac{\pi}{L}\right]$. The union of all admissible values of λ forms the Lax spectrum of the double-periodic solutions.

With λ defined in the Lax spectrum, the spectral problem (5.112) determines the spectral parameter Ω from the periodicity conditions $\chi(t+T, x) = \chi(t,x)$ for every $x \in \mathbb{R}$. The relation between $\varphi = \varphi(t,x)$ and the solution $v = v(t,x)$ of the linearized NLS equation (5.2) is given by Lemma 5.1. □

REMARK 5.57. If $\mathrm{Re}(\Omega) > 0$ for λ in the Lax spectrum, then the double-periodic wave $u = U(t,x)e^{-2ibt}$ is linearly unstable with the instability growth rate given by $\mathrm{Re}(\Lambda) = 2\,\mathrm{Re}(\Omega)$, which follows from (5.3) and (5.4).

EXAMPLE 5.10. For the double-periodic wave with the profile (5.95) for $k = 0.85$ (top) and $k = 0.95$ (bottom), Figure 5.15 shows the Lax spectrum of the spectral problem (5.111) with $\theta \in \left[-\frac{\pi}{L}, \frac{\pi}{L}\right]$ on the λ-plane (left) and the stability spectrum on the Λ-plane (right). The unstable spectrum is located at the boundary $\mathrm{Im}(\Lambda) = \pm\frac{\pi}{T}$ of the strip for every $k \in (0,1)$. The double-periodic wave (5.95) is linearly unstable.

EXAMPLE 5.11. For the double-periodic wave with the profile (5.96) for $k = 0.2$ (top) and $k = 0.6$ (bottom), Figure 5.16 shows the same as Figure 5.15. The Lax spectrum on the λ-plane has three bands, two of which are connected either across the imaginary axis (top) or across the real axis (bottom), the third band is located on the real axis. The unstable spectrum on the Λ-plane includes the figure-eight band and bands located near the boundary $\mathrm{Im}(\Lambda) = \pm\frac{\pi}{T}$.

REMARK 5.58. Rogue waves are spatially and temporally localized waves which may exist on a nonzero background. These waves appear from nowhere and disappear without a trace. Existence of rogue waves is related to the linear instability of the nonzero background. Rogue waves for modulationally unstable standing periodic waves were studied for the NLS equation in [**46, 94**] and for the derivative NLS equation in [**43, 44**]. Rogue waves on the double-periodic wave background were studied for the NLS equation in [**45**].

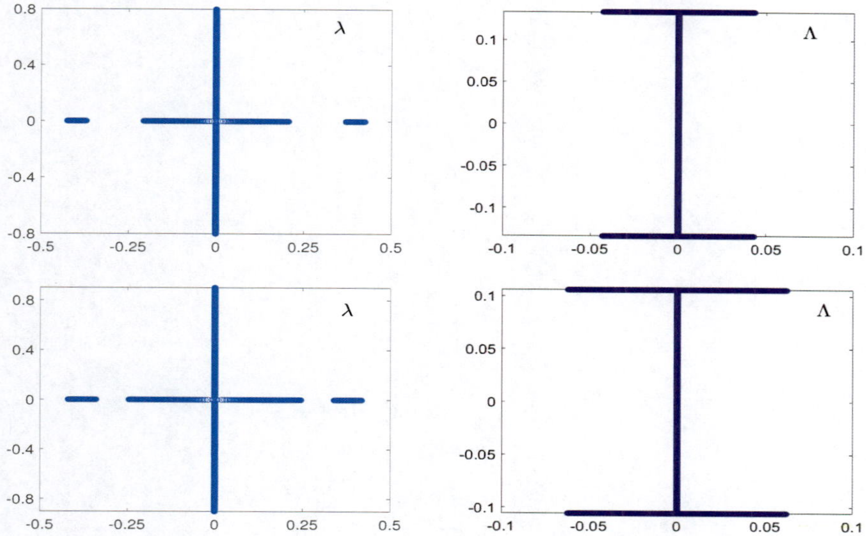

FIGURE 5.15. Lax spectrum on the λ-plane (left) and stability spectrum on the Λ-plane (right) for the double-periodic wave with the profile (5.95) for $k = 0.85$ (top) and $k = 0.95$ (bottom).

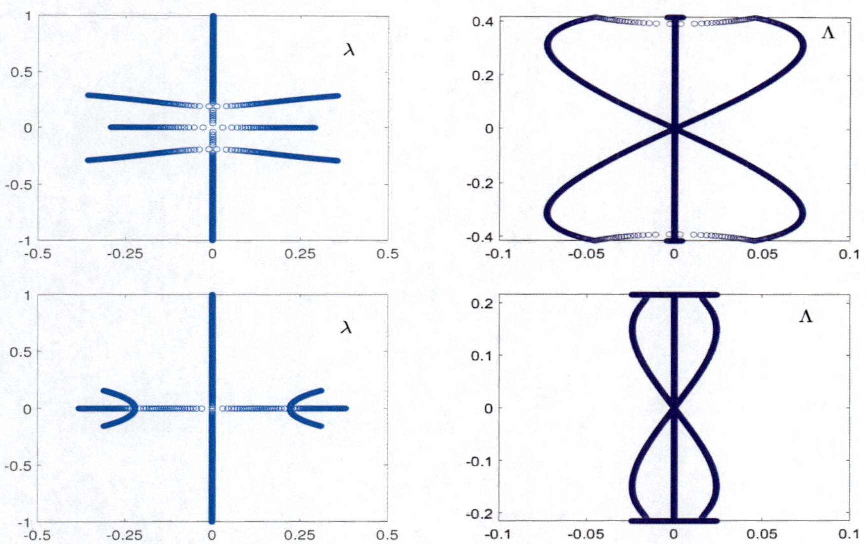

FIGURE 5.16. The same as Figure 5.15 but for the double-periodic wave with the profile (5.96) for $k = 0.2$ (top) and $k = 0.6$ (bottom).

CHAPTER 6

Stability of peaked waves

The phenomenon of peaked waves and wave breaking may occur in any body of water and can be observed in regions near the shore as well as in the open sea. The term **peak** refers to a corner at the crest of the wave, i.e., the wave profile $u(t, x)$ is not smooth at its maximum at $x = \xi(t)$ but exhibits a finite jump discontinuity in the derivative:

$$\lim_{x \to \xi(t)^+} u_x(t, x) \neq \lim_{x \to \xi(t)^-} u_x(t, x).$$

To mathematically model the dynamics of such peaked waves, one can resort again to infinite-dimensional Hamiltonian systems written in evolution form. Compared to the theory we developed for smooth (at least C^1) solutions in Chapters 2, 3, 4, and 5, several challenges arise when studying peaked waves. For instance, recall that the translation symmetry present in Hamiltonian evolution equations implies that the kernel of the linearized Hamiltonian is generated by the symmetry mode given by the spatial derivative of the wave profile. This derivative is easy to compute when the wave is smooth, but becomes problematic for peaked waves whose profiles are only piecewise C^1 with finite jumps of the first derivative. Other difficulties that occur in the analysis of systems with peaked solutions are the lack of well-posedness results, nonstandard Hamiltonian formulations, and the failure of the spectral gap assumption.

In Section 6.1 we will explain how to deal with these challenges in the stability analysis of peaked traveling waves using two main examples: the b-CH equation and the reduced Ostrovsky equation, which we write in the form of a perturbed inviscid Burgers equation. For this particular class of models, we will argue in Section 6.2 that the peaked waves and the time evolution problem are well defined in the Banach space $X = H^1(\mathbb{R}) \cap W^{1,\infty}(\mathbb{R})$. This will allow us to adapt the linearization method to the presence of peaked singularities. Using the linearized equations we will address linear instability in Section 6.3 and spectral instability in Section 6.4. Regarding nonlinear instability, we will show in Section 6.5 that the energy method is applicable if conserved quantities are available but might give misleading (conditional) stability predictions for peaked solitary waves if the energy space does not coincide with the space in which the equation is locally well-posed. We show that nonlinear instability of peaked waves in X can be proven using the method of characteristics.

6.1. Hamiltonian systems with peaked waves

We will start with three examples of Hamiltonian systems which admit traveling peaked waves. These examples will allow us to explain the three main challenges which commonly occur in the stability analysis of peaked waves in Hamiltonian systems. Each example can be considered as a variation of the Korteweg–de Vries (KdV) equation (Example 3.11).

EXAMPLE 6.1. Peaked solitary waves, called **peakons**, arise in the b-family of the Camassa–Holm equations known as the b-CH model:

$$u_t - u_{xxt} + (b+1)uu_x = bu_x u_{xx} + uu_{xxx}, \tag{6.1}$$

where $(t,x) \in \mathbb{R} \times \mathbb{R}$, $u \in \mathbb{R}$ and $b \in \mathbb{R}$ is a parameter, see [**84, 87, 92**]. This equation is completely integrable when $b = 2$ (Camassa–Holm equation) and $b = 3$ (Degasperis–Procesi equation), in which case the equation models the unidirectional propagation of shallow water waves over a flat bottom and u can be interpreted as the horizontal fluid velocity, see [**62, 142**] and Example 2.7.

The b-CH model (6.1) can be rewritten in the momentum variable as

$$m_t + um_x + bu_x m = 0, \tag{6.2}$$

where $m := u - u_{xx}$ denotes the fluid's momentum. Let $\phi(x) = e^{-|x|}$ be Green's function satisfying $(1-\partial_x^2)\phi = 2\delta$ with δ being the Dirac delta distribution centered at 0. Since the equation (6.1) is equivalent to

$$(1-\partial_x^2)(u_t + uu_x) + buu_x + (3-b)u_x u_{xx} = 0,$$

Green's function ϕ allows us to write the b-CH model (6.1) in the evolution form

$$u_t + uu_x + \frac{1}{4}\phi' * \left[bu^2 + (3-b)u_x^2\right] = 0, \tag{6.3}$$

where $*$ denotes the convolution of two functions $(f*g)(x) := \int_{\mathbb{R}} f(x-y)g(y)dy$. It can be checked with explicit computations of the convolution integrals that $\phi(x) = e^{-|x|}$ satisfies the integral equation

$$-\phi + \frac{1}{2}\phi^2 + \frac{1}{4}\phi * \left[b\phi^2 + (3-b)(\phi')^2\right] = 0, \quad x \in \mathbb{R},$$

piecewise on both sides of the peak of ϕ at $x = 0$ and continuously at $x = 0$. Consequently, the C^1-piecewise function $u(t,x) = \phi(x-t)$ is a traveling peakon of the b-CH equation (6.3), see Figure 6.1. Note that if $u(t,x)$ is a solution of (6.3), so is $\gamma u(\gamma t, x)$ for every $\gamma \in \mathbb{R}$. Hence, $u(t,x) = \gamma\phi(x-\gamma t)$ is a one-parameter family of traveling peakons. This C^1-piecewise function $u(t,x) = \gamma\phi(x-\gamma t)$ corresponds to the δ-singular solution $m(t,x) = 2\gamma\delta(x-\gamma t)$ to the b-CH equation in the momentum form (6.2). In view of this scaling invariance, one can take $\gamma = 1$ without loss of generality.

EXAMPLE 6.2. The b-CH equation (6.1) can be considered in the symmetric periodic domain $(t,x) \in \mathbb{R} \times [-T,T]$ for some $T > 0$. The same evolution form (6.3) holds with the convolution integral taken over the periodic domain $[-T,T]$ and ϕ stands for the periodic extension of Green's function given by

$$\phi(x) = \sum_{n\in\mathbb{Z}} \phi_0(x-2nT), \quad \phi_0(x) := \frac{\cosh(T-|x|)}{\sinh(T)}\chi_{[-T,T]}(x),$$

where $\chi_S(x)$ denotes the characteristic function given by $\chi_S(x) = 1$ if $x \in S$ and $\chi_S(x) = 0$ if $x \notin S$. Note that the even function ϕ_0 satisfies $(1-\partial_x^2)\phi_0 = 2\delta$ on $[-T,T]$ together with the boundary conditions $\phi_0'(-T) = \phi_0'(T) = 0$ and the jump

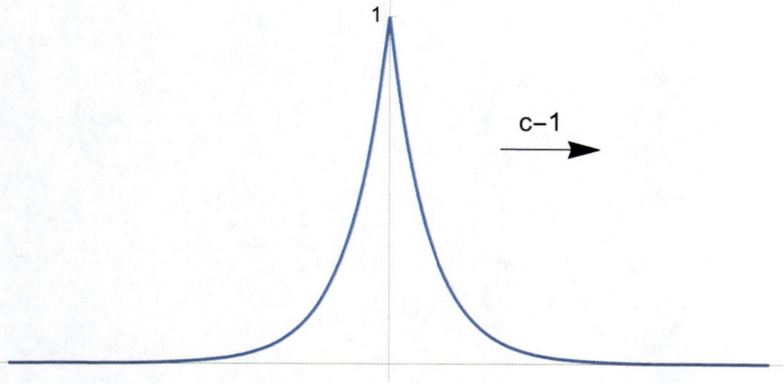

FIGURE 6.1. The graph of the peakon solution $u(t,x) = \phi(x-t)$ of the b-CH equation (6.3) with wave speed $c = 1$.

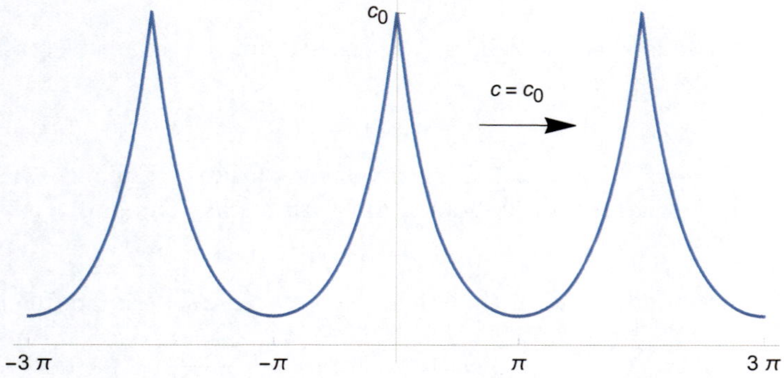

FIGURE 6.2. The graph of the peaked 2π-periodic wave solution $u(t,x) = \phi(x - c_0 t)$ of the b-CH equation (6.3) with wave speed $c = c_0 > 0$.

conditions at the δ-singularity $\phi_0'(0^\pm) = \mp 1$. It follows again from explicit computations of the convolution integral that Green's function ϕ satisfies the integral equation

$$-c_0\phi + \frac{1}{2}\phi^2 + \frac{1}{4}\phi * \left[b\phi^2 + (3-b)(\phi')^2\right] = d_0,$$

where

$$c_0 = \max_{x\in[-T,T]} \phi_0(x) = \coth(T)$$

and

$$d_0 = \frac{1}{2}(b-1)\left(\min_{x\in[-T,T]} \phi_0(x)\right)^2 = \frac{1}{2}(b-1)\,\mathrm{csch}^2(L),$$

see [**194**]. Consequently, the C^1-piecewise $2T$-periodic function $u(t,x) = \phi(x - c_0 t)$ is a traveling peaked periodic wave of the b-CH equation (6.3) with speed c_0, see Figure 6.2. Due to the scaling invariance, $u(t,x) = \gamma\phi(x - \gamma c_0 t)$ is a one-parameter family of traveling peaked periodic waves of the b-CH equation (6.3) with wave speed γc_0 for every $\gamma \in \mathbb{R}$.

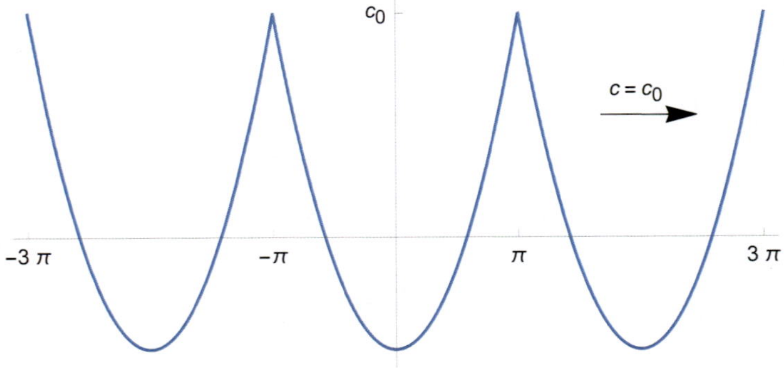

FIGURE 6.3. The graph of the peaked 2π-periodic wave solution $u(t,x) = \phi(x - c_0 t)$ of the reduced Ostrovsky equation (6.4) with speed $c = c_0 > 0$.

EXAMPLE 6.3. Peaked periodic waves also arise in the reduced Ostrovsky equation

$$(u_t + uu_x)_x = u, \tag{6.4}$$

where $(t,x) \in \mathbb{R} \times [-T,T]$ and $u \in \mathbb{R}$ is the velocity variable. The reduced Ostrovsky equation can be rewritten in the momentum variable $m := 1 - 3u_{xx}$ as

$$m_t + um_x + 3u_x m = 0. \tag{6.5}$$

It follows by direct substitution that the C^1-piecewise $2T$-periodic function

$$\phi(x) = \sum_{n \in \mathbb{Z}} \phi_0(x - 2nT), \quad \phi_0(x) := \frac{3x^2 - T^2}{18} \chi_{[-T,T]}(x) \tag{6.6}$$

satisfies

$$(\phi - c_0)\phi'' + (\phi')^2 = \phi, \qquad c_0 := \max_{x \in [-T,T]} \phi_0(x) = \frac{T^2}{9},$$

piecewise on both sides of the peaks of ϕ at $x \in \{(2n+1)T\}_{n \in \mathbb{Z}}$. Consequently, the C^1-piecewise $2T$-periodic function $u(t,x) = \phi(x - c_0 t)$ is a traveling peaked periodic wave of the reduced Ostrovsky equation (6.4) with speed c_0, see Figure 6.3. Since $\phi_0'(\pm T) = \pm\frac{T}{3}$, this C^1-piecewise function corresponds to the δ-singular solution

$$m(t,x) = 2T \sum_{n \in \mathbb{Z}} \delta(x - (2n+1)T - ct)$$

of the reduced Ostrovsky equation in the momentum form (6.5). For every period $T > 0$, the speed c_0 of the traveling peaked periodic wave is uniquely determined [**105**].

REMARK 6.1. The reduced Ostrovsky equation (6.4) also admits smooth periodic waves with zero mean, orbital stability of which can be handled with the methods of Chapter 3, see [**104, 121**]. Similarly, orbital stability of smooth periodic waves can be studied in the Camassa–Holm equation (6.1) with $b = 2$ [**108**], $b = 3$ [**107**], and for every $b > 1$ [**91**]. Orbital stability of smooth solitary waves at the constant nonzero background can be handled with the methods of Chapter 2 for the b-CH model (6.1) for every $b > 1$ [**65, 168, 177**], see also Example 2.7.

The first challenge occurring in the analysis of the Hamiltonian systems with peaked waves is the lack of local and global well-posedness results in the energy space (Definition 2.1). Assumptions 2.1 and 3.1 were taken for granted in the previous chapters because the local and global well-posedness theory in the energy space is well developed for standard examples such as the nonlinear wave equation, the nonlinear Schrödinger equation, and the generalized Korteweg–de Vries equation [**181**].

Corresponding results for equations that admit peaked solutions are often only partially available and much more recent. For the b-CH equation with peaked solitary waves (Example 6.1), the integral formulation (6.3) suggests that $H^1(\mathbb{R})$ might be a suitable space for the weak solutions in Definition 2.1. However, the initial-value problem for the evolution equation (6.3) was shown to be well-posed in $H^s(\mathbb{R})$ for $s > \frac{3}{2}$ and ill-posed in $H^s(\mathbb{R})$ for $s < \frac{3}{2}$ [**129, 211**] due to the lack of continuous dependence and norm inflation. It was shown that the initial-value problem is locally well-posed in the space $H^1(\mathbb{R}) \cap W^{1,\infty}(\mathbb{R})$ [**182**]. Since $\phi(x) = e^{-|x|}$ belongs to this space, it is still a suitable space for the stability analysis of peaked waves.

When a weak solution evolves in time, it may blow up in finite time in the form of **wave breaking** as follows, see for instance [**59, 245**].

DEFINITION 6.1. We say that a weak solution $u(t,x)$ breaks in a finite time τ_0 if

$$\lim_{t\uparrow\tau_0} \|u(t,\cdot)\|_{L^\infty} < \infty, \quad \lim_{t\uparrow\tau_0} \|u(t,\cdot)\|_{W^{1,\infty}} = \infty. \tag{6.7}$$

REMARK 6.2. Definition 6.1 implies that the wave profile remains bounded in the supremum norm but its gradient (the wave slope) becomes unbounded in the supremum norm as the solution approaches its maximum time of existence τ_0. Peaked waves with bounded profile and finite slope can be thought of as bordering states between the global (in time) weak solutions on the one hand and the local weak solutions that break as $t \uparrow \tau_0$ on the other hand.

REMARK 6.3. There exists another type of singular traveling wave solution of the b-CH equation (6.3) in $H^1(\mathbb{R})$ called **cusped wave** which has an unbounded first derivative at the crest, see for instance [**174, 175**]. Looking at the wave breaking criterion (6.7), it seems natural that such waves with infinite slopes should exist. Unfortunately, the initial-value problem for the cusped waves in $H^1(\mathbb{R})$ is not locally well-posed and therefore their stability cannot be studied.

The second challenge arising in the analysis of Hamiltonian systems with peaked waves is the lack of standard Hamiltonian structure. For the b-CH equation, a variety of different Hamiltonian structures is available only for certain values of b. This makes the stability analysis of peaked waves for other values of b more difficult since the stability analysis relies on the Hamiltonian formulation of the evolution problem.

For the CH equation, which corresponds to $b = 2$, three Hamiltonian formulations are available. The first (standard) Hamiltonian formulation is similar to the Hamiltonian form (2.1) and is given by

$$\frac{du}{dt} = J_1 \nabla H(u), \quad J_1 := -(1-\partial_x^2)^{-1}\partial_x, \quad H(u) := \frac{1}{2}\int_{\mathbb{R}} (u^3 + u u_x^2) dx. \tag{6.8}$$

The second (nonstandard) Hamiltonian formulation is given in terms of the momentum variable $m := u - u_{xx}$ by

$$\frac{dm}{dt} = J_2 \nabla_m E(u), \quad J_2 := -(m\partial_x + \partial_x m), \quad E(u) = \frac{1}{2}\int_{\mathbb{R}} (u^2 + u_x^2) dx, \tag{6.9}$$

where $\nabla_m E(u) = (1 - \partial_x^2)^{-1} m = u$ is obtained by expressing $E(u)$ in the variable m as

$$E(u) = \frac{1}{2}\int_{\mathbb{R}} (u^2 + u_x^2) dx = \frac{1}{2}\int_{\mathbb{R}} um dx = \frac{1}{2}\int_{\mathbb{R}} m(1 - \partial_x^2)^{-1} m dx.$$

The third (nonstandard) Hamiltonian formulation exists in the form

$$\frac{dm}{dt} = J_3 \nabla_m M(u), \quad M(u) = \int_{\mathbb{R}} u dx, \tag{6.10}$$

with

$$J_3 := -(2m\partial_x + m_x)(1 - \partial_x^2)^{-1}\partial_x^{-1}(2m\partial_x + m_x),$$

where $\nabla_m M(u) = 1$ since $M(u) = M(m)$. The functionals $H(u)$ in (6.8), $E(u)$ in (6.9), and $M(u)$ in (6.10) are the first three conserved quantities the infinite hierarchy of infinite conservation laws of CH due to its integrability, see [**58**].

For $b \neq 2$, only the third (nonstandard) Hamiltonian formulation (6.10) can be extended to other values of b as the following Hamiltonian structure for the b-CH equation

$$\frac{dm}{dt} = J_b \nabla_m M(u), \quad M(u) = \int_{\mathbb{R}} u dx, \tag{6.11}$$

where for $b \neq 1$ the skew-symmetric operator is given by

$$J_b := -\frac{1}{b-1}(bm\partial_x + m_x)(1 - \partial_x^2)^{-1}\partial_x^{-1}(bm\partial_x + (b-1)m_x).$$

The Hamiltonian formulation (6.11) holds also for $b = 1$ but with a different expression for $J_{b=1}$ [**84**].

REMARK 6.4. Choosing $b = 3$ in the b-CH equation yields the Degasperis–Procesi equation. Two Hamiltonian formulations are available in this case. The first Hamiltonian formulation is given by

$$\frac{du}{dt} = J\nabla H(u), \quad J := -\frac{1}{2}(4 - \partial_x^2)(1 - \partial_x^2)^{-1}\partial_x, \quad H(u) := \frac{1}{3}\int_{\mathbb{R}} u^2 dx. \tag{6.12}$$

The second Hamiltonian formulation is given by (6.11) for $b = 3$. Besides the conserved quantities $H(u)$ and $M(u)$, there exists another conserved quantity given by

$$E(u) := \frac{1}{2}\int_{\mathbb{R}} u(1 - \partial_x^2)(4 - \partial_x^2)^{-1} u dx. \tag{6.13}$$

The third challenge in the analysis of Hamiltonian systems with peaked waves is the violation of the spectral gap condition (2.50) for the linearized operator at the traveling peaked wave. To illustrate how the spectral gap condition (2.50) fails to hold, let us consider the reduced Ostrovsky equation (Example 6.3) on the normalized periodic domain $[-\pi, \pi]$. The standard Hamiltonian formulation is given by

$$\frac{du}{dt} = J\nabla H(u), \quad J := \partial_x, \quad H(u) := -\frac{1}{2}\oint ((\partial_x^{-1}u)^2 + \frac{1}{3}u^3) dx, \tag{6.14}$$

where ∂_x^{-1} is defined by means of the Fourier series on the 2π-periodic functions with zero mean:

$$f(x) = \sum_{k\in\mathbb{Z}\backslash\{0\}} f_k e^{ikx}, \quad [\partial_x^{-1} f](x) = \sum_{k\in\mathbb{Z}\backslash\{0\}} (ik)^{-1} f_k e^{ikx}.$$

To simplify notation we denote the subspace of 2π-periodic functions in L^2_{per} with zero mean by $\dot{L}^2_{\text{per}}$. Besides the Hamiltonian $H(u)$, there exist two other conserved quantities of the reduced Ostrovsky equation (6.4) bearing the names mass and momentum, which are given by

$$M(u) = \oint u dx, \quad P(u) = \frac{1}{2}\oint u^2 dx. \tag{6.15}$$

Traveling waves of the reduced Ostrovsky equation with profile $u(t,x) = u_{c,b}(x-ct)$ satisfy the differential equation

$$(u-c)u'' + (u')^2 = u,$$

which after integrating twice becomes

$$\frac{1}{2}u^2 - cu = \partial_x^{-2} u + b,$$

where b is a constant of integration. This is the Euler–Lagrange equation for the augmented energy

$$\Lambda_{c,b}(u) := H(u) + cP(u) + bM(u).$$

The associated linearized operator $\mathcal{L}_{c,b} : \dot{L}^2_{\text{per}} \mapsto \dot{L}^2_{\text{per}}$ obtained from linearizing the equation at the traveling wave $u_{c,b}$ is given by

$$\mathcal{L}_{c,b} := \partial_x^{-2} + c - u_{c,b}.$$

The peaked 2π-periodic wave (6.6) with $T = \pi$ is defined for

$$c = c_0 := \frac{\pi^2}{9} \quad \text{and} \quad b = b_0 := -\frac{c_0^2}{2} = -\frac{\pi^4}{162}.$$

Let us now consider the spectrum of $\mathcal{L}_{c,b}$, see Definition B.2. Since $\partial_x^{-2} : \dot{L}^2_{\text{per}} \mapsto \dot{L}^2_{\text{per}}$ is a compact operator in the trace class, Kato's theorem (see [**154**] and also Lemma 6.5) implies that the spectrum of $\mathcal{L}_{c,b}$ consists of the continuous spectrum of the multiplicative operator $c - u_{c,b}$ which coincides with the range of the function $x \mapsto c - u_{c,b}(x)$, and a finite number of isolated eigenvalues outside the continuous spectrum. For the smooth periodic waves, we have that $c - u_{c,b}(x) > 0$ for every $x \in [-\pi, \pi]$ and hence the spectral gap condition (2.50) holds. However, for peaked periodic wave (6.6) with $c = c_0$, $b = b_0$ and $\phi_0 = u_{c_0,b_0}$, we have that $c_0 - \phi_0(x) \geq 0$ for every $x \in [-\pi, \pi]$ with $\phi_0(\pm\pi) = \frac{\pi^2}{9} = c_0$. Therefore, the continuous spectrum of $\mathcal{L}_0 := \mathcal{L}_{c_0,b_0}$ touches the zero eigenvalue and hence the spectral gap condition (2.50) fails to hold. Consequently, Lemma 2.4, Theorem 2.7, Theorem 2.8, and other related results of Chapters 2 and 3 are not applicable to the peaked waves of the reduced Ostrovsky equation. Similar obstacles exist for the peaked waves in other examples of relevant Hamiltonian systems.

In the following we will show how to develop the stability analysis of peaked traveling waves in spite of the three challenges explained above.

6.2. Peaked waves and wave breaking

We consider weak solutions satisfying Definition 2.1 in the function space $X = H^1(\mathbb{R}) \cap W^{1,\infty}(\mathbb{R})$, where the wave slope is allowed to have finite jump discontinuities. In order to explain the relevance of this space for the local well-posedness analysis, we consider an abstract initial-value problem for the perturbed inviscid Burgers equation:

$$\text{(6.16)} \qquad \begin{cases} u_t + uu_x = Q, & t > 0, \\ u|_{t=0} = u_0, \end{cases}$$

where $u = u(t,x)$, u_0 is the initial data and $Q = Q(t,x,u(t,x))$. The initial-value problem (6.16) is locally well-posed in Sobolev spaces $H^s(\mathbb{R})$ of sufficiently high regularity $s > \frac{3}{2}$, provided that $Q \in C^0(\mathbb{R}, H^s(\mathbb{R}))$ with the same s. This is well studied since the work of T. Kato [**155**]. By Sobolev's embedding theorem, if $f \in H^s(\mathbb{R})$ with $s > \frac{3}{2}$, then $f \in C^1(\mathbb{R}) \cap W^{1,\infty}(\mathbb{R})$ with f and f' decaying to zero at infinity. However, peaked waves with jump discontinuities of their slopes do not belong to these spaces but belongs to the space $H^1(\mathbb{R}) \cap W^{1,\infty}(\mathbb{R})$.

The following lemma gives the local well-posedness result for (6.16) in $H^1(\mathbb{R}) \cap W^{1,\infty}(\mathbb{R})$.

LEMMA 6.1. *Assume that $Q \in C^0(\mathbb{R}, H^1(\mathbb{R}) \cap W^{1,\infty}(\mathbb{R}))$. For every $u_0 \in H^1(\mathbb{R}) \cap W^{1,\infty}(\mathbb{R})$, there exist the maximal existence time $\tau_0 > 0$ and the unique solution*

$$u \in C^0([0,\tau_0), H^1(\mathbb{R}) \cap W^{1,\infty}(\mathbb{R}))$$

to the initial-value problem (6.16)*, which depends continuously on the initial data u_0. Moreover, $\tau_0 < \infty$ if and only if $\lim_{t \uparrow \tau_0} \|u_x(t,\cdot)\|_{L^\infty} = \infty$, that is, wave breaking occurs in the sense of Definition* 6.1.

PROOF. A robust way of solving the initial-value problem (6.16) is the method of characteristic curves, which reduces the evolution problem to a system of ordinary differential equations with a parameter. Let $x = X(t,s)$ be the family of characteristic curves satisfying the initial-value problem

$$\text{(6.17)} \qquad \begin{cases} \frac{d}{dt} X = u(t,X), & t > 0, \\ X|_{t=0} = s. \end{cases}$$

Let $U(t,s) = u(t,X(t,s))$ be defined along the characteristic curves. It follows from (6.16) and (6.17) that U is found from the initial-value problem

$$\text{(6.18)} \qquad \begin{cases} \frac{d}{dt} U = Q(t,X), & t > 0, \\ U|_{t=0} = u_0(s). \end{cases}$$

Recall that if $f \in H^1(\mathbb{R}) \cap W^{1,\infty}(\mathbb{R})$, then f is globally Lipschitz. By the existence and uniqueness theory for differential equations, there exist unique solutions $X(\cdot,s) \in C^1([0,\tau_0))$ and $U(\cdot,s) \in C^1([0,\tau_0))$ to the initial-value problems (6.17) and (6.18), respectively, for every fixed $s \in \mathbb{R}$ as long as $u \in C^0([0,\tau_0), H^1(\mathbb{R}) \cap W^{1,\infty}(\mathbb{R}))$. Moreover, it follows from the continuous dependence of solutions on initial data that $U(t,\cdot) \in H^1(\mathbb{R}) \cap W^{1,\infty}(\mathbb{R})$ for $t \in [0,\tau_0)$ since $u_0 \in H^1(\mathbb{R}) \cap W^{1,\infty}(\mathbb{R})$. It remains to ensure that $u \in C^0([0,\tau_0), H^1(\mathbb{R}) \cap W^{1,\infty}(\mathbb{R}))$ from the relations

$$U(t,s) = u(t.X(t,s)) \quad \text{and} \quad U_s(t,s) = u_x(t,X(t,s))X_s(t,s).$$

Differentiating (6.17) in s and solving the linear differential equation gives

$$X_s(t,s) = e^{\int_0^t u_x(t',X(t',s))dt'}.$$

The condition that $u(t,\cdot) \in H^1(\mathbb{R}) \cap W^{1,\infty}(\mathbb{R})$ is equivalent to the condition that $U(t,\cdot) \in H^1(\mathbb{R}) \cap W^{1,\infty}(\mathbb{R})$ for $t \in [0,\tau_0)$ if and only if the mapping $s \mapsto X(t,s)$ is an increasing diffeomorphism and there exist $0 < C_- \leq C_+ < \infty$ such that

$$C_- \leq X_s(t,s) \leq C_+, \quad t \in [0,\tau_0), \quad s \in \mathbb{R}. \tag{6.19}$$

If the maximal existence time τ_0 is finite, then assuming that $\lim_{t\uparrow\tau_0} \|u_x(t,\cdot)\|_{L^\infty} < \infty$ implies that the solutions $X(\cdot,s) \in C^1([0,\tau_0))$ and $U(\cdot,s) \in C^1([0,\tau_0))$ can be continued beyond τ_0 with the property (6.19). This is a contradiction with τ_0 being the maximal time and hence $\lim_{t\uparrow\tau_0} \|u_x(t,\cdot)\|_{L^\infty} = \infty$. On the other hand, if $\lim_{t\uparrow\tau_0} \|u_x(t,\cdot)\|_{L^\infty} = \infty$, then τ_0 is the finite maximal time for the solution $u \in C^0([0,\tau_0), H^1(\mathbb{R}) \cap W^{1,\infty}(\mathbb{R}))$. Combining both implications, we conclude that $\tau_0 < \infty$ if and only if $\lim_{t\uparrow\tau_0} \|u_x(t,\cdot)\|_{L^\infty} = \infty$. □

REMARK 6.5. Although the statement of Lemma 6.1 was formulated and proven in $H^1(\mathbb{R}) \cap W^{1,\infty}(\mathbb{R})$, exactly the same result can be obtained in $H^1_{\text{per}} \cap W^{1,\infty}$ for the periodic domain of any fixed period.

EXAMPLE 6.4. For the b-CH equation (6.3) (Example 6.1), we define

$$Q[u] := -\frac{1}{4}\phi' * \left[bu^2 + (3-b)u_x^2\right],$$

where $\phi(x) = e^{-|x|}$. Since $\phi - \phi'' = 2\delta$, we obtain

$$\partial_x Q[u] := \frac{1}{2}\left[bu^2 + (3-b)u_x^2\right] - \frac{1}{4}\phi * \left[bu^2 + (3-b)u_x^2\right].$$

Since $\phi \in H^1(\mathbb{R}) \cap W^{1,\infty}(\mathbb{R})$, Young's convolution inequality $\|f*g\|_{L^p} \leq \|f\|_{L^p}\|g\|_{L^1}$ yields

$$\|Q(u)\|_{L^2\cap L^\infty} \leq \frac{1}{4}\|\phi'\|_{L^2\cap L^\infty}\left(|b|\|u\|^2_{L^2} + |b-3|\|u_x\|^2_{L^2}\right)$$

and

$$\begin{aligned}\|\partial_x Q(u)\|_{L^2\cap L^\infty} \leq\ & \frac{1}{4}\|\phi\|_{L^2\cap L^\infty}\left(|b|\|u\|^2_{L^2} + |b-3|\|u_x\|^2_{L^2}\right) \\ & + \frac{1}{2}\left(|b|\|u\|_{L^\infty}\|u\|_{L^2\cap L^\infty} + |b-3|\|u_x\|_{L^\infty}\|u_x\|_{L^2\cap L^\infty}\right).\end{aligned}$$

Hence the mapping $Q(u) : H^1(\mathbb{R}) \cap W^{1,\infty}(\mathbb{R}) \mapsto H^1(\mathbb{R}) \cap W^{1,\infty}(\mathbb{R})$ is continuous and we can apply Lemma 6.1 to conclude that the b-CH equation has a unique weak solution $u \in C^0([0,\tau_0), H^1(\mathbb{R}) \cap W^{1,\infty}(\mathbb{R}))$ and the maximal existence time is finite only in the case of wave breaking.

EXAMPLE 6.5. For the reduced Ostrovsky equation (6.4), we define

$$Q[u] := \partial_x^{-1} u,$$

where $\partial_x^{-1} : \dot{L}^2_{\text{per}} \mapsto \dot{L}^2_{\text{per}}$ is well-defined on the space of periodic functions with zero mean. If $u \in \dot{H}^1_{\text{per}} \cap W^{1,\infty}$, then $Q[u] \in \dot{H}^2_{\text{per}} \cap W^{2,\infty} \subset \dot{H}^1_{\text{per}} \cap W^{1,\infty}$. Hence, the mapping $Q(u) : H^1_{\text{per}} \cap W^{1,\infty} \mapsto H^1_{\text{per}} \cap W^{1,\infty}$ is continuous and the assumptions of Lemma 6.1 are satisfied for the solution $u \in C^0([0,\tau_0), H^1_{\text{per}} \cap W^{1,\infty})$ in the periodic setting.

EXAMPLE 6.6. Solutions of the initial-value problem for the inviscid Burgers equation

$$\begin{cases} u_t + uu_x = 0, & t > 0, \\ u|_{t=0} = u_0, \end{cases}$$

always break in finite time if $u_0 \in H^1(\mathbb{R}) \cap W^{1,\infty}(\mathbb{R})$. Indeed, it follows from the method of characteristics that the solutions of (6.17) and (6.18) with $Q = 0$ are given by

$$X(t, s) = s + tu_0(s) \quad \text{and} \quad U(t, s) = u_0(s).$$

Therefore,

$$X_s(t, s) = 1 + tu_0'(s) \quad \text{and} \quad u_0'(s) = u_x(t, s + tu_0(s))X_s(t, s).$$

Since there exist $s \in \mathbb{R}$ such that $u_0'(s) < 0$ if $u_0 \in H^1(\mathbb{R}) \cap W^{1,\infty}(\mathbb{R})$, $X_s(t, s)$ vanishes for some $s \in \mathbb{R}$ at the maximal existence time

$$\tau_0 = \frac{1}{\sup_{s\in\mathbb{R}:\, u_0'(s)<0} |u_0'(s)|} > 0$$

so that the solution satisfies $\|u(t, \cdot)\|_{L^\infty} = \|u_0\|_{L^\infty} < \infty$ and $\|u_x(t, \cdot)\|_{L^\infty} \to \infty$ as $t \uparrow \tau_0$.

By using apriori energy estimates, one can obtain the following precise blow-up criterion for solutions of the initial-value problem (6.16). This criterion is in line with the fact that the wave profile of solutions of the inviscid Burgers equation (Example 6.6) has a negative slope at breaking time, a so-called plunging breaker, see for instance [**59**].

LEMMA 6.2. *Let $u \in C^0([0, \tau_0), H^1(\mathbb{R}) \cap W^{1,\infty}(\mathbb{R}))$ be the solution of* (6.16) *in Lemma* 6.1 *with $\tau_0 < \infty$. Then, $\lim_{t\uparrow\tau_0} \|u(t, \cdot)\|_{H^1\cap W^{1,\infty}} = \infty$ if and only if $\lim_{t\uparrow\tau_0} \inf_{x\in\mathbb{R}} u_x(t, x) = -\infty$.*

PROOF. If $\inf_{x\in\mathbb{R}} u_x(t, x) \to -\infty$ as $t \uparrow \tau_0$ then $\lim_{t\uparrow\tau_0} \|u(t, \cdot)\|_{W^{1,\infty}} = \infty$ which implies the finite-time blowup in $H^1(\mathbb{R}) \cap W^{1,\infty}(\mathbb{R})$. To see that the other implication holds we assume that $\lim_{t\uparrow\tau_0} \|u(t, \cdot)\|_{H^1\cap W^{1,\infty}} = \infty$ and there is $M > 0$ such that

$$\inf_{x\in\mathbb{R}} u_x(t, x) \geq -M, \quad \forall t \in [0, \tau_0). \tag{6.20}$$

We will show that $\|u(t, \cdot)\|_{H^1\cap W^{1,\infty}}$ remains bounded as $t \uparrow \tau_0$ in contradiction with the assumption of $\lim_{t\uparrow\tau_0} \|u(t, \cdot)\|_{H^1\cap W^{1,\infty}} = \infty$.

Applying density arguments, we can approximate the initial value $u_0 \in H^1(\mathbb{R}) \cap W^{1,\infty}(\mathbb{R})$ by a sequence of functions $\{u_0^n\}_{n\in\mathbb{N}} \in H^2(\mathbb{R})$ such that $\lim_{n\to\infty} u_0^n = u_0$ in $H^1(\mathbb{R}) \cap W^{1,\infty}(\mathbb{R})$. Furthermore, let $u^n \in C^0([0, \tau_0^n), H^2(\mathbb{R}))$ be the unique solution of the initial-value problem (6.16) with initial data $u_0^n \in H^2(\mathbb{R})$. By uniqueness of solutions, we have $\lim_{n\to\infty} \tau_0^n = \tau_0$. By Sobolev's embedding, if $u^n \in C^0([0, \tau_0^n), H^2(\mathbb{R}))$, then $u^n \in C^0([0, \tau_0^n), C^1(\mathbb{R}) \cap W^{1,\infty}(\mathbb{R}))$ for any $t \in [0, \tau_0^n)$ such that $|u^n(t, x)| + |u_x^n(t, x)| \to 0$ as $|x| \to \infty$. Integrating by parts yields

$$\frac{d}{dt}\frac{1}{2}\int_{\mathbb{R}} (u^n)^2 dx = \int_{\mathbb{R}} u^n Q dx \leq \|u^n\|_{L^2} \|Q\|_{L^2}$$

and

$$\frac{d}{dt}\frac{1}{2}\int_{\mathbb{R}} (u_x^n)^2 dx = -\frac{1}{2}\int_{\mathbb{R}} (u_x^n)^3 dx + \int_{\mathbb{R}} u_x^n Q_x dx \leq \frac{M}{2}\|u_x^n\|_{L^2}^2 + \|u_x^n\|_{L^2} \|Q_x\|_{L^2}.$$

where we have used (6.20). We obtain

$$\|u^n(t,\cdot)\|_{L^2} \leq \|u_0^n\|_{L^2} + \int_0^t \|Q(t',\cdot)\|_{L^2} dt'$$

and

$$\|u_x^n(t,\cdot)\|_{L^2} \leq \left(\|(u_0^n)_x\|_{L^2} + \int_0^t \|Q_x(t',\cdot)\|_{L^2} dt' \right) e^{\frac{M}{2}t},$$

where we have used the integral formulation of Gronwall's inequality in the second inequality. Since $\|u_0^n\|_{H^1}$ converges to $\|u_0\|_{H^1}$ as $n \to \infty$, the bounds above imply that the $\|u(t,\cdot)\|_{H^1}$ in Lemma 6.1 remains bounded as $t \uparrow \tau$.

It remains to show that $\sup_{x\in\mathbb{R}} u_x(t,x)$ remains bounded as $t \uparrow \tau$ in contradiction with the assumption of $\lim_{t\uparrow\tau_0} \|u(t,\cdot)\|_{H^1\cap W^{1,\infty}} = \infty$. Differentiating of (6.16) in x yields

$$u_{tx} + uu_{xx} + u_x^2 = Q_x$$

Define $V(t,s) := u_x(t, X(t,s))$ along the characteristic curves. Then, it satisfies

$$\begin{cases} V_t = -V^2 + Q_x(t,X) \leq Q_x(t,X), & t > 0, \\ V|_{t=0} = u_0'(s), \end{cases}$$

so that $V(t,s) \leq u_0'(s) + \int_0^t Q_x(t', X(t',s))dt'$ yields the bound

$$\sup_{t'\in[0,t]} u_x(t',\cdot) \leq \|u_0'\|_{L^\infty} + \int_0^t \sup_{t'\in[0,t]} \|Q_x(t',\cdot)\|_{L^\infty} dt'.$$

This expression remains bounded for every $t > 0$ since $Q \in W^{1,\infty}$, and therefore, if $\lim_{t\uparrow\tau_0} \|u(t,\cdot)\|_{H^1\cap W^{1,\infty}} = \infty$, then $\lim_{t\uparrow\tau_0} \inf_{x\in\mathbb{R}} u_x(t,x) = -\infty$. ☐

REMARK 6.6. The energy estimates of Lemma 6.2 can be extended to $H^s(\mathbb{R})$ for every integer $s \geq 2$. It follows from these estimates that the smooth solutions in $H^s(\mathbb{R})$ blow up in finite time if and only if $\lim_{t\to\tau_0^+} \inf_{x\in\mathbb{R}} u_x(t,x) = -\infty$. Hence, smooth solutions to the initial-value problem (6.16) can be continued in time as long as the wave slope $u_x(t,x)$ remains finite and bounded from below.

For the dispersive generalizations of the inviscid Burgers equation with nonzero Q, wave breaking occurs if the initial data has a wave profile with sufficiently steep negative slope. This was shown for instance in [**60, 61**] for the Camassa–Holm equation (Example 6.4) and in [**185**] for the reduced Ostrovsky equation (Example 6.5). Other dispersive equations generalizing the inviscid Burgers equation for which wave breaking has been studied include the fractional Korteweg–de Vries equation [**160**], the Whitham equations [**134**] and their coupled systems [**135**], the Fornberg–Whitham equation [**127**] and the equation for the free surface elevation corresponding to the CH equation [**62**].

The following theorem gives a sufficient condition for wave breaking in the reduced Ostrovsky equation (Example 6.5, [**185**]). The proof is based on the method of characteristics.

THEOREM 6.1. *Let $u \in C^0([0,\tau_0), \dot{H}^1_{\text{per}} \cap W^{1,\infty})$ be the unique solution of the reduced Ostrovsky equation* (6.4) *on the normalized periodic domain $\mathbb{T} = \mathbb{R}/\{2\pi\mathbb{Z}\}$. If $u_0 \in \dot{H}^1_{\text{per}} \cap W^{1,\infty}$ satisfies*

$$u_0'(s_0) \leq -2\sqrt{\|u_0\|_{L^\infty} + \sqrt{2\pi}\,\tau_1\,\|u_0\|_{L^2}},$$

for some $s_0 \in \mathbb{T}$ *and* $0 < \tau_1 < \infty$, *then wave breaking occurs in finite time* $\tau_0 \in (0, \tau_1)$, *and the solution blows up according to*

$$\lim_{t \uparrow \tau_0} (\tau_0 - t) \inf_{x \in \mathbb{T}} u_x(t, x) = -1. \tag{6.21}$$

PROOF. The proof consists of three steps. In the first step, we derive a bound on the supremum norm of the solution, which does not blow up in finite time. In the second step, we specify the sufficient condition on u_0 such that $\tau_0 < \infty$. In the third step, we prove the precise blow-up rate given by (6.21).

STEP 1. The solution $u \in C^0([0, \tau_0), \dot{H}^1_{\rm per} \cap W^{1,\infty})$ satisfies the bound

$$\sup_{t' \in [0,t]} \|u(t', \cdot)\|_{L^\infty} \leq \|u_0\|_{L^\infty} + \sqrt{2\pi}\, t\, \|u_0\|_{L^2}, \quad \forall t \in [0, \tau_0). \tag{6.22}$$

PROOF OF STEP 1. It follows from the proof of Lemma 6.1 that the mapping $\mathbb{T} \ni s \mapsto X(t, s) \in \mathbb{T}$ is invertible for any $t \in [0, \tau_0)$, since X is an increasing diffeomorphism. Therefore we have for $u(t, X(t, s)) = U(t, s)$ that

$$\sup_{t' \in [0,t]} \sup_{x \in \mathbb{T}} |u(t', x)| = \sup_{t' \in [0,t]} \sup_{s \in \mathbb{T}} |U(t', s)|, \quad t \in [0, \tau_0).$$

Since $Q = \partial_x^{-1} u(t, x) \in C([0, \tau_0); \dot{H}^2_{\rm per})$ is a mean-zero periodic function of x for each $t \in [0, \tau_0)$ (Example 6.5), there exists a $\xi(t) \in \mathbb{T}$ such that $\partial_x^{-1} u(t, \xi(t)) = 0$. Then for any $x \in \mathbb{T}$ and $t \in [0, \tau_0)$, we have

$$|\partial_x^{-1} u(t, x)| \leq \left| \int_{\xi(t)}^x u(t, x)\, dx \right| \leq \int_{\mathbb{T}} |u(t, x)| dx \leq \sqrt{2\pi} \|u_0\|_{L^2},$$

where we have used the Cauchy–Schwarz inequality and the momentum conservation $\|u(t, \cdot)\|^2_{L^2} = \|u_0\|^2_{L^2}$ in (6.15). Using the integral equation

$$U(t, s) = u_0(s) + \int_0^t Q(t', s) dt', \quad t \in [0, \tau_0),$$

we obtain

$$\begin{aligned} \sup_{t' \in [0,t]} \sup_{x \in \mathbb{T}} |u(t', x)| &\leq \|u_0\|_{L^\infty} + t \sup_{t' \in [0,t]} \sup_{x \in \mathbb{T}} \left|\partial_x^{-1} u(t', x)\right| \\ &\leq \|u_0\|_{L^\infty} + \sqrt{2\pi}\, t\, \|u_0\|_{L^2}, \end{aligned}$$

and the bound (6.22) is proved. □

STEP 2. Let τ_1 be the smallest positive root of

$$2\tau_1 h(\tau_1) = \log(3) \tag{6.23}$$

where

$$h(t) := \sqrt{\|u_0\|_{L^\infty} + \sqrt{2\pi}\, t\, \|u_0\|_{L^2}}$$

and assume that there is a $s_0 \in \mathbb{T}$ such that

$$u_0'(s_0) \leq -2h(\tau_1). \tag{6.24}$$

Then, the maximal existence time τ_0 belongs to $(0, \tau_1)$.

PROOF OF STEP 2. Define $V(t,s) := u_x(t, X(t,s))$. Since $u_{tx} + uu_{xx} = u - u_x^2$, this variable satisfies the initial-value problem

$$\begin{cases} \dot{V} = U - V^2, & t > 0, \\ V|_{t=0} = u_0', \end{cases}$$

almost everywhere on $\mathbb{T}$, where $\dot{}$ denotes differentiation with respect to t. It follows from (6.22) that V satisfies the following differential inequality almost everywhere on $\mathbb{T}$:

$$\dot{V} \le -V^2 + \|u_0\|_{L^\infty} + \sqrt{2\pi} t \|u_0\|_{L^2}, \quad t \in (0, \tau_0).$$

It follows from (6.23) and (6.24) that $V_0(t) := V(t, s_0)$ satisfies

$$\begin{cases} \dot{V}_0(t) \le -V_0^2(t) + h^2(\tau_1), & t \in [0, \tau_1] \cap [0, \tau_0), \\ V_0(0) \le -2h(\tau_1). \end{cases}$$

By the comparison principle for differential equations, we have

$$V_0(t) \le V_+(t) < 0, \quad t \in [0, \tau_1] \cap [0, \tau_0),$$

where $V_+(t)$ solves the equation

$$\begin{cases} \dot{V}_+(t) = -V_+^2(t) + h^2(\tau_1), & t \in [0, \tau_1), \\ V_+(0) = -2h(\tau_1). \end{cases}$$

This equation admits an implicit solution

$$\frac{V_+(t) + h(\tau_1)}{V_+(t) - h(\tau_1)} = \frac{V_+(0) + h(\tau_1)}{V_+(0) - h(\tau_1)} e^{2h(\tau_1)t}, \quad t \in [0, \tau_1),$$

which satisfies $\dot{V}_+(t) < 0$. In view of (6.23), this implies that

$$\frac{V_+(t) + h(\tau_1)}{V_+(t) - h(\tau_1)} = \frac{1}{3} e^{2h(\tau_1)t} \to 1, \quad \text{as} \quad t \uparrow \tau_1,$$

so that $\lim_{t \uparrow \tau_1} V_+(t) = -\infty$. Since $V_+(t)$ is a supersolution for $V_0(t)$ for $t \in [0, \tau_1)$, there is $\tau_0 \in (0, \tau_1)$ such that $\lim_{t \uparrow \tau_0} \inf_{x \in \mathbb{T}} u_x(t, x) = -\infty$. □

STEP 3. If $\tau_0 < \infty$, then the solution satisfies the limits (6.21).

PROOF OF STEP 3. Let $m(t) := \inf_{x \in \mathbb{T}} u_x(t, x)$ and $\xi(t)$ be the argument of the infimum. In view of the assumption on $\tau_0 < \infty$ and Step 2, we have $\lim_{t \uparrow \tau_0} m(t) = -\infty$. For every $t \in [0, \tau_0)$, $m(t) = u_x(t, \xi(t))$ satisfies

$$\dot{m}(t) = -m^2(t) + u(t, \xi(t)) \quad \text{a.e.} \quad t \in (0, \tau_0).$$

By using the bound (6.22), we obtain

$$-m^2(t) - h^2(t) \le \dot{m}(t) \le -m^2(t) + h^2(t) \quad \text{a.e.} \quad t \in (0, \tau_0),$$

where $h(t) \le h(\tau_0)$ for every $t \in (0, \tau_0)$. Since $\lim_{t \uparrow \tau_0} m(t) = -\infty$, we have that for every $\varepsilon \in (0, 1)$ there exists $t_0 \in [0, \tau_0)$ such that

$$m(t_0) < -\frac{\sqrt{1+\varepsilon}}{\sqrt{\varepsilon}} h(\tau_0).$$

It follows by the continuation of solutions that m is decreasing on $[t_0, \tau_0)$ so that

$$m(t) \le m(t_0) < -\frac{\sqrt{1+\varepsilon}}{\sqrt{\varepsilon}} h(\tau_0), \quad t \in [t_0, \tau_0)$$

and

$$1-\frac{\varepsilon}{1+\varepsilon}\le\frac{d}{dt}\left(\frac{1}{m(t)}\right)\le 1+\frac{\varepsilon}{1+\varepsilon},\qquad t\in[t_0,\tau_0].$$

Integrating the above relation on (t,τ_0) with $t\in[t_0,\tau_0)$ and noticing that

$$\lim_{t\uparrow\tau_0} m(t)=-\infty,$$

we deduce that

$$\frac{1}{1+\varepsilon}(\tau_0-t)\le-\frac{1}{m(t)}\le\frac{1+2\varepsilon}{1+\varepsilon}(\tau_0-t).$$

Since $\varepsilon\in(0,1)$ is arbitrary, in view of the definition of $m(t)$, the above inequality in the limit $\varepsilon\downarrow 0$ implies (6.21). □

REMARK 6.7. We can also show that $\sup_{x\in\mathbb{T}} u_x(t,x)$ remains bounded as $t\uparrow\tau_0$. Let $M(t):=\sup_{x\in\mathbb{T}} u_x(t,x)$ and $\eta(t)$ be the argument of the supremum. For every $t\in[0,\tau_0)$, $M(t)=u_x(t,\eta(t))$ satisfies

$$\frac{d}{dt}M(t)=-M^2(t)+u(t,\eta(t)),\quad \text{a.e.}\quad t\in(0,\tau_0).$$

Integrating and using the bound (6.22) yields

$$M(t)\le\sup_{x\in\mathbb{T}} u_0'(x)+\tau_0\|u_0\|_{L^\infty}+\frac{\sqrt{2\pi}}{2}\tau_0^2\|u_0\|_{L^2}<\infty.$$

Since $u(t,x)$ is periodic on $\mathbb{T}$ for all $t\in[0,T)$, there exists $\eta_0(t)\in\mathbb{T}$ for every $t\in[0,\tau_0)$ such that $u_x(t,\eta_0(t))=0$. Therefore, $M(t)\ge u_x(t,\eta_0(t))=0$ for all $t\in[0,T)$, and hence $M(t)$ remains bounded as $t\uparrow\tau_0$.

Theorem 6.1 only gives a sufficient condition for wave breaking. In general, solutions of dispersive generalizations of the inviscid Burgers equation do not have to blow up in a finite time. This is the case for the Camassa–Holm equation [**60**,**61**] and for the reduced Ostrovsky equation [**118**], where sufficient conditions for the existence of global weak solutions have been found.

The following result gives a sufficient condition for the local solution to the b-CH equation (Example 6.4) to be continued as a global solution. It relies on the equivalent formulation (6.2) of the b-CH equation (6.1) for the momentum density $m:=u-u_{xx}$.

THEOREM 6.2. *Let $u\in C^0([0,\tau_0),H^1(\mathbb{R})\cap W^{1,\infty}(\mathbb{R}))$ be the unique solution of the b-CH equation* (6.1). *Assume that $m_0:=u_0-u_0''\in H^{-1}(\mathbb{R})$ is strictly positive in the sense of Radon measure and that there exists $C>0$ such that*

$$|b|\|u(t,\cdot)\|_{L^2}^2+|3-b|\|u_x(t,\cdot)\|_{L^2}^2\le C,\quad t\in[0,\tau_0). \tag{6.25}$$

Then, $\tau_0=\infty$.

PROOF. The proof consists of two steps. In the first step, we prove that the positivity of m is preserved in the time evolution of (6.2). In the second step, we prove that this property implies that the negative slope of u remains bounded according to (6.20). By Lemma 6.2, this implies that $\tau_0=\infty$.

STEP 1. If $m_0>0$, then $m(t,\cdot)>0$ for every $t\in[0,\tau_0)$ in the sense of Radon measure.

PROOF OF STEP 1. Applying density arguments, we approximate initial value $u_0 \in H^1(\mathbb{R}) \cap W^{1,\infty}(\mathbb{R})$ by a sequence of functions $\{u_0^n\}_{n\in\mathbb{N}} \in H^2(\mathbb{R})$ such that $\lim_{n\to\infty} u_0^n = u_0$ in $H^1(\mathbb{R}) \cap W^{1,\infty}(\mathbb{R})$. Consider the unique solution

$$u^n \in C^0([0, \tau_0^n), H^2(\mathbb{R}))$$

of the initial-value problem and define $m^n := (1 - \partial_x^2)u^n \in C^0([0, \tau_0^n), L^2(\mathbb{R}))$ such that $\lim_{n\to\infty} m_0^n = m_0$ in $H^{-1}(\mathbb{R})$. By uniqueness of solutions, we have $\lim_{n\to\infty} \tau_0^n = \tau_0$. By strict positivity of m_0 in the sense of Radon measure and convergence of $\{m_0^n\}_{n\in\mathbb{N}}$ to m_0 in $H^{-1}(\mathbb{R})$, we have strict positivity $m_0^n(x) > 0$ a.e. for $x \in \mathbb{R}$, if n is sufficiently large. It follows from the evolution equation (6.2) and the method of characteristics that $M^n(t, s) := m^n(t, X(t, s))$ satisfies a.e. for $s \in \mathbb{R}$

$$\begin{cases} \dot{M}^n = -bu_x^n(t, X)M^n, & t > 0, \\ M^n|_{t=0} = m_0^n, \end{cases}$$

which admits the exact solution

$$M^n(t, s) = m_0^n(s)e^{-b\int_0^t u_x^n(t', X(t,s))dt'}, \quad t \in [0, \tau_0^n).$$

Since $m_0^n(x) > 0$ a.e. for $x \in \mathbb{R}$ and $u^n \in C^0([0, \tau_0^n), H^2(\mathbb{R}))$, then $M^n(t, s) > 0$ a.e. for $s \in \mathbb{R}$ and for every $t \in [0, \tau_0^n)$. Since the mapping $s \mapsto X(t, s)$ is an increasing diffeomorphism for $t \in [0, \tau_0^n)$, this implies that $m^n(t, x) > 0$ a.e. for $x \in \mathbb{R}$ and for every $t \in [0, \tau_0^n)$. By taking the limit $n \to \infty$ in $H^{-1}(\mathbb{R})$, we obtain that $m(t, \cdot) = \lim_{n\to\infty} m^n(t, \cdot)$ is strictly positive in the sense of Radon measure for every $t \in [0, \tau_0)$. □

STEP 2. If $m(t, \cdot) > 0$ for $t \in [0, \tau_0)$, then there exists $M > 0$ such that

$$\inf_{x\in\mathbb{R}} u_x(t, x) \geq -M, \quad t \in [0, \tau_0).$$

PROOF OF STEP 2. Inverting $m = (1 - \partial_x^2)u$, we can write

$$u(t, x) = \frac{1}{2}\phi * m,$$

where $\phi(x) = e^{-|x|} \in H^1(\mathbb{R}) \cap W^{1,\infty}(\mathbb{R})$ and $m(t, \cdot) \in H^{-1}(\mathbb{R})$. Since $\phi \pm \phi' \geq 0$, we deduce from strict positivity of m that $u + u_x \geq 0$ and $u - u_x \geq 0$. Therefore, $u_x(t, x) \geq -\|u(t, \cdot)\|_{L^\infty}$ for every $t \in [0, \tau_0)$.

It remains to control the L^∞ norm of the unique solution $u \in C^0([0, \tau_0), H^1(\mathbb{R}) \cap W^{1,\infty}(\mathbb{R}))$. This follows from the evolution of U along the characteristic curves in (6.18) with

$$Q[u] = -\frac{1}{4}\phi' * \left[bu^2 + (3 - b)u_x^2\right].$$

Since $\|\phi'\|_{L^\infty} = 1$, we obtain

$$\begin{aligned} \sup_{t'\in[0,t]} \|u(t, \cdot)\|_{L^\infty} &\leq \|u_0\|_{L^\infty} + \frac{1}{4}t \sup_{t'\in[0,t]} \left(|b|\|u(t', \cdot)\|_{L^2}^2 + |3 - b|\|u_x(t', \cdot)\|_{L^2}^2\right) \\ &\leq \|u_0\|_{L^\infty} + \frac{1}{4}Ct, \qquad t \in [0, \tau_0), \end{aligned}$$

where the second inequality follows from (6.25). For every finite $\tau_0 > 0$, $\|u(t, \cdot)\|_{L^\infty}$ is bounded from above and hence $\inf_{x\in\mathbb{R}} u_x(t, x)$ is bounded from below by some constant. □

REMARK 6.8. Condition (6.25) in Theorem 6.2 is satisfied for $b = 2$ and $b = 3$ in view of the conserved quantity $E(u)$ in (6.9) and (6.13), respectively.

REMARK 6.9. In some other dispersive generalizations of the inviscid Burgers equation, smooth solutions exist only for finite time for any initial data. This was proven for instance for the Burgers–Hilbert equation

$$u_t + uu_x = H(u),$$

where $H(u)$ is the Hilbert transform of u, see [**133**].

REMARK 6.10. In Examples 6.1, 6.2, and 6.3 we have seen that the momentum variable $m = u - u_{xx}$ for the peaked traveling waves is zero everywhere except at the location of the peak, where m is proportional to the Dirac delta function. In this sense, one could say that the peaked waves are at the border between the wave breaking result of Theorem 6.1 and the global solutions of Theorem 6.2.

In order to study the stability of the peaked waves, we need to understand how the location of the peak of a weak solution $u \in C^0([0, \tau_0), H^1(\mathbb{R}) \cap W^{1,\infty}(\mathbb{R}))$ moves in the time evolution of (6.16). We simplify the analysis by restricting to solutions with a single peak positioned at the point $\xi(t)$ so that $u(t, \cdot + \xi(t)) \in C^1(\mathbb{R}\backslash\{0\})$ for every $t \in [0, \tau_0)$. The following lemma implies that the peak at $\xi(t)$ moves with local characteristic speed given by (6.17). To simplify the exposition, we introduce the notation

$$[v]_-^+ := \lim_{x \to \xi^+} v(t, x) - \lim_{x \to \xi^-} v(t, x)$$

for the jump of a function v across the peak's location at $x = \xi(t)$.

LEMMA 6.3. *Let $u \in C^0([0, \tau_0), H^1(\mathbb{R}) \cap W^{1,\infty}(\mathbb{R}))$ be the unique solution in Lemma 6.1 such that $u(t, \cdot + \xi(t)) \in C^1(\mathbb{R}\backslash\{0\})$ for every $t \in [0, \tau_0)$ with*

$$[u_x]_-^+ \neq 0.$$

Then, $\xi \in C^1(0, \tau_0)$ satisfies

$$\frac{d\xi}{dt} = u(t, \xi(t)), \quad t \in (0, \tau_0). \tag{6.26}$$

PROOF. We consider the weak solution $u \in C^0([0, \tau_0), H^1(\mathbb{R}) \cap W^{1,\infty}(\mathbb{R}))$ of the initial-value problem (6.16) in the integral form

$$\int_0^{\tau_0} \int_{\mathbb{R}} \left(u\varphi_t + \frac{1}{2}u^2\varphi_x + Q\varphi \right) dxdt + \int_{\mathbb{R}} u_0(x)\varphi(0, x)dx = 0, \tag{6.27}$$

where $\varphi \in C^0([0, \tau_0), H^1(\mathbb{R})) \cap C^1([0, \tau_0), L^2(\mathbb{R}))$ is the test function such that $\varphi|_{t=\tau_0} = 0$. Integrating (6.27) by parts separately on $(-\infty, \xi(t))$ and $(\xi(t), \infty)$ and using the fact that $u(t, \cdot) \in C^0([0, \tau_0), H^1(\mathbb{R}) \cap W^{1,\infty}(\mathbb{R}))$ and $u(t, \cdot + \xi(t)) \in C^1(\mathbb{R}\backslash\{0\})$ for every $t \in [0, \tau_0)$, we obtain

$$u_t(t, x) + u(t, x)u_x(t, x) = Q(t, x), \quad \text{a.e. } \pm (x - \xi(t)) > 0, \quad t \in (0, \tau_0). \tag{6.28}$$

If $Q \in C^0([0, \tau_0), H^1(\mathbb{R}) \cap W^{1,\infty}(\mathbb{R}))$, then Q is a continuous function of $(t, x) \in (0, \tau_0) \times \mathbb{R}$ due to the Sobolev embedding of $H^1(\mathbb{R})$ into the space of continuous functions. Therefore, it follows from (6.28) that

$$[u_t]_-^+ + u(t, \xi(t))[u_x]_-^+ = 0, \quad \text{a.e. } t \in (0, \tau_0). \tag{6.29}$$

On the other hand, since $u(t, \cdot + \xi(t)) \in C^1(\mathbb{R}\backslash\{0\})$ for $t \in [0, \tau_0)$, we differentiate $u(t, \xi(t))$ on both sides of $\xi(t)$. Since $u(t, \xi(t))$ is continuous for $t \in [0, \tau_0)$ we obtain that

$$\lim_{x\to\xi(t)^+} \left(u_t + \frac{d\xi}{dt} u_x \right) = \lim_{x\to\xi(t)^-} \left(u_t + \frac{d\xi}{dt} u_x \right), \quad \text{a.e. } t \in (0, \tau_0).$$

Therefore,

$$[u_t]_-^+ + \frac{d\xi}{dt}[u_x]_-^+ = 0, \quad \text{a.e. } t \in (0, \tau_0) \tag{6.30}$$

Since $[u_x]_-^+ \neq 0$, it follows from (6.29) and (6.30) that $\xi(t)$ satisfies (6.26) a.e. for $t \in (0, \tau_0)$. Since $u \in C^0((0, \tau_0) \times \mathbb{R})$, then $\xi \in C^1(0, \tau_0)$ and equation (6.26) holds for every $t \in (0, \tau_0)$. □

COROLLARY 6.1. *Let $u(t, x) = \phi(x - ct)$ be the peaked traveling wave solution of (6.16) with $\phi \in H^1(\mathbb{R}) \cap W^{1,\infty}(\mathbb{R})$, $\phi \in C^1(\mathbb{R}\backslash\{0\})$, and jump $[\phi']_-^+ \neq 0$ at 0. Then, $c = \phi(0)$.*

PROOF. For $\xi(t) = ct$ we obtain from (6.26) that $c = u(t, ct) = \phi(0)$. □

REMARK 6.11. The result of Lemma 6.3 is well known in the context of quasilinear hyperbolic systems with continuous and piecewise C^1 solutions. In the same context, **shock waves** with finite jump discontinuities of the wave profile have also been studied in quasilinear hyperbolic equations. Shock waves propagate with a speed different from the local characteristic speed given by (6.26). This speed is found from the Rankine–Hugoniot conditions [**226**]. Shock waves for dispersive perturbations of the inviscid Burgers equation (6.16) are not considered in this monograph.

6.3. Linear instability of peaked waves

The standard method to study the stability of nonlinear waves consists of adding a perturbation to the nonlinear wave, substituting it into the evolution equation and exploring the Hamiltonian structure of the linearized equation. After separation of the temporal and spatial variables, the linearized evolution can be studied by means of spectral stability theory. If the nonlinear wave is a constrained minimizer of energy and the spectral gap assumption is satisfied, both the linear and spectral stability can be proven from the linearized equations as in Section 3.5 and the orbital stability can be studied as in Section 3.3.

For peaked waves, all of the above becomes problematic: the linearized equation may not exhibit the required Hamiltonian structure and the spectral gap assumption is typically violated. The formal linearization of the evolution equation at the peaked waves may become inconsistent to the dynamics of peaks prescribed by Lemma 6.3, see Remark 6.12. We will show how to overcome these challenges in the framework of the perturbed Burgers equation (6.16), for which we will study the cases of the b-CH equation (Example 6.4) and the reduced Ostrovsky equation (Example 6.5).

We will use the local well-posedness result of Lemma 6.1 for the weak solution $u \in C^0([0, \tau_0), H^1(\mathbb{R}) \cap W^{1,\infty}(\mathbb{R}))$ of the perturbed Burgers equation (6.16). We assume that $Q = Q[u]$ depends on $u = u(t, x)$ and is independent of (t, x). We decompose the weak solution into the peaked wave $\phi(x - ct)$ with $c = \phi(0)$, see Corollary 6.1, and the perturbation $v(t, x - ct)$. However, this formal decomposition might break down already at the level of the linearized equations. According to Lemma 6.3, if the peak is located at $x = ct$, then the perturbation would need to satisfy the constraint $v(t,0) = 0$ which is in general not preserved in the time evolution of the perturbed Burgers equation (6.16). In order to overcome this problem, we introduce the deviation $a(t)$ of the peak's position from its unperturbed coordinate moving at speed c and consider the decomposition

$$u(t,x) = \phi(x - ct - a(t)) + v(t, x - ct - a(t)). \tag{6.31}$$

It follows from the peak's equation (6.26) for $\xi(t) = ct + a(t)$ with $c = \phi(0)$ that $a(t)$ is defined by the differential equation $a'(t) = v(t,0)$. The perturbation satisfies the evolution equation

$$v_t = (c - \phi + v|_{x=0} - v)v_x + \phi'(v|_{x=0} - v) + Q[\phi + v] - Q[\phi], \tag{6.32}$$

where x now denotes the moving frame coordinate $x - ct - a(t)$ and $v|_{x=0} = v(t,0)$. Formal truncation of (6.32) at linear order in v gives the linearized equation

$$v_t = (c - \phi)v_x + \phi'(v|_{x=0} - v) + Q'[\phi]v. \tag{6.33}$$

By Lemma 6.1, the initial-value problem for the linearized equation (6.33) is locally well-posed for $v|_{t=0} = v_0 \in H^1(\mathbb{R}) \cap W^{1,\infty}(\mathbb{R})$. Since $Q'[\phi]$ is t-independent, the unique local solution $v \in C^0([0,\tau_0), H^1(\mathbb{R}) \cap W^{1,\infty}(\mathbb{R}))$ to the linearized equation (6.33) can be extended globally to $v \in C^0(\mathbb{R}, H^1(\mathbb{R}) \cap W^{1,\infty}(\mathbb{R}))$. The linearized equation (6.33) is the starting point of the linear stability analysis of the peaked wave with the profile ϕ.

REMARK 6.12. If we consider the formal linearization without the deviation $a(t)$, we would obtain (6.33) without the term $\phi' v|_{x=0}$. This formal linearization does not possess a global solution $v \in C^0(\mathbb{R}, H^1(\mathbb{R}) \cap W^{1,\infty}(\mathbb{R}))$. Indeed, since $\phi(0) = c$ and $\phi'(0^+) \neq \phi'(0^-)$, the linearized evolution (6.33) without the term $\phi' v|_{x=0}$, with $Q'[\phi] \in C^0(\mathbb{R}, H^1(\mathbb{R}) \cap W^{1,\infty}(\mathbb{R}))$ and $v_0(0) \neq 0$ leads to the discontinuity $v(t,0^+) \neq v(t,0^-)$ for every $t > 0$, which is not possible for solutions $v(t,\cdot)$ in $H^1(\mathbb{R})$. Thus, the inclusion of the deviation $a(t)$ in the decomposition (6.31) is necessary for the derivation of the proper linearized equation (6.33).

The following definition gives the meaning of the linear stability and instability of peaked traveling waves with the profile ϕ in a Banach space X.

DEFINITION 6.2. A peaked traveling wave solution with the profile ϕ is said to be **linearly stable** if for every $v_0 \in X \subset L^2(\mathbb{R})$, there exists a positive constant C and a unique global solution $v \in C^0([0,\infty), X)$ to the linearized equation (6.33) with $v|_{t=0} = v_0$ such that

$$\|v(t,\cdot)\|_{L^2} \leq C\|v_0\|_{L^2}, \quad t > 0.$$

Otherwise, it is said to be linearly unstable.

Unfortunately, there are only very few examples of equations for which we can solve the linearized equation (6.33) in $X = H^1(\mathbb{R}) \cap W^{1,\infty}(\mathbb{R})$. Moreover, even when we can solve it explicitly, it is still difficult to prove whether the peaked wave is linearly stable or unstable in the sense of Definition 6.2. We illustrate this in the following example, which shows the potential complexity of the explicit solution of the linearized equation (6.33).

EXAMPLE 6.7. For the Camassa–Holm equation (6.3) with $b = 2$, we obtain

$$Q'[\phi]v = -\frac{1}{2}\phi' * [2\phi v + \phi' v_x],$$

where $\phi(x) = e^{-|x|}$. The following elementary identities hold for every $v \in H^1(\mathbb{R})$:

$$\phi' * (\phi' v') = \phi * (\phi' v) - \phi' * (\phi v) + 2(v|_{x=0} - v)\phi' \tag{6.34}$$

and

$$\phi * (\phi' v) + \phi' * (\phi v) = -2\phi \int_0^x v(x')dx'. \tag{6.35}$$

As a result, the linearized equation (6.33) with $c = \phi(0) = 1$ can be cast in the equivalent form:

$$v_t = (1 - \phi)v_x + \phi \int_0^x v(t, x')dx'. \tag{6.36}$$

Integrating this equation in x and introducing the variable $\mathfrak{v}(t, x) := \int_0^x v(t, x')dx'$ satisfying the boundary conditions $\mathfrak{v}(t, 0) = 0$ yields

$$\mathfrak{v}_t = (1 - \phi)\mathfrak{v}_x + \phi' \mathfrak{v}$$

on both sides of $x > 0$ and $x < 0$. This linear equation can be solved by using the method of characteristics piece for $x > 0$ and $x < 0$. The family of characteristic curves $X(t, s)$ satisfying the initial-value problem

$$\begin{cases} \frac{dX}{dt} = \phi(X) - 1, \\ X|_{t=0} = s, \end{cases} \tag{6.37}$$

are uniquely defined in the exact form

$$X(t, s) = \begin{cases} \log\left[1 + (e^s - 1)e^{-t}\right], & s > 0, \\ -\log\left[1 + (e^{-s} - 1)e^t\right], & s < 0, \end{cases} \tag{6.38}$$

such that $\lim_{s \to 0^\pm} X(t, s) = 0$ for every $t \in \mathbb{R}$. Along each characteristic curve satisfying (6.37), $\mathfrak{V}(t, s) := \mathfrak{v}(t, X(t, s))$ satisfies the initial-value problem

$$\begin{cases} \frac{d\mathfrak{V}}{dt} = \phi'(X(t, s))\mathfrak{V}, \\ \mathfrak{V}|_{t=0} = \int_0^s v_0(s')ds', \end{cases}$$

which can be solved uniquely in the exact form

$$\mathfrak{V}(t, s) = X_s(t, s) \int_0^s v_0(s')ds', \quad X_s(t, s) = \begin{cases} \frac{1}{1 + (e^t - 1)e^{-s}}, & s > 0, \\ \frac{1}{1 + (e^{-t} - 1)e^s}, & s < 0, \end{cases}$$

such that $\lim_{s \to 0^\pm} \mathfrak{V}(t, s) = 0$ for every $t \in \mathbb{R}$. Finally, along each characteristic curve satisfying (6.37), $V(t, s) := v(t, X(t, s))$ satisfies the initial-value problem

$$\begin{cases} \frac{dV}{dt} = \phi(X(t, s))\mathfrak{V}(t, s), \\ V|_{t=0} = v_0(s), \end{cases}$$

which can be solved uniquely in the exact form

$$V(t, s) = v_0(s) + Y(t, s) \int_0^s v_0(s')ds', \quad Y(t, s) = \begin{cases} \frac{(e^t - 1)e^{-s}}{1 + (e^t - 1)e^{-s}}, & s > 0, \\ \frac{(1 - e^{-t})e^s}{1 + (e^{-t} - 1)e^s}, & s < 0, \end{cases}$$

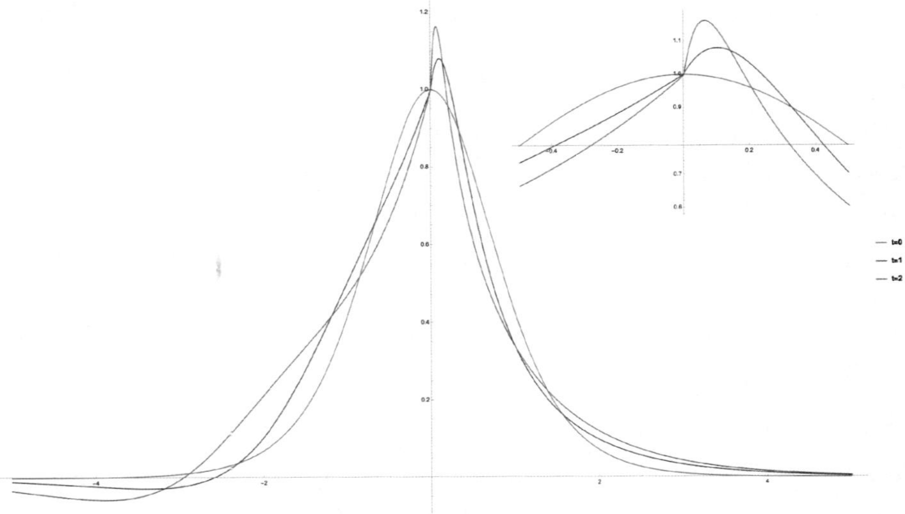

FIGURE 6.4. A plot of $v(t,x)$ versus x for some values of t in the case of $v_0(x) = \text{sech}^2(x)$.

with $\lim_{s\to 0^\pm} V(t,s) = v_0(0)$ for every $t \in \mathbb{R}$. Figure 6.4 illustrates the solution $v(t,x)$ in the parametric form $x = X(t,s)$ and $v(t, X(t,s)) = V(t,s)$ for $v_0(s) = \text{sech}^2(s)$. By using the chain rule, we can compute the L^2 norm of the perturbation after integration by parts and find that

$$\|v(t,\cdot)\|^2_{L^2(0,\infty)} = \int_0^\infty \frac{[v_0(s)]^2}{1+(e^t-1)e^{-s}} ds + \int_0^\infty \frac{(e^t-1)e^{-s}}{[1+(e^t-1)e^{-s}]^3} \left(\int_0^s v_0(s')ds' \right)^2 ds$$

and

$$\|v(t,\cdot)\|^2_{L^2(-\infty,0)} = \int_{-\infty}^0 \frac{[v_0(s)]^2}{1+(e^{-t}-1)e^{s}} ds + \int_{-\infty}^0 \frac{(e^{-t}-1)e^{s}}{[1+(e^{-t}-1)e^{s}]^3} \left(\int_0^s v_0(s')ds' \right)^2 ds.$$

Although these formulas are exact, it is difficult to conclude if the peaked wave is linearly stable or unstable in the sense of Definition 6.2. For instance, the formal limit $t \to +\infty$ under the integration sign depends on the behavior of $v_0(s)$ as $s \to 0^\pm$ and $s \to \pm\infty$.

REMARK 6.13. For the Camassa–Holm equation (Example 6.7), it was shown in [**205**] that if $v_0 \in H^1(\mathbb{R}) \cap W^{1,\infty}(\mathbb{R})$, then the solution $v \in C^0(\mathbb{R}, H^1(\mathbb{R}) \cap W^{1,\infty}(\mathbb{R}))$ satisfies

$$\|v(t,\cdot)\|^2_{H^1(0,\infty)} = \|v_0\|^2_{H^1(0,\infty)} + 2(e^t-1) \int_0^\infty \phi(s) \left([v_0(s)]^2 + \frac{1}{2}[v_0'(s)]^2 \right) ds$$

and

$$\|v(t,\cdot)\|^2_{H^1(-\infty,0)} = \|v_0\|^2_{H^1(-\infty,0)} + 2(e^{-t}-1)\int_{-\infty}^{0} \phi(s)\left([v_0(s)]^2 + \frac{1}{2}[v_0'(s)]^2\right) ds.$$

Hence, the perturbation grows in the $H^1(\mathbb{R})$ norm from the right side of the peak located at $x = 0$, see Figure 6.4. In addition, the general solution satisfies the following limits:

$$v_x(t,0^+) = v_0'(0^+)e^t + v_0(0)(e^t - 1)$$

and

$$v_x(t,0^-) = v_0'(0^-)e^{-t} + v_0(0)(1 - e^{-t}).$$

Hence, the perturbation also grows in the $W^{1,\infty}(\mathbb{R})$ norm on the right side of the peak located at $x = 0$, see Figure 6.4. Similar exact solutions were found for the linearized equation at the peaked periodic wave (Example 6.2 with $b = 2$) in [**194**].

Because it is difficult to solve the linearized equation (6.33) and to analyze the solution $v \in C^0(\mathbb{R}, H^1(\mathbb{R}) \cap W^{1,\infty}(\mathbb{R}))$ even if it can be expressed in explicit form, we will rewrite the linear evolution defined in space $H^1(\mathbb{R}) \cap W^{1,\infty}(\mathbb{R}) \subset L^2(\mathbb{R})$ in the form

$$w_t = \mathcal{A}w, \tag{6.39}$$

where the linear operator $\mathcal{A} : \mathrm{Dom}(\mathcal{A}) \subset L^2(\mathbb{R}) \to L^2(\mathbb{R})$ acts on the dense domain

$$\mathrm{Dom}(\mathcal{A}) = \left\{ w \in L^2(\mathbb{R}) : \ (c - \phi)\partial_x w \in L^2(\mathbb{R}) \right\}.$$

The initial-value problem is well-posed for $w|_{t=0} = w_0 \in \mathrm{Dom}(\mathcal{A})$ if the linearized equation (6.39) has a unique solution $w \in C^0(\mathbb{R}, \mathrm{Dom}(\mathcal{A}))$. The existence of such local solutions can be obtained from the spectral properties of the linear operator $\mathcal{A}$.

Remark 6.14. If $c > 0$ and $\phi \in [0, c]$, it follows from the bound

$$\|(c - \phi)\partial_x w\|_{L^2} \le c\|\partial_x w\|_{L^2}$$

that $H^1(\mathbb{R})$ is continuously embedded into $\mathrm{Dom}(\mathcal{A})$. However, $H^1(\mathbb{R})$ is not equivalent to $\mathrm{Dom}(\mathcal{A})$: functions $w \in \mathrm{Dom}(\mathcal{A})$ are allowed to have finite jump discontinuities across the peak at $x = 0$ so that $\partial_x w|_{x=0}$ is not defined, in which case $w \notin H^1(\mathbb{R})$. The relevance of $\mathrm{Dom}(\mathcal{A})$ for the linearized equation (6.33) is particularly apparent for the homogeneous problems for which $Q'[\phi]$ is independent of (t, x). The translational symmetry implies that $\mathcal{A}\phi' = 0$, where $\phi' \in \mathrm{Dom}(\mathcal{A})$ but $\phi' \notin H^1(\mathbb{R})$ since $x\delta = 0$ in $L^2(\mathbb{R})$ but $\delta \notin L^2(\mathbb{R})$, where the Dirac singularity δ arises in $\phi - \phi'' = 2\delta$.

The stability and instability of the peaked waves of the linearized equation (6.39) in $\mathrm{Dom}(\mathcal{A}) \subset L^2(\mathbb{R})$ can be studied using spectral properties of the linear operator $\mathcal{A}$ in $L^2(\mathbb{R})$, see Section 6.4. We adopt the following definition of spectral stability of the peaked wave with the profile ϕ.

Definition 6.3. A peaked traveling wave solution with profile ϕ is said to be **spectrally stable** if the spectrum $\sigma(\mathcal{A})$ of the linear operator $\mathcal{A}$ in $L^2(\mathbb{R})$ satisfies $\sigma(\mathcal{A}) \subset i\mathbb{R}$. Otherwise, it is said to be spectrally unstable.

Remark 6.15. Spectral stability is weaker than linear stability and spectral instability is stronger than linear instability. Since the peaked waves are spectrally unstable for the two basic examples (Examples 6.4 and 6.5), the linear instability of the peaked waves according to Definition 6.2 with $X = \mathrm{Dom}(\mathcal{A})$ can be concluded from the spectral instability. However, in some other models, the peaked wave may be spectrally stable and linearly unstable.

We will show for the two main examples how the linearized equation (6.39) can be obtained from the linearized equation (6.33).

EXAMPLE 6.8. For the b-CH equation (6.3) of Example 6.1, we obtain

$$Q'[\phi]v = -\frac{1}{2}\phi' * [b\phi v + (3-b)\phi' v_x],$$

where $\phi(x) = e^{-|x|}$. With the help of (6.34), the linearized equation (6.33) with $c = \phi(0) = 1$ can be cast in the equivalent form

$$v_t = (1-\phi)v_x + (b-2)(v|_{x=0} - v)\phi' + \frac{1}{2}(b-3)\phi * (\phi' v) - \frac{1}{2}(2b-3)\phi' * (\phi v). \tag{6.40}$$

This suggests to define the linear operator $\mathcal{A} : \mathrm{Dom}(\mathcal{A}) \subset L^2 \to L^2$ by

$$\mathcal{A} := (1-\phi)\partial_x + (2-b)\phi' + \frac{1}{2}(b-3)\phi * (\phi' \cdot\,) - \frac{1}{2}(2b-3)\phi' * (\phi \cdot\,). \tag{6.41}$$

Explicit computations yield

$$\mathcal{A}\phi = (2-b)\phi' \text{ and } \mathcal{A}\phi' = 0. \tag{6.42}$$

The decomposition $v(t,x) = v|_{x=0}\phi(x) + w(t,x)$ substituted into the linearized equation (6.40) satisfies

$$\left(\frac{d}{dt} v|_{x=0}\right)\phi + w_t = \mathcal{A}w.$$

It follows from (6.40) that

$$\frac{d}{dt} v|_{x=0} = \frac{3}{2}(b-2)\langle \phi\phi', w\rangle,$$

where $\langle \cdot, \cdot \rangle$ denotes the inner product in $L^2(\mathbb{R})$. Thus, we obtain the linearized equation in the form

$$w_t = \mathcal{A}w + \frac{3}{2}(2-b)\langle \phi\phi', w\rangle \phi. \tag{6.43}$$

The linearized equation (6.43) is equally defined for $w(t,\cdot) \in H^1(\mathbb{R}) \cap W^{1,\infty}(\mathbb{R})$ for which $w(t,0) = 0$ and for $w(t,\cdot) \in \mathrm{Dom}(\mathcal{A})$ for which $w(t,0)$ may not exist. If $w(t,\cdot) \in \mathrm{Dom}(\mathcal{A})$, we can transform (6.43) into the homogeneous form (6.39) by applying yet another decomposition $w(t,x) = \alpha(t)\phi(x) + \beta(t)\phi'(x) + \mathfrak{w}(t,x)$, which satisfies

$$\frac{d\alpha}{dt}\phi + \frac{d\beta}{dt}\phi' + \mathfrak{w}_t = (2-b)\alpha\phi' + \mathcal{A}\mathfrak{w} + (2-b)\beta\phi + \frac{3}{2}(2-b)\langle \phi\phi', \mathfrak{w}\rangle \phi,$$

where we have used (6.42), $\langle \phi\phi', \phi\rangle = 0$, and $\langle \phi, (\phi')^2\rangle = \frac{2}{3}$. The linearized equation (6.39) is obtained for $\mathfrak{w}$ if α and β satisfy the system of two first-order differential equations

$$\frac{d\alpha}{dt} = (2-b)\beta + \frac{3}{2}(2-b)\langle \phi\phi', \mathfrak{w}\rangle, \quad \frac{d\beta}{dt} = (2-b)\alpha. \tag{6.44}$$

If $\mathfrak{w} = 0$ and $b \neq 2$, then

$$\alpha(t) = c_+ e^{(2-b)t} + c_- e^{-(2-b)t}, \quad \beta(t) = c_+ e^{(2-b)t} - c_- e^{-(2-b)t} \tag{6.45}$$

with some constant c_+ and c_-. Therefore, the peaked wave is linearly unstable in the sense of Definition 6.2 for $b \neq 2$ in $\mathrm{Dom}(\mathcal{A}) \subset L^2(\mathbb{R})$. For $b = 2$, α and β

are constant in time and we have to study the homogeneous equation (6.39) with $\mathfrak{w} \neq 0$. This is done in Theorem 6.3 and Lemma 6.6 for all values of b.

EXAMPLE 6.9. For the reduced Ostrovsky equation (6.4) of Example 6.3, we obtain

$$Q'[\phi]v = \partial_x^{-1} v,$$

where $\partial_x^{-1} : \dot{L}^2_{\text{per}} \mapsto \dot{L}^2_{\text{per}}$ is defined for periodic L^2 functions with zero mean. We consider 2π-periodic functions, for which $c_0 = \pi^2/9$ and

$$\phi_0(x) = \frac{3x^2 - \pi^2}{18} \chi_{[-\pi,\pi]}(x).$$

The linearized equation (6.33) then reads

$$v_t = (c_0 - \phi)v_x + (v|_{x=\pm\pi} - v)\phi' + \partial_x^{-1} v, \tag{6.46}$$

where $\phi(x) = \sum_{n\in\mathbb{Z}} \phi_0(x - 2n\pi)$ so that the peaks of the peaked periodic wave with the profile ϕ are located at $x \in \{(2n+1)\pi\}_{n\in\mathbb{Z}}$. The unique solution $v \in C^0(\mathbb{R}, \dot{H}^1_{\text{per}} \cap W^{1,\infty})$ of the initial-value problem for the linearized equation (6.46) is well-defined for initial data $v_0 \in \dot{H}^1_{\text{per}} \cap W^{1,\infty}$. It follows from (6.46) that

$$\frac{d}{dt} v|_{x=\pm\pi} = \partial_x^{-1} v|_{x=\pm\pi},$$

since $\phi_0(\pm\pi) = c_0$. We define $\partial_x^{-1} : \dot{L}^2_{\text{per}} \to \dot{L}^2_{\text{per}}$ pointwise by

$$(\partial_x^{-1} v)(x) := \int_{-\pi}^{x} v(x')dx' - \frac{1}{2\pi}\int_{-\pi}^{\pi}\int_{-\pi}^{x} v(x')dx'dx, \quad x \in [-\pi, \pi],$$

and note that since $\phi'(x) = \frac{x}{3}$,

$$-\frac{1}{2\pi}\int_{-\pi}^{\pi}\int_{-\pi}^{x} v(x')dx'dx = -\frac{1}{2\pi}\int_{-\pi}^{\pi}(\pi - x)v(x)dx = -\frac{1}{2}\int_{-\pi}^{\pi} v(x)dx + \frac{3}{2\pi}\langle \phi', v\rangle.$$

In view of the fact that v has zero mean we obtain that

$$\frac{d}{dt} v|_{x=\pm\pi} = \frac{3}{2\pi}\langle \phi', v\rangle. \tag{6.47}$$

In order to transform (6.46) to the homogeneous form (6.39), we define

$$\mathcal{A} := (c_0 - \phi)\partial_x - \phi' + \partial_x^{-1}, \tag{6.48}$$

with the domain

$$\text{Dom}(\mathcal{A}) = \left\{ w \in \dot{L}^2_{\text{per}} : \ \partial_x[(c_0 - \phi)w] \in \dot{L}^2_{\text{per}} \right\}. \tag{6.49}$$

It follows from (6.48) that $\mathcal{A} = \partial_x \mathcal{L}$, where $\mathcal{L} : \dot{L}^2_{\text{per}} \mapsto \dot{L}^2_{\text{per}}$ is a self-adjoint operator given by

$$\mathcal{L} := c_0 - \phi + \partial_x^{-2}. \tag{6.50}$$

Since $\mathcal{L}\phi' = 0$ with $\phi' \in \dot{L}^2_{\text{per}}$, it follows from (6.46) that $\langle \phi, v(t,\cdot)\rangle$ is independent of t. This property is related to the conservation of the momentum functional $\|u\|^2_{L^2_{\text{per}}}$ in (6.15). Therefore, we can consider the linear stability of the peaked periodic wave in $\dot{L}^2_{\text{per}}$ with the following constraint on the perturbation:

$$\langle \phi, v(t,\cdot)\rangle = 0, \quad t \in \mathbb{R}, \tag{6.51}$$

which represents the vanishing linear variation of the momentum as in Section 2.4. It follows from (6.46) after integration by parts that

$$\begin{aligned}\frac{d}{dt}\langle\phi', v\rangle &= -\langle(c_0-\phi)\phi'' - (\phi')^2 + \phi, v\rangle + \langle(\phi')^2, v|_{x=\pm\pi} - v\rangle \\ &= -\frac{2}{3}\langle\phi, v\rangle + \frac{2\pi^3}{27} v|_{x=\pm\pi},\end{aligned} \tag{6.52}$$

where we have used the derivative of the stationary equation $(c_0-\phi)\phi' + \partial_x^{-1}\phi = 0$ on $x \in (-\pi, \pi)$ and the fact that $(\phi')^2 = \frac{2}{3}\phi + \frac{\pi^2}{27}$ together with the zero mean-value property of ϕ and v. It follows from (6.47), (6.51), and (6.52) that the time evolution of the linearized equation (6.46) is closed in the function space $\dot{H}^1_{\rm per} \cap W^{1,\infty} \cap L^2_c$ with two constraints:

$$L^2_c = \{v \in \dot{L}^2_{\rm per} : \quad \langle\phi, v\rangle = 0, \quad \langle\phi', v\rangle = 0\}, \tag{6.53}$$

which ensure that $v|_{x=\pm\pi} = 0$ independently of time t. Therefore, the linearized equation (6.46) can be reduced to the homogeneous form (6.39) in $\dot{H}^1_{\rm per} \cap W^{1,\infty} \cap L^2_c$, i.e., $w_t = \mathcal{A}w$ with $\mathcal{A} = \partial_x\mathcal{L}$ and $\mathcal{L}$ defined in (6.50). Subsequently, the same homogeneous equation can be considered on $\mathrm{Dom}(\mathcal{A}) \subset L^2_{\rm per}$.

Remark 6.16. If a dynamical system with peaked waves has no Hamiltonian structure then the linear operator $\mathcal{A}$ can not be expressed as $J\mathcal{L}$, where J is skew-adjoint and $\mathcal{L}$ is self-adjoint. This can be seen in Example 6.8 with $b \neq 2$ and $b \neq 3$. On the other hand, if the dynamical system has a Hamiltonian structure as for instance in (6.14) for the reduced Ostrovsky equation in Example 6.9, then $\mathcal{A} = J\mathcal{L}$ with $J = \partial_x$ and $\mathcal{L}$ defined in (6.50). In the latter case, we can use the concept of linear orbital stability of the peaked periodic wave due to the conservation of the momentum $P(u)$ in (6.15), which allows us to define the constrained subspace $L^2_c \subset \dot{L}^2_{\rm per}$ in (6.53).

6.4. Spectral instability of peaked waves

Compared to the spectral stability of smooth waves in linear Hamiltonian systems (Section 3.5), the spectral instability of peaked waves is not defined by the negative index of the Hamiltonian–Krein theorem (Theorems 3.10 and 3.11). One needs to study the entire spectrum of the linear operator $\mathcal{A} : \mathrm{Dom}(\mathcal{A}) \subset L^2(\mathbb{R}) \mapsto L^2(\mathbb{R})$ linearized at the peaked wave. Hence, we recall the general definition (see Definition B.2) for the spectrum of an abstract linear operator A on a Banach space X.

Definition 6.4. Let A be a linear operator on a Banach space X with a dense domain $\mathrm{Dom}(A) \subset X$. The complex plane $\mathbb{C}$ is decomposed into the following two sets:

(1) The resolvent set

$$\rho(A) = \left\{ \lambda \in \mathbb{C} : \begin{array}{l} \mathrm{Ker}(A - \lambda I) = \{0\}, \\ \mathrm{Ran}(A - \lambda I) = X, \\ (A - \lambda I)^{-1} : X \to X \text{ is bounded} \end{array} \right\}.$$

(2) The spectrum

$$\sigma(A) = \mathbb{C} \setminus \rho(A),$$

which is further decomposed into the following three disjoint sets:

(a) the point spectrum

$$\sigma_p(A) = \{\lambda \in \sigma(A) : \ \mathrm{Ker}(A - \lambda I) \neq \{0\}\},$$

(b) the residual spectrum

$$\sigma_r(A) = \left\{\lambda \in \sigma(A) : \begin{array}{l} \mathrm{Ker}(A - \lambda I) = \{0\}, \\ \mathrm{Ran}(A - \lambda I) \neq X \end{array}\right\},$$

(c) the continuous spectrum

$$\sigma_c(A) = \left\{\lambda \in \sigma(A) : \begin{array}{l} \mathrm{Ker}(A - \lambda I) = \{0\}, \\ \mathrm{Ran}(A - \lambda I) = X, \\ (A - \lambda I)^{-1} : X \to X \text{ is unbounded} \end{array}\right\}.$$

When the Banach space X is also a Hilbert space equipped with the standard inner product $\langle \cdot, \cdot \rangle$ and the induced norm $\| \cdot \|$, two useful results enable us to distinguish between different parts of the spectrum of A.

LEMMA 6.4. *Let X be the Hilbert space and A^* be the adjoint operator to A with respect to $\langle \cdot, \cdot \rangle$ with a dense domain* $\mathrm{Dom}(A^*) \subset X$. *If $\sigma_c(A) = \sigma_c(A^*)$, then*

$$\sigma_p(A) \subseteq \sigma_r(A^*) \cup \sigma_p(A^*) \quad \text{and} \quad \sigma_r(A) \subseteq \sigma_p(A^*)$$

and

$$\sigma_p(A^*) \subseteq \sigma_r(A) \cup \sigma_p(A) \quad \text{and} \quad \sigma_r(A^*) \subseteq \sigma_p(A).$$

If $A = A^*$ and $\mathrm{Dom}(A) = \mathrm{Dom}(A^*)$, *then $\sigma_r(A)$ is an empty set.*

PROOF. Assume $\lambda_0 \in \sigma_p(A^*)$ so that there is $v_0 \in \mathrm{Dom}(A^*)$ such that

$$\langle (A^* - \lambda_0 I)v_0, u\rangle = 0, \qquad \text{for every } \ u \in \mathrm{Dom}(A) \subset X.$$

Hence, $\langle v_0, (A - \lambda_0 I)u\rangle = 0$ for every $u \in \mathrm{Dom}(A)$ so that either $\lambda_0 \in \sigma_p(A)$ if there exists $u_0 \in \mathrm{Dom}(A)$ such that $(A - \lambda_0 I)u_0 = 0$ or $\lambda_0 \in \sigma_r(A)$. Hence, $\sigma_p(A^*) \subseteq \sigma_r(A) \cup \sigma_p(A)$.

Assume $\lambda_0 \in \sigma_r(A)$. Since X is a Hilbert space, there exists $v \in X$ such that

$$\langle v, (A - \lambda_0 I)u\rangle = 0 \qquad \text{for every } \ u \in \mathrm{Dom}(A) \subset X.$$

Since $\mathrm{Dom}(A^*) \subset X$ is dense in X, for every $v \in X$ there is a sequence $\{v_n\}_{n\in\mathbb{N}} \in \mathrm{Dom}(A^*)$ such that $v_n \to v$ in X as $n \to \infty$. Without loss of generality we may assume the both sequences are normalized such that $\|v\| = \|v_n\| = 1$. Then,

$$0 = \lim_{n\to\infty} \langle v_n, (A - \lambda_0 I)u\rangle = \lim_{n\to\infty} \langle (A^* - \lambda_0 I)v_n, u\rangle, \qquad \text{for every } \ u \in \mathrm{Dom}(A).$$

Since $\mathrm{Dom}(A) \subset X$ is also dense in X, this implies that $\lim_{n\to\infty} \|(A^* - \lambda_0 I)v_n\| = 0$ so that either $\lambda_0 \in \sigma_p(A^*)$ or $\lambda_0 \in \sigma_c(A^*)$. However, since $\sigma_c(A) = \sigma_c(A^*)$ and the three parts of the spectra in Definition B.2 are disjoint, we conclude that $\lambda \in \sigma_p(A^*)$ so that $\sigma_r(A) \subseteq \sigma_p(A^*)$. Repeating these arguments with of A and A^* interchanged, we also obtain $\sigma_p(A) \subseteq \sigma_r(A^*) \cup \sigma_p(A^*)$ and $\sigma_r(A^*) \subseteq \sigma_p(A)$.

If $A = A^*$ and $\mathrm{Dom}(A) = \mathrm{Dom}(A^*)$, then we have $\sigma_r(A) \subseteq \sigma_p(A^*) = \sigma_p(A)$. However, the three parts of the spectrum in Definition B.2 are disjoint and the condition $\mathrm{Ker}(A - \lambda_0 I) \neq \{0\}$ is checked first, hence $\sigma_r(A)$ is an empty set. ☐

The following result is a variant of Kato's theorem formulated for self-adjoint operators A and A_0 that ensures that $\sigma_c(A) = \sigma_c(A_0)$ if $A - A_0 = K$ is compact in the trace class [**154**]. This particular version of Kato's theorem was proven in [**106**].

LEMMA 6.5. *Let $A : \mathrm{Dom}(A) \subset X \to X$ and $A_0 : \mathrm{Dom}(A_0) \subset X \to X$ be linear operators on Hilbert space X with the same domain $\mathrm{Dom}(A_0) = \mathrm{Dom}(A)$ such that $A - A_0 = K$ is a compact operator in X. Assume that the intersections $\sigma_p(A) \cap \rho(A_0)$ and $\sigma_p(A_0) \cap \rho(A)$ are empty. Then, $\sigma(A) = \sigma(A_0)$.*

PROOF. Assume that $\lambda \in \sigma(A_0)$ but $\lambda \in \rho(A)$. Hence, for every $f \in \mathrm{Dom}(A)$, we can write

$$f = (A - \lambda I)^{-1}(K + A_0 - \lambda I)f, \tag{6.54}$$

where $(A - \lambda I)^{-1} : X \to X$ is a bounded operator. The operator $(A - \lambda I)^{-1}K : X \to X$ is compact as a composition of bounded and compact operators. Therefore, the spectrum of $I - (A - \lambda I)^{-1}K$ in X consists of eigenvalues accumulating at 1. By the Fredholm alternative theorem, either (i) this operator is invertible for this λ with a bounded inverse or (ii) there exists $f_0 \in \mathrm{Dom}(A)$, $f_0 \neq 0$ such that $f_0 = (A - \lambda I)^{-1}Kf_0$.

In the case (i), we can rewrite (6.54) for every $f \in \mathrm{Dom}(A)$ in the form

$$f = (I - (A - \lambda I)^{-1}K)^{-1}(A - \lambda I)^{-1}(A_0 - \lambda I)f, \tag{6.55}$$

from which we obtain a contradiction against the assumption $\lambda \in \sigma(A_0)$. Indeed,

- If $\lambda \in \sigma_p(A_0)$, then there exists $f_0 \in \mathrm{Dom}(A_0)$, $f_0 \neq 0$ such that $(A_0 - \lambda I)f_0 = 0$, in which case equation (6.55) yields that $f_0 = 0$, a contradiction.
- If $\lambda \in \sigma_r(A_0)$, then there exists $g_0 \in X$ such that $g_0 \notin \mathrm{Ran}(A_0 - \lambda I)$. This is in contradiction with (6.55) since for every $g_0 \in X$, there exists a unique $f_0 \in \mathrm{Dom}(A)$ such that
$$(A - \lambda I)(I - (A - \lambda I)^{-1}K)f_0 = g_0 = (A_0 - \lambda I)f_0.$$
- If $\lambda \in \sigma_c(A_0)$, then for $f \in \mathrm{Dom}(A_0)$ we let $g := (A_0 - \lambda I)f \in X$ and obtain from (6.55) that
$$\|f\|_X = \|(I - (A - \lambda I)^{-1}K)^{-1}(A - \lambda I)^{-1}g\|_X \leq C\|g\|_X, \tag{6.56}$$
for some $C > 0$. Since $\lambda \in \sigma_c(A_0)$, we have $\mathrm{Ran}(A_0 - \lambda I) = X$ for this λ and since $f \in \mathrm{Dom}(A_0)$ is arbitrary, the bound (6.56) implies that for every $g \in X$,
$$\|(A_0 - \lambda I)^{-1}g\|_X \leq C\|g\|_X,$$
in contradiction with the assumption $\lambda \in \sigma_c(A_0)$.

In the case (ii), there exists $f_0 \in \mathrm{Dom}(A)$, $f_0 \neq 0$, such that

$$f_0 = (A - \lambda I)^{-1}Kf_0,$$

and hence we can rewrite (6.54) for this f_0 as

$$(A - \lambda I)^{-1}(A_0 - \lambda I)f_0 = 0.$$

Therefore, we have $(A_0 - \lambda I)f_0 = 0$, and hence $\lambda \in \sigma_p(A_0)$, in contradiction with the assumption that the intersection $\sigma_p(A_0) \cap \rho(A)$ is empty.

Thus, if $\lambda \in \sigma(A_0)$, then $\lambda \in \sigma(A)$. Since $A_0 - A = -K$ and the previous argument does not depend on the sign of K, the reverse statement is true. Hence, $\sigma(A) = \sigma(A_0)$. ☐

We will next analyze the spectral instability of peaked waves in the two main examples of the b-CH equation and the reduced Ostrovsky equation.

6.4.1. Peaked solitary waves in the b-CH equation. The spectral instability of the peaked solitary waves in the b-CH equation (6.3) of Example 6.8 follows from the next theorem, which was proven in [**169**].

THEOREM 6.3. *For every $b \in \mathbb{R}$, the spectrum of the linear operator $\mathcal{A}$ defined by (6.41) with the domain* $\mathrm{Dom}(\mathcal{A}) = \{w \in L^2(\mathbb{R}) : (1-\phi)\partial_x w \in L^2(\mathbb{R})\}$ *is given by*

$$\sigma(\mathcal{A}) = \left\{ \lambda \in \mathbb{C} : \; |\,\mathrm{Re}(\lambda)\,| \leq \left| \frac{5}{2} - b \right| \right\}.$$

Moreover, as illustrated in Figure 6.5,

- *the point spectrum contains a double eigenvalue $\lambda = 0$ for every b,*
- *the point spectrum also includes the open strip $0 < |\,\mathrm{Re}(\lambda)\,| < \frac{5}{2} - b$ if $b < \frac{5}{2}$,*
- *the residual spectrum is located inside $0 < |\,\mathrm{Re}(\lambda)\,| < b - \frac{5}{2}$ if $b > \frac{5}{2}$,*
- *the continuous spectrum is located on $\lambda \in i\mathbb{R}\backslash\{0\}$ and on $\mathrm{Re}(\lambda) = \pm\left|\frac{5}{2} - b\right|$.*

PROOF. The idea of the proof is to decompose the operator $\mathcal{A}$ as in Lemma 6.5, such that $\mathcal{A} = \mathcal{A}_0 + \mathcal{K}$, where

$$\mathcal{A}_0 := (1-\phi)\partial_x + (2-b)\phi', \quad \mathcal{K} := \frac{1}{2}(b-3)\phi * (\phi' \cdot\,) - \frac{1}{2}(2b-3)\phi' * (\phi \cdot\,).$$

Note that $\mathcal{K} : L^2(\mathbb{R}) \mapsto L^2(\mathbb{R})$ is a (compact) Hilbert–Schmidt operator

$$(\mathcal{K}v)(x) = \int_{\mathbb{R}} k(x,y) v(y) dy$$

with square integrable kernel

$$k(x,y) = \frac{1}{2}(3-b)\,\mathrm{sgn}(y) e^{-|y|-|x-y|} + \frac{1}{2}(2b-3)\,\mathrm{sgn}(x-y) e^{-|y|-|x-y|},$$

so that $\|k\|_{L^2(\mathbb{R}\times\mathbb{R})} < \infty$. The proof is based on three steps, after which Lemma 6.5 is applied. The first step identifies the three parts of the spectrum of $\mathcal{A}_0$. The second step gives explicit estimates for the resolvent operator in the resolvent set of $\mathcal{A}_0$ and confirms that $\rho(\mathcal{A}_0) = \mathbb{C}\backslash\sigma(\mathcal{A}_0)$. The third step gives the point and residual spectra of $\mathcal{A}$.

STEP 1. The point, residual, and continuous spectra of $\mathcal{A}_0$ with $\mathrm{Dom}(\mathcal{A}_0) = \mathrm{Dom}(\mathcal{A})$ in $L^2(\mathbb{R})$ are given by

$$\sigma_p(\mathcal{A}_0) = \left\{ \lambda \in \mathbb{C} : \; 0 < |\,\mathrm{Re}(\lambda)\,| < \frac{5}{2} - b \right\} \quad \text{if } b < \frac{5}{2},$$

$$\sigma_r(\mathcal{A}_0) = \left\{ \lambda \in \mathbb{C} : \; 0 < |\,\mathrm{Re}(\lambda)\,| < b - \frac{5}{2} \right\} \quad \text{if } b > \frac{5}{2},$$

$$\sigma_c(\mathcal{A}_0) = \left\{ \lambda \in \mathbb{C} : \; \mathrm{Re}(\lambda) = 0 \;\; \text{and} \;\; |\,\mathrm{Re}(\lambda)\,| = \left| b - \frac{5}{2} \right| \right\}.$$

PROOF OF STEP 1. For $\sigma_p(\mathcal{A}_0)$, we consider the differential equation

$$(1-\phi)\frac{dv}{dx} + (2-b)\phi' v = \lambda v, \quad x \in \mathbb{R}. \tag{6.57}$$

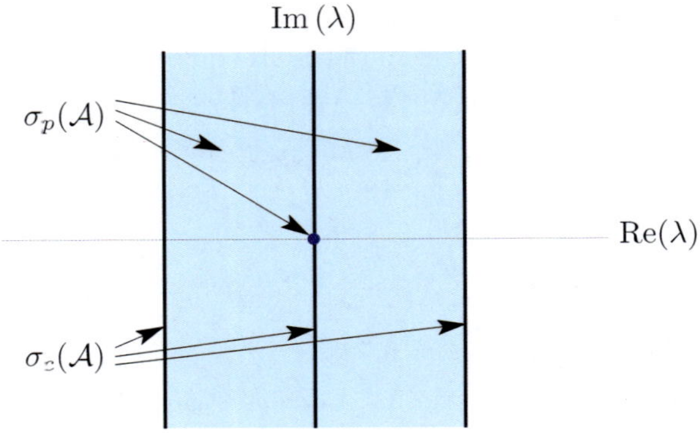

(a) $b < \frac{5}{2}$

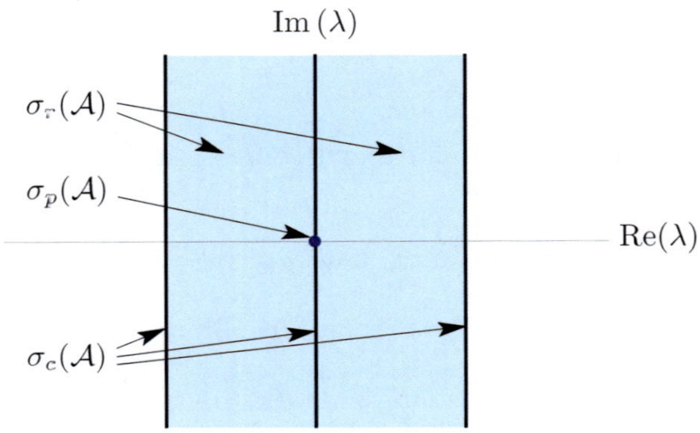

(b) $b > \frac{5}{2}$

FIGURE 6.5. Sketches of the vertical strip in the complex plane containing the different parts of the spectrum of the linear operator $\mathcal{A}$ in Theorem 6.3.

The differential equation (6.57) is solved separately for $x > 0$ and $x < 0$, and has the general solution

$$v(x) = \begin{cases} v_+ e^{\lambda x}(1 - e^{-x})^{2+\lambda-b}, & x > 0, \\ v_- e^{\lambda x}(1 - e^{x})^{2-\lambda-b}, & x < 0, \end{cases} \tag{6.58}$$

where v_+ and v_- are arbitrary constants. The following symmetry exists for (6.57) because ϕ is even and ϕ' is odd.

> If $\lambda = \lambda_0$ is an eigenvalue with the eigenfunction $v = v_0(x)$, then $\lambda = -\lambda_0$ is an eigenvalue with the eigenfunction $v = v_0(-x)$.

Due to this symmetry, it is sufficient to consider eigenvalues λ with $\mathrm{Re}(\lambda) \geq 0$. Since $v(x) \sim v_+ e^{\lambda x}$ as $x \to +\infty$, $v \in L^2(\mathbb{R})$ only if $v_+ = 0$. Since $v(x) \sim v_- e^{\lambda x}$ as $x \to -\infty$, $v \in L^2(\mathbb{R})$ is satisfied for arbitrary v_- if and only if $\mathrm{Re}(\lambda) > 0$. Since $v(x) \sim v_- |x|^{2-b-\lambda}$ as $x \to 0^-$, $v \in L^2(\mathbb{R})$ is satisfied for arbitrary v_- if and only if $\mathrm{Re}(\lambda) < \frac{5}{2} - b$. In all other cases, we have to set $v_- = 0$ for $v \in L^2(\mathbb{R})$ so that $v = 0$ is the only solution of (6.57) in $L^2(\mathbb{R})$. Summarizing and using the symmetry above, $\sigma_p(\mathcal{A}_0)$ is nonempty if $b < \frac{5}{2}$ and it is located in the strip determined by $0 < |\operatorname{Re}(\lambda)| < \frac{5}{2} - b$.

For $\sigma_c(\mathcal{A}_0)$, we consider the resolvent equation $\mathcal{A}_0 v - \lambda v = f$ for $f \in L^2(\mathbb{R})$ separately for $x > 0$ and $x < 0$. By Theorem 4 in [**88**, p. 1438], if $\mathcal{A}_0$ is a differential operator on $\mathbb{R} = (-\infty, 0) \cup (0, \infty)$, and $\mathcal{A}_0^{\pm}$ are restrictions of $\mathcal{A}_0$ on $(-\infty, 0)$ and $(0, \infty)$, respectively, then

$$\sigma_c(\mathcal{A}_0) = \sigma_c(\mathcal{A}_0^+) \cup \sigma_c(\mathcal{A}_0^-).$$

For $x > 0$, we use the transformation $(0, \infty) \ni x \mapsto z = \log(e^x - 1) \in \mathbb{R}$ and obtain the resolvent equation $\mathfrak{A}_0^+ \mathfrak{v} - \lambda \mathfrak{v} = \mathfrak{f}$, where $\mathfrak{v}(z) := v(x)$, $\mathfrak{A}_0^+ := \partial_z + (b-2)(1+e^z)^{-1}$ and $\mathfrak{f}(z) := f(x)$. Since

$$\int_0^\infty v(x)^2 dx = \int_{-\infty}^0 e^z \frac{\mathfrak{v}(z)^2 dz}{1+e^z} + \int_0^\infty \frac{\mathfrak{v}(z)^2 dz}{1+e^{-z}},$$

$\sigma_c(\mathcal{A}_0^+)$ is the union of $\sigma_c(\mathfrak{A}_0^+|_{L^2(\mathbb{R}_+)})$ in $L^2(\mathbb{R}_+)$ and $\sigma_c(\mathfrak{A}_0^+|_{L^2(\mathbb{R}_-;e^z)})$ in the exponential weighted space $L^2(\mathbb{R}_-; e^z)$. Since $\mathfrak{A}_0^+ = \partial_z + e^{-z}(b-2)(1+e^{-z})^{-1}$ and $e^{-z}(1+e^{-z})^{-1}$ is bounded and decaying exponentially to 0 as $z \to +\infty$, $\sigma_c(\mathfrak{A}_0^+|_{L^2(\mathbb{R}_+)})$ is located on the line determined by $\operatorname{Re}(\lambda) = 0$. Since $\mathfrak{A}_0^+ = \partial_z + (b-2) - (b-2)e^z(1+e^z)^{-1}$ and $e^z(1+e^z)^{-1}$ is bounded and decaying exponentially to 0 as $z \to -\infty$, $\sigma_c(\mathfrak{A}_0^+|_{L^2(\mathbb{R}_-;e^z)})$ is located on the line determined by $\operatorname{Re}(\lambda) = b - \frac{5}{2}$. Thus,

$$\sigma_c(\mathcal{A}_0^+) = \left\{ \lambda \in \mathbb{C} : \operatorname{Re}(\lambda) = 0 \text{ and } \operatorname{Re}(\lambda) = b - \frac{5}{2} \right\}.$$

Similarly, for $x < 0$, we use the transformation $(-\infty, 0) \ni x \mapsto z = -\log(e^{-x} - 1) \in \mathbb{R}$ and obtain the resolvent equation $\mathfrak{A}_0^- \mathfrak{v} - \lambda \mathfrak{v} = \mathfrak{f}$, where $\mathfrak{v}(z) := v(x)$, $\mathfrak{f}(z) := f(x)$ and $\mathfrak{A}_0^- := \partial_z + (2-b)(1+e^{-z})^{-1}$. Since

$$\int_{-\infty}^0 v(x)^2 dx = \int_{-\infty}^0 \frac{\mathfrak{v}(z)^2 dz}{1+e^z} + \int_0^\infty e^{-z} \frac{\mathfrak{v}(z)^2 dz}{1+e^{-z}},$$

$\sigma_c(\mathcal{A}_0^-)$ is the union of $\sigma_c(\mathfrak{A}_0^-|_{L^2(\mathbb{R}_-)})$ in $L^2(\mathbb{R}_-)$ and $\sigma_c(\mathfrak{A}_0^-|_{L^2(\mathbb{R}_+;e^{-z})})$ in $L^2(\mathbb{R}_+; e^{-z})$. Similar to the previous arguments, the continuous spectrum of $\mathfrak{A}_0^- = \partial_z + e^z(2-b)(1+e^z)^{-1}$ in $L^2(\mathbb{R}_-)$ is located on the line determined by $\operatorname{Re}(\lambda) = 0$ and the continuous spectrum of $\mathfrak{A}_0^- = \partial_z + (2-b) - (2-b)e^{-z}(1+e^{-z})^{-1}$ in $L^2(\mathbb{R}_+; e^{-z})$ is located on the line determined by $\operatorname{Re}(\lambda) = \frac{5}{2} - b$. Thus,

$$\sigma_c(\mathcal{A}_0^-) = \left\{ \lambda \in \mathbb{C} : \operatorname{Re}(\lambda) = 0 \text{ and } \operatorname{Re}(\lambda) = \frac{5}{2} - b \right\}.$$

For $\sigma_r(\mathcal{A}_0)$, recall from Lemma 6.4 that if $\sigma_p(\mathcal{A}_0)$ is empty, then $\sigma_r(\mathcal{A}_0) = \sigma_p(\mathcal{A}_0^*)$, which can be determined by considering the differential equation

$$-(1-\phi)\frac{dv}{dx} + (3-b)\phi' v = \lambda v, \quad x \in \mathbb{R}. \tag{6.59}$$

Since (6.59) is obtained from (6.57) using the transformation $\lambda \mapsto -\lambda$ and $b - 2 \mapsto 3 - b$, we can obtain its general solution by applying the same transformation to (6.58). Proceeding similarly with the limits $x \to \pm\infty$ and $x \to 0^{\pm}$ shows that nonzero solutions of (6.59) exist in $L^2(\mathbb{R})$ if and only if $v_- = 0$, $b > \frac{5}{2}$, and

$0 < |\operatorname{Re}(\lambda)| < b - \frac{5}{2}$, which yields the location of $\sigma_p(\mathcal{A}_0^*)$. Since $\sigma_p(\mathcal{A}_0)$ is an empty set for $b > \frac{5}{2}$, we obtain $\sigma_r(\mathcal{A}_0) = \sigma_p(\mathcal{A}_0^*)$. □

STEP 2. The resolvent set of $\mathcal{A}_0$ in $L^2(\mathbb{R})$ is located in the strip determined by

$$\rho(\mathcal{A}_0) = \left\{ \lambda \in \mathbb{C} : \ |\operatorname{Re}(\lambda)| > |b - \frac{5}{2}| \right\}.$$

PROOF OF STEP 2. Let us consider the resolvent equation

$$\mathcal{A}_0 v - \lambda v = f, \tag{6.60}$$

where $f \in L^2(\mathbb{R})$ is arbitrary and $\operatorname{Re}(\lambda) \geq 0$ is assumed without loss of generality. We multiply both sides of (6.60) by $\bar{v}$ and integrate over $\mathbb{R}$ to obtain

$$\langle ((1-\phi)v)', v \rangle + (3-b)\langle \phi' v, v \rangle - \lambda \|v\|^2 = \langle f, v \rangle. \tag{6.61}$$

Integrating by parts and using that $\lim_{x\to\pm\infty} v(x) = 0$ for $v \in \operatorname{Dom}(\mathcal{A})$, we have that

$$\langle ((1-\phi)v)', v \rangle = -\langle v, ((1-\phi)v)' \rangle - \langle \phi' v, v \rangle,$$

which yields

$$\operatorname{Re}(\langle ((1-\phi)v)', v \rangle) = -\frac{1}{2}\langle \phi' v, v \rangle.$$

Using this equation in the real part of (6.61), we obtain

$$\left(b - \frac{5}{2}\right)\langle \phi' v, v \rangle + \operatorname{Re}(\lambda)\|v\|^2 = -\operatorname{Re}(\langle f, v \rangle). \tag{6.62}$$

Since $-\|v\|^2 \leq \langle \phi' v, v \rangle \leq \|v\|^2$, we obtain from (6.62) in the case $b \leq \frac{5}{2}$ that

$$\left(\operatorname{Re}(\lambda) + b - \frac{5}{2}\right)\|v\|^2 \leq |\operatorname{Re}(\langle f, v \rangle)| \leq \|f\|\|v\|,$$

where the Cauchy–Schwarz inequality has been used. Hence, for every eigenvalue λ with $\operatorname{Re}(\lambda) > \frac{5}{2} - b$ with $b \leq \frac{5}{2}$, there exists C_λ such that $\|v\| \leq C_\lambda \|f\|$ so that this λ belongs to $\rho(\mathcal{A}_C)$. In a very similar way, we obtain from (6.62) in the case $b \geq \frac{5}{2}$ that

$$\left(\operatorname{Re}(\lambda) - b + \frac{5}{2}\right)\|v\|^2 \leq |\operatorname{Re}(\langle f, v \rangle)| \leq \|f\|\|v\|.$$

Therefore, every λ with $\operatorname{Re}(\lambda) > b - \frac{5}{2}$ and $b \geq \frac{5}{2}$ belongs to $\rho(\mathcal{A})$. □

STEP 3. $\sigma_p(\mathcal{A})$ and $\sigma_r(\mathcal{A})$ coincide with $\sigma_p(\mathcal{A}_0)$ and $\sigma_r(\mathcal{A}_0)$, respectively, except that $\sigma_p(\mathcal{A})$ also includes the double zero eigenvalue for every value of b.

PROOF OF STEP 3. For $\sigma_p(\mathcal{A})$, it follows from (6.42) that the zero eigenvalue is double due to the two-dimensional invariant subspace $\{\phi, \phi'\}$. Let us look for other eigenvalues of $\sigma_p(\mathcal{A})$ for $\operatorname{Re}(\lambda) \geq 0$. Applying the operator $1 - \partial_x^2$ to $\mathcal{A}v = \lambda v$ separately for $x < 0$ and $x > 0$ yields the differential equation

$$(1-\phi)(v' - v''') - b\phi'(v - v'') = \lambda(v - v''). \tag{6.63}$$

Let $m := v - v''$. The differential equation (6.63) becomes the first-order equation

$$(1-\phi)\frac{dm}{dx} - b\phi' m = \lambda m, \tag{6.64}$$

which coincides with the differential equation (6.57) after replacing b by $b-2$. Therefore, the exact solution is given by the corresponding transformation of the solution (6.58), namely

$$m(x) = \begin{cases} m_+ e^{\lambda x}(1-e^{-x})^{\lambda-b}, & x>0, \\ m_- e^{\lambda x}(1-e^{x})^{-\lambda-b}, & x<0, \end{cases} \tag{6.65}$$

where m_+ and m_- are arbitrary constants. The eigenfunction v is recovered from solutions of the second-order equation

$$v(x) - v''(x) = \begin{cases} m_+ e^{\lambda x}(1-e^{-x})^{\lambda-b}, & x>0, \\ m_- e^{\lambda x}(1-e^{x})^{-\lambda-b}, & x<0. \end{cases} \tag{6.66}$$

If $v \in L^2(\mathbb{R})$, then $m \in H^{-2}(\mathbb{R})$, the dual space of $H^2(\mathbb{R})$. Since $m(x) \sim m_+ e^{\lambda x}$ as $x \to +\infty$, then $v \in L^2(\mathbb{R})$ for $\mathrm{Re}(\lambda) \geq 0$ exists if and only if $m_+ = 0$ and $v(x) = c_+ e^{-x}$, where c_+ is arbitrary. Similarly, $m(x) \sim m_- e^{\lambda x}$ as $x \to -\infty$. If $\mathrm{Re}(\lambda) = 0$, then $v \in L^2(\mathbb{R})$ exists if and only if $m_- = 0$ and $v(x) = c_- e^{x}$, where c_- is arbitrary. In this case, $v = c_1\phi + c_2\phi'$ with $c_1 \pm c_2 = c_{\mp}$, which corresponds to the double eigenvalue $\lambda = 0$ in $\sigma_p(L)$. If $\mathrm{Re}(\lambda) > 0$, we set $m_+ = 0$, $m_- = 1$, and rewrite the solution of the differential equation (6.66) for $x<0$ in the form

$$v(x) = e^{\lambda x} f(x)(1-e^{x})^{2-\lambda-b}, \quad x<0, \tag{6.67}$$

where $f(x)$ satisfies the second-order differential equation

$$\begin{aligned} &(1-e^x)^2(f'' + 2(2-b)f' + (b-1)(b-3)f) \\ &+ (\lambda+b-2)(1-e^x)(2f' + (3-2b)f) + (\lambda+b-2)(\lambda+b-1)f = -1. \end{aligned} \tag{6.68}$$

The homogeneous part of the second-order equation (6.68) has the regular singular point at $x=0$, which is associated with the indicial equation

$$\sigma^2 + (2\lambda + 2b - 5)\sigma + (\lambda+b-2)(\lambda+b-1) = 0$$

for power solutions $f(x) \sim x^{\sigma}$. If $\lambda + b \neq \{1,2\}$, then 0 is not among the roots of the indicial equation, whereas if $\lambda + b = \{1,2\}$, then 0 is a simple root of the indicial equation. By the Frobenius theory, there exists a particular solution to the differential equation (6.68) with the following behavior near the regular singular point $x=0$:

$$f_p(x) \sim \begin{cases} 1 + \mathcal{O}(|x|), & \lambda + b \neq \{1,2\}, \\ \log|x| + \mathcal{O}(|x|\log|x|), & \lambda+b = \{1,2\}, \end{cases} \quad \text{as} \quad x \to 0^-,$$

which yields the corresponding behavior of $v_p(x)$ found from (6.67):

$$v_p(x) \sim \begin{cases} |x|^{2-b-\lambda}, & \lambda+b \neq \{1,2\}, \\ \log|x|, & \lambda+b = 2, \\ |x|\log|x|, & \lambda+b=1, \end{cases} \quad \text{as} \quad x \to 0^-.$$

Hence, (6.66) has a solution $v \in L^2(\mathbb{R})$ if and only if $0 < \mathrm{Re}(\lambda) < \frac{5}{2} - b$. This yields $\sigma_p(\mathcal{A}) = \{\lambda \in \mathbb{C} : \ 0 < |\,\mathrm{Re}(\lambda)\,| < \frac{5}{2} - b\} \cup \{0\}$ if $b < \frac{5}{2}$.

For $\sigma_r(\mathcal{A})$, we again first look for $\sigma_p(\mathcal{A}^*)$, where

$$\mathcal{A}^* := (\phi - 1)\partial_x + (3-b)\phi' + \frac{1}{2}(b-3)\phi'(\phi * \cdot) + \frac{1}{2}(2b-3)\phi(\phi' * \cdot).$$

By making the substitution $v = (1-\partial_x^2)k$ in $\mathcal{A}^*v = \lambda v$ and integrating by parts under the assumption that k and k' are bounded and continuous functions, we obtain

$$(1-\phi)k''' + (b-3)\phi' k'' + (2(b-1)\phi - 1)k' = \lambda(k - k''). \tag{6.69}$$

The operator $(1-\partial_x^2)$ can be factored out for both cases $x>0$ and $x<0$ as

$$(1-\partial_x^2)\left[(\phi-1)k' + (1-b)\phi' k - \lambda k\right] = 0. \tag{6.70}$$

The general solution of (6.70) is

$$(\phi-1)k' + (1-b)\phi' k - \lambda k = \begin{cases} K_+ e^{-x} + M_+ e^{x}, & x>0,\\ K_- e^{x} + M_- e^{-x}, & x<0, \end{cases} \tag{6.71}$$

where $K_\pm$ and $M_\pm$ are arbitrary constants. The coefficients $M_\pm$ give unbounded solutions which can be normalized to be $k(x) \sim e^{|x|}$ if $\lambda \neq 1$ and $k(x) \sim xe^{|x|}$ if $\lambda = 1$. Since $v = (1-\partial_x^2)k$, we obtain from (6.69) for $\lambda \neq 1$ that

$$\lim_{|x|\to\infty} (v' + \lambda v) = \lim_{|x|\to\infty} e^{-|x|}\left(-k''' - (b-3)\,\mathrm{sgn}(x)k'' + 2(b-1)k'\right) = b. \tag{6.72}$$

This implies for $b \neq 0$ and $\lambda \neq 0$ that $v(x) \to b/\lambda$ as $|x| \to \infty$, hence $v \notin L^2(\mathbb{R})$. For $b \neq 0$ and $\lambda = 1$ (the exceptional case), the limit (6.72) is unbounded, hence $v \notin L^2(\mathbb{R})$. For $b = 0$ and any $\lambda \neq 0$, equation (6.69) is satisfied with $k = e^{|x|}$ for which $v \equiv 0$. Hence, for any b and $\lambda \neq 0$, we have to set $M_+ = M_- = 0$ to ensure that $v \in L^2(\mathbb{R})$.

It follows from (6.71) with $M_+ = M_- = 0$ that k and k' are continuous across $x = 0$ if and only if

$$\begin{cases} K_\pm = (\pm(b-1) - \lambda)k(0),\\ \mp K_\pm = (\pm(b-2) - \lambda)k'(0) + (1-b)k(0), \end{cases}$$

where we have used $\phi'(0^\pm) = \mp 1$ and $\phi(0^\pm) = 1$. If $\lambda \neq 0$, this system yields $K_+ = K_- = 0$. Setting all constants to zero in (6.71), we obtain the first-order equation

$$(\phi - 1)k' + (1-b)\phi' k = \lambda k.$$

This equation follows from (6.64) after the transformation: $\lambda \mapsto -\lambda$ and $b \mapsto 1-b$. Therefore, we obtain the following general solution by applying this transformation to (6.65):

$$k(x) = \begin{cases} k_+ e^{-\lambda x}(1-e^{-x})^{b-\lambda-1}, & x>0,\\ k_- e^{-\lambda x}(1-e^{x})^{b+\lambda-1}, & x<0, \end{cases}$$

where k_+ and k_- are arbitrary constants. Proceeding similarly with the limits $x \to \pm\infty$ and $x \to 0^\pm$ for $\mathrm{Re}(\lambda) > 0$ shows that the corresponding nonzero function $v = (1-\partial_x^2)k$ is in $L^2(\mathbb{R})$ if and only if $k_- = 0$ and $0 < \mathrm{Re}(\lambda) < b - \frac{5}{2}$ if $b > \frac{5}{2}$.

In order to show that $0 \in \sigma_p(\mathcal{A}^*)$, we obtain two exact solutions of the differential equation (6.69) for $\lambda = 0$:

$$k_1 = 1, \quad k_2 = (1-\phi)^{b-1} - 1,$$

whereas the third solution k_3 cannot be computed explicitly. However, it follows from (6.71) and (6.72) for $\lambda = 0$ that the third solution diverges like $k_3(x) \sim e^{|x|}$ and gives either the divergent solution $v(x) \sim bx$ as $|x| \to \infty$ for $b \neq 0$ or the zero solution $v \equiv 0$ if $b = 0$. At the same time, $k_1 \notin L^2(\mathbb{R})$, whereas $v = (1-\partial_x^2)k_2$

belong to $\mathrm{Dom}(\mathcal{A}) \subset L^2(\mathbb{R})$ provided that $b > \frac{5}{2}$. We conclude that $0 \in \sigma_p(\mathcal{A}^*)$ if and only if $b > \frac{5}{2}$, that is, $\sigma_p(\mathcal{A}^*) = \{\lambda \in \mathbb{C} : 0 < |\,\mathrm{Re}(\lambda)\,| < b - \frac{5}{2}\} \cup \{0\}$ if $b > \frac{5}{2}$.

Since $\sigma_c(\mathcal{A}) = \sigma_c(\mathcal{A}^*)$ by a similar behavior of the differential equations (6.63) and (6.69), Lemma 6.4 implies that $\sigma_p(\mathcal{A}^*) \subseteq \sigma_r(\mathcal{A}) \cup \sigma_p(\mathcal{A})$ and $\sigma_r(\mathcal{A}) \subseteq \sigma_p(\mathcal{A}^*)$. Since $\sigma_p(\mathcal{A}) = \{0\}$ for $b > \frac{5}{2}$, we obtain $\sigma_r(\mathcal{A}) = \sigma_p(\mathcal{A}^*)\backslash\{0\} = \{\lambda \in \mathbb{C} : 0 < |\,\mathrm{Re}(\lambda)\,| < b - \frac{5}{2}\}$ if $b > \frac{5}{2}$. ☐

End of proof of Theorem 6.3. Combining the information gathered in the three steps, we can apply Lemma 6.5 to finish the proof. For $b < \frac{5}{2}$, $\sigma_p(\mathcal{A}_0)$ consists of the bands $0 < |\,\mathrm{Re}(\lambda)\,| < \frac{5}{2} - b$, whereas $\sigma_p(\mathcal{A})$ consists of the same bands and an additional double zero eigenvalue. Since the resolvent set $\rho(\mathcal{A}_0)$ consists of the bands $|\,\mathrm{Re}(\lambda)\,| > \frac{5}{2} - b$, $\sigma_p(\mathcal{A}) \cap \rho(\mathcal{A}_0)$ is an empty set. Since $\sigma_p(\mathcal{A}_0) \subset \sigma_p(\mathcal{A})$, $\sigma_p(\mathcal{A}_0) \cap \rho(\mathcal{A})$ is also an empty set. Hence, $\sigma(\mathcal{A}) = \sigma(\mathcal{A}_0)$ for $b < \frac{5}{2}$. For $b \geq \frac{5}{2}$, $\sigma_p(\mathcal{A}_0)$ is an empty set, hence $\sigma_p(\mathcal{A}_0) \cap \rho(\mathcal{A})$ is an empty set, whereas $\sigma_p(\mathcal{A}) = \{0\}$ does not belong to $\rho(\mathcal{A}_0)$, hence $\sigma_p(\mathcal{A}) \cap \rho(\mathcal{A}_0)$ is also an empty set. Hence, $\sigma(\mathcal{A}) = \sigma(\mathcal{A}_0)$.

Since the three subsets of the spectrum in Definition B.2 are disjoint, we have that $\sigma_c(\mathcal{A}) = \sigma(\mathcal{A})\backslash\sigma_p(\mathcal{A}) \cup \sigma_r(\mathcal{A})$ so that $\sigma_c(\mathcal{A})$ is located on $\lambda \in i\,\mathbb{R}\backslash\{0\}$ and on $\mathrm{Re}(\lambda) = \pm\left|\frac{5}{2} - b\right|$. ☐

Remark 6.17. The only difference between $\sigma(\mathcal{A})$ and $\sigma(\mathcal{A}_0)$ in Theorem 6.3 is the double zero eigenvalue in the point spectrum of $\mathcal{A}$, which is embedded into $\sigma_c(\mathcal{A})$.

Theorem 6.3 gives the spectral instability of the peaked waves in the sense of Definition 6.3. The following lemma shows that this result implies the linear instability of peaked waves in the sense of Definition 6.2 in $\mathrm{Dom}(\mathcal{A}) \subset L^2(\mathbb{R})$.

Lemma 6.6. *The spectral instability of peaked waves in Theorem* 6.3 *for* $b \neq \frac{5}{2}$ *implies the linear instability of peaked waves in* $\mathrm{Dom}(\mathcal{A}) \subset L^2(\mathbb{R})$.

Proof. Let $\lambda_0 > 0$ be a real eigenvalue in $\sigma_p(\mathcal{A})$ for $b < \frac{5}{2}$ and let $w_0 \in \mathrm{Dom}(\mathcal{A}) \subset L^2(\mathbb{R})$ be the corresponding eigenfunction of $\mathcal{A}$. Then the linearized equation (6.39) has the exact solution $w(t,x) = e^{\lambda_0 t} w_0(x)$ with exponential growth of $\|w(t,\cdot)\|_{L^2}$. This implies the linear instability of peaked waves in the sense of Definition 6.2 in $\mathrm{Dom}(\mathcal{A}) \subset L^2(\mathbb{R})$.

Let $\lambda_0 > 0$ be a real eigenvalue in $\sigma_r(\mathcal{A}) = \sigma_p(\mathcal{A}^*)\backslash\{0\}$ for $b > \frac{5}{2}$ and let $w_0 \in \mathrm{Dom}(\mathcal{A}^*) \subset L^2(\mathbb{R})$ be the corresponding eigenfunction of $\mathcal{A}^*$. Since $\mathrm{Dom}(\mathcal{A}^*) = \mathrm{Dom}(\mathcal{A})$, we consider the decomposition $w(t,x) = a(t)w_0(x) + \tilde{w}(t,x)$, where $a(t)$ is uniquely determined by the orthogonality condition $\langle w_0, \tilde{w}(t,\cdot)\rangle = 0$. Both $a(t)$ and $\tilde{w}(t,x)$ are found from (6.39) which reads

$$\frac{da}{dt} w_0 + \frac{d\tilde{w}}{dt} = a\mathcal{A}w_0 + \mathcal{A}\tilde{w}.$$

Projecting to w_0 yields $\frac{da}{dt} = \lambda_0 a$ with exponential growth of $a(t)$ since $\lambda_0 > 0$. This also implies the exponential growth of $\|w(t,\cdot)\|_{L^2}$ due to the orthogonality $\langle w_0, \tilde{w}(t,\cdot)\rangle = 0$, which implies the linear instability of peaked waves in the sense of Definition 6.2. ☐

Remark 6.18. For $b = \frac{5}{2}$, Theorem 6.3 does not imply spectral instability of the peaked waves. However, projections to $\{\phi, \phi'\}$ are linearly unstable for every $b \neq 2$ according to the explicit solution (6.45) of the first-order system (6.44) with

$\mathfrak{w} = 0$ (Example 6.8). Hence, the peaked waves of the b-CH model are linearly unstable for every b.

REMARK 6.19. For the Camassa–Holm equation ($b = 2$), the differential equation (6.66) with $\lambda \neq \{-1, 0, 1\}$ can be solved exactly in the form

$$v(x) = \frac{1}{\lambda(1-\lambda^2)} \begin{cases} m_+(\lambda + e^{-x})(e^x - 1)^{\lambda}, & x > 0, \\ m_-(\lambda - e^x)(e^{-x} - 1)^{-\lambda}, & x < 0. \end{cases} \tag{6.73}$$

One can verify that this eigenfunction is a solution of the spectral problem

$$(1-\phi)v'(x) + \phi \int_0^x v(x')dx' = \lambda v(x),$$

which follows from the separation of variables in the linearized equation (6.36). It follows from the exact solution (6.73) that $\sigma_p(\mathcal{A}) = \{\lambda \in \mathbb{C} : \ 0 < |\operatorname{Re}(\lambda)| < \frac{1}{2}\} \cup \{0\}$ in agreement with Theorem 6.3 for $b = 2$. By Lemma 6.6, the spectral instability gives the linear instability of peaked waves in the linearized equation (6.36). Recall that the exact solution of the linearized equation for $b = 2$ has been obtained with the method of characteristics (Example 6.7) but it was difficult to analyze the behavior of $\|v(t,\cdot)\|_{L^2}$ if $v_0 \in H^1(\mathbb{R}) \cap W^{1,\infty}(\mathbb{R})$. On the other hand, $\|v(t,\cdot)\|_{L^2}$ grows exponentially if $v_0 \in \operatorname{Dom}(\mathcal{A}) \subset L^2(\mathbb{R})$ is the eigenfunction of $\mathcal{A}$ with weak singularity at $x \to 0$ as in (6.73).

The following example illustrates the linear instability of peaked waves with the profile ϕ for the b-CH equation in Example 6.8. Compared to Lemma 6.6, we replace $\mathcal{A}$ by $\mathcal{A}_0$ and show that perturbations to the peaked waves grow differently due to contributions from the point spectrum of $\mathcal{A}_0$ and those from the residual spectrum of $\mathcal{A}_0$.

EXAMPLE 6.10. Recall the truncated operator $\mathcal{A}_0$ in Theorem 6.3 with

$$\operatorname{Dom}(\mathcal{A}_0) = \operatorname{Dom}(\mathcal{A})$$

and consider the corresponding initial-value problem $v_t = \mathcal{A}_0 v$ given by

$$\begin{cases} v_t = (1-\phi)v_x + (2-b)\phi' v, & t > 0, \\ v|_{t=0} = v_0, \end{cases} \tag{6.74}$$

where $v_0 \in \operatorname{Dom}(\mathcal{A})$. The initial-value problem (6.74) can be solved exactly by using the method of characteristics, like in Example 6.7 for the CH equation with $b = 2$. The characteristic curves $x = X(t,s)$ are given by the exact solution (6.38). Along the characteristic curves, the function $V(t,s) := v(t, x = X(t,s))$ satisfies

$$\begin{cases} V_t = (2-b)\phi'(X(t,s))V, & t > 0, \\ V|_{t=0} = v_0(s), \end{cases}$$

which is uniquely solved separately for $s > 0$ and $s < 0$ by

$$V(t,s) = v_0(s)\,[X_s(t,s)]^{2-b} = \begin{cases} v_0(s)\,[1 + (e^t - 1)e^{-s}]^{b-2}, & s > 0, \\ v_0(s)\,[1 + (e^{-t} - 1)e^{s}]^{b-2}, & s < 0. \end{cases}$$

By the chain rule, we obtain

$$\|v(t,\cdot)\|^2_{L^2(0,\infty)} = \int_0^\infty v_0(s)|^2 \left[1 + (e^t - 1)e^{-s}\right]^{2b-5} ds \tag{6.75}$$

and

$$\|v(t,\cdot)\|^2_{L^2(-\infty,0)} = \int_{-\infty}^0 |v_0(s)|^2 \left[1 + (e^{-t} - 1)e^s\right]^{2b-5} ds. \tag{6.76}$$

We can now analyze the expressions (6.75) and (6.76) in three different cases.

- If $b = \frac{5}{2}$, the linear evolution is L^2-preserving in time so that $\|v(t,\cdot)\|_{L^2} = \|v_0\|_{L^2}$. Preservation of the L^2 norm of the solution $v \in C^0(\mathbb{R}, \mathrm{Dom}(\mathcal{A}))$ is related to $\sigma(\mathcal{A}_0) = \{i\mathbb{R}\}$ in Step 1 of Theorem 6.3.
- If $b > \frac{5}{2}$, then it follows from (6.75) and (6.76) that
$$\|v(t,\cdot)\|^2_{L^2(0,\infty)} \geq (e^t - 1)^{2b-5} \int_0^\infty |v_0(s)|^2 e^{-(2b-5)s} ds$$
and
$$\|v(t,\cdot)\|^2_{L^2(-\infty,0)} \leq \int_{-\infty}^0 |v_0(s)|^2 ds.$$
This proves that the L^2 norm of the perturbation grows exponentially in time on the right side of the peak, while it stays bounded in time on the left side of the peak. The growth rate $e^{(b-\frac{5}{2})t}$ of the L^2 norm of the solution $v \in C^0(\mathbb{R}, \mathrm{Dom}(\mathcal{A}))$ agrees with the width of the unstable spectrum in $\sigma(\mathcal{A}_0)$ located in the strip determined by $0 < |\,\mathrm{Re}(\lambda)\,| \leq b - \frac{5}{2}$ in Step 1 of Theorem 6.3.
- If $b < \frac{5}{2}$, then it follows from (6.75) that
$$\|v(t,\cdot)\|^2_{L^2(0,\infty)} \leq \int_0^\infty |v_0(s)|^2 ds,$$
so that the L^2 norm of the perturbation stays bounded in time on the right side of the peak. If v_0 is the eigenfunction of the point spectrum of $\mathcal{A}_0$ for the eigenvalue $\lambda_0 \in (0, \frac{5}{2} - b)$, which is obtained from (6.58), that is,
$$v_0(s) = \frac{e^{\lambda_0 s}}{(1 - e^s)^{\lambda_0 + b - 2}}, \quad s < 0. \tag{6.77}$$
Then, after elementary transformations, we obtain for $s < 0$ that
$$V(t,s) = v_0(s)\left[1 + (e^{-t} - 1)e^s\right]^{b-2} = \frac{e^{\lambda_0 t + \lambda_0 X(t,s)}}{(1 - e^{X(t,s)})^{\lambda_0 + b - 2}} = v_0(X(t,s))e^{\lambda_0 t},$$
so that $\|v(t,\cdot)\|_{L^2(-\infty,0)} = \|v_0\|_{L^2(-\infty,0)} e^{\lambda_0 t}$ grows exponentially in time. Therefore, linear instability is obtained from the eigenfunction (6.77) corresponding to the eigenvalue of the point spectrum in $\sigma(\mathcal{A}_0)$ located for $0 < |\,\mathrm{Re}(\lambda)\,| < \frac{5}{2} - b$ in Step 1 of Theorem 6.3.

REMARK 6.20. For $b < \frac{5}{2}$, it may first seem from (6.76) that the L^2 norm on the left side of the peak stays bounded with the formal limit

$$\lim_{t \to +\infty} \|v(t,\cdot)\|^2_{L^2(-\infty,0)} = \int_{-\infty}^0 v_0(s)|^2 (1 - e^s)^{2b-5} ds. \tag{6.78}$$

However, the limit may not exist if $v_0(0) \neq 0$. For the eigenfunction in (6.77), the integral of $|v_0(s)|^2(1-e^s)^{2b-5}$ in (6.78) diverges due to the weak singularity of $v_0(s)$ as $s \to 0^-$. The weak singularity is admitted if $v_0 \in \mathrm{Dom}(\mathcal{A})$ and $v_0 \notin H^1(\mathbb{R}) \cap W^{1,\infty}(\mathbb{R})$. On the other hand, if $v_0 \in H^1(\mathbb{R}) \cap W^{1,\infty}(\mathbb{R})$ satisfies $v_0(0) = 0$, the integral of $|v_0(s)|^2(1-e^s)^{2b-5}$ in (6.78) converges and the L^2 norm of the solution $v \in C^0(\mathbb{R}, H^1(\mathbb{R}) \cap W^{1,\infty}(\mathbb{R}))$ is bounded for every $t \in \mathbb{R}$.

6.4.2. Peaked periodic waves in the reduced Ostrovsky equation. In the periodic setting, similar instability results can be obtained for the b-CH equation in Example 6.2 and the reduced Ostrovsky equation in Example 6.3, see [**194**] and [**105, 106**], respectively. However, the proofs of these results require an extra step: one needs to use a coordinate transformation to map the compact domain to the infinite line. We will explain this step by reviewing the following result obtained in [**106**], which implies the spectral instability of the peaked periodic wave in the reduced Ostrovsky equation, using the linearized equations obtained in Example 6.9.

THEOREM 6.4. *The spectrum of the linear operator* $\mathcal{A}$ *defined by* (6.48) *with the domain* $\mathrm{Dom}(\mathcal{A}) = \{w \in L^2(\mathbb{R}) : \partial_x[(1-\phi)w] \in \dot{L}^2_{\mathrm{per}}(\mathbb{R})\}$ *in* (6.49) *is given by*

$$\sigma(\mathcal{A}) = \left\{\lambda \in \mathbb{C} : \; |\,\mathrm{Re}(\lambda)\,| \leq \frac{\pi}{6}\right\}, \tag{6.79}$$

where the point spectrum $\sigma_p(\mathcal{A})$ *consists only of the simple eigenvalue* $\lambda = 0$.

PROOF. The structure of the proof is similar as the one for Theorem 6.3. With the intention of applying Lemma 6.5, we decompose $\mathcal{A} = \mathcal{A}_0 + \mathcal{K}$ with

$$\mathcal{A}_0 := (c_0 - \phi)\partial_x - \phi', \quad \mathcal{K} := \partial_x^{-1},$$

where $\mathcal{K} : \dot{L}^2_{\mathrm{per}} \mapsto \dot{L}^2_{\mathrm{per}}$ is the (compact) Hilbert–Schmidt operator with square summable eigenvalues and spectrum

$$\sigma(\mathcal{K}) \equiv \sigma_p(\mathcal{K}) = \left\{\frac{1}{in}, \quad n \in \mathbb{Z}\backslash\{0\}\right\}.$$

To be able to carry out some explicit calculations in the spectral analysis below, we recall from Example 6.9 that $c_0 = \frac{\pi^2}{9}$ and $\phi_0(x) = \frac{1}{18}(3x^2 - \pi^2)$, so that

$$\mathcal{A}_0 = \frac{1}{6}(\pi^2 - x^2)\partial_x - \frac{x}{3}, \quad x \in (-\pi, \pi).$$

The proof is based on the following three steps after which we apply Lemma 6.5. The first step identifies the point spectrum of $\mathcal{A}$. The second step uses a coordinate transformation to map the compact domain to the infinite line. The third step gives the spectrum of $\mathcal{A}_0$.

STEP 1. $\sigma_p(\mathcal{A}) = \{0\}$.

PROOF OF STEP 1. First we note that if $f \in \mathrm{Dom}(\mathcal{A})$, then $f \in H^1(-\pi,\pi)$ so that $f \in C^0(-\pi,\pi)$ by Sobolev embedding. Bootstrapping arguments for $\mathcal{A}f = \lambda f$ immediately yield that $f \in C^\infty(-\pi,\pi)$, hence the spectral problem $\mathcal{A}f = \lambda f$ for $f \in \mathrm{Dom}(\mathcal{A})$ can be differentiated once in x to yield the second-order differential equation

$$(\pi^2 - x^2)f''(x) - 4xf'(x) + 4f(x) = 6\lambda f'(x), \quad x \in (-\pi, \pi).$$

One solution is available in closed form: $f_1(x) = 2x + 3\lambda$. In order to obtain the second linearly independent solution, we write $f_2(x) = (2x + 3\lambda)g(x)$ and derive the following equation for g:

$$(\pi^2 - x^2)(2x + 3\lambda)g''(x) + 2\left[2(\pi^2 - x^2) - (2x + 3\lambda)^2\right] g'(x) = 0, \quad x \in (-\pi, \pi).$$

This equation can be integrated once to obtain

$$g'(x) = \frac{g_0}{(\pi^2 - x^2)^2(2x + 3\lambda)^2} \left(\frac{\pi + x}{\pi - x}\right)^{\frac{3\lambda}{\pi}}, \quad x \in (-\pi, \pi),$$

where g_0 is a constant of integration. Computing the limits $x \to \pm\pi$ shows that if $\pm 2\pi + 3\lambda \neq 0$, then

$$g'(x) \sim \begin{cases} (\pi - x)^{-\frac{3\lambda}{\pi} - 2}, & x \to \pi, \\ (\pi + x)^{\frac{3\lambda}{\pi} - 2}, & x \to -\pi. \end{cases}$$

This sharp asymptotical behavior shows that $(\pi^2 - x^2)g'(x) \notin L^2(-\pi, \pi)$, even if $g \in L^2(-\pi, \pi)$. Therefore, for every $\lambda \in \mathbb{C}$ with $\pm 2\pi + 3\lambda \neq 0$, the second solution $f_2(x)$ does not belong to $\mathrm{Dom}(\mathcal{A}) \subset \dot{L}^2_{\mathrm{per}}$ because it diverges as $x \to \pm\pi$. For $\pm 2\pi + 3\lambda = 0$, the explicit expression for $g(x)$ yields

$$g'(x) = \frac{g_0}{4(\pi^2 - x^2)^2(\pi \pm x)^2}, \quad x \in (-\pi, \pi),$$

which still implies that f_2 does not belong to $\mathrm{Dom}(\mathcal{A}) \subset \dot{L}^2_{\mathrm{per}}$. Hence, for every $\lambda \in \mathbb{C}$, if $f \in \mathrm{Dom}(\mathcal{A}) \subset \dot{L}^2_{\mathrm{per}}$ is a solution to $\mathcal{A}f = \lambda f$, then f is proportional to $f_1(x) = 2x + 3\lambda$ only. The zero mean constraint $\int_{-\pi}^{\pi} f_1(x)dx = 0$ yields $\lambda = 0$, so that $f_1(x) = 2x$ is proportional to $\phi'(x)$. No other $\lambda \in \mathbb{C}$ such that a nonzero solution f of $\mathcal{A}f = \lambda f$ belongs to $\mathrm{Dom}(\mathcal{A}) \subset \dot{L}^2_{\mathrm{per}}$ exists. ☐

STEP 2. The spectral problem $\mathcal{A}_0 v = \lambda v$ on $(-\pi, \pi)$ is equivalent to the spectral problem $\mathcal{B}_0 w = \mu w$ on $\mathbb{R}$ with $\mu := \frac{6}{\pi}\lambda$, where $\mathcal{B}_0 : H^1(\mathbb{R})|_{\{\varphi\}^\perp} \to L^2(\mathbb{R})|_{\{\varphi\}^\perp}$ is the linear operator given by

$$(\mathcal{B}_0 w)(y) := \partial_y w(y) - \tanh(y) w(y), \quad y \in \mathbb{R}$$

and $\varphi(y) := \mathrm{sech}(y)$.

PROOF OF STEP 2. In view of the structure of the operator

$$(\mathcal{A}_0 v)(x) = \frac{1}{6}\partial_x \left[(\pi^2 - x^2)v(x)\right]$$

for $x \in (-\pi, \pi)$, we consider the differential equation

$$\frac{dx}{d\xi} = \frac{1}{6}(\pi^2 - x^2),$$

whose solution provides us with the coordinate transformation

$$x = \pi \tanh(y) \quad \text{with } y := \frac{\pi\xi}{6},$$

where the constant of integration is defined without loss of generality from the condition that $x = 0$ at $\xi = 0$. Setting

$$v(x) := \cosh(y) w(y),$$

we see that

$$\int_{-\pi}^{\pi} v^2(x)dx = \pi \int_{-\infty}^{\infty} v^2(\pi \tanh y) \operatorname{sech}^2(y)dy = \pi \int_{-\infty}^{\infty} w^2(y)dy.$$

Therefore $v \in L^2(-\pi,\pi)$ if and only if $w \in L^2(\mathbb{R})$. The zero-mean constraint in $\dot{L}^2_{\rm per}$ is transformed to $\langle w, \varphi \rangle = 0$ since

$$0 = \int_{-\pi}^{\pi} v(x)dx = \pi \int_{-\infty}^{\infty} v(\pi \tanh y) \operatorname{sech}^2(y)dy = \pi \int_{-\infty}^{\infty} w(y) \operatorname{sech}(y)dy.$$

Furthermore, we see that $\partial_x\big[(\pi^2 - x^2)v\big] \in L^2(-\pi,\pi)$ if and only if $\partial_y w - \tanh(y) w \in L^2(\mathbb{R})$. Therefore, $w \in H^1(\mathbb{R})$ and we obtain that the orthogonality constraint holds in view of

$$\langle \mathcal{B}_0 w, \varphi \rangle = \int_{\mathbb{R}} [w'(y) - \tanh(y)w(y)] \operatorname{sech}(y)dy = \int_{\mathbb{R}} \frac{d}{dy} [w(y) \operatorname{sech}(y)]\, dy = 0.$$

Hence, $\operatorname{Dom}(\mathcal{B}_0) = H^1(\mathbb{R})|_{\{\varphi\}^\perp}$ and $\operatorname{Ran}(\mathcal{B}_0) = L^2(\mathbb{R})|_{\{\varphi\}^\perp}$. The coordinate transformation between the spectral problems $\mathcal{A}_0 v = \lambda v$ and $\mathcal{B}_0 w = \mu w$ implies the relation $\mu = \frac{6}{\pi}\lambda$. □

STEP 3. The spectrum $\sigma(\mathcal{B}_0)$ consists of the following parts:

$$\sigma_p(\mathcal{B}_0) = \emptyset, \quad \sigma_r(\mathcal{B}_0) = \{\mu \in \mathbb{C} : |\operatorname{Re}(\mu)| < 1\}, \quad \sigma_c(\mathcal{B}_0) = \{\mu \in \mathbb{C} : |\operatorname{Re}(\mu)| = 1\}.$$

PROOF OF STEP 3. Let $\mu \in \mathbb{C}$ and consider $w \in \operatorname{Ker}(\mathcal{B}_0 - \mu I)$, i.e., w satisfies the differential equation

$$\frac{dw}{dy} = \mu w(y) + \tanh(y) w(y)$$

with solution $w(y) = C\cosh(y)e^{\mu y}$, where C is arbitrary. We have $w(y) \sim e^{(\mu \pm 1)y}$ as $y \to \pm\infty$ and hence the two exponential functions decay to zero as $y \to \pm\infty$ in two disjoint sets for μ in $\mathbb{C}$. Hence, $w \in L^2(\mathbb{R})$ if and only if $C = 0$ for every $\mu \in \mathbb{C}$ so that $\sigma_p(\mathcal{B}_0) = \emptyset$.

Let $f \in L^2(\mathbb{R})|_{\{\varphi\}^\perp}$ and consider the resolvent equation for $w \in H^1(\mathbb{R})|_{\{\varphi\}^\perp}$ given by

$$\frac{dw}{dy} - \tanh(y)w(y) - \mu w(y) = f(y). \tag{6.80}$$

Since the spectrum $\sigma(\mathcal{B}_0)$ is invariant under translations along the imaginary axis, due to the transformation $f \mapsto fe^{iky}$ and $w \mapsto we^{iky}$ with $k \in \mathbb{R}$, it suffices to study (6.80) for $\mu \in \mathbb{R}$. Solving the first-order inhomogeneous equation (6.80) by variation of parameters yields

$$w(y) = \cosh(y)e^{\mu y}\left[C + \int_0^y e^{-\mu y'} \operatorname{sech}(y') f(y') dy'\right], \tag{6.81}$$

from which we infer that $w \in H^1_{\rm loc}(\mathbb{R})$. However, we also need to consider the behavior of $w(y)$ as $y \to \pm\infty$ to ensure that $w \in \operatorname{Dom}(\mathcal{B}_0)$.

If $\mu \neq 0$ and $w \in H^1(\mathbb{R})$ is a solution to (6.80), then the constraint $\langle f, \varphi \rangle = 0$ implies $\langle w, \varphi \rangle = 0$, so that $w \in H^1(\mathbb{R})$ implies $w \in H^1(\mathbb{R})|_{\{\varphi\}^\perp}$. If $\mu = 0$ and $w \in H^1(\mathbb{R})$ is a solution to (6.80), then the constraint $\langle w, \varphi \rangle = 0$ is an additional condition to be checked. Hence we require different estimates for different values of $\mu \in \mathbb{R}$.

CASE $|\mu| > 1$. We show that $I_+ := \{\mu \in \mathbb{R} : \mu > 1\}$ belongs to the resolvent set of $\mathcal{B}_0$. Since $e^{(\mu+1)y}$ diverges as $y \to +\infty$ for every $\mu \in I_+$, we define C in (6.81) by

$$C := -\int_0^\infty e^{-\mu y'} \operatorname{sech}(y') f(y') dy', \tag{6.82}$$

so that the unique solution (6.81) can be rewritten as

$$w(y) = \int_{+\infty}^{y} e^{\mu(y-y')} \frac{\cosh(y)}{\cosh(y')} f(y') dy'. \tag{6.83}$$

The following two equivalent representations will be useful in the estimates below:

$$\frac{\cosh(y)}{\cosh(y')} = \frac{1+e^{2y}}{1+e^{2y'}} e^{y'-y} \tag{6.84}$$

$$= \frac{1+e^{-2y}}{1+e^{-2y'}} e^{y-y'}. \tag{6.85}$$

Let $f = f\chi_{\{y>0\}} + f\chi_{\{y<0\}}$, where χ_S is the characteristic function on the set $S \subset \mathbb{R}$, and define $w_\pm$ by (6.83) with f replaced by $f\chi_{\{\pm y>0\}}$ so that $w = w_+ + w_-$. Using (6.84) for $y < 0$ and (6.85) for $y > 0$, we obtain

$$|w_+(y)| \le \int_0^{+\infty} e^{-(\mu-1)(y'-y)} |f(y')| dy' \quad \text{for } y < 0,$$

and

$$|w_+(y)| \le 2 \int_y^{+\infty} e^{-(\mu+1)(y'-y)} |f(y')| dy' \quad \text{for } y > 0.$$

By the Cauchy–Schwarz inequality for $y < 0$ and by the generalized Young's inequality for $y > 0$, we obtain

$$\|w_+\|_{L^2(\mathbb{R}^-)} \le \|e^{-(\mu-1)\cdot}\|_{L^2(\mathbb{R}^+)} \|f\|_{L^2(\mathbb{R}^+)} \|e^{(\mu-1)\cdot}\|_{L^2(\mathbb{R}^-)} \le \frac{1}{2(\mu-1)} \|f\|_{L^2(\mathbb{R}^+)}$$

and

$$\|w_+\|_{L^2(\mathbb{R}^+)} \le 2\|e^{-(\mu+1)\cdot}\|_{L^1(\mathbb{R}^+)} \|f\|_{L^2(\mathbb{R}^+)} \le \frac{2}{\mu+1} \|f\|_{L^2(\mathbb{R}^+)}.$$

On the other hand, $w_-(y) = 0$ for $y > 0$ and

$$y < 0: \quad |w_-(y)| \le 2 \int_y^0 e^{-(\mu-1)(y'-y)} |f(y')| dy',$$

where the representation (6.84) has been used. By the generalized Young's inequality, we obtain

$$\|w_-\|_{L^2(\mathbb{R})} \le 2\|e^{-(\mu-1)\cdot}\|_{L^1(\mathbb{R}^+)} \|f\|_{L^2(\mathbb{R}^-)} \le \frac{2}{\mu-1} \|f\|_{L^2(\mathbb{R}^-)}.$$

Putting these bounds together yields

$$\|w\|_{L^2(\mathbb{R})} \le C_\mu \|f\|_{L^2(\mathbb{R})}, \tag{6.86}$$

where the constant $C_\mu > 0$ depends on μ and is bounded for every $\mu > 1$. Thus, we have shown that $I_+ \subset \rho(\mathcal{B}_0)$. Similarly, one can show that $I_- := \{\mu \in \mathbb{R} : \mu < -1\}$ also belongs to the resolvent set of $\mathcal{B}_0$ due to the same bound (6.86) for every $\mu \in I_-$. Hence, $I_+ \cup I_- \subset \rho(\mathcal{B}_0)$ and by the invariance of $\rho(\mathcal{B}_0)$ along the imaginary axis, we have established that $\rho(\mathcal{B}_0)$ is given by the two half-planes defined by $|\operatorname{Re}(\mu)| > 1$.

CASE $|\mu| < 1$. We show that $(-1,1)$ belongs to $\sigma_r(\mathcal{B}_0)$. To do so, we use again the explicit solution $w \in H^1_{\rm loc}(\mathbb{R})$ given in (6.81). If $\mu \in (-1,1)$, then the exponential functions $e^{(\mu+1)y}$ and $e^{(\mu-1)y}$ do not decay to zero as $y \to +\infty$ and $y \to -\infty$, respectively. Therefore, to ensure decay of $w(y)$ as $y \to \pm\infty$, the constant C in (6.81) would have to be defined twice

$$C = -\int_0^\infty e^{-\mu y'} \operatorname{sech}(y') f(y') dy' = \int_{-\infty}^0 e^{-\mu y'} \operatorname{sech}(y') f(y') dy'.$$

This implies that $f \in L^2(\mathbb{R})|_{\{\varphi\}^\perp}$ would have to satisfy an additional constraint

$$\int_{\mathbb{R}} e^{-\mu y'} \operatorname{sech}(y') f(y') dy' = 0, \tag{6.87}$$

which is different from $\langle f, \varphi \rangle = 0$ if $\mu \neq 0$. Hence, if $f \in L^2(\mathbb{R})|_{\{\varphi\}^\perp}$ does not satisfy (6.87), then there exists no solution $w \in \operatorname{Dom}(\mathcal{B}_0)$ to the resolvent equation (6.80) because the solution $w \in H^1_{\rm loc}(\mathbb{R})$ diverges at infinity. This implies that $\operatorname{Ran}(\mathcal{B}_0 - \mu I) \subsetneq L^2(\mathbb{R})|_{\{\varphi\}^\perp}$ so that $(-1,1)\backslash\{0\}$ belongs to $\sigma_r(\mathcal{B}_0)$.

In the special case $\mu = 0$, the constraint (6.87) coincides with $\langle f, \varphi \rangle = 0$. The unique solution (6.81) with C as in (6.82) can be rewritten for $\mu = 0$ as

$$w(y) = \int_\infty^y \frac{\cosh(y)}{\cosh(y')} f(y') dy'. \tag{6.88}$$

If $\langle f, \varphi \rangle = 0$, then the solution (6.88) belongs to $H^1(\mathbb{R})$. The constraint $\langle w, \varphi \rangle = 0$, however, is satisfied only under the additional constraint

$$\int_{\mathbb{R}} \int_\infty^y \operatorname{sech}(y') f(y') dy' dy = 0. \tag{6.89}$$

Therefore, for $\mu = 0$, there exists no solution $w \in \operatorname{Dom}(\mathcal{B}_0)$ to the resolvent equation (6.80) unless $f \in L^2(\mathbb{R})|_{\{\varphi\}^\perp}$ satisfies (6.89). This implies again that $\operatorname{Ran}(\mathcal{B}_0) \subsetneq L^2(\mathbb{R})|_{\{\varphi\}^\perp}$ and so $0 \in \sigma_r(\mathcal{B}_0)$. By using the invariance of $\sigma_r(\mathcal{B}_0)$ along the imaginary axis, we have established that $\sigma_r(\mathcal{B}_0)$ is given by the strip determined by $|\operatorname{Re}(\mu)| < 1$.

CASE $|\mu| = 1$. We show that $\pm 1 \in \sigma_c(\mathcal{B}_0)$. One of the two exponential functions $e^{(\mu+1)y}$ and $e^{(\mu-1)y}$ in (6.81) does not decay to zero both as $y \to +\infty$ and $y \to -\infty$ if $\mu = \pm 1$. Moreover, the improper integral in (6.81) does not converge for $f \in L^2(\mathbb{R})$, $f \notin L^1(\mathbb{R})$ because $e^{\pm y'} \operatorname{sech}(y') \to 2$ as $y' \to \pm\infty$. Therefore, the solution w in (6.81) does not decay to zero and does not belong to $\operatorname{Dom}(\mathcal{B}_0)$ independently on the constraint on C and hence

$$(\mathcal{B}_0 - \mu I)^{-1} : L^2(\mathbb{R})|_{\{\varphi\}^\perp} \to L^2(\mathbb{R})|_{\{\varphi\}^\perp}$$

is unbounded. Hence $\pm 1 \in \sigma_c(\mathcal{B}_0)$. Due to the invariance of $\sigma_c(\mathcal{B}_0)$ along the imaginary axis, we conclude that $\sigma_c(\mathcal{B}_0)$ is given by the two lines determined by $|\operatorname{Re}(\mu)| = 1$. ∎

END OF PROOF OF THEOREM 6.4. It follows from Steps 2 and 3 that the point spectrum of $\mathcal{A}_0$ is an empty set, the residual spectrum of $\mathcal{A}_0$ is located in the strip determined by $|\operatorname{Re}(\lambda)| < \frac{\pi}{6}$ and the continuous spectrum of $\mathcal{A}_0$ is located on the line determined by $|\operatorname{Re}(\lambda)| = \frac{\pi}{6}$. It also follows from Step 1 that $\sigma_p(\mathcal{A}) = \{0\}$. Since $\rho(\mathcal{A}) \cap \sigma_p(\mathcal{A}_0) = \emptyset$ and $\rho(\mathcal{A}_0) \cap \sigma_p(\mathcal{A}) = \emptyset$, we have $\sigma(\mathcal{A}) = \sigma(\mathcal{A}_0)$ by Lemma 6.5. ∎

REMARK 6.21. Compared to the proof of Theorem 6.3, we obtained the residual spectrum of $\mathcal{A}_0$ directly instead of considering the point spectrum of the adjoint operator $\mathcal{A}_0^*$. This is because the spectrum of $\mathcal{A}_0$ is defined in $\dot{L}^2_{\rm per}$ with the zero-mean constraint, which gives the orthogonality constraint $\langle \cdot, \varphi \rangle = 0$ and defines the spectrum of $\mathcal{B}_0$ in $L^2(\mathbb{R})|_{\{\varphi\}^\perp}$. If we were to consider the adjoint operator $\mathcal{B}_0$, then we would need to incorporate the constraint into the computations. Another related difference is that we do not distinguish between the residual and the continuous spectrum of $\mathcal{A}$ in Theorem 6.4.

REMARK 6.22. From Example 6.9, we know that the spectrum of $\mathcal{A}$ should be defined in the constrained space $L^2_c \subset \dot{L}^2_{\rm per}$ given by (6.53) with two constraints. However, the result of Theorem 6.4 remains true in L^2_c, because a finite number of constraints cannot change the location of the vertical strip of $\sigma(\mathcal{A}) \subset \mathbb{C}$.

Because the residual and continuous spectra of $\mathcal{A}$ are not distinguished in Theorem 6.4, we cannot use an analogue of Lemma 6.6 to show that linear instability of peaked waves follows from their spectral instability. However, this can be shown with direct estimates of solutions to the linearized equation $w_t = \mathcal{A}w$ for $w \in L^2_c \subset \dot{L}^2_{\rm per}$ given by (6.53), see [**105**]. The following example illustrates how the linear instability of the peaked periodic waves with the profile ϕ given in Example 6.9 unfolds in the time evolution generated by $\mathcal{A}_0 : \mathrm{Dom}(\mathcal{A}_0) \subset \dot{L}^2_{\rm per} \to \dot{L}^2_{\rm per}$, where $\mathrm{Dom}(\mathcal{A}_0) = \mathrm{Dom}(\mathcal{A})$.

EXAMPLE 6.11. Recall the truncated operator $\mathcal{A}_0$ in Theorem 6.4 with

$$\mathrm{Dom}(\mathcal{A}_0) = \mathrm{Dom}(\mathcal{A})$$

and consider the corresponding initial-value problem $v_t = \mathcal{A}_0 v$ given by

$$\begin{cases} v_t = (c_0 - \phi)v_x - \phi' v, & t > 0, \\ v|_{t=0} = v_0, \end{cases} \tag{6.90}$$

where $v_0 \in \mathrm{Dom}(\mathcal{A}) \subset \dot{L}^2_{\rm per}$. The initial-value problem (6.90) can be solved by using the method of characteristics. The characteristic curve $x = X(t,s)$ is found from

$$\begin{cases} X_t = \dfrac{1}{6}\left[X(s,t)^2 - \pi^2\right], & t > 0, \\ X(0,s) = s, \end{cases}$$

where $s \in [-\pi, \pi]$ and $c_0 - \phi(x) = \frac{1}{6}(\pi^2 - x^2)$ for $x \in [-\pi, \pi]$. The exact solution for the characteristic curves is given by

$$X(t,s) = \pi \frac{s\cosh(\pi t/6) - \pi \sinh(\pi t/6)}{\pi \cosh(\pi t/6) - s \sinh(\pi t/6)}, \quad s \in [-\pi, \pi], \quad t \in \mathbb{R}. \tag{6.91}$$

Along the characteristic curves, the function $V(t,s) := v(t, X(t,s))$ satisfies

$$\begin{cases} V_t = -\dfrac{1}{3}XV, & t > 0, \\ V(0,s) = v_0(s), \end{cases}$$

which is uniquely solved by

$$V(t,s) = \frac{1}{\pi^2}[\pi\cosh(\pi t/6) - s\sinh(\pi t/6)]^2 v_0(s), \quad s \in [-\pi, \pi], \quad t \in \mathbb{R}. \tag{6.92}$$

The zero-mean constraint is preserved by the time evolution:

$$\int_{-\pi}^{\pi} v(t,x)dx = \int_{-\pi}^{\pi} V(t,s)X_s(t,s)ds = \int_{-\pi}^{\pi} v_0(s)ds = 0.$$

By the chain rule, we obtain

$$\begin{aligned}\int_{-\pi}^{\pi} v(t,x)^2 dx &= \int_{-\pi}^{\pi} V(t,s)^2 X_s(t,s)ds \\ &= \frac{1}{\pi^2}\int_{-\pi}^{\pi} [\pi\cosh(\pi t/6) - s\sinh(\pi t/6)]^2 v_0(s)^2 ds,\end{aligned}$$

which provides the upper bound

$$\|v(t,\cdot)\|_{L^2_{\rm per}} \leq e^{\frac{\pi}{6}t}\|v_0\|_{L^2_{\rm per}}.$$

If the initial condition $v_0 \in \mathrm{Dom}(\mathcal{A})$ satisfies the additional constraint

$$\int_{-\pi}^{\pi} s v_0(s)^2 ds = 0,$$

then we obtain the lower bound

$$\|v(t,\cdot)\|^2_{L^2_{\rm per}} = \cosh(\pi t/6)^2\|v_0\|^2_{L^2_{\rm per}} + \frac{1}{\pi^2}\sinh(\pi t/6)^2\|sv_0\|^2_{L^2_{\rm per}} \geq \frac{1}{4}e^{\frac{\pi}{3}t}\|v_0\|^2_{L^2_{\rm per}},$$

which shows that $\|v(t,\cdot)\|_{L^2_{\rm per}}$ grows exponentially as $e^{\frac{\pi}{6}t}$ in agreement with the width of the unstable strip in (6.79). Figure 6.6 illustrates the solution $v(t,x)$ versus x in the polar form $x = X(t,s)$ and $v(t,X(t,s)) = V(t,s)$ for

$$v_0(x) = x(\pi^2 - x^2)(1 - \frac{7}{3\pi^2}x^2).$$

The coefficients of v_0 are computed to satisfy all the constraints of the theory:

$$\langle 1, v_0\rangle = \langle \phi, v_0\rangle = \langle \partial_x\phi, v_0\rangle = \langle x v_0, v_0\rangle = 0.$$

REMARK 6.23. It follows from the exact solution (6.92) in Example 6.11 that if $v_0 \in \dot{H}^1_{\rm per}$ with $v_0(\pm\pi) \neq 0$, then $v(t,\cdot) \notin \dot{H}^1_{\rm per}$ for every $t \neq 0$, since

$$v(t,\pm\pi) = V(t,\pm\pi) = e^{\mp\frac{\pi}{3}t}v_0(\pm\pi).$$

On the other hand, if $v_0(\pm\pi) = 0$, then $v|_{x=\pm\pi} = 0$ for every $t > 0$. The constraints were discussed in Example 6.9 for the time evolution of the linearized equation (6.46) in $\dot{H}^1_{\rm per} \cap W^{1,\infty}$. Compared to the evolution in $\dot{H}^1_{\rm per} \cap W^{1,\infty}$, $\lim_{x\to\pm\pi} v(t,x)$ do not have to be defined in $\mathrm{Dom}(\mathcal{A})$ since $x = \pm\pi$ are the locations of the peaks in the peaked periodic wave. It follows from (6.91) and (6.92) that

$$(c_0 - \phi)v_x \in C^0(\mathbb{R}, L^2(-\pi,\pi)) \quad \text{if} \quad (c_0 - \phi)v_0' \in C^0(\mathbb{R}, L^2(-\pi,\pi))$$

so that the time evolution of the initial-value problem (6.90) is well-defined in $\mathrm{Dom}(\mathcal{A}) \subset \dot{L}^2_{\rm per}$ with the unique solution $v \in C^0(\mathbb{R}, \mathrm{Dom}(\mathcal{A}))$.

REMARK 6.24. Spectral instability can be shown for other modifications of the perturbed Burgers equation (6.16), e.g., for the modified reduced Ostrovsky equation

$$(u_t + u^2 u_x)_x = u,$$

which can be rewritten for the momentum variable $m := 1 - 2(u_x)^2$ as

$$m_t + u^2 m_x + 4uu_x m = 0.$$

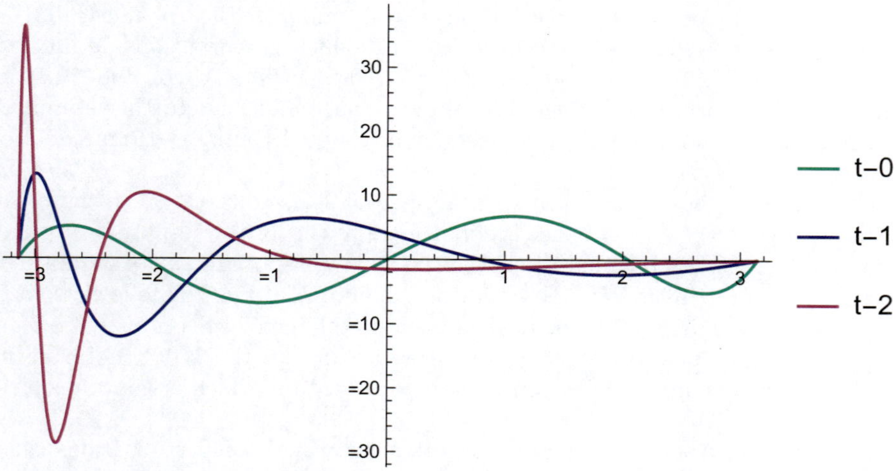

FIGURE 6.6. A plot of the solution $v(t, x)$ of the initial-value problem (6.90) in Example 6.11 as a function of x at three instances of time t illustrating the exponential growth of perturbation near $x = -\pi$.

The C^1-piecewise 2π-periodic function

$$\phi(x) = \sum_{n\in\mathbb{Z}} \phi_0(x - 2\pi n), \quad \phi_0(x) := \frac{x}{\sqrt{2}}\chi_{[-\pi,\pi]}(x)$$

satisfies

$$(\phi^2 - c_0)\phi'' + 2\phi(\phi')^2 = \phi, \qquad c_0 := \max_{x\in[-\pi,\pi]} \phi_0(x) = \frac{\pi}{\sqrt{2}},$$

piecewise on both sides of the peaks of ϕ at $x \in \{(2n+1)\pi n\}_{n\in\mathbb{Z}}$. Consequently, the C^1-piecewise 2π-periodic function $u(t, x) = \phi(x - c_0 t)$ is the peaked periodic traveling wave solution for the particular speed c_0. The linearized operator $\mathcal{A}$ is given by

$$\mathcal{A} := \partial_x(c_0 - \phi)^2 + \partial_x^{-1}$$

with

$$\mathrm{Dom}(\mathcal{A}) := \left\{ w \in \dot{L}^2_{\mathrm{per}} : \ \partial_x(c_0 - \phi)^2 w \in \dot{L}^2_{\mathrm{per}} \right\}.$$

It is shown in [**106**] with similar computations to those in Theorem 6.4 that the spectrum of $\mathcal{A}$ in $\dot{L}^2_{\mathrm{per}}$ is located in the vertical strip for

$$\sigma(\mathcal{A}) = \left\{ \lambda \in \mathbb{C} : \ |\,\mathrm{Re}(\lambda)\,| \le \frac{\pi}{4} \right\}.$$

Consequently, the peaked wave with the profile ϕ is spectrally unstable in

$$\mathrm{Dom}(\mathcal{A}) \subset \dot{L}^2_{\mathrm{per}}.$$

6.5. Nonlinear instability of peaked waves

In Hamiltonian systems with peaked waves for which conserved energy functionals are available, it is sometimes possible to show that perturbations of the peaked waves do not grow in some suitable norms. Typical examples are the Camassa–Holm equation ($b = 2$), for which perturbations do not grow in the H^1 norm [**63**, **64**], and the Degasperis–Procesi equation ($b = 3$) for which perturbations do not grow in the L^2 norm [**179**]. These findings allowed the authors to conclude that the peaked waves are orbitally stable in the energy space and have further stimulated much research on more complicated systems of coupled Camassa–Holm equations [**47**, **176**].

These results should be taken with a grain of salt, however, since the orbital stability theory in the energy space is conditional to the local well-posedness theory in that same space. As Lemma 6.1 shows, the local well-posedness of the perturbed Burgers equation holds in $H^1 \cap W^{1,\infty}$ rather than in H^1 or L^2. It follows from the spectral and linear instability of peaked waves explained in Sections 6.3 and 6.4 that perturbations of the peaked waves grow in time in the $W^{1,\infty}$ norm even if their H^1 norm remains bounded on the time interval for which a smooth solution exists.

The following theorem explains the intrinsic growth of perturbations of the peaked solitary wave in the time evolution of the b-CH equation (6.3) for every $b > 1$, which implies its nonlinear instability in $H^1(\mathbb{R}) \cap W^{1,\infty}(\mathbb{R})$. It generalizes the results obtained in [**205**] for $b = 2$.

THEOREM 6.5. *Fix $b > 1$ in the b-CH equation (6.3) and assume that the peaked solitary wave with profile $\phi(x) = e^{-|x|}$ is orbitally stable in $H^1(\mathbb{R})$. For every $\delta > 0$, there exists initial data $u_0 \in H^1(\mathbb{R}) \cap C^1(\mathbb{R} \setminus \{0\})$ satisfying*

$$\|u_0 - \phi\|_{H^1} + \|u_0' - \phi'\|_{L^\infty} < \delta,$$

and $t_0 \in (0, \tau_0)$ such that the local solution $u \in C^0([0,\tau_0), H^1(\mathbb{R}) \cap W^{1,\infty}(\mathbb{R}))$ with $u(0,\cdot) = u_0$ and maximal existence time $\tau_0 > 0$ satisfies

$$\|u_x(t_0,\cdot) - \phi'(\cdot - \xi(t_0))\|_{L^\infty} > 1, \tag{6.93}$$

where $\xi(t) \in \mathbb{R}$ is the location of the single peak of $u(t,\cdot)$ for $t \in [0,\tau_0)$ with $\xi(0) = 0$. Moreover, there exists u_0 such that $\tau_0 < \infty$.

PROOF. The proof consists of the following four steps. In the first step, the evolution problem for the perturbation of the peaked solitary wave is derived. In the second step, the method of characteristics is used to analyze the evolution problem for the first spatial derivative of the perturbation at the peak. In the third step, the proof of existence of u_0 and t_0 in the instability bound (6.93) is established by using Gronwall's inequality. In the fourth step, the proof of existence of u_0 sufficient for wave breaking in finite time $\tau_0 < \infty$ is developed based on Riccati's equation and the comparison principle.

STEP 1. The unique local solution $u \in C^0([0,\tau_0), H^1(\mathbb{R}) \cap W^{1,\infty}(\mathbb{R}))$ can be written in the form

$$u(t,x) = \phi(x - t - a(t)) + v(t, x - t - a(t))$$

with $a'(t) = v(t,0)$ and $v(t,x)$ satisfying the initial-value problem:

$$\begin{cases} v_t = (1-\phi)v_x + (v|_{x=0} - v)\phi' - \dfrac{1}{2}\phi' * [b\phi v + (3-b)\phi' v_x] \\ \qquad\quad + (v|_{x=0} - v)v_x - \dfrac{1}{4}\phi' * \left[bv^2 + (3-b)v_x^2\right], \\ v|_{t=0} = v_0, \end{cases} \tag{6.94}$$

where $v_0 := u_0 - \phi$. If in addition, $v_0 \in C^1(\mathbb{R} \setminus \{0\})$, then $v(t,\cdot) \in C^1(\mathbb{R} \setminus \{0\})$.

PROOF OF STEP 1. By using (6.32) with $Q[u]$ as in Example 6.8 for $c = 1$, we obtain (6.94). By Lemma 6.1 and the method of characteristics, for every $v_0 \in H^1(\mathbb{R}) \cap C^1(\mathbb{R}\backslash\{0\})$, there exists a unique solution $v \in C^0([0, \tau_0), H^1(\mathbb{R}) \cap W^{1,\infty}(\mathbb{R}))$ of (6.94) such that $v(t, \cdot) \in C^1(\mathbb{R}\backslash\{0\})$. ☐

STEP 2. The limit $U_0^+(t) := \lim_{x\downarrow 0} v_x(t, x)$ on the right side of the peak satisfies the evolution equation

$$\begin{aligned}\frac{dU_0^+}{dt} &= (b-1)U_0^+ + bV_0 - \frac{1}{2}\langle \phi, b\phi v + (3-b)\phi' v_x\rangle \\ &\quad + \frac{1}{2}bV_0^2 + \frac{1}{2}(1-b)(U_0^+)^2 - \frac{1}{4}\langle \phi, bv^2 + (3-b)v_x^2\rangle,\end{aligned} \tag{6.95}$$

where $V_0(t) := \lim_{x\to 0} v(t, x)$.

PROOF OF STEP 2. Related to (6.95), we define the family of characteristic curves $x = X(t, s)$ satisfying the initial-value problem

$$\begin{cases} X_t = \phi(X) - 1 + v(t, X) - v(t, 0), \\ X|_{t=0} = s, \end{cases} \tag{6.96}$$

where ϕ and $v(t, \cdot)$ are Lipschitz continuous across the equilibrium point $X = 0$ for $s = 0$ as long as $v \in C^0([0, \tau_0), H^1(\mathbb{R}) \cap W^{1,\infty}(\mathbb{R}))$. This implies that the peak's location at $x = 0$ is preserved along the time evolution of the initial-value problem (6.94).

Since $v(t, \cdot) \in C^1(\mathbb{R}\backslash\{0\})$ for $t \in [0, \tau_0)$, we can differentiate (6.94) in x separately for $x > 0$ and $x < 0$ and obtain

$$\begin{cases} v_{xt} = (1-\phi)v_{xx} + (1-b)\phi' v_x + b\phi v + (v|_{x=0} - v)\phi - \frac{1}{2}\phi * [b\phi v + (3-b)\phi' v_x] \\ \qquad + (v|_{x=0} - v)v_{xx} + \frac{1}{2}bv^2 + \frac{1}{2}(1-b)v_x^2 - \frac{1}{4}\phi * \left[bv^2 + (3-b)v_x^2\right], \\ v_x|_{t=0} = v_0', \end{cases}$$

where $\phi - \phi'' = 2\delta$ has been used. Along the characteristic curves,

$$V(t, s) := v(t, X(t, s)) \text{ and } U(t, s) := v_x(t, X(t, s))$$

satisfy the initial-value problems

$$\begin{cases} V_t = (V|_{x=0} - V)\phi' - \frac{1}{2}\phi' * [b\phi v + (3-b)\phi' v_x] - \frac{1}{4}\phi' * \left[bv^2 + (3-b)v_x^2\right], \\ V|_{t=0} = v_0, \end{cases}$$

and

$$\begin{cases} U_t = (1-b)\phi' U + b\phi V + (V|_{x=0} - V)\phi - \frac{1}{2}\phi * [b\phi v + (3-b)\phi' v_x] \\ \qquad + \frac{1}{2}bV^2 + \frac{1}{2}(1-b)U^2 - \frac{1}{4}\phi * \left[bv^2 + (3-b)v_x^2\right], \\ U|_{t=0} = v_0', \end{cases}$$

The evolution equation (6.95) follows from the latter equation in the limit $s \downarrow 0$ since $\phi(0) = 1$ and $\phi'(0^+) = -1$. ☐

STEP 3. For every $\delta > 0$, there exists $v_0 \in H^1(\mathbb{R}) \cap C^1(\mathbb{R} \setminus \{0\})$ satisfying

$$\|v_0\|_{H^1} + \|v_0'\|_{L^\infty} < \delta$$

and $t_0 \in (0, \tau_0)$ such that $\|v_x(t_0, \cdot)\|_{L^\infty} > 1$.

PROOF OF STEP 3. Let $\delta > 0$. By the assumption of orbital stability of peakons in $H^1(\mathbb{R})$, for every small $\varepsilon > 0$ there exists $\nu(\varepsilon) > 0$ such that if $\|v_0\|_{H^1} < \nu(\varepsilon)$, then $\|v(t,\cdot)\|_{H^1} < \varepsilon$ for every $t \in [0,\tau_0)$. By Sobolev's embedding of $H^1(\mathbb{R})$ into $L^\infty(\mathbb{R})$, there is a positive constant C such that

$$|V_0(t)| \leq \|v(t,\cdot)\|_{L^\infty} \leq C\|v(t,\cdot)\|_{H^1} < C\varepsilon.$$

By the Cauchy–Schwarz inequality, we obtain

$$|\langle \phi, b\phi v + (3-b)\phi' v_x \rangle| \leq C(|b|\|v\|_{L^2} + |b-3|\|v_x\|_{L^2}),$$
$$|\langle \phi, bv^2 + (3-b)(v_x)^2 \rangle| \leq C(|b|\|v\|_{L^2}^2 + |b-3|\|v_x\|_{L^2}^2).$$

These bounds are sufficient to control all terms in the evolution equation (6.95) except the U_0^+ terms, according to the differential inequality

$$\frac{dU_0^+}{dt} \leq (b-1)U_0^+ + \frac{1}{2}(1-b)(U_0^+)^2 + C_b\varepsilon, \tag{6.97}$$

where the positive constant C_b may depend on b. Choose $\varepsilon > 0$ small enough such that

$$\nu(\varepsilon) + \frac{2C_b\varepsilon}{b-1} < \delta$$

and assume that the initial data $v_0 \in H^1(\mathbb{R}) \cap C^1(\mathbb{R}\setminus\{0\})$ satisfies

$$\lim_{x\to 0^+} v_0'(\xi) = -\|v_0'\|_{L^\infty} = -\frac{2C_b\varepsilon}{b-1}, \tag{6.98}$$

such that $\|v_0\|_{H^1} + \|v_0'\|_{L^\infty} < \delta$. By integrating (6.97), dropping the quadratic term in $(U_0^+)^2$ for $b > 1$, and using (6.98), we obtain

$$U_0^+(t) \leq e^{(b-1)t}\left[U_0^+(0) + \frac{C_b\varepsilon}{b-1}\right] = -\frac{C_b\varepsilon}{b-1}e^{(b-1)t}, \quad t \in [0,\tau_0).$$

Hence, there exists

$$t_1 := -\frac{\log(C_b\varepsilon) - \log(b-1)}{b-1} > 0,$$

such that $U_0^+(t_1) \leq -1$. If $t_1 < \tau_0$, the bound $\|v_x(t_0,\cdot)\|_{L^\infty} > 1$ is true with some $t_0 \in (0,t_1]$ since $\|v_x(t,\cdot)\|_{L^\infty} \geq |U_0^+(t)|$ for $t \in [0,\tau_0)$. If $t_1 \geq \tau_0$, then τ_0 is the finite maximal existence time for the solution along the characteristic curves. It follows from (6.96), see Lemma 6.2, that if $\tau_0 < \infty$, then $\lim_{t\uparrow\tau_0} \|v_x(t,\cdot)\|_{L^\infty} = \infty$. Hence, there exists $t_0 \in (0,\tau_0)$ such that the bound $\|v_x(t_0,\cdot)\|_{L^\infty} > 1$ is true. In both cases, the claim is proven. □

STEP 4. There exists $v_0 \in H^1(\mathbb{R}) \cap C^1(\mathbb{R}\setminus\{0\})$ such that $\tau_0 < \infty$.

PROOF OF STEP 4. Let $U_b(\epsilon)$ be the negative root of the quadratic equation

$$C_b\varepsilon + (b-1)U - \frac{1}{2}(b-1)U^2 = 0$$

It is clear that $U_b(\varepsilon) \in (-\frac{C_b\varepsilon}{b-1}, 0)$. Choose $\varepsilon > 0$ small enough such that

$$\nu(\varepsilon) + 2|U_b(\varepsilon)| < \delta$$

and assume that the initial data $v_0 \in H^1(\mathbb{R}) \cap C^1(\mathbb{R}\setminus\{0\})$ satisfies

$$\lim_{x\to 0^+} v_0'(x) = -\|v_0'\|_{L^\infty} = 2U_b(\varepsilon).$$

If $U_0^+(0) = 2U_b(\varepsilon)$, then the differential equation (6.95) for U_0^+ implies that

$$\frac{d}{dt}U_0^+(0) < 0,$$

hence the map $t \to U_0^+(t)$ is monotonically decreasing.

Let $\overline{U}$ be the supersolution that satisfies the Ricatti differential equation

$$\frac{d\overline{U}}{dt} = (b-1)\overline{U} - \frac{1}{2}(b-1)\overline{U}^2 + C_b\varepsilon.$$

with $\overline{U}(0) = 2U_b(\varepsilon)$. It follows from the comparison theory for differential equations that $U_0^+(t) \le \overline{U}(t)$ for every $t > 0$ for which $U_0^+(t)$ exists. Since there exists a finite $\overline{\tau}_0 > 0$ such that $\overline{U}(t) \to -\infty$ as $t \uparrow \overline{\tau}_0$, there exists $\tau_0 \in (0, \overline{\tau}_0]$ such that $U_0^+(t) \to -\infty$ as $t \uparrow \tau_0$. The maximal existence time τ_0 of the solution $v \in C^0([0,\tau_0), H^1 \cap C^1(\mathbb{R}\backslash\{0\}))$ is hence finite. □

Steps 3 and 4 imply the weak and strong instability of the peaked waves with the profile ϕ formulated in Theorem 6.5. □

REMARK 6.25. The arguments of Theorem 6.5 can be extended to $b < 1$ by considering $U_0^-(t) := \lim_{x\uparrow 0} v_x(t,x)$ on the left side of the peak. This quantity satisfies the differential inequality

$$\frac{dU_0^-}{dt} \ge (1-b)U_0^- + \frac{1}{2}(1-b)(U_0^-)^2 - C_b\varepsilon,$$

from which one can prove very similarly to Steps 3 and 4 that there exists t_0 such that $U_0^-(t_0) > 1$ and that there exists $v_0 \in H^1(\mathbb{R}) \cap C^1(\mathbb{R}\setminus\{0\})$ such that $\tau_0 < \infty$.

The assumption of orbital stability of the peaked solitary wave with profile ϕ in H^1 in Theorem 6.5 was shown to be true for $b = 2$ in [**63, 64**] using the conserved quantities $H(u)$ and $E(u)$ given in (6.8) and (6.9). We review the result in the following lemma to clarify in what sense the peaked solitary wave is orbitally stable in $H^1(\mathbb{R})$.

LEMMA 6.7. *For every $\delta > 0$, let $u \in C^0([0,\tau_0), H^1(\mathbb{R}))$ be a solution of the Camassa–Holm equation with $b = 2$ such that $u(0,\cdot) = u_0$, $\|u_0 - \phi\|_{H^1} < \delta$, and $\tau_0 > 0$ is the maximal existence time. There exists $C > 0$ such that*

$$\|u(t,\cdot) - \phi(\cdot - \xi(t))\|_{H^1} < C\delta^{1/4}, \quad t \in [0,\tau_0),$$

where $\xi(t)$ is the location of the single peak of $u(t,\cdot)$ for $t \in [0,\tau_0)$.

PROOF. It follows from the decomposition $u(t,x) = \phi(x-\xi(t)) + v(t,x-\xi(t))$ by the translational symmetry that

$$\begin{aligned} E(u) &= E(\phi) + \langle v, \phi\rangle + \langle v_x, \phi'\rangle + E(v) \\ &= E(\phi) + 2v(t,0) + E(v), \end{aligned}$$

which implies that

$$\|v(t,\cdot)\|_{H^1}^2 = 2E(u(t,\cdot)) - 2E(\phi) - 4v(t,0).$$

Due to the energy conservation and Cauchy–Schwarz inequality, the bound $\|u_0 - \phi\|_{H^1} < \delta$ on the initial data $u_0 \in H^1(\mathbb{R}) \cap W^{1,\infty}(\mathbb{R}))$ implies

$$\|v(t,\cdot)\|_{H^1}^2 \le C_0(\delta + |v(t,0)|), \quad t \in [0,\tau_0)$$

as long as the local solution exists in $H^1(\mathbb{R}) \cap W^{1,\infty}(\mathbb{R})$ for $t \in [0, \tau_0)$. Thus, controlling $|v(t,0)|$ gives the H^1 orbital stability of the peaked solitary wave.

Assume now that the solution consists of a single peak such that the location of the peak at $\xi(t)$ coincides with the location of the single maximum $M(t) := \max_{x\in\mathbb{R}} u(t,x) = u(t,\xi(t))$. Dropping the time t from the arguments, we obtain

$$H(u) \leq ME(u) - \frac{1}{3}M^3$$

from

$$\begin{aligned} H(u) - \frac{2}{3}M^3 &= \frac{1}{2}\int_{-\infty}^{\xi} u(u-u_x)^2 dx + \frac{1}{2}\int_{\xi}^{\infty} u(u+u_x)^2 dx \\ &\leq \frac{1}{2}M\int_{-\infty}^{\xi} (u-u_x)^2 dx + \frac{1}{2}M\int_{\xi}^{\infty} (u+u_x)^2 dx \\ &= ME(u) - M^3. \end{aligned}$$

Hence, $F(M,u) := M^3 - 3ME(u) + 3H(u)$ satisfies $F(M,u) \leq 0$. For $u = \phi$ with $E(\phi) = 1$ and $H(\phi) = \frac{2}{3}$, $F(M,\phi) = M^3 - 3M + 2 = (M+2)(M-1)^2$ has a simple root at -2 and a double root at 1, the latter one being the relevant one since $M_\varphi := \max_{x\in\mathbb{R}} \phi(x) = 1$. Since $E(u)$ and $H(u)$ are conserved and $\|u_0 - \phi\|_{H^1} < \delta$, there is a δ-independent constant $C > 0$ such that

$$|E(u) - 1| + |H(u) - \frac{2}{3}| \leq C\delta$$

for every small $\delta > 0$. Thus, we can write

$$F(M,u) = (M-1)^2(M+2) - 3M(E(u)-1) + 3\left(H(u) - \frac{2}{3}\right)$$

In view of $F(M,u) \leq 0$, the perturbation theory near the double root $M = 1$ implies that

$$|M(t) - 1| \leq C\delta^{1/2}, \quad t \in [0, \tau_0)$$

for another δ-independent constant $C > 0$. If $\xi(t)$ is the location of the maximum $M(t)$, we thus have $|v(t,0)| \leq C\delta^{1/2}$, which concludes the proof of the bound

$$\|v(t,\cdot)\|_{H^1} \leq C\delta^{1/4}, \quad t \in [0, \tau_0)$$

for another δ-independent constant $C > 0$. □

REMARK 6.26. Note that Lemma 6.7 implies only the **energetic stability** of the peaked solitary waves in $H^1(\mathbb{R})$, also known as **conditional orbital stability**, where the condition is the local well-posedness of the Camassa–Holm equation in $H^1(\mathbb{R})$. However, after the energetic stability results in [**63, 64**] were published it was shown in [**129**] that the Camassa–Holm equation is ill-posed in $H^1(\mathbb{R})$.

The assumption of orbital stability in Theorem 6.5 is not satisfied for $b = 3$ because the peaked solitary waves are energetically stable in $L^2(\mathbb{R})$ rather than in $H^1(\mathbb{R})$. The L^2-stability result was obtained in [**179**] by using the conserved quantities $H(u)$ and $E(u)$ defined in (6.12) and (6.13). The following result extends the proof of Theorem 6.5 to this case with minor modifications of the arguments.

THEOREM 6.6. *Consider the Degasperis–Procesi equation* ($b = 3$). *For every* $\delta > 0$, *there exists* $u_0 \in H^1(\mathbb{R}) \cap C^1(\mathbb{R} \setminus \{0\})$ *satisfying*

$$\|u_0 - \phi\|_{H^1} + \|u_0' - \phi'\|_{L^\infty} < \delta,$$

and $t_0 \in (0, \tau_0)$ *such that the local solution* $u \in C^0([0, \tau_0), H^1(\mathbb{R}) \cap W^{1,\infty}(\mathbb{R}))$ *with* $u(0, \cdot) = u_0$ *and with the maximal existence time* $\tau_0 > 0$ *satisfies*

$$\|u_x(t_0, \cdot) - \phi'(\cdot - \xi(t_0))\|_{L^\infty} > 1,$$

where $\xi(t) \in \mathbb{R}$ *is the location of a single peak of* $u(t, \cdot)$ *for* $t \in [0, \tau_0)$ *with* $\xi(0) = 0$. *Moreover, there exist* u_0 *such that* $\tau_0 < \infty$.

PROOF. The proof of Theorem 6.5 is modified as follows. Steps 1 and 2 are the same. For Step 3, although $V_0(t)$ can no longer be controlled, there is no need to control the H^1 norm of the perturbation v because the nonlocal terms in the initial-value problem (6.94) do not contain v_x terms if $b = 3$. Consequently, we rewrite the main equations of Step 2 as

$$\begin{cases} V_t = (V|_{x=0} - V)\phi' - \frac{3}{2}\phi' * (\phi v) - \frac{3}{4}\phi' * v^2, \\ V|_{t=0} = v_0, \end{cases}$$

and

$$\begin{cases} U_t = -2\phi' U + 3\phi V + (V|_{x=0} - V)\phi - \frac{3}{2}\phi * (\phi v) + \frac{3}{2}V^2 - U^2 - \frac{3}{4}\phi * v^2, \\ U|_{t=0} = v_0'. \end{cases}$$

By using the Cauchy–Schwartz inequality and the L^2-orbital stability result with the bound $\|v(t, \cdot)\|_{L^2} < \varepsilon$ for every $t \in [0, \tau_0)$, we find from the first system as $X \to 0$ that there is a constant $C_1 > 0$ such that $|V_0'(t)| \leq C_1\varepsilon$ for small $\varepsilon > 0$. Combining this with the initial bound $\|v_0\|_{H^1} < \nu(\varepsilon)$ yields at most linear growth of $|V_0(t)|$:

$$|V_0(t)| \leq \nu(\varepsilon) + C_1\varepsilon t, \qquad t \in [0, \tau_0).$$

Similarly, we find from the second system as $X \to 0^+$ that

$$\frac{dU_0^+}{dt} = 2U_0^+ + 3V_0 - \frac{3}{2}\langle \phi, \phi v\rangle + \frac{3}{2}V_0^2 - (U_0^+)^2 - \frac{3}{4}\langle \phi, v^2\rangle.$$

hence there is another constant $C_2 > 0$ such that

$$\frac{dU_0^+}{dt} \leq 2U_0^+ - (U_0^+)^2 + C_2\varepsilon + 3|V_0(t)| + \frac{3}{2}V_0(t)^2, \tag{6.99}$$

for small $\varepsilon > 0$. The bound on $U_0^+(t)$ is obtained from integration of

$$\frac{d}{dt}(e^{-2t}U_0^+) \leq \left(C_2\varepsilon + 3|V_0(t)| + \frac{3}{2}V_0(t)^2\right) e^{-2t},$$

to the form

$$U_0^+(t) \leq e^{2t}\left[U_0^+(0) + \int_0^t \left(C_2\varepsilon + 3|V_0(t)| + \frac{3}{2}V_0(t)^2\right) e^{-2t} dt\right].$$

By using the bounds

$$\int_0^t te^{-2t} dt \leq \frac{1}{4}, \qquad \int_0^t t^2 e^{-2t} dt \leq \frac{1}{4},$$

we obtain from $|V_0(t)| \leq \nu(\varepsilon) + C_1 \varepsilon t$ that

$$U_0^+(t) \leq (U_0^+(0) + C_3\varepsilon)e^{2t}$$

with another constant $C_3 > 0$ for small $\varepsilon > 0$. This yields the weak instability result. Step 4 for the strong instability is obtained from the differential inequality (6.99) with the polynomially growing $V_0(t)$ and $V_0(t)^2$ similarly to Step 2 in Theorem 6.1. $\square$

EXAMPLE 6.12. Peaked solitary waves exist in the cubic modification of the Camassa–Holm equation known as the Novikov equation

$$u_t - u_{xxt} + 4u^2 u_x = 3uu_x u_{xx} + u^2 u_{xxx}.$$

In terms of the momentum density $m := u - u_{xx}$, this equation is written in the form:

$$m_t + u^2 m_x + 3uu_x m = 0,$$

which shows that the local characteristic speed for the peak at $x = \xi(t)$ is given by

$$\xi'(t) = u^2(t, \xi(t)).$$

The peaked solitary wave with unit speed $u(t,x) = \phi(x-t)$ is given again by the function $\phi(x) := e^{-|x|}$. Decomposing with the perturbation $v(t,x) := u(t, x + t + a(t)) - \phi(x)$, we find that $a'(t) = 2v(t,0) + v(t,0)^2$ and obtain the evolution equation in the form

$$v_t + \mathcal{N}_1(v) + \mathcal{N}_2(v) + \mathcal{N}_3(v) = 0,$$

where the linear, quadratic, and cubic terms are given by

$$\begin{aligned}
\mathcal{N}_1(v) &= (\phi^2 - 1)v_x + 2\phi'(\phi v - v|_{x=0}) \\
&\quad - \frac{3}{2}\phi' * (\phi^2 v + \frac{1}{2}(\phi')^2 v + \phi\phi' v_x) + \frac{3}{4}\phi * ((\phi')^2 v_x), \\
\mathcal{N}_2(v) &= 2(\phi v - v|_{x=0})v_x + \phi'(v^2 - v^2|_{x=0}) \\
&\quad + \frac{1}{2}\phi' * (\frac{3}{2}\phi v_x^2 + 3\phi' v v_x + 3\phi v^2) + \frac{3}{4}\phi * (\phi' v_x^2), \\
\mathcal{N}_3(v) &= (v^2 - v^2|_{x=0})v_x + \frac{1}{2}\phi' * (\frac{3}{2}v v_x^2 + v^3) + \frac{1}{4}\phi * v_x^3.
\end{aligned}$$

With the help of the identities (6.34) and (6.35), the linearized equation $v_t + \mathcal{N}_1(v) = 0$ can be rewritten in the equivalent form

$$v_t + (\phi^2 - 1)v_x + \phi'(v|_{x=0} - \phi v) = 0.$$

This linearized equation can be solved with the method of characteristics similar to Example 6.7. The characterstic curves at $x = X(t,s)$ are given by

$$X(t,s) = \begin{cases} \frac{1}{2}\log\left(1 + (e^{2s} - 1)e^{-2t}\right), & s > 0, \\ -\frac{1}{2}\log\left(1 + (e^{-2s} - 1)e^{2t}\right), & s < 0, \end{cases}$$

where the solution $V(t,s) := v(t, X(t,s))$ along the characteristic curves is given by

$$V(t,s) = \begin{cases} \dfrac{v_0(s) + v_0(0)(e^t - 1)e^{-s}}{\sqrt{1 + (e^{2t} - 1)e^{-2s}}}, & s > 0, \\ \dfrac{v_0(s) - v_0(0)(1 - e^{-t})e^{s}}{\sqrt{1 - (1 - e^{-2t})e^{2s}}}, & s < 0. \end{cases}$$

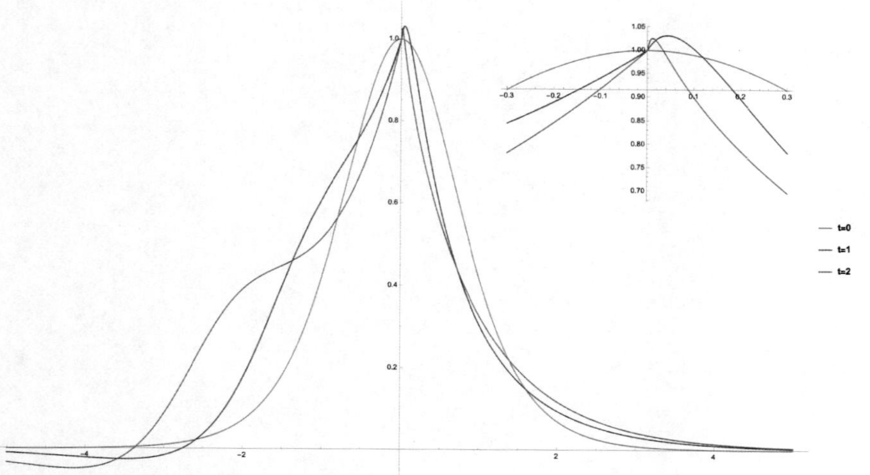

FIGURE 6.7. A plot of the solution $v(t,x)$ of linearized equation in Example 6.12 as a function of x for the initial condition $v_0(x) = \text{sech}^2(x)$ at three instances of time t illustrating exponential growth.

It was proven in [**48**] that for every initial data $v_0 \in H^1(\mathbb{R}) \cap C^1(\mathbb{R}\backslash\{0\})$, the unique global solution of the linearized Novikov equation with $v(0,\cdot) = v_0$ preserves the H^1 norm separately for $x > 0$ and $x < 0$ as $\|v(t,\cdot)\|^2_{H^1(\mathbb{R}^\pm)} = \|v_0\|^2_{H^1(\mathbb{R}^\pm)}$ with the $W^{1,\infty}$ norm growing in time at the right side of the peak:

$$\|v_x(t,\cdot)\|_{L^\infty(\mathbb{R}^+)} \geq |v_0(0) + v_0'(0^+)|e^t - |v_0(0)|, \quad t > 0.$$

Figure 6.7 illustrates the solution $v(t,x)$ for $v_0(x) = \text{sech}^2(x)$. The linear $W^{1,\infty}$ instability of the peaked solitary wave persists as the nonlinear instability and the result similar to Theorem 6.5 was proven in [**48**].

REMARK 6.27. Although Example 6.12 shows that the peaked solitary wave is linearly stable in $L^2(\mathbb{R})$ if the initial data is in $H^1(\mathbb{R}) \cap C^1(\mathbb{R}\backslash\{0\})$, this result does not eliminate the linear instability in $L^2(\mathbb{R})$ if the initial data is defined in the domain of the linearized operator. The spectral instability for the Novikov equation was recently proven in [**167**].

REMARK 6.28. Peaked solitary waves were considered in many other equations similar to the Camassa–Holm equations [**47, 176**]. Although the H^1-orbital stability was routinely proven in these models based on the energy conservation, it was missed that the peaked waves are unstable in the $W^{1,\infty}$ norm due to finite time wave breaking in the nonlinear evolution of these models. Due to the mismatch between the spaces where local well-posedness holds and where the energy functions are defined, these stability results give only energetic stability of peaked solitary waves. Only recently, it was proven that the peaked waves in these equations are linearly and nonlinearly unstable using similar methods as the ones described in this section, see [**53, 85**].

CHAPTER 7

Stability of domain walls and black solitons

Domain walls are examples of **topological solitary waves** which commonly occur in nonlinear dispersive partial differential equations. Other examples are given by kinks (Example 2.3) and black solitons (Example 2.4). Analysis of stability of the topological solitary waves is complicated due to nonzero boundary conditions at infinity. Topological solitary waves do not belong to the Hilbert space $\mathcal{H} = L^2(\mathbb{R})$ and have to be defined in the appropriate Hilbert space where the energy functions are bounded and smooth.

Although the main strategy of the orbital stability analysis for the topological solitary waves follow the same algorithm as the one in Chapters 2, 3, and, in the particular case of integrable systems, in Chapter 4, we have to develop more elaborate estimates to obtain coercivity of the energy function near the wave profile and to control the remainder terms in the expansion of the augmented energy. Details of the stability analysis is a subject of this chapter, where we explain the estimates for several study cases. We also describe some practically useful methods of the spectral stability analysis.

The study cases with domain walls and black solitons are introduced in Section 7.1. Orbital stability of the domain walls as the energy minimizers is obtained in Section 7.2. Orbital stability of the black solitons as the constrained energy minimizers is obtained in Section 7.3. Criteria for spectral stability of the black solitons are given in Section 7.4.

7.1. Hamiltonian systems with domain walls

Domain walls can be studied by using the following system of two coupled nonlinear Schrödinger (NLS) equations:

$$\begin{cases} i\partial_t\psi_1 = -\partial_x^2\psi_1 + (|\psi_1|^2 + \gamma|\psi_2|^2)\psi_1, \\ i\partial_t\psi_2 = -\partial_x^2\psi_2 + (\gamma|\psi_1|^2 + |\psi_2|^2)\psi_2, \end{cases} \tag{7.1}$$

where $t \in \mathbb{R}$, $x \in \mathbb{R}$, $(\psi_1, \psi_2) \in \mathbb{C}^2$, and γ is a real coupling parameter. Other parameters of the model have been scaled to unity. Similar to other models, the coupled NLS system (7.1) is an example of the Hamiltonian system with formally conserved energy in the form

$$H(\psi_1, \psi_2) = \int_{\mathbb{R}} [|\partial_x\psi_1|^2 + |\partial_x\psi_2|^2 + P_\gamma(|\psi_1|^2, |\psi_2|^2)]dx, \tag{7.2}$$

where

$$P_\gamma(|\psi_1|^2, |\psi_2|^2) = \frac{1}{2}(|\psi_1|^2 + |\psi_2|^2 - 1)^2 + (\gamma - 1)|\psi_1|^2|\psi_2|^2. \tag{7.3}$$

The energy space for the wave function $(\psi_1, \psi_2) \in \mathbb{C}^2$ for which the energy $H(\psi_1, \psi_2)$ is bounded and smooth depends on the boundary conditions of (ψ_1, ψ_2) at infinity. Topological solitary waves correspond to nonzero boundary conditions and, for normalization purposes, the choice of P_γ in (7.3) suggests that suitable solutions of the coupled NLS system (7.1) satisfy

$$(7.4) \qquad |\psi_1(t,x)|^2 + |\psi_2(t,x)|^2 \to 1, \quad |\psi_1(t,x)|^2|\psi_2(t,x)|^2 \to 0 \quad \text{as} \quad |x| \to \infty,$$

for every $t \in \mathbb{R}$. The decay rate at infinity is assumed to be sufficiently fast such that (ψ_1, ψ_2) is defined in the energy space X for which

$$(7.5) \qquad \partial_x \psi_1, \partial_x \psi_2 \in L^2(\mathbb{R}), \quad |\psi_1|^2 + |\psi_2|^2 - 1 \in L^2(\mathbb{R}), \quad \psi_1 \psi_2 \in L^2(\mathbb{R}).$$

The energy function $H(\psi_1, \psi_2)$ in (7.2) is bounded and smooth for $(\psi_1, \psi_2) \in X$. Critical points of $H(\psi_1, \psi_2)$ in X satisfy the Euler–Lagrange equations

$$(7.6) \qquad \begin{cases} -u_1'' + (|u_1|^2 + \gamma|u_2|^2 - 1)u_1 = 0, \\ -u_2'' + (\gamma|u_1|^2 + |u_2|^2 - 1)u_2 = 0, \end{cases}$$

which correspond to the standing wave solutions $\psi_{1,2}(t,x) = e^{-it}u_{1,2}(x)$ of the coupled NLS system (7.1).

Compared to the systems of differential equations in Chapters 2, 3, and 4, the system (7.6) defines a dynamical system in $\mathbb{C}^4$, bounded solutions of which could be extremely complicated. Hence, before proceeding with the stability analysis, we have to identify a suitable class of the topological solitary waves and verify their existence for some values of the parameter γ. Similar to the questions of well-posedness of PDEs, existence of bounded solutions in systems of ODEs is a subject of a different monograph. Hence we will avoid details of the asymptotic and variational methods for the domain walls in the system (7.6), which have been used in [**2, 21, 112, 230**] and [**4, 240**], respectively. Instead, we will define **domain walls** as well as **black solitons** as suitable bounded solutions of the system (7.6) and explain how these solutions arise.

We will start with a simple result on the constant (real and positive) solutions of the system (7.6).

LEMMA 7.1. *For every $\gamma > -1$, there exist three constant (real and positive) solutions of the system* (7.6)*:*

$$S_+ := \begin{bmatrix} 1 \\ 0 \end{bmatrix}, \quad S_- := \begin{bmatrix} 0 \\ 1 \end{bmatrix}, \quad S_0 := \frac{1}{\sqrt{1+\gamma}} \begin{bmatrix} 1 \\ 1 \end{bmatrix},$$

in addition to the trivial (zero) solution. The solution S_0 is a global minimizer of P_γ if $\gamma \in (-1,1)$ and a saddle point of P_γ if $\gamma \in (1,\infty)$, whereas the solutions $S_\pm$ are saddle points of P_γ if $\gamma \in (-1,1)$ and global minimizers of P_γ if $\gamma \in (1,\infty)$. At $\gamma = 1$, all three solutions belong to a continuous family of constant (real and positive) solutions of the system (7.6) *given by*

$$S_\theta := \begin{bmatrix} \cos\theta \\ \sin\theta \end{bmatrix}, \quad \theta \in \left[0, \frac{\pi}{2}\right].$$

These solutions are degenerate minimizers of $P_{\gamma=1}$.

PROOF. Let us rewrite P_γ in (7.3) as a quadratic form in variables $x_{1,2} = |\psi_{1,2}|^2 \geq 0$:

$$P_\gamma(x_1, x_2) = \frac{1}{2}(x_1 + x_2 - 1)^2 + (\gamma - 1)x_1 x_2$$
$$= \frac{1}{2}(x_1^2 + x_2^2 + 2\gamma x_1 x_2 - 2x_1 - 2x_2 + 1).$$

The only critical point of P_γ in the first quadrant for $\gamma \neq 1$ is

$$(x_1, x_2) = \left(\frac{1}{1+\gamma}, \frac{1}{1+\gamma}\right)$$

for which

$$P_\gamma\left(\frac{1}{1+\gamma}, \frac{1}{1+\gamma}\right) = -\frac{(1-\gamma)}{2(1+\gamma)}.$$

Convexity of P_γ at the critical point is defined by the Hessian matrix

$$\begin{bmatrix} 1 & \gamma \\ \gamma & 1 \end{bmatrix},$$

which is strictly positive for $\gamma \in (-1, 1)$ and sign-indefinite for $\gamma \in (1, \infty)$. Hence, the constant solution S_0 of the system (7.6) represented by $(x_1, x_2) = ((1+\gamma)^{-1}, (1+\gamma)^{-1})$ is a global minimizer of P_γ for $\gamma \in (-1, 1)$ and a saddle point of P_γ for $\gamma \in (1, \infty)$.

At the two boundaries of the first quadrant, the only critical points occur at $(x_1, x_2) = (1, 0)$ and $(x_1, x_2) = (0, 1)$, for which

$$P_\gamma(1, 0) = P_\gamma(0, 1) = 0.$$

Each point is a local minimizer of P_γ at the corresponding boundaries. Combined with the behavior of P_γ inside the first quadrant, we conclude that the constant solutions $S_\pm$ of the system (7.6) represented by $(x_1, x_2) = (1, 0)$ and $(x_1, x_2) = (0, 1)$ are saddle points of P_γ for $\gamma \in (-1, 1)$ and global minimizers of P_γ for $\gamma \in (1, \infty)$.

It remains to consider the borderline case $\gamma = 1$, for which $P_{\gamma=1}$ is given by

$$P_{\gamma=1}(x_1, x_2) = \frac{1}{2}(x_1 + x_2 - 1)^2.$$

The critical points occur on the line segment $x_1 + x_2 = 1$ in the first quadrant and correspond to the degenerate minimizers of P_γ. The line segment represents the constant solutions S_θ of the system (7.6) with $\gamma = 1$ which connects S_0 and $S_\pm$ at $\theta = 0$, $\theta = \frac{\pi}{4}$, and $\theta = \frac{\pi}{2}$. □

REMARK 7.1. The constant solutions $S_\pm$ belong to X but correspond to the reduction of the coupled NLS system (7.1) to a scalar NLS equation (Example 3.16). The constant solution S_0 does not belong to X for $\gamma \neq 1$ because

$$|\psi_1|^2 + |\psi_2|^2 - 1 = \frac{1-\gamma}{1+\gamma} \neq 0.$$

EXAMPLE 7.1. If $\psi_2 = 0$, the system (7.1) reduces to the scalar defocusing NLS equation for $\psi \equiv \psi_1$:

$$i\partial_t \psi = -\partial_x^2 \psi + |\psi|^2 \psi.$$

It admits the following family of traveling wave solutions satisfying the boundary conditions (7.4):

$$\psi(t,x) = e^{-it}\left[\kappa \tanh\left(\frac{\kappa}{\sqrt{2}}(x-ct)\right) + \frac{ic}{\sqrt{2}}\right], \quad \kappa = \sqrt{1-\frac{c^2}{2}},$$

where $c \in (-\sqrt{2}, \sqrt{2})$. The traveling wave solutions are known as **dark solitons** as they correspond to the dip of the squared amplitudes:

$$|\psi(t,x)|^2 = 1 - \kappa^2 \operatorname{sech}^2\left(\frac{\kappa}{\sqrt{2}}(x-ct)\right),$$

with $\kappa \in (0,1]$. In the case of the standing wave solutions with $c = 0$ and $\kappa = 1$, the family of dark solitons recovers the black soliton in the form

$$\psi(t,x) = e^{-it} \tanh\left(\frac{x}{\sqrt{2}}\right).$$

There exist symmetrically reflected dark and black solitons by replacing ψ by $-\psi$.

In addition to the constant solutions in Lemma 7.1 and the black solitons in Example 7.1, we define the domain walls of the system (7.6) by using the following definition.

DEFINITION 7.1. We say that $(u_1, u_2) \in X$ is the domain wall solution of the system (7.6) if it satisfies the boundary conditions

$$\begin{bmatrix} u_1(x) \\ u_2(x) \end{bmatrix} \to S_{\pm} \quad \text{as either } x \to \pm\infty \quad \text{or } x \to \mp\infty. \tag{7.7}$$

EXAMPLE 7.2. There exists a particular solution of the system (7.6) with $\gamma = 3$ for the domain wall solutions of Definition 7.1:

$$u_1(x) = \frac{1}{2}\left[1 + \tanh\left(\frac{x}{\sqrt{2}}\right)\right], \quad u_2(x) = \frac{1}{2}\left[1 - \tanh\left(\frac{x}{\sqrt{2}}\right)\right],$$

which satisfies (7.7) as $x \to \pm\infty$. The domain wall solution satisfying (7.7) as $x \to \mp\infty$ is obtained by reflection of x to $-x$. The domain wall solutions can be translated in x with an arbitrary parameter $\xi \in \mathbb{R}$ as the orbit $\{u_1(\cdot - \xi), u_2(\cdot - \xi)\}_{\xi\in\mathbb{R}}$. Without the translation, the solution satisfies the following symmetry:

$$u_1(x) = u_2(-x), \quad x \in \mathbb{R}. \tag{7.8}$$

It follows from the behavior of the system of differential equations (7.6) near the critical points $S_{\pm}$ that no exponential convergence of solutions to $S_{\pm}$ as either $x \to \pm\infty$ or $x \to \mp\infty$ is possible for $|\gamma| \le 1$. This restricts consideration of the domain wall solutions to $\gamma \in (1, \infty)$. Their existence for $\gamma \in (1,\infty)$ follows from the variational characterization of domain walls as minimizers of energy $H(u_1, u_2)$ in the energy space X subject to the boundary conditions (7.7). Moreover, the variational methods are also useful to clarify properties of the spatial profiles (u_1, u_2) versus x. The following lemma proven in Theorems 2.1, 2.4, and 3.1 of [4] states the existence of the domain wall solutions of the system (7.6).

LEMMA 7.2. *For every $\gamma > 1$, there exists an orbit $\{u_1(\cdot - \xi)e^{-i\alpha_1}, u_2(\cdot - \xi)e^{-i\alpha_2}\}_{\xi,\alpha_1,\alpha_2\in\mathbb{R}}$ of minimizers of energy (7.2) with (7.3) in the energy space X subject to the boundary conditions (7.7). The real-valued spatial profiles (u_1, u_2) are uniquely defined by the symmetry (7.8) and satisfy*

- $u_1^2(x) + u_2^2(x) \leq 1$ *for all* $x \in \mathbb{R}$,
- *either* $u_1'(x) > 0$, $u_2'(x) < 0$ *or* $u_1'(x) < 0$, $u_2'(x) > 0$ *for all* $x \in \mathbb{R}$,
- *the convergence rate in* (7.7) *is exponential:* $e^{-\sqrt{2}|x|}$ *near* 0 *and* $e^{-\sqrt{\gamma-1}|x|}$ *near* 1.

REMARK 7.2. Uniqueness of domain wall solutions with properties in Lemma 7.2 was proven in [**2**, Theorem 1.3] and in a more general setting in [**93**].

REMARK 7.3. The existence and bifurcations of the domain wall solutions in the coupled NLS system (7.1) with the harmonic potential x^2 were studied in [**69**]. In this case, the family of standing wave solutions has additional parameter ω in

$$\psi_{1,2}(t,x) = e^{-i\omega t} u_{1,2}(x)$$

and the translational invariance is broken. The definition of the domain wall solutions has to be revised compared to Definition 7.1 because all standing wave solutions in the harmonic potential are spatially decaying. The domain wall solutions were defined in [**69**] to satisfy the symmetry (7.8) and to be different from the symmetric solutions satisfying also $u_1(x) = u_2(x)$ for $x \in \mathbb{R}$.

REMARK 7.4. The system of coupled NLS equation (7.1) admits formally conserved masses in each component:

$$Q_1(\psi) = \int_{\mathbb{R}} |\psi_1|^2 dx, \qquad Q_2(\psi) = \int_{\mathbb{R}} |\psi_2|^2 dx.$$

However, these quantities are not useful in the stability analysis of the domain walls as they are infinite even if we subtract any constant value from the integrand. Additionally, the system has the conserved momentum in the form:

$$P(\psi_1, \psi_2) = i \int_{\mathbb{R}} (\bar{\psi}_1 \partial_x \psi_1 + \bar{\psi}_2 \partial_x \psi_2 - \psi_1 \partial_x \bar{\psi}_1 - \psi_2 \partial_x \bar{\psi}_2) dx.$$

For the stability analysis of the domain walls, we do not need any help from additional conserved quantities as the domain walls in Lemma 7.2 are minimizers (rather than constrained minimizers) of the conserved energy $H(\psi_1, \psi_2)$.

REMARK 7.5. Similar to Example 7.1, one can define coupled black solitons of the system (7.6) which are exponentially decaying to the constant solutions $\pm S_0$ as either $x \to \pm\infty$ or $x \to \mp\infty$. The function P_γ in the energy function $H(\psi_1, \psi_2)$ has to be renormalized by subtracting a constant value to ensure that the integrals converge at infinity. Both the existence and stability for such coupled black solitons are open problems. The stability of the black solitons in the scalar NLS equation (Example 7.1) is addressed in Section 7.3.

7.2. Minimizers of energy

The initial-value problem for the coupled NLS equations (7.1) is locally well-posed for perturbations to the domain walls and black solitons defined in a subset of $L^2(\mathbb{R})$ [**99**, **102**]. However, the main difficulty in developing their orbital stability analysis is that the energy is not coercive at the domain walls and black solitons due to nonzero boundary conditions.

We will explain this main difficulty by working with the domain wall $U = (u_1, u_2) \in \mathbb{R}^2$ of Lemma 7.2 and the energy function $H(\psi_1, \psi_2)$ defined in (7.2) and (7.3). By adding a perturbation to U and separating its real and imaginary parts as $W + iV$, we obtain the following decomposition of energy:

$$H(U+W+iV) - H(U) = \langle L_+ W, W\rangle + \langle L_- V, V\rangle + R(W,V), \tag{7.9}$$

where $L_\pm : H^2(\mathbb{R}) \subset L^2(\mathbb{R}) \to L^2(\mathbb{R})$ are the linear self-adjoint operators given by

$$L_+ = \begin{bmatrix} -\partial_x^2 + 3u_1^2 + \gamma u_2^2 - 1 & 2\gamma u_1 u_2 \\ 2\gamma u_1 u_2 & -\partial_x^2 + \gamma u_1^2 + 3u_2^2 - 1 \end{bmatrix}$$

and

$$L_- = \begin{bmatrix} -\partial_x^2 + u_1^2 + \gamma u_2^2 - 1 & 0 \\ 0 & -\partial_x^2 + \gamma u_1^2 + u_2^2 - 1 \end{bmatrix},$$

whereas $R(W,V) = \int_{\mathbb{R}} r(W,V)dx$ is the remainder term with the density $r(W,V)$ given by the cubic and quartic terms:

$$\begin{aligned} r(W,V) = {} & 2(u_1w_1 + u_2w_2)(w_1^2 + v_1^2 + w_2^2 + v_2^2) + 2(\gamma-1)u_1w_1(w_2^2 + v_2^2) \\ & + 2(\gamma-1)u_2w_2(w_1^2 + v_1^2) + \frac{1}{2}(w_1^2 + v_1^2 + w_2^2 + v_2^2)^2 \\ & + (\gamma-1)(w_1^2 + v_1^2)(w_2^2 + v_2^2). \end{aligned}$$

As previously, $\langle \cdot, \cdot \rangle$ indicate the inner product in $L^2(\mathbb{R})$ and $\langle L_\pm \cdot, \cdot\rangle$ denote the quadratic form of $L_\pm$ defined in $H^1(\mathbb{R})$. If small perturbations are defined in $H^1(\mathbb{R})$, then

$$R(W,V) = \mathcal{O}(\|W + iV\|_{H^1}^3)$$

is cubically small because $U \in L^\infty(\mathbb{R})$ and $H^1(\mathbb{R})$ is a Banach algebra with respect to pointwise multiplication.

The following lemma specifies the properties of the linear operators $L_\pm$.

LEMMA 7.3. *Let $U = (u_1,u_2) \in \mathbb{R}^2$ be the domain wall solution of Lemma 7.2. Then, the quadratic forms for L_+ and L_- are positive semi-definite on $H^1(\mathbb{R})$. Zero is a simple eigenvalue of L_+, with associated eigenfunction $\partial_x U$ and there exists $C_+ > 0$ such that*

$$\langle L_+ W, W\rangle \geq C_+ \|W\|_{H^1}^2 \quad \text{for every} \quad W \in H^1(\mathbb{R}) : \quad \langle W, \partial_x U\rangle = 0. \tag{7.10}$$

Zero is the bottom of the essential spectrum of L_- such that

$$L_- U_1 = 0, \quad L_- U_2 = 0 \tag{7.11}$$

with $U_1 = (u_1, 0) \notin L^2(\mathbb{R})$ and $U_2 = (0, u_2) \notin L^2(\mathbb{R})$.

PROOF. Positivity of the quadratic form for L_+ and L_- in $H^1(\mathbb{R})$ follows from the fact that the domain wall with the profile $U = (u_1,u_2) \in \mathbb{R}^2$ is a minimizer of the energy $H(\psi_1, \psi_2)$ by Lemma 7.2.

The essential spectrum of L_+ and L_- is found from the boundary conditions (7.7) due to the exponential convergence of the domain walls to the constant solutions $S_\pm$, see Lemma 7.2. Since L_+ converges to the diagonal operator with entries $-\partial_x^2 + 2$ and $-\partial_x^2 + \gamma - 1$ with $\gamma > 1$, the essential spectrum of L_+ is located at $[\min(\gamma - 1, 2), \infty)$ and is bounded away from 0. Since L_- converges to to the diagonal operator with entries $-\partial_x^2$ and $-\partial_x^2 + \gamma - 1$, the essential spectrum of L_- is located at $[0, \infty)$. We can now analyze the other properties of L_+ and L_-.

For L_+, the translational invariance of the coupled NLS system (7.1) implies that $L_+ \partial_x U = 0$ with $\partial_x U \in H^2(\mathbb{R})$ due to smoothness and decay of the domain wall solutions with the profile U. Hence 0 is an eigenvalue of L_+. To show that it is a simple eigenvalue, we write

$$W_1(x) = A_1(x)u_1'(x), \quad W_2(x) = A_2(x)u_2'(x)$$

with real $(A_1, A_2) \in C^1(\mathbb{R})$ and transform the quadratic form for L_+ in $H^1(\mathbb{R})$ to the form

$$\langle L_+W, W\rangle = \int_{\mathbb{R}} \left[(A_1')^2(u_1')^2 + (A_2')^2(u_2')^2 - 2\gamma u_1 u_2 u_1' u_2'(A_1 - A_2)^2\right] dx,$$

where $u_1(x)u_2(x)u_1'(x)u_2'(x) < 0$ for all $x \in \mathbb{R}$ by Lemma 7.2. Thus, $\langle L_+W, W\rangle \geq 0$ and attain zero if and only if $A_1 = A_2$ for which W is proportional to $\partial_x U$. Hence, 0 is a simple eigenvalue of L_+. Since it is isolated from the rest of the spectrum of L_+, the coercivity bound (7.10) follows from the spectral gap and the standard estimates, as in the proof of Theorem 2.2.

For L_-, the rotational invariance of the coupled NLS system (7.1) implies (7.11), where bounded eigenfunctions $U_1 = (u_1, 0)$ and $U_2 = (0, u_2)$ are not in $L^2(\mathbb{R})$ since they decay to nonzero boundary conditions at one infinity. We can also confirm that 0 is at the bottom of the spectrum of L_- by writing

$$V_1(x) = B_1(x)u_1(x), \quad V_2(x) = B_2(x)u_2(x)$$

with real $(V_1, V_2) \in C^1(\mathbb{R})$ and transforming the quadratic form for L_- in $H^1(\mathbb{R})$ to the form

$$\langle L_-V, V\rangle = \int_{\mathbb{R}} \left[(B_1')^2(u_1)^2 + (B_2')^2(u_2)^2\right] dx.$$

Hence, zero is the bottom of the spectrum of L_- but it is not isolated from the essential spectrum of L_- located for $[0, \infty)$. $\square$

REMARK 7.6. Since zero is not separated from the essential spectrum of L_- by a spectral gap, the quadratic form $\langle L_-V, V\rangle$ is not coercive in $H^1(\mathbb{R})$ so that the cubic and quartic remainder terms in $R(W, V)$ cannot be controlled in $H^1(\mathbb{R})$ from the energy conservation. Hence the stability analysis of the domain walls is not developed by considering their perturbations in $H^1(\mathbb{R})$ contrary to the stability analysis of the kinks in Theorem 2.2 and Corollary 2.1 (Section 2.2).

In order to obtain coercivity of the quadratic form $\langle L_-V, V\rangle$ and to control the remainder terms $R(W, V)$, we introduce the exponentially weighted Hilbert space. The weighted Hilbert space denoted by $\mathcal{H}$ is introduced according to the following inner product for $\Psi = (\psi_1, \psi_2)$ and $\Phi = (\phi_1, \phi_2)$,

$$\langle \Psi, \Phi\rangle_{\mathcal{H}} := \int_{\mathbb{R}} \left[\psi_1'\bar{\phi}_1' + \psi_2'\bar{\phi}_2' + (\gamma - 1)(1 - u_1^2)\psi_1\bar{\phi}_1 + (\gamma - 1)(1 - u_2^2)\psi_2\bar{\phi}_2\right] dx,$$

and the induced norm $\|\Psi\|_{\mathcal{H}} := \sqrt{\langle \Psi, \Psi\rangle_{\mathcal{H}}}$. Positivity of the weights is ensured by $\gamma > 1$ and properties $u_1^2 + u_2^2 \leq 1$ of the domain walls with the profile $U = (u_1, u_2)$ in Lemma 7.2.

REMARK 7.7. The Sobolev space $H^1(\mathbb{R})$ is continuously embedded into the weighted space $\mathcal{H}$ because there is a positive constant C_H such that

$$\|\Psi\|_{\mathcal{H}} \leq C_H \|\Psi\|_{H^1}, \quad \text{for every } \Psi \in H^1(\mathbb{R}).$$

Functions in $\mathcal{H}$ can apriori be slowly growing at infinity due to the exponential weights in the norm of $\mathcal{H}$. This is not acceptable since the domain wall solutions with the profile $U = (u_1, u_2)$ are bounded at infinity. Following the ideas from [**113**] introduced for black solitons of the cubic defocusing NLS equation (Example 7.1), we can control perturbations at infinity by using new variables

$$\eta_1 := |u_1 + w_1 + iv_1|^2 - u_1^2 = 2u_1 w_1 + w_1^2 + v_1^2,$$
$$\eta_2 := |u_2 + w_2 + iv_2|^2 - u_2^2 = 2u_2 w_2 + w_2^2 + v_2^2,$$

which are defined consistently with the energy space (7.5) for the domain wall solutions with the profile $U = (u_1, u_2)$. If we denote $\Upsilon := (\eta_1, \eta_2)$, the decomposition of energy in (7.9) can be rewritten in a compact form as the sum of three quadratic forms:

$$H(U + W + iV) - H(U) = \langle L_- W, W\rangle + \langle L_- V, V\rangle + \frac{1}{2}\langle M\Upsilon, \Upsilon\rangle, \tag{7.12}$$

where

$$M = \begin{bmatrix} 1 & \gamma \\ \gamma & 1 \end{bmatrix}.$$

This follows from (7.9) since

$$2u_1^2 w_1^2 + 2u_2^2 w_2^2 + 4\gamma u_1 u_2 w_1 w_2 + r(W, V) = \frac{1}{2}(\eta_1 + \eta_2)^2 + (\gamma - 1)\eta_1\eta_2.$$

REMARK 7.8. The third term in the energy decomposition (7.12) is sign-indefinite since $\gamma > 1$. This creates a complication of the stability analysis of the domain walls and requires us to further modify the decomposition of energy into different parts considered in the weighted Hilbert space $\mathcal{H}$.

By Lemma 7.3, the first two terms in the energy decomposition (7.12) are positive semi-definite in $H^1(\mathbb{R})$ but not coercive. By using the weighted Hilbert space $\mathcal{H}$, we can obtain the desired coercivity of the quadratic form $\langle L_- V, V\rangle$ subject to two orthogonality conditions, according to the following result.

LEMMA 7.4. *There exists $C_- > 0$ such that*

$$\langle L_- V, V\rangle \geq C_- \|V\|_{\mathcal{H}}^2 \quad \text{for every } V \in \mathcal{H}: \quad \langle TV, U_1\rangle_{\mathcal{H}} = \langle TV, U_2\rangle_{\mathcal{H}} = 0, \tag{7.13}$$

where $U_1 = (u_1, 0) \in \mathcal{H}$, $U_2 = (0, u_2) \in \mathcal{H}$, and the linear operator $T : \mathcal{H} \to \mathcal{H}$ is defined by the bilinear form

$$\langle T\Psi, \Phi\rangle_{\mathcal{H}} := \int_{\mathbb{R}} \left(1 - u_1^2 - u_2^2\right)\left(\psi_1\bar{\phi}_1 + \psi_2\bar{\phi}_2\right) dx.$$

PROOF. Since $1 - u_1^2 - u_2^2$ decays to zero at both infinities exponentially fast, the operator T is compact in $\mathcal{H}$ and its spectrum in $\mathcal{H}$ is purely discrete. Since

$$\langle L_- V, V\rangle = \|V\|_{\mathcal{H}}^2 - \gamma\langle TV, V\rangle_{\mathcal{H}},$$

the spectrum of the operator $\mathcal{L}_- := I - \gamma T$ in $\mathcal{H}$ consists of isolated eigenvalues $\{\lambda_n\}_{n\in\mathbb{N}}$ accumulating to 1. Moreover, $\lambda_n < 1$ since $1 - u_1^2 - u_2^2 \geq 0$ by Lemma 7.2.

By Lemma 7.3, the zero eigenvalue of $\mathcal{L}_-$ is at least double with eigenfunctions $U_1, U_2 \in \mathcal{H}$. To show that $U_{1,2}$ belong to $\mathcal{H}$, we note that $\partial_x u_{1,2}$ decays exponentially to zero at both infinities, whereas $u_{1,2}$ decays exponentially to zero at the same infinity where $1 - u_{1,2}^2$ is nonzero and vice versa. Therefore, $U_{1,2} \in \mathcal{H}$.

To study eigenvalues $\{\lambda_n\}_{n\in\mathbb{N}}$ of $\mathcal{L}_-$ in $\mathcal{H}$, we write the generalized spectral problem

$$L_-\Psi = \lambda K\Psi, \quad \Psi \in \mathcal{H},$$

with K being given by

$$K = \begin{bmatrix} -\partial_x^2 + (\gamma-1)(1-u_1^2) & 0 \\ 0 & -\partial_x^2 + (\gamma-1)(1-u_2^2) \end{bmatrix},$$

since $\|\Psi\|_{\mathcal{H}}^2 = \langle K\Psi, \Psi\rangle$. Both L_- and K are diagonal Schrödinger operators, hence the generalized spectral problem can be written as two separate equations:

$$\left[-\partial_x^2 + u_1^2 + \gamma u_2^2 - 1\right]\psi_1 = \lambda\left[-\partial_x^2 + (\gamma-1)(1-u_1^2)\right]\psi_1$$

and

$$\left[-\partial_x^2 + \gamma u_1^2 + u_2^2 - 1\right]\psi_2 = \lambda\left[-\partial_x^2 + (\gamma-1)(1-u_2^2)\right]\psi_2.$$

Each equation with $\lambda = 0$ has one bounded and one unbounded linearly independent solutions. The unbounded solution grows exponentially at the same infinity where the bounded solution decays exponentially because the Wronskian of the two linearly independent solutions is constant and nonzero. Since the exponentially growing solutions do not belong to the space $\mathcal{H}$, the kernel of $\mathcal{L}_-$ in $\mathcal{H}$ is exactly two-dimensional, spanned by U_1 and U_2.

No solutions of the generalized spectral problem $L_-\Psi = \lambda K\Psi$ exists in $H^1(\mathbb{R})$ for $\lambda < 0$ since both $\langle L_-V, V\rangle$ and $\langle KV, V\rangle$ are positive semi-definite in $H^1(\mathbb{R})$. To check the same in $\mathcal{H}$, we rewrite the first scalar spectral problem in the equivalent form:

$$\left[-\partial_x^2 + P(x;\lambda)\right]\psi_1 = 0, \quad P(\cdot;\lambda) := \frac{u_1^2 + \gamma u_2^2 - 1 + \lambda(\gamma-1)(u_1^2-1)}{1-\lambda}.$$

Since the mapping of $\lambda \mapsto P(x;\lambda)$ is decreasing for every $x \in \mathbb{R}$, which follows from

$$\partial_\lambda P(x;\lambda) = -\frac{\gamma}{(1-\lambda)^2}\left[1 - u_1(x)^2 - u_2(x)^2\right] \le 0,$$

Sturm's Comparison Theorem (Theorem B.10 in Appendix B) implies that bounded solutions exists for a fixed value of $\lambda = \lambda_1$ only if the Schrödinger operator $-\partial_x^2 + P(x;\lambda_2) : H^2(\mathbb{R}) \to L^2(\mathbb{R})$ with $\lambda_2 > \lambda_1$ has a negative eigenvalue associated with a decaying eigenfunction at infinity. Since $-\partial_x^2 + P(x;0) : H^2(\mathbb{R}) \to L^2(\mathbb{R})$ has no negative eigenvalues, no bounded solutions exist for $\lambda < 0$. A similar argument applies to the second scalar spectral problem for $\lambda < 0$.

Thus, the generalized spectral problem $L_-\Psi = \lambda K\Psi$ has no negative eigenvalues λ in $\mathcal{H}$, whereas the zero eigenvalue is double and isolated from the rest of the purely discrete spectrum of $\mathcal{L}_-$. Since the orthogonality conditions

$$0 = \langle \Psi, U_{1,2}\rangle_{\mathcal{H}} = \gamma \int_{\mathbb{R}} (1 - u_1^2 - u_2^2) u_{1,2}\psi_{1,2}\,dx = \gamma\langle T\Psi, U_{1,2}\rangle_{\mathcal{H}}$$

remove projections to the eigenfunctions U_1 and U_2, the spectral theorem in $\mathcal{H}$ yields the coercivity bound (7.13) with C_- being the first positive eigenvalue of $\mathcal{L}_-$ in $\mathcal{H}$. □

REMARK 7.9. Lemma 7.4 gives the desired coercivity bound for the quadratic form $\langle L_-V, V\rangle$ in (7.12) associated with the imaginary part of perturbations V. The two orthogonality conditions in (7.13) can be preserved in the time evolution of the perturbation V by using the rotational invariance of the coupled NLS system

(7.1) with two complex phases for each component of $\Psi = (\psi_1, \psi_2)$. However, the coercivity bound (7.13) canot be used for the quadratic form $\langle L_- W, W\rangle$ in (7.12) associated with the real part of perturbations W since only one orthogonality condition $\langle W, \partial_x U\rangle = 0$ as in (7.10) can be preserved in the time evolution of the perturbation W by using the translational invariance of the coupled NLS system (7.1). It is also insufficient for the stability argument to use that $\langle L_- W, W\rangle \geq 0$ since the third term in (7.12) is sign-indefinite if $\gamma > 1$, see Remark 7.8, and it provides no control for the real part of perturbations W.

Due to the difficulties in the stability analysis explained in Remarks 7.8 and 7.9, we divide the line into three regions,

$$\mathbb{R} = (-\infty, -R] \cup [-R, R] \cup [R, \infty) \tag{7.14}$$

for a fixed $R > 0$. We will control the real part of perturbations W separately inside and outside the compact support $[-R, R]$. Hence, we equip the weighted Hilbert space $\mathcal{H}$ with an additional metric which takes into account the variable $\Upsilon = (\eta_1, \eta_2)$ and the partition of the line:

$$\rho_R(\Psi, \Phi) := \left\|\Psi - \Phi\right\|_{\mathcal{H}} + \left\||\psi_1|^2 - |\phi_1|^2\right\|_{L^2(|x|\geq R)} + \left\||\psi_2|^2 - |\phi_2|^2\right\|_{L^2(|x|\geq R)}. \tag{7.15}$$

Coercivity of the energy is obtained in the metric ρ_R for sufficiently large $R > 0$. This leads to the following orbital stability result for the domain walls for every $\gamma > 1$. The theorem was proven in [**67**].

THEOREM 7.1. *Let $\Psi_0 \in X \cap L^\infty(\mathbb{R})$. There exist $R_0 > 0$ and $\varepsilon_0 > 0$ such that for every $R \in (R_0, \infty)$ and for every $\varepsilon \in (0, \varepsilon_0)$, there exist $\delta > 0$ and real functions $\xi(t), \alpha_1(t), \alpha_2(t)$ such that if*

$$\rho_R(\Psi_0, U) < \delta,$$

then

$$\rho_R(\Psi(t,\cdot), U_{\xi(t),\alpha_1(t),\alpha_2(t)}) < \varepsilon, \quad t \geq 0,$$

where $U_{\xi(t),\alpha_1(t),\alpha_2(t)} = (e^{-i\alpha_1(t)} u_1(\cdot - \xi(t)), e^{-i\alpha_2(t)} u_2(\cdot - \xi(t)))$ is an orbit of domain walls with the profile $U = (u_1, u_2)$.

PROOF. We start with the improved energy decomposition, which will be used for the proof of the orbital stability result in the metric ρ_R. This decomposition combines the two decompositions (7.9) and (7.12) on the partition (7.14):

$$\begin{aligned} H(U + W + iV) - H(U) = {} & \langle L_R W, W\rangle + \langle L_- V, V\rangle \\ & + \int_{-R}^{R} r(W,V)dx + \frac{1}{2}\left(\int_{-\infty}^{-R} + \int_{R}^{\infty}\right)(\eta_1^2 + \eta_2^2)dx \\ & + \gamma \int_{-\infty}^{-R} \eta_2(2u_1 w_1 + w_1^2 + v_1^2)dx \\ & + \gamma \int_{R}^{\infty} \eta_1(2u_2 w_2 + w_2^2 + v_2^2)dx, \end{aligned}$$

where L_R is a linear operator that interpolates between L_- as $R \to 0$ and L_+ as $R \to \infty$:

$$L_R = L_- + 2 \begin{bmatrix} u_1^2 & \gamma u_1 u_2 \\ \gamma u_1 u_2 & u_2^2 \end{bmatrix} \chi_{[-R,R]}$$
$$= L_+ - 2 \begin{bmatrix} u_1^2 & \gamma u_1 u_2 \\ \gamma u_1 u_2 & u_2^2 \end{bmatrix} \chi_{(-\infty,-R)\cup(R,\infty)},$$

with χ being the characteristic function. The proof of Theorem 7.1 can be broken into three steps. In the first step, we show that the quadratic form $\langle L_R W, W\rangle$ is coercive in $\mathcal{H}$ under only one orthogonality condition $\langle W, \partial_x U\rangle_{\mathcal{H}} = 0$ if $R > 0$ is sufficiently large. In the second step, we obtain the lower bound on the energy difference $H(U+W+iV) - H(U)$. In the third step, we provide the decomposition of a point in $X \cap L^\infty(\mathbb{R})$ near an orbit of the domain wall with the profile U to satisfy the three orthogonality conditions on the perturbations with the profile $W + iV$. The three steps are combined together in the proof of the orbital stability of the domain wall in ρ_R with suitable choosen δ in terms of $\varepsilon > 0$ and $R > 0$.

STEP 1. There exist $R_0 > 0$ and $C_+ > 0$ such that for every $R > R_0$, we have

$$\langle L_R W, W\rangle \geq C_+ \|W\|_{\mathcal{H}}^2 \quad \text{for every} \quad W \in \mathcal{H}: \quad \langle W, \partial_x U\rangle_{\mathcal{H}} = 0. \tag{7.16}$$

PROOF OF STEP 1. By using the explicit formula for L_R, we obtain

$$\langle L_R W, W\rangle = \|W\|_{\mathcal{H}}^2 - \langle T_R W, W\rangle_{\mathcal{H}},$$

where the linear operator $T_R : \mathcal{H} \to \mathcal{H}$ is defined by the bilinear form

$$\langle T_R \Psi, \Phi\rangle_{\mathcal{H}} := \gamma \int_{\mathbb{R}} \left(1 - u_1^2 - u_2^2\right)\left(\psi_1 \bar{\varphi}_1 + \psi_2 \bar{\varphi}_2\right) dx$$
$$- 2 \int_{-R}^{R} \left(u_1^2 \psi_1 \bar{\varphi}_1 + \gamma u_1 u_2 (\psi_1 \bar{\varphi}_2 + \psi_2 \bar{\varphi}_1) + u_2^2 \psi_2 \bar{\varphi}_2\right) dx.$$

Since $1 - u_1^2 - u_2^2$ decays to zero at both infinities exponentially fast and the support of the second integral is compact, the operator T_R is compact in $\mathcal{H}$ for any fixed $R > 0$. Hence the spectrum of the operator $\mathcal{L}_R := I - T_R$ in $\mathcal{H}$ consists of isolated eigenvalues $\{\lambda_n(R)\}_{n\in\mathbb{N}}$ accumulating to 1 as $n \to \infty$. These eigenvalues are given by the generalized spectral problem

$$L_R \Psi = \lambda(R) K \Psi, \quad \Psi \in \mathcal{H},$$

with the same diagonal operator K as in the proof of Lemma 7.4. We enumerate these eigenvalues in the increasing order:

$$\lambda_1(R) \leq \lambda_2(R) \leq \cdots.$$

The formal limiting operator T_∞ defined in the limit of $R \to \infty$ is not compact in $\mathcal{H}$. The spectrum of $\mathcal{L}_+ := I - T_\infty$ is only defined in $H^1(\mathbb{R})$ from the generalized spectral problem

$$L_+ \Psi = \lambda K \Psi, \quad \Psi \in H^1(\mathbb{R}).$$

It follows from the asymptotic values of the potentials of L_+ and K at infinity that the spectrum of $\mathcal{L}_+$ in $H^1(\mathbb{R})$ includes the essential spectrum for $\lambda \in [1, \infty)$ and a finite number of isolated eigenvalues for $\lambda \in [0, 1)$, which we denote by N, where the first eigenvalue is located at $\lambda = 0$ with the eigenfunction $\Psi = \partial_x U$ and the second eigenvalue is strictly positive.

It was shown in [**67**, Appendix A] that the first N eigenvalues $\{\lambda_n(R)\}_{n=1}^N$ of $\mathcal{L}_R$ in $\mathcal{H}$ are continuous with respect to R for sufficiently large R and converge to the N isolated eigenvalues of $\mathcal{L}_+$ in $H^1(\mathbb{R})$ as $R \to \infty$. Moreover, the associated eigenfunctions of $\mathcal{L}_R$ converges weakly in $\mathcal{H}$ to the associated eigenfunctions of $\mathcal{L}_+$ defined in $H^1(\mathbb{R})$. The coercivity bound (7.16) for a sufficiently large fixed R follows from the coercivity bound (7.10) and the continuity and convergence of eigenvalues and eigenfunctions of $\mathcal{L}_R$ in $\mathcal{H}$ to those of $\mathcal{L}_+$ in $H^1(\mathbb{R})$ as $R \to \infty$. $\square$

STEP 2. Assume that W satisfies (7.16) and V satisfies (7.13). For every $R > R_0$ in Step 1, there exist positive constants $C > 0$ and $\kappa > 0$ which depend on γ such that the energy difference can be estimated by

$$\begin{aligned}
&H(U + W + iV) - H(U) \\
&\geq C_+\|W\|_{\mathcal{H}}^2 + C_-\|V\|_{\mathcal{H}}^2 + \frac{1}{2}\|\eta_1\|_{L^2(|x|\geq R)}^2 + \frac{1}{2}\|\eta_2\|_{L^2(|x|\geq R)}^2 \\
&\quad - Ce^{-\kappa R}\|W + iV\|_{\mathcal{H}}\left(\|\eta_1\|_{L^2(|x|\geq R)} + \|\eta_2\|_{L^2(|x|\geq R)}\right) \\
&\quad - C\|W + iV\|_{\mathcal{H}}^2\left(\|\eta_1\|_{L^2(|x|\geq R)} + \|\eta_2\|_{L^2(|x|\geq R)}\right) \\
&\quad - Ce^{3\kappa R}\|W + iV\|_{\mathcal{H}}^3. \qquad (7.17)
\end{aligned}$$

PROOF OF STEP 2. Let us estimate the energy decomposition on the semi-infinite interval $[R, \infty)$. Since $u_2(x)$ is exponentially small as $x \to +\infty$, it follows from the definition of $\mathcal{H}$ that

$$\|w_2 + iv_2\|_{H^1(R,\infty)} \leq C\|W + iV\|_{\mathcal{H}},$$

for some positive constant C that depends on $\gamma > 1$. By Sobolev's embedding, we have $w_2(x) + iv_2(x) \to 0$ as $x \to +\infty$ and

$$\|w_2 + iv_2\|_{L^\infty(R,\infty)} \leq C\|W + iV\|_{\mathcal{H}},$$

where the constant C can change from one line to another line. Since $u_2(x)$ is exponentially small as $x \to +\infty$, we can control the last term in the energy difference:

$$\begin{aligned}
&\gamma\left|\int_R^\infty \eta_1(2u_2w_2 + w_2^2 + v_2^2)dx\right| \\
&\qquad \leq C\left(e^{-\kappa R}\|W + iV\|_{\mathcal{H}} + \|W + iV\|_{\mathcal{H}}^2\right)\|\eta_1\|_{L^2(|x|\geq R)}.
\end{aligned}$$

Since $u_1(x)$ is exponentially small as $x \to -\infty$, similar estimates are obtained on $(-\infty, -R]$ for the other term in the energy difference:

$$\begin{aligned}
&\gamma\left|\int_{-\infty}^{-R} \eta_2(2u_1w_1 + w_1^2 + v_1^2)dx\right| \\
&\qquad \leq C\left(e^{-\kappa R}\|W + iV\|_{\mathcal{H}} + \|W + iV\|_{\mathcal{H}}^2\right)\|\eta_2\|_{L^2(|x|\geq R)}.
\end{aligned}$$

It follows from (7.13), (7.16), and (7.17) that all terms of the expansions are controlled except for the nonlinear terms $\int_{-R}^R r(W,V)dx$, where $r(W,V)$ includes the cubic and quartic terms in (W,V). The quartic terms are positive and hence are controlled from below by zero. The cubic terms are bounded by $C\|V+iW\|_{H^1(-R,R)}^3$ for some positive constant C. However, since $1 - u_2^2(x)$ is exponentially small as $x \to -\infty$ and $1 - u_1^2(x)$ is exponentially small as $x \to +\infty$, it follows from the definition of $\mathcal{H}$ that

$$\|W + iV\|_{H^1(-R,R)} \leq Ce^{\kappa R}\|V + iW\|_{\mathcal{H}},$$

for some positive constants C and κ. By combining all estimates together, we obtain the estimate (7.17). $\square$

STEP 3. Let

$$\Psi_{\xi,\alpha_1,\alpha_2} := (e^{i\alpha_1}\psi_1(\cdot+\xi), e^{i\alpha_2}\psi_2(\cdot+\xi))$$

for $\Psi = (\psi_1, \psi_2)$. There exists $\varepsilon_0 > 0$ such that, for any $\Psi \in X \cap L^\infty(\mathbb{R})$ satisfying

$$\varepsilon := \inf_{\xi,\alpha_1,\alpha_2 \in \mathbb{R}} \|\Psi_{\xi,\alpha_1,\alpha_2} - U\|_{\mathcal{H}} \le \varepsilon_0,$$

there exist $\xi \in \mathbb{R}$, $\alpha_1 \in \mathbb{R}/(2\pi\mathbb{Z})$, and $\alpha_2 \in \mathbb{R}/(2\pi\mathbb{Z})$ such that

$$\Psi_{\xi,\alpha_1,\alpha_2} = U + W + iV,$$

where the real-valued functions W and V satisfy the orthogonality conditions

$$\langle W, \partial_x U\rangle_{\mathcal{H}} = 0, \qquad \langle V, U_1\rangle_{\mathcal{H}} = \langle V, U_2\rangle_{\mathcal{H}} = 0.$$

Moreover, $\xi \in \mathbb{R}$, $\alpha_1 \in \mathbb{R}/(2\pi\mathbb{Z})$, and $\alpha_2 \in \mathbb{R}/(2\pi\mathbb{Z})$ depend continuously on Ψ in $\mathcal{H}$.

PROOF OF STEP 3. Given $\Psi \in X \cap L^\infty(\mathbb{R})$, we consider the smooth vector field $\mathbf{f} : \mathbb{R}^3 \to \mathbb{R}^3$ defined by

$$\mathbf{f}(\xi,\alpha_1,\alpha_2) := \begin{pmatrix} \langle \operatorname{Re}\Psi_{\xi,\alpha_1,\alpha_2}, \partial_x U\rangle_{\mathcal{H}} \\ \langle \operatorname{Im}\Psi_{\xi,\alpha_1,\alpha_2}, U_1\rangle_{\mathcal{H}} \\ \langle \operatorname{Im}\Psi_{\xi,\alpha_1,\alpha_2}, U_2\rangle_{\mathcal{H}} \end{pmatrix}, \qquad (\xi,\alpha_1,\alpha_2) \in \mathbb{R}^3.$$

Since $\langle U, \partial_x U\rangle_{\mathcal{H}} = 0$, the construction implies that $\mathbf{f}(\xi,\alpha_1,\alpha_2) = \mathbf{0}$ if and only if Ψ can be represented as $\Psi_{\xi,\alpha_1,\alpha_2} = U + W + iV$ for some real-valued functions W and V satisfying the three orthogonality conditions.

Without loss of generality, we assume that

$$\varepsilon := \inf_{\xi,\alpha_1,\alpha_2 \in \mathbb{R}} \|\Psi_{\xi,\alpha_1,\alpha_2} - U\|_{\mathcal{H}} \le \varepsilon_0,$$

is attained at $(\xi,\alpha_1,\alpha_2) = (0,0,0)$. By Cauchy–Schwarz inequality, since $\partial_x U, U_1$, $U_2 \in \mathcal{H}$, we have $\|\mathbf{f}(0,0,0)\| \le C\varepsilon$ for some positive ε-independent constant C. Furthermore, the Jacobian matrix of the function $\mathbf{f}$ at $(0,0,0)$ is given by

$$D\mathbf{f}(0,0,0) = \begin{pmatrix} \|\partial_x U\|^2_{\mathcal{H}} & 0 & 0 \\ 0 & \|U_1\|^2_{\mathcal{H}} & 0 \\ 0 & 0 & \|U_2\|^2_{\mathcal{H}} \end{pmatrix}$$
$$+ \begin{pmatrix} \langle \operatorname{Re}\partial_x(\Psi - U), \partial_x U\rangle_{\mathcal{H}} & -\langle \operatorname{Im}(\Psi - U), \partial_x U_1\rangle_{\mathcal{H}} & -\langle \operatorname{Im}(\Psi - U), \partial_x U_2\rangle_{\mathcal{H}} \\ \langle \operatorname{Im}\partial_x(\Psi - U), U_1\rangle_{\mathcal{H}} & \langle \operatorname{Re}(\Psi - U), U_1\rangle_{\mathcal{H}} & 0 \\ \langle \operatorname{Im}\partial_x(\Psi - U), U_2\rangle_{\mathcal{H}} & 0 & \langle \operatorname{Re}(\Psi - U), U_2\rangle_{\mathcal{H}}. \end{pmatrix}.$$

The first term in $D\mathbf{f}(0,0,0)$ is a fixed invertible matrix. The second term in $D\mathbf{f}(0,0,0)$ is bounded in the matrix norm by $C\varepsilon$ for another positive ε-independent constant C. Indeed, for the second and third columns, these bounds follow by the Cauchy–Schwarz inequality. For the first column, before applying the Cauchy–Schwarz inequality, the x derivative can be moved from $\Psi - U$ to $\partial_x U$, U_1, and U_2 by integration by parts, with the use of smoothness and decay of $\partial_x U$, U_1, and U_2. Hence $D\mathbf{f}(0,0,0)$ is invertible if ε is small enough and the norm of the inverse of $D\mathbf{f}(0,0,0)$ is bounded by a constant independent of ε. By the local inverse mapping theorem, there exists a unique triple (ξ,α_1,α_2) in the size $\mathcal{O}(\varepsilon)$ neighborhood of $(0,0,0)$ such that $\mathbf{f}(\xi,\alpha_1,\alpha_2) = \mathbf{0}$. This justifies both the decomposition and the continuity of the modulation parameters (ξ,α_1,α_2) on Ψ in $\mathcal{H}$. $\square$

END OF PROOF OF THEOREM 7.1. Let $\Psi(t,\cdot) \in X \cap L^\infty(\mathbb{R})$ be a solution of the coupled NLS system (7.1) starting with the initial data $\Psi_0 \in X \cap L^\infty(\mathbb{R})$. This local solution always exists due to the local well-posedness of the coupled NLS system (7.1) in $X \cap L^\infty(\mathbb{R})$ [**253**]. We assume that $\rho_R(\Psi_0, U) < \delta$ and introduce the modulation parameters according to the decomposition

$$\begin{aligned}\Psi_{\xi(t),\alpha_1(t),\alpha_2(t)}(t,x) &:= (e^{it+i\alpha_1(t)}\psi_1(t,x+\xi(t)), e^{it+i\alpha_2(t)}\psi_2(t,x+\xi(t))) \\ &= U(x) + W(t,x) + iV(t,x),\end{aligned}$$

where the perturbations W and V are real-valued and satisfy the orthogonality conditions

$$\langle W(t,\cdot), \partial_x U\rangle_{\mathcal{H}} = 0, \qquad \langle V(t,\cdot), U_1\rangle_{\mathcal{H}} = \langle V(t,\cdot), U_2\rangle_{\mathcal{H}} = 0, \qquad t \in \mathbb{R}.$$

The decomposition exists at least for small values of time t by Step 3.

Let $\nu > 0$ be a small number that defines radius of a ball in $\mathcal{H}$ for the perturbation terms such that

$$\|W + iV\|_{\mathcal{H}} + \|\eta_1\|_{L^2(|x|\geq R)} + \|\eta_2\|_{L^2(|x|\geq R)} \leq \nu e^{-3\kappa R}. \tag{7.18}$$

By the translational and rotational invariancies of the coupled NLS system (7.1) we have the energy function:

$$H(\Psi(t,\cdot)) = H(\Psi_{\xi(t),\alpha_1(t),\alpha_2(t)}(t,\cdot)) = H(U + W(t,\cdot) + iV(t,\cdot)).$$

It follows from Step 2 that for sufficiently small $\nu > 0$ and sufficiently large $R > 0$, the perturbation term can be controlled the energy difference from (7.17) which yields

$$\|W + iV\|^2_{\mathcal{H}} + \|\eta_1\|^2_{L^2(|x|\geq R)} + \|\eta_2\|^2_{L^2(|x|\geq R)} \leq C\,[H(U + W + iV) - H(U)].$$

On the other hand, we obtain from the initial data and the energy conservation in time that

$$\begin{aligned}H(U + W(t,\cdot) + iV(t,\cdot)) &= H(U + W_0 + iV_0) \\ &\leq \|W_0 + iV_0\|^2_{\mathcal{H}} + \|\eta_1\|^2_{L^2(|x|\geq R)} \\ &\quad + \|\eta_2\|^2_{L^2(|x|\geq R)} + Ce^{4\kappa R}\|W_0 + iV_0\|^4_{\mathcal{H}} \\ &\leq C\delta^2(1 + e^{4\kappa R}\delta^2) \\ &\leq \varepsilon^2.\end{aligned}$$

Since $\varepsilon := \nu e^{-3\kappa R}$, we can choose $\delta = \mathcal{O}(\nu e^{-3\kappa R})$ for $R > R_0$ and obtain

$$\rho_R(\Psi(t,\cdot), U_{\xi(t),\alpha_1(t),\alpha_2(t)}) < \varepsilon,$$

locally in time. However, if $\varepsilon \in (0, \varepsilon_0)$ with ε_0 defined in Step 3, we can extend the construction to all times $t \geq 0$ and thus obtain the orbital stability of the domain walls with the profile U in metric ρ_R. □

REMARK 7.10. Modulation parameters ξ, α_1, and α_2 in Step 3 of Theorem 7.1 are uniquely determined by the projections in the weighted space $\mathcal{H}$. Their time evolution is controlled from the modulation equations in the form:

$$|\xi(t)| + |\alpha_1(t)| + |\alpha_2(t)| \leq C\varepsilon \max\{1, t\}, \quad t \geq 0,$$

with some ε-independent constant $C > 0$. See Lemma 5.2 in [**67**] for the projections in $\mathcal{H}$ applied to the system of modulation equations for $\xi(t)$, $\alpha_1(t)$, and $\alpha_2(t)$.

REMARK 7.11. As $R \to 0$, $\mathcal{L}_R$ converges to $\mathcal{L}_- = I - \gamma T$ in the norm of $\mathcal{H}$, where the compact operator T is defined in Lemma 7.4. Hence, the spectrum of $\mathcal{L}_R$ in $\mathcal{H}$ contains two small eigenvalues converging to the double zero eigenvalue as $R \to 0$. The coercivity bound (7.16) with a single orthogonality condition only holds for large $R > 0$ and clearly fails as $R \to 0$. This is why $R > R_0$ with R_0 being sufficiently large in Step 1 of Theorem 7.1.

REMARK 7.12. The ball (7.18) with the size of $\nu e^{-3\kappa R}$ is exponentially small in terms of large parameter R. Perturbations in the ball are controlled according to the distance ρ_R defined by (7.15).

REMARK 7.13. Theorem 7.1 is an improved orbital stability result for the domain walls compared to [**4**, Theorems 1.4 and 1.5]. The latter results are inspired by similar results for the black solitons in [**22**]. The phase parameters $\alpha_1(t)$ and $\alpha_2(t)$ are not defined precisely in [**4**] and their time evolution is not controlled in the coupled NLS system (7.1).

7.3. Constrained minimizers of energy

Compared to domain walls, black solitons are saddle points of energy. Additional conserved quantities must be used to characterize the black solitons as constrained minimizers of energy. To use the conserved quantities, we again need to take into account the nonzero boundary conditions at infinity, which might force us to renormalize the conserved quantities appropriately.

We will use the cubic defocusing NLS equation (Example 7.1) as the simplest example to explain the main ingredients of the orbital stability theory of the black solitons. The stability theory is developed similarly to Section 7.2 but with the account of the constrained minimization properties and the renormalization of the conserved quantities. Only if the black solitons and their perturbations are restricted to the spatially odd functions, the unconstrained minimization properties can be used for the proof of their orbital stability. In addition, integrability of the cubic defocusing NLS equation (Section 4.2) can be used to prove their stability from higher-order conserved energy.

7.3.1. Orbital stability due to the constrained energy minimization. We consider the cubic defocusing NLS equation in the form

$$i\partial_t\psi + \partial_x^2\psi - |\psi|^2\psi = 0, \tag{7.19}$$

for the class of solutions satisfying the boundary conditions $|\psi(t,x)|^2 \to 1$ as $|x| \to \infty$ for every $t \in \mathbb{R}$. The NLS equation (7.19) admits the Hamiltonian formulation with the conserved energy

$$H(\psi) = \int_{\mathbb{R}} \left[|\partial_x\psi|^2 + \frac{1}{2}(1 - |\psi|^2)^2 \right] dx \tag{7.20}$$

as well as the formally conserved mass and momentum:

$$Q(\psi) = \int_{\mathbb{R}} \left[|\psi|^2 - 1 \right] dx, \qquad P(\psi) = \frac{i}{2}\int_{\mathbb{R}} \left[\bar{\psi}\partial_x\psi - \psi\partial_x\bar{\psi} \right] dx. \tag{7.21}$$

The energy space X is given by

$$X = \left\{ \psi \in H^1_{\rm loc}(\mathbb{R}) : \quad \partial_x\psi \in L^2(\mathbb{R}), \ |\psi|^2 - 1 \in L^2(\mathbb{R}) \right\}. \tag{7.22}$$

As we can see from (7.21), both the mass and momentum are not defined in the energy space X and this is the first difficulty in the orbital stability analysis of black solitons where the momentum conservation is needed.

Let us show why we need to use the momentum conservation for the black soliton

$$\psi(t,x) = e^{-it}u(x), \qquad u(x) = \tanh\left(\frac{x}{\sqrt{2}}\right).$$

By adding a perturbation to the black soliton with the profile u and separating its real and imaginary parts as $w + iv$, we obtain from (7.20) the following decomposition of energy:

$$H(u + w + iv) - H(u) = \langle L_+ w, w\rangle + \langle L_- v, v\rangle + \int_{\mathbb{R}} r(w,v)dx,$$

where $L_\pm : H^2(\mathbb{R}) \subset L^2(\mathbb{R}) \to L^2(\mathbb{R})$ are the linear self-adjoint operators given by

$$L_+ = -\partial_x^2 + 3u^2 - 1,$$
$$L_- = -\partial_x^2 + u^2 - 1,$$

whereas the density $r(w,v)$ of the remainder term is given by

$$r(w,v) = 2uw(w^2+v^2) + \frac{1}{2}(w^2+v^2)^2.$$

The following lemma presents the spectral properties of the linear operators $L_\pm$ which are standard for the black solitons in other NLS models.

LEMMA 7.5. *The quadratic form for L_+ is positive semi-definite on $H^1(\mathbb{R})$. Zero is a simple eigenvalue of L_+ with the associated eigenfunction $\partial_x u \in H^2(\mathbb{R})$ and there exists $C_+ > 0$ such that*

$$\langle L_+ w, w\rangle \geq C_+\|w\|_{H^1}^2 \quad \textit{for every} \quad w \in H^1(\mathbb{R}) : \quad \langle w, \partial_x u\rangle = 0.$$

On the other hand, L_- admits exactly one simple negative eigenvalue and zero is the bottom of the essential spectrum of L_- with the associated eigenfunction $u \in L^\infty(\mathbb{R})$, $u \notin L^2(\mathbb{R})$.

PROOF. Since L_+ and L_- are Schrödinger operators, we apply Sturm's oscillation theory (Theorems B.7, B.9, and B.10). Since $u(x) \to \pm 1$ as $x \to \pm\infty$ exponentially fast, the essential spectrum of L_+ and L_- are located on $[2,\infty)$ and $[0,\infty)$, respectively.

Due to the translational invariance, we have $L_+\partial_x u = 0$ with $\partial_x u \in H^2(\mathbb{R})$. Since $\partial_x u(x) > 0$ for all $x \in \mathbb{R}$, 0 is the smallest (simple) eigenvalue of L_+ isolated from the rest of the spectrum of L_+. The coercivity bound for $\langle Lw, w\rangle$ follows from the spectral gap and the Gärdner's estimate, similar to the proof of Theorem 2.2.

Due to the rotational invariance, we have $L_- u = 0$, where $u \in L^\infty(\mathbb{R})$ but $u \notin L^2(\mathbb{R})$, since $u(x) \to \pm 1$ as $x \to \pm\infty$. Thus, 0 is the bottom of the essential spectrum of L_- located on $[0,\infty)$. There exists a single point $x_0 = 0$ where $u(x_0) = 0$, hence there exists exactly one (simple) negative eigenvalue of L_-. □

EXAMPLE 7.3. The spectrum of the Schrödinger operator

$$L_- = -\partial_x^2 - \operatorname{sech}^2\left(\frac{x}{\sqrt{2}}\right)$$

in $L^2(\mathbb{R})$ is available explicitly. The only negative eigenvalue is located at $-1/2$ with the associated eigenfunction $\operatorname{sech}(x/\sqrt{2})$.

We follow the methods of Section 7.2 and obtain coercivity of the quadratic form $\langle L_- v, v\rangle$ in the exponentially weighted Hilbert space $\mathcal{H}$ introduced by means of the following inner product

$$\langle \psi, \phi \rangle_{\mathcal{H}} := \int_{\mathbb{R}} \left[\partial_x \psi \, \partial_x \bar{\phi} + (1 - u^2) \psi \bar{\phi} \right] dx,$$

and the induced norm $\|\psi\|_{\mathcal{H}} := \sqrt{\langle \psi, \psi \rangle_{\mathcal{H}}}$, where the weight

$$1 - u(x)^2 = \operatorname{sech}^2\left(\frac{x}{\sqrt{2}} \right)$$

is positive and exponentially decaying at infinity. To control boundedness of functions in the exponentially weighted Hilbert space $\mathcal{H}$, we also intoduce

$$\eta := |u + w + iv|^2 - u^2 = 2uw + w^2 + v^2,$$

consistently with the energy space X. We obtain from (7.20) an alternative decomposition of energy:

$$H(u + w + iv) - H(u) = \langle L_- w, w\rangle + \langle L_- v, v\rangle + \frac{1}{2}\|\eta\|^2_{L^2}. \tag{7.23}$$

By Lemma 7.5, $\langle L_- v, v\rangle$ is both sign-indefinite and noncoercive in $H^1(\mathbb{R})$. However, if the quadratic form $\langle L_- v, v\rangle$ is considered in the weighted space $\mathcal{H}$, then it is coercive under two orthogonality conditions, as in the following lemma.

LEMMA 7.6. *There exists $C_- > 0$ such that*

$$\langle L_- v, v\rangle \geq C_- \|v\|^2_{\mathcal{H}} \quad \text{for every } v \in \mathcal{H}: \quad \langle 1, v\rangle_{\mathcal{H}} = \langle u, v\rangle_{\mathcal{H}} = 0.$$

PROOF. We can write

$$\langle L_- v, v\rangle = \|v\|^2_{\mathcal{H}} - 2\langle Tv, v\rangle_{\mathcal{H}},$$

where the linear operator $T : \mathcal{H} \to \mathcal{H}$ is defined by the bilinear form

$$\langle T\psi, \phi\rangle_{\mathcal{H}} := \int_{\mathbb{R}} (1 - u^2)\, \psi \bar{\phi} dx.$$

Since $1 - u^2$ is positive and exponentially decaying at infinity, the operator T is compact in $\mathcal{H}$ and its spectrum in $\mathcal{H}$ is purely discrete and positive. Hence the spectrum of the operator $\mathcal{L}_- := I - 2T$ in $\mathcal{H}$ consists of isolated eigenvalues $\{\lambda_n\}_{n\in\mathbb{N}}$ accumulating to 1 such that $\lambda_n < 1$ for every $n \in \mathbb{N}$. We order the eigenvalues in the increasing order:

$$\lambda_1 \leq \lambda_2 \leq \lambda_3 \leq \cdots.$$

To study eigenvalues $\{\lambda_n\}_{n\in\mathbb{N}}$ of $\mathcal{L}_-$ in $\mathcal{H}$, we use the generalized spectral problem

$$L_- \psi = \lambda K \psi, \quad \psi \in \mathcal{H},$$

with $K := -\partial_x^2 + 1 - u^2$. By Sylvester's theorem (Theorem B.11), it follows from Lemma 7.5 that $\lambda_1 < 0$, $\lambda_2 = 0$, and $\lambda_3 > 0$. Moreover, we can find explicitly that

$$\lambda_1 = -1, \quad \psi_1 = 1,$$
$$\lambda_2 = 0, \quad \psi_2 = u.$$

Since the orthogonality conditions $\langle 1, v\rangle_{\mathcal{H}} = \langle u, v\rangle_{\mathcal{H}} = 0$ remove projections to the eigenfunctions for the negative and zero eigenvalues of $\mathcal{L}_-$, the spectral theorem in $\mathcal{H}$ yields the coercivity bound for $\langle L_- v, v\rangle$ with $C_- = \lambda_3 > 0$. □

COROLLARY 7.1. *It follows that*

$$\langle L_- v, v\rangle \geq 0 \quad \textit{for every } v \in \mathcal{H}: \quad \langle \partial_x u, v\rangle = 0.$$

PROOF. We compute explicitly that

$$\langle 1, v\rangle_{\mathcal{H}} = \langle T1, v\rangle_{\mathcal{H}} = \sqrt{2}\langle \partial_x u, v\rangle,$$
$$\langle u, v\rangle_{\mathcal{H}} = 2\langle Tu, v\rangle_{\mathcal{H}} = -2\langle \partial_x^2 u, v\rangle.$$

Thus, the orthogonality condition $\langle 1, v\rangle_{\mathcal{H}} = 0$ is equivalent to $\langle \partial_x u, v\rangle = 0$. Since it removes the projection to the eigenfunction for the only negative eigenvalue of $\mathcal{L}_-$, the result follows from Lemma 7.6 and the spectral theorem in $\mathcal{H}$. □

REMARK 7.14. For the proof of orbital stability of black solitons, we are going to use the decomposition of energy in (7.23). The coercivity bound for $\langle L_- v, v\rangle$ from Lemma 7.6 can be used if the two orthogonality conditions can be preserved in the time evolution of the NLS equation (7.19). The second orthogonality condition $\langle u, v\rangle_{\mathcal{H}} = 0$, or equivalently $\langle \partial_x^2 u, v\rangle = 0$ can be satisfied by choosing the phase α in the orbit $\{u(x-\xi)e^{-i\alpha}\}_{\xi\in\mathbb{R},\alpha\in\mathbb{R}}$ generated by the black soliton with the profile u. To satisfy the first orthogonality condition $\langle 1, v\rangle_{\mathcal{H}} = 0$, or equivalently $\langle \partial_x u, v\rangle = 0$, we can consider the expansion of the momentum $P(\psi)$ in (7.21):

$$P(u + w + iv) = \int_{\mathbb{R}} (v\partial_x u - u\partial_x v)dx + P(w + iv).$$

If the second term is integrated by parts, then the first orthogonality condition $\langle 1, v\rangle_{\mathcal{H}} = 0$ is satisfied in the tangent approximation when the quadratic term $P(w+iv)$ is neglected. The first orthogonality condition can be preserved in time by extending the black soliton into the family of dark solitons with the additiional parameter $c \in (-\sqrt{2}, \sqrt{2})$ (Example 7.1). Nevertheless, since the integral of $u\partial_x v$ is not well-defined in X given by (7.22), the argument based on the momentum P in (7.21) has a flaw. Therefore, we need to use the renormalized version of the momentum

$$\hat{P}(\psi) = -\frac{i}{2}\int_{\mathbb{R}} \frac{1-|\psi|^2}{|\psi|^2}[\bar{\psi}\partial_x\psi - \psi\partial_x\bar{\psi}]dx, \tag{7.24}$$

which is also a conserved quantity (see Remark 7.15). Integration by parts is now justified due to the exponential decay of $1-u^2$ to zero at infinity,

$$\int_{\mathbb{R}} \frac{1-u^2}{u^2}(v\partial_x u - u\partial_x v)dx = \frac{(1-u^2)v}{u}\Big|_{x\to-\infty}^{x\to+\infty} + 2\int_{\mathbb{R}} v\partial_x u dx,$$

but we need to pay attention to the singularity of $u(x)^{-1}$ at $x = 0$ and to the quadratic and higher-order terms in the expansion of $\hat{P}(u+w+iv)$ with respect to $w + iv$.

REMARK 7.15. The renormalized momentum $\hat{P}(\psi)$ defined by (7.24) can be written in the form

$$\hat{P}(\psi) = P(\psi) + \arg(\psi)|_{x\to-\infty}^{x\to+\infty}, \tag{7.25}$$

where $P(\psi)$ is defined by (7.21). Compared to (7.24), the representation (7.25) avoids singularity at zeros of ψ at finite values of x. The renormalized momentum written in the form (7.25) is a conserved quantity in the time evolution of the NLS equation (7.19) since $\arg(\psi)|_{x\to-\infty}^{x\to+\infty}$ is independent of time. The difference

$\hat{P}(\psi) - P(\psi)$ is the Casimir functional in the sense of Definition 3.1, which does not contribute to the Euler–Lagrange equations of the augmented energy

$$\hat{\Lambda}_c(\psi) := H(\psi) + c\hat{P}(\psi).$$

The Euler–Lagrange equation for $\hat{\Lambda}_c$ is satisfied by the profile u_c of the dark solitons (Example 7.1). The same Casimir functional was mentioned in Example 3.1 in the context of the traveling periodic wave solutions of the generalized NLS equation.

REMARK 7.16. We need to balance between two terms $\langle L_+ w, w\rangle$ and $\|\eta\|^2_{L^2}$ in the decomposition of energy in (7.23) in order to control the real part of perturbations w. We can only use one orthogonality condition $\langle \partial_x u, w\rangle = 0$ as in Corollary 7.1. The orthogonality condition $\langle \partial_x u, w\rangle = 0$ can be satisfied by choosing the phase parameter ξ in the orbit $\{u(x-\xi)e^{-i\alpha}\}_{\xi\in\mathbb{R},\alpha\in\mathbb{R}}$ generated by the black soliton with the profile u.

We can now consider the orbital stability of the black solitons in the metric

$$\rho(\psi,\phi) := \|\psi - \phi\|_{\mathcal{H}} + \||\psi|^2 - |\phi|^2\|_{L^2}, \tag{7.26}$$

which is consistent with the energy space X and with the choice of the variable η in the decomposition (7.23). Functions in the exponentially weighted space $\mathcal{H}$ do not have to be bounded at infinity. The additional requirement $\eta \in L^2(\mathbb{R})$ enforces the perturbation with the profile $(w + iv)$ to be bounded in $L^\infty(\mathbb{R})$ as in the following lemma.

LEMMA 7.7. *There exists $\varepsilon_0 > 0$ and $C_0 > 0$ such that for every $\varepsilon \in (0,\varepsilon_0)$ if*

$$\rho(\psi, u + w + iv) < \varepsilon$$

then

$$\|w + iv\|_{L^\infty} \le C_0, \quad \|\eta\|_{L^\infty} \le C_0\varepsilon, \tag{7.27}$$

where $\eta := |u + w + iv|^2 - u^2 = 2uw + w^2 + v^2$.

PROOF. Since $(u + w)^2 + v^2 = u^2 + \eta$, it is clear that there exists $C > 0$ such that

$$\|w + iv\|_{L^\infty} \le C(1 + \|\eta\|_{L^\infty}).$$

By using the bound

$$\|\eta\|^2_{L^\infty} \le \|\eta\|_{L^2}\|\eta'\|_{L^2},$$

it follows from the product rule for η and the triangle inequality that

$$\begin{aligned}\|\eta\|^2_{L^\infty} &\le 2\|\eta\|_{L^2}\left(\|u'w + uw'\|_{L^2} + \|ww'\|_{L^2} + \|vv'\|_{L^2}\right)\\ &\le C\|\eta\|_{L^2}\left(\|w\|_{\mathcal{H}} + \|w\|_{L^\infty}\|w'\|_{L^2} + \|v\|_{L^\infty}\|v'\|_{L^2}\right)\\ &\le C\left(1 + \|w + iv\|_{L^\infty}\right)\left(\|w + iv\|^2_{\mathcal{H}} + \|\eta\|^2_{L^2}\right).\end{aligned}$$

Since $\rho(\psi, u + w + iv) < \varepsilon$, we obtain

$$\|\eta\|^2_{L^\infty} \le C(1 + \|\eta\|_{L^\infty})\varepsilon^2.$$

Hence, there exists a sufficiently small $\varepsilon_0 > 0$ and a fixed $C_0 > 0$ such that for every $\varepsilon \in (0,\varepsilon_0)$, we have $\|\eta\|_{L^\infty} \le C_0\varepsilon$, which yields (7.27). □

We can now prove the orbital stability of the black soliton with the profile u by using its constrained minimization properties. The following theorem was proven in [**113**].

THEOREM 7.2. *Let $\psi_0 \in X \cap L^\infty(\mathbb{R})$ and $\psi \in C^0(\mathbb{R}_+, X \cap L^\infty(\mathbb{R}))$ be the corresponding solution of the NLS equation (7.19). There exists $\varepsilon_0 > 0$ such that for every $\varepsilon \in (0, \varepsilon_0)$, there exist $\delta > 0$ and real functions $\xi(t), \alpha(t)$ such that if $\rho(\psi_0, u) < \delta$, then*

$$\rho(\psi(t,\cdot), e^{-i\alpha(t)} u(\cdot - \xi(t))) < \varepsilon, \quad t \geq 0.$$

PROOF. The proof can be divided into three steps. In the first step, we obtain the lower bound on the energy difference by using the decomposition (7.23). In the second step, we provide the decomposition of a point in $\mathcal{H}$ near an orbit of the dark solitons $\{e^{-i\alpha} u_c(\cdot - \xi)\}_{\xi,\alpha\in\mathbb{R}}$ with small speed $c \in (-\sqrt{2}, \sqrt{2})$, where

$$u_c(x) = \kappa \tanh\left(\frac{\kappa x}{\sqrt{2}}\right) + \frac{ic}{\sqrt{2}}, \quad \kappa = \sqrt{1 - \frac{c^2}{2}}, \tag{7.28}$$

see Example 7.1. The additional parameter $c \in (-\sqrt{2}, \sqrt{2})$ to the two translational parameters $\xi, \alpha \in \mathbb{R}$ is necessary to satisfy the three orthogonality conditions on the perturbations of the black soliton u which corresponds to $c = 0$. In the third step, we estimate the difference of the renormalized momentum $\hat{P}$ at the perturbation with the profile $(w + iv)$ to the dark solition with the profile u_c. The three steps are combined together in the proof of the orbital stability of the black soliton in ρ with suitable choosen δ in terms of $\varepsilon > 0$.

STEP 1. Assume that $w + iv \in \mathcal{H}$ satisfy the orthogonality conditions

$$\langle 1, w\rangle_{\mathcal{H}} = 0, \quad \langle 1, v\rangle_{\mathcal{H}} = 0, \quad \langle u, v\rangle_{\mathcal{H}} = 0.$$

There exist constants $C_1 > 0$ and $C_2 > 0$ such that the energy difference (7.23) can be estimated by

$$H(u + w + iv) - H(u) \geq C_1 \left(\|w + iv\|_{\mathcal{H}}^2 + \|\eta\|_{L^2}^2\right) - C_2 \|w + iv\|_{\mathcal{H}}^3. \tag{7.29}$$

PROOF OF STEP 1. By Lemma 7.6 and Corollary 7.1, we already have

$$H(u + w + iv) - H(u) \geq C_- \|v\|_{\mathcal{H}}^2 + \frac{1}{2}\|\eta\|_{L^2}^2.$$

Hence we only need to control coercivity with respect to the real part of perturbations w. Since w satisfy the orthogonality condition $\langle 1, w\rangle_{\mathcal{H}} = 0$, we can use the orthogonal decomposition

$$w = \frac{\langle u, w\rangle_{\mathcal{H}}}{\|u\|_{\mathcal{H}}^2} u + \tilde{w}, \quad \langle u, \tilde{w}\rangle_{\mathcal{H}} = 0$$

and obtain from Lemma 7.6 and from $\mathcal{L}_- u = 0$ that

$$\begin{aligned}
\langle L_- w, w\rangle &= \langle \mathcal{L}_- w, w\rangle_{\mathcal{H}} \\
&= \langle \mathcal{L}_- \tilde{w}, \tilde{w}\rangle_{\mathcal{H}} \\
&\geq C_1 \|\tilde{w}\|_{\mathcal{H}}^2 \\
&= C_1 \left\| w - \frac{\langle u, w\rangle_{\mathcal{H}}}{\|u\|_{\mathcal{H}}^2} u \right\|_{\mathcal{H}}^2 \\
&= C_1 \|w\|_{\mathcal{H}}^2 - C_1 \frac{|\langle u, w\rangle_{\mathcal{H}}|^2}{\|u\|_{\mathcal{H}}^2},
\end{aligned}$$

for some sufficiently small $C_1 \in (0, C_-)$. The energy difference is hence estimated from below by

$$H(u+w+iv) - H(u) \geq C_1 \|w+iv\|_{\mathcal{H}}^2 + \frac{1}{2}\|\eta\|_{L^2}^2 - C_1 \frac{|\langle u, w\rangle_{\mathcal{H}}|^2}{\|u\|_{\mathcal{H}}^2}.$$

To compensate the negative term, we note that

$$\begin{aligned}\int_{\mathbb{R}} (1-u^2)\eta^2 dx &= \int_{\mathbb{R}} (1-u^2)(4u^2w^2 + 4uw(w^2+v^2) + (w^2+v^2)^2) dx \\ &\geq 4\int_{\mathbb{R}} (1-u^2)u^2w^2 dx + 4\int_{\mathbb{R}} u(1-u^2)w(w^2+v^2) dx.\end{aligned}$$

By using the Cauchy–Schwarz inequality and the Sobolev embedding of $H^1(\mathbb{R})$ into $L^\infty(\mathbb{R})$, we obtain

$$\|(1-u^2)^{1/2}(w+iv)\|_{L^\infty} \leq C\|(1-u^2)^{1/2}(w+iv)\|_{H^1} \leq C\|w+iv\|_{\mathcal{H}},$$

from which it follows that the cubic term in (w, v) is cubically small in $\mathcal{H}$:

$$\begin{aligned}\int_{\mathbb{R}} u(1-u^2)w(w^2+v^2) dx &= -\int_{\mathbb{R}} u''w(w^2+v^2) dx \\ &= \int_{\mathbb{R}} u'(3w^2+v^2)w' dx + 2\int_{\mathbb{R}} u'wvv' dx \\ &\leq C\|(1-u^2)^{1/2}(w+iv)\|_{L^\infty}\|w+iv\|_{\mathcal{H}}^2 \\ &\leq C\|w+iv\|_{\mathcal{H}}^3,\end{aligned}$$

for some generic positive constant C. On the other hand, the quadratic term allows us to estimate

$$\begin{aligned}|\langle u, w\rangle_{\mathcal{H}}|^2 &= 4\left|\int_{\mathbb{R}} (1-u^2)uw dx\right|^2 \\ &\leq 4\left(\int_{\mathbb{R}} (1-u^2) dx\right)\left(\int_{\mathbb{R}} (1-u^2)u^2w^2 dx, \right),\end{aligned}$$

by using the Cauchy–Schwarz inequality. Combining all estimates together yields

$$\begin{aligned}&H(u+w+iv) - H(u) \\ &\quad\geq C_1\left(\|w+iv\|_{\mathcal{H}}^2 + \|\eta\|_{L^2}^2\right) \\ &\qquad + \left(\frac{1}{2} - C_1\right)\int_{\mathbb{R}} (1-u^2)\eta^2 dx - C_1\frac{|\langle u, w\rangle_{\mathcal{H}}|^2}{\|u\|_{\mathcal{H}}^2} \\ &\quad\geq C_1\left(\|w+iv\|_{\mathcal{H}}^2 + \|\eta\|_{L^2}^2\right) - C_2\|w+iv\|_{\mathcal{H}}^3 \\ &\qquad + \left(2 - 4C_1 - \frac{4C_1}{\|u\|_{\mathcal{H}}^2}\int_{\mathbb{R}} (1-u^2) dx\right)\left(\int_{\mathbb{R}} (1-u^2)u^2w^2 dx\right).\end{aligned}$$

The last term is strictly positive if C_1 is sufficiently small so that the estimate yields (7.29). □

STEP 2. There exists $\varepsilon_0 > 0$ such that, for any $\psi \in \mathcal{H}$ satisfying

$$\|e^{i\alpha_0}\psi(\cdot + \xi_0) - u\|_{\mathcal{H}} := \inf_{\xi, \alpha \in \mathbb{R}} \|e^{i\alpha}\psi(\cdot + \xi) - u\|_{\mathcal{H}} \leq \varepsilon_0,$$

there exist $\xi \in \mathbb{R}$, $\alpha \in \mathbb{R}/(2\pi\mathbb{Z})$, and $c \in (-\sqrt{2}, \sqrt{2})$ such that

$$e^{i\alpha}\psi(\cdot + \xi) = u_c + w + iv$$

where the real-valued functions w, v satisfy the orthogonality conditions

$$\langle 1, w\rangle_{\mathcal{H}} = 0, \qquad \langle 1, v\rangle_{\mathcal{H}} = \langle u, v\rangle_{\mathcal{H}} = 0.$$

Moreover, $\xi \in \mathbb{R}$, $\alpha \in \mathbb{R}/(2\pi\mathbb{Z})$, and $c \in (-\sqrt{2}, \sqrt{2})$ depend continuously on ψ in $\mathcal{H}$.

PROOF OF STEP 2. For small $c \in (-\sqrt{2}, \sqrt{2})$, we have the following expansion for the profile u_c of the dark soliton (7.28):

$$u_c = u + \frac{ic}{\sqrt{2}} + \mathcal{O}_{\mathcal{H}}(c^2),$$

where u is the black soliton. We define the vector function

$$\vec{F}(c, \alpha, \xi; \psi) : (-\sqrt{2}, \sqrt{2}) \times \mathbb{R} \times \mathbb{R} \times \mathcal{H} \mapsto \mathbb{R}^3$$

representing the three constraints:

$$\vec{F}(c, \alpha, \xi; \psi) := \begin{pmatrix} \langle 1, \mathrm{Re}(e^{i\alpha}\psi(\cdot + \xi) - u_c)\rangle_{\mathcal{H}} \\ \langle 1, \mathrm{Im}(e^{i\alpha}\psi(\cdot + \xi) - u_c)\rangle_{\mathcal{H}} \\ \langle u, \mathrm{Im}(e^{i\alpha}\psi(\cdot + \xi) - u_c)\rangle_{\mathcal{H}} \end{pmatrix}.$$

The infimum $\inf_{\xi,\alpha\in\mathbb{R}} \|e^{i\alpha}\psi(\cdot + \xi) - u\|_{\mathcal{H}}$ is attained for $(\alpha, \xi) = (\alpha_0, \xi_0)$ for the given $\psi \in \mathcal{H}$. Since $u \in \mathcal{H}$, the Cauchy–Schwarz inequality implies that there exists $C > 0$ such that

$$|\vec{F}(0, \alpha_0, \xi_0; \psi)| \leq C\varepsilon,$$

where $|\cdot|$ denotes the standard Euclidean norm for vectors in $\mathbb{C}^3$. We can write

$$\vec{F}(0, \alpha_0, \xi_0; \psi) = \mathcal{O}(\varepsilon)$$

to indicate the remainder term for $\psi = e^{i\alpha_0}u(\cdot + \xi_0) + \mathcal{O}_{\mathcal{H}}(\varepsilon)$.

The function $\vec{F}$ is C^1 with respect to its arguments and we can compute the Jacobian of $\vec{F}$ with respect to (c, α, ξ) at $(0, \alpha_0, \xi_0)$ and for fixed $\psi \in \mathcal{H}$. The Jacobian evaluated at $(c, \alpha, \xi) = (0, \alpha_0, \xi_0)$ is given by the following matrix

$$\begin{bmatrix} 0 & 0 & \langle 1, u'\rangle_{\mathcal{H}} \\ -\langle 1, 1\rangle_{\mathcal{H}} & 0 & 0 \\ 0 & \langle u, u\rangle_{\mathcal{H}} & 0 \end{bmatrix} + \mathcal{O}(\varepsilon),$$

where we have used that $\langle 1, u\rangle_{\mathcal{H}} = 0$ due to different polarities. It is clear that the entries $\langle 1, 1\rangle_{\mathcal{H}}$, $\langle u, u\rangle_{\mathcal{H}}$, and $\langle 1, u'\rangle_{\mathcal{H}}$ are nonzero. By choosing ε_0 small enough, the Jacobian for $\vec{F}$ is invertible. By the local inverse mapping theorem, for any $\psi \in \mathcal{H}$ there exists a unique solution $(c, \alpha, \xi) \in (-\sqrt{2}, \sqrt{2}) \times \mathbb{R} \times \mathbb{R}$ of $\vec{F}(c, \alpha, \xi; \psi) = 0$ satisfying

$$|c| + |\alpha - \alpha_0| + |\xi - \xi_0| \leq C\varepsilon,$$

for some ε-independent $C > 0$. This justifies the decomposition $e^{i\alpha}\psi(\cdot + \xi) = u_c + w + iv$ and the continuous dependence of parameters on ψ in $\mathcal{H}$. ☐

STEP 3. For every $\psi \in \mathcal{H}$ decomposed in Step 2, there is $C > 0$ such that

$$\hat{P}(u_c + w + iv) - \hat{P}(u_c) = 2\int_{\mathbb{R}} [\mathrm{Re}(u_c')v - \mathrm{Im}(u_c')w]dx + \hat{R}(w, v), \tag{7.30}$$

with

$$|\hat{R}(w, v)| \leq C(\|w + iv\|_{\mathcal{H}}^2 + \|\eta_c\|_{L^2}^2),$$

where

$$\eta_c := |u_c + w + iv|^2 - |u_c|^2 = 2\,\mathrm{Re}(u_c)w + 2\,\mathrm{Im}(u_c)v + w^2 + v^2.$$

PROOF OF STEP 3. We fix $R > 0$ and split the integral in $\hat{P}(u_c + w + iv)$ on $(-\infty, -R] \cup [-R, R] \cup [R, \infty)$.

The outer integrals on $(-\infty, -R] \cup [R, \infty)$ are treated based on the representation (7.24):

$$P_R(\psi) := \frac{1}{2i}\left(\int_{-\infty}^{-R} + \int_R^{\infty}\right) \frac{(1 - |\psi|^2)}{|\psi|^2}(\bar{\psi}\partial_x\psi - \psi\partial_x\bar{\psi})dx.$$

The integrand is nonsingular because $|u_c + w + iv|^2 \geq |u_c|^2 - |\eta_c|$ is bounded away from zero on $(-\infty, -R] \cup [R, \infty)$. This follows from the boundedness of $|u_c|^2$ away from zero and smallness of $\|\eta_c\|_{L^\infty}$ according to the bound (7.27) in Lemma 7.7. We expand

$$\bar{\psi}\partial_x\psi - \psi\partial_x\bar{\psi} = \bar{u}_c u_c' - \bar{u}_c' u_c + \nu_c + 2i(wv' - w'v),$$

where $\nu_c := 2i\,\mathrm{Im}(u_c')w - 2i\,\mathrm{Re}(u_c')v - 2i\,\mathrm{Im}(u_c)w' + 2i\,\mathrm{Re}(u_c)v'$, so that

$$P_R(u_c + w + iv) = P_R(u_c) + P_1 + P_2,$$

where

$$P_1 := \frac{1}{2i}\int\left[\frac{(1 - |u_c|^2)}{|u_c|^2}\nu_c - \frac{1}{|u_c|^4}(\bar{u}_c u_c' - \bar{u}_c' u_c)\eta_c\right]dx$$

and

$$\begin{aligned} P_2 := &\int \frac{(1 - |u_c|^2 - \eta_c)}{|u_c|^2 + \eta_c}(wv' - w'v)dx \\ &+ \frac{1}{2i}\int\left[\frac{(1 - |u_c|^2 - \eta_c)}{|u_c|^2 + \eta_c} - \frac{(1 - |u_c|^2)}{|u_c|^2}\right]\nu_c dx, \\ &+ \frac{1}{2i}\int\left[\frac{(1 - |u_c|^2 - \eta_c)}{|u_c|^2 + \eta_c} - \frac{(1 - |u_c|^2)}{|u_c|^2} + \frac{1}{|u_c|^4}\eta_c\right](\bar{u}_c u_c' - \bar{u}_c' u_c)dx, \end{aligned}$$

and the integration is performed over $(-\infty, -R] \cup [R, \infty)$. Substituting ν_c and η_c into P_1 and integrating w' and v' by parts yield

$$\begin{aligned} P_1 = &\frac{(1 - |u_c|^2)}{|u_c|^2}(-\mathrm{Im}(u_c)w + \mathrm{Re}(u_c)v)\left(\Big|_{x=R}^{x\to+\infty} + \Big|_{x\to-\infty}^{x=-R}\right) \\ &+ 2\int(-\mathrm{Im}(u_c')w + \mathrm{Re}(u_c')v)dx + \frac{i}{2}\int\frac{1}{|u_c|^4}(\bar{u}_c u_c' - \bar{u}_c' u_c)(w^2 + v^2)dx, \end{aligned}$$

The first term in P_1 is zero in the limits $x \to \pm\infty$ due the boundnedness of u and v in (7.27) in Lemma 7.7. The limits $x = \pm R$ will be combined with the inner integrals on $[-R, R]$. The second term in P_1 combined with a similar term on $[-R, R]$ gives the linear term in the expansion (7.30). The third term in P_1 is bounded by $\|w + iv\|_{\mathcal{H}}^2$.

The terms in P_2 are analyzed by using the facts that $|u_c|^2$ is bounded away from zero on $(-\infty, -R] \cup [R, \infty)$, $\|\eta_c\|_{L^\infty}$ is small, and the perturbation $w + iv$ is bounded according to (7.27). The first term in P_2 is estimated by using Cauchy–Schwarz inequality:

$$\begin{aligned}
&\int \frac{(1-|u_c|^2-\eta_c)}{|u_c|^2+\eta_c}(wv'-w'v)dx \\
&\le C\int (1-|U_{c,\omega}|^2+|\eta|)(|w||v'|+|w'||v|)dx \\
&\le C\left(\|w+iv\|_{\mathcal{H}}\|w'+iv'\|_{L^2}+\|w+iv\|_{L^\infty}\|w'+iv'\|_{L^2}\|\eta_c\|_{L^2}\right) \\
&\le C\left(\|w+iv\|_{\mathcal{H}}^2+\|\eta_c\|_{L^2}^2\right).
\end{aligned}$$

The second and third terms in P_2 are estimated similarly:

$$\left|\frac{1}{2i}\int \frac{\eta_c\nu_c}{|u_c|^2(|u_c|^2+\eta_c)}dx\right| \le C\|w+iv\|_{\mathcal{H}}\|\eta_c\|_{L^2}$$

and

$$\frac{1}{2i}\int \frac{\eta_c^2}{|u_c|^4(|u_c|^2+\eta_c)}(\bar{u}_c u_c'-\bar{u}_c' u_c)dx \le C\|\eta_c\|_{L^2}^2.$$

The inner integrals on $[-R,R]$ is treated based on the representation (7.25):

$$\hat{P}_R(\psi) := \frac{i}{2}\int_{-R}^{R}(\bar{\psi}\partial_x\psi-\psi\partial_x\bar{\psi})dx+\alpha(R)-\alpha(-R),$$

where $\alpha(x) := \arg(\psi(x))$. Expanding $\psi = u_c + w + iv$ and $\alpha = \arg(u_c)+\alpha_1+\alpha_2$ with

$$\alpha_1 = \frac{\mathrm{Re}(u_c)v-\mathrm{Im}(u_c)w}{|u_c|^2}$$

and

$$\alpha_2 = \arg(u_c+w+iv)-\arg(u_c)-\frac{\mathrm{Re}(u_c)v-\mathrm{Im}(u_c)w}{|u_c|^2}$$

yields the expansion

$$\hat{P}_R(u_c+w+iv) = \hat{P}_R(u_c)+\hat{P}_1+\hat{P}_2,$$

where

$$\hat{P}_1 := -\int [\mathrm{Im}(u_c')w-\mathrm{Re}(u_c')v-\mathrm{Im}(u_c)w'+\mathrm{Re}(u_c)v'\;dx+\alpha_1(R)-\alpha_1(-R)$$

and

$$\hat{P}_2 := -\int (wv'-w'v)dx+\alpha_2(R)-\alpha_2(-R),$$

and the integration is performed over $[-R,R]$. Integrating w' and v' by parts yields

$$\begin{aligned}
\hat{P}_1 &= (\mathrm{Im}(u_c)w-\mathrm{Re}(u_c)v)\Big|_{x=-R}^{x=R} \\
&\quad +2\int(-\mathrm{Im}(u_c')w+\mathrm{Re}(u_c')v)dx+\alpha_1(R)-\alpha_1(-R).
\end{aligned}$$

The terms at $x=\pm R$ are combined together in P_1 and $\hat{P}_1$ to yield zero, where we have used the explicit expressions for $\alpha_1(\pm R)$. The linear integral term in $\hat{P}_1$ is combined together with a similar term in P_1 to yield

$$2\int_{\mathbb{R}}(-\mathrm{Im}(u_c')w+\mathrm{Re}(u_c')v)dx,$$

which is linear term in the expansion (7.30). The integral term in $\hat{P}_2$ is bounded by $\|w + iv\|_{\mathcal{H}}^2$ because integration on $[-R, R]$ is bounded by integration on $\mathbb{R}$ with the weight $(1 - u^2)$. Similarly, by using the Sobolev embedding of $H^1([-R, R])$ to $L^\infty([-R, R])$ and the Taylor expansion with $|u_c(\pm R)|^2$ being bounded away from zero, we obtain

$$|\alpha_2(\pm R)| \leq C_R(|u(\pm R)|^2 + |v(\pm R)|^2) \leq C_R\|u + iv\|_{H^1([-R,R])}^2,$$

which is bounded by $\|u + iv\|_{\mathcal{H}}^2$ since $[-R, R]$ is compact and $1 - u^2$ is bounded away from zero on $[-R, R]$. Combining all estimates together yields the expansion (7.30). □

END OF PROOF OF THEOREM 7.2. Let $\psi_0 \in X \cap L^\infty(\mathbb{R})$ satisfy $\rho(\psi_0, u) < \delta$ for some small $\delta \in (0, \varepsilon_0)$, where ρ is defined by (7.26) and ε_0 is defined in Step 2. By the local well-posedness of the NLS equation (7.19) in $X \cap L^\infty(\mathbb{R})$ from [**99, 102, 253**], there exists a local solution $\psi \in C^0([0, \tau_0], X \cap L^\infty(\mathbb{R}))$ for some small $\tau_0 > 0$ such that $\psi(0, \cdot) = \psi_0$. Since $\psi_0 \in \mathcal{H}$, then $\psi(t, \cdot) \in \mathcal{H}$ for $t \in [0, \tau_0]$. Therefore, the orthogonal decomposition of Step 2 holds true for $t \in [0, \tau_0]$ at least for small $\tau_0 > 0$.

By using the expression for the dark solitons (7.28), we can compute explicitly

$$\begin{aligned}\hat{P}(u_c) &= \pi + c\sqrt{2 - c^2} + 2\arctan\frac{c}{\sqrt{2 - c^2}}\\ &= \pi + 2\sqrt{2}c + \mathcal{O}(c^3).\end{aligned}$$

Due to conservation of energy $H(\psi)$ given by (7.20) and the renormalized momentum $\hat{P}(\psi)$ given by (7.25), there is $C > 0$ such that

$$\begin{cases} |H(\psi) - H(u)| = |H(\psi_0) - H(u)| \leq C\delta, \\ |\hat{P}(\psi) - \hat{P}(u)| = |\hat{P}(\psi_0) - \hat{P}(u)| \leq C\delta, \end{cases}$$

We construct the Lyapunov functional for the orbital stability of dark solitons:

$$\hat{\Lambda}_c(w, v) := H(u_c + w + iv) - H(u) + c\left[\hat{P}(u_c + w + iv) - \hat{P}(u)\right].$$

By using Step 3, we have the decomposition:

$$\hat{\Lambda}_c(w, v) = \Delta(c) + \langle L_-(c)w, w\rangle + \langle L_-(c)v, v\rangle + \frac{1}{2}\|\eta\|_{L^2}^2 + c\hat{R}(w, v),$$

where

$$\Delta(c) := H(u_c) - H(u) + c\left[\hat{P}(u_c) - \hat{P}(u)\right]$$

and

$$L_-(c) := -\partial_x^2 + |u_c|^2 - 1.$$

There is a constant $C_0 > 0$ such that for every $f \in \mathcal{H}$ and every $c \in (-\sqrt{2}, \sqrt{2})$, it is true that

$$\left|\int_{\mathbb{R}}(1 - |u_c|^2)|f|^2dx - \int_{\mathbb{R}}(1 - u^2)|f|^2dx\right| \leq C_0c^2\|f\|_{\mathcal{H}}^2.$$

As a result, all estimates in Step 1 are valid for the functions $w, v \in \mathcal{H}$ satisfying the three orthogonality conditions and for small $|c|$. In addition, $\hat{R}(w, v)$ is quadratic with respect to $w, v \in \mathcal{H}$ in the $\mathcal{H}$ norm. Hence we obtain the lower energy bound:

$$\hat{\Lambda}_c(w,v) \geq \Delta(c) + C_1(\|w+iv\|^2_{\mathcal{H}} + \|\eta\|^2_{L^2}) - C_2\|w+iv\|^3_{\mathcal{H}} - C_3|c|\|w+iv\|^2_{\mathcal{H}},$$

for some constants $C_1, C_2, C_3 > 0$. We have the expansion

$$\Delta(c) = \sqrt{2}c^2 + \mathcal{O}(c^4), \tag{7.31}$$

which follows from $\Delta(0) = 0$, $\Delta'(0) = 0$, and $\Delta''(c) = \frac{d}{dc}\hat{P}(u_c)$. By using Young's inequality, we have

$$|c|\|w+iv\|^2_{\mathcal{H}} \leq \frac{1}{3}|c|^3 + \frac{2}{3}\|w+iv\|^3_{\mathcal{H}}$$

so that the coercivity of the Lyapunov functional and the initial bound us to control the perturbation terms in the time evolution:

$$|c(t)|^2 + \|w(t,\cdot)+iv(t,\cdot)\|^2_{\mathcal{H}} + \|\eta(t,\cdot)\|^2_{L^2} \leq C\delta, \quad t \in [0,\tau_0].$$

By the triangle inequality and the orthogonal decomposition in Step 2, we obtain

$$\rho(\psi(t,\cdot), e^{-i\alpha(t)}, u(\cdot - \xi(t))) < C\delta^{1/2}, \quad t \in [0,\tau_0],$$

for some real functions $\xi(t)$, $\alpha(t)$. Thus for every $\varepsilon \in (0,\varepsilon_0)$ there is $\delta < \min(\varepsilon^2/C^2, \varepsilon_0)$ such that the orthogonal decomposition in Step 2 can be extended beyond the time $t = \tau_0$ with the same uniform estimate. Hence, the local solution $\psi \in C^0([0,\tau_0], X \cap L^\infty(\mathbb{R}))$ near the orbit $\{e^{i\alpha}u(\cdot+\xi)\}_{\xi,\alpha\in\mathbb{R}}$ is extended globally for every $t \in \mathbb{R}$. ☐

REMARK 7.17. It was used in the end of the proof of Theorem 7.2 that u_c is a critical point of $\Lambda_c(\psi) = E(\psi) + c\hat{P}(\psi)$ so that the first variation in the expansion of $\hat{\Lambda}_c(w,v)$ vanishes.

REMARK 7.18. The method used in the proof of Theorem 7.2 was developed in [**113**]. This is the improvement of the earlier results in [**22**] where the phase parameter $\alpha(t)$ was not controlled in the decomposition near the orbit of the black soliton. Asymptotic stability of the black solitons was also obtained in [**113**] independently of the integrability of the defocusing cubic NLS equation. The proof of orbital and asymptotic stability of the black and dark solitons can be developed based on the integrability methods as in [**73, 103**]. The method of [**113**] was used for the proof of orbital stability of the black soliton in nonintegrable NLS equations such as the defocusing quintic NLS equation in [**6**] and the NLS equation with intensity-dependent dispersion in [**218**].

Readers may suggest that the method of the proof of Theorem 7.2 is too complicated. It is nice to know that the proof can be greatly simplified if we restrict our attention to the spatially odd perturbations which are preserved in the time evolution of the NLS equation (7.19). This restriction allows us to simplify the three steps in the proof of Theorem 7.2 and obtain the following corollary.

COROLLARY 7.2. *Let $\psi_0 \in X \cap L^\infty(\mathbb{R})$ be odd on $\mathbb{R}$ and $\psi \in C^0(\mathbb{R}_+, X \cap L^\infty(\mathbb{R}))$ be the unique solution of the NLS equaiton* (7.19). *There exists $\varepsilon_0 > 0$ such that for every $\varepsilon \in (0,\varepsilon_0)$, there exist $\delta > 0$ and real function $\alpha(t)$ such that if $\rho(\psi_0, u) < \delta$, then*

$$\rho(\psi(t,\cdot), e^{-i\alpha(t)}u) < \varepsilon, \quad t \geq 0.$$

PROOF. If $\psi_0 \in\in X \cap L^\infty(\mathbb{R})$ is odd, then $|\psi_0|^2$ is even and $\psi_0'' + |\psi_0|^2\psi_0$ is odd, so that the suitable solution $\psi \in C^0(\mathbb{R}_+, X \cap L^\infty(\mathbb{R}))$ to the NLS equation (7.19) with $\psi(0,\cdot) = \psi_0$ remains odd in x for all times $t \geq 0$. Since the black soliton u is odd, the perturbation $w + iv$ is also odd in x for all times $t \geq 0$.

The coercivity bound (7.29) in Step 1 is satisfied with the only orthogonality condition

$$\langle u, v\rangle_{\mathcal{H}} = 0,$$

since the other two orthogonality conditions are satisfied due to the odd parity of the perturbation $w + iv$. As a result, in Step 2 we only need to introduce the decomposition

$$e^{i\alpha}\psi = u + w + iv$$

and choose only one parameter $\alpha \in \mathbb{R}/(2\pi\mathbb{Z})$ from the only orthogonality condition. The renormalized momentum $\hat{P}$ in Step 3 is not needed because the black soliton with the profile u is now the minimizer of energy $H(\psi)$ with respect to odd functions in X. Indeed, the spectral properties of L_+ and L_- are changed due to the restriction to odd perturbations as follows. The spectrum of L_+ in $L^2(\mathbb{R})$ is strictly positive and bounded away from zero, see Lemma 7.5, and the spectrum of L_- in $\mathcal{H}$ is nonnegative and admits a simple zero eigenvalue bounded away from positive eigenvalues, see Lemma 7.6. Hence, Step 3 becomes redundant and we complete the proof with the Lyapunov functional:

$$\begin{aligned}\Lambda(w,v) &:= H(u + w + iv) - H(u)\\ &= \langle L_- w, w\rangle + \langle L_- v, v\rangle + \frac{1}{2}\|\eta\|_{L^2}^2\\ &\geq C_1(\|w + iv\|_{\mathcal{H}}^2 + \|\eta\|_{L^2}^2) - C_2\|w + iv\|_{\mathcal{H}}^3\end{aligned}$$

from which we obtain

$$\|w(t,\cdot) + iv(t,\cdot)\|_{\mathcal{H}}^2 + \|\eta(t,\cdot)\|_{L^2}^2 \leq C\delta, \quad t \in [0, \tau_0],$$

which suggests the orbital stability of the black soliton in the metric ρ by the same continuation argument as in the proof of Theorem 7.2. $\square$

7.3.2. Orbital stability due to the higher-order energy minimization. The approach is based on the existence of the higher-order energy of the defocusing NLS equation (7.19) which exists due to its integrability (Section 4.1) and which has been used in the proof of orbital stability of the periodic waves (Section 4.2). The black soliton with the profile u is a critical point of the conserved energy $H(\psi)$ given by (7.20). In addition, it is also a critical point of the conserved higher-order energy $S(\psi)$ given by

(7.32)
$$S(\psi) = \int_{\mathbb{R}} [|\partial_x^2\psi|^2 + 3|\psi|^2|\partial_x\psi|^2 + \frac{1}{2}(\bar{\psi}\partial_x\psi + \psi\partial_x\bar{\psi})^2 + \frac{1}{2}(1 - |\psi|^2)^2(2 + |\psi|^2)]dx,$$

The energy $S(\psi)$ is constructed in Section 4.2 as a linear combination of the higher-order energies of the integrable NLS equation (Example 4.1).

The energy space $\mathcal{X}$ for the higher-order energy (7.32) is defined by

$$\mathcal{X} = \left\{\psi \in H^2_{\text{loc}}(\mathbb{R}) : \quad \psi_x \in H^1(\mathbb{R}),\ 1 - |\psi|^2 \in L^2(\mathbb{R})\right\}. \tag{7.33}$$

If $\psi \in \mathcal{X}$, then $\zeta := 1 - |\psi|$ belongs to $H^1(\mathbb{R})$, because $|\zeta| \leq |1 - |\psi|^2| \in L^2(\mathbb{R})$ and $\partial_x\zeta = -\partial_x|\psi| \in L^2(\mathbb{R})$. By Sobolev's embedding of $H^1(\mathbb{R})$ into $L^\infty(\mathbb{R})$, we thus have $|\psi| = 1 - \zeta \in L^\infty(\mathbb{R})$, and from the definitions (7.32) and (7.33) it follows easily that $S(\psi) < \infty$.

Since u', u'', and $1-u^2$ decay exponentially to zero as $|x| \to \infty$, it is clear that the functional (7.32) is well defined for H^2 perturbations of the black soliton with the profile u. This allows us to define the differential of S at u, and a direct calculation reveals that u is a critical point of S, in the sense that $\nabla S(u) = 0$. Note that computations in Example 4.1 also suggest that the profile u of the black soliton satisfies $\nabla S(u) = 0$.

Similar to $H(\psi)$ in (7.20), $S(\psi)$ in (7.32) does not have a positive second variation at the black soliton with the profile u, hence u is a saddle point of both $H(\psi)$ and $S(\psi)$. Nevertheless, it is possible to construct a linear combination of $H(\psi)$ and $S(\psi)$ with a nonnegative second variation at u with perturbations in $H^2(\mathbb{R})$. Expanding $H(\psi)$ and $S(\psi)$ at $\psi = u$ up to the second variation yields

$$\begin{aligned} H(u+w+iv) - H(u) &= \langle L_+ w, w\rangle + \langle L_- v, v\rangle + \mathcal{O}(\|w+iv\|_{H^1}^3), \\ S(u+w+iv) - S(u) &= \langle M_+ w, w\rangle + \langle M_- v, v\rangle + \mathcal{O}(\|w+iv\|_{H^2}^3), \end{aligned}$$

where $L_\pm : H^2(\mathbb{R}) \subset L^2(\mathbb{R}) \to L^2(\mathbb{R})$ are same as before and $M_\pm : H^4(\mathbb{R}) \subset L^2(\mathbb{R}) \to L^2(\mathbb{R})$ are given by

$$\begin{aligned} M_+ &:= \partial_x^4 - 5\partial_x u^2 \partial_x - 5u^4 + 15u^2 - 4, \\ M_- &:= \partial_x^4 - 3\partial_x u^2 \partial_x + u^2 - 1. \end{aligned}$$

We will construct the following linear combination of the two energies:

$$\Lambda(\psi) :=, S(\psi) - 2H(\psi), \qquad \psi \in \mathcal{X}.$$

The second variation of $\Lambda(\psi)$ at $\psi = u$ is given by the quadratic forms

$$\begin{aligned} \langle K_+ w, w\rangle &= \int_{\mathbb{R}} [w_{xx}^2 + (5u^2-2)w_x^2 + (9u^2 - 5u^4 - 2)w^2]dx, \\ \langle K_- v, v\rangle &= \int_{\mathbb{R}} [v_{xx}^2 + (3u^2-2)v_x^2 + (1-u^2)v^2]dx, \end{aligned}$$

associated with the linear operators $K_\pm = M_\pm - 2L_\pm : H^4(\mathbb{R}) \subset L^2(\mathbb{R}) \to L^2(\mathbb{R})$.

We will show in the following two lemmas that the quadratic forms for K_+ and K_- are nonnegative on $H^2(\mathbb{R})$. Due to translation invariance of the NLS equation (7.19), we have

$$L_+ \partial_x u = M_+ \partial_x u = 0 \quad \Rightarrow \quad K_+ \partial_x u = 0.$$

As $\partial_x u \in H^2(\mathbb{R})$, this shows that the quadratic form for K_+ is not strictly positive due to a simple zero eigenvalue. The quadratic form for K_+ becomes coercive under a single constraint $\langle \partial_x u, w\rangle = 0$. The situation is slightly different for K_-. Due to rotational invariance of the NLS equation (7.19), we have

$$L_- u = M_- u = 0 \quad \Rightarrow \quad K_- u = 0,$$

but of course $u \notin H^2(\mathbb{R})$. This is related to the essential spectrum of K_- which extends all the way to 0. We will show that the quadratic form associated with K_- is strictly positive in a weaker topology than $H^2(\mathbb{R})$.

LEMMA 7.8. *For any $w \in H^2(\mathbb{R})$, we have*

$$\langle K_+ w, w\rangle = \|\partial_x \tilde{w}\|_{L^2}^2 + \|\tilde{w}\|_{L^2}^2 \geq 0, \tag{7.34}$$

where $\tilde{w} := \partial_x w + \sqrt{2} u w$.

PROOF. Integrating by parts and using the differential equation $\sqrt{2}\partial_x u = 1 - u^2$, we easily obtain

$$\int_{\mathbb{R}} \tilde{w}^2 dx = \int_{\mathbb{R}} [w_x^2 + 2\sqrt{2}uww_x + 2u^2w^2]dx = \int_{\mathbb{R}} [w_x^2 + (3u^2 - 1)w^2]dx.$$

and

$$\begin{aligned}\int_{\mathbb{R}} \tilde{w}_x^2 dx &= \int_{\mathbb{R}} [w_{xx}^2 + 2\sqrt{2}uw_x w_{xx} + 2u^2w_x^2 + 2\sqrt{2}u'ww_{xx} + 4uu'ww_x + 2u'^2w^2]dx \\ &= \int_{\mathbb{R}} [w_{xx}^2 + (5u^2 - 3)w_x^2 + 8uu'ww_x + 2(u')^2w^2]dx \\ &= \int_{\mathbb{R}} [w_{xx}^2 + (5u^2 - 3)w_x^2 + (1 - u^2)(5u^2 - 1)w^2]dx,\end{aligned}$$

because $2(u')^2 - 4(uu')' = (1 - u^2)(5u^2 - 1)$. Adding the two computations together proves (7.34). □

REMARK 7.19. The right-hand side of (7.34) vanishes if and only if $\tilde{w} = 0$, which is equivalent to $w = C\partial_x u$ for some constant C. Zero is hence a simple eigenvalue of K_+ in $L^2(\mathbb{R})$. Moreover, since $u(x) \to \pm 1$ as $x \to \pm\infty$, it is clear that the essential spectrum of K_+ is the interval $[2, \infty)$. Thus if we restrict ourselves to the orthogonal complement of $\partial_x u$ with respect to the scalar product $\langle \cdot , \cdot \rangle_{L^2}$, the spectrum of K_+ is bounded from below by a strictly positive constant, and the corresponding quadratic form is thus coercive in the topology of $H^2(\mathbb{R})$, see Lemma 7.10.

LEMMA 7.9. *For any* $v \in H^2(\mathbb{R})$, *we have*

$$\langle K_- v, v \rangle = \|L_- v\|_{L^2}^2 + \|u\partial_x v - v\partial_x u\|_{L^2}^2 \geq 0, \tag{7.35}$$

where $L_- = -\partial_x^2 + u^2 - 1$.

PROOF. Integrating by parts we obtain

$$\begin{aligned}\int_{\mathbb{R}} (L_- v)^2 dx &= \int_{\mathbb{R}} [v_{xx}^2 + 2(1 - u^2)vv_{xx} + (1 - u^2)^2v^2]dx \\ &= \int_{\mathbb{R}} [v_{xx}^2 + 2(u^2 - 1)v_x^2 - 2(uu')'v^2 + (1 - u^2)^2v^2]dx.\end{aligned}$$

Similarly, we have

$$\int_{\mathbb{R}} [u\partial_x v - v\partial_x u]^2 dx = \int_{\mathbb{R}} [u^2v_x^2 + (uu')'v^2 + u'^2v^2]dx.$$

It follows that

$$\|L_- v\|_{L^2}^2 + \|u\partial_x v - v\partial_x u\|_{L^2}^2 = \int_{\mathbb{R}} [v_{xx}^2 + (3u^2 - 2)v_x^2 + [(1 - u^2)^2 - uu'']v^2]dx,$$

which proves (7.35) since $(1 - u^2)^2 - uu'' = 1 - u^2$. □

REMARK 7.20. The right-hand side of (7.35) vanishes if and only if $L_- v = 0$ and $u\partial_x v - v\partial_x u = 0$, namely if $v = Cu$ for some constant C. As $u \notin H^2(\mathbb{R})$, this shows that $\langle K_- v, v \rangle > 0$ for any nonzero $v \in H^2(\mathbb{R})$. However, since $|u(x)| \to 1$ as $|x| \to \infty$, it is clear that zero belongs to the essential spectrum of the operator K_-, hence the associated quadratic form is not coercive in the topology of $H^2(\mathbb{R})$. Nevertheless, some weaker coercivity property will be established below, see Lemma 7.11.

The following two lemmas establish coercivity of the quadratic forms for K_+ and K_- under some constraints from Lemmas 7.8 and 7.9.

LEMMA 7.10. *There exists a constant* $C_+ > 0$ *such that*

$$\langle K_+ w, w\rangle \geq C_+ \|w\|^2_{H^2}, \tag{7.36}$$

for every $w \in H^2(\mathbb{R})$ *satisfying* $\langle \partial_x u, w\rangle = 0$.

PROOF. The result follows from (7.34) if we can show that there exists a positive constant C such that

$$\|w\|_{H^2} \leq C\|\tilde{w}\|_{H^1}, \quad \tilde{w} := \partial_x w + \sqrt{2}uw, \tag{7.37}$$

for every $w \in H^2(\mathbb{R})$ satisfying $\langle \partial_x u, w\rangle = 0$. To show (7.37), we solve the linear differential equation $\partial_x w + \sqrt{2}uw = \tilde{w}$ by the variation of constant formula and find $w = A\partial_x u + \hat{K}(\tilde{w})$ for some $A \in \mathbb{R}$, where

$$\hat{K}(\tilde{w}) := \int_0^x K(x,y)\tilde{w}(y)dy, \qquad K(x,y) = \frac{\cosh^2(y/\sqrt{2})}{\cosh^2(x/\sqrt{2})}.$$

The constant A is uniquely determined by the orthogonality condition $\langle \partial_x u, w\rangle = 0$, which implies that $A\|\partial_x u\|^2_{L^2} + \langle \partial_x u, \hat{K}(\tilde{w})\rangle = 0$. We obtain

$$\begin{aligned}
\langle \partial_x u, \hat{K}(\tilde{w})\rangle_{L^2} &= \int_{-\infty}^{\infty} \left\{ \int_0^x K(x,y)\tilde{w}(y)dy \right\} u'(x)dx \\
&= \int_0^{\infty} \left\{ \int_y^{\infty} K(x,y)u'(x)dx \right\} \Big(\tilde{w}(y) - \tilde{w}(-y)\Big)dy \\
&= \frac{1}{3}\int_0^{\infty} e^{-\sqrt{2}y}\,\frac{3+e^{-\sqrt{2}y}}{1+e^{-\sqrt{2}y}} \Big(\tilde{w}(y) - \tilde{w}(-y)\Big)dy,
\end{aligned}$$

hence $|\langle \partial_x u, \hat{K}(\tilde{w})\rangle| \leq C\|\tilde{w}\|_{L^2}$ for some $C > 0$. As a result, we have

$$|A| \leq C\|\tilde{w}\|_{L^2}/\|\partial_x u\|^2_{L^2}.$$

In order to treat the term $\hat{K}(\tilde{w})$, we use the fact that $\hat{K}$ is a bounded operator from $L^\infty(\mathbb{R})$ to $L^\infty(\mathbb{R})$ with norm

$$K_\infty = \sup_{x\in\mathbb{R}} \int_0^{|x|} K(x,y)dy = \frac{1}{\sqrt{2}} \sup_{x\in\mathbb{R}} \frac{1 + 2\sqrt{2}|x|e^{-\sqrt{2}|x|} - e^{-2\sqrt{2}|x|}}{1 + 2e^{-\sqrt{2}|x|} + e^{-2\sqrt{2}|x|}} < \infty,$$

as well as a bounded operator from $L^1(\mathbb{R})$ to $L^1(\mathbb{R})$ with norm

$$K_1 = \sup_{y\in\mathbb{R}} \int_{|y|}^{\infty} K(x,y)dx = \frac{1}{\sqrt{2}} \sup_{y\in\mathbb{R}} \Big(1 + e^{-\sqrt{2}|y|}\Big) = \sqrt{2}.$$

By the Riesz–Thorin interpolation theorem, it follows that $\hat{K}$ is a bounded operator from $L^2(\mathbb{R})$ to $L^2(\mathbb{R})$, and we have the estimate $\|\hat{K}(\tilde{w})\|_{L^2} \leq (K_1 K_\infty)^{1/2}\|\tilde{w}\|_{L^2}$.

Summarizing, we have shown that $\|w\|_{L^2} \leq |A|\|\partial_x u\|_{L^2} + \|\hat{K}(\tilde{w})\|_{L^2} \leq C\|\tilde{w}\|_{L^2}$ for some $C > 0$. Since $\tilde{w} = \partial_x w + \sqrt{2}uw$, we also have

$$\|w_x\|_{L^2} \leq \|\tilde{w}\|_{L^2} + \sqrt{2}\|w\|_{L^2}$$

and (after differentiating)

$$\|w_{xx}\|_{L^2} \leq \|\tilde{w}_x\|_{L^2} + \sqrt{2}\|w_x\|_{L^2} + \|w\|_{L^2}.$$

This proves the bound (7.37) and hence the coercivity bound (7.36). □

LEMMA 7.11. *There exists a constant $C_- > 0$ such that*

$$\langle K_- v, v\rangle \geq C_-\Big(\|\partial_x v\|_{H^1}^2 + |v(0)|^2\Big), \tag{7.38}$$

for every $v \in H^2_{\rm loc}(\mathbb{R})$ satisfying $\partial_x v \in H^1(\mathbb{R})$ and $\langle \partial_x^2 u, v\rangle = 0$.

PROOF. The result follows from (7.35) if we can show that there exists a positive constant C such that

$$\|\partial_x v\|_{H^1} + |v(0)| \leq C(\|p\|_{L^2} + \|q\|_{L^2}), \tag{7.39}$$

where

$$p := u\partial_x v - v\partial_x u,$$
$$q := \partial_x^2 v + (1-u^2)v,$$

for every $v \in H^2_{\rm loc}(\mathbb{R})$ satisfying $\partial_x v \in H^1(\mathbb{R})$ and $\langle \partial_x^2 u, v\rangle = 0$. We have

$$\partial_x p = u\partial_x^2 v - v\partial_x^2 u = u(\partial_x^2 v + (1-u^2)v) = uq.$$

Any solution of the linear differential equation $u\partial_x v - v\partial_x u = p$ can be written in the form $v = Bu + Z$ for some $B \in \mathbb{R}$, where

$$Z(x) = u(x)\int_0^x [p(y) + \sqrt{2}q(y)]dy - \sqrt{2}p(x).$$

Indeed, we have

$$\partial_x v(x) = \partial u(x)\left(B + \int_0^x [p(y) + \sqrt{2}q(y)]dy\right) + u(x)p(x),$$

so that $u\partial_x v - v\partial_x u = (u^2 + \sqrt{2}\partial_x u)p = p$. The constant B is uniquely determined by the orthogonality condition $\langle \partial_x^2 u, v\rangle = 0$, which implies that $B\|\partial_x u\|_{L^2}^2 = \langle \partial_x^2 u, Z\rangle$.

Since $p \in L^2(\mathbb{R})$ and $\partial_x p = uq \in L^2(\mathbb{R})$, we have $p \in H^1(\mathbb{R})$ and hence $p \in L^\infty(\mathbb{R})$ by Sobolev's embedding, which implies the bound

$$\|p\|_{L^\infty}^2 \leq \|p\|_{L^2}\|\partial_x p\|_{L^2} \leq \|p\|_{L^2}\|q\|_{L^2}.$$

By Hölder's inequality, we deduce that

$$|Z(x)| \leq \sqrt{2}(|x|^{1/2} + 1)(\|p\|_{L^2} + \|q\|_{L^2}), \quad x \in \mathbb{R}.$$

This moderate growth of Z is compensated for by the exponential decay of $u''(x) \to 0$ as $|x| \to \infty$, and we obtain for some $C > 0$:

$$|\langle \partial_x^2 u, Z\rangle| \leq C(\|p\|_{L^2} + \|q\|_{L^2}) \quad \Rightarrow \quad |B| \leq C(\|p\|_{L^2} + \|q\|_{L^2}).$$

In the same way, we have $\|\partial_x v\|_{L^2} \leq C(\|p\|_{L^2} + \|q\|_{L^2})$. A similar estimate holds for $\|\partial_x^2 v\|_{L^2}$ because $\partial_x^2 v = q - (1-u^2)v$ and $1-u^2$ has the exponential decay to zero at infinity. Finally, since $v(0) = -\sqrt{2}p(0)$, we also have $|v(0)| \leq C(\|p\|_{L^2} + \|q\|_{L^2})$. This proves the bound (7.39) and hence the coercivity bound (7.38). □

REMARK 7.21. The orthogonality condition $\langle \partial_x u, w\rangle = 0$ needed for the coercivity bound (7.36) is very similar to the one in Lemma 7.5. The orthogonality condition $\langle \partial_x^2 u, v\rangle = 0$ needed for the coercivity bound (7.38) is very similar to the orthogonality condition $\langle u, v\rangle_{\mathcal{H}} = 0$ in Lemma 7.6. We do not need the other orthogonality condition $\langle 1, v\rangle_{\mathcal{H}} = 0$ in Lemma 7.6 because K_- is nonnegative in $L^2(\mathbb{R})$ compared to L_- which has a simple negative eigenvalue in $L^2(\mathbb{R})$.

REMARK 7.22. The second variation of $\Lambda(\psi)$ at $\psi = u$ is positive except for degeneracies due to the two symmetries: K_+ has a simple zero eigenvalue which is due to translation invariance and the essential spectrum of K_- extends all the way to 0 due to rotational invariance. As a consequence, perturbations in $H^2(\mathbb{R})$ can include slow modulations of the phase of the black soliton far away from the origin, which do not increase the functional Λ. Lemmas 7.10 and 7.11 show that the orthogonality conditions can be satisfied by the modulation parameters along the translation and rotational symmetries but the coercivity of Λ is obtained at a weaker topology than $H^2(\mathbb{R})$. The black soliton is a minimizer of Λ in this topology, but it is a degenerate minimizer in the topology of $H^2(\mathbb{R})$.

To obtain the orbital stability of the black soliton in a weaker topology of $H^2(\mathbb{R})$ (Remark 7.22), we equip the space $\mathcal{X}$ with the distance

$$\rho_R(\psi, \phi) := \|\partial_x(\psi - \phi)\|_{H^1(\mathbb{R})} + \||\psi|^2 - |\phi|^2\|_{L^2(\mathbb{R})} + \|\psi - \phi\|_{L^2(-R,R)}, \tag{7.40}$$

where $R \geq 1$ is a fixed parameter. A function $\psi \in H^2_{\text{loc}}(\mathbb{R})$ belongs to $\mathcal{X}$ if and only if $\rho_R(\psi, u) < \infty$. Moreover, different choices of R give equivalent distances on $\mathcal{X}$. If $\|\psi - u\|_{H^2}$ is small, then so is $\rho_R(\psi, u)$ but the converse is not true because $\rho_R(\psi, u)$ does not control the L^2 norm of the difference $\psi - u$ on the whole real line. The following theorem proven in [**98**] gives the orbital stability of the black soliton with the profile u in the metric ρ_R in (7.40).

THEOREM 7.3. *Fix $R \geq 1$. Given any $\epsilon > 0$, there exists $\delta > 0$ such that, for any $\psi_0 \in \mathcal{X}$ satisfying $\rho_R(\psi_0, u) \leq \delta$, the global solution $\psi \in C^0([0,\infty), \mathcal{X})$ of the NLS equation* (7.19) *has the following property. For every $t \geq 0$, there exist $\xi(t) \in \mathbb{R}$ and $\alpha(t) \in \mathbb{R}/(2\pi\,\mathbb{Z})$ such that*

$$\rho_R(\psi(t,\cdot), e^{-i\alpha(t)} u(\cdot - \xi(t))) \leq \epsilon.$$

PROOF. The theorem is proven based on the following four steps. The first step is a standard decomposition of the solution near the two-dimensional orbit of the black soliton with the profile u generated by the two symmetries to satisfy the two orthogonality conditions needed for coercivity of the higher-order energy. The second step provides an equivalent distance to ρ_R which is better entitled to measure the distance for the perturbation to the black soliton with the profile u. The third step gives the coercivity of the energy difference with respect to the new distance in $\mathcal{X}$. The fourth step transfers the coercivity of the energy difference with respect to the same distance ρ_R in (7.40).

STEP 1. Fix $R \geq 1$. There exists $\epsilon_0 > 0$ such that, for any $\psi \in \mathcal{X}$ satisfying

$$\inf_{\xi,\alpha \in \mathbb{R}} \rho_R(e^{i\alpha}\psi(\cdot + \xi), u) \leq \epsilon_0,$$

there exist $\xi \in \mathbb{R}$ and $\alpha \in \mathbb{R}/(2\pi\,\mathbb{Z})$ such that

$$e^{i\alpha}\psi(x + \xi) = u(x) + w(x) + iv(x), \quad x \in \mathbb{R}, \tag{7.41}$$

where the real-valued functions w and v satisfy the orthogonality conditions

$$\langle \partial_x u, w \rangle = 0 \text{ and } \langle \partial_x^2 u, v \rangle = 0.$$

Moreover, the modulation parameters $\xi \in \mathbb{R}$ and $\alpha \in \mathbb{R}/(2\pi\,\mathbb{Z})$ depend continuously on ψ in the topology defined by the distance ρ_R.

PROOF OF STEP 1. It is sufficient to prove the decomposition for any $\psi \in \mathcal{X}$ such that $\epsilon := \rho_R(\psi, u)$ is sufficiently small. To do so, we consider the smooth function $\mathbf{f} : \mathbb{R}^2 \to \mathbb{R}^2$ defined by

$$\mathbf{f}(\xi, \alpha) = \begin{pmatrix} \langle \partial_x u(\cdot - \xi), \mathrm{Re}(e^{i\alpha}\psi)\rangle \\ \langle \partial_x^2 u(\cdot - \xi), \mathrm{Im}(e^{i\alpha}\psi)\rangle \end{pmatrix}, \qquad (\xi, \alpha) \in \mathbb{R}^2 .$$

By construction, we have $\mathbf{f}(\xi, \alpha) = \mathbf{0}$ if and only if ψ can be represented by the orthogonal decomposition (7.41) with the two orthogonality conditions $\langle \partial_x u, w\rangle = 0$ and $\langle \partial_x^2 u, v\rangle = 0$ because $\langle \partial_x u, u\rangle = 0$. As in the proof of Lemma 7.11, we observe that

$$|u(x)| \le C\Big(\|u\|_{L^2(-1,1)} + (1 + |x|^{1/2})\|\partial_x u\|_{L^2(\mathbb{R})}\Big) \le C(1 + |x|^{1/2})\rho_R(\psi, u).$$

If $\epsilon = \rho_R(\psi, u)$ is sufficiently small, then

$$|\langle \partial_x u, \mathrm{Re}(\psi)\rangle| \le C\rho_R(\psi, u) \text{ and } |\langle \partial_x^2 u, \mathrm{Im}(\psi)\rangle| \le C\rho_R(\psi, u).$$

This shows that $\|\mathbf{f}(0, 0)\| \le C\epsilon$ for some positive constant C independent of ϵ. The Jacobian matrix of the function $\mathbf{f}$ at $(0, 0)$ is given by

$$D\mathbf{f}(0, 0) = \begin{pmatrix} \|\partial_x u\|_{L^2}^2 & 0 \\ 0 & -\|\partial_x u\|_{L^2}^2 \end{pmatrix} + \begin{pmatrix} -\langle \partial_x^2 u, \mathrm{Re}(\psi - u)\rangle & -\langle \partial_x u, \mathrm{Im}(\psi - u)\rangle \\ -\langle \partial_x^3 u, \mathrm{Im}(\psi - u)\rangle & \langle \partial_x^2 u, \mathrm{Re}(\psi - u)\rangle \end{pmatrix}.$$

The first term in the right-hand side is a fixed invertible matrix and the second term is bounded in norm by $C\epsilon$, hence $D\mathbf{f}(0, 0)$ is invertible if ϵ is small enough. In addition, the norm of the inverse of $D\mathbf{f}(0, 0)$ is bounded by a constant independent of ϵ. Finally, it is straightforward to verify that the second-order derivatives of $\mathbf{f}$ are uniformly bounded if $\epsilon \le 1$. These observations together imply that there exists a unique pair (ξ, α), in a neighborhood of size $\mathcal{O}(\epsilon)$ of $(0, 0)$, such that $\mathbf{f}(\xi, \alpha) = \mathbf{0}$. Thus the orthogonal decomposition holds for these values of (ξ, α). In addition, the above argument shows that the modulation parameters ξ, α depend continuously on $\psi \in X$ in the topology defined by the distance ρ_R. ☐

STEP 2. Fix $R \ge 1$, and assume that $\psi = u + w + iv$. Define $\hat{\rho}_R(u, v, \eta)$ by

$$\hat{\rho}_R(w, v, \eta) := \sqrt{\int_{\mathbb{R}} \Big(w_{xx}^2 + v_{xx}^2 + w_x^2 + v_x^2\Big)dx + \int_{|x|\le R} \Big(w^2 + R^{-2}v^2\Big)dx + \int_{|x|\ge R} \Big(\eta_x^2 + \eta^2\Big)dx},$$

where

$$\eta := |u + w + iv|^2 - u^2 = 2uw + w^2 + v^2.$$

Then, $\rho_R(\psi, u) < \infty$ if and only if $\hat{\rho}_R(w, v, \eta) < \infty$. Moreover, there exists a constant $C_0 \ge 1$ (independent of R) such that, if $\rho_R(\psi, u) \le 1$ or if $R^{1/2}\hat{\rho}_R(w, v, \eta) \le 1$, then

$$C_0^{-1}\hat{\rho}_R(w, v, \eta) \le \rho_R(\psi, u) \le C_0 R\hat{\rho}_R(w, v, \eta).$$

PROOF OF STEP 2. Assume first that $\rho_R(\psi, u) < \infty$ so that $w_x, v_x \in H^1(\mathbb{R})$, $w, v \in L^2(-R, R)$, and $\eta \in L^2(\mathbb{R})$. We claim that $w, v \in L^\infty(\mathbb{R})$ and that

$$K := \|w\|_{L^\infty(\mathbb{R})} + \|v\|_{L^\infty(\mathbb{R})} \le C[1 + \rho_R(\psi, u)], \tag{7.42}$$

for some universal constant $C > 0$. Indeed, if $f = |\psi| - |u|$, we observe that

$$[\rho_R(\psi, u)]^2 \ge \int_{\mathbb{R}} \eta^2 dx \ge \int_{|x|\ge 1} (|\psi| - |u|)^2(|\psi| + |u|)^2 dx \ge C\int_{|x|\ge 1} f^2 dx,$$

hence $f \in L^2(I)$, where $I = \{x \in \mathbb{R} : |x| \geq 1\}$, and $\|f\|_{L^2(I)} \leq C\rho_R(\psi, u)$. Moreover, we have $|\partial_x f| \leq 2\partial_x u + |\partial_x w| + |\partial_x v|$ almost everywhere, hence $\partial_x f \in L^2(\mathbb{R})$ and

$$\|\partial_x f\|_{L^2(\mathbb{R})} \leq C[1 + \rho_R(\psi, u)].$$

By Sobolev embedding of $H^1(\mathbb{R})$ into $L^\infty(\mathbb{R})$, this implies that $f \in L^\infty(I)$, hence also $w, v \in L^\infty(I)$, and we have the bound

$$| w\|_{L^\infty(I)} + \|v\|_{L^\infty(I)} \leq C[1 + \rho_R(\psi, u)].$$

Finally, since $\|\partial_x w\|_{L^2(\mathbb{R})} + \|\partial_x v\|_{L^2(\mathbb{R})} \leq C\rho_R(\psi, u)$, we conclude that $w, v \in L^\infty(\mathbb{R})$ and that (7.42) holds.

Next, we assume that $\hat{\rho}_R(w, v, \eta) < \infty$, so that $\partial_x w, \partial_x v \in H^1(\mathbb{R})$, $w, v \in L^2(-R, R)$, and $\eta \in H^1(I_R)$, where $I_R = \{x \in \mathbb{R} : |x| \geq R\}$. We claim that $w, v \in L^\infty(\mathbb{R})$ and that

$$K := \|w\|_{L^\infty(\mathbb{R})} + \|v\|_{L^\infty(\mathbb{R})} \leq C[1 + R^{1/2}\hat{\rho}_R(w, v, \eta)], \tag{7.43}$$

for some universal constant $C > 0$. Indeed, we know that $\eta \in L^\infty(I_R)$ with

$$\|\eta\|_{L^\infty(I_R)} \leq C\hat{\rho}_R(w, v, \eta).$$

This implies that $\psi \in L^\infty(I_R)$, hence also $w, v \in L^\infty(I_R)$, and that

$$\|w\|_{L^\infty(I_R)} + \|v\|_{L^\infty(I_R)} \leq C[1 + \hat{\rho}_R(w, v, \eta)]^{1/2}.$$

On the other hand, we know that $\|w\|_{L^\infty(-R,R)} \leq C\|w\|_{H^1(-R,R)} \leq C\hat{\rho}_R(w, v, \eta)$ and that

$$\begin{aligned}\|v\|_{L^\infty(-R,R)} &\leq C\Big(\frac{\|v\|_{L^2(-R,R)}}{R^{1/2}} + \|v\|^{1/2}_{L^2(-R,R)}\|\partial_x v\|^{1/2}_{L^2(-R,R)}\Big)\\ &\leq CR^{1/2}\hat{\rho}_R(w, v, \eta),\end{aligned}$$

because $\|v\|_{L^2(-R,R)} \leq R\hat{\rho}_R(w, v, \eta)$ and $\|\partial_x v\|_{L^2(-R,R)} \leq \hat{\rho}_R(w, v, \eta)$. Thus we conclude that $w, v \in L^\infty(\mathbb{R})$ and that (7.43) holds.

Finally we assume that $K = \|w\|_{L^\infty(\mathbb{R})} + \|v\|_{L^\infty(\mathbb{R})} < \infty$, which is the case if either $\rho_R(\psi, u) < \infty$ or $\hat{\rho}_R(w, v, \eta) < \infty$. As $\eta = 2uw + w^2 + v^2$, we find

$$\begin{aligned}\|\eta\|_{L^2(-R,R)} &\leq C(1 + K)\Big(\|w\|_{L^2(-R,R)} + \|v\|_{L^2(-R,R)}\Big)\\ &\leq C(1 + K)R\hat{\rho}_R(w, v, \eta),\end{aligned}$$

because $\|w\|_{L^2(-R,R)} \leq \hat{\rho}_R(w, v, \eta)$ and $\|v\|_{L^2(-R,R)} \leq R\hat{\rho}_R(w, v, \eta)$. This shows that, if $\hat{\rho}_R(w, v, \eta) < \infty$, then $\eta \in L^2(\mathbb{R})$, so that $\rho_R(\psi, u) < \infty$, and we have the bound $\rho_R(\psi, u) \leq C(1 + K)R\hat{\rho}_R(w, v, \eta)$. Conversely, since

$$\partial_x \hat{\rho}_R(w, v, \eta) = 2(w\partial_x u + u\partial_x w + w\partial_x w + v\partial_x v),$$

we obtain

$$\begin{aligned}\|\partial_x \eta\|_{L^2(\mathbb{R})} &\leq C(1 + K)\Big(\|w\|_{L^2(-1,1)} + \|\partial_x w\|_{L^2(\mathbb{R})} + \|\partial_x v\|_{L^2(\mathbb{R})}\Big)\\ &\leq C(1 + K)\rho_R(\psi, u),\end{aligned}$$

where to estimate $w\partial_x u$ we used the fact that $|w(x)| \leq C(\|w\|_{L^2(-1,1)} + (1 + |x|)^{1/2}\|\partial_x w\|_{L^2(\mathbb{R})})$. This shows that, if $\rho_R(\psi, u) < \infty$, then $\partial_x \eta \in L^2(\mathbb{R})$, so that $\hat{\rho}_R(w, v, \eta) < \infty$, and we have the bound $\hat{\rho}_R(w, v, \eta) \leq C(1 + K)\rho_R(\psi, u)$. □

STEP 3. If $R \geq 1$ is sufficiently large, there exists a constant $C_0 > 0$ such that, if $\psi = u + w + iv$ satisfies $\rho_R(\psi, u) \leq 1$ and if $\langle \partial_x u, w \rangle = \langle \partial_x^2 u, v \rangle = 0$, then

$$\Lambda(\psi) - \Lambda(u) \geq C_0 \hat{\rho}_R(w, v, \eta)^2 + \mathcal{O}(R^3 \hat{\rho}_R(w, v, \eta)^3),$$

where the estimate is uniform in R.

PROOF OF STEP 3. We first show that if $\psi = u+w+iv$ satisfies $\rho_R(\psi, u) < \infty$, then

$$\begin{aligned}\Lambda(\psi) - \Lambda(u) = \int_{\mathbb{R}} \Big(& w_{xx}^2 + v_{xx}^2 + (3u^2 - 2)(w_x^2 + v_x^2) + (1 - u^2)(w^2 + v^2) \\ & - 3(1 - u^2)(1 - 3u^2)w^2 + \frac{1}{2}\eta_x^2 + \frac{1}{2}(3u^2 - 2)\eta^2 \\ & + \frac{1}{2}\eta^3 + 3\eta(w_x^2 + v_x^2) + 6u'(w^2 + v^2)w_x \Big) dx.\end{aligned}$$

In order to prove the decomposition, we observe that $|\psi|^2 = u^2 + \eta$ and

$$\bar{\psi}\partial_x\psi + \psi\partial_x\bar{\psi} = 2u\partial_x u + \partial_x \eta.$$

If the integrand in the functional $\Lambda = S - 2E$ is denoted by

$$A(\psi) := |\psi_{xx}|^2 + |\psi_x|^2(3|\psi|^2 - 2) + \frac{1}{2}(\bar{\psi}\psi_x + \psi\bar{\psi}_x)^2 + \frac{1}{2}|\psi|^2(1 - |\psi|^2)^2,$$

a direct calculation shows that

$$\begin{aligned}A(\psi) - A(u) = {} & \mathcal{L}(w, \eta) + 6\eta u' w_x + w_{xx}^2 + v_{xx}^2 + (3u^2 - 2)(w_x^2 + v_x^2) \\ & + \frac{1}{2}\eta_x^2 + \frac{1}{2}(3u^2 - 2)\eta^2 + \frac{1}{2}\eta^3 + 3\eta(w_x^2 + v_x^2),\end{aligned}$$

where

$$\mathcal{L}(w, \eta) := 2u'' w_{xx} + 2(3u^2 - 2)u' w_x + 2uu' \eta_x + \eta(1 - u^2)(2 - 3u^2).$$

We now integrate the right-hand side over $x \in \mathbb{R}$, starting with the terms $\mathcal{L}(w, \eta)$ which are linear in w and η. Using the identities $u'' + u - u^3 = 0$ and $u'''' + (1 - 3u^2)u'' - 6u(u')^2 = 0$, we find

$$\begin{aligned}2\int_{\mathbb{R}} \Big(u'' w_{xx} + (3u^2 - 2)u' w_x \Big) dx &= 2\int_{\mathbb{R}} \Big(u'''' - (3u^2 - 2)u'' - 6u(u')^2 \Big) w dx \\ &= 2\int_{\mathbb{R}} u'' w dx \\ &= -2\int_{\mathbb{R}} (1 - u^2) u w dx.\end{aligned}$$

Similarly, as $2(uu')' = (1 - u^2)(1 - 3u^2)$, we have

$$2\int_{\mathbb{R}} uu' \eta_x dx = -2\int_{\mathbb{R}} (uu')' \eta dx = -\int_{\mathbb{R}} (1 - u^2)(1 - 3u^2)\eta dx.$$

We conclude that

$$\int_{\mathbb{R}} \mathcal{L}(w, \eta) dx = \int_{\mathbb{R}} (1 - u^2)(\eta - 2uw) dx = \int_{\mathbb{R}} (1 - u^2)(w^2 + v^2) dx.$$

This expansion is quadratic in w and v since u is a critical point of the functional Λ. We next consider the quadratic term $6\eta u' w_x$ which has no definite sign. Using the definition of η, we find $6\eta u' w_x = 12uu'ww_x + 6u'(w^2+v^2)w_x$, and integrating by parts, we obtain

$$6\int_{\mathbb{R}} \eta u' w_x dx = -3\int_{\mathbb{R}} (1-u^2)(1-3u^2)w^2 dx + 6\int_{\mathbb{R}} u'(w^2+v^2)w_x dx.$$

Combining the previous expansions yields the explicit expression for $\Lambda(\psi) - \Lambda(u)$.

To proceed further, we simplify the notations and define

$$\begin{aligned} B_0(w) &:= w_{xx}^2 + (5u^2-2)w_x^2 + (9u^2-5u^4-2)w^2, \\ B_1(w) &:= w_{xx}^2 + (3u^2-2)w_x^2 + (1-u^2)w^2 - 3(1-u^2)(1-3u^2)w^2, \\ B_2(v) &:= v_{xx}^2 + (3u^2-2)v_x^2 + (1-u^2)v^2, \\ B_3(\eta) &:= \frac{1}{2}\eta_x^2 + \frac{1}{2}(3u^2-2)\eta^2. \end{aligned}$$

We also use the following estimates:

$$\|w\|_{L^2(-2R,2R)} \leq C_1[\hat{\rho}_R(w,v,\eta) + R^{3/2}\hat{\rho}_R(w,v,\eta)^2], \tag{7.44}$$

$$\|w\|_{L^\infty(-2R,2R)} + \|v\|_{L^\infty(-2R,2R)} \leq C_1 R^{1/2}\hat{\rho}_R(w,v,\eta), \tag{7.45}$$

The bound (7.45) is obtained from the estimate

$$|f(x)| \leq C(R^{-1/2}\|f\|_{L^2(-R,R)} + (|x|+R)^{1/2}\|f_x\|_{L^2(\mathbb{R})}),$$

where f is either w or v. To prove estimate (7.44), we recall that $\|w\|_{L^2(-R,R)} \leq \hat{\rho}_R(w,v,\eta)$, so we only need to control $w(x)$ for $R \leq |x| \leq 2R$. In that region we have $|w| \leq C(|\eta| + w^2 + v^2)$, hence using the bound (7.45) and the fact that $\|\eta\|_{L^2(|x|\geq R)} \leq \hat{\rho}_R(2,v,\eta)$ we obtain (7.44).

It is easy to estimate the cubic terms in the expansion for $\Lambda(u+w+iv) - \Lambda(u)$ in terms of $\hat{\rho}_R(w,v,\eta)$ using bounds (7.42) and (7.45). We thus find

$$\Lambda(\psi) - \Lambda(u) = Q(u,v,\eta) + \mathcal{O}(R^3\rho(w,v,\eta)^3),$$

where

$$Q(w,v,\eta) = \int_{\mathbb{R}} [B_1(w) + B_2(v) + B_3(\eta)]dx.$$

$Q(w,v,\eta)$ contains $\langle K_- v, v\rangle \equiv \int_{\mathbb{R}} B_2(v)dx$ but does not contain

$$\langle K_+ w, w\rangle \equiv \int_{\mathbb{R}} B_0(w)dx.$$

Instead, $\langle K_+ w, w\rangle$ is obtained from $\int_{\mathbb{R}} B_1(w)dx$ and a part of $\int_{\mathbb{R}} B_3(\eta)dx$ since $\eta = 2uw + w^2 + v^2$. Due to the relation between w and η, it is not obvious that each quadratic term in $Q(w,v,\eta)$ is positive independently of the others. To avoid that difficulty, we fix some $R \geq 1$ and split the integration domain into two regions. When $|x| \leq R$, we replace η by $2uw + w^2 + v^2$ and apply the same arguments as in Lemmas 7.10 and 7.11 to prove positivity of the quadratic terms in $Q(w,v,\eta)$. In the outer region $|x| > R$, the analysis is much simpler, because the expressions $B_1(w)$, $B_2(v)$, and $B_3(\eta)$ are obviously positive if R is large enough.

Thus we split the integral using the partition of unity $1 = \chi_R + (1-\chi_R)$ and we use the estimate

$$\begin{aligned} \int_{\mathbb{R}} [B_1(w) + B_3(\eta)]\chi_R(x)dx = &\int_{\mathbb{R}} B_0(w)\chi_R(x)dx \\ &+ \mathcal{O}(R^3\hat{\rho}_R(w,v,\eta)^3 + e^{-R}\hat{\rho}_R(w,v,\eta)^2), \end{aligned}$$

uniformly for $R \geq 1$. Indeed, we have

$$B_3(\eta) = 2(u')^2 w^2 + 2u^2 w_x^2 + 4uu'ww_x + 2(3u^2 - 2)u^2 w^2 + \tilde{N}(w,v),$$

where

$$\begin{aligned}\tilde{N}(w,v) := &\, 4(ww_x + vv_x)(u'w + uw_x) + 2(ww_x + vv_x)^2 \\ &+ 2(3u^2 - 2)uw(w^2 + v^2) + \frac{1}{2}(3u^2 - 2)(w^2 + v^2)^2.\end{aligned}$$

This implies that

$$B_1(w) + B_3(\eta) \;=\; B_0(w) + (2uu'w^2)_x + \tilde{N}(w,v),$$

which yields the estimate after we multiply both sides by $\chi_R(x)$, integrate over $x \in \mathbb{R}$, and use (7.42) and (7.45) to obtain

$$\begin{aligned}-2\int_{\mathbb{R}} uu'w^2 \chi_R'(x)dx \;&=\; \mathcal{O}(e^{-R}\hat{\rho}_R(w,v,\eta)^2), \\ \int_{\mathbb{R}} \tilde{N}(w,v)\chi_R(x)dx \;&=\; \mathcal{O}(R^3 \hat{\rho}_R(w,v,\eta)^3).\end{aligned}$$

Thus we obtain

$$\begin{aligned}Q(w,v,\eta) \;=&\; \int_{\mathbb{R}} B_2(v)dx + \int_{\mathbb{R}} B_0(w)\chi_R(x)dx \\ &+ \int_{\mathbb{R}} [B_1(w) + B_3(\eta)](1 - \chi_R(x))dx \\ &+ \mathcal{O}(R^3 \hat{\rho}_R(w,v,\eta)^3 + e^{-R}\hat{\rho}_R(w,v,\eta)^2).\end{aligned}$$

As $\langle \partial_x^2 u, v\rangle = 0$, we know from Lemma 7.11 that

$$\int_{\mathbb{R}} B_2(v)dx \;\geq\; C\int_{\mathbb{R}} (v_{xx}^2 + v_x^2)dx + \frac{C}{R^2}\int_{|x|\leq R} v^2 dx,$$

where the last term in the right-hand side follows from the bound

$$|v(x)| \leq |v(0)| + |x|^{1/2}\|v_x\|_{L^2},$$

which implies

$$\int_{|x|\leq R} v^2 dx \;\leq\; 4R|v(0)|^2 + 2R^2 \int_{\mathbb{R}} v_x^2 dx \;\leq\; CR^2 \int_{\mathbb{R}} B_2(v)dx.$$

If $R \geq 1$ is large enough so that $3u^2 - 2 \geq \frac{1}{2}$ for $|x| \geq R$, it is clear that

$$\int_{\mathbb{R}} [B_1(w) + B_3(\eta)](1 - \chi_R(x))dx \;\geq\; C\int_{|x|\geq R} (w_{xx}^2 + w_x^2 + \eta_x^2 + \eta^2)dx.$$

Finally, we estimate from below the term $\int_{\mathbb{R}} B_0(w)\chi_R(x)dx$ under the orthogonality condition $\langle \partial_x u, w\rangle = 0$. Arguing as in Lemmas 7.8 and 7.9, we introduce the auxiliary variable $\tilde{w} := w_x + \sqrt{2}uw$. After integrating by parts, we obtain the identity

$$\begin{aligned}&\int_{\mathbb{R}} B_0(w)\chi_R(x)dx \\ &= \int_{\mathbb{R}} [\tilde{w}_x^2 + \tilde{w}^2]\chi_R(x)dx + \int_{\mathbb{R}} [\sqrt{2}uw_x^2 \\ &+ 2\sqrt{2}u'ww_x + (2uu' - \sqrt{2}u'')w^2 + \sqrt{2}u^2 w^2]\chi_R'(x)dx.\end{aligned}$$

Since $\chi_R'(x) = R^{-1}\chi'(R^{-1}x)$, we use (7.44) and obtain the following estimate for the second term:

$$\begin{aligned}
&\left|\int_{\mathbb{R}}[\sqrt{2}uw_x^2 + 2\sqrt{2}u'ww_x + (2uu' - \sqrt{2}u'')w^2 + \sqrt{2}u^2w^2]\chi_R'(x)dx\right| \\
&\quad \le \frac{C}{R}\int_{|x|\le 3R/2}[w_x^2 + w^2]dx \\
&\quad \le \frac{C}{R}\hat{\rho}_R(w,v,\eta)^2 + \mathcal{O}(R^2\hat{\rho}_R(w,v,\eta)^4),
\end{aligned}$$

where $C > 0$ is independent of R. Moreover, proceeding as in the proof of Lemma 7.10, we find

$$\int_{|x|\le R}[w_{xx}^2 + w_x^2 + w^2]dx \le C\int_{|x|\le R}[\tilde{w}_x^2 + \tilde{w}^2]dx + \mathcal{O}(e^{-R}\hat{\rho}_R(w,v,\eta)^2). \tag{7.46}$$

Indeed, we have the representation $w = A\partial_x u + \hat{K}(\tilde{w})$, where the constant A is fixed by the orthogonality condition $\langle \partial_x u, w\rangle = 0$. The proof of Lemma 7.10 shows that $\|\hat{K}(\tilde{w})\|_{L^2(|x|\le R)} \le C\|\tilde{w}\|_{L^2(|x|\le R)}$. From the orthogonality relation

$$0 = \int_{|x|\le R} u'(x)[Au'(x) + \hat{K}(\tilde{w})(x)]dx + \int_{|x|\ge R} u'(x)w(x)dx,$$

we easily obtain the bound $|A| \le C\|\hat{K}(\tilde{w})\|_{L^2(|x|\le R)} + \mathcal{O}(e^{-R}\hat{\rho}_R(w,v,\eta))$. This shows that

$$\|w\|_{L^2(|x|\le R)} \le C\|\tilde{w}\|_{L^2(|x|\le R)} + \mathcal{O}(e^{-R}\hat{\rho}_R(w,v,\eta)),$$

and since $w_x = \tilde{w} - \sqrt{2}uw$ we obtain similar estimates for the derivatives w_x and w_{xx}, which altogether give the upper bound (7.46). Summarizing, we have shown

$$\begin{aligned}
\int_{\mathbb{R}} B_0(w)\chi_R(x)dx \ge C\int_{|x|\le R}[w_{xx}^2 + w_x^2 + w^2]dx - \frac{C}{R}\hat{\rho}_R(w,v,\eta)^2 \\
+ \mathcal{O}(R^3\hat{\rho}_R(w,v,\eta)^3 + e^{-R}\hat{\rho}_R(w,v,\eta)^2),
\end{aligned}$$

where we have used $\hat{\rho}_R(w,v,\eta) \le C_0\rho_R(\psi,u) \le C_0$ from Step 2. Now, combining the previous estimates and taking $R \ge 1$ sufficiently large, we arrive at the lower bound for $\Lambda(\psi) - \Lambda(u)$. □

STEP 4. Fix $R \ge 1$. There exist $\epsilon_0 \in (0,1)$ and $C_0 \ge 1$ such that, if $\psi = u + w + iv$ satisfies $\rho_R(\psi,u) \le \epsilon_0$ and if $\langle \partial_x u, w\rangle = \langle \partial_x^2 u, v\rangle = 0$, then

$$C_0^{-1}\rho_R(\psi,u)^2 \le \Lambda(\psi) - \Lambda(u) \le C_0\rho_R(\psi,u)^2. \tag{7.47}$$

PROOF OF STEP 4. Choose $R \ge 1$ large enough so that the conclusion of Step 3 holds, and $\rho_0 > 0$ small enough so that $R^3\rho_0 \ll C_0$, where C_0 is defined in Step 3. Take $\epsilon_0 \in (0,1)$ such that $C_0\epsilon_0 \le \rho_0$, where C_0 is defined in Step 2. If $\psi = u+w+iv$ satisfies $\rho_R(\psi,u) \le \epsilon_0$ and $\langle \partial_x u, w\rangle = \langle \partial_x^2 u, v\rangle = 0$, it follows from Step 2 that the quantity $\hat{\rho}_R(w,v,\eta)$ satisfies $\hat{\rho}_R(w,v,\eta) \le \rho_0$. By Step 3, we thus have

$$\frac{1}{2}C_0\hat{\rho}_R(w,v,\eta)^2 \le \Lambda(\psi) - \Lambda(u) \le C_1\hat{\rho}_R(w,v,\eta)^2,$$

where the upper bound can be established by a much simpler argument (which does not use orthogonality conditions). Since $\hat{\rho}_R(w,v,\eta)$ is equivalent to $\rho_R(\psi,u_0)$ by Step 2, we obtain (7.47). The estimate holds for any $R \ge 1$ because different values of R give equivalent distances ρ_R on $\mathcal{X}$. □

END OF THE PROOF. The initial-value problem for the NLS equation (7.19) is globally well-posed in the space $\mathcal{X}$ [**102, 253**]. Given any $\epsilon > 0$, we take

$$\delta = \frac{1}{2C_0} \min(2\epsilon, \epsilon_0),$$

where $C_0 \geq 1$ is defined in Step 4 and $\epsilon_0 > 0$ is the smallest between Steps 1 and 4. If $\psi_0 \in \mathcal{X}$ satisfies $\rho_R(\psi_0, u) \leq \delta$, then $\Lambda(\psi_0) - \Lambda(u) \leq C_0\delta^2$ by the upper bound in (7.47), which does not require orthogonality conditions. Since Λ is a conserved quantity, we deduce that the solution $\psi(t, \cdot)$ of the cubic NLS equation (7.19) with initial data ψ_0 satisfies $\Lambda(\psi(t, \cdot)) - \Lambda(u) \leq C_0\delta^2$ for all $t \geq 0$. We claim that, for all $t \geq 0$, we have

$$\inf_{\xi,\alpha\in\mathbb{R}} \rho_R\Big(e^{i\alpha}\psi(t, \cdot + \xi), u\Big) \leq 2C_0\delta \leq \epsilon_0. \tag{7.48}$$

Indeed, the bound (7.48) holds for $t = 0$ by assumption. Since the solutions of the cubic NLS equation (7.19) with initial data $\psi_0 \in \mathcal{X}$ depend continuously on time with respect to the distance $\rho_R(\psi, u)$ [**102, 253**], the left-hand side of the bound (7.48) is a continuous function of t so that the estimate holds for $t \in [0, t_0]$ with some $t_0 > 0$. By Step 1 we can find $\xi(t), \alpha(t) \in \mathbb{R}$ for $t \in [0, t_0]$ such that the solution $\psi(t, \cdot)$ can be decomposed as

$$e^{i\alpha(t)}\psi(t, x + \xi(t)) = u(x) + w(t, x) + iv(t, x), \qquad (t, x) \in [0, t_0] \times \mathbb{R},$$

where the perturbations w, v are real-valued and satisfy the orthogonality conditions

$$\langle \partial_x u, w(t, \cdot)\rangle = 0, \qquad \langle \partial_x^2 u, v(t, \cdot)\rangle = 0, \qquad t \in [0, t_0].$$

Applying Step 4 we deduce that

$$\begin{aligned} C_0^{-1}\rho_R(e^{i\alpha(t)}\psi(t, x + \xi(t)), u)^2 &\leq \Lambda(e^{i\alpha(t)}\psi(t, x + \xi(t))) - \Lambda(u) \\ &= \Lambda(\psi_0) - \Lambda(u) \leq C_0\delta^2, \end{aligned}$$

so that $\rho_R(e^{i\alpha(t)}\psi(t, x + \xi(t)), u_0) \leq C_0\delta$. Using again a continuity argument, we conclude that $[0, t_0]$ can be extended to $[0, \infty)$ so that the bound (7.48) holds for all $t \geq 0$. Using Step 1, we thus obtain modulations parameters $\xi(t)$, $\alpha(t)$ such that

$$\rho_R(e^{i\alpha(t)}\psi(t, \cdot + \xi(t))\,, u) \leq C_0\delta \leq \epsilon, \qquad t \geq 0,$$

which concludes the proof. □

REMARK 7.23. Except for the use of a different distance ρ_R, which controls the perturbations in the topology of $H^2_{\text{loc}}(\mathbb{R})$, Theorem 7.3 is the exact analogue of the orbital stability result of Theorem 7.2. However, the proof is quite different, and in some sense simpler, because the profile u of the black soliton is an unconstrained local minimizer of the higher-order functional Λ.

REMARK 7.24. Since the black soliton with the profile u is the critical point of the functional Λ, we have the following expansion for the perturbation $w + iv \in H^2(\mathbb{R})$:

$$\Lambda(u + w + iv) - \Lambda(u) = \langle K_+w, w\rangle + \langle K_-v, v\rangle + N(w, v),$$

where $N(w, v)$ collects all terms that are at least cubic in w and v in $H^2(\mathbb{R})$. However, this decomposition is not sufficient to prove the orbital stability of the black soliton because the quadratic terms are nonnegative, but they are degenerate in the sense that they do not control the $L^2(\mathbb{R})$ norm of v, as can be seen from the lower bound (7.38). This is due to the fact that the operator K_- has essential spectrum touching the origin, with generalized eigenfunctions corresponding to slow modulations of the phase of the black soliton. As is clear from the proof of Lemma 7.11, one cannot even prove that $v \in L^\infty(\mathbb{R})$ if we only know that $\langle K_- v, v\rangle < \infty$. This in turn makes it impossible to control the nonlinearity $N(u,v)$ in terms of the quadratic terms $\langle K_+ w, w\rangle + \langle K_- v, v\rangle$. There are good reasons to believe that the above problem is not just a technical one, and that the H^2 topology for the perturbations $w + iv$ is not appropriate to prove orbital stability of the black soliton. Indeed, a small, spatially localized perturbation of the black soliton can lead to the formation of the dark soliton (7.28) with a small nonzero speed $c \in (-\sqrt{2}, \sqrt{2})$. If this happens, the perturbations $w + iv$ cannot stay bounded in $L^2(\mathbb{R})$ for all times because $u_c - u \notin L^2(\mathbb{R})$ if v is nonzero.

REMARK 7.25. Similar to the case of the periodic waves in Lemma 4.3, the modulation parameters $\xi(t), \alpha(t)$ in the decomposition

$$e^{i\alpha(t)}\psi(t, x + \xi(t)) = u(x) + w(t,x) + iv(t,x), \qquad (t,x) \in \mathbb{R}_+ \times \mathbb{R},$$

subject to the orthogonality conditions

$$\langle \partial_x u, w(t,\cdot)\rangle = 0, \qquad \langle \partial_x^2 u, v(t,\cdot)\rangle = 0, \qquad t \in \mathbb{R}_+,$$

are continuously differentiable functions of $t \in \mathbb{R}_+$ satisfying

$$|\dot{\xi}(t)| + |\dot{\alpha}(t)| \leq C\epsilon, \quad t \in \mathbb{R}_+,$$

for some positive constant C.

7.4. Spectral stability and instability

A practical way to escape delicacies of the functional-analytic framework and to provide robust criteria for stability or instability of the black solitons is to consider the concept of the spectral stability which relies on the linearized equations of motion. In the spectral stability case, the results would provide the first step towards the nonlinear stability analysis since the spectral stability does not imply the nonlinear (orbital) stability. In the spectral instability case, it is quite possible that the nonlinear instability of the black solitons can be deduced from the linear instability similar to Theorem 2.4.

Since the domain walls are energy minimizers, they are spectrally stable. On the other hand, the black solitons are saddle points of energy. They can be orbitally stable if they are constrained energy minimizers, as in the case of the cubic defocusing NLS equation (7.19), see Theorem 7.2. They can also be saddle points of energy under the constraint of fixed renormalized momentum, in which case they are expected to be spectrally unstable.

REMARK 7.26. The criterion of Theorem 2.7 is expected to play the role in order to distinguish the spectral stability and instability of the black solitons, where the slope condition (2.58) is to be computed from the renormalized momentum. This slope condition may have been missed in the proof of Theorem 7.2 for the particular NLS equation (7.19). However, the same criterion plays the role in getting $\Delta(c) = \sqrt{2}c^2 + \mathcal{O}(c^4)$ and $\Delta''(c) = \frac{d}{dc}\hat{P}(u_c)$, see (7.31).

Here we relate the criterion for spectral stability of the black solitons with the constrained energy minimization under fixed renormalized momentum. To explain this concept in more generality, we shall consider the generalized NLS equation

$$i\psi_t = -\psi_{xx} + F'(|\psi|^2)\psi, \tag{7.49}$$

where $F'(|\psi|^2) : [0,\infty) \to \mathbb{R}$ and $F \in C^2([0,\infty),\mathbb{R})$. The black soliton with the spatial profile $u(x) : \mathbb{R} \to \mathbb{R}$ satisfying $u(x) \to \pm 1$ as $x \to \pm\infty$ is a solution of the form $\psi(t,x) = u(x)e^{-itF'(1)}$. From Example 2.4, the black soliton with the profile u exists if $W(u^2) := F(u^2) - u^2F'(1)$ satisfies the following properties:

- $W(1) = 0$, $W'(1) = 0$, $W''(1) > 0$,
- $W(u^2) > 0$ for $u \in (-1,1)$.

By using the perturbation in the form

$$\psi(t,x) = e^{-itF'(1)}\left[u(x) + w(t,x) + iv(t,x)\right] \tag{7.50}$$

and separating variables as

$$w(t,x) = W(x)e^{\lambda t}, \quad v(t,x) = V(x)e^{\lambda t},$$

we obtain the following spectral stability problem

$$L_- V = \lambda W, \quad L_+ W = -\lambda V, \tag{7.51}$$

where $\lambda \in \mathbb{C}$ is the spectral parameter and $L_\pm : \mathrm{Dom}(L_\pm) \subset L^2(\mathbb{R}) \to L^2(\mathbb{R})$ are self-adjoint operators in the form

$$\begin{aligned} L_+ &= -\partial_x^2 + F'(u^2) + 2u^2F''(u^2) - F'(1), \\ L_- &= -\partial_x^2 + F'(u^2) - F'(1). \end{aligned}$$

The self-adjoint operators are associated with the energy

$$H(\psi) = \int_{\mathbb{R}} \left[|\psi_x|^2 + F(|\psi|^2) + F'(1)(1 - |\psi|^2)\right] dx$$

defined in the energy space X and formally expanded at the perturbation (7.50) up to the quadratic terms:

$$H(\psi) - H(u) = \langle L_+ w, w\rangle + \langle L_- v, v\rangle + R(w,v),$$

where $R(w,v)$ is the remainder term beyond the quadratic terms.

Definition 7.2. We say that the black soliton is spectrally unstable if the spectral problem (7.51) admits an isolated eigenvalue $\lambda_0 \in \mathbb{C}$ such that $\mathrm{Re}(\lambda_0) > 0$ with the associated eigenfunction $(W,V) \in \mathrm{Dom}(L_+)\times\mathrm{Dom}(L_-)$. The black soliton is considered to be spectrally stable if no such eigenvalue λ exists.

In the case of the cubic defocusing NLS equation (7.19), we get correspondence with the general case (7.49) for $F'(|\psi|^2) = |\psi|^2$ so that we can use

$$F(|\psi|^2) = \frac{1}{2}(|\psi|^4 - 1) \quad \Rightarrow \quad F(|\psi|^2) + F'(1)(1 - |\psi|^2) = \frac{1}{2}(1 - |\psi|^2)^2.$$

The black soliton has the profile

$$u(x) := \tanh\left(\frac{x}{\sqrt{2}}\right)$$

and the spectral stability problem (7.51) is defined with the two operators:

$$\begin{aligned} L_+ &= -\partial_x^2 + 3u^2 - 1, \\ L_- &= -\partial_x^2 + u^2 - 1. \end{aligned}$$

For practical purposes, as in Lemma 7.5, we can assume that L_+ is a non-negative operator with a simple zero eigenvalue isolated from the strictly positive spectrum due to the translational mode $\text{Ker}(L_+) = \text{Span}\{\partial_x u\}$, whereas L_- is a sign-indefinite operator with only one simple negative eigenvalue and the essential spectrum located on $[0,\infty)$ with $L_- u = 0$ for $u \in L^\infty(\mathbb{R})$ and $u \notin L^2(\mathbb{R})$. The following theorem gives the criterion for spectral stability and instability of black solitons based on analysis of the spectral stability problem (7.51). It was proven in [**86**] and in [**217**].

THEOREM 7.4. *Assume that L_+ and L_- are the same as in Lemma 7.5. The black soliton is spectrally stable if $\langle \partial_x u, V_u\rangle \leq 0$ and spectrally unstable if $\langle \partial_x u, V_u\rangle > 0$, where $V_u \in L^\infty(\mathbb{R})$ is a solution of the linear inhomogeneous equation $L_- V_u = \partial_x u$.*

PROOF. By Fredholm's theory, if there exists a solution (W,V) of the spectral problem (7.51) in $\text{Dom}(L_+)\times\text{Dom}(L_-)$ for $\lambda \neq 0$, then the component V satisfies the orthogonality condition $\langle \partial_x u, V\rangle = 0$. We will show that an eigenvalue $\lambda_0 \in \mathbb{C}\backslash i\mathbb{R}$ exists if and only if $\langle \partial_x u, V_u\rangle > 0$. Since, for every eigenvalue $\lambda_0 \in \mathbb{C}$ with the eigenvector (W,V), $-\lambda_0 \in \mathbb{C}$ is also an eigenvalue with the eigenvector $(W,-V)$, this will prove the theorem.

For any eigenvalue $\lambda_0 \neq 0$, the corresponding eigenvector $(W,V) \in \text{Dom}(L_+)\times \text{Dom}(L_-)$ must satisfy $\langle \partial_x u, V\rangle = 0$. One can select the component $W \in \text{Dom}(L_+)$ uniquely by requiring that $\langle \partial_x u, W\rangle = 0$ after adding a multiple of $\partial_x u$ to W. With this convention, the second equation of the system (7.51) can be rewritten in the form

$$W = -\lambda_0 (L_+|_{\{\partial_x u\}^\perp})^{-1} V, \quad L_+|_{\{\partial_x u\}^\perp} : \text{Dom}(L_+) \subset L^2(\mathbb{R})|_{\{\partial_x u\}^\perp} \to L^2(\mathbb{R})|_{\{\partial_x u\}^\perp}.$$

Substituting this formula into the first equation of the system (7.51) yields

$$L_- V = -\lambda_0^2 (L_+|_{\{\partial_x u\}^\perp})^{-1} V.$$

Since this generalized eigenvalue problem is self-adjoint and

$$(L_+|_{\{\partial_x u\}^\perp})^{-1} : L^2(\mathbb{R}) \to L^2(\mathbb{R})$$

is a strictly positive operator by the assumption, we obtain $\lambda_0^2 \in \mathbb{R}$, i.e., $\lambda_0 \in \mathbb{R}\cup i\mathbb{R}$. Therefore, λ_0 is an eigenvalue in $\mathbb{C}\backslash i\mathbb{R}$ if and only if $-\lambda_0^2 < 0$. Consequently, such an eigenvalue exists if and only if

$$\inf_{\substack{V \in X\backslash\{0\} \\ \langle \partial_x u, V\rangle = 0}} \frac{\langle L_- V, V\rangle}{\langle (L_+|_{\{\partial_x u\}^\perp})^{-1} V, V\rangle} < 0, \tag{7.52}$$

where X is the energy space which is the form domain for the operator $L_- : \text{Dom}(L_-) \subset L^2(\mathbb{R}) \to L^2(\mathbb{R})$. Since the denominator is strictly positive, the sign of the left-hand side of (7.52) is determined by the sign of $-\mu_0^2$ in a simpler variational problem

$$-\mu_0^2 = \inf_{\substack{V \in X\backslash\{0\} \\ \langle \partial_x u, V\rangle = 0}} \frac{\langle L_- V, V\rangle}{\|V\|^2}. \tag{7.53}$$

As in Theorem 2.7, the sign of $-\mu_0^2$ in (7.53) depends on the sign of

$$\begin{aligned}\lim_{\lambda\to 0^-} ((L_- - \lambda I)^{-1}\partial_x u, \partial_x u) &= \lim_{\lambda \to 0^-} ((L_- - \lambda I)^{-1} L_- V_u, \partial_x u) \\ &= \langle V_u, \partial_x u\rangle.\end{aligned}$$

Hence we have the following dichotom:

(i) $-\mu_0^2 \geq 0$ if $\langle \partial_x u, V_u \rangle \leq 0$,
(ii) $-\mu_0^2 < 0$ if $\langle \partial_x u, V_u \rangle > 0$.

In case (i), the condition (7.52) is not satisfied so that no isolated eigenvalue λ_0 with $\text{Re}(\lambda_0) > 0$ of the spectral problem (7.51) exists. This is the spectral stability case which corresponds to $\langle \partial_x u, V_u \rangle \leq 0$. In case (ii), the condition (7.52) is satisfied so that there exists an isolated positive eigenvalue $\lambda_0 \in \mathbb{R}$. This is the spectral instability case which corresponds to $\langle \partial_x u, V_u \rangle > 0$. □

EXAMPLE 7.4. An example where the spectral stability and instability of the black soliton in Theorem 7.4 can change was given in [**219**] in the framework of the cubic defocusing NLS equation with the regularized dispersion:

$$i(1 - \epsilon^2 \partial_x^2)\psi_t + \psi_{xx} + 2(1 - |\psi|^2)\psi = 0. \tag{7.54}$$

The black soliton exists in the form $\psi(t, x) = u(x) := \tanh(x)$. The spectral stability problem can be derived by the substitution $\psi(t, x) = u(x) + w(t, x) + iv(t, x)$ and separation of variables. Compared to (7.51), the spectral stability problem is modified to the form:

$$L_- V = \lambda(1 - \epsilon^2 \partial_x^2) W, \qquad L_+ W = -(1 - \epsilon^2 \partial_x^2) V, \tag{7.55}$$

with

$$\begin{aligned} L_+ &= -\partial_x^2 + 4 - 6\,\text{sech}^2(x), \\ L_- &= -\partial_x^2 - 2\,\text{sech}^2(x), \end{aligned}$$

which are the same linearized operators as above, up to the scaling transformation. Due to the positive operator $(1 - \epsilon^2 \partial_x^2)$ in (7.55), we shall use the inner product

$$\langle f, g \rangle_\epsilon := \int_{\mathbb{R}} \left[\bar{f}(x) g(x) + \epsilon^2 f'(x) g'(x) \right] dx$$

instead of the $L^2(\mathbb{R})$ inner product. The stability criterion of Theorem 7.4 is applied with $\langle \partial_x u, V_u \rangle_\epsilon$, where V_u is a bounded solution of the linear inhomoegeneous equation

$$L_- V_u = (1 - \epsilon^2 \partial_x^2) \partial_x u.$$

A general solution V_u can be found in the explicit form by substitutions as

$$V_u(x) = -\frac{1}{2}(1 + 2\epsilon^2) + \frac{3}{2}\epsilon^2\,\text{sech}^2(x) + c_1 \tanh(x) + c_2 \left[x \tanh(x) - 1 \right],$$

where c_1 and c_2 are arbitrary constants. If V_u is required to be bounded, then $c_2 = 0$. The constant c_1 does not play any role due to the odd parity of u. Although $V_u \notin L^2(\mathbb{R})$, the inner product $\langle \partial_x u, V_u \rangle_\epsilon$ makes sense due to the exponential decay of $u'(x) \to 0$ as $|x| \to \infty$ and can be computed explicitly:

$$\begin{aligned} \langle \partial_x u, V_u \rangle_\epsilon &= \frac{1}{2} \int_{\mathbb{R}} \left[(1 - 4\epsilon^2)\,\text{sech}^2(x) + 6\epsilon^2\,\text{sech}^4(x) \right] \left[-(1 + 2\epsilon^2) + 3\epsilon^2\,\text{sech}^2(x) \right] dx \\ &= -(1 + 2\epsilon^2)(1 - 4\epsilon^2) + 2\epsilon^2(1 - 4\epsilon^2) - 4\epsilon^2(1 + 2\epsilon^2) + \frac{48}{5}\epsilon^4 \\ &= -1 + \frac{8}{5}\epsilon^4. \end{aligned}$$

Hence $\langle \partial_x u, V_u \rangle_\epsilon \leq 0$ if $\epsilon \in [0, \epsilon_0]$ and $\langle \partial_x u, V_u \rangle_\epsilon > 0$ if $\epsilon \in (\epsilon_0, \infty)$, where $\epsilon_0 := (5/8)^{1/4}$. By Theorem 7.4, the black soliton is spectrally stable if $\epsilon \in [0, \epsilon_0]$ and spectrally unstable if $\epsilon \in (\epsilon_0, \infty)$.

REMARK 7.27. Depending on the properties of the nonlinear function $F'(|\psi|^2)$ in (7.49), there may exist a different standing wave solution $\psi(t,x) = u(x)e^{-itF'(1)}$ satisfying the boundary conditions $u(x) \to +1$ for $|x| \to \infty$. Such a solution is usally called the bubble, not the black soliton. The existence of the bubble follows from the ODE theory of the Newton equation (see Lemma 2.3) for $W(u^2) := F(u^2) - u^2F'(1)$ to satisfy the properties

- $W(1) = 1$, $W'(1) = 0$, $W''(1) > 0$,
- there exists $u_0 \in (0,1)$ such that $W(u_0^2) = 0$ and $W'(u_0^2) > 0$,
- $W(u^2) > 0$ for $u \in (u_0, 1)$.

The spectral stability problem is still given by (7.51) but $L_+ \partial_x u = 0$ with $\partial_x u(x)$ having exactly one zero on $\mathbb{R}$ suggests that L_+ has exactly one (simple) negative eigenvalue. At the same time $L_- u = 0$ with $u(x)$ having no zeros on $\mathbb{R}$ suggests that L_- has the essential spectrum on $[0,\infty)$ and no negative eigenvalues. The constraint $\langle \partial_x u, V \rangle = 0$ for every eigenvector $(W,V) \in \mathrm{Dom}(L_+) \times \mathrm{Dom}(L_-)$ with $\lambda \neq 0$ does not affect the number of negative eigenvalues of L_+. Thus, there exists exactly one real positive eigenvalue λ of the spectral stability problem (7.51) for the bubbles by a generalization of the Hamiltonian–Krein theorem (Theorem 1.8) developed in [**180**]. Hence, the bubble is spectrally unstable. This conclusion was derived in [**77**] and confirmed in [**178**, **217**].

REMARK 7.28. The case $\epsilon = 0$ in Example 7.4 corresponds to the cubic defocusing NLS equation (7.19) up to the scaling transformation. Since $\langle \partial_x u, V_u \rangle < 0$ for $\epsilon = 0$, the black soliton is spectrally stable, in agreement with the orbital stability result of Theorem 7.2. We can further relate the criterion of $\langle \partial_x u, V_u \rangle < 0$ with the slope condition for the renormalized momentum (7.24). Recall that the traveling dark soliton solution with the profile u_c in (7.28) is a solution of the differential equation

$$u_c''(x) - icu_c'(x) - |u_c(x)|^2 u_c(x) = 0. \tag{7.56}$$

Since the mapping $(-\sqrt{2}, \sqrt{2}) \ni c \mapsto u_c \in X$ is differentiable in c and $u_{c=0} = u$ is real, where X is given by (7.22), we obtain by differentiating of (7.56) with respect to c that

$$L_- \partial_c \operatorname{Im}(u_c)|_{c=0} = -\partial_x \operatorname{Re}(u_c)|_{c=0}.$$

Hence $V_u := -\partial_c \operatorname{Im}(u_c)|_{c=0}$ so that

$$\begin{aligned}\langle \partial_x u, V_u \rangle &= -\langle \partial_x \operatorname{Re}(u_c)|_{c=0}, \partial_c \operatorname{Im}(u_c)|_{c=0} \\ &= -\frac{1}{2}\partial_c \hat{P}(u_c)|_{c=0},\end{aligned}$$

where we have used

$$\hat{P}(u_c) = \int_{\mathbb{R}} \frac{1-|u_c|^2}{|u_c|^2} \left[\operatorname{Re}(u_c)\partial_x \operatorname{Im}(u_c) - \operatorname{Im}(u_c)\partial_x \operatorname{Re}(u_c)\right] dx$$

and obtained by integration by parts that

$$\partial_c \hat{P}(u_c) = -2\int_{\mathbb{R}} \left[\partial_c \operatorname{Re}(u_c)\partial_x \operatorname{Im}(u_c) - \partial_c \operatorname{Im}(u_c)\partial_x \operatorname{Re}(u_c)\right] dx$$

with $\partial_c \mathrm{Re}(u_c)|_{c=0} = 0$. The integration by parts is justified in the renormalized momentum $\hat{P}(u_c)$ since $|u_c(x)|^2 \to 1$ as $|x| \to \infty$ exponentially fast. The stability criterion $\langle \partial_x u, V_u \rangle < 0$ of the black soliton is equivalent to the slope condition $\partial_c \hat{P}(u_c)|_{c=0} > 0$, which has been used to obtain (7.31) in the proof of Theorem 7.2.

Remark 7.28 points out the relation between the spectral stability criterion of the black soliton in Theorem 7.4 and the slope condition for the renormalized momentum $\hat{P}(u_c)$ computed at the family of the dark solitons in the limit $c \to 0$ and used in Theorem 7.2. This correspondence of the stability criterion and the slope condition can be extended to the entire family of dark solitons for every c defined in some existence interval.

In order to define the family of dark solitons in the generalized NLS equation (7.49), we consider traveling wave solutions of the form $\psi(t,x) = u_c(x-ct)e^{-iF'(1)t}$ with the spatial profile $u_c(x) : \mathbb{R} \to \mathbb{C}$ satisfying $|u_c(x)| \to 1$ as $x \to \pm\infty$. The profile is found from the differential equation

$$-u_c''(x) + \left[F'(|u_c(x)|^2) - F'(1)\right] u_c(x) + icu_c'(x) = 0, \quad x \in \mathbb{R}, \tag{7.57}$$

and we assume that the family of dark solitons exist in some interval $\mathcal{I} \subset \mathbb{R}$ which contains the point $c = 0$ with the black soliton profile $u_{c=0} \equiv u$. The profile u_c is found from the Euler–Lagrange equation for the augmented energy $\Lambda_c(\psi) := H(\psi) + c\hat{P}(\psi)$, where $H(\psi)$ and $\hat{P}(\psi)$ are the conserved energy and renormalized momentum of the generalized NLS equation (7.49). The renormalized momentum $\hat{P}(\psi)$ exists in the same form

$$\begin{aligned} \hat{P}(\psi) &= -\frac{i}{2}\int_{\mathbb{R}} \frac{1-|\psi|^2}{|\psi|^2}[\bar{\psi}\partial_x\psi - \psi\partial_x\bar{\psi}]dx \\ &= \frac{i}{2}\int_{\mathbb{R}} [\bar{\psi}\partial_x\psi - \psi\partial_x\bar{\psi}]dx + \arg(\psi)|_{x\to-\infty}^{x\to+\infty}, \end{aligned} \tag{7.58}$$

as the one given in (7.24) and (7.25) for the cubic defocusing NLS equation (7.19). By using the perturbation in the form

$$\psi(t,x) = e^{-itF'(1)}\left[u_c(x-ct) + \varphi(t,x-ct)\right]$$

and separating variables as

$$\varphi(t,x) = \phi_1(x)e^{\lambda t}, \quad \bar{\varphi}(t,x) = \phi_2(x)e^{\lambda t},$$

we obtain the following spectral stability problem

$$L_c \begin{bmatrix} \phi_1 \\ \phi_2 \end{bmatrix} = i\lambda\sigma_3 \begin{bmatrix} \phi_1 \\ \phi_2 \end{bmatrix} \tag{7.59}$$

where $\lambda \in \mathbb{C}$ is the spectral parameter, $\sigma_3 = \mathrm{diag}(1,-1)$, and $L_c : \mathrm{Dom}(L_c) \subset L^2(\mathbb{R}) \to L^2(\mathbb{R})$ is the self-adjoint operator in the form

$$\begin{aligned} L_c = &\left[-\partial_x^2 + F'(|u_c|^2) - F'(1)\right]\begin{bmatrix} 1 & 0 \\ 0 & 1 \end{bmatrix} \\ &+ \begin{bmatrix} ic\partial_x + |u_c|^2F''(|u_c|^2) & u_c^2F''(|u_c|^2) \\ \bar{u}_c^2F''(|u_c|^2) & -ic\partial_x + |u_c|^2F''(|u_c|^2) \end{bmatrix}. \end{aligned}$$

Using $u \equiv u_{c=0}$ at $c = 0$, we can see that $L_{c=0}$ can be diagonalized to the linear operators $L_\pm : \mathrm{Dom}(L_\pm) \subset L^2(\mathbb{R}) \to L^2(\mathbb{R})$, for which the result of Lemma 7.5 suggests that there exists only one (simple) negative eigenvalue of $L_{c=0}$ isolated from the essential spectrum located on $[0,\infty)$. Since $|u_c(x)| \to 1$ as $|x| \to \infty$

exponentially fast for every $c \in \mathcal{I} \subset \mathbb{R}$, the essential spectrum of L_c is located at the union of the following two closed sets

$$\overline{\{F''(1) + k^2 + \sqrt{[F''(1)]^2 + c^2k^2}, \ k \in \mathbb{R}\}} \\ \cup \overline{\{F''(1) + k^2 - \sqrt{[F''(1)]^2 + c^2k^2}, \ k \in \mathbb{R}\}}$$

which yields $[0, \infty)$ for every $c \in \mathcal{I} \subset \mathbb{R}$. It is also natural to assume that there exists only one (simple) negative eigenvalue of L_c for every $c \in \mathcal{I} \subset \mathbb{R}$. Under these conditions, the following theorem gives the spectral stability criterion for the dark solitons with the profile u_c in the spectral stability problem (7.59).

THEOREM 7.5. *Assume that the mapping $\mathcal{I} \ni c \mapsto u_c \in X$ is C^1. Assume that there exists only one (simple) negative eigenvalue of L_c isolated from the essential spectrum located on $[0, \infty)$ for every $c \in \mathcal{I}$. The dark soliton with the profile u_c is spectrally stable if $\partial_c \hat{P}(u_c) \geq 0$ and is spectrally unstable if $\partial_c \hat{P}(u_c) < 0$, where $\hat{P}$ is the renormalized momentum given by* (7.58).

PROOF. It follows from the translational symmetry that

$$L_c \begin{bmatrix} \partial_x u_c \\ \partial_x \bar{u}_c \end{bmatrix} = \begin{bmatrix} 0 \\ 0 \end{bmatrix},$$

where $(\partial_x u_c, \partial_x \bar{u}_c) \in \text{Dom}(L_c) \subset L^2(\mathbb{R})$. Hence every solution $(\phi_1, \phi_2) \in \text{Dom}(L_c)$ to the spectral stability problem (7.59) with $\lambda \neq 0$ satisfies the orthogonality constraint

$$\langle \partial_x u_c, i\phi_1 \rangle - \langle \partial_x \bar{u}_c, i\phi_2 \rangle = 0.$$

Differentiating (7.57) in c, we confirm that

$$L_c \begin{bmatrix} \partial_c u_c \\ \partial_c \bar{u}_c \end{bmatrix} = i\sigma_3 \begin{bmatrix} \partial_x u_c \\ \partial_x \bar{u}_c \end{bmatrix},$$

where $\partial_c u_c \in L^\infty(\mathbb{R})$ since the mapping $\mathcal{I} \ni c \mapsto u_c \in X$ is assumed to be C^1 and $\sigma_3 = \text{diag}(1, -1)$.. Using the exponential decay of $\partial_x u_c(x) \to 0$ as $|x| \to \infty$, we define

$$\begin{aligned}
&\lim_{\lambda \to 0^-} \langle (L_c - \lambda I)^{-1} i\sigma_3(\partial_x u_c, \partial_x \bar{u}_c), i\sigma_3(\partial_x u_c, \partial_x \bar{u}_c) \rangle \\
&\quad = \lim_{\lambda \to 0^-} \langle (L_c - \lambda I)^{-1} L_c(\partial_c u_c, \partial_c \bar{u}_c), i\sigma_3(\partial_x u_c, \partial_x \bar{u}_c) \rangle \\
&\quad = \langle \partial_c u_c, i\partial_x u_c \rangle - \langle \partial_c \bar{u}_c, i\partial_x \bar{u}_c \rangle \\
&\quad = -\frac{1}{2} \partial_c \hat{P}(u_c),
\end{aligned}$$

where the last identity can be confirmed from (7.58) after integration by parts.

Similarly to the proof of Theorem 7.4, the Rayleigh quotient for the linear operator L_c tells us that if $\partial_c \hat{P}(u_c) \geq 0$, then $L_c|_{\{i\sigma_3(\partial_x u_c, \partial_x \bar{u}_c)\}^\perp} \geq 0$. Since the quadratic form associated with $L_c|_{\{i\sigma_3(\partial_x u_c, \partial_x \bar{u}_c)\}^\perp}$ is a constant of motion for the time-evolution linearized system, the spectral stability problem (7.59) does not have any eigenvalue $\lambda \in \mathbb{C}$ with $\text{Re}(\lambda) \neq 0$ as the existence of such eigenvalue is in contradiction with the conservation of the positive quadratic form. Hence, the dark soliton is spectrally stable in this case. On the other hand, if $\partial_c \hat{P}(u_c) < 0$, then $L_c|_{\{i\sigma_3(\partial_x u_c, \partial_x \bar{u}_c)\}^\perp}$ admits exactly one (simple) negative eigenvalue. By a generalization of the Hamiltonian–Krein theorem (Theorem 1.7) developed in [**180**], there exists an eigenvalue of the spectral stability problem (7.59) with a real (positive) λ. Hence, the dark soliton is spectrally unstable in this case. $\square$

EXAMPLE 7.5. For the cubic defocusing NLS equation (7.19), we have obtained the exact expression for $\hat{P}(u_c)$ in the proof of Theorem 7.2. This expression is given by

$$\hat{P}(u_c) = \pi + c\sqrt{2-c^2} + 2\arctan\frac{c}{\sqrt{2-c^2}}.$$

Consequently, we compute

$$\begin{aligned}\partial_c \hat{P}(u_c) &= \frac{2(1-c^2)}{\sqrt{2-c^2}} + \frac{2}{\sqrt{2-c^2}}\\ &= 2\sqrt{2-c^2} > 0.\end{aligned}$$

The stability criterion of Theorem 7.5 is satisfied for every $c \in (-\sqrt{2}, \sqrt{2})$. By Theorem 7.5, the dark solitons of the cubic defocusing NLS equation (7.19) are spectrally stable for every $c \in (-\sqrt{2}, \sqrt{2})$.

REMARK 7.29. The assumption of Theorem 7.5 on the existence of only one (simple) negative eigenvalue of L_c isolated from the essential spectrum is not so easy to show for the generalized NLS equation (7.49). The linear operator L_c is not diagonalizable into two uncoupled Schrödinger operators for $c \neq 0$ compared to the case of black solitons for $c = 0$.

REMARK 7.30. Spectral stability of the dark solitons with the profile u_c can be extended to the orbital stability result in the energy space for the generalized NLS equation (7.49) under some assumptions. The first result of this kind was obtained in [**178**]. It was extended in [**22, 23, 52, 86**] and in the particular setting of the cubic defocusing NLS equation (7.19) in [**73, 103**].

REMARK 7.31. For the regularized NLS equation (7.54), the two conserved quantities of energy and momentum are given by

$$H(\psi) = \int_{\mathbb{R}} \left[|\psi_x|^2 + (1-|\psi|^2)^2\right] dx$$

and

$$P(\psi) = i\int_{\mathbb{R}} \left[(\bar{\psi}\psi_x - \bar{\psi}_x\psi) + \epsilon^2(\bar{\psi}_x\psi_{xx} - \bar{\psi}_{xx}\psi_x)\right] d\xi.$$

The energy $H(\psi)$ is well-defined for the perturbations $w(t,\cdot) + iv(t,\cdot)$ to the black soliton $u(x) = \tanh(x)$ in the Sobolev space $H^1(\mathbb{R})$ but the momentum $P(\psi)$ is only defined for the perturbations in the Sobolev space $H^2(\mathbb{R})$. The coercivity of the operators $L_\pm$ in the orbital stability analysis of Theorem 7.2 can only provide the bounds on the perturbations in the $H^1(\mathbb{R})$ norm. This controls the perturbations in $H^1(\mathbb{R})$ globally in time. However, the momentum is defined for the perturbations in $H^2(\mathbb{R})$ and hence, this argument does not prove the orbital stability of the black solitons in the regularized NLS equation (7.54) [**219**]. Similarly, there is no orbital stability proof of the dark solitons in the regularized NLS equation (7.54).

APPENDIX A

Jacobi elliptic functions and integrals

We introduce the elliptic integrals of the first and second kind, respectively:

$$F(\varphi;k) := \int_0^{\varphi} \frac{d\theta}{\sqrt{1-k^2\sin^2\theta}}, \quad E(\varphi;k) := \int_0^{\varphi} \sqrt{1-k^2\sin^2\theta}d\theta, \tag{A.1}$$

where $k \in (0,1)$ is the elliptic modulus. The complete elliptic integrals of the first and second kind are given, respectively, by

$$K(k) := F\left(\frac{\pi}{2};k\right) = \int_0^{\frac{\pi}{2}} \frac{d\theta}{\sqrt{1-k^2\sin^2\theta}} \tag{A.2}$$

and

$$E(k) := E\left(\frac{\pi}{2};k\right) = \int_0^{\frac{\pi}{2}} \sqrt{1-k^2\sin^2\theta}d\theta. \tag{A.3}$$

Jacobi elliptic functions $\mathrm{sn}(u;k) = \sin\varphi$ and $\mathrm{cn}(u;k) = \cos\varphi$ generalize the trigonometric sine and cosine functions and satisfy the fundamental relation

$$\mathrm{sn}^2(u;k) + \mathrm{cn}^2(u;k) = 1. \tag{A.4}$$

These functions are derived from the inversion of the elliptic integral of the first kind with $u = F(\varphi;k)$. Considered on $\mathbb{R}$, these functions are smooth and periodic with the period equal to $4K(k)$.

Jacobi elliptic function $\mathrm{dn}(u;k) = \sqrt{1-k^2\sin^2\varphi}$ is strictly positive and satisfies the fundamental relation

$$\mathrm{dn}^2(u;k) + k^2\,\mathrm{sn}^2(u;k) = 1. \tag{A.5}$$

Considered on $\mathbb{R}$, this function is smooth and periodic with the period equal to $2K(k)$. It follows from (A.1) with $u = F(\varphi;k)$ that $d\varphi = \mathrm{dn}(u;k)du$. Hence the elliptic integral of the second kind in (A.1) can be expressed as

$$E(\varphi;k) = \int_0^u \mathrm{dn}^2(s;k)ds,$$

from which it follows that $E(\varphi + K(k);k) = E(\varphi;k) + E(k)$.

The derivatives of the Jacobi elliptic functions are given by

$$\frac{d}{du}\begin{bmatrix}\mathrm{sn}(u;k)\\ \mathrm{cn}(u;k)\\ \mathrm{dn}(u;k)\end{bmatrix} = \begin{bmatrix}\mathrm{cn}(u;k)\ \mathrm{dn}(u;k)\\ -\mathrm{sn}(u;k)\ \mathrm{dn}(u;k)\\ -k^2\,\mathrm{sn}(u;k)\ \mathrm{cn}(u;k)\end{bmatrix}.$$

The second derivatives of the Jacobi elliptic functions are closed in view of relations (A.4) and (A.5), from which the Jacobi elliptic functions arise as particular periodic solutions of the following second-order differential equations:

$$\begin{cases} y(x) = \mathrm{sn}(x;k): & y'' + (1+k^2)y - 2k^2y^3 = 0, \\ y(x) = \mathrm{cn}(x;k): & y'' + (1-2k^2)y + 2k^2y^3 = 0, \\ y(x) = \mathrm{dn}(x;k): & y'' + (k^2-1)y + 2y^3 = 0. \end{cases}$$

In the limit $k \to 0$, the Jacobi elliptic functions degenerate into the trigonometric functions according to the asymptotic expansion:

$$\begin{cases} \mathrm{sn}(u;k) = \sin(u) + \frac{1}{16}\left[\sin(u) + \sin(3u) - 4u\cos(u)\right]k^2 + \mathcal{O}(k^4), \\ \mathrm{cn}(u;k) = \cos(u) + \frac{1}{16}\left[-\cos(u) + \cos(3u) + 4u\sin(u)\right]k^2 + \mathcal{O}(k^4), \\ \mathrm{dn}(u;k) = 1 - \frac{1}{2}k^2\sin^2(u) + \mathcal{O}(k^4). \end{cases}$$

In the limit $k \to 1$, the Jacobi elliptic functions degenerate into the hyperbolic functions according to the asymptotic expansion:

$$\begin{cases} \mathrm{sn}(u;k) = \tanh(u) + \frac{1}{4}\left[\sinh(u)\cosh(u) - u\right]\mathrm{sech}^2(u)(1-k^2) \\ \qquad + \mathcal{O}((1-k^2)^2), \\ \mathrm{cn}(u;k) = \mathrm{sech}(u) - \frac{1}{4}\left[\sinh(u)\cosh(u) - u\right]\tanh(u)\,\mathrm{sech}(u)(1-k^2) \\ \qquad + \mathcal{O}((1-k^2)^2), \\ \mathrm{dn}(u;k) = \mathrm{sech}(u) + \frac{1}{4}\left[\sinh(u)\cosh(u) + u\right]\tanh(u)\,\mathrm{sech}(u)(1-k^2) \\ \qquad + \mathcal{O}((1-k^2)^2). \end{cases}$$

The complete elliptic integrals $K(k)$ and $E(k)$ have similar expansions as $k \to 0$

$$K(k) = \frac{\pi}{2}\left[1 + \frac{1}{4}k^2 + \mathcal{O}(k^4)\right], \quad E(k) = \frac{\pi}{2}\left[1 - \frac{1}{4}k^2 + \mathcal{O}(k^4)\right], \tag{A.6}$$

and as $k \to 1$

$$K(k) = \log\frac{4}{1-k^2} + \frac{1}{4}\left[\log\frac{4}{1-k^2} - 1\right](1-k^2) + \mathcal{O}(|\log(1-k^2)|(1-k^2)^2),$$

$$E(k) = 1 + \frac{1}{2}\left[\log\frac{4}{1-k^2} - \frac{1}{2}\right](1-k^2) + \mathcal{O}(|\log(1-k^2)|(1-k^2)^2),$$

Furthermore, the derivatives of the complete elliptic integrals are given by

$$\frac{d}{dk}E(k) = \frac{E(k) - K(k)}{k}, \quad \frac{d}{dk}K(k) = \frac{E(k)}{k(1-k^2)} - \frac{K(k)}{k}. \tag{A.7}$$

It follows from (A.2) and (A.3) that the derivative of $K(k)$ is positive and the derivative of $E(k)$ is negative for $k \in (0,1)$. Hence the complete elliptic integrals satisfy the following bounds:

$$(1-k^2)K(k) < E(k) < K(k), \tag{A.8}$$

for every $k \in (0,1)$.

APPENDIX B

Spectral theory for linear operators

Let H be a Hilbert space equipped with the inner product $\langle\cdot,\cdot\rangle$ and the induced norm $\|\cdot\|$. A linear operator is a linear mapping $L : \mathrm{Dom}(L) \subseteq H \to H$, where $\mathrm{Dom}(L) = \{f \in H : Lf \in H\}$ is the domain of L. The **kernel** of L is defined by

$$\mathrm{Ker}(L) := \{f \in \mathrm{Dom}(L) : Lf = 0\} \subseteq H$$

and the **range** of L is defined by

$$\mathrm{Ran}(L) := \{Lf : f \in \mathrm{Dom}(L)\} \subseteq H.$$

The operator L is called **bounded** if

$$\|L\| := \sup_{\|f\|_H=1} \|Lf\|_H < \infty.$$

The operator L is **closed** if $\forall \{f_n\}_{n\in\mathbb{N}} \in \mathrm{Dom}(L)$ such that $f_n \to f$ in H and $Lf_n \to g$ in H for some $f, g \in H$, it follows that $f \in \mathrm{Dom}(L)$ and $Lf = g$.

The **adjoint operator** $L^* : \mathrm{Dom}(L^*) \subseteq H \to H$ for an unbounded densely defined operator L and its domain $\mathrm{Dom}(L^*)$ are defined by

$$\mathrm{Dom}(L^*) = \{g \in H : \langle Lf, g\rangle = \langle f, L^*g\rangle \ \forall f \in \mathrm{Dom}(L)\}.$$

The operator L is called **self-adjoint** if $L^* = L$. Moreover, since L^* is a closed operator, every self-adjoint operator $L = L^*$ is closed. Finally, we say that an operator L is **densely defined** if $\mathrm{Dom}(L)$ is dense in H.

DEFINITION B.1. We say that L is a Fredholm operator of index zero if

$$\dim(\mathrm{Ker}(L)) = \dim(\mathrm{Ker}(L^*)) < \infty$$

and $\mathrm{Ran}(L)$ is closed.

The following theorem gives a practical criterion to check if a given self-adjoint operator is the Fredholm operator of index zero.

THEOREM B.1 ([**216**, Appendix B.4]). *If L is a densely defined self-adjoint operator in H and the continuous spectrum of L is bounded away from zero, then L is a Fredholm operator of index zero.*

The following Fredholm Alternative Theorem gives the necessary and sufficient condition of existence of solutions of the inhomogeneous equation $Lg = f$ for a given $f \in H$ if L is a Fredholm operator on H.

THEOREM B.2 ([**216**, Appendix B.4]). *Let $L : \mathrm{Dom}(L) \subset H \to H$ be a Fredholm operator of index zero. There exists a solution $u \in \mathrm{Dom}(L)$ of the inhomogeneous equation $Lu = f$ for a given $f \in H$ if and only if $\langle f, v_0\rangle_H = 0$ for all $v_0 \in \mathrm{Ker}(L^*)$. Moreover, the solution is unique if $\langle u, w_0\rangle_H = 0$ for all $w_0 \in \mathrm{Ker}(L)$.*

Next we introduce the **spectrum** of a linear operator in Hilbert space H.

DEFINITION B.2. Let L be a linear operator on a Hilbert space H with dense domain $\mathrm{Dom}(L) \subset H$. The complex plane $\mathbb{C}$ is decomposed into the following two sets:

(1) The resolvent set $\rho(L)$, defined as

$$\begin{aligned}\rho(L) := \{\lambda \in \mathbb{C} : \ & \mathrm{Ker}(L - \lambda I) = \{0\}, \\ & \mathrm{Ran}(L - \lambda I) = H, \\ & (L - \lambda I)^{-1} : H \to H \text{ is bounded}\}.\end{aligned}$$

(2) The spectrum $\sigma(L)$, defined as $\sigma(L) := \mathbb{C} \setminus \rho(L)$, which is further decomposed into the following three disjoint sets:
   (a) the point spectrum

$$\sigma_p(L) := \{\lambda \in \sigma(A) : \ \mathrm{Ker}(L - \lambda I) \neq \{0\}\},$$

   (b) the residual spectrum

$$\sigma_r(L) := \{\lambda \in \sigma(A) : \ \mathrm{Ker}(L - \lambda I) = \{0\}, \ \mathrm{Ran}(A - \lambda I) \neq H\},$$

   (c) the continuous spectrum

$$\begin{aligned}\sigma_c(L) := \{\lambda \in \sigma(L) : \ & \mathrm{Ker}(L - \lambda I) = \{0\}, \\ & \mathrm{Ran}(L - \lambda I) = H, \\ & (L - \lambda I)^{-1} : H \to H \text{ is unbounded}\}.\end{aligned}$$

In the case if $L = L^*$ is also self-adjoint, then the following alternative definition for the spectrum of L is often used.

DEFINITION B.3. Let L be a linear, self-adjoint operator on a Hilbert space H with dense domain $\mathrm{Dom}(L) \subset H$. The spectrum $\sigma(L)$ can be alternatively decomposed into the two disjoint sets $\sigma_d(L)$ and $\sigma_{\mathrm{ess}}(L)$, which are defined as follows:

(1) discrete spectrum:

$$\sigma_d(L) := \{\lambda \in \sigma(L) : \quad \mathrm{Ker}(L - \lambda I) \neq \{0\}, \quad \dim \mathrm{Ker}(L - \lambda I) < \infty\}.$$

The elements $\lambda \in \sigma_d(L)$ are called eigenvalues of L and $u \in \mathrm{Ker}(L - \lambda I), u \neq 0$ is called eigenvector of L for λ and satisfies $Lu = \lambda u$.

(2) essential spectrum: $\sigma_{\mathrm{ess}}(L) := \sigma(L) \setminus \sigma_d(L)$.

REMARK B.1. The essential spectrum σ_{ess} in Definition B.3 includes the continuous spectrum σ_c and point eigenvalues in σ_p with infinite multiplicity in Definition B.2. The residual spectrum $\sigma_r(L)$ of a self-adjoint operator L is empty.

The following two theorems give useful properties of the linear self-adjoint operators L and its associated quadratic form $\langle Lu, u\rangle$. The first result is known as the **Spectral Theorem** and the second result is the **Minimax Principle**.

THEOREM B.3 ([**237**, Theorem 2.19 and Lemma 2.21]). *Let $L : \mathrm{Dom}(L) \subset H \to H$ be a linear, self-adjoint, and densely defined operator. Then the spectrum $\sigma(L)$ of L is real and the eigenvectors of L are orthogonal to each other. Moreover, in this case, the operator satisfies*

$$\langle Lf, f\rangle \geq \lambda_- \|f\|^2, \quad \forall f \in \mathrm{Dom}(L)$$

if and only if $\sigma(L) \subseteq [\lambda_-, \infty)$ *and we have that*

$$\|(L-\lambda I)^{-1} f\| \leq |\lambda - \lambda_-|^{-1} \|f\|, \quad \forall f \in H, \quad \forall \lambda < \lambda_-.$$

THEOREM B.4 ([**237**, Theorem 4.12]). *Let* $L : \mathrm{Dom}(L) \subset H \to H$ *be a linear, self-adjoint, and densely defined operator and let* $\lambda_1 \leq \lambda_2 \leq \ldots$ *be the eigenvalues of L below the essential spectrum and set* $\lambda_j = \inf \sigma_{\mathrm{ess}}(L)$ *for* $j > N$ *if the number of eigenvalues* N *is finite. Then*

$$\lambda_n(L) = \sup_{\psi_1, \ldots, \psi_{n-1}} \inf_{u \in U} \frac{\langle Lu, u \rangle}{\langle u, u \rangle},$$

where $U := \{u \in \mathrm{Dom}(L) : u \neq 0, u \in \mathrm{Span}\{\psi_1, \ldots, \psi_{n-1}\}^\perp\}$.

As long as λ_n is an eigenvalue, then the sup and inf in Theorem B.4 are in fact max and min. As a consequence, one has the following result which is **Rayleigh–Ritz Theorem**.

THEOREM B.5 ([**227**, Theorem XIII.3]). *Let* $L : \mathrm{Dom}(L) \subset H \to H$ *be a linear, self-adjoint, and densely defined operator with the spectrum* $\sigma(L)$ *bounded from below. Let* $Y \subset \mathrm{Dom}(L)$ *be an* n*-dimensional subspace. Let* P *be the orthogonal projection onto* Y *such that* $L|_Y = PLP$. *Let* $\{\lambda_j\}_j$ *be the eigenvalues of* $L|_Y$ *and let* $\{\mu_j\}_j$ *be the eigenvalues of* L, *in increasing order. Then*

$$\sigma(L|_Y) \geq \sigma(L)$$

and more specifically

$$\mu_1 \leq \lambda_1 \leq \mu_2 \leq \lambda_2 \leq \ldots.$$

This is the infinite dimensional analogue of **Cauchy interlacing theorem** (also known as **Poincaré separation theorem**).

THEOREM B.6 ([**20**]). *Let* A *be a symmetric* $n \times n$ *matrix and let* $B = PAP^*$, *where* P *is an orthogonal projection onto a subspace of dimension* $m \leq n$. *Denote by* $\{\alpha_1, \ldots, \alpha_n\}$ *the eigenvalues of* A, *and by* $\{\beta_1, \ldots, \beta_m\}$ *the eigenvalues of* B. *Then we have for all* $j \leq m$ *that*

$$\alpha_j \leq \beta_j \leq \alpha_{n-m+j}.$$

If $m = n-1$, *then* $\alpha_j \leq \beta_j \leq \alpha_{j+1}$.

The following result characterizes the essential spectrum of a self-adjoint operator and is referred as **Weyl's Theorem**.

THEOREM B.7 ([**216**, Theorem B.14]). *Let* $L : \mathrm{Dom}(L) \subset H \to H$ *be a linear, self-adjoint, and densely defined operator. Then,* $\lambda \in \sigma_{\mathrm{ess}}(L)$ *if and only if there exists a sequence* $\{u_n\}_{n \geq 1} \in \mathrm{Dom}(L)$ *such that* $\|u_n\|_H = 1$ *for all* $n \geq 1$, $u_n \rightharpoonup 0$ *in* H *as* $n \to \infty$, *and* $\lim_{n\to\infty} \|(L - \lambda I) u_n\|_H = 0$.

Theorem B.7 can be used to find the essential spectrum of a Schrödinger operator with an exponentially decaying potential.

THEOREM B.8 ([**216**, Lemma B.5]). *Let* $L : H^2(\mathbb{R}) \subset L^2(\mathbb{R}) \to L^2(\mathbb{R})$ *be the Schrödinger operator given by*

$$L := -\partial_x^2 + V(x),$$

where the potential $V \in L^\infty(\mathbb{R})$ *decays exponentially according to*

$$\lim_{|x|\to\infty} e^{k|x|}|V(x) - c| = C,$$

for some $c, k, C > 0$. *Then*

$$\sigma_{\text{ess}}(L) = [c, \infty).$$

Moreover, we have the following result on the point spectrum of a Schrödinger operator with an exponentially decaying potential.

THEOREM B.9 ([**216**, Corollary 4.1]). *Let* $L : H^2(\mathbb{R}) \subset L^2(\mathbb{R}) \to L^2(\mathbb{R})$ *be the same as in Theorem* B.8. *If* $\lambda \in \sigma_p(L)$, *then* λ *is simple. Moreover, if* λ *is the* n*th eigenvalue sorted in increasing order, then the eigenfunction has exactly* $n-1$ *zeros on* $\mathbb{R}$.

More generally, solutions of the spectral problem for a Schrödinger operator enjoys the following property referred to as **Sturm comparison theorem**.

THEOREM B.10 ([**216**, Theorem B.10]). *Let* $L : H^2(\mathbb{R}) \subset L^2(\mathbb{R}) \to L^2(\mathbb{R})$ *be the same as in Theorem* B.8. *Let* $u(x; V)$ *be a solution of* $Lu = \lambda u$ *such that*

$$\lim_{x\to-\infty} u(x; V) = 1.$$

Assume that $V_1(x) > V_2(x)$ *for all* $x \in \mathbb{R}$ *and* $u(x; V_2)$ *has* n *zeros on* $\mathbb{R}$. *Then,* $u(x; V_1)$ *has at most* n *zeros on* $\mathbb{R}$.

Finally, we give two versions of **Sylvester's law of inertia**, one for matrices and another one for linear operators.

THEOREM B.11 ([**234**]). *Two symmetric square matrices* A *and* B *of the same size have the same number of positive, negative and zero eigenvalues if and only if they are congruent, i.e., if there exists a nonsingular matrix* S *such that* $B = SAS^T$.

THEOREM B.12 ([**188**]). *Let* $L : \text{Dom}(L) \subset H \to H$ *be a linear, self-adjoint, and densely defined operator with the essential spectrum* $\sigma_{\text{ess}}(L)$ *bounded from below by a positive number. Let* $S : H \to H$ *be a bounded invertible operator. Then,* S^*LS *and* L *have the same dimensions of the negative and null invariant subspaces of* H.

Bibliography

[1] R. Adami, E. Serra, and P. Tilli, *NLS ground states on graphs*, Calc. Var. Partial Differential Equations **54** (2015), no. 1, 743–761, DOI 10.1007/s00526-014-0804-z. MR3385179

[2] A. Aftalion and C. Sourdis, *Interface layer of a two-component Bose-Einstein condensate*, Commun. Contemp. Math. **19** (2017), no. 5, 1650052, 46, DOI 10.1142/S0219199716500528. MR3670791

[3] N. N. Akhmediev, V. M. Eleonskiĭ, and N. E. Kulagin, *First-order exact solutions of the nonlinear Schrödinger equation* (Russian, with English summary), Teoret. Mat. Fiz. **72** (1987), no. 2, 183–196. MR915545

[4] S. Alama, L. Bronsard, A. Contreras, and D. E. Pelinovsky, *Domain walls in the coupled Gross-Pitaevskii equations*, Arch. Ration. Mech. Anal. **215** (2015), no. 2, 579–610, DOI 10.1007/s00205-014-0789-y. MR3294411

[5] M. A. Alejo, *Nonlinear stability of Gardner breathers*, J. Differential Equations **264** (2018), no. 2, 1192–1230, DOI 10.1016/j.jde.2017.09.035. MR3720842

[6] M. Á. Alejo and A. J. Corcho, *Orbital stability of the black soliton for the quintic Gross-Pitaevskii equation*, Rev. Mat. Iberoam. **40** (2024), no. 5, 1731–1780, DOI 10.4171/rmi/1467. MR4792547

[7] M. A. Alejo, L. Fanelli, and C. Muñoz, *The Akhmediev breather is unstable*, São Paulo J. Math. Sci. **13** (2019), no. 2, 391–401, DOI 10.1007/s40863-019-00145-4. MR4025573

[8] M. A. Alejo, L. Fanelli, and C. Muñoz, *Stability and instability of breathers in the $U(1)$ Sasa-Satsuma and nonlinear Schrödinger models*, Nonlinearity **34** (2021), no. 5, 3429–3484, DOI 10.1088/1361-6544/abea6c. MR4260801

[9] M. A. Alejo and C. Muñoz, *Nonlinear stability of MKdV breathers*, Comm. Math. Phys. **324** (2013), no. 1, 233–262, DOI 10.1007/s00220-013-1792-0. MR3116324

[10] M. A. Alejo and C. Muñoz, *Dynamics of complex-valued modified KdV solitons with applications to the stability of breathers*, Anal. PDE **8** (2015), no. 3, 629–674, DOI 10.2140/apde.2015.8.629. MR3353827

[11] G. Alves, F. Natali, and A. Pastor, *Sufficient conditions for orbital stability of periodic traveling waves*, J. Differential Equations **267** (2019), no. 2, 879–901, DOI 10.1016/j.jde.2019.01.029. MR3957975

[12] S. Amaral, H. Borluk, G. M. Muslu, F. Natali, and G. Oruc, *On the existence, uniqueness, and stability of periodic waves for the fractional Benjamin-Bona-Mahony equation*, Stud. Appl. Math. **148** (2022), no. 1, 62–98, DOI 10.1111/sapm.12428. MR4429038

[13] V. Ambrosio, *On some convergence results for fractional periodic Sobolev spaces*, Opuscula Math. **40** (2020), no. 1, 5–20, DOI 10.7494/opmath.2020.40.1.5. MR4075889

[14] J. Angulo Pava, *Nonlinear dispersive equations: Existence and stability of solitary and periodic traveling wave solutions*, Mathematical Surveys and Monographs, vol. 156, American Mathematical Society, Providence, RI, 2009, DOI 10.1090/surv/156. MR2567568

[15] J. Angulo Pava, *Nonlinear stability of periodic traveling wave solutions to the Schrödinger and the modified Korteweg-de Vries equations*, J. Differential Equations **235** (2007), no. 1, 1–30, DOI 10.1016/j.jde.2007.01.003. MR2309564

[16] J. A. Pava, *Stability properties of solitary waves for fractional KdV and BBM equations*, Nonlinearity **31** (2018), no. 3, 920–956, DOI 10.1088/1361-6544/aa99a2. MR3784994

[17] J. Angulo Pava, J. L. Bona, and M. Scialom, *Stability of cnoidal waves*, Adv. Differential Equations **11** (2006), no. 12, 1321–1374. MR2276856

[18] J. Angulo Pava and F. Natali, *On the instability of periodic waves for dispersive equations*, Differential Integral Equations **29** (2016), no. 9-10, 837–874. MR3513583

[19] L. Barreira and C. Valls, *Ordinary differential equations: Qualitative theory*, Graduate Studies in Mathematics, vol. 137, American Mathematical Society, Providence, RI, 2012. Translated from the 2010 Portuguese original by the authors, DOI 10.1090/gsm/137. MR2931599

[20] R. Bellman, *Introduction to matrix analysis*, Classics in Applied Mathematics, vol. 19, Society for Industrial and Applied Mathematics (SIAM), Philadelphia, PA, 1997. Reprint of the second (1970) edition; With a foreword by Gene Golub, DOI 10.1137/1.9781611971170. MR1455129

[21] H. Berestycki, T.-C. Lin, J. Wei, and C. Zhao, *On phase-separation models: asymptotics and qualitative properties*, Arch. Ration. Mech. Anal. **208** (2013), no. 1, 163–200, DOI 10.1007/s00205-012-0595-3. MR3021546

[22] F. Béthuel, P. Gravejat, J.-C. Saut, and D. Smets, *Orbital stability of the black soliton for the Gross-Pitaevskii equation*, Indiana Univ. Math. J. **57** (2008), no. 6, 2611–2642, DOI 10.1512/iumj.2008.57.3632. MR2482993

[23] F. Bethuel, P. Gravejat, and D. Smets, *Asymptotic stability in the energy space for dark solitons of the Gross-Pitaevskii equation* (English, with English and French summaries), Ann. Sci. Éc. Norm. Supér. (4) **48** (2015), no. 6, 1327–1381, DOI 10.24033/asens.2271. MR3429470

[24] D. Bilman and P. D. Miller, *A robust inverse scattering transform for the focusing nonlinear Schrödinger equation*, Comm. Pure Appl. Math. **72** (2019), no. 8, 1722–1805, DOI 10.1002/cpa.21819. MR3974953

[25] B. Birnir, H. P. McKean, and A. Weinstein, *The rigidity of sine-Gordon breathers*, Comm. Pure Appl. Math. **47** (1994), no. 8, 1043–1051, DOI 10.1002/cpa.3160470803. MR1288631

[26] J. L. Bona, Y. Liu, and N. V. Nguyen, *Stability of solitary waves in higher-order Sobolev spaces*, Commun. Math. Sci. **2** (2004), no. 1, 35–52, DOI 10.4310/cms.2004.v2.n1.a3. MR2082818

[27] J. L. Bona, P. E. Souganidis, and W. A. Strauss, *Stability and instability of solitary waves of Korteweg-de Vries type*, Proc. Roy. Soc. London Ser. A **411** (1987), no. 1841, 395–412. MR897729

[28] N. Bottman and B. Deconinck, *KdV cnoidal waves are spectrally stable*, Discrete Contin. Dyn. Syst. **25** (2009), no. 4, 1163–1180, DOI 10.3934/dcds.2009.25.1163. MR2552133

[29] N. Bottman, B. Deconinck, and M. Nivala, *Elliptic solutions of the defocusing NLS equation are stable*, J. Phys. A **44** (2011), no. 28, 285201, 24, DOI 10.1088/1751-8113/44/28/285201. MR2812335

[30] J. Bourgain, *Global solutions of nonlinear Schrödinger equations*, American Mathematical Society Colloquium Publications, vol. 46, American Mathematical Society, Providence, RI, 1999, DOI 10.1090/coll/046. MR1691575

[31] N. Boussaïd and A. Comech, *Nonlinear Dirac equation: Spectral stability of solitary waves*, Mathematical Surveys and Monographs, vol. 244, American Mathematical Society, Providence, RI, [2019] ©2019, DOI 10.1090/surv/244. MR3971006

[32] J. C. Bronski, M. A. Johnson, and T. Kapitula, *An index theorem for the stability of periodic traveling waves of Korteweg-de Vries type*, Proc. Roy. Soc. Edinburgh Sect. A **141** (2011), no. 6, 1141–1173, DOI 10.1017/S0308210510001216. MR2855892

[33] J. Bronski, M. A. Johnson, and T. Kapitula, *An instability index theory for quadratic pencils and applications*, Comm. Math. Phys. **327** (2014), no. 2, 521–550, DOI 10.1007/s00220-014-1949-5. MR3183408

[34] J. C. Bronski, V. M. Hur, and M. A. Johnson, *Modulational instability in equations of KdV type*, New approaches to nonlinear waves, Lecture Notes in Phys., vol. 908, Springer, Cham, 2016, pp. 83–133, DOI 10.1007/978-3-319-20690-5_4. MR3408757

[35] T. Buckmaster and H. Koch, *The Korteweg–de Vries equation at H^{-1} regularity*, Ann. Inst. H. Poincaré C Anal. Non Linéaire **32** (2015), no. 5, 1071–1098, DOI 10.1016/j.anihpc.2014.05.004. MR3400442

[36] T. Bühler and D. A. Salamon, *Functional analysis*, Graduate Studies in Mathematics, vol. 191, American Mathematical Society, Providence, RI, 2018, DOI 10.1090/gsm/191. MR3823238

[37] A. Calini and C. M. Schober, *Dynamical criteria for rogue waves in nonlinear Schrödinger models*, Nonlinearity **25** (2012), no. 12, R99–R116, DOI 10.1088/0951-7715/25/12/R99. MR2997697

[38] T. Candy, *Global existence for an L^2 critical nonlinear Dirac equation in one dimension*, Adv. Differential Equations **16** (2011), no. 7-8, 643–666. MR2829499

[39] R. Carles and D. Pelinovsky, *On the orbital stability of Gaussian solitary waves in the log-KdV equation*, Nonlinearity **27** (2014), no. 12, 3185–3202, DOI 10.1088/0951-7715/27/12/3185. MR3291147

[40] T. Cazenave, *Semilinear Schrödinger equations*, Courant Lecture Notes in Mathematics, vol. 10, New York University, Courant Institute of Mathematical Sciences, New York; American Mathematical Society, Providence, RI, 2003, DOI 10.1090/cln/010. MR2002047

[41] J. Chen and D. E. Pelinovsky, *Rogue periodic waves of the modified KdV equation*, Nonlinearity **31** (2018), no. 5, 1955–1980, DOI 10.1088/1361-6544/aaa2da. MR3816661

[42] J. Chen and D. E. Pelinovsky, *Periodic traveling waves of the modified KdV equation and rogue waves on the periodic background*, J. Nonlinear Sci. **29** (2019), no. 6, 2797–2843, DOI 10.1007/s00332-019-09559-y. MR4030403

[43] J. Chen and D. E. Pelinovsky, *Rogue waves on the background of periodic standing waves in the derivative nonlinear Schrödinger equation*, Phys. Rev. E **103** (2021), no. 6, Paper No. 062206, 25, DOI 10.1103/physreve.103.062206. MR4286591

[44] J. Chen, D. E. Pelinovsky, and J. Upsal, *Modulational instability of periodic standing waves in the derivative NLS equation*, J. Nonlinear Sci. **31** (2021), no. 3, Paper No. 58, 32, DOI 10.1007/s00332-021-09713-5. MR4255683

[45] J. Chen, D. E. Pelinovsky, and R. E. White, *Rogue waves on the double-periodic background in the focusing nonlinear Schrödinger equation*, Phys. Rev. E **100** (2019), no. 5, 052219, 18, DOI 10.1103/physreve.100.052219. MR4041411

[46] J. Chen, D. E. Pelinovsky, and R. E. White, *Periodic standing waves in the focusing nonlinear Schrödinger equation: rogue waves and modulation instability*, Phys. D **405** (2020), 132378, 13, DOI 10.1016/j.physd.2020.132378. MR4066770

[47] R. M. Chen, X. Liu, Y. Liu, and C. Qu, *Stability of the Camassa-Holm peakons in the dynamics of a shallow-water-type system*, Calc. Var. Partial Differential Equations **55** (2016), no. 2, Art. 34, 22, DOI 10.1007/s00526-016-0972-0. MR3472673

[48] R. M. Chen and D. E. Pelinovsky, *$W^{1,\infty}$ instability of H^1-stable peakons in the Novikov equation*, Dyn. Partial Differ. Equ. **18** (2021), no. 3, 173–197, DOI 10.4310/DPDE.2021.v18.n3.a1. MR4291309

[49] G. Chen, J. Liu, and B. Lu, *Long time asymptotics and stability for the sine-Gordon equation*, arXiv:2009.04260, 2020.

[50] C. Chicone, *The monotonicity of the period function for planar Hamiltonian vector fields*, J. Differential Equations **69** (1987), no. 3, 310–321, DOI 10.1016/0022-0396(87)90122-7. MR903390

[51] C. Chicone, *Ordinary differential equations with applications*, Texts in Applied Mathematics, vol. 34, Springer-Verlag, New York, 1999. MR1707333

[52] D. Chiron, *Stability and instability for subsonic traveling waves of the nonlinear Schrödinger equation in dimension one*, Anal. PDE **6** (2013), no. 6, 1327–1420, DOI 10.2140/apde.2013.6.1327. MR3148057

[53] G. Chong, Y. Fu, and H. Wang, *Instability of H^1-stable periodic peakons for the higher-order μ-Camassa-Holm equation*, J. Math. Anal. Appl. **531** (2024), no. 1, Paper No. 127830, 40, DOI 10.1016/j.jmaa.2023.127830. MR4654781

[54] S.-N. Chow and J. A. Sanders, *On the number of critical points of the period*, J. Differential Equations **64** (1986), no. 1, 51–66, DOI 10.1016/0022-0396(86)90071-9. MR849664

[55] M. Chugunova and D. Pelinovsky, *Count of eigenvalues in the generalized eigenvalue problem*, J. Math. Phys. **51** (2010), no. 5, 052901, 19, DOI 10.1063/1.3406252. MR2666968

[56] K. M. Claassen and M. A. Johnson, *Nondegeneracy and stability of antiperiodic bound states for fractional nonlinear Schrödinger equations*, J. Differential Equations **266** (2019), no. 9, 5664–5712, DOI 10.1016/j.jde.2018.10.033. MR3912763

[57] A. Comech and D. Pelinovsky, *Purely nonlinear instability of standing waves with minimal energy*, Comm. Pure Appl. Math. **56** (2003), no. 11, 1565–1607, DOI 10.1002/cpa.10104. MR1995870

[58] A. Constantin, *The Hamiltonian structure of the Camassa-Holm equation*, Exposition. Math. **15** (1997), no. 1, 53–85. MR1438436

[59] A. Constantin, *Nonlinear water waves with applications to wave-current interactions and tsunamis*, CBMS-NSF Regional Conference Series in Applied Mathematics, vol. 81,

Society for Industrial and Applied Mathematics (SIAM), Philadelphia, PA, 2011, DOI 10.1137/1.9781611971873. MR2867413

[60] A. Constantin and J. Escher, *Well-posedness, global existence, and blowup phenomena for a periodic quasi-linear hyperbolic equation*, Comm. Pure Appl. Math. **51** (1998), no. 5, 475–504, DOI 10.1002/(SICI)1097-0312(199805)51:5¡475::AID-CPA2¿3.0.CO;2-5. MR1604278

[61] A. Constantin and J. Escher, *Wave breaking for nonlinear nonlocal shallow water equations*, Acta Math. **181** (1998), no. 2, 229–243, DOI 10.1007/BF02392586. MR1668586

[62] A. Constantin and D. Lannes, *The hydrodynamical relevance of the Camassa-Holm and Degasperis-Procesi equations*, Arch. Ration. Mech. Anal. **192** (2009), no. 1, 165–186, DOI 10.1007/s00205-008-0128-2. MR2481064

[63] A. Constantin and L. Molinet, *Orbital stability of solitary waves for a shallow water equation*, Phys. D **157** (2001), no. 1-2, 75–89, DOI 10.1016/S0167-2789(01)00298-6. MR1854962

[64] A. Constantin and W. A. Strauss, *Stability of peakons*, Comm. Pure Appl. Math. **53** (2000), no. 5, 603–610, DOI 10.1002/(sici)1097-0312(200005)53:5¡603::aid-cpa3¿3.0.co;2-l. MR1737505

[65] A. Constantin and W. A. Strauss, *Stability of the Camassa-Holm solitons*, J. Nonlinear Sci. **12** (2002), no. 4, 415–422, DOI 10.1007/s00332-002-0517-x. MR1915943

[66] A. Contreras and D. Pelinovsky, *Stability of multi-solitons in the cubic NLS equation*, J. Hyperbolic Differ. Equ. **11** (2014), no. 2, 329–353, DOI 10.1142/S0219891614500106. MR3214610

[67] A. Contreras, D. E. Pelinovsky, and M. Plum, *Orbital stability of domain walls in coupled Gross-Pitaevskii systems*, SIAM J. Math. Anal. **50** (2018), no. 1, 810–833, DOI 10.1137/17M1114892. MR3757105

[68] A. Contreras, D. E. Pelinovsky, and Y. Shimabukuro, *L^2 orbital stability of Dirac solitons in the massive Thirring model*, Comm. Partial Differential Equations **41** (2016), no. 2, 227–255, DOI 10.1080/03605302.2015.1123272. MR3462129

[69] A. Contreras, D. E. Pelinovsky, and V. Slastikov, *Domain walls in the coupled Gross-Pitaevskii equations with the harmonic potential*, Calc. Var. Partial Differential Equations **61** (2022), no. 5, Paper No. 164, 28, DOI 10.1007/s00526-022-02277-6. MR4444177

[70] W. A. Coppel and L. Gavrilov, *The period function of a Hamiltonian quadratic system*, Differential Integral Equations **6** (1993), no. 6, 1357–1365. MR1235199

[71] S. Cuccagna, *On asymptotic stability in energy space of ground states of NLS in 1D*, J. Differential Equations **245** (2008), no. 3, 653–691, DOI 10.1016/j.jde.2008.02.042. MR2422523

[72] S. Cuccagna, *On instability of excited states of the nonlinear Schrödinger equation*, Phys. D **238** (2009), no. 1, 38–54, DOI 10.1016/j.physd.2008.08.010. MR2571965

[73] S. Cuccagna and R. Jenkins, *On the asymptotic stability of N-soliton solutions of the defocusing nonlinear Schrödinger equation*, Comm. Math. Phys. **343** (2016), no. 3, 921–969, DOI 10.1007/s00220-016-2617-8. MR3488549

[74] S. Cuccagna and M. Maeda, *On orbital instability of spectrally stable vortices of the NLS in the plane*, J. Nonlinear Sci. **26** (2016), no. 6, 1851–1894, DOI 10.1007/s00332-016-9322-9. MR3562397

[75] S. Cuccagna and D. E. Pelinovsky, *The asymptotic stability of solitons in the cubic NLS equation on the line*, Appl. Anal. **93** (2014), no. 4, 791–822, DOI 10.1080/00036811.2013.866227. MR3180019

[76] S. Cuccagna, D. Pelinovsky, and V. Vougalter, *Spectra of positive and negative energies in the linearized NLS problem*, Comm. Pure Appl. Math. **58** (2005), no. 1, 1–29, DOI 10.1002/cpa.20050. MR2094265

[77] A. De Bouard, *Instability of stationary bubbles*, SIAM J. Math. Anal. **26** (1995), no. 3, 566–582, DOI 10.1137/S0036141092237029. MR1325903

[78] J. Denzler, *Nonpersistence of breather families for the perturbed sine Gordon equation*, Comm. Math. Phys. **158** (1993), no. 2, 397–430. MR1249601

[79] T. Kapitula and B. Deconinck, *On the spectral and orbital stability of spatially periodic stationary solutions of generalized Korteweg–de Vries equations*, Hamiltonian partial differential equations and applications, Fields Inst. Commun., vol. 75, Fields Inst. Res. Math. Sci., Toronto, ON, 2015, pp. 285–322, DOI 10.1007/978-1-4939-2950-4_10. MR3445506

[80] B. Deconinck and B. L. Segal, *The stability spectrum for elliptic solutions to the focusing NLS equation*, Phys. D **346** (2017), 1–19, DOI 10.1016/j.physd.2017.01.004. MR3624690

[81] B. Deconinck, P. McGill, and B. L. Segal, *The stability spectrum for elliptic solutions to the sine-Gordon equation*, Phys. D **360** (2017), 17–35, DOI 10.1016/j.physd.2017.08.010. MR3710734

[82] J. Upsal and B. Deconinck, *Real Lax spectrum implies spectral stability*, Stud. Appl. Math. **145** (2020), no. 4, 765–790, DOI 10.1111/sapm.12335. MR4174173

[83] J.-M. Delort and N. Masmoudi, *Long-time dispersive estimates for perturbations of a kink solution of one-dimensional cubic wave equations*, Memoirs of the European Mathematical Society, vol. 1, EMS Press, Berlin, 2022, DOI 10.4171/mems/1. MR4529848

[84] A. Degasperis, D. D. Holm, and A. N. W. Hone, *Integrable and non-integrable equations with peakons*, Nonlinear physics: theory and experiment, II (Gallipoli, 2002), World Sci. Publ., River Edge, NJ, 2003, pp. 37–43, DOI 10.1142/9789812704467_0005. MR2028761

[85] X. Deng and A. Chen, *Instability of H^1-stable periodic peakons for the μ-Camassa-Holm equation*, J. Dynam. Differential Equations **36** (2024), no. 1, 515–534, DOI 10.1007/s10884-022-10165-y. MR4710788

[86] L. Di Menza and C. Gallo, *The black solitons of one-dimensional NLS equations*, Nonlinearity **20** (2007), no. 2, 461–496, DOI 10.1088/0951-7715/20/2/010. MR2290470

[87] H. R. Dullin, G. A. Gottwald, and D. D. Holm, *An integrable shallow water equation with linear and nonlinear dispersion*, Phys. Rev. Lett. **87** (2001), 194501.

[88] N. Dunford and J. T. Schwartz, *Linear operators. Part II: Spectral theory. Self adjoint operators in Hilbert space*, Interscience Publishers John Wiley & Sons, New York-London, 1963. With the assistance of William G. Bade and Robert G. Bartle. MR188745

[89] N. Duruk Mutlubaş and A. Geyer, *Orbital stability of solitary waves of moderate amplitude in shallow water*, J. Differential Equations **255** (2013), no. 2, 254–263, DOI 10.1016/j.jde.2013.04.010. MR3047403

[90] M. S. P. Eastham, *The spectral theory of periodic differential equations*, Texts in Mathematics (Edinburgh), Scottish Academic Press, Edinburgh; Hafner Press, New York, 1973. MR3075381

[91] B. Ehrman and M. A. Johnson, *Orbital stability of periodic traveling waves in the b-Camassa-Holm equation*, Phys. D **461** (2024), Paper No. 134105, 11, DOI 10.1016/j.physd.2024.134105. MR4711186

[92] J. Escher and Z. Yin, *Well-posedness, blow-up phenomena, and global solutions for the b-equation*, J. Reine Angew. Math. **624** (2008), 51–80, DOI 10.1515/CRELLE.2008.080. MR2456624

[93] A. Farina, B. Sciunzi, and N. Soave, *Monotonicity and rigidity of solutions to some elliptic systems with uniform limits*, Commun. Contemp. Math. **22** (2020), no. 5, 1950044, 24, DOI 10.1142/S0219199719500445. MR4123442

[94] B.-F. Feng, L. Ling, and D. A. Takahashi, *Multi-breather and high-order rogue waves for the nonlinear Schrödinger equation on the elliptic function background*, Stud. Appl. Math. **144** (2020), no. 1, 46–101, DOI 10.1111/sapm.12287. MR4057933

[95] R. L. Frank and E. Lenzmann, *Uniqueness of non-linear ground states for fractional Laplacians in $\mathbb{R}$*, Acta Math. **210** (2013), no. 2, 261–318, DOI 10.1007/s11511-013-0095-9. MR3070568

[96] T. Gallay and M. Hărăgus, *Orbital stability of periodic waves for the nonlinear Schrödinger equation*, J. Dynam. Differential Equations **19** (2007), no. 4, 825–865, DOI 10.1007/s10884-007-9071-4. MR2357528

[97] T. Gallay and D. Pelinovsky, *Orbital stability in the cubic defocusing NLS equation: I. Cnoidal periodic waves*, J. Differential Equations **258** (2015), no. 10, 3607–3638, DOI 10.1016/j.jde.2015.01.018. MR3319432

[98] T. Gallay and D. Pelinovsky, *Orbital stability in the cubic defocusing NLS equation: II. The black soliton*, J. Differential Equations **258** (2015), no. 10, 3639–3660, DOI 10.1016/j.jde.2015.01.019. MR3319433

[99] C. Gallo, *The Cauchy problem for defocusing nonlinear Schrödinger equations with non-vanishing initial data at infinity*, Comm. Partial Differential Equations **33** (2008), no. 4-6, 729–771, DOI 10.1080/03605300802031614. MR2424376

[100] L. Gavrilov, *Remark on the number of critical points of the period*, J. Differential Equations **101** (1993), no. 1, 58–65, DOI 10.1006/jdeq.1993.1004. MR1199482

[101] P. Germain and F. Pusateri, *Quadratic Klein-Gordon equations with a potential in one dimension*, Forum Math. Pi **10** (2022), Paper No. e17, 172, DOI 10.1017/fmp.2022.9. MR4451296

[102] P. Gérard, *The Cauchy problem for the Gross-Pitaevskii equation* (English, with English and French summaries), Ann. Inst. H. Poincaré C Anal. Non Linéaire **23** (2006), no. 5, 765–779, DOI 10.1016/j.anihpc.2005.09.004. MR2259616

[103] P. Gérard and Z. Zhang, *Orbital stability of traveling waves for the one-dimensional Gross-Pitaevskii equation* (English, with English and French summaries), J. Math. Pures Appl. (9) **91** (2009), no. 2, 178–210, DOI 10.1016/j.matpur.2008.09.009. MR2498754

[104] A. Geyer and D. E. Pelinovsky, *Spectral stability of periodic waves in the generalized reduced Ostrovsky equation*, Lett. Math. Phys. **107** (2017), no. 7, 1293–1314, DOI 10.1007/s11005-017-0941-3. MR3685174

[105] A. Geyer and D. Pelinovsky, *Linear instability and uniqueness of the peaked periodic wave in the reduced Ostrovsky equation*, SIAM J. Math. Anal. **51** (2019), no. 2, 1188–1208, DOI 10.1137/18M117978X. MR3936896

[106] A. Geyer and D. E. Pelinovsky, *Spectral instability of the peaked periodic wave in the reduced Ostrovsky equations*, Proc. Amer. Math. Soc. **148** (2020), no. 12, 5109–5125, DOI 10.1090/proc/14937. MR4163826

[107] A. Geyer and D. E. Pelinovsky, *Stability of smooth periodic traveling waves in the Degasperis-Procesi equation*, J. Differential Equations **404** (2024), 354–390, DOI 10.1016/j.jde.2024.05.047. MR4756908

[108] A. Geyer, R. H. Martins, F. Natali, and D. E. Pelinovsky, *Stability of smooth periodic traveling waves in the Camassa-Holm equation*, Stud. Appl. Math. **148** (2022), no. 1, 27–61, DOI 10.1111/sapm.12430. MR4429037

[109] A. Geyer and J. Villadelprat, *On the wave length of smooth periodic traveling waves of the Camassa-Holm equation*, J. Differential Equations **259** (2015), no. 6, 2317–2332, DOI 10.1016/j.jde.2015.03.027. MR3353646

[110] D. Gilbarg and N. S. Trudinger, *Elliptic partial differential equations of second order*, Classics in Mathematics, Springer-Verlag, Berlin, 2001. Reprint of the 1998 edition. MR1814364

[111] I. C. Gohberg and M. G. Kreĭn, *Introduction to the theory of linear nonselfadjoint operators*, Translations of Mathematical Monographs, Vol. 18, American Mathematical Society, Providence, RI, 1969. Translated from the Russian by A. Feinstein. MR246142

[112] M. Goldman and B. Merlet, *Phase segregation for binary mixtures of Bose-Einstein condensates*, SIAM J. Math. Anal. **49** (2017), no. 3, 1947–1981, DOI 10.1137/15M1051105. MR3662031

[113] P. Gravejat and D. Smets, *Asymptotic stability of the black soliton for the Gross-Pitaevskii equation*, Proc. Lond. Math. Soc. (3) **111** (2015), no. 2, 305–353, DOI 10.1112/plms/pdv025. MR3384514

[114] M. Grillakis, *Linearized instability for nonlinear Schrödinger and Klein-Gordon equations*, Comm. Pure Appl. Math. **41** (1988), no. 6, 747–774, DOI 10.1002/cpa.3160410602. MR948770

[115] M. Grillakis, *Analysis of the linearization around a critical point of an infinite-dimensional Hamiltonian system*, Comm. Pure Appl. Math. **43** (1990), no. 3, 299–333, DOI 10.1002/cpa.3160430302. MR1040143

[116] M. Grillakis, J. Shatah, and W. Strauss, *Stability theory of solitary waves in the presence of symmetry. I*, J. Funct. Anal. **74** (1987), no. 1, 160–197, DOI 10.1016/0022-1236(87)90044-9. MR901236

[117] M. Grillakis, J. Shatah, and W. Strauss, *Stability theory of solitary waves in the presence of symmetry. II*, J. Funct. Anal. **94** (1990), no. 2, 308–348, DOI 10.1016/0022-1236(90)90016-E. MR1081647

[118] R. Grimshaw and D. Pelinovsky, *Global existence of small-norm solutions in the reduced Ostrovsky equation*, Discrete Contin. Dyn. Syst. **34** (2014), no. 2, 557–566, DOI 10.3934/dcds.2014.34.557. MR3094592

[119] P. G. Grinevich and P. M. Santini, *The linear and nonlinear instability of the Akhmediev breather*, Nonlinearity **34** (2021), no. 12, 8331–8358, DOI 10.1088/1361-6544/ac3143. MR4344718

[120] S. Hakkaev and A. G. Stefanov, *Stability of periodic waves for the fractional KdV and NLS equations*, Proc. Roy. Soc. Edinburgh Sect. A **151** (2021), no. 4, 1171–1203, DOI 10.1017/prm.2020.54. MR4284516

[121] S. Hakkaev, M. Stanislavova, and A. Stefanov, *Spectral stability for classical periodic waves of the Ostrovsky and short pulse models*, Stud. Appl. Math. **139** (2017), no. 3, 405–433, DOI 10.1111/sapm.12166. MR3708844

[122] M. Hărăgus and G. Iooss, *Local bifurcations, center manifolds, and normal forms in infinite-dimensional dynamical systems*, Universitext, Springer-Verlag London, Ltd., London; EDP Sciences, Les Ulis, 2011, DOI 10.1007/978-0-85729-112-7. MR2759609

[123] M. Hărăguş and T. Kapitula, *On the spectra of periodic waves for infinite-dimensional Hamiltonian systems*, Phys. D **237** (2008), no. 20, 2649–2671, DOI 10.1016/j.physd.2008.03.050. MR2514124

[124] M. Hărăgus, J. Li, and D. E. Pelinovsky, *Counting unstable eigenvalues in Hamiltonian spectral problems via commuting operators*, Comm. Math. Phys. **354** (2017), no. 1, 247–268, DOI 10.1007/s00220-017-2898-6. MR3656518

[125] M. Hărăgus and D. E. Pelinovsky, *Linear instability of breathers for the focusing nonlinear Schrödinger equation*, J. Nonlinear Sci. **32** (2022), no. 5, Paper No. 66, 40, DOI 10.1007/s00332-022-09819-4. MR4450127

[126] B. Harrop-Griffiths, R. Killip, and M. Vişan, *Large-data equicontinuity for the derivative NLS*, Int. Math. Res. Not. IMRN **6** (2023), 4601–4642, DOI 10.1093/imrn/rnab374. MR4565673

[127] S. V. Haziot, *Wave breaking for the Fornberg–Whitham equation*, J. Differential Equations **263** (2017), no. 12, 8178–8185, DOI 10.1016/j.jde.2017.08.037. MR3710682

[128] D. B. Henry, J. F. Perez, and W. F. Wreszinski, *Stability theory for solitary-wave solutions of scalar field equations*, Comm. Math. Phys. **85** (1982), no. 3, 351–361. MR678151

[129] A. A. Himonas, K. Grayshan, and C. Holliman, *Ill-posedness for the b-family of equations*, J. Nonlinear Sci. **26** (2016), no. 5, 1175–1190, DOI 10.1007/s00332-016-9302-0. MR3551270

[130] A. Hoffman and C. E. Wayne, *Orbital stability of localized structures via Bäcklund transformations*, Differential Integral Equations **26** (2013), no. 3-4, 303–320. MR3059166

[131] J. Holmer and Q. Lin, *Phase-driven interaction of widely separated nonlinear Schrödinger solitons*, J. Hyperbolic Differ. Equ. **9** (2012), no. 3, 511–543, DOI 10.1142/S0219891612500166. MR2974768

[132] H. Huh and B. Moon, *Low regularity well-posedness for Gross-Neveu equations*, Commun. Pure Appl. Anal. **14** (2015), no. 5, 1903–1913, DOI 10.3934/cpaa.2015.14.1903. MR3359550

[133] J. K. Hunter and M. Ifrim, *Enhanced life span of smooth solutions of a Burgers-Hilbert equation*, SIAM J. Math. Anal. **44** (2012), no. 3, 2039–2052, DOI 10.1137/110849791. MR2982741

[134] V. M. Hur, *Wave breaking in the Whitham equation*, Adv. Math. **317** (2017), 410–437, DOI 10.1016/j.aim.2017.07.006. MR3682673

[135] V. M. Hur and L. Tao, *Wave breaking in a shallow water model*, SIAM J. Math. Anal. **50** (2018), no. 1, 354–380, DOI 10.1137/15M1053281. MR3749383

[136] V. M. Hur and M. A. Johnson, *Stability of periodic traveling waves for nonlinear dispersive equations*, SIAM J. Math. Anal. **47** (2015), no. 5, 3528–3554, DOI 10.1137/12090215X. MR3397429

[137] I. S. Iohvidov, M. G. Kreĭn, and H. Langer, *Introduction to the spectral theory of operators in spaces with an indefinite metric*, Mathematical Research, vol. 9, Akademie-Verlag, Berlin, 1982. MR691137

[138] G. Iooss and M. Adelmeyer, *Topics in bifurcation theory and applications*, 2nd ed., Advanced Series in Nonlinear Dynamics, vol. 3, World Scientific Publishing Co., Inc., River Edge, NJ, 1998. MR1695170

[139] R. Jenkins, J. Liu, P. A. Perry, and C. Sulem, *Global well-posedness for the derivative non-linear Schrödinger equation*, Comm. Partial Differential Equations **43** (2018), no. 8, 1151–1195, DOI 10.1080/03605302.2018.1475489. MR3913998

[140] M. A. Johnson, *Nonlinear stability of periodic traveling wave solutions of the generalized Korteweg-de Vries equation*, SIAM J. Math. Anal. **41** (2009), no. 5, 1921–1947, DOI 10.1137/090752249. MR2564200

[141] E. R. Johnson and D. E. Pelinovsky, *Orbital stability of periodic waves in the class of reduced Ostrovsky equations*, J. Differential Equations **261** (2016), no. 6, 3268–3304, DOI 10.1016/j.jde.2016.05.026. MR3527630

[142] R. S. Johnson, *Camassa-Holm, Korteweg-de Vries and related models for water waves*, J. Fluid Mech. **455** (2002), 63–82, DOI 10.1017/S0022112001007224. MR1894796

[143] C. K. R. T. Jones, *An instability mechanism for radially symmetric standing waves of a nonlinear Schrödinger equation*, J. Differential Equations **71** (1988), no. 1, 34–62, DOI 10.1016/0022-0396(88)90037-X. MR922197

[144] C. K. R. T. Jones, *Instability of standing waves for nonlinear Schrödinger-type equations*, Ergodic Theory Dynam. Systems **8*** (1988), no. Charles Conley Memorial Issue, 119–138, DOI 10.1017/S014338570000938X. MR967634

[145] A. Kairzhan and D. E. Pelinovsky, *Nonlinear instability of half-solitons on star graphs*, J. Differential Equations **264** (2018), no. 12, 7357–7383, DOI 10.1016/j.jde.2018.02.020. MR3779640

[146] A. Kairzhan and D. E. Pelinovsky, *Spectral stability of shifted states on star graphs*, J. Phys. A **51** (2018), no. 9, 095203, 23, DOI 10.1088/1751-8121/aaa89f. MR3766212

[147] A. Kairzhan, D. E. Pelinovsky, and R. H. Goodman, *Drift of spectrally stable shifted states on star graphs*, SIAM J. Appl. Dyn. Syst. **18** (2019), no. 4, 1723–1755, DOI 10.1137/19M1246146. MR4013916

[148] A. M. Kamchatnov, *On improving the effectiveness of periodic solutions of the NLS and DNLS equations*, J. Phys. A **23** (1990), no. 13, 2945–2960. MR1062995

[149] T. Kapitula, *On the stability of N-solitons in integrable systems*, Nonlinearity **20** (2007), no. 4, 879–907, DOI 10.1088/0951-7715/20/4/005. MR2307885

[150] T. Kapitula, P. G. Kevrekidis, and B. Sandstede, *Counting eigenvalues via the Krein signature in infinite-dimensional Hamiltonian systems*, Phys. D **195** (2004), no. 3-4, 263–282, DOI 10.1016/j.physd.2004.03.018. MR2089513

[151] T. Kapitula and K. Promislow, *Stability indices for constrained self-adjoint operators*, Proc. Amer. Math. Soc. **140** (2012), no. 3, 865–880, DOI 10.1090/S0002-9939-2011-10943-2. MR2869071

[152] T. Kapitula and K. Promislow, *Spectral and dynamical stability of nonlinear waves*, Applied Mathematical Sciences, vol. 185, Springer, New York, 2013. With a foreword by Christopher K. R. T. Jones, DOI 10.1007/978-1-4614-6995-7. MR3100266

[153] T. Kapitula and A. Stefanov, *A Hamiltonian-Krein (instability) index theory for solitary waves to KdV-like eigenvalue problems*, Stud. Appl. Math. **132** (2014), no. 3, 183–211, DOI 10.1111/sapm.12031. MR3180620

[154] T. Kato, *Perturbation of continuous spectra by trace class operators*, Proc. Japan Acad. **33** (1957), 260–264. MR92133

[155] T. Kato, *On the Korteweg-de Vries equation*, Manuscripta Math. **28** (1979), no. 1-3, 89–99, DOI 10.1007/BF01647967. MR535697

[156] D. J. Kaup, *Closure of the squared Zakharov-Shabat eigenstates*, J. Math. Anal. Appl. **54** (1976), no. 3, 849–864, DOI 10.1016/0022-247X(76)90201-8. MR413877

[157] C. E. Kenig, G. Ponce, and L. Vega, *Well-posedness and scattering results for the generalized Korteweg-de Vries equation via the contraction principle*, Comm. Pure Appl. Math. **46** (1993), no. 4, 527–620, DOI 10.1002/cpa.3160460405. MR1211741

[158] R. Killip and M. Vişan, *KdV is well-posed in H^{-1}*, Ann. of Math. (2) **190** (2019), no. 1, 249–305, DOI 10.4007/annals.2019.190.1.4. MR3990604

[159] R. Killip, M. Vişan, and X. Zhang, *Low regularity conservation laws for integrable PDE*, Geom. Funct. Anal. **28** (2018), no. 4, 1062–1090, DOI 10.1007/s00039-018-0444-0. MR3820439

[160] C. Klein, J.-C. Saut, and Y. Wang, *On the modified fractional Korteweg–de Vries and related equations*, Nonlinearity **35** (2022), no. 3, 1170–1212, DOI 10.1088/1361-6544/ac4814. MR4374000

[161] H. Koch and D. Tataru, *Conserved energies for the cubic nonlinear Schrödinger equation in one dimension*, Duke Math. J. **167** (2018), no. 17, 3207–3313, DOI 10.1215/00127094-2018-0033. MR3874652

[162] H. Koch and D. Tataru, *Multisolitons for the cubic NLS in 1-D and their stability*, Publ. Math. Inst. Hautes Études Sci. **140** (2024), 155–270, DOI 10.1007/s10240-024-00148-8. MR4824748

[163] H. Koch and X. Liao, *Conserved energies for the one dimensional Gross-Pitaevskii equation*, Adv. Math. **377** (2021), Paper No. 107467, 83, DOI 10.1016/j.aim.2020.107467. MR4186010

[164] M. Kowalczyk, Y. Martel, and C. Muñoz, *Kink dynamics in the ϕ^4 model: asymptotic stability for odd perturbations in the energy space*, J. Amer. Math. Soc. **30** (2017), no. 3, 769–798, DOI 10.1090/jams/870. MR3630087

[165] M. Kowalczyk, Y. Martel, C. Muñoz, and H. Van Den Bosch, *A sufficient condition for asymptotic stability of kinks in general* $(1+1)$*-scalar field models*, Ann. PDE **7** (2021), no. 1, Paper No. 10, 98, DOI 10.1007/s40818-021-00098-y. MR4242134

[166] E. A. Kuznetsov, *Solitons in a parametrically unstable plasma*, Sov. Phys. Dokl. **22** (1977), 507–508.

[167] S. Lafortune, *Spectral and linear stability of peakons in the Novikov equation*, Stud. Appl. Math. **152** (2024), no. 4, 1404–1424. MR4752670

[168] S. Lafortune and D. E. Pelinovsky, *Stability of smooth solitary waves in the b-Camassa-Holm equation*, Phys. D **440** (2022), Paper No. 133477, 10, DOI 10.1016/j.physd.2022.133477. MR4467706

[169] S. Lafortune and D. E. Pelinovsky, *Spectral instability of peakons in the b-family of the Camassa-Holm equations*, SIAM J. Math. Anal. **54** (2022), no. 4, 4572–4590, DOI 10.1137/21M1458776. MR4458895

[170] P. D. Lax, *Integrals of nonlinear equations of evolution and solitary waves*, Comm. Pure Appl. Math. **21** (1968), 467–490, DOI 10.1002/cpa.3160210503. MR235310

[171] P. D. Lax, *Periodic solutions of the KdV equation*, Comm. Pure Appl. Math. **28** (1975), 141–188, DOI 10.1002/cpa.3160280105. MR369963

[172] U. Le and D. E. Pelinovsky, *Periodic waves of the modified KdV equation as minimizers of a new variational problem*, SIAM J. Appl. Dyn. Syst. **21** (2022), no. 4, 2518–2534, DOI 10.1137/21M1465329. MR4518501

[173] S. Le Coz and Z. Wang, *Stability of the multi-solitons of the modified Korteweg–de Vries equation*, Nonlinearity **34** (2021), no. 10, 7109–7143, DOI 10.1088/1361-6544/ac20a7. MR4313706

[174] J. Lenells, *Traveling wave solutions of the Camassa-Holm equation*, J. Differential Equations **217** (2005), no. 2, 393–430, DOI 10.1016/j.jde.2004.09.007. MR2168830

[175] J. Lenells, *Traveling wave solutions of the Degasperis-Procesi equation*, J. Math. Anal. Appl. **306** (2005), no. 1, 72–82, DOI 10.1016/j.jmaa.2004.11.038. MR2132889

[176] J. Li and Y. Liu, *Stability of solitary waves for the modified Camassa-Holm equation*, Ann. PDE **7** (2021), no. 2, Paper No. 14, 35, DOI 10.1007/s40818-021-00104-3. MR4270851

[177] J. Li, Y. Liu, and Q. Wu, *Spectral stability of smooth solitary waves for the Degasperis-Procesi equation* (English, with English and French summaries), J. Math. Pures Appl. (9) **142** (2020), 298–314, DOI 10.1016/j.matpur.2020.08.003. MR4149693

[178] Z. Lin, *Stability and instability of traveling solitonic bubbles*, Adv. Differential Equations **7** (2002), no. 8, 897–918. MR1895111

[179] Z. Lin and Y. Liu, *Stability of peakons for the Degasperis-Procesi equation*, Comm. Pure Appl. Math. **62** (2009), no. 1, 125–146, DOI 10.1002/cpa.20239. MR2460268

[180] Z. Lin and C. Zeng, *Instability, index theorem, and exponential trichotomy for linear Hamiltonian PDEs*, Mem. Amer. Math. Soc. **275** (2022), no. 1347, v+136, DOI 10.1090/memo/1347. MR4352468

[181] F. Linares and G. Ponce, *Introduction to nonlinear dispersive equations*, Universitext, Springer, New York, 2009. MR2492151

[182] F. Linares, G. Ponce, and T. C. Sideris, *Properties of solutions to the Camassa-Holm equation on the line in a class containing the peakons*, Asymptotic analysis for nonlinear dispersive and wave equations, Adv. Stud. Pure Math., vol. 81, Math. Soc. Japan, Tokyo, [2019] ©2019, pp. 197–246, DOI 10.2969/aspm/08110197. MR4388135

[183] P.-L. Lions, *The concentration-compactness principle in the calculus of variations. The locally compact case. I* (English, with French summary), Ann. Inst. H. Poincaré Anal. Non Linéaire **1** (1984), no. 2, 109–145. MR778970

[184] P.-L. Lions, *The concentration-compactness principle in the calculus of variations. The locally compact case. II* (English, with French summary), Ann. Inst. H. Poincaré Anal. Non Linéaire **1** (1984), no. 4, 223–283. MR778974

[185] Y. Liu, D. Pelinovsky, and A. Sakovich, *Wave breaking in the Ostrovsky-Hunter equation*, SIAM J. Math. Anal. **42** (2010), no. 5, 1967–1985, DOI 10.1137/09075799X. MR2684307

[186] J. Liu, P. A. Perry, and C. Sulem, *Global existence for the derivative nonlinear Schrödinger equation by the method of inverse scattering*, Comm. Partial Differential Equations **41** (2016), no. 11, 1692–1760, DOI 10.1080/03605302.2016.1227337. MR3563476

[187] T. Long and C. Liu, *Orbital stability of smooth solitary waves for the b-family of Camassa-Holm equations*, Phys. D **446** (2023), Paper No. 133680, 7, DOI 10.1016/j.physd.2023.133680. MR4547334

[188] O. Lopes, *A class of isoinertial one parameter families of selfadjoint operators*, Nonlinear equations: methods, models and applications (Bergamo, 2001), Progr. Nonlinear Differential Equations Appl., vol. 54, Birkhäuser, Basel, 2003, pp. 191–195. MR2023242

[189] D. G. Luenberger and Y. Ye, *Linear and nonlinear programming*, 4th ed., International Series in Operations Research & Management Science, vol. 228, Springer, Cham, 2016, DOI 10.1007/978-3-319-18842-3. MR3363684

[190] A. M. Lyapunov, *The general problem of the stability of motion*, Internat. J. Control **55** (1992), no. 3, 521–790, DOI 10.1080/00207179208934253. Translated by A. T. Fuller from Édouard Davaux's French translation (1907) of the 1892 Russian original; With an editorial (historical introduction) by Fuller, a biography of Lyapunov by V. I. Smirnov, and the bibliography of Lyapunov's works collected by J. F. Barrett; Lyapunov centenary issue. MR1154209

[191] Y. C. Ma, *The perturbed plane-wave solutions of the cubic Schrödinger equation*, Stud. Appl. Math. **60** (1979), no. 1, 43–58, DOI 10.1002/sapm197960143. MR517602

[192] R. S. MacKay, *Stability of equilibria of Hamiltonian systems*, Nonlinear phenomena and chaos (Malvern, 1985), Malvern Phys. Ser., Hilger, Bristol, 1986, pp. 254–270. MR854706

[193] J. H. Maddocks and R. L. Sachs, *On the stability of KdV multi-solitons*, Comm. Pure Appl. Math. **46** (1993), no. 6, 867–901, DOI 10.1002/cpa.3160460604. MR1220540

[194] A. Madiyeva and D. E. Pelinovsky, *Growth of perturbations to the peaked periodic waves in the Camassa-Holm equation*, SIAM J. Math. Anal. **53** (2021), no. 3, 3016–3039, DOI 10.1137/20M1347474. MR4261112

[195] V. G. Makhankov, Y. P. Rybakov, and V. I. Sanyuk, *Localised nontopological structures: construction of solutions and stability properties*, Phys. Usp. **37** (1994), 113–137.

[196] M. Maeda, *Stability of bound states of Hamiltonian PDEs in the degenerate cases*, J. Funct. Anal. **263** (2012), no. 2, 511–528, DOI 10.1016/j.jfa.2012.04.006. MR2923422

[197] W. Magnus and S. Winkler, *Hill's equation*, Dover Publications, Inc., New York, 1979. Corrected reprint of the 1966 edition. MR559928

[198] F. Mañosas and J. Villadelprat, *Criteria to bound the number of critical periods*, J. Differential Equations **246** (2009), no. 6, 2415–2433, DOI 10.1016/j.jde.2008.07.002. MR2498846

[199] Y. Martel, F. Merle, and T.-P. Tsai, *Stability in H^1 of the sum of K solitary waves for some nonlinear Schrödinger equations*, Duke Math. J. **133** (2006), no. 3, 405–466, DOI 10.1215/S0012-7094-06-13331-8. MR2228459

[200] F. Merle and L. Vega, *L^2 stability of solitons for KdV equation*, Int. Math. Res. Not. **13** (2003), 735–753, DOI 10.1155/S1073792803208060. MR1949297

[201] T. Mizumachi, *Asymptotic stability of small solitary waves to 1D nonlinear Schrödinger equations with potential*, J. Math. Kyoto Univ. **48** (2008), no. 3, 471–497, DOI 10.1215/kjm/1250271380. MR2511047

[202] T. Mizumachi and R. L. Pego, *Asymptotic stability of Toda lattice solitons*, Nonlinearity **21** (2008), no. 9, 2099–2111, DOI 10.1088/0951-7715/21/9/011. MR2430663

[203] T. Mizumachi and D. Pelinovsky, *Bäcklund transformation and L^2-stability of NLS solitons*, Int. Math. Res. Not. IMRN **9** (2012), 2034–2067. MR2920823

[204] L. Molinet, D. Pilod, and S. Vento, *On well-posedness for some dispersive perturbations of Burgers' equation*, Ann. Inst. H. Poincaré C Anal. Non Linéaire **35** (2018), no. 7, 1719–1756, DOI 10.1016/j.anihpc.2017.12.004. MR3906854

[205] F. Natali and D. E. Pelinovsky, *Instability of H^1-stable peakons in the Camassa-Holm equation*, J. Differential Equations **268** (2020), no. 12, 7342–7363, DOI 10.1016/j.jde.2019.11.059. MR4079004

[206] F. Natali, U. Le, and D. E. Pelinovsky, *New variational characterization of periodic waves in the fractional Korteweg–de Vries equation*, Nonlinearity **33** (2020), no. 4, 1956–1986, DOI 10.1088/1361-6544/ab6a79. MR4075882

[207] F. Natali, U. Le, and D. E. Pelinovsky, *Periodic waves in the fractional modified Korteweg-de Vries equation*, J. Dynam. Differential Equations **34** (2022), no. 2, 1601–1640, DOI 10.1007/s10884-021-10000-w. MR4413398

[208] A. Neves, *Floquet's theorem and stability of periodic solitary waves*, J. Dynam. Differential Equations **21** (2009), no. 3, 555–565, DOI 10.1007/s10884-009-9143-8. MR2538951

[209] D. Noja and D. E. Pelinovsky, *Standing waves of the quintic NLS equation on the tadpole graph*, Calc. Var. Partial Differential Equations **59** (2020), no. 5, Paper No. 173, 31, DOI 10.1007/s00526-020-01832-3. MR4150875

[210] S. P. Novikov, *A periodic problem for the Korteweg-de Vries equation. I* (Russian), Funkcional. Anal. i Priložen. **8** (1974), no. 3, 54–66. MR382878

[211] E. Novruzov, *Construction of peakon-antipeakon solutions and ill-posedness for the b-family of equations*, J. Differential Equations **272** (2021), 544–559, DOI 10.1016/j.jde.2020.10.013. MR4163105

[212] M. Ohta, *Instability of bound states for abstract nonlinear Schrödinger equations*, J. Funct. Anal. **261** (2011), no. 1, 90–110, DOI 10.1016/j.jfa.2011.03.010. MR2785894

[213] D. E. Pelinovsky, *Inertia law for spectral stability of solitary waves in coupled nonlinear Schrödinger equations*, Proc. R. Soc. Lond. Ser. A Math. Phys. Eng. Sci. **461** (2005), no. 2055, 783–812, DOI 10.1098/rspa.2004.1345. MR2121936

[214] D. Pelinovsky, *Survey on global existence in the nonlinear Dirac equations in one spatial dimension*, Harmonic analysis and nonlinear partial differential equations, RIMS Kôkyûroku Bessatsu, B26, Res. Inst. Math. Sci. (RIMS), Kyoto, 2011, pp. 37–50. MR2883845

[215] D. E. Pelinovsky, *Spectral stability on nonlinear waves in KdV-type evolution equations*, Nonlinear physical systems, Mech. Eng. Solid Mech. Ser., Wiley, Hoboken, NJ, 2014, pp. 377–400. MR3330810

[216] D. E. Pelinovsky, *Localization in periodic potentials: From Schrödinger operators to the Gross-Pitaevskii equation*, London Mathematical Society Lecture Note Series, vol. 390, Cambridge University Press, Cambridge, 2011, DOI 10.1017/CBO9780511997754. MR2894529

[217] D. E. Pelinovsky and P. G. Kevrekidis, *Dark solitons in external potentials*, Z. Angew. Math. Phys. **59** (2008), no. 4, 559–599, DOI 10.1007/s00033-007-6120-0. MR2417380

[218] D. E. Pelinovsky and M. Plum, *Stability of black solitons in optical systems with intensity-dependent dispersion*, SIAM J. Math. Anal. **56** (2024), no. 2, 2521–2568, DOI 10.1137/23M1552838. MR4721062

[219] D. E. Pelinovsky and M. Plum, *Dynamics of the black soliton in a regularized nonlinear Schrödinger equation*, Proc. Amer. Math. Soc. **152** (2024), no. 3, 1217–1231, DOI 10.1090/proc/16633. MR4693678

[220] D. E. Pelinovsky, A. Saalmann, and Y. Shimabukuro, *The derivative NLS equation: global existence with solitons*, Dyn. Partial Differ. Equ. **14** (2017), no. 3, 271–294, DOI 10.4310/DPDE.2017.v14.n3.a3. MR3702542

[221] D. E. Pelinovsky and Y. Shimabukuro, *Orbital stability of Dirac solitons*, Lett. Math. Phys. **104** (2014), no. 1, 21–41, DOI 10.1007/s11005-013-0650-5. MR3147657

[222] D. E. Pelinovsky and Y. Shimabukuro, *Existence of global solutions to the derivative NLS equation with the inverse scattering transform method*, Int. Math. Res. Not. IMRN **18** (2018), 5663–5728, DOI 10.1093/imrn/rnx051. MR3862117

[223] G. Perelman, *Asymptotic stability of multi-soliton solutions for nonlinear Schrödinger equations*, Comm. Partial Differential Equations **29** (2004), no. 7-8, 1051–1095, DOI 10.1081/PDE-200033754. MR2097576

[224] D. H. Peregrine, *Water waves, nonlinear Schrödinger equations and their solutions*, J. Austral. Math. Soc. Ser. B **25** (1983), no. 1, 16–43, DOI 10.1017/S0334270000003891. MR702914

[225] L. Pontrjagin, *Hermitian operators in spaces with indefinite metric* (Russian, with English summary), Bull. Acad. Sci. URSS. Sér. Math. [Izvestia Akad. Nauk SSSR] **8** (1944), 243–280. MR12200

[226] M. Renardy and R. C. Rogers, *An introduction to partial differential equations*, 2nd ed., Texts in Applied Mathematics, vol. 13, Springer-Verlag, New York, 2004. MR2028503

[227] M. Reed and B. Simon, *Methods of modern mathematical physics. IV. Analysis of operators*, Academic Press [Harcourt Brace Jovanovich, Publishers], New York-London, 1978. MR493421

[228] J. Shatah and W. Strauss, *Instability of nonlinear bound states*, Comm. Math. Phys. **100** (1985), no. 2, 173–190. MR804458

[229] J. Shatah and W. Strauss, *Spectral condition for instability*, Nonlinear PDE's, dynamics and continuum physics (South Hadley, MA, 1998), Contemp. Math., vol. 255, Amer. Math. Soc., Providence, RI, 2000, pp. 189–198, DOI 10.1090/conm/255/03982. MR1752509

[230] C. Sourdis, *Weak separation limit of a two-component Bose-Einstein condensate*, Electron. J. Differential Equations (2018), Paper No. 40, 12. MR3762827

[231] M. Stanislavova and A. Stefanov, *Linear stability analysis for traveling waves of second order in time PDE's*, Nonlinearity **25** (2012), no. 9, 2625–2654, DOI 10.1088/0951-7715/25/9/2625. MR2967119

[232] M. Stanislavova and A. Stefanov, *On the spectral problem $\mathcal{L}u = \lambda u'$ and applications*, Comm. Math. Phys. **343** (2016), no. 2, 361–391, DOI 10.1007/s00220-015-2542-2. MR3477342

[233] S. Selberg and A. Tesfahun, *Low regularity well-posedness for some nonlinear Dirac equations in one space dimension*, Differential Integral Equations **23** (2010), no. 3-4, 265–278. MR2588476

[234] J. Sylvester, *A demonstration of the theorem that every homogeneous quadratic polynomial is reducible by real orthogonal substitutions to the form of a sum of positive and negative squares*, Philosophical Magazine **4** (1852), 138–142.

[235] M. Tajiri and Y. Watanabe, *Breather solutions to the focusing nonlinear Schrödinger equation*, Phys. Rev. E (3) **57** (1998), no. 3, 3510–3519, DOI 10.1103/PhysRevE.57.3510. MR1611665

[236] G. Teschl, *Ordinary differential equations and dynamical systems*, Graduate Studies in Mathematics, vol. 140, American Mathematical Society, Providence, RI, 2012, DOI 10.1090/gsm/140. MR2961944

[237] G. Teschl, *Mathematical methods in quantum mechanics: With applications to Schrödinger operators*, 2nd ed., Graduate Studies in Mathematics, vol. 157, American Mathematical Society, Providence, RI, 2014, DOI 10.1090/gsm/157. MR3243083

[238] G. Z. Tu, *The trace identity, a powerful tool for constructing the Hamiltonian structure of integrable systems*, J. Math. Phys. **30** (1989), no. 2, 330–338, DOI 10.1063/1.528449. MR978545

[239] N. G. Vakhitov and A. A. Kolokolov, *Stationary solutions of the wave equation in a medium with nonlinearity saturation*, Radiophys. Quantum Electron **16** (1973) 783–789.

[240] G. J. B. van den Berg and R. C. A. M. van der Vorst, *A domain-wall between single-mode and bimodal states*, Differential Integral Equations **13** (2000), no. 1-3, 369–400. MR1811963

[241] J. Villadelprat, *On the reversible quadratic centers with monotonic period function*, Proc. Amer. Math. Soc. **135** (2007), no. 8, 2555–2565, DOI 10.1090/S0002-9939-07-08749-7. MR2302576

[242] Z. Wang, *Isoinertial operators around the KdV multi-solitons*, Nonlinear Anal. **219** (2022), Paper No. 112820, 24, DOI 10.1016/j.na.2022.112820. MR4385208

[243] M. I. Weinstein, *Nonlinear Schrödinger equations and sharp interpolation estimates*, Comm. Math. Phys. **87** (1982/83), no. 4, 567–576. MR691044

[244] M. I. Weinstein, *Lyapunov stability of ground states of nonlinear dispersive evolution equations*, Comm. Pure Appl. Math. **39** (1986), no. 1, 51–67, DOI 10.1002/cpa.3160390103. MR820338

[245] G. B. Whitham, *Linear and nonlinear waves*, Pure and Applied Mathematics, Wiley-Interscience [John Wiley & Sons], New York-London-Sydney, 1974. MR483954

[246] J. Yang, *Classification of solitary wave bifurcations in generalized nonlinear Schrödinger equations*, Stud. Appl. Math. **129** (2012), no. 2, 133–162, DOI 10.1111/j.1467-9590.2012.00549.x. MR2959901

[247] K. Yagasaki, *Monotonicity of the period function for $u'' - u + u^p = 0$ with $p \in \mathbb{R}$ and $p > 1$*, J. Differential Equations **255** (2013), no. 7, 1988–2001, DOI 10.1016/j.jde.2013.06.002. MR3072678

[248] K. Yagasaki and S. Yamazoe, *Bifurcation of relative equilibria in infinite-dimensional Hamiltonian systems*, SIAM J. Appl. Dyn. Syst. **18** (2019), no. 1, 393–417, DOI 10.1137/18M1187738. MR3917921

[249] J. Yang, *Nonlinear waves in integrable and nonintegrable systems*, Mathematical Modeling and Computation, vol. 16, Society for Industrial and Applied Mathematics (SIAM), Philadelphia, PA, 2010, DOI 10.1137/1.9780898719680. MR2721855

[250] V. E. Zakharov and A. B. Shabat, *Exact theory of two-dimensional self-focusing and one-dimensional self-modulation of waves in nonlinear media* (Russian, with English summary),

Ž. Èksper. Teoret. Fiz. **61** (1971), no. 1, 118–134; English transl., Soviet Physics JETP **34** (1972), no. 1, 62–69. MR406174

[251] V. E. Zaharov and A. B. Shabat, *A plan for integrating the nonlinear equations of mathematical physics by the method of the inverse scattering problem. I* (Russian), Funkcional. Anal. i Priložen. **8** (1974), no. 3, 43–53. MR481668

[252] V. E. Zaharov and A. B. Shabat, *Integration of the nonlinear equations of mathematical physics by the method of the inverse scattering problem. II* (Russian), Funktsional. Anal. i Prilozhen. **13** (1979), no. 3, 13–22. MR545363

[253] P. E. Zhidkov, *Korteweg-de Vries and nonlinear Schrödinger equations: qualitative theory*, Lecture Notes in Mathematics, vol. 1756, Springer-Verlag, Berlin, 2001. MR1831831

[254] R. Zhou, *Nonlinearizations of spectral problems of the nonlinear Schrödinger equation and the real-valued modified Korteweg-de Vries equation*, J. Math. Phys. **48** (2007), no. 1, 013510, 9, DOI 10.1063/1.2424554. MR2292625

[255] R. Zhou, *Finite-dimensional integrable Hamiltonian systems related to the nonlinear Schrödinger equation*, Stud. Appl. Math. **123** (2009), no. 4, 311–335, DOI 10.1111/j.1467-9590.2009.00452.x. MR2561079

[256] J. Zweck, Y. Latushkin, J. L. Marzuola, and C. K. R. T. Jones, *The essential spectrum of periodically stationary solutions of the complex Ginzburg-Landau equation*, J. Evol. Equ. **21** (2021), no. 3, 3313–3329, DOI 10.1007/s00028-020-00640-8. MR4350275

Index

Selected Published Titles in This Series

288 **Anna Geyer and Dmitry E. Pelinovsky,** Stability of Nonlinear Waves in Hamiltonian Dynamical Systems, 2025

287 **Jacques Sauloy,** Basic Modern Theory of Linear Complex Analytic q-Difference Equations, 2024

286 **Zalman Balanov, Wieslaw Krawcewicz, Dmitrii Rachinskii, Hao-Pin Wu, and Jianshe Yu,** Degree Theory and Symmetric Equations Assisted by GAP System, 2025

285 **Niles Johnson and Donald Yau,** Bimonoidal Categories, E_n-Monoidal Categories, and Algebraic K-Theory, 2024

284 **Donald Yau,** Bimonoidal Categories, E_n-Monoidal Categories, and Algebraic K-Theory, 2024

283 **Donald Yau,** Bimonoidal Categories, E_n-Monoidal Categories, and Algebraic K-Theory, 2024

282 **Michael Kapovich and Pranab Sardar,** Trees of Hyperbolic Spaces, 2024

281 **Horst R. Thieme,** Discrete-Time Dynamics of Structured Populations and Homogeneous Order-Preserving Operators, 2024

280 **Tadashi Ochiai,** Iwasawa Theory and Its Perspective, Volume 2, 2024

279 **Ellen Elizabeth Eischen, Wee Teck Gan, Aaron Pollack, and Zhiwei Yun,** Automorphic Forms Beyond GL_2, 2024

278 **Delaram Kahrobaei, Ramón Flores, Marialaura Noce, Maggie E. Habeeb, and Christopher Battarbee,** Applications of Group Theory in Cryptography, 2024

277 **George Metcalfe, Francesco Paoli, and Constantine Tsinakis,** Residuated Structures in Algebra and Logic, 2023

276 **Balázs Bárány, Károly Simon, and Boris Solomyak,** Self-similar and Self-affine Sets and Measures, 2023

275 **Alekos Vidras and Alain Yger,** Multidimensional Residue Theory and Applications, 2023

274 **Willi Freeden and M. Zuhair Nashed,** Recovery Methodologies: Regularization and Sampling, 2023

273 **Gennadiy Feldman,** Characterization of Probability Distributions on Locally Compact Abelian Groups, 2023

272 **Tadashi Ochiai,** Iwasawa Theory and Its Perspective, Volume 1, 2023

271 **Hideto Asashiba,** Categories and Representation Theory, 2022

270 **Leslie Hogben, Jephian C.-H. Lin, and Bryan L. Shader,** Inverse Problems and Zero Forcing for Graphs, 2022

269 **Ralph S. Freese, Ralph N. McKenzie, George F. McNulty, and Walter F. Taylor,** Algebras, Lattices, Varieties, 2022

268 **Ralph S. Freese, Ralph N. McKenzie, George F. McNulty, and Walter F. Taylor,** Algebras, Lattices, Varieties, 2022

267 **Dragana S. Cvetković Ilić,** Completion Problems on Operator Matrices, 2022

266 **Kate Juschenko,** Amenability of Discrete Groups by Examples, 2022

265 **Nabil H. Mustafa,** Sampling in Combinatorial and Geometric Set Systems, 2022

264 **J. Scott Carter and Seiichi Kamada,** Diagrammatic Algebra, 2021

263 **Vugar E. Ismailov,** Ridge Functions and Applications in Neural Networks, 2021

262 **Ragnar-Olaf Buchweitz,** Maximal Cohen–Macaulay Modules and Tate Cohomology, 2021

261 **Shiri Artstein-Avidan, Apostolos Giannopoulos, and Vitali D. Milman,** Asymptotic Geometric Analysis, Part II, 2021

For a complete list of titles in this series, visit the
AMS Bookstore at **www.ams.org/bookstore/survseries/**.

FSC
www.fsc.org
MIX
Papier | Fördert
gute Waldnutzung
FSC® C083411

Kolibri 360 GmbH
Ferdinand-Jühlke-Straße 7
99095 Erfurt, Deutschland
produktsicherheit@kolibri360.de